Sustainable Approaches to Controlling Plant Pathogenic Bacteria

Sustainable Approaches to Controlling Plant Pathogenic Bacteria

Edited by

Velu Rajesh Kannan

Bharathidasan University
Tiruchirappalli, India

Kubilay Kurtulus Bastas

Selçuk University
Konya, Turkey

CRC Press
Taylor & Francis Group
Boca Raton London New York

CRC Press is an imprint of the
Taylor & Francis Group, an **informa** business

CRC Press
Taylor & Francis Group
6000 Broken Sound Parkway NW, Suite 300
Boca Raton, FL 33487-2742

Printed on acid-free paper
Version Date: 20150624

International Standard Book Number-13: 978-1-4822-4053-5 (Hardback)

Library of Congress Cataloging-in-Publication Data

Sustainable approaches to controlling plant pathogenic bacteria / editors: Velu Rajesh Kannan and Kubilay Kurtulus Bastas.
pages cm
Includes bibliographical references and index.
ISBN 978-1-4822-4053-5 (alk. paper)
1. Phytopathogenic bacteria--Control. 2. Bacterial diseases of plants. I. Rajesh Kannan, Velu, editor. II. Bastas, Kubilay Kurtulus, editor.

SB734.S97 2016
632'.32--dc23 2015007950

Visit the Taylor & Francis Web site at
http://www.taylorandfrancis.com

and the CRC Press Web site at
http://www.crcpress.com

Contents

Foreword

Biological science research has made significant and tremendous developments during the first two decades of the twenty-first century; affiliated researchers incessantly provide required solutions to complex biological systems every day. However, some of the biological activities remain in unanswered categories and their effects are simply exposed in other biological systems including human health and economic status. The primary producers of plant systems are linked with all other biological systems, such as positive and negative effects in plants, which affect the food systems of the herbivores as well as the carnivores. Plant diseases caused by many biotic and abiotic particles, particularly biotic particles of virus, bacteria, phytoplasma, fungi, nematode, and insects, are devastating crop yields globally at massive levels. These diseases reduce yields, lower the quality of human life, and the nutritional value of plant products; the plant pathogenic bacterial role is crucial among plant pathogens.

Globally, many plant pathological researchers take this kind of challenging task on their shoulders to provide solid solutions. In this context, I am proud to introduce this book entitled *Sustainable Approaches on Controlling Plant Pathogenic Bacteria*, edited by Dr. Velu Rajesh Kannan (Bharathidasan University, India) and Dr. Kubilay Kurtulus Bastas (Selçuk University, Turkey), published by CRC Press, Taylor & Francis Group, USA, to researchers and readers. Both editors deserve high praise and their extensive contributions to agricultural research since the turn of the millenium, particularly in plant growth and development highlighting microbial sources, are exemplary. In fact, this book will be yet another feather in their cap.

This book contains twenty chapters. Each chapter focuses on different aspects in the exploration of plant pathogenic bacteria, such as pathogenesis, epidemiology, and forecasting, various mechanisms and controlling strategies, application methods, and their toxic molecules impact on consumers. I strongly believe that this book will be a keystone of plant pathology in agricultural fields. I congratulate Dr. Rajesh Kannan and Dr. Bastas, and all the contributors for their arduous efforts in producing this volume.

Dr. V. M. Muthukumar
Vice-Chancellor
Bharathidasan University
Tiruchirappalli, India

Preface

Whether the reader is a plant pathology graduate student, doctoral fellow, professor, scientist, or agriculturalist, he or she will find useful information in this book, whose content covers many aspects of the activities of pathogenic bacteria that interact with plants. These 20 chapters focus on pathogenesis; epidemiology; forecasting systems; control measures including diagnosis, quarantine, and eradication; adoption of agro-traditional practices; some tools for the control of antibacterial polypeptides; nutrient supplements; metabolic substances from other organisms; mechanisms of siderophores; host resistances; quorum sensing and quenching; seed and foliar applications; and the impact of plant pathogens on scientific and economic levels. The chapters are arranged progressively and linked thematically.

The idea for this book originated during our interactions and deliberations concerning various biological issues. Some of these occurred at scientific meetings that we attended, particularly the Theoretical and Practical Course on Quorum Sensing in Plant-Associated Bacteria, which was held May 23–26, 2011, at the International Centre for Genetic Engineering and Biotechnology (ICGEB), Trieste, Italy. Our approach combines our efforts with a broader perspective, which includes modern trends in ecology that approach plant pathogenic bacterial control from all angles. We hope that combining these approaches into a single platform that addresses plant pathogenic bacterial epidemiology via numerous application strategies will be helpful to the full range of plant pathogenic bacterial researchers, from students to policy makers as well as agriculturists.

We wish to express deep appreciation to the contributors and are particularly indebted to Dr. M. Murugesan, professor of English, for his constant encouragement and constructive suggestions. Special thanks go to Randy Brehm, Dave Fausel, and Ashley Weinstein at CRC Press. We are also grateful to CRC Press (Taylor & Francis Group, USA), Bharathidasan University (India), and Selcuk University (Turkey), for their solid cooperation and encouragement.

Velu Rajesh Kannan
Kubilay Kurtulus Bastas

Editors

Dr. Velu Rajesh Kannan, currently an assistant professor of microbiology at Bharathidasan University, Tiruchirappalli, Tamilnadu, India, is a botany graduate of Madurai Kamaraj University. He earned his master's and doctoral degrees in plant sciences from Bharathiar University, Coimbatore, Tamilnadu, India. During his 15 years as a faculty member, Dr. Rajesh Kannan has received several awards, including the CSIR and DST Young Scientist fellowships. He has supervised 8 PhDs and 30 MPhils, is presently guiding 8 PhD and MPhil candidates, and has completed several major research projects (more are in progress). His professional service includes membership on India's Board of Studies for Microbiology, Biotechnology, Botany and Life Sciences, as well as editorial board membership on several scientific publications and journals. His research specializations are rhizosphere biology and bioremediation.

Dr. Rajesh Kannan has served on the review board of more than 50 internationally reputed scientific journals and has published more than 80 research papers and several review articles in international refereed journals; his edited volume, *Microbiological Research in Agroecosystem Management*, was published in 2013 by Springer-Verlag. In addition, he has deposited nearly 100 microbial rRNA sequences and 9 pyrosequences in the NCBI GenBank. He has organized 2 national scientific conferences, delivered the keynote address at an international conference in Taiwan, and given several talks on All India Radio. He is an active member of the Indian Science Congress Association, the International Mycorrhizae Research Society, the International Plant-Microbes Interactions Association, the International Ecological Association, the Indian Association of Biomedical Scientists, and the Association of Microbiologists of India. He has also visited several Asian and European countries to present papers and participate in deliberations.

Dr. Kubilay Kurtulus Bastas is an assistant professor of plant protection at Selcuk University, Turkey. He completed his graduate work on plant protection and postgraduate work on phytopathology-bacteriology at Ankara University, Turkey; earned his PhD from Selcuk University in Turkey; and earned a postdoctoral degree in molecular phytopathology from Warwick University, UK. He has worked on *Erwinia amylovora* and onion diseases in Professor S. V. Beer's bacteriology laboratory at Cornell University, USA. His research areas of interest are plant bacterial diseases, fire blight, seed pathology, and molecular plant bacteriology. His 10 years of teaching and 20 years of research have earned him several awards and scholarships from national (TUBITAK, TMMOB, and YOK) and international (British Council and ERASMUS) foundations. He has produced 8 MPhils and is presently guiding 4 PhDs and 9 MPhils. He completed several major research projects (more are in progress). His professional service extends to membership on the review boards of more than 25 reputed scientific journals and membership on the editorial boards of several scientific publications.

Dr. Bastas has published nearly 100 research papers and several review articles in international refereed journals. He has reported incidences in Turkey of several new bacterial pathogens in cultivations of apples, grapes, beans, berries, some ornamentals, rice, rosehips, and kiwi. He is an active member in professional bodies such as the Turkish Phytopathology Association, the National Fire Blight Working Group, the American Society for Horticultural Science (ASHS), the American Phytopathological Society (APS), the Pathogen and Disease Detection Committee, Bacteriology Committee, and Teaching Committee. He has also visited and participated in scientific deliberations in several countries.

Contributors

Robert Antony
Department of Microbiology
Rhizosphere Biology Laboratory
Bharathidasan University
Tamilnadu, India

Yesim Aysan
Department of Plant Protection
Cukurova University
Adana, Turkey

Pudukadu Munusamy Ayyasamy
Department of Microbiology
Periyar University
Tamilnadu, India

Yakup Bakir
Department of Biology
Marmara University
Istanbul, Turkey.

Natesan Balasubramanian
CREM & Department of Life Sciences
New University of Lisbon
Caparica, Portugal

Kubilay Kurtulus Bastas
Department of Plant Protection
Selçuk University
Konya, Turkey.

Ömür Baysal
Department of Molecular Biology and Genetics
Muğla Sitki Koçman University
Muğla, Turkey

Vahap Eldem
Department of Biology
Cankiri Karatekin University
Cankiri, Turkey.

Arakalagud Nanjundaiah Ganeshamurthy
Division of Soil Science and Agricultural Chemistry
Soil Microbiology Laboratory
Indian Institute of Horticultural Research
Karnataka, India

Hariharan Harikrishnan
Department of Microbial Technology
School of Biological Sciences
Madurai Kamaraj University
Tamilnadu, India

Sumer Horuz
Department of Plant Protection
Cukurova University
Adana, Turkey

Mani Jayaprakashvel
Department of Marine Biotechnology
AMET University
Tamilnadu, India

Kanthaiah Kannan
Department of Microbiology
Rhizosphere Biology Laboratory
Bharathidasan University
Tamilnadu, India

Velu Rajesh Kannan
Department of Microbiology
Bharathidasan University
Tamilnadu, India

Kotamraju Vijay Krishna Kumar
College of Agriculture
ANGR Agricultural University
Andhra Pradesh, India

Koothala Jyothirmai Madhavi
Fruit Research Station
Dr. YSR Horticultural University
Andhra Pradesh, India

Bhekie Brilliance Mamba
Department of Applied Chemistry
University of Johannesburg
Johannesburg, South Africa

Uppala Naga Mangala
International Crops Research Institute for the Semi-Arid Tropics (ICRISAT)
Andhra Pradesh, India

Mustafa Mirik
Department of Plant Protection
Namik Kemal University
Tekirdag, Turkey

Titus Alfred Makudali Msagati
Department of Applied Chemistry
University of Johannesburg
Johannesburg, South Africa

Gangatharan Muralitharan
Department of Microbiology
Center of Excellence in Life Sciences
Bharathidasan University
Tamilnadu, India

Thangavelu Muthukumar
Department of Botany
Root and Soil Biology Laboratory
Bharathiar University
Tamilnadu, India

Karmegham Nithya
Department of Microbial Technology
School of Biological Sciences
Madurai Kamaraj University
Tamilnadu, India

Duraisamy Nivas
Department of Microbiology
Rhizosphere Biology Laboratory
Bharathidasan University
Tamilnadu, India

Sezer Okay
Department of Biology
Istanbul University
Istanbul, Turkey

Cansu Öksel
Department of Plant Protection
Namik Kemal University
Tekirdag, Turkey

Periyasamy Panneerselvam
Division of Soil Science and Agricultural Chemistry
Soil Microbiology Laboratory
Indian Institute of Horticultural Research
Bangalore, India

Mylsamy Prabhakaran
PG and Research Department of Botany
Pachaiyappa's College
Tamilnadu, India

Perumalsamy Priyadharsini
Department of Botany
Root and Soil Biology Laboratory
Bharathiar University
Tamilnadu, India

Sundaram Rajakumar
Department of Marine Biotechnology
Bharathidasan University
Tamilnadu, India

Karunakaran Rojamala
Department of Botany
Root and Soil Biology Laboratory
Bharathiar University
Tamilnadu, India

Rathinagiri Sakthivel
Department of Geology
Centre for Remote Sensing
Bharathidasan University
Tamilnadu, India

Rajendran Sangeethadevi
Department of Microbiology
Rhizosphere Biology Laboratory
Bharathidasan University
Tamilnadu, India

Boya Saritha
Division of Soil Science and Agricultural Chemistry
Soil Microbiology Laboratory
Indian Institute of Horticultural Research
Karnataka, India

Muthukrishnan Sathiyabama
Department of Plant Science
Bharathidasan University
Tamilnadu, India

Govindan Selvakumar
Division of Soil Science and Agricultural Chemistry
Soil Microbiology Laboratory
Indian Institute of Horticultural Research
Karnataka, India

Vellasamy Shanmugaiah
Department of Microbial Technology
School of Biological Sciences
Madurai Kamaraj University
Tamilnadu, India

Sumathy Shunmugam
Department of Microbiology
Bharathidasan University
Tamilnadu, India

Venkataraman Sivasankar
PG and Research Department of Chemistry
Pachaiyappa's College
Tamilnadu, India

Nooruddin Thajuddin
Department of Microbiology
Center of Excellence in Life Sciences
Bharathidasan University
Tamilnadu, India

Subramanian Umadevi
Department of Marine Biotechnology
Bharathidasan University
Tamilnadu, India

Turgay Unver
Department of Biology
Cankiri Karatekin University
Cankiri, Turkey

Penumatsa Kishore Varma
College of Agriculture
ANGR Agricultural University
Andhra Pradesh, India

1 Plant Pathogenic Bacteria
An Overview

Velu Rajesh Kannan, Kubilay Kurtulus Bastas, and Robert Antony

CONTENTS

ABSTRACT Phytopathology is an interdisciplinary biological science that comprises botany, microbiology, crop science, soil science, ecology, genetics, biochemistry, molecular biology, and physiology. In general, plant diseases are caused by living organisms (called pathogens) and by nonliving agents. Plant diseases and changes in existing pathogens remain a constant threat to our forests, food and fiber crops, and landscape plants. However, many economically important pathosystems are largely unexplored and biologically relevant life stages of even familiar systems remain poorly understood. Development of new and innovative ways to control plant diseases is a constant challenge for plant pathologists. The modern mindset in this context favors a broad perspective that considers multifaceted approaches to plant pathogenic bacterial control. This chapter, which discusses impact of plant pathogenic bacterial pathogenesis on scientific and economic levels, introduces the various mechanisms, measuring tools, and controlling strategies that will be elaborated in the following chapters.

KEYWORDS: plant pathogen, phytopathology, plant bacterial disease, pathosystem, disease-controlling strategy

1.1 INTRODUCTION

The immense diversity of plant pathogens, which include viruses, bacteria, fungi, nematodes, and insects, approximates 7100 species. Among these, roughly 150 are bacterial species that cause diseases to plants. The major ways that bacterial pathogens cause plant diseases are by obtaining nutrients one or more host plants for their own growth; using specific mechanisms to secrete proteins and other molecules to locations on, in, and near their hosts; and by exploiting these proteins and other molecules modulate or avoid plant defense circuitry to enable parasitic colonization (Chisholm et al., 2006; Davis et al., 2008). Bacterial plant diseases are most frequent and severe in tropical and subtropical places, where warm and humid conditions are ideal for bacterial growth. Indeed, consistent annual crop losses are recorded in all countries.

The problem of plant diseases, particularly in developing countries, is aggravated by the paucity of resources devoted to pathological studies. This gap in the literature may be the result of an inability to quantify plant diseases, followed closely by an inability to relate information about plant diseases to the failure of crops to reach manageable yields. In general, plant pathogenic bacterial species belonging to Xanthomonadaceae, Pseudomonadaceae, and Enterobacteriaceae families target all types of plants that can supply them with appropriate food and shelter. The most devastating plant pathogens belong to genera such as *Erwinia*, *Pectobacterium*, *Pantoea*, *Agrobacterium*, *Pseudomonas*, *Ralstonia*, *Burkholderia*, *Acidovorax*, *Xanthomonas*, *Clavibacter*, *Streptomyces*, *Xylella*, *Spiroplasma*, and *Phytoplasma*.

1.2 PATHOGENESIS OF PLANT PATHOGENIC BACTERIA

Plant bacterial diseases are generally characterized by plant morphological symptoms such as leaf and fruit spots, cankers, blights, vascular wilts, rots, and tumors. Phytopathogenic bacteria provoke diseases in plants by penetrating into host tissues (Buonaurio, 2008). Microbial pathogenicity has often been defined as the biochemical mechanisms whereby pathogenic microorganisms cause disease in a host organism (Fuchs, 1998). Microbial virulence is defined as the degree or measure of pathogenicity shown by one or more plants. Pathogenicity and/or virulence of Gram-negative plant pathogenic bacteria are strictly dependent on the presence of secretion apparatuses in host cells, through which they secrete proteins or nucleoproteins involved in their virulence within the apoplast or inject these substances into host cells (Buonaurio, 2008).

Bacterial pathogens contain several classes of genes, called virulence genes, that are essential for causing disease or for increasing virulence in one or more hosts. Pathogenicity factors that are encoded by pathogenicity genes (*pat*) and disease-specific genes (*dsp*) are crucially involved in the establishment of diseases. Some of these genes are essential for the recognition of a host by a pathogen, attachment of a pathogen to a plant's surface, formation of infection structures on or within the host, penetration of the host, and colonization of host tissue. The pathogenicity genes that are involved in the synthesis and modification of the lipopolysaccharide cell wall of Gram-negative bacteria may help condition the host range of a bacterium.

Plant pathogenic bacterial virulence factors are associated with the bacterial surface or secreted into the surrounding environment. Proteins secreted by bacteria are transported via molecular systems out of bacterial cells; unrelated virulence factors often share the same secretion mechanism (Fuchs, 1998). Bacterial pathogenicity depends upon bacterial secretion systems (types i–iv), quorum sensing (QS), plant cell-wall-degrading enzymes, toxins, hormones, polysaccharides, proteinases, siderophores, and melanin. All of these systems and substances, which are essential for pathogenic infection and virulence, are produced by pathogens during bacterial pathogen–plant interactions (Agrios, 2005). They are all regulated by proteins through signal transduction systems.

A case study of the infection process of *Rhizobium radiobacter* (*Agrobacterium tumefaciens*), the causal agent of crown gall in many fruit and forest trees, vegetables, and herbaceous dicotyledonous plants from 90 families, showed that *R. radiobacter* is able to genetically transform healthy host cells into tumoral cells by inserting T-DNA from its tumor-inducing (Ti) plasmid into the plant genome. Binding of *R. radiobacter* to plant surfaces, an essential stage for establishing long-term interaction with host plants, is dependent upon plant factors such as bacterial polysaccharides and bacterial cellulose synthesis. In addition, infected host plants release signals that include specific phytochemicals in combination with acidic pH and temperatures.

Cultivar-specific resistance, which generally conforms to the gene-for-gene hypothesis and is genetically determined, may generate unfavorable conditions for bacterial multiplication within the apoplast. Phytopathogenic bacterial strains in which any virulence factor mutates more or less reduce their virulence, but their pathogenicity remains unchanged. Bacterial infection can, in fact, induce structural and biochemical defenses in plants; in turn these defenses can contribute to disease resistance (see Chapter 2).

1.3 EPIDEMIOLOGY AND FORECASTING SYSTEMS ON PLANT PATHOGENIC BACTERIA

Plant productivity and corresponding yields are facing greater losses each year due to pathogenic bacteria. Chapter 3 discusses the epidemiology and forecasting of plant pathogenic bacterial diseases and mathematical models for measurement (e.g., multiple regression, nonlinear models, and growth models) that have enabled forecasts to be made of various plant diseases using parametric variables. Plant pathogenic bacterial disease epidemics can cause such huge losses in crop yields that species extinction can occur. Triangulation of epidemic elements include susceptible host, pathogen, and favorable environment; each of these is vital in the occurrence of a disease and the duration of an epidemic. Epidemics can be monocyclic (typically, soil-borne) or polycyclic (typically, airborne). Polyetic epidemics can be caused by both monocyclic and polycyclic pathogens. In such cases, an initial infestation by inocula continues over several growing seasons, causing the same disease each time, because bacteria are able to travel to previously uninfected areas.

Plant disease forecasting is a management system that attempts to accurately predict the occurrence of or changes in severity of plant diseases as well as the resultant economic losses. These systems, which help growers make economic decisions about disease management (Agrios, 2004). Although some plant disease forecasts focus on avoiding initial inocula and reducing the seasonal rate of an epidemic, all forecasting systems emphasize plant diseases (Campbell and Madden, 1990; Agrios, 2004). A forecast management system must measure initial inocula or disease, favorable weather conditions for the development of secondary inocula, and the characteristics of both initial and secondary inocula.

Plant disease modelers and researchers seek to improve producers' profitability through model validations based on cost quantifications that involve false-positive and/or false-negative predictions. One example is the multiple disease and/or pest forecasting system called EPIPRE (epidemiology, prediction, prevention), developed for winter season wheat, that focuses on multiple pathogens (Reinink, 1986). The proper validation of a developed model affirms its accuracy and reliability as a plant-disease forecasting system, but false-positive predictions will undermine both of those attributes. False predictions can arise when a forecast is made for a disease at a location in which it cannot possibly develop (false positive), or when a forecast predicts that a disease will not occur yet the disease becomes established; such errors may have different economic effects for producers (Madden, 2006). Certain mathematical models could forecast diseases based on parameters such as temperature, relative humidity, and precipitation. The progress of a disease over time, which is measured by a progress curve, has been modeled and well demonstrated with various bacterial organisms that, along with biotic and abiotic factors, frequently and/or seasonally affect plants and trees.

1.4 CONTROLLING STRATEGIES TO PLANT PATHOGENIC BACTERIA

Chapter 4 begins with certain facts: that plant bacterial pathogens cause considerable loss to the productivity of major crop plants worldwide, that bacterial pathogens cannot be restricted effectively through agrochemicals, and that alternative methods can be both user-friendly and effective. These include early detection in seeds, contamination-free planting materials, and ensuring non-infectious planting materials through rapid diagnostics. Traditional detection protocols, which are based on cultural and biochemical characteristics, can still be used to confirm the exact identity of a causal pathogenic bacterium (Mondal and Shanmugam, 2013). In addition to various diagnostic methods that are grounded in symptomatology, biological assays and molecular detections can be used for accurate and effective control of bacterial diseases at all levels of cultivation (Romeiro et al., 1999; Yamada et al., 2007; Mondal, 2011; Narayanasamy, 2011; De Boer and Lopez, 2012).

Various preventive measures for controlling plant pathogenic bacteria, in research conditions as well as in fields, include quarantine, sanitation, and eradication. One of the most important preventive measures to control bacterial diseases is the identical sample inspection method. In the laboratory, an evaluation of the representative sample is made by a certification agency to determine germination percentage, moisture content, weed–seed mixing levels, admixture, purity, and the presence of seed-borne pathogens. The same procedure has been adopted for propagation materials (e.g., seeds, cuttings, rootstocks, buds, tubers, rhizomes, bulbs, and corms). Other preventive practices include producing propagation materials in disease-free or new areas, and inspecting such materials carefully before transporting them into a disease-free new area.

This chapter also contains an overview of quarantine and eradication methods to control fire blight disease caused by *Erwinia amylovora*, a very destructive bacterial pathogen that targets the Rosaceae family in hosts of apple, pear, and quince (Bastas, 2011; OEPP/EPPO, 2014). The chapter concludes with current trends, such as EPPO protocols for plant pathogen detection that combine conventional, serological, and molecular techniques in integrated approaches.

1.5 AGRO-TRADITIONAL PRACTICES OF PLANT PATHOGEN CONTROL

Chapter 5 discusses practices that have been developed and adopted over millennia to control plant diseases. These include direct mechanical methods, smoking out and flushing out spore networks, cultivation methods based on religious practices, land use in crop rotation, regular periods of planting and harvesting, sanitation procedures, tillage, and mulching methods (Galindo et al., 1983; Campbell, 1989; Djumalieva and Vassilev, 1993; Sharma, 2008).

Effective control of some crop diseases is achieved through solarization, a physical method; in addition, agrochemicals are still used worldwide. In India, organic farming is based on the traditional approaches of *vrikshayurvedha* (the *ayurveda* of plants) and *panchagavya* (mixtures of animal-produced substances to promote plant growth, immunity, etc.) (Narayanasamy, 2002; Firake et al., 2013). Although twenty-first-century agricultural research provides scientific knowledge, some traditional agricultural practices are still being effectively used; in general, such methods are very simple, feasible, low-cost, and easily adopted (Srivastava et al., 1988; Kotkar et al., 2001; Kiruba et al., 2008). They can maintain soil fertility without damaging the environment, and thereby eliminate the use of synthetic pesticides.

1.6 ANTIMICROBIAL PEPTIDES IN PLANT PATHOGENIC BACTERIA CONTROL

Chapter 6 considers antimicrobial peptides (AMPs), which are synthesized as defense molecules by all types of organisms. Peptides are short in length, contain fewer than 50 amino acids, and have broad antimicrobial spectra. AMPs have been ascertained to aid plant protection against phytopathogenic agents (Kessler, 2011) that lead to decreases in crop quality and agricultural losses. In conducive environments, plants contain systematic and structural antibiotic barriers that include chemical antimicrobial compounds such as phenolics, terpenoids, and nitrogen-containing compounds, as well as small peptides; all of these can be secreted to infected areas (Castro and Fontes, 2005; Kessler, 2011).

Several types of AMPs such as thionins, defensins, lipid-transfer proteins, cyclotides, and snakins have been isolated from diverse plant species, and their AMP structures, mechanisms of action, and activities for the regulation of plant disease resistance have been discovered. Plant protection against phytopathogens via AMPs is based on the expression of AMP-encoding genes by diverse types of organisms. Designing and constructing new synthetic peptides are also used in transgenic research, for the enhancement of plant disease resistance. Understanding the structural bases of AMPs will help to generate new synthetic versions, as well as fusion proteins, to regulate and enhance plant disease resistance against a broad spectrum of phytopathogens.

1.7 NUTRIENT SUPPLEMENTS IN PLANT PATHOGENIC BACTERIAL CONTROL

Because phytopathogens cause such enormous economic damage in agriculture, it is necessary to achieve comprehensive understanding of the factors responsible for flare-ups of plant bacterial diseases in economically important plants. Chapter 7 highlights the major and minor nutrient sources that infect and establish pathogenic bacteria in host plants. Bacterial phytopathogens exhibit specific nutrient preferences for initiating disease (Goto, 1992). Also, the types of available nutrient supplements often dictate virulence factors such as toxin production levels (Gallarato et al., 2012), exopolysaccharides (Kumar et al., 2003), and pathogen-related enzymes and proteins (Smith, 1958), as well as on the function of type III secretion systems (Tang et al., 2006). Proper nutritional supplements also enhance QS mechanisms, which are the key factors that regulates the production of virulence factors such as extracellular polysaccharides cell-wall-degrading enzymes, antibiotic production of competitive saprophytic ability, iron-chelating agents of siderophores, pigments, and the *hrp* gene expression that regulates disease development in a susceptible host (Arlat et al., 1991; Salmond, 1994; Sunish Kumar and Sakthivel, 2001; Dellagi et al., 2009; Raaijmakers and Mazzola, 2012).

Modifications of the nutrient compositions of the majority of plant pathogenic bacteria genetically target the sites of plant exudates in the rhizoplane or phylloplane. This approach aims to modify surface nutrient composition to favor recognition and regulation of a pathogen, rather than inducing its virulence. Other areas of current knowledge that can contribute to devising management strategies against plant pathogenic bacteria include dose-dependent and need-based nutrient supplements that simultaneously promote plant growth and suppress pathogen virulence.

1.8 SIDEROPHORES AGAINST PLANT PATHOGENIC BACTERIA

Chapter 8 provides an overview of the significance of siderophores in the inhibition of plant pathogenic bacteria. Siderophores, which are chelating compounds with a high affinity to iron, have mechanisms that can be used in biological control to suppress plant diseases. In general, siderophores are produced in iron-limited conditions to sequester the less-available iron from the environment (Hider and Kong, 2010); depriving a pathogen of iron ultimately leads to its inhibition (Kloepper et al., 1980a,b). Siderophores are divided into three main groups; hydroxymates, catecholates, and mixed. Hydroxymate siderophores are formed from aceylated or formylated hydroxylamines that are usually derived from lysine or ornithine (Miethke and Marahiel, 2007). The catecholate class comprises phenol catecholates, which contain a mono or dihydroxybenzoic acid group to chelate iron (Neilands, 1981; Meneely, 2007); siderophores in this group are derived from salicylate or dihydroxybenzoic acid. Mixed siderophores use multiple functional groups to chelate iron (Meneely, 2007). An example of a mixed siderophore would be the mycobactins produced by *Mycobacterium* sp., which consist of hydroxymate and phenol-catecholates and are highly lipid-soluble.

Siderophore production is influenced by fluorescent pseudomonads that involve a great variety of factors such as concentration of iron (Kloepper et al., 1980a,b), nature and concentration of carbon and nitrogen sources (Park et al., 1988), level of phosphates (Barbhaiya and Rao, 1985), pH and light (Greppin and Gouda, 1965), temperature (Weisbeek et al., 1986), degree of aeration (Lenhoff, 1963), presence of trace elements such as magnesium (Georgia and Poe, 1931), zinc (Chakrabarty and Roy, 1964), and molybdenum (Lenhoff et al., 1956). Siderophores also improve plant growth-promoting rhizobacteria and exclude other microorganisms from the ecological niche provided by the rhizosphere (Haas and Défago, 2005), Siderophores that include salicylic acid, pyochelin, and pyoverdine, which chelate iron and other metals, confer a competitive advantage to biocontrol agents for the limited supply of essential trace minerals in natural habitats (Höfte et al., 1992). The stability of the siderophore iron complex is an important factor for the molecule's efficiency.

1.9 PLANT METABOLIC SUBSTANCES IN PLANT PATHOGENIC BACTERIAL CONTROL

Most plant metabolic substances, particularly secondary metabolic substances, are inhibited by microorganisms. Chapter 9 discusses various plant metabolic compounds, including proteins, that recognize pathogen avirulence determinants and in turn trigger signal transduction cascades that lead to rapid defense mobilization (Hannond-Kosack and Parker, 2003). Secondary metabolites were generally viewed as waste products that are of little importance to plant metabolism and growth, but it has become clear that many secondary products are key components of active and potent defense mechanisms (Bennett and Wallsgrove, 1994).

Secondary metabolites comprise three chemically distinct groups (terpenes, phenolics, nitrogen, and/or sulfur-containing compounds) that defend plants against a variety of herbivores, pathogenic microorganisms, and abiotic stresses (Mazid et al., 2011); they also reduce bacterial growth (Tsuchiya et al., 1996; Da Carvalho and Da Fonseca, 2006; Burow et al., 2008; Freeman and Beattie, 2008). Some plant transformation techniques and their contributions to the creation of disease-resistant plants are discussed as well (Mourgues et al., 1998).

1.10 HOST RESISTANCE: SAR AND ISR IN PLANT PATHOGENIC BACTERIAL CONTROL

Chapter 10 considers plant–bacteria interactions that result from the activation, by infection, of various mechanisms in plants. This resistance phase in response to bacteria is associated with cell signaling, during which a set of localized responses are elicited by the pathogens themselves in and around the infected plant cells. These localized responses are the primary indicator of incompatible interactions; they include an oxidative burst followed by cell death, depending on the resistance stimulants (Lamb and Dixon, 1997; Baysal et al., 2008). Cell death traps the bacterial pathogen at sites of early infection and prevents the bacteria from spreading.

A plant's initial defense response consists of a pathogen-associated molecular pattern that triggers its immunity. Host plants have also evolved several other defense strategies to prevent bacterial invasions. Studies have shown that bacteria can be inhibited by two types of resistance: one is constitutive, and the other is briefly initiated by chemical and biological stimulants. The key players in this case are bacterial effector proteins, which are delivered through the type III secretion system and have the capability to suppress basal defenses (Alfano and Collmer, 2004).

In host plants, varietal resistance to disease is based on the recognition of effectors by the products of resistance (*r*) genes. When recognized, the effector (in this case, an avirulence or Avr protein) triggers a hypersensitive resistance (HR) reaction, which generates antimicrobial compounds that can inhibit the pathogenic properties. However, this type of gene-for-gene resistance usually fails due to the emergence of virulent strains of the pathogen that no longer trigger the hypersensitive resistance reaction. In turn, plants respond to the emergence of virulent strains by developing nonhost resistance, which is a broad-spectrum plant defense that provides sufficient immunity to all members of a plant species against all isolates of a microorganism that is pathogenic to other plant species (Nomura et al., 2005).

1.11 QUORUM SENSING IN PLANT PATHOGENIC BACTERIA

Pathogenic bacteria use small signaling molecules to determine their local population density, an activity known as quorum sensing (QS). The characteristics of certain phenotypes also coordinate bacterial activities that allow pathogens to function as multicellular systems in plant pathogenicity properties in detail is described in Chapter 11 provides a comprehensive overview of QS in bacteria, including mechanisms, autoinducers, quorum quenching (QQ) and its mechanisms, and applications of QS in biological control efforts. Plant pathogenic bacteria use QS signals to regulate genes for epiphytic fitness, such as motility in *Ralstonia solanacearum*, antibiosis in *Erwinia carotovora*, and UV-light resistance in *Xanthomonas campestris*. QS signals are also used by plant pathogenic bacteria to initiate major pathogenicity factors, including exopolysaccharide in *Pseudomonas stewartii*, *X. campestris*, and *R. solanacearum*, type III secretion systems in *Ps. stewartii* and *E. carotovora*, and exoenzyme production in *E. carotovora*, *X. campestris*, and *R. solanacearum* (von Bodman et al., 2003).

This chapter also discusses QQ as a potential method of plant pathogenic bacterial control (Dong et al., 2000; Uroz et al., 2003; Ma et al., 2013).

1.12 CYANOBACTERIA AND ALGAE IN PHYTOPATHOGENIC BACTERIA

Chapter 12 includes detailed evidence on the use of cyanobacteria and macroalgal activities to control plant pathogenic bacteria. Many cyanobacteria and macroalgae naturally exhibit such properties (Ali et al., 2008). Although this chapter sheds some light on the beneficial roles of algae and cyanobacteria in the biological control of phytopathogens, and takes into account several promising results, it also acknowledges that much work remains to be done not only on the exact chemical nature of the bioactive principle but also about the mechanisms of microbicidal/static activity on targeted phytopathogenic organisms (Arunkumar and Rengasamy, 2000a,b; Arunkumar et al., 2005; Kumar et al., 2008).

1.13 ARBUSCULAR MYCORRHIZA FUNGI IN PLANT PATHOGENIC BACTERIAL CONTROL

This chapter is focused on arbuscular mycorrhizal fungal symbiosis, which can alleviate the negative effects of plant pathogens and is the preferred choice for control of plant pathogens. Because the effect of mycorrhizae on pathogenic bacteria varies with the host plant, there are several mechanisms by which mycorrhizae control bacterial plant pathogens. In addition, the arbuscular mycorrhizal fungal association induces resistivity in host plant against various pathogens (Lingua et al., 2002; Ismail and Hijri, 2010; Kempel et al., 2010).

This chapter emphasizes mycorrhizal mechanisms that could be promoted in host plants, such as nutrition improvement, root-damage compensation, reduced competition for photosynthates, changes in the anatomy or morphology of the root system, changes in mycorrhizosphere microbial populations, and activation of plant defense mechanisms (Azcón-Aguilar and Barea, 1996). The primary limitation in exploiting the beneficial effects of arbuscular mycorrhizal fungi in the control of bacterial diseases of plants is the obligate nature of the symbiont. To date, studies on the control of plant pathogenic bacteria by mycorrhizae have only considered fungi and nematodes. Much more work is needed on the mechanisms involved in the mycorrhizae fungal-mediated control of bacterial plant diseases.

1.14 PLANT GROWTH-PROMOTING RHIZOBACTERIA ON PLANT PATHOGENIC BACTERIAL CONTROL

Chapter 14 discusses the role of plant growth in promoting rhizobacteria to improve plant growth and development through mechanisms such as nutrition improvement, protection from pathogens, induction of plant host defense mechanisms, and others (Kloepper and Schroth, 1978). Microorganisms that are responsible for the activities of the genera *Azotobacter*, *Azospirillum*, *Bacillus*, *Burkholderia*, *Gluconoacetobacter*, *Pseudomonas*, and *Paenibacillus* include some of the Enterobacteriaceae (e.g., yeasts and actinobacteria).

Plant growth-promoting rhizobacteria are exhibited in several mechanisms of biocontrol, most of which involve competition and the production of secondary metabolites that affect the pathogen directly; examples of the latter include antibiotics, cell-wall-degrading enzymes, siderophores, and hydrogen cyanide (Kloepper, 1983; Weller, 1988; Enebak et al., 1998). This chapter presents the broad spectrum of plant-growth-promoting rhizobacteria, including antibiotics, siderophores, bacteriocin, and induction of ISR. All of these have been reported to be effective, economical, and practical ways to protect plants from bacterial diseases. Globally, many findings have also established that the synergistic activities of numerous mixtures of plant-growth-promoting rhizobacterial isolates have induced strong plant protection against many diseases (Stockwell et al., 2011).

1.15 BACTERIOPHAGES IN PLANT PATHOGENIC BACTERIAL CONTROL

Chapter 15, which deals with plant pathogenic bacterial control by bacteriophage, considers existing plant disease control methods, phages as an approach to disease control, and the early development of this approach and its modern usage. Bacterial resistance to antibiotics complicates this disease control strategy, however. Some research has tested the use of lytic phages for the control of phytopathogens.*

Conversely, the therapeutic use of phages was started in the year of 1896 by Ernest Hankin, which was against the *Vibrio cholera* for the causative agent of cholera threats to humans. In 1915, Fredric Twort hypothesized the virus (phage), but the bacteriophages were discovered by Fe'lix d'He'relle in 1917 (Hermoso et al., 2007). Though, the phage therapy has been found to be an effective tool for the control of several phytopathogenic bacteria such as: *Xanthomonas* spp. (Ceverolo, 1973), *Pseudomonas* spp. (Kim et al., 2011), *Erwinia* spp. (Nagy et al., 2012), *Pantoea* spp. (Thomas, 1935), *Ralsotnia* spp. (Fujiwara et al., 2011), *Streptomyces* spp. (Goyer, 2005), *Dickeya* spp. (Adriaenssens et al., 2012), and *Pectobacterium* spp. (Lim et al., 2013). In conclusion, the success of phage therapy in controlling plant bacterial disease is pathogen-specific. Therefore, a complete understanding of phage biology is necessary if effective therapy practices are to be developed.

1.16 PLANT DEFENSE ENZYMES IN PLANT PATHOGENIC BACTERIAL CONTROL

Chapter 16 provides information on various defense enzymes in the control of plant pathogenic bacteria, particularly the role of defense enzymes induced by pathogenic/nonpathogenic bacteria and the hypothetical defense mechanisms by which plants directly counteract phytopathogenic bacteria. In general, plant pathogens reveal themselves to a host's immune system through molecules called pathogen-associated/microbe-associated molecular patterns (PAMP or MAMP), such as flagellins or bacterial lipopolysaccharides. Chitinases, peroxidases, and polyphenol oxidases are other

* This chapter authors focused to usage of phages for the control of plant pathogens, the existing plant disease control methods, the usage of phages in disease control, the early development and its modern usage in agriculture for controlling the plant pathogens and the complete analysis about disease control strategies of phages in plant disease control.

defense enzymes that are induced in plants during pathogenesis and participate in plant resistance to phytopathogens (Chen et al., 2000).

Abundant proteins are found in plants; their physiological function is strong antimicrobial activity, but they also display lysozymal activity (Majeau et al., 1990; Mauch-Mani and Métraux, 1998) and thus may be involved in conferring resistance to plants against bacterial pathogens. The peroxidases, a group of heme-containing glycosylated proteins that are activated in response to pathogen attacks, are attributed to plant peroxidases in host–pathogen interactions (Chittoor et al., 1999; Do et al., 2003; Saikia et al., 2006), specifically in the oxidation of phenols (Schmid and Feucht, 1980) and plant protection (Hammerschmidt et al., 1982). Polyphenol oxidases, which are nuclear-encoded enzymes of almost ubiquitous distribution in plants (Mayer and Harel, 1979; Mayer, 1987), use molecular oxygen to oxidize common ortho-diphenolic compounds (e.g., caffeic acid and catechol) to their respective quinones. The reactive polyphenol oxidase-generated quinines have been suggested to participate in plant defenses against pests and pathogens. Although the study of plant defense mechanisms against pathogens has made remarkable progress, further efforts are necessary to understand the mechanisms that are responsible for the induction of defence enzymes and the restriction of bacterial growth during resistance.

1.17 PLANT PATHOGENIC BACTERIAL CONTROL THROUGH SEED APPLICATION

Chapter 17 covers plant pathogenic bacterial control strategies through seed treatment. Seeds are seldom treated because the risk of damage to them exceeds the expected benefits. Moreover, pesticide registrations are difficult to obtain for extremely limited applications (Taylor and Harman, 1990). The fact that some seeds (e.g., hybrid flower seeds) are very expensive and produced in small quantities is also important. Nonetheless, seed treatments have been used for centuries and for decades have been used widely on a commercial basis. Hypothetically, seed treatments may be profitable for seeds in one crop but may not be economically reasonable for the seeds of another crop.

Cleaning seeds before using them can help to reduce pathogen inoculum because seeds can retain soil fragments, plant debris, and pathogen particles after they have been stored or sieved under controlled conditions. Various seed-treatment practices include physical techniques (i.e., hot-water treatments, dry-heat treatments, vapor-heat treatments, and radiation treatments), fermentation processes, and chemical techniques. Several reports have confirmed potentially valuable microorganisms for biological control, including organic amendments, but the developmental process to bring these into commercial use is long and arduous (Cook, 1993). To date, biological control is considered much more variable and less effective than pesticide use. Moreover, biological agents may not be able to survive or compete with other living organisms (Harman, 1991).

1.18 PLANT PATHOGENIC BACTERIAL CONTROL THROUGH FOLIAR APPLICATIONS

Chapter 18 focuses on plant pathogenic bacterial control through effective foliar applications. Although the biological control or biocontrol of pathogens by introduced microorganisms has been studied since the 1940s, it is not yet commercially feasible. Plant surfaces provide habitat for epi- and saprophytic pathogens that cause foliar disease. According to the relationship between biocontrol agents and pathogens or other microorganisms, mechanisms such as substrate competition and niche exclusion, siderophores, antibiotics, and induced resistance (IR) may be initiated. These mechanisms are dependent on the microclimatic conditions at the plant surface as well as on the chemical environment.

Synergistic phenomena have been observed in integrated control approaches that utilize fungicides; in addition, biocontrol agents may be more efficient and longer-lasting than biocontrol agents or fungicides alone (Shoda, 2000; Paulitz and Belanger, 2001). Future research will allow essential traits to be altered to make new strains of biocontrol agents more effective.

1.19 MODERN TRENDS IN PLANT PATHOGENIC BACTERIAL CONTROL

Chapter 19, which discusses plant pathogenic bacterial control through modern practices, also discusses conventional control methods such as avoidance, exclusion, eradication, protection, resistance, and therapeutic applications. Modern agricultural scenarios propose sustainable agriculture and related scientific and technological tools (Huang and Wu, 2009). Because sustainable agriculture results may resolve environmental and ecological issues that continue to challenge agriculture, modern technologies developed for crop production must be economically feasible, ecologically sound, environmentally safe, and socially acceptable. New disease-management strategies could include one or more of the following: PGPR (Chernin, 2011), mycorrhizae (Bastas et al., 2006), essential oil and extracts from aromatic plants (McManus et al., 2002), and bacteriophages (Ravensdale et al., 2007) are capable of producing volatile compounds and quorum quenching or QQ (Dong et al., 2007).

Plant pathogenic bacteria are also controlled through chemicals (bactericides). These include bronopol, copper hydroxide, copper sulfate, copper oxychloride, cresol, dichlorophen, dipyrithione, dodicin, fenaminosulf, formaldehyde, 8-hydroxyquinoline sulfate, kasugamycin, nitrapyrin, octhilinone, oxolinic acid, oxytetracycline, probenazole, streptomycin, tecloftalam, and thiomersal. These chemicals, which are regularly and repeatedly used on plants to manage disease, cause serious damage to agricultural and natural ecosystems (Gottlieb et al., 2002). The use of genetic engineering to manage plant diseases targets two major immunity systems used by plants to defend themselves against pathogens; these include pathogen-associated molecular patterns and pattern recognition receptors, both of which lead to the activation of defense responses in plants (De Wit, 1997). Integrated strategies for sustainable management of crop diseases include the use of natural toxic substances from plants, soil management techniques such as biofumigation and organic soil amendment, and crop management, as well as the use of pathogen-free seeds or other planting materials, disease-resistant cultivars, biological control agents, and cultural practices such as field sanitation and crop rotation. The development of less toxic and more stable compounds by improving preparative syntheses and biotechnological applications will be crucial for decreasing production costs and thereby encouraging wider use.

1.20 SCIENTIFIC AND ECONOMIC IMPACT OF PLANT PATHOGENIC BACTERIA

Chapter 20 considers plant pathogenic bacterial impact in terms of host plant morphology, growth, and development (Jones and Dangl, 2006), as well as yield impacts both economically (Stefani, 2010) and nutritionally. Plant pathogenic bacteria are opportunistic to organisms worldwide. One example, *Pantoea* spp., is a plant pathogen but is also pathogenic and even deadly to humans, who encounter it on contaminated closures of infusion fluid bottles (Maki et al., 1976), in intravenous fluid, parenteral nutrition or PN (intravenous feeding), blood products, propofol, and transference tubes (Bicudo et al., 2007; Kirzinger et al., 2011). Various wild and cultivated plant species have been reported to host opportunistic human pathogens in their rhizospheres (Berg et al., 2005). In contrast to these methods of invasion of animal hosts, enteric bacteria appear to reside mostly in the apoplastic spaces of plant hosts (Holden et al., 2009).

A wide range of plant pathogenic bacteria secrete proteins and other molecules to different cellular compartments of their hosts. Sometimes, these secretions are introduced via the production of a wide spectrum of nonhost-specific phytotoxins to nontarget hosts that offer necessary resources (Buonaurio, 2008). Scientific understanding of the mechanisms of bacterial toxic molecules and the actions of phytotoxins has increased enormously (Schiavo and van der Goot, 2001). However, greater understanding and control of plant bacterial disease is of urgent importance for the simple reason that phytopathogen-infested plants produce toxic molecules which, if spread to herbivores and carnivores through the food chain, can collapse the performance of entire ecosystems (Szabòo et al., 2002).

1.21 CONCLUSION AND FUTURE PERSPECTIVES

Plant pathogenic bacteria produce diseases most frequently and severely in tropical and subtropical places, where warm and humid conditions are ideal for bacterial growth. All countries record annual and consistent crop losses from such diseases. Such problems are particularly severe in developing countries, where they are aggravated by the paucity of resources devoted to pathological studies. Contributing factors to this literature gap include poor ability to quantify plant diseases, from those that occur during cultivation to those that manifest in storage, and relating such information to the failure of crops to reach manageable yields. Another important fact is that in day-to-day life, humans and animals spread bacterial disease by cultivating, consuming, and excreting low-quality plant-based foods. This overview chapter has provided an overview of plant pathogenic bacterial diseases. We hope that the discussions and reviews that comprise the rest of this book, which cover a wide range of aspects of plant pathogenic bacterial pathogenicity, epidemiology, and impact on the food chain as well as strategies for control, are helpful to students, plant pathogenic bacterial researchers, and policy makers.

ACKNOWLEDGMENT

The authors are grateful to all of the contributors to this book.

REFERENCES

Abedon, S.T., Abedon, C.T., Thomas, A., Mazure. H. 2011. Bacteriophage prehistory. Is or is not Hankin, 1896, a phage reference? *Bacteriophage* 1(3): 174–178.

Adriaenssens, E.M., Vaerenbergh, J.V., Vandenheuvel, D., Dunon, V., Ceyssens, P.J., Proft, M.D., Kropinski, A.M. et al. 2012. T4-related bacteriophage LIMEstone isolates for the control of soft rot on potato caused by '*Dickeya solani*'. *PLoS One* 7(3): 1–10.

Agrios, G.N. 2004. *Plant Pathology*, 5th ed. San Diego: Academic Press.

Agrios, G.N. 2005. *Plant Pathology*, 5th ed. Amsterdam, The Netherlands: Elsevier/Academic, 948p.

Alfano, J.R., Collmer A. 2004. Type III secretion system effector proteins: Double agents in bacterial disease and plant defense. *Annual Review of Phytopathology* 42: 385–414.

Ali, D.M., Kumar, T.V., Thajuddin, N. 2008. Screening of some selected hypersaline cyanobacterial isolates for biochemical and antibacterial activity. *Indian Hydrobiology* 11: 241–246.

Arlat, M., Gough, C., Barber, C.E., Boucher, C., Daniels, M. 1991. *Xanthomonas campestris* contains a cluster of *hrp* genes related to the larger *hrp* cluster of *Pseudomonas solanacearum*. *Molecular Plant-Microbe Interactions* 4: 593–601.

Arunkumar, K., Rengasamy, R. 2000a. Antibacterial activities of seaweed extracts/fractions obtained through a TLC profile against phytopathogenic bacterium *Xanthomonas oryzae* sp *oryzae*. *Botanica Marina* 43: 417–421.

Arunkumar, K., Rengasamy, R. 2000b. Evaluation of antibacterial potential of seaweeds occurring along the coast of Tamilnadu, India against the plant pathogenic bacterium *Xanthomonas oryzae* pv. *oryzae* (Ishiyama) dye. *Botanica Marina* 43: 409–415.

Arunkumar, K., Selvapalam, N., Rengasamy, R. 2005. The antibacterial compound sulphoglycerolipid 1–0 palmitoyl-3–0 (6'-sulpho-α-quinovopyranosyl)-glycerol from *Sargassum wightii*. *Botanica Marina* 48: 441–445.

Azcón-Aguilar, C., Barea, J.M. 1996. Arbuscular mycorrhizas and biological control of soil-borne plant pathogens—an overview of the mechanisms involved. *Mycorrhiza* 6: 457–464.

Barbhaiya, H.B., Rao, K.K. 1985. Production of pyoverdine, the fluorescent pigment of *Pseudomonas aeruginosa* PAO1. *FEMS Microbiology Letters* 27: 233–235.

Bastas, K.K. 2011. An Integrated Management Program for Fire Blight Disease on Apples. ASHS Annual Conference, *HortScience* 46(9): S129.

Bastas, K.K., Akay, A., Maden, S. 2006. A new approach to fire blight control: Mycorrhiza. *HortScience* 41(5): 1309–1312.

Baysal, O., Calıskan, M., Yesilova, O. 2008. An inhibitory effect of a new *Bacillus subtilis* strain (EU07) against *Fusarium oxysporum* f. sp. *radicis lycopersici*. *Physiological and Molecular Plant Pathology* 73: 25–32.

Bennett, R.N., Wallsgrove, R.M. 1994. Secondary metabolites in plant defence mechanisms. *New Phytologist* 127: 617–633.

Berg, G., Eberl, L., Hartmann, A. 2005. The rhizosphere as a reservoir for opportunistic human pathogenic bacteria. *Environmental Microbiology* 7: 1673–1685.

Bicudo, E.L., Macedo, V.O., Carrara, M.A., Castro, F.F., Rage, R.I. 2007. Nosocomial outbreak of *Pantoea agglomerans* in a pediatric urgent care center. *Brazilian Journal of Infectious Diseases* 2007 11: 281–284.

Buonaurio, R. 2008. Infection and plant defense responses during plant–bacterial interaction. *Plant-Microbe Interactions* 169–197.

Burow, M., Wittstock, U., Jonathan, G. 2008. Sulfur-containing secondary metabolites and their role in plant defense. *Advances in Photosynthesis and Respiration* 27: 201–222.

Campbell, R. 1989. *Biological Control of Microbial Plant Pathogens*. Cambridge: Cambridge University Press, p. 41.

Campbell, C.L., Madden, L.V. 1990. *Introduction to Plant Disease Epidemiology*. New York: John Wiley and Sons.

Castro, M.S., Fontes, W. 2005. Plant defense and antimicrobial peptides. *Protein and Peptide Letters* 12(1): 11–16.

Ceverolo, E.L. 1973. Characterization of *Xanthomonas pruni* bacteriophages to bacterial spot disease in prunus. *Phytopathology* 63: 1279–1284.

Chakrabarty, A.M., Roy, S.C. 1964. Effects of trace elements on the production of pigments by a pseudomonad. *Biochemical Journal* 93: 228–231.

Chen, C., Belanger, R.R., Benhamou, N., Paulitz, T. 2000. Defense enzymes induced in cucumber roots by treatment with plant-growth promoting rhizobacteria (PGPR) and *Pythium aphanidermatum*. *Physiological and Molecular Plant Pathology* 56: 13–23.

Chernin, L. 2011. Quorum-sensing signals as mediators of PGPRs' beneficial traits. In *Bacteria in Agrobiology: Plant Nutrient Management*, Vol. 3. Maheshwari, D.K. (Ed.). Berlin, Germany: Springer-Verlag, pp. 209–236.

Chisholm, S.T., Coaker, G., Day, B., Staskawicz, B.J. 2006. Host-microbe interactions: Shaping the evolution of the plant immune response. *Cell* 124: 803 – 814.

Chittoor, J.M., Leach, J.E., White, E.F. 1999. Differential induction of a peroxidise during defense against pathogens. In *Molecular Biology Intelligence Unit: Pathogeneis-Related Proteins in Plants*. Datta, S.K., S.M. Muthukrishnan, S.M. (Eds.). New York: CRC Press, pp. 171–193.

Cook, R.J. 1993. Making greater use of introduced microorganisms for biological control of plant pathogens. *Annual Review of Phytopathology* 31: 53–80.

Da Carvalho, C.C.C.R., Da Fonseca, M.R.M. 2006. Biotransformation of terpenes. *Biotechnology Advances* 24(2): 134–142.

Davis, E.L., Hussey, R.S., Mitchum, M.G., Baum, T.J. 2008. Parasitism proteins in nematode–plant interactions. *Current Opinion in Plant Biology* 11: 360–366.

De Boer, S.H., Lopez, M.M. 2012. New grower-friendly methods for plant pathogen monitoring. *Annual Review of Phytopathology* 50: 197–218.

De Wit, P.J.G.M. 1997. Pathogen avirulence and plant resistance: A key role for recognition. *Trends in Plant Science* 2: 452–458.

Dellagi, A., Segond, D., Rigault, M., Fagard, M., Simon, C., Saindrenan, P., Expert, D. 2009. Microbial siderophores exert a subtle role in arabidopsis during infection by manipulating the immune response and the iron status. *Plant Physiology* 150: 1687–1696.

Djumalieva D., Vassilev, A. 1993. *Cropping Systems in Intensive Agriculture*. New Delhi: MD Publications.

Do, H.M., Hong, J.K., Jung, H.W. 2003. Expression of peroxidise-like genes, H_2O_2 production and peroxidise activity during the hypersensitive responses to *Xanthomonas campestris* pv. *vesicatoria* in *Capsicum annum*. *Molecular Plant–Microbe Interaction* 16: 196–205.

Dong, Y.H., Wang, L.H., Zhang, L.H. 2007. Quorum-quenching microbial infections: Mechanisms and implications. *Philosophical Transactions of the Royal Society B* 362: 1201–1211.

Dong, Y.H., Xu, J.L., Li, X.Z., Zhang, L.H. 2000. AiiA, an enzyme that inactivates the acylhomoserine lactone quorum-sensing signal and attenuates the virulence of *Erwinia carotovora*. *Proceedings of the National Academy of Sciences USA* 97(7): 3526–3531.

Enebak, S.A., Wei, G., Kloepper, J.W. 1998. Effects of plant growth-promoting rhizobacteria on loblolly and slash pine seedlings. *Forest Science* 44: 139–144.

Firake, D M., Lytan, D., Thubru, D.P., Behere, G.T., Firale, P.D., Thakur, N.S.A. 2013. Traditional pest management practices and beliefs of different ethnic tribes of Meghalaya, North Eastern Himalaya. *Indian Journal of Hill Farming* 26(1): 58–61.

Freeman, B.C., Beattie, G.A. 2008. An overview of plant defenses against pathogens and herbivores. *The Plant Health Instructor*. DOI: 10.1094/PHI-I-2008-0226-01.

Fuchs, T.M. 1998. Molecular mechanisms of bacterial pathogenicity. *Naturwissenschaften* 85: 99–108.

Fujiwara, A., Fujisawa, M., Hamasaki, R., Kawasaki, T., Fujie, M., Yamada, T. 2011. Biocontrol of *Ralstonia solanacearum* by treatment with lytic bacteriophages. *Applied and Environmental Microbiology* 77(12): 4155–4162.

Galindo, J.J., Abawi, G.S.,Thurston, H.D., Galvez, G. 1983. Source of inoculam and development of bean web blight in Costa Rica. *Plant Diseases* 67: 1016–1021.

Gallarato, L.A., Primo, E.D., Lisa, A.T., Garrido, M.N. 2012. Choline promotes growth and tabtoxin production in a *Pseudomonas syringae* strain. *Advances in Microbiology* 2: 327–331.

Georgia, F.R., Poe, C.P. 1931. Study of bacterial fluorescence in various media. Inorganic substances necessary for bacterial fluorescence. *Journal of Bacteriology* 22: 349–361.

Goto, M. 1992. *Fundamentals of Bacterial Plant Pathology*. San Diego, CA: Academic Press, p. 342.

Gottlieb, O.R., Borin, M.R., Brito, N.R. 2002. Integration of ethnobotany and phytochemistry: Dream or reality? *Phytochemistry* 60: 145–152.

Goyer, C. 2005. Isolation and characterization of phages Stsc1 and Stsc3 infecting *Sterptomyces scabiei* and their potential as biocontrol agents. *Canadian Journal of Plant Pathology* 27: 210–216.

Greppin, H., Gouda, S. 1965. Action de la lumiere sur le pigment de *Pseudomonas fluorescens* Migula. *Archives Des Sciences (Geneve)* 18: 721–725.

Haas, D., Défago, G. 2005. Biological control of soil-borne pathogens by fluorescent pseudomonads. *Nature Reviews Microbiology* 3: 307–319.

Hammerschmidt, R., Nucckles, R., Kuc, J. 1982. Association of enhance peroxidase activity with induced systemic resistance of cucumber to *Colletotrichum largenarium. Physiological Plant Pathology* 20: 73–82.

Hannond-Kosack, K.E., Parker, J.E. 2003. Deciphering plant-pathogen communication: Fresh perspectives for molecular resistance breeding. *Current Opinion in Biotechnology* 14(2): 177–193.

Harman, G.E. 1991. Seed treatments for biological control of plant disease. *Crop Protection* 10: 166–171.

Hermoso, J.A., Garcia, J.L., Garcia, P. 2007. Taking aim on bacterial pathogens: From phage therapy to enzybiotics. *Current Opinion in Microbiology* 10(5): 461–472.

Hider, R.C., Kong, X. 2010. Chemistry and biology of siderophores. *Natural Product Reports* 27: 637–657.

Höfte, M., Boelens, J., Verstraete, W. 1992. Survival and root colonization of mutants of plant growth-promoting pseudomonads affected in siderophore biosynthesis or regulation of siderophore production. *Journal of Plant Nutrition* 15: 2253–2262.

Holden, N., Pritchard, L., Toth, I. 2009. Colonization out with the colon: Plants as an alternative environmental reservoir for human pathogenic enterobacteria. *FEMS Microbiology Reviews* 33: 689–703.

Huang, H.C., Wu, M.T. 2009. Plant disease management in the era of energy conservation. *Plant Pathology Bulletin* 18: 1–12.

Ismail, Y., Hijri, M. 2010. Induced resistance in plants and the role of arbuscular mycorrhizal fungi. In *Mycorrhizal Biotechnology*. Thangadurai, D., Busso, C.A., Hijri. M. (Eds). Enfield, NH: Science, pp. 77–99.

Jones, J.D., Dangl, J.L. 2006. The plant immune system. *Nature*. 444: 323–329.

Kempel, A., Schmidt, A.K., Brandl, R., Schädler, M. 2010. Support from the underground: Induced plant resistance depends on arbuscular mycorrhizal fungi. *Functional Ecology* 24: 293–300.

Kessler, A. 2011. Plant defense: Warding off attack by pathogens, herbivores, and parasitic plants, by Dale R. Walters. *The Quarterly Review of Biology* 86(4): 356–357.

Kim, M.H., Park, S.W., Kim, Y.K. 2011. Bacteriophages of *Pseudomonas tolaassi* for the biological control of brown blotch disease. *Journal of the Korean Society for Applied Biological Chemistry* 54(1): 99–104.

Kiruba, S., Jeeva, S., Kanagappan, M., Stalin, I.S., Das, S.S.M. 2008. Ethnic storage strategies adopted by farmers of Tirunelveli district of Tamil Nadu, Southern Peninsular India. *Journal of Agricultural Technology* 4(1): 1–10.

Kirzinger, M.W.B., Nadarasah, G., Stavrinides, J. 2011. Insights into cross-kingdom plant pathogenic bacteria. *Genes* 2: 980–997.

Kloepper, J.W. 1983. Effect of seed piece inoculation with plant growth-promoting rhizobacteria on populations of *Erwinia carotovora* on potato roots and daughter tubers. *Phytopathology* 73: 217–219.

Kloepper, J.W., Leong, J., Teintze, M., Schroth, M.N. 1980a. Enhancing plant growth by siderophores produced by plant growth-promoting rhizobacteria. *Nature* 286: 885–886.
Kloepper, J.W., Leong, J., Teintze, M., Schroth, M.N. 1980b. *Pseudomonas* siderophores: A mechanism explaining disease suppressive soils. *Current Microbiology* 4: 317–320.
Kloepper, J.W., Schroth, M.N. 1978. Plant growth promoting rhizobacteria on radishes. In *Proceedings of the Fourth International Conference on Plant Pathogenic Bacteria*. Angers, France, pp. 879–882.
Kotkar, H.M., Mendki, P.S., Sadan S.V.G.S., Jha, S.R., Upasani, S.M., Maheshwari, V.L. 2001. Antimicrobial and pesticidal activity of partially purified flavonoids of *Annona squamosa*. *Pest Management Science* 58(1): 33–37.
Kumar, A., Sunish Kumar, R., Sakthivel, N. 2003. Compositional difference of the exopolysaccharides produced by the virulent and virulence-deficient strains of *Xanthomonas oryzae* pv. *oryzae*. *Current Microbiology* 46: 251–255.
Kumar, C.S., Sarada, D.V.L., Rengasamy, R. 2008. Seaweed extracts control the leaf spot disease of the medicinal plant *Gymnema sylvestre*. *Indian Journal of Science and Technology* 1: 1–5.
Lamb, C., Dixon, R.A. 1997. The oxidative burst in plant disease resistance. *Annual Review of Plant Physiology and Plant Molecular Biology* 48: 251–275.
Lenhoff, H.M. 1963. An inverse relationship of the effects of oxygen and iron on the production of fluorescin and cytochrome C by *Pseudomonas fluorescens*. *Nature* 199: 601–602.
Lenhoff, H.M., Nicholas, D.J.D., Kaplan, N.O. 1956. Effects of oxygen, iron and molybdenum on routes of electron transfer in *Pseudomonas fluorescens*. *Journal of Biological Chemistry* 220: 983–995.
Lim, J.A., Jee, S., Lee, D.H., Roh, E., Jung, K., Oh, C., Heu, S. 2013. Biocontrol of *Pectobacterium carotovorum* subsp. *carotovorum* using bacteriophage PP1. *Journal of Microbiology and Biotechnology* 23(8): 1147–1153.
Lingua, G., D'Agostino, G., Massa, N., Antosiano, M., Berta, G. 2002. Mycorrhiza-induced differential response to a yellows disease in tomato. *Mycorrhiza* 12: 191–198.
Ma, L., Liu, X., Liang, H., Che, Y., Chen, C., Dai, H. 2013. Effects of 14-alpha lipoylandrographolide on quorum sensing in *Pseudomonas aeruginosa*. *Antimicrobial Agents and Chemotherapy* 56: 6088–6094.
Madden, L.V. 2006. Botanical epidemiology: Some key advances and its continuing role in disease management. *European Journal of Plant Pathology* 115: 3–23.
Majeau, N., Trudel, J., Asselin, A. 1990. Diversity of cucumber chitinase isoforms and characterization of one seed basic chitinase with lysozyme activity. *Plant Science* 68: 9–16.
Maki, D.G., Rhame, F.S., Mackel, D.C., Bennett, J.V. 1976. Nationwide epidemic of septicemia caused by contaminated intravenous products. I. Epidemiologic and clinical features. *American Journal of Medicine* 60: 471–485.
Mauch-Mani, B., Métraux, J.-P. 1998. Salicylic acid and systemic acquired resistance to pathogen attack. *Annals of Botany* 82: 535–540.
Mayer, A.M. 1987. Polyphenol oxidases in plants—Recent progress. *Phytochemistry* 26: 11–20.
Mayer, A.M., Harel, E. 1979. Polyphenol oxidase in plant. *Phytochemistry* 18: 193–215.
Mazid, M., Khan, T.A., Mohammad, F. 2011. Role of secondary metabolites in defense mechanisms of plants. *Biology and Medicine* 3(2): 232–249.
McManus, P.S., Stockwell, V.O., Sundin, G.W., Jones, A.L. 2002. Antibiotic use in plant agriculture. *Annual Review of Phytopathology* 40: 443–465.
Meneely, K.M. 2007. The biochemistry of siderophore biosynthesis. PhD Thesis. University of Kansas, USA.
Miethke, M., Marahiel, M.A. 2007. Siderophore-based iron acquisition and pathogen control. *Microbiology and Molecular Biology Review* 71: 413–445.
Mondal, K.K. 2011. *Plant Bacteriology*, New Delhi: Kalyani Publishers, ISBN: 978-93-272-1631-8, p. 190.
Mondal, K.K., Shanmugam, V. 2013. Advancements in the diagnosis of bacterial plant pathogens: An overview. *Biotechnology and Molecular Biology Review* 8(1): 1–11.
Mourgues, F., Brisset, M., Chevreau, E. 1998. Strategies to improve plant resistance to bacterial diseases through genetic engineering. *Trends in Biotechnology* 16(5): 203–210.
Nagy, J.K., Kiraly, L., Schwariczinger, I. 2012. Phage therapy for plant disease control with a focus on fire blight. *Central European Journal of Biology* 7(1): 1–12.
Narayanasamy, P. 2002. Traditional pest control: A retrospection. *Indian Journal of Traditional Knowledge* 1(1): 40–50.
Narayanasamy, P. 2011. *Microbial Plant Pathogens-Detection and Disease Diagnosis: Bacterial and Phytoplasmal Pathogens*, Vol. 2. Springer, the Netherlands, pp. 5–169.
Neilands, J.B. 1981. Microbial iron compounds. *Annual Review of Biochemistry* 50: 715–731.
Nomura, K., Melotto, M., He, S.Y. 2005. Suppression of host defense in compatible plant *Pseudomonas syringae* interactions. *Current Opinion in Plant Biology* 8: 361–368.

OEPP/EPPO. 2014. EPPO A1 and A2 lists of quarantine pests (version 2014-09).
Park, C.S., Paulitz, T.C., Baker, R. 1988. Biocontrol of *Fusarium* wilt of cucumber resulting from interactions between *Pseudomonas putida* and non-pathogenic isolates of *Fusarium oxysporum*. *Phytopathology* 78: 190–194.
Paulitz, T.C. Belanger, R.R. 2001. Biological control in greenhouse systems. *Annual Review of Phytopathology* 39: 103–133.
Raaijmakers, J.S., Mazzola, M. 2012. Diversity and natural functions of antibiotics produced by beneficial and plant pathogenic bacteria. *Annual Review of Phytopathology* 50: 403–424.
Ravensdale, M., Blom, T.J., Gracia-Garza, J.A., Svircev, A.M., Smith, R.J. 2007. Bacteriophages and the control of *Erwinia carotovora* subsp. *carotovora*. *Canadian Journal of Plant Pathology* 29: 121–130.
Reinink, K. 1986. Experimental verification and development of EPIPRE, a supervised disease and pest management system for wheat. *The Netherland Journal of Plant Pathology* 92(1): 3–14.
Romeiro, R.S., Silva, H.S.A., Beriam, L.O.S., Rodrigues Neto, J., de Carvalho, M.G. 1999. A bioassay for the detection of *Agrobacterium tumefaciens* in soil and plant material. *Summa Phytopathologica* 25: 359–362.
Saikia, R., Kumar, R., Arora, D.K. 2006. *Pseudomonas aeruginosa* inducing rice resistance against *Rhizoctonia solani*: Production of salicylic acid and peroxidases. *Folia Microbiologica* 51: 375–380.
Salmond, G.P.C. 1994. Secretion of extracellular virulence factors by plant pathogenic bacteria. *Annual Review of Phytopathology* 32: 181–200.
Schiavo, G., van der Goot, E.G. 2001. The bacterial toxin toolkit. *Nature Reviews Molecular Cell Biology* 2: 530–537.
Schmid, P.S., Feucht, W. 1980. Tissues-specific oxidation browning of polyphenols by peroxidase in cherry shoots. *Gartenbauwisenschaft* 45: 68–73.
Sharma, R.N. 2008. *Origin and Development of Agriculture Science*. Delhi: Vista International Publishing House, pp. 1–19.
Shoda, M. 2000. Bacterial control of plant diseases. *Journal of Bioscience and Bioengineering* 89: 515–521.
Smith, W.K. 1958. A survey of the production of pectic enzymes by plant pathogenic and other bacteria. *Microbiology* 18(1): 33–41.
Srivastava, S., Gupta, K.C., Agrawal, A. 1988. Effect of plant product on *Callosobruchus chinensis* L. infection on red gram. *Seed Research* 16(1): 98–101.
Stefani, E. 2010. Economic significance and control of bacterial spot/canker of stone fruits caused by *Xanthomonas arboricola* pv. *Pruni*. *Journal of Plant Pathology* 92: 99–103.
Stockwell, V.O., Johnson, K. B., Sugar, D., Loper, J.E. 2011. Mechanistically compatible mixtures of bacterial antagonists improve biological control of fire blight of pear. *Biological Control* 101(1): 113–123.
Sunish Kumar, R., Sakthivel, N. 2001. Exopolysaccharides of *Xanthomonas* pathovar strains that infect rice and wheat crops. *Applied Microbiology and Biotechnology* 55: 782–786.
Szabò Z., Gróf, P., Schagina, L.V., Gurnev, P.A., Takemoto, J.Y., Mátyus, E., K. Blaskò, K. 2002. Syringotoxin pore formation and inactivation in human red blood cell and model bilayer lipid membranes. *Biochimica et Biophysica Acta* 1567: 143–149.
Tang, X., Xiao, Y., Zhou, J. 2006. Regulation of the type III secretion system in phytopathogenic bacteria. *Molecular Plant Microbe Interactions* 19(11): 1159–1166.
Taylor, A.G., Harman, G.E. 1990. Concepts and technologies of selected seed treatments. *Annual Review of Phytopathology* 28: 321–339.
Thomas, R.C. 1935. A bacteriophage in relation to Stewart's disease of corn. *Phytopathology* 25: 371–372.
Tsuchiya, H., Sato, M., Miyazaki, T., Fujiwara, S., Tanigaki, S., Ohyama, M., Tanaka, T., Iinuma, M. 1996. Comparative study on the antibacterial activity of phytochemical flavanones against methicillin-resistant *Staphylococcus aureus*. *Journal of Ethnopharmacology* 50: 27–34.
Uroz, S., Dangelo, C., Carlier, A., Faure, D., Petit, A., Oger, P., Sicot, C., Dessaux, Y. 2003. Novel bacteria degrading *N*-acyl homoserine lactones and their use as quenchers of quorum-sensing regulated functions of plant pathogenic bacteria. *Microbiology* 149: 1981–1989.
von Bodman, S.B., Dietz Bauer, W., Coplin, D.L. 2003. Quorum sensing in plant-pathogenic bacteria. *Annual Review of Phytopathology* 41: 455–482.
Weisbeek, P.J., Van der Hofstad, G.A.J.M., Schippers, B., Marugg, J.D. 1986. Genetic analysis of the iron uptake system of two plant growth promoting *Pseudomonas* strains. *NATO ASI Series A* 117: 299–313.
Weller, D.M. 1988. Biological control of soilborne plant pathogens in the rhizosphere with bacteria. *Annual Review of Phytopathology* 26: 379–407.
Yamada, T., Kawasaki, T., Nagata, S., Fujiwara, A., Usami, S., Fujie, M. 2007. New bacteriophages that infect the phytopathogen *Ralstonia solanacearum*. *Microbiology* 153: 2630–2639.

2 Pathogenesis of Plant Pathogenic Bacteria

Kubilay Kurtulus Bastas and Velu Rajesh Kannan

CONTENTS

ABSTRACT Plant pathogenic bacteria stimulate disease symptoms in plants using various pathways. Pathogenicity factors encoded by pathogenicity genes and disease-specific genes are involved in steps that are crucial for the establishment of disease. Many plant bacterial pathogens require different types of secretion systems to mount successful invasions of susceptible host-plant species. Quorum sensing is known to be involved in the regulation of important physiological functions of bacteria such as symbiosis, conjugation, and virulence. The expression of the extracellular polysaccharides in host cells is known to play a central role by providing resistance to oxidative stress during pathogenesis. Basal resistance is elicited in plants by pathogen-associated or microbial-associated molecular patterns. The overproduction of reactive oxygen species is an early event that characterizes the HR (Hypersensitive response) induced by bacteria; this induction, at least in part, may generate unfavorable conditions for bacterial multiplication in the apoplast. Aside from hydrogen peroxide, plant antibacterial molecules could be responsible for the restriction of bacterial growth observed during HR. Induced plant defenses (SAR and ISR) against pathogens are regulated by networks of interconnecting signaling pathways in which the primary components are the plant-signal molecules (salicylic acid, jasmonic acid, ethylene, and nitric oxide). In many host–pathogen interactions, plants react to attack by pathogens with enhanced production of these substances; at the same time, a distinct set of gene-to-gene resistance comprised of defense-related genes is activated and attempts to block the infection. The severity of disease symptoms caused by bacteria in plants is determined by the production of a number of virulence factors such as phytotoxins, plant cell wall-degrading enzymes, extracellular polysaccharides, and phytohormones. In general, phytopathogenic bacterial strains that mutate in any virulence factor more or less reduce their virulence, whereas their pathogenicity remains unchanged. Structural and biochemical defenses induced in plants by bacterial infection can contribute to disease resistance.

KEYWORDS: plant pathogenic bacteria, pathogenesis, phytohormones, virulence, disease resistance

2.1 INTRODUCTION

Plant bacterial diseases are most frequent and severe in tropical and subtropical countries, where warmth and humidity create ideal conditions for bacterial growth. Plant bacterial diseases are generally characterized by symptoms of leaf and fruit spots, cankers, blights, vascular wilts, rots, and tumors. Phytopathogenic bacteria provoke diseases in plants by penetrating into host tissues (Buonaurio, 2008). Bacteria enter plants mostly through wounds and less frequently through natural openings (i.e., stomata, hydathodes, nectarthodes, and lenticels), but never directly through undamaged cell walls. Laceration or death of tissues may be the result of environmental factors such as wind breakage and hail; animal feeding (e.g., by insects and large animals); human cultural practices such as pruning, transplanting, and harvesting; self-inflicted injuries such as leaf scars; and, finally, wounds or lesions caused by other pathogens. If water soaking occurs, bacteria present in a film of water over a stoma can easily swim through the stoma and into the substomatal cavity, where they can multiply and start infection.

Air dissemination of some bacteria occurs infrequently, and only under special conditions or indirectly. For example, bacteria causing fire blight of apple and pear produce fine strands of dried bacterial exudate that contain viable bacteria; such strands may be broken off and disseminated by wind. Bacteria present in soil may be blown away along with plant debris or soil particles in dust. Wind, which also helps in the dissemination of bacteria by carrying rain-splash droplets that contain bacterial pathogens, also carries insects that may contain or are smeared with bacteria.

Water is important in disseminating pathogens in three ways: (i) rain or irrigation water that moves on or through the soil's surface disseminate bacteria in soil; (ii) all bacteria, being exuded in sticky liquid, depend on rain or (overhead) irrigation water that disseminate them either by washing them downward or splashing them about; and (iii) raindrops or drops from overhead irrigation pick up bacteria that are present in the air and wash them downward, where some may land on susceptible plants. Hundreds of insect species and almost all animals are by far the most important vectors of bacterial pathogens (e.g., bacterial wilt of cucurbits and bacterial soft rots). Many bacterial pathogens are present on or in seeds, transplants, budwood, or nursery stock, and are disseminated by them as they are transported from field to field or are sold and transported to other areas near and far. Such types of transmission are important because they bring pathogens into areas where they may never have existed before (Agrios, 2005).

Bacteria colonize plant apoplast (e.g., intercellular spaces or xylem vessels), in which they respectively cause parenchymatous and vascular diseases or parenchymatous vascular diseases. Aside from the endophytic habitat, some bacterial species have the capacity to survive as epiphytes on plant surfaces (phylloplane, rhizoplane, carpoplane, etc.). Once they are inside plant tissues, bacteria may implement two main attack strategies to exploit the host plant nutrients: biotrophy, in which the plant cells are kept alive as long as possible and the bacteria extract nutrients from live cells; and necrotrophy, in which the bacteria kill plant cells and extract nutrients from dead cells. Infections caused by fastidious xylem- or phloem-inhabiting bacteria are systemic; that is, from one initial point in a plant, the pathogen spreads and invades most or all of the susceptible cells and tissues throughout the plant. Bacteria and mollicutes reproduce by fission in which one mature individual splits into two equal, smaller individuals. Under optimum nutritional and environmental conditions (in culture), bacteria divide (i.e., double their numbers) every 20–30 min; presumably, they also multiply at that rate in a susceptible plant for as long as nutrients and space are available and the temperature is favorable. Because millions of bacteria may be present in a single drop of infected plant sap, the numbers of bacteria per plant must be astronomical. Fastidious bacteria appear to reproduce more slowly than typical bacteria, although they spread systemically throughout the plant's vascular system (Buonaurio, 2008).

The fitness of a bacterial pathogen can be quantified by measuring its reproductive rate, rate of multiplication, efficiency of infection, and the amount of disease caused by its aggressiveness. The presence of excess genes for virulence, however, imposes a fitness penalty upon the pathogen. Therefore, a mutation from avirulence to virulence occurs only if it is needed to overcome an *R* gene for resistance; that is, only if the mutation is absolutely necessary for the pathogen to survive. New biotypes of bacteria seem to arise with varying frequency by means of at least three sex-like processes: (i) conjugation, which occurs when two compatible bacteria come into contact and a small portion of the chromosome or plasmid from one is transferred to the other through a conjugation bridge or pilus; (ii) genetic transformation, which occurs when bacterial cells absorb and incorporate genetic material that is secreted by other bacteria or released during their rupture; and (iii) transduction, which occurs when a bacterial virus (bacteriophage) transfers genetic material from the bacterium in which the phage was produced to the bacterium it will infect next (Agrios, 2005).

2.2 PATHOGENICITY

Microbial pathogenicity has often been defined as the biochemical mechanisms whereby microorganisms cause disease in a host organism (Fuchs, 1998). Virulence is the degree or measure of pathogenicity of a pathogen. The pathogenicity and/or virulence of Gram-negative plant pathogenic bacteria strictly depends upon the presence of secretion apparatuses in their cells, through which they secrete proteins or nucleoproteins that contribute to their virulence in the apoplast or are injected into the host cell (Buonaurio, 2008).

Bacterial pathogens include several classes of genes that are essential for causing disease (pathogenicity genes or pat) or for increasing virulence in one or a few hosts (virulence genes). Pathogenicity factors encoded by pat and by disease-specific genes (dsp) are involved in steps that are crucial for the establishment of disease. Such genes include those that are essential for recognition of the host by the pathogen, attachment of the pathogen to the plant surface, formation of infection structures on the plant surface, penetration of the host, and colonization of the host tissue. Genes involved in the synthesis and modification of the lipopolysaccharide (LPS) cell wall of Gram-negative bacteria may help condition the host range of the bacteria. Some plant cell wall-degrading enzymes, toxins, hormones, polysaccharides, proteinases, and siderophores, as well as melanin, are produced by pathogens during pathogen–plant interactions in which they are essential for the pathogen to infect its host and cause disease (Agrios, 2005).

Both pathogenic and nonpathogenic bacteria take the first step toward successful colonization when they are able to maintain close proximity to a plant surface by adhesion. One of the major groups of bacterial adhesins is the fimbriae or pili, which share a common feature in terms of the molecular machinery they need for pilus biogenesis and assembly onto the bacterial surface. The family of P pili encoded by *pap* genes (pyelonephritis-associated pili) is a well-known example of pilus assembly (Marklund et al., 1992; Fuchs, 1998).

Most bacteria do not need adhesion mechanisms except perhaps when they are moving through the xylem and phloem. The crown gall bacterium *Rhizobium radiobacter* (formerly *Agrobacterium tumefaciens*), however, requires attachment to plant-surface receptors as the first step in the transfer of T-DNA and development of disease symptoms. The attachment requires three components: a glucan molecule, the synthesis and export of which requires three genes; genes for the synthesis of cellulose; and the *att* region of the bacterial genome, which contains several genes for attachment. In addition to these genes, *Agrobacterium* contains numerous other genes that have homology to genes of mammalian pathogens for adhesins and for pilus biosynthesis. Several other plant pathogenic bacteria also have genes that encode proteins likely to be involved in attachment and aggregation (Agrios, 2005).

The interactions between plants and microbial pathogens, which are specific, complex, and dynamic, often result in either disease (compatible interaction) or resistance (incompatible interaction). In an incompatible interaction, a plant mounts a rapid response to a virulent pathogen that leads to localized cell death at the site of infection; in turn, the localized cell death prevents further pathogen infection. Incompatible interactions are usually associated with a hypersensitive response (HR) such as localized cell death and the appearance of necrotic flecks at the site of infection (Huang et al., 2005). There are four requirements of pathogenicity: (i) entry into host tissues through these surfaces; (ii) *in vivo* multiplication in the environment; (iii) interference with the host's defense mechanisms; and (iv) damage to the host (Fuchs, 1998).

Prior to the initiation of infection by pathogenic bacterial species, stable biofilm-population size must be established. Epiphytic multiplication is influenced by environmental factors and/or host physiology. Localization in protected sites and clustering in structures (e.g., biofilms) that lead to enhanced stress resistance are considered to be adaptive traits of epiphytic bacteria. The most common sites of bacterial colonization have been shown to be stomates, the bases of trichomes, and depressions along veins. The biofilms in these environments provide protection to the microbial pathogens from external environmental conditions such as desiccation, and thus is a favorable mileu for multiplication. For instance, *Xanthomonas axonopodis* pv. *phaseoli* could survive on leaf surfaces and also endophytically. The occurrence of *Xa. axonopodis* pv. *phaseoli* population as biofilms on bean leaves was examined during three field experiments on plots established with naturally contaminated bean seeds (Weller and Saettler, 1980).

2.2.1 Bacterial Secretion Systems

Because most bacterial virulence factors are associated with the bacterial surface or secreted into the surrounding environment, the transport of these molecular systems are known to move proteins out of the bacterial cell. Unrelated virulence factors often share the same secretion mechanism (Fuchs, 1998).

2.2.1.1 Type I

A type I secretion system is present in almost all plant pathogenic bacteria. It carries out the secretion of toxins such as hemolysins, cyclolysin, and rhizobiocin. These toxins, which consist of ATP-binding cassette proteins, are involved in the export and import of a variety of compounds through energy provided by the hydrolysis of ATP (Agrios, 2005). Production of macromolecule-like enzymes is regulated by four major secretion pathways. Some exoenzymes and toxins are secreted through the type I pathway, which is *sec*-independent. The proteases produced by *Pectobacterium chrysanthemi*, which cause soft rot, are secreted directly into the extracellular environment. This one-step secretion is governed by the type I secretory apparatus, which consists of three accessory proteins: PrtD, PrtE, and PrtF (Pugsley, 1993).

2.2.1.2 Type II

The type II secretion system is common in Gram-negative bacteria. It is involved in the export of various proteins, enzymes, toxins, and virulence factors. Proteins are exported in a two-step process: first as unfolded proteins to the periplasm via the *sec* pathway across the inner membrane, and second as processed and folded proteins through the periplasm and across the outer membrane via an apparatus that consists of 12–14 proteins encoded by a cluster of genes. *Ralstonia* and *Xanthomonas*, which have two type II systems per cell, use them for secretion outside the bacterium of virulence factors such as pectinolytic and cellulolytic enzymes. *Xylella* and *Rhizobium* (*Agrobacterium*) have one type II per cell; in fact, *Rhizobium* has genes for only the first step of protein transport across the inner membrane and uses type IV for the other steps (Agrios, 2005).

In these two *sec*-dependent steps, type II controls the secretion of pectinases and cellulases through what is known as the general secretory pathway (GSP). Some of the enzymes shown to be virulence factors are secreted through GSP by *Pe. chrysanthemi* and *Pectobacterium carotovorum* ssp. *carotovorum*. The pectates lyase, polygalacturonase, and cellulase are exported to the periplasm, but are unable to move through the outer membrane. In the second step, the periplasmic form of these enzymes moves through the outer membrane into the extracellular environment. Secretion of major pectate lyases, polygalacturonases, and cellulases is controlled by the *out* gene cluster; the number of *out* genes may vary with *Pectobacterium* spp. (Lindberg and Collmer, 1992; Reeves et al., 1993). The purified pectate lyases from *Pe. chrysanthemi* (i.e., PelA, PelB, PelC, and PelE) are important virulence factors that are involved in the degradation of plant-cell walls (Preston et al., 1992; Normura et al., 1998; Noueiry et al., 1999). The cell wall-degrading enzyme from the principal virulence determinants of *Pe. carotovorum* ssp. *carotovorum* and their synthesis is coordinately regulated by a complex network (Eriksson et al., 1998). The aggressiveness of several *Pe. chrysanthemi* strains on potato has been shown to be negatively correlated with their osmotic tolerance.

In *Xanthomonas oryzae* pv. *oryzae*, several proteins that include a xylanase required for virulence are secreted through the type II secretion system. The *xynB* gene encoding for secreted xylanase and a paralog *xynA* were cloned. These genes, *xynA* and *xynB*, are adjacent in *Xa. oryzae* pv. *oryzae*, as well as in *Xa. axonopodis* pv. *citri*. There seems to be functional redundancy among the type II-secreted proteins *Xa. oryzae* pv. *oryzae* in promoting virulence on rice (Rajeswari et al., 2005). *Xylella fastidiosa* systemically colonizes the xylem elements of infected grapevine plants by breaching the pit-pore membranes and consequently separating the xylem vessels. Based on the detection of genes involved in the plant cell-wall degradation in *Xy. fastidiosa* genome, it was thought that the cell wall-grading enzymes produced by *Xy. fastidiosa* may be involved in the cell-wall degradation. Several beta-1, 4-endoglucanases xylanases and xylosidases and one PG-encoding gene were identified in the *Xy. fastidiosa* genome. PG is required for successful infection of grapevine by *Xy. fastidiosa* and for its function as a critical virulence factor for *Xy. fastidiosa* pathogenesis in grapevine (Roper et al., 2007).

2.2.1.3 Type III

Bacterial plant pathogens colonize the apoplastic space between plant cells. As extracellular pathogens, these organisms deploy an arsenal of secreted virulence factors to modulate host-cell processes from outside plant cells. These factors include: (i) low molecular-weight phytotoxins, plant hormones, and hormone analogs that are secreted into the apoplast, many of which presumably enter or are taken up by plant cells; (ii) protein-virulence factors (effectors) that are delivered directly into the plant-cell cytosol via a specialized type III secretion system; and (iii) plant cell wall-degrading enzymes that are secreted through a *sec*-dependent type II secretion system (Alfano and Collmer, 1997; Sandkvist, 2001) and function to degrade or remodel the plant-cell wall. Type III-delivered effector molecules are proposed to function inside plant cells to modulate host-cell physiology, thus rendering host tissue suitable for pathogen growth and disease development. The activities of these molecules may include suppression of plant-defense responses, stimulation of the release of nutrients and water into the apoplast, promotion of disease-symptom development, facilitation of pathogen release from infected tissue, and hence pathogen transmission (Kunkel and Chen, 2006) (Figure 2.1A).

Many plant-bacterial pathogens require type III secretion systems to mount successful invasions of susceptible host-plant species. Type III was first discovered in Gram-negative bacterial pathogens of plants (Lindgren, 1997; He, 1998). The type III system acts as a syringe through which virulence proteins are injected by the bacteria into the host cells. The type III effector proteins are

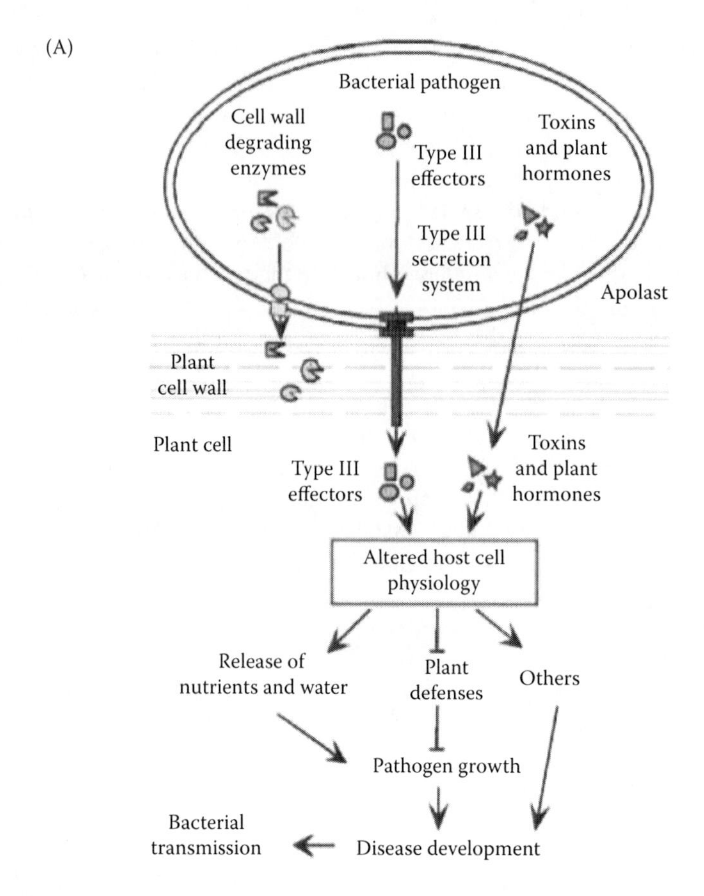
(A)
Bacterial pathogen
Cell wall degrading enzymes
Type III effectors
Toxins and plant hormones
Type III secretion system
Apolast
Plant cell wall
Plant cell
Type III effectors
Toxins and plant hormones
Altered host cell physiology
Release of nutrients and water
Plant defenses
Others
Pathogen growth
Bacterial transmission
Disease development

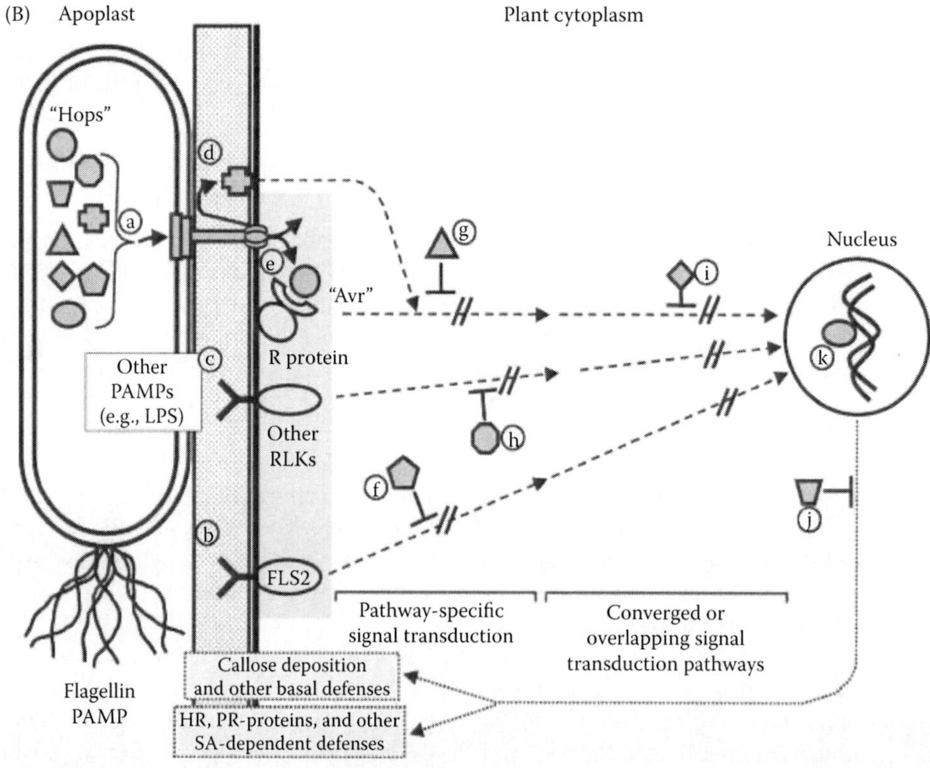
(B) Apoplast
Plant cytoplasm
"Hops"
a
d
e
g
"Avr"
i
Nucleus
k
R protein
c
Other PAMPs (e.g., LPS)
Other RLKs
h
f
b
FLS2
j
Pathway-specific signal transduction
Converged or overlapping signal transduction pathways
Callose deposition and other basal defenses
HR, PR-proteins, and other SA-dependent defenses
Flagellin PAMP

thought to facilitate disease development by altering the normal physiology of the plant in favor of the pathogen. The type III bacterial pathogens belonging to the genera *Erwinia, Pseudomonas,* and *Xanthomonas* possess *hrp* (HR and pathogenicity) and *hrc* (HR and conserved) genes. The *hrp* genes are involved both in the functions of pathogenicity and in the elicitation of resistance by bacterial pathogens such as *Erwinia amylovora*, *Pseudomonas syringae* pv. *tomato*, and *Xanthomonas vesicatoria* that induce necrosis; these genes have also been found to be homologous (Van Gijsegem et al., 1995a, b). Virulence of bacterial pathogens is frequently abolished/reduced if mutations occur that affect type III protein secretion. When the pathogens come into contact with host cells, type III is activated *in vivo*; the virulence proteins are thought to accelerate the leakage of plant nutrients in the extracellular space (apoplast) in plant tissues infected by bacteria (Lindgren, 1997; He, 1998). Host-specific symptoms may be induced by the expression of single genes, as in the case of *Xa. axonopodis* pv. *citri* (citrus canker). The *pthA* gene controls division, enlargement, and death of host cells (Duan et al., 1999).

Type III effectors may be required to promote bacterial growth in apoplast by defeating host defenses and releasing nutrients from plant cells. Basal defenses are triggered from the outsides of plant cells by nonpathogens, type III mutants, heat-killed bacteria, and factors that bear pathogen-associated molecular patterns (PAMPs) (Medzhitov and Janeway, 2002; Nurnberger and Brunner, 2002). These factors are also known as general elicitors, which include LPSs (Dow et al., 2000), flagellin (Gomez-Gomez and Boller, 2002), and cold shock proteins (Felix and Boller, 2003). Basal resistance to nonpathogenic bacteria is indicated by the production of a callose-rich papilla beneath the plant-cell wall at the site of bacterial contact, localized induction of reactive oxygen species (ROS), and enhancement of the expression of defense-related genes (Jones and Takemoto, 2004). Basal defences are triggered early in the plant–pathogen interaction in response to the perception by plant pathogen recognition receptors (PRRs) of extracellular PAMPs. Plants have evolved PRRs that can recognize PAMPs, which are important molecules for the microbial lifestyle. For their part, PAMPs contain a conserved structural feature that is recognized by a PRR. Recognition of a PAMP accelerates several early front-line defenses against bacterial pathogens (Gomez-Gomez and Boller, 2002). By contrast, the second level of host defenses, HR, is elicited inside plant cells by many bacterial pathogen effectors during gene-for-gene interactions that involve matching *avr* and *R* genes.

Plants respond to an array of microbe-associated molecular patterns (MAMPs) from both pathogenic and nonpathogenic microbes (Ausubel, 2005). In the absence of specialized immune systems (as in animals), plant cells seem to have developed the ability to respond to MAMPs such as flagellin, harpin (HrpZ), LPS, chitin, and necrosis-inducing Phytophthora proteins (NPP) and activate defense-gene transcription and MAPK signaling (Asai et al., 2002; Fellbrich et al., 2002; Navarro et al., 2004; Ramonell et al., 2005).

FIGURE 2.1 (A) Secreted virulence factors deployed by bacterial plant pathogens. Bacterial plant pathogens colonize the apoplastic space between plant cells. As extracellular pathogens, these organisms deploy an arsenal of secreted virulence factors to modulate host cell processes from outside plant cells. (From Springer Science+Business Media: *The Prokaryotes*, 2006, Kunkel, B.N., Z. Chen.) (B) Model depicting sites of action for bacterial type III effectors as both elicitors and suppressors of plant defense: (a) the effectors may be injected into host cells by the type III of *Pseudomonas syringae* pv. *tomato* DC3000, (b) *P. syringae* flagellin activates basal defense following recognition by the FLS2, (c) bacterial PAMPs are recognized by RLKs, (d) the type III-secreted harpins act in the apoplast and can trigger HR/SAR defense responses from outside of plant cells, (e) Avr proteins are recognized inside plant cells by R proteins and activate HR defenses, (f) and (h) type III effectors include pathway-specific components for basal defenses, (g) HR defenses, (i) additional sites of action may be at points where signal transduction pathways, (j) antimicrobial responses activated by defense-signaling pathways or may be inhibited, (k) AvrBs3 family members act to alter host transcription within the nucleus. (Adapted from Alfano, J.R., Collmer, A. 2004. *Annual Reviews of Phyopathology* 42: 385–414.)

The leucine-rich receptor (LRR) kinase perceives a conserved 22-amino acid peptide (flg22) from bacterial flagellin and activates both MAPK cascades and an WRKY transcription factor in *Arabidopsis* (Asai et al., 2002; Gomez-Gomez and Boller, 2002). Activation of the flg22-mediated MAPK cascade confers resistance to both bacterial and fungal pathogens (Asai et al., 2002). Many type III effectors (e.g., AvrPto, AvrRpt2, AvrRpm1, and HopAI1) suppress defense responses elicited by either TTSS-defective mutants or flg22 (Hauck et al., 2003; Kim et al., 2005; Li et al., 2005a,b).

Programmed cell death (PCD) is a result of recognition and signal-transduction events; in addition, the surrounding plant tissues exhibit resistance following complex physiological activities by the host tissues (Alfano and Collmer, 2004). Many of the type III effectors seem to be acquired by horizontal gene transfer and are commonly associated with mobile genetic elements (Kim and Alfano, 2002; Arnold et al., 2003). The twin-arginine translocation (Tat) pathway operates in the inner membrane of many Gram-negative bacteria (Berks et al., 2003). The Tat substrates are synthesized with cleavable N-terminal signal peptides that are characterized by a highly conserved twin-arginine motif in the positively charged N-terminal region, a weakly hydrophobic core region, and a positively charged *sec* pathway-avoidance signal in the C-terminal region (Cristobal et al., 1999). The Tat system has a remarkable ability to transport proteins that have already folded within the cytoplasm (Berks et al., 2000). Inactivation of the Tat system in *Ps. syringae* pv. *tomato* DC 3000 led to multiple complex phenotypes, including loss of motility on soft agar plates, deficiency in siderophore synthesis and iron acquisition, sensitivity to copper, loss of extracellular phospholipase activity, and attenuation in host-plant leaves (Kloek et al., 2000; Bronstein et al., 2005).

The *hrp* (HR and pathogenicity) genes, to date found only in Gram-negative bacteria, are additional bacterial genes that seem to be essential for the ability of some bacteria to cause visible disease on a host plant, to induce a HR on certain plants that are normally not infected by the bacteria, and to enable bacteria to multiply and reach high numbers in a susceptible host. Harpin, a protein that has a major role in virulence, is exported *in vitro* via type III machinery. This protein could not be detected inside the host cells by using a specific antiserum, but it was found to be associated with the pathogen cells and secreted. The extracellular localization of harpin is in agreement with the physiological effects induced by purified harpin applied as an exogenous elicitor (Banar et al., 1994; Perino et al., 1999). The *dsp* (disease-specific) region located next to the *hrp* gene cluster is essential for *E. amylovora* (Figure 2.2), but not for elicitation of HR. The operon *dspEF* contains genes *dspE* and *dspF* and is positively regulated by the *hrpL* gene. Similarity in the *dspE* to a partial sequence of the *avrE* locus of *Ps. syringae* pv. *tomato* was observed.

When the *dspEF* was introduced into a plasmid, *Ps. syringae* pv. *glycinea* race 4 became avirulent on soybean (Bogdanove et al., 1998). Three proteins secreted by *E. amylovora* through a functional Hrp-secretion pathway were identified: Harpin (HrpN), a glycine-rich, heat-stable protein capable of eliciting HR on tobacco; DspA/E, which is homologous and functionally equivalent to AvrEl; and HrpW, which has structural similarity to harpin and is homologous to Class III pectate lyases. These three proteins were purified and their functions were determined by examining the defective mutants in *hrpN*, *dspA*, and *hrpW*. HrpN was shown to be a virulence factor rather than a pathogenicity factor. DspA was shown to be an essential pathogenicity factor. HrpW was not found to be required for pathogenicity of *E. amylovora*. It was further shown that although harpin was the main HR elicitor, DspA also participated in HR elicitation on tomato. HrpW acted as a negative factor because hrpW mutants could induce stronger HR than the wild-type strain could (Barny et al., 1999). DspA/E interacted physically and specifically with four similar putative LRR-like serine/threonine kinases (RLK) from apple, an important host of *E. amylovora*. The genes encoding those four DspA/E-interacting proteins of *Malus* × *domestica* (DIPM 1–4) are conserved in a wide range of cultivars that represent highly susceptible to highly resistant categories. All four DIPMs were expressed constitutively in host plants. These proteins may act as susceptibility factors required for disease progression (Meng et al., 2006). Wide variations in the expression of *dspE* and *PelD* in culture and in plants were observed (Oh et al., 2005; Peng et al., 2006).

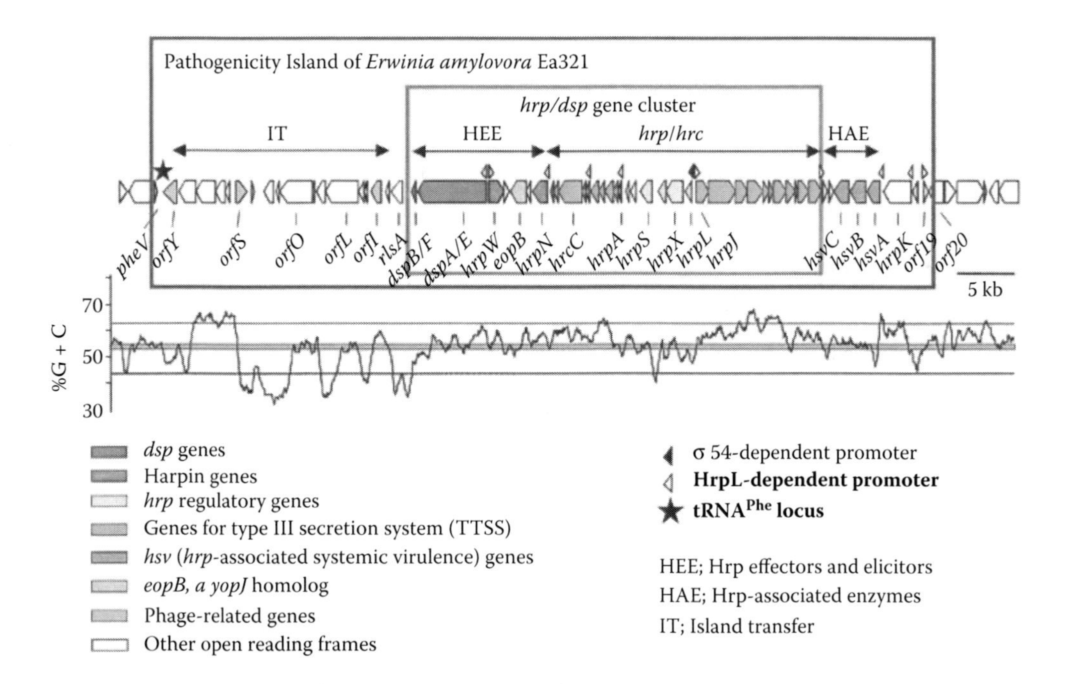

FIGURE 2.2 The Hrp pathogenicity island of *Erwinia amylovora* strain Ea321. The *hrp/dsp* gene cluster includes the *hrp/hrc* region and the HEE region. (Adapted from Oh, C.S., Beer, S.V. 2005. *FEMS Microbiology Letters* 253: 185–192.)

The proteins secreted via type III and the T3 secretome include effector and helper proteins. Effectors are virulence proteins that interfere with host metabolism and defense, whereas helper proteins were predicted to function extracellularly in effector delivery (Collmer et al., 2002). Six effector-like or helper-like proteins (Eop1, Eop2, Eop3, Eop4 [*Erwinia* outer protein], HrpJ, and HrpK) were identified. These proteins had been detected earlier as proteins secreted by type III of *E. amylovora*. In addition, HrpJ was required for the extracellular accumulation of harpins for pathogenicity and for wild-type elicitation of HR (Nissinen et al., 2007). A total of 12 proteins secreted through type III of *E. amylovora* in a defined and inducing minimal medium were identified. Only four of these proteins (Eop1, Eop3, Eop4, and DspA/E) exhibited similarity to known effectors. Analysis of the secretome of a nonpolar *hrpJ* mutant revealed that HrpJ was required for the accumulation of wild-type levels of secreted harpins. HrpJ was found to be essential for pathogenesis and also to have an important role in the elicitation of HR in tobacco (Nissinen et al., 2007).

Necrotrophic bacterial plant pathogens, such as *Pe. chrysanthemi*, produce pectinolytic enzymes that play a key role in inducing soft-rot symptoms. These factors include HrpN, which is a heat-stable, glycine-rich hydrophilic protein secreted by type III (Nasser et al., 2005). In addition to translocating effector proteins, *Pseudomonas syringae* type III secretes several extracellular accessory or helper proteins within extracellular spaces of plant leaves (i.e., the apoplast); these proteins aid the translocation of effectors and are designated as HOP proteins (Hrp outer proteins) (Alfano and Collmer, 2004). Many type III substrates utilize intracellular accessory proteins known as type III chaperones (TTCs) that facilitate their secretion and translocation via type III (Feldman and Cornelis, 2003; Alfano and Collmer, 2004).

The pathogenicity of *Xanthomonas* spp. depends on a specialized type III secretion system that secretes some helper and accessory proteins into the extracellular milieu that supports the injection of effector proteins, which comprise the major substrate class of the type III system, into the host cell. The specialized protein system is encoded by a 23-kb hrp chromosomal *hrp* gene cluster that contains six operons, *hrpA* to *hrpF*. HrpF is essential for pathogenicity but not for secretion in *Xa. vesicatoria* (Rossier et al., 2000). The *hrpF* mutants of *Xa. oryzae* pv. *oryzae* were reduced only in virulence (Sugio et al., 2005). The other type III-secreted proteins, HrpE and AvrBs3, are also in close contact with Hrp pilus during and/or after their secretion. HrpE protein is the predominant component of the pilus preparations from *Xa. vesicatoria* (Weber et al., 2005). The gene *hrpE* is unique to the genus *Xanthomonas* and exhibits no sequence similarity to other pili genes.

Bacteria with mutations in *hrpE* are unable to either cause disease or elicit HR in plants. In addition, HrpE is essential for all of the type III used in secretions of HrpF, AvrBs1, and AvrBs3. The Hrp pilus is therefore an indispensable component of a functional type III (Weber and Koebnik, 2005). Mutations in type III caused loss of the ability to induce disease, which indicates that the effectors are essential agents of *Ps. syringae* pathogenesis (Collmer et al., 2000). Many effectors have been isolated based on their ability to trigger host immunity. The concept of plant immunity based on the gene-for-gene model is in turn based on the recognition of a pathogen avirulence (Avr) effector protein by a plant resistance (R) protein (Cohn and Martin, 2003). Necrosis associated with disease was increased when both AvrPto and AvrPtoB were delivered by type III of *Ps. syringae* pv. *tomato* DC 3000 (Lin and Martin, 2005). The conservation of AvrPtoB type III effector protein among diverse genera of plant-bacterial pathogens suggests that it plays an important role in pathogenesis.

More than 30 effectors may be injected into host cells by the type III of the model pathogen *Ps. syringae* pv. *tomato* DC3000. *Pseudomonas syringae* flagellin activates basal-defense pathways after recognition by the FLS2 RLK. Other bacterial PAMPs may be recognized by additional RLKs. The type III-secreted harpins act in the apoplast and can trigger HR/SAR defense responses from outside the plant cells. A subset of type III effectors (i.e., Avr proteins) are recognized inside plant cells by R proteins; according to the guard hypothesis, these detect the activities of effectors on "guarded" susceptibility targets and activate HR defenses. The shaded box that encompasses the R-protein and basal defense receptors denotes the potential relationship among guarded effector targets, the basal defense system, and the plant membrane, possibly in recognition complexes. Putative sites of action for defense-suppressive type III effectors include pathway-specific components for basal defenses and HR defenses. Additional sites of action may exist at points where signal transduction pathways converge. Antimicrobial responses activated by defense-signaling pathways (or involving preformed factors) may also be inhibited post-transcriptionally. Finally, type III effectors such as AvrBs3 family members may act within the nucleus to alter host transcription (Alfano and Collmer, 2004) (Figure 2.1B).

Xanthomonas oryzae pv. *oryzae* incitant of rice bacterial blight disease produces AvrXa7 protein, the product of the pathogen gene *avrXa7*; this protein is a virulence factor. The gene *avrXa7* is included in the *avrBs* avirulence gene family, which encodes proteins targeted to plant cells by the type III secretion system (Yang et al., 2000). AvrXa7 is the only known type III secretion-system effector from *Xa. oryzae* pv. *oryzae* that makes a major contribution to bacterial growth and lesion formation in bacterial blight disease (Yang and White, 2004; Makino et al., 2006). A second system employs disease-resistance (R) proteins to recognize type III effector proteins that are delivered into the plant cell by bacterial type III. RIN4 from *Arabidopsis* is a host target of type III effectors that is guarded by R proteins. The effector AvrRpt2, a protease, targets RIN4 and induces its post-transcriptional disappearance. The disappearance of RIN4 then activates RPS2, the cognate R protein of AvrRpt2 (Axtell and Staskawicz, 2003; Mackey et al., 2003). RIN4 is a protein respectively targeted by AvrRpt2 and AvrRmp1 for degradation and phosphorylation. RIN4 itself was found to be a regulator of PAMP signaling. The effector-induced perturbations of RIN4 were sensed by the R proteins RPS2 and RPM1. RIN4 negatively regulates PAMP signaling, and PAMP-induced defense responses are inhibited or enhanced in plants that, respectively, overexpress or lack RIN4 (Kim et al., 2005).

AvrRpt2, an effector protein from *Pseudomonas syringae pv. tomato*, is able to function as a virulence factor that activates resistance in *Arabidopsis thaliana* lines that express the resistance gene *RPS2*. In addition, AvrRpt2 can enhance pathogen fitness by promoting the ability of the bacteria to grow and cause disease on susceptible *A. thaliana* lines that lack functional *RPS2* (Leister and Katagiri, 2000). AvrRpt2 has been shown to induce disappearance of the *A. thaliana* protein RIN4 (Axtell and Staskawicz, 2003; Mackey et al., 2003). The effector protein VirPphA has been shown to allow *Pseudomonas savastanoi* pv. *phaseolicola* to evade HR-based immunity in bean (Jackson et al., 1999). In addition, the other effectors of *Ps. savastanoi* pv. *Phaseolicola*, including AvrPphC and AvrPphF, help the pathogen to avoid triggering host immunity (Tsiamis et al., 2000). AvrPto was reported to suppress a cell wall-based defense in transgenic *Arabidopsis* lines (Hauck et al., 2003). Several effector proteins, such as AvrPtoB, are also thought to act as general suppressors of host PCD; such suppression is the hallmark of HR-based immunity in plants, depending on the genetically controlled and regulated process (Abramovitch et al., 2003; Jamir et al., 2004).

2.2.1.4 Type IV

The type IV secretion system transports macromolecules from the bacterium to the host cell. The transferred proteins are very similar to those responsible for the mobilization of plasmids among bacteria. The *Rh. radiobacter vir*B operon encodes 11 proteins that form an organized structure and are involved in the transfer of the T-DNA strand from the bacterium to the plant cell cytoplasm. The transporting structure stretches from the bacterial inner membrane through the outer membrane and terminates in a pilus-like structure that protrudes from the bacterial cell (Agrios, 2005).

Successful infection by bacterial pathogens commences with the attachment of pathogen cells to the host tissue surface. Many Gram-negative bacteria have type IV fimbriae for adhesion. The presence of type IV fimbriae genes distributed in several loci on the genome of *Xa. oryzae* pv. *oryzae* has been observed (Ray et al., 2002; Ochiai et al., 2005). *Ralstonia solanacearum*, *Pseudomonas*, *Xanthomonas*, and *Xylella* have as many as 35 genes that are homologous to type IV pili genes. In *Xanthomonas* and *Pseudomonas*, these genes are involved in cell-to-cell aggregation and protection from environmental stress, whereas in *Xylella* type IV, pili are necessary for the establishment of an aggregated bacterial population in the turbulent environment of the xylem via adherence to the vessels in conjunction with components such as polysaccharides (Agrios, 2005).

2.2.2 Quorum Sensing

Bacteria commonly control expression of gene circuits in a population-dependent manner via a regulatory mechanism known as quorum sensing (QS) (Bellemann et al., 1994; Whitehead et al., 2001). Cell-to-cell communication in diverse species of Gram-negative bacteria, which is achieved by QS gene regulation, is characterized by the presence of specific proteins that function as transcriptional regulators (LuxR homologs) and *N*-acetylhomoserine lactone (AHL) synthases (LuxI homologs) (Whitehead et al., 2001). The QS process has been shown to be involved in the regulation of many important physiological functions of bacteria (e.g., symbiosis, conjugation, and virulence). AHL-mediated QS has been found to govern a myriad of vital processes in pathogenic and beneficial bacteria (Pierson et al., 1997; Miller and Bassler, 2001; Wisniewski-Dye and Downie, 2002; von Bodman et al., 2003). AHL signals are required for conjugal transfer of the tumor-inducing (Ti) plasmid in phytopathogenic *Rh. radiobacter*, for antibiotic biosynthesis in plant-beneficial *Pseudomonas chlororaphis*, for nodulation factors in *Rhizobium leguminosarum*, and for synthesis of exoenzyme virulence factors in *Erwinia carotovora*, to name just a few of these processes (Chen et al., 2003; Heurlier et al., 2003; Whistler and Pierson, 2003). The *luxR* homolog *aviR* in *Rhizobium vitis* strain F2/5 was associated with induction of an HR on tobacco and necrosis on grapevine, which indicated that the responses were regulated by QS (Hao et al., 2005). Many soil-borne plant-pathogenic bacteria, including *Ra. solanacearum*, have swimming motility that makes an important quantitative contribution to bacterial-wilt virulence in the early stages of host invasion and colonization (Trans-Kersten et al., 2001).

Chemotaxis is an essential trait required for virulence and pathogenic fitness in *Ra. solanacearum* (Yao and Allen, 2006). Two major plant-virulence traits, namely production of extracellular polysaccharides (EPSs; amylovoran and levan) and tolerance to oxygen free radicals, were controlled in a bacterial cell density-dependent manner. The bacterial populations attached to a surface are referred to as biofilms; these bacteria are thought to be resistant to adverse conditions such as desiccation and extreme temperature. They may also function as an important virulence factor. *Pantoea stewartii* ssp. *stewartii* produces stewartan exo/capsular polysaccharide (EPS) in a cell density-dependent manner governed by an EsaI/EsaR QS system (Koutsoudis et al., 2006). In a further study, the role of flagella of *Pe. carotovorum* ssp. *carotovorum* in biofilm formation was critically examined. The biofilm-forming ability of bacteria may help the bacterial pathogens in host colonization, and flagellar motility may be indirectly involved in pathogenicity (Hossain et al., 2005; Hossain and Tsuyumu, 2006).

Burkholderia glumae that cause seedling and grain rot in rice also produce the phytotoxin toxoflavin, which functions as a major virulence factor. The expression of genes involved in the biosynthesis and transport of toxoflavin was shown to be regulated by *N*-octanoyl homoserine lactone (C8-HSL) (Kim et al., 2004). Interspecies communication through the use of autoinducers has been inferred as a possible mechanism by which the pathogenicity of certain virulent bacterial species such as *Burkholderia cepacia* may be enhanced (McKenny et al., 1995).

2.2.3 Bacterial Toxins

Toxins have long been known to play a central role in parasitism and in the pathogenesis of plants by several plant-pathogenic bacteria. *Pseudomonas syringae*, *Ps. syringae* pv. *tomato*, and *Ps. syringae* pv. *maculicola* are primarily associated with production of the phytotoxin coronatine (COR). COR functions primarily by suppressing the induction of defense-related genes; however, as with most bacterial phytotoxins, it does not seem to be essential for pathogenicity by all strains. The bacterium *Ps. syringae*, along with its pathovars, produces several pathotoxins, including syringomycin (Agrios, 2005).

The expression of the *syrB* gene that controls synthesis of syringomycin, a nonhost-specific phytotoxin produced by *Ps. syringae* pv. *syringae* (bacterial canker) was determined *in vitro* by using aqueous extracts of peach-bark tissue. The expression of *syrB* was significantly less in plants receiving N fertilization compared with nonfertilized plants. Nitrogen fertilization appeared to decrease host susceptibility to *Ps. syringae* pv. *syringae* by producing or reducing compounds that may induce or antagonize syrB expression (Cao et al., 2005). The *salA* gene in *Ps. syringae* pv. *syringae*, a key regulatory element for the production of syringomycin, encodes a member of the LuxR regulatory protein family. In addition, *salA*, a member of the GacS/GacA signal transduction system, has been shown to be essential for bacterial virulence, syringomycin production, and expression of the *syrB1* synthetase gene. The subgenomic oligonucleotide microarray has potential for use as a power tool for defining the *SalA* regulon; its relationship to other genes is important to plant pathogenesis. Likewise, syringopeptin production is activated by plant-signal molecules and GacS, SalA, and SyrF regulatory pathways mediate transmission of plant molecules to the *syr-syp* biosynthesis apparatus. The synthetase genes for syringomycin (*syrB1* and *syrE*) and syringopeptin (*sypA, sypB*, and *sypC*), as well as four regulatory genes (salA, syrF, syrG, and syrP) and nine putative secretion genes dedicated to the production of two toxins were activated by the phenolic plant-signal molecule arbutin (Lu et al., 2005; Wang et al., 2006).

Xanthomonas albilineans causing leaf scald disease in sugar cane is a systemic, xylem-invading bacterial pathogen. Albicidins, produced by *Xa. albilineans*, block the replication of prokaryotic DNA and the development of plastids, thereby causing chlorosis in emerging leaves. Albicidins interfere with host-defense mechanisms and thereby the bacteria gain systemic invasion of the host plant (Agrios, 2005). Albicidin, a major toxic compound produced by the pathogen, has a critical role in pathogenicity. Albicidin-deficient mutants cannot induce disease symptoms; for example, transgenic sugar cane plants that express an albicidin-detoxifying gene also exhibit resistance to leaf scald (Zhang et al., 1999).

2.2.4 Extracellular Polysaccharides

Extracellular polysaccharides play an important role in the pathogenesis of many bacteria both by direct intervention with host cells and by providing resistance to oxidative stress (Agrios, 2005). EPSs may be associated with the bacterial cell as a capsule, be released as fluidal slime, or be present in both forms (Denny, 1999). EPSs are important pathogenicity or virulence factors, particularly for bacteria with a vascular habitat. Large amounts of EPS are produced by many bacterial pathogens both *in vivo* and *in vitro*. The *eps* genes govern the synthesis of EPS, and the ability to produce EPS is positively correlated to the virulence of the strains of *E. amylovora*, *Pe. chrysanthemi*, *Burkholderia solanacearum*, *Clavibacter michiganensis*, and *Xanthomonas campestris* (Narayanasamy, 2002).

The EPSs amylovoran and levan are, respectively, pathogenicity and virulence factors of *E. amylovora*, the causal agent of fire blight on some rosaceous plants (Oh and Beer, 2005). Amylovoran is an acidic heteropolysaccharide composed of a pentasaccharide repeating unit of four differently linked galactose molecules and a glucuronic acid residue, which are decorated with pyruvate and acetate groups; levan is a neutral homopolysaccharide composed of 2,6-linked fructose (Jahr, 1999). The gene clusters *cms* and *cps* that control the biosynthesis of levan and amylovoran have been sequenced (Bernhard et al., 1993). Proteins with transport functions were tagged by fusions with β-lactamase. Expression of levan sucrose is influenced by *rlsA* located in the *dsp/hrp* region. Although no signal peptide could be detected in the amino acid sequence of levan sucrose, its fusion with β-lactamase produced antibiotic-resistant cells, which confirmed its transport in *E. amylovora*. A protease-deficient strain was retarded in the colonization of host plants (Geider et al., 1999). Amylovoran affects plants primarily by plugging the vascular tissue, thus inducing shoot wilt, and is considered a pathogenicity factor because amylovoran-deficient mutants are not pathogenic (Bellemann and Geider, 1992). The biosynthesis of amylovoran requires the *ams* operon, which consists of 12 genes whose expression is controlled by the regulatory proteins RcsA and RcsB, which in turn are able to bind the promoter region of the *ams* operon (Oh and Beer, 2005). Levan is synthesized by levansucrase and encoded by the *lsc* gene, mutation of which results in slow symptom formation on host-plant shoots (Geier and Geider, 1993).

As a general response to environmental changes, the current global regulation of bacterial transcription depends upon many factors, including histone-like proteins such as H-NS (Dorman and Deighan, 2003). H-NS is mainly a multifunctional gene regulator, and is predominantly negative, but it may also activate genes by repressing a repressor (Dorman, 2004); in small form it is known to have attenuated virulence of animal pathogens. An *hns* homolog from *E. amylovora* was identified by complementing an *Escherichia coli hns*-mutant strain with a cosmid library from *E. amylovora*. The functions of the two *hns*-like genes in *E. amylovora* differed in terms of their production of EPS amylovoran. Levan production was significantly increased by *hns* mutations; however, synthesis of the capsular EPS amylovoran and of levan were reduced when *hns* from *E. amylovora* plasmid was overexpressed. Increase in amylovoran synthesis is attributed to a mutation in the chromosomal *hns* of *E. amylovora*. Both mutations adversely affected development of symptoms on immature pear fruit (Hildebrand et al., 2006).

In *Pe. chrysanthemi*, the *eps* genes are clustered on the chromosome and are repressed by a regulator of pectate lyase synthesis (PecT). Reduction in the efficiency of tissue maceration in an *eps* mutant indicated that full expression of virulence in *Pe. chrysanthemi* required the production of EPS (Condemine et al., 1999). In *B.* (*Pseudomonas*) *solanacearum*, biosynthesis and transport of the acid EPS (high MW) are encoded by at least nine structural genes located in the *eps* operon. Inactivation of the *eps* operon by transposon tagging resulted in reduction in its ability to produce

EPS in plants; consequently, the EPS mutants showed reduced virulence. Acidic EPS is an essential wilt-inducing factor produced by *B. solanacearum* (Denny and Schell, 1994).

In *Clavibacter michiganensis*, which produces two distinct types of EPS (A and B), no difference could be discerned in the virulence of EPS mutants. Therefore, wilting was not attributed to EPS (Bermpohl et al., 1996). After the pathogenicity locus *pat-1* located in the plasmid pCM2 was shown to be essential for the virulence of *C. michiganensis* ssp. *michiganensis*, the endophytic plasmid-free isolates of *Cmm* became virulent following the introduction of the *pat-1* region (Dreier et al., 1997).

Ralstonia solanacearum, which causes the wilting of several hundred plant species (e.g., potato, tomato, tobacco, peanut, and banana), is another phytopathogenic bacterium whose virulence mainly depends on EPS production. Among the EPSs it produces, EPS1, an acidic high-molecular-mass heteropolysaccharide, is the bacterium's single most important virulence factor. This is so because *eps* mutants are severely reduced during the systemic colonization of tomato plants that are inoculated via unwounded roots; even when introduced directly into stem wounds, they do not cause typical wilt symptoms. Along with their involvement in wilting, EPSs might help to shield bacteria from toxic plant compounds, reduce contact with plant cells to minimize host defense responses, promote multiplication by prolonging water-soaking of tissues, or otherwise aid invasion or systemic colonization (Denny and Schell, 1994).

Yun et al., (2006a,b) have demonstrated that xanthan, the major exopolysaccharide secreted by *Xanthomonas* spp., plays an important role in *X. campestris* pv. *campestris* pathogenesis. They also demonstrated that xanthan suppresses callose deposition in plant-cell walls, which is a basal form of resistance to bacterial colonization. Xanthan is produced abundantly at later stages of pathogenesis in tissues that are undergoing necrosis (Vojnov et al., 2001). Xanthan polymers have been implicated in several symptoms, including wilting induced by vascular pathogens and the water-soaking symptoms associated with foliar pathogens (Denny, 1995). The *pigB* gene has been shown to govern the production of EPS xanthomonadin pigements and the diffusible signal molecule diffusible factor (DF). Following the extracellular application of DF, *pigB* mutants were able to synthesize EPS and xanthomonadin. They could infect cauliflower through wounds but not through natural openings (e.g., hydathodes). A functional *pigB* gene may be essential for epiphytic survival and natural host infection (Poplawsky et al., 1998; Poplawsky and Chun, 1998). Production of extracellular enzymes (including proteases, pectinases and cellulases, and EPS) by *Xa. campestris* pv. *campestris* was shown to be regulated by the products of the *rpfABFCHG* gene cluster (Barber et al., 1997; Slater et al., 2000; Dow et al., 2003; He et al., 2006).

2.3 REGULATION OF PATHOGENICITY FACTORS

During a complete life-cycle, pathogenic bacteria must survive within strikingly different environments; therefore, they must be able to sense whether they are staying outside their host, entering an organism, or occupying a specialized compartment inside a host organ or a specific cell type. Virulence factors to be expressed, it is also linked to various environmental signals; these include host responses such as increases in temperature, acidification, oxygen level, osmolarity, iron concentration, urea, and carbon sources (Fuchs, 1998).

Two-component signal transduction (TCST) systems are a large family of proteins that are involved in signal transduction and gene regulation in bacteria. They are typically composed of a sensor protein located within the cell membrane, and a cytoplasmic regulatory protein. The conserved cytoplasmic domain of the sensor contains a histidine kinase activity that transforms the environmental stimulus into a cellular signal via phosphoryl group transfer. An autophosphorylation event at histidine residues of the sensor is followed by a phosphate transfer to an aspartate residue of the regulator. The covalent modification of the regulator modulates its DNA-binding activity and thus the transcription from dependent virulence-promoters (Gross, 1993). Two-component systems control, for example, tumor formation in *Rh. radiobacter* (VirA/VirG).

In addition to the two-component family, other groups of regulatory factors influence virulence expression. Members of the LysR family also contain a DNA-binding motif that regulates transcription of virulence genes from, for example, *Ps. solanacearum* and *Rh. radiobacter.* TCST systems in *E. amylovora* played major roles in virulence on immature pear fruit and in the regulation of amylovoran biosynthesis and swarming motility (Hoch and Silhavy, 1995; Stock et al., 2000; Wolanin et al., 2002; Zhao et al., 2009). The xylem-limited *Xy. fastidiosa* has the fewest TCST genes; the relatively adaptable *Ps. syringae* and *Xanthomonas* spp. have the largest numbers (Lavin et al., 2007; Qian et al., 2008).

Two-component signal transduction systems are used by pathogenic bacteria to control expression of the virulence factors required for infection. The *Rh. radiobacter* VirA/VirG system, the GacA/GacS of both *Pseudomonas* spp. and *Pe. carotovorum*, and the RpfCG of *Xanthomonas* spp. and *Xy. fastidiosa* are probably the most well known and studied TCST systems involved in virulence gene expression in plant pathogens (Brencic and Winans, 2005; Beier and Gross, 2006; Lavin et al., 2007; Mole et al., 2007; Qian et al., 2008).

Among nutrients that are essential for bacterial growth and survival, iron is the most important. Iron is sequestered in plant by ferritins, which are multimeric iron-storage proteins. Bacterial pathogenesis is affected by the milieu of the available iron in the apoplastic fluids of plants. Ferric citrate is the major iron carrier in the conducting elements of plants.

Pectobacterium chrysanthemi produces a catechol-type siderophore, chrysobactin, that aids in the sequestration of free iron, as well as two receptors that are capable of interacting with bacterial iron-carrier molecules. Pel activity registers significant increases under low iron conditions. A regulatory *cbrAB* positively regulates the expression of *pelB*, *pelC*, and *pelE* (Sauvage and Expert, 1994). The iron-transport pathway is mediated by desferrioxamine (DFO), which is required for iron utilization by *Pe. chrysanthemi* during pathogenesis. DFO may also have a role in the oxidative burst elicited by the bacteria (Dellagi et al., 1998).

Chlorosis, the visible result of the loss of chlorophyll, may occur by a variety of mechanisms, including response to the activity of toxins; this results in cell/tissue collapse and, ultimately, plant death. In tomato, *Xa. vesicatoria* causes bacterial spot disease which induces chlorosis as the early visible symptom; later, the infected tissues become necrotic. A novel locus from *Xa. vesicatoria* that induces early chlorosis in tomato and several nonhosts has been identified and characterized. The gene, chlorosis factor (*ecf*), encodes a hydrophobic protein that is similar to four other proteins in plants, including HolPsyAE (Morales et al., 2005).

Damage to plant tissues due to biotic or abiotic stress may result in the release of chlorophyll from the thylakoid membranes. The chlorophylls must be degraded rapidly to prevent cellular damage by the photodynamic action (Takamiya et al., 2000). In the absence of chlorophyll degradation, the amount of ROS produced may exceed the antioxidant capacity of the plant system. In such cases, the toxic molecules may cause cell death. The direct correlation between the absence of callose deposition and the production of disease symptoms suggests a major role for callose in the pathogenesis of *Xa. campestris* pv. *campestris* (Yun et al., 2006a,b).

Virulence and pathogenicity genes of *Ra. solanacearum* are regulated by a complex network whose core is the phenotype conversion (Phc) system. This system consists of gene *PhcA*, a lysine-rich type transcriptional regulator, and the products of the operon *phcBRSQ*, which control levels of active *PhcA* depending on cell density or crowding. Cells that contain high levels of active *PhcA* produce large amounts of major virulence factors (e.g., EPS1 and some exoenzymes) and are very virulent. When *PhcA* is inactivated, the bacterial cells become quite avirulent and produce almost no EPS1 and exoproteins; instead, they activate genes that produce polygalacturonase, siderophores, the Hrp secretion apparatus, and swimming motility (Agrios, 2005).

The proteins encoded by *R* genes, which are quite similar, are classified according to certain structural characteristics and their localization in the plant cell. All but two R proteins contain a domain that is rich in the amino acid leucine (leucine-rich repeats or LRR) and is thought to participate in protein–protein (e.g., elicitor–receptor) interactions. Depending on where in the plant cell the

R protein LRR reside, they have either cytoplasmic LRRs or extracytoplasmic LRRs. The R proteins that have a cytoplasmic LRR domain also have a nucleotide-binding site; some have a zipper-like domain of leucine molecules known as coiled coil, or a domain of Toll/interleukin 1 receptors.

Induced defenses of plants against pathogens are regulated by networks of interconnecting signaling pathways in which the primary components are the plant-signal molecules salicylic acid (SA), jasmonic acid (JA), ethylene (ET), and probably nitric oxide (NO). In many host–pathogen interactions, plants react to pathogen attack with enhanced production of these substances while a distinct set of gene-to-gene resistance defense-related genes is also activated and attempts to block the infection. In addition, an exogenous application of SA, JA, ET, or NO to the plant often results in a higher level of resistance. SA reacts with several plant proteins, including the two major H_2O_2-scavenging enzymes, catalase and ascorbate peroxidase, as well as with a chloroplast SA-binding protein that has antioxidant activity. The main components of the SA-mediated pathway that leads to disease resistance appear to be constitutively expressed genes that encode pathogenesis-related (PR) proteins.

Some of these genes also activate the JA- and ET-mediated pathways, which results in the induction of the gene- encoding defensin. NO synthase activity also increases dramatically upon inoculation of resistant plants (but not susceptible ones). NO induces the expression of PR-1 and the early defense gene phenylalanine lyase (PAL). Production of SA occurs within the NO-mediated pathway downstream from NO. As with SA, NO reacts with and inhibits the activity of the enzymes aconitase, catalase, and ascorbate peroxidase. SA is not generally required for resistance genes *R* to determine resistance at the infection site; however, in at least some plants, SA is required at the primary infection site and in distal secondary tissues for the establishment and maintenance of SAR (Agrios, 2005).

2.4 PATHOGENICITY FACTORS RELATED WITH PLANTS

The severity of disease symptoms caused by bacteria in plants is determined by the production of a number of virulence factors such as phytotoxins, cell wall-degrading enzymes, EPSs, and phytohormones. In general, phytopathogenic bacterial strains mutated by any virulence factor more or less reduce their virulence, whereas their pathogenicity remains unchanged.

2.4.1 Phytotoxins

Phytopathogenic bacteria of the *Pseudomonas* genus, especially *Ps. syringae*, produce a wide spectrum of nonhost-specific phytotoxins. These are toxic compounds that cause symptoms in many plants independently of whether they can or cannot be infected by the toxin-producing bacterium. On the basis of the symptoms they induce in plants, phytotoxins of *Pseudomonas* spp. have been grouped as necrosis-inducing and chlorosis-inducing (Buonaurio, 2008).

Pseudomonas syringae pv. *syringae*, the causal agent of many diseases and types of symptoms in herbaceous and woody plants, produces necrosis-inducing phytotoxins called lipodepsipeptides. Based on their amino acid chain length, these are usually divided into two groups: mycins (e.g., syringomycins) and peptins (e.g., syringopeptins) (Melotto et al., 2006). Syringomycins and syringopeptins are synthesized by modular nonribosomal peptide synthases (NRPS), whose chromosomal genes are present, respectively, in the *syr* and *syp* clusters. Both types of phytotoxins induce necrosis in plant tissues and form pores in plant plasma membranes, thereby promoting transmembrane ion flux and cell death (Bender et al., 1999).

Tissue necrosis is the primary symptom induced by a phytotoxin (Barta et al., 1992). Chlorosis-inducing phytotoxins include COR, produced by *Ps. syringae* pvs. *atropurpurea*, *glycinea*, *maculicola*, *morsprunorum*, and *tomato*; phaseolotoxin, produced by *Ps. savastanoi* pv. *phaseolicola* and *Ps. syringae* pv. *Actinidiae*; and tabtoxin, produced by the pvs. *tabaci*, *coronafaciens*, and *garcae* of *Ps. syringae*. COR, which structurally resembles a polyketide, consists of two distinct moieties, coronafacic acid and coronamic acid, which function as intermediates in the biosynthetic pathway to COR and are fused by an amide bond (Bender, 1999). COR is also known to induce hypertrophy of storage tissue, compression of thylakoids, thickening of plant-cell walls, accumulation of protease inhibitors, inhibition of root elongation, and stimulation of ET production in some plant species (Weingart et al., 2003).

Pseudomonas syringae pv. *Syringae*, which causes mango apical necrosis disease, elaborates an antimetabolite mangotoxin, production of which is controlled by the chromosomal region of 11.1 kb. Six complete ORFs, including a large gene (ORF5) with a modular architecture characteristic of NRPS (*mgoA*), were identified in this chromosomal region. The involvement of *mgoA* in the virulence of the *Pss* strain that infects mango has been confirmed (Arrebola et al., 2007). A neutral leucine aminopeptidase (LAP-N) and an acidic LAP (LAP-A) are expressed constitutively in tomato plants during floral development and in leaves in response to infection by *Ps. syringae* pv. *tomato* (Pautot et al., 2001). *Burkholderia glumae* (which causes grain rot disease in rice seedlings) elaborates a phytotoxin, toxoflavin, that retards the growth of leaves and roots and also inducts chlorotic symptoms on panicles in the grain-rot phase. Toxoflavin might be synthesized in part through a biosynthetic pathway common to the synthesis of riboflavin, with which the toxin exhibited certain physical and chemical similarities (Yoneyama et al., 1998; Suzuki et al., 2004).

2.4.2 Plant Cell Wall-Degrading Enzymes

Extracellular enzymes that are able to degrade plant-cell walls are essential virulence factors for necrotrophic soft rot bacteria such as the soft rot erwiniae, now belonging to the *Pectobacterium* genus (e.g., *Pe. carotovorum* ssp. *carotovorum*, *Pectobacterium atrosepticum*, *Pe. chrysanthemi*). A combination of extracellular enzymes (i.e., pectate lyases, pectin methylesterases, pectin lyases, polygalacturonases, cellulase, and proteases) are involved in the depolymerization process of plant-cell walls provoked by these bacteria (Toth et al., 2003). Proteases are secreted by the type I secretion system, whereas the rest of the abovementioned enzymes are produced by the type II secretion system (Jha et al., 2005; Preston et al., 2005). Pectinases are the most important element of pathogenesis because they cause tissue maceration by degrading the pectic substances in the middle lamella and thus, indirectly, also cause cell death. Four main types of pectin-degrading enzymes are produced: three (pectate lyase, pectin lyase, and pectin methyl esterase) with a low alkaline pH optimum, and one, polygalacturonase, with a pH optimum about 6. All are present in many forms or isoenzymes, each encoded by independent genes (Agrios, 1997).

Among these enzymes, pectate lyases (Pels) are mainly involved in the virulence of soft rot *Pectobacterium* species (Toth et al., 2003). *Pectobacterium chrysanthemi* has five major Pel isoenzymes. These are encoded by the *pelA*, *pelB*, *pelC*, *pelD,* and *pelE* genes, which are organized in two clusters: *pelADE* and *pelBC* (Hugouvieux-Cotte-Pattat et al., 1996). Mutations in individual major *pel* genes do not bring about any significant changes in *Pe. chrysanthemi* virulence (Ried and Collmer, 1988; Beaulieu et al., 1993). In addition, deletion of all major *pel* genes not only failed to eliminate tissue maceration activity but also showed a set of secondary Pel isoenzymes (e.g., pelL, pelZ, and pelI) whose activities are only expressed in plants and appear to have an important role in both infection and host specificity (Ried and Collmer, 1988; Beaulieu et al., 1993; Lojkowska et al., 1995; Pissavin et al., 1996; Shevchik et al., 1997). Bauer et al. (1994) have demonstrated that an *hrp/hrc* cluster is also present in the genome of *Pe. chrysanthemi* and that mutations in this cluster provoke a slight virulence reduction in susceptible hosts.

2.4.3 Phytohormones

Phytohormones appear to have dual functions in plant pathogenesis. They are involved in both plant-defense response and disease progression (O'Donnell et al., 2003). The interactions of three phytohormones (ET, SA, and JA) are believed to regulate the specificity of plant-defense responses (Dong, 1998; Reymond and Farmer, 1998).

The gaseous hormone ET is a critical component of plant responses to pathogen attack, in addition to being essential for developmental processes such as fruit ripening and senescence. ET regulates a variety of growth and developmental processes, but it also mediates responses to a range of biotic and abiotic stresses in higher plants (Bleecker and Kende, 2000; Ciardi and Klee, 2001). Although ET is an important signaling component in plant–pathogen interactions, its role in pathogenesis and resistance is occasionally ambiguous (Knoester et al., 2001). ET, which increases disease-symptom development, is produced by various microorganisms, including plant pathogenic bacteria (Weingart and Völksch, 1997), and can be considered a virulence factor for some of them. *Pseudomonas syringae* pv. *glycinea* and *Ps. savastanoi* pv. *Phaseolicola*, which produce ET very efficiently, generate it by using 2-oxoglutarate as the substrate and the ET-forming enzyme (Weingart et al., 1999). In tomato, ET perception was needed for later stages of symptom development following inoculation with *Ps. syringae* pv. *tomato* and *Xanthomonas* spp., as characterized by extensive necrosis of infected tissue after initial lesion formation (Lund et al., 1998). In addition, ET has been reported to promote SA production in tomato in response to *Xa. campestris* pv. *Vesicatoria*, which is critical for the development of necrosis induced by this pathogen (O'Donnell et al., 2003).

Production of the auxins (e.g., indole-3-acetic acid, IAA) and cytokinins are important virulence factors for the gall-forming phytopathogenic bacteria *Pantoea agglomerans* pv. *gypsophilae*, the causal agent of crown and root gall disease of *Gypsophila paniculata*, and the pvs. *savastanoi* and *nerii* of *Ps. savastanoi*, which respectively incite olive and oleander knot diseases. Starting from L-tryptophan, these bacteria synthesize IAA by the indole-3-acetamide (IAM) route, respectively mediated by tryptophan-2-monooxygenase and indole-3-acetamide hydrolase and respectively encoded by *iaaM* and *iaaH* genes (Caponero et al., 1995; Brandl and Lindow, 1996; Manulis et al., 1998). A group of T-DNA genes in *Rh. radiobacter* directs the synthesis of plant-growth hormones that are required for the enlargement and proliferation of transformed plant cells, which results in the production of characteristic galls or tumors in the infected plants. The *iaaM* and *iaaH* gene products are involved in the conversion of tryptophan via indoleacetamide to indole acetic acid (auxin) (Bins and Costantino, 1998). Two other T-DNA genes, gene 5 and gene 6b (tml), are also thought to play ancillary roles in tumorigenesis (Hooykaas et al., 1988).

2.4.4 Plant Defense Responses against Bacterial Attacks

Plants are continually exposed to a vast number of potential phytopathogenic bacteria, against which they try to defend themselves through a multilayered system of passive and active defense mechanisms. Nonhost resistance has been overcome by individual phytopathogenic strains of a given bacterial species through the acquisition of virulence factors that enabled them to either evade or suppress plant-defense mechanisms. As a result of co-evolution between host plants and host bacteria, individual plant genotypes have evolved resistance genes that specifically recognize bacterial strains or race-specific factors and allow a plant to resist the infection of that particular race (e.g., race-specific resistance or cultivar-specific resistance) (Nürnberger et al., 2004). In general, cultivar-specific resistance conforms to the gene-for-gene hypothesis and is genetically determined by complementary pairs of bacterium-encoded avirulence (*Avr*) genes (Bonas and Van den Ackerveken, 1999; Mudgett, 2005) and plant resistance (*R*) genes (Martin et al., 2003).

Nonhost resistance of the type II and cultivar-specific resistances to phytopathogenic bacteria are typically associated with HR, which is characterized by the localized, rapid death of host cells at the infection site and contributes to limiting the growth and spread of the invading bacterium. However, some plant–bacterium interactions are characterized by nonhost resistance that does not involve the HR (e.g., type I nonhost resistance) (Mysore and Ryu, 2004). Instead, they mainly rely on basal resistance or innate immunity (Jones and Takemoto, 2004; Da Cunha et al., 2006). Basal resistance is elicited in plants by PAMPs or, more precisely, by MAMPs; basal resistance to bacteria is elicted by LPS (Dow et al., 2000), flagellin (Gomez-Gomez and Boller, 2002), and cold-shock protein (Felix and Boller, 2003). Nonpathogenic bacteria were shown to localize production of ROS and increase the expression of phenylpropanoid pathway genes (Alfano and Collmer, 2004; Jones and Takemoto, 2004; Soylu et al., 2005).

Preformed defenses, both structural and chemical, can discourage entry of bacteria into plant tissue and restrict their growth when ingress has been gained (Anderson, 1982). Particular morphological features of plant natural openings may contribute to plant resistance during the bacterium-entry stage of the infection process. Ramos et al. (1992) showed a relationship between frequency of stomata, stomatal size, some morphological leaf characteristics of tomato plants, and resistance to bacterial leaf spot disease caused by *Xa. campestris* pv. *vesicatoria*. Zinsou et al. (2006) postulated that the number of adaxial leaf stomata, together with leaf surface wax, contribute to the resistance of cassava to *Xa. axonopodis* pv. *manihotis*. Horino (1984) reported that the openings of the hydathode water pores of *Leersia japonica*, which is resistant to *Xa. oryzae* pv. *oryzae*, are narrower than those of the susceptible rice leaves.

When phytopathogenic bacteria are inside host tissues, preexisting structural and chemical plant defenses can restrict their spread and growth *in planta*. The movement in the vascular tissue of the xylem-inhabiting bacterium *C. michiganensis* ssp. *insidiosus*, the agent of bacterial wilt of alfalfa, is reduced in resistant alfalfa cultivars because they have fewer vascular bundles, shorter vessel elements, and thicker cortexes than susceptible cultivars do (Cho et al., 1973). The resistance of pear plants to *E. amylovora* seems to be associated with a high level of arbutin-hydroquinone, the antibacterial compound present in the exterior parts of the blossoms, where the plant is most susceptible to the bacterium (Schroth and Hildebrand, 1965). This compound is released from the glucoside arbutin through the action of β-glucosidase, an activity that is particularly elevated at these locations.

Structural and biochemical defenses induced *in planta* by bacterial infection can contribute to disease resistance. The formation of tylose and gum occlusions in the xylem vessels of grapevine seems to be associated with resistance to Pierce's disease, which is caused by *Xy. fastidiosa* (Fry and Milholland, 1990). However, further research is necessary to verify this assumption in light of contrasting results (Krivanek et al., 2005).

The overproduction of ROS, the so-called oxidative burst, is an early event that characterizes the HR induced by bacteria; it may, at least in part, generate unfavorable conditions for bacterial multiplication in the apoplast (Baker and Orlandi, 1995). This adverse environment for bacterial life could be due to the antimicrobial activity of hydrogen peroxide, which is strongly increased around bacterial cells, and to the oxidative cross-linking of the cell wall, which is driven by the rapid accumulation of hydrogen peroxide at plant-cell walls that are adjacent to attached bacteria (Bestwick et al., 1997). The oxidative burst associated with bacterially induced HR is generated by the plasma membrane NADPH oxidase and by apoplastic peroxidases. Aside from hydrogen peroxide, plant antibacterial molecules could be responsible for the restriction of bacterial growth observed during HR.

Phenolics have been proposed to modify cell-wall polysaccharides to resist the action of lytic enzymes (Matern et al., 1995). It is known that chitinases have a bifunctional activity of both chitinase and lysozyme and therefore may hydrolyze bacterial cell walls (Graham and Sticklen, 1994). Transcripts of an extracellular pepper class II basic chitinase, CAChi2, are highly expressed in pepper plants that are undergoing HR in response to *Xa. campestris* pv. *vesicatoria* (Hong et al., 2000). Lignin accumulation (Reimers and Leach, 1991; Lee et al., 2001) and changes in apoplastic water potential (Wright and Beattie, 2004) are additional possible post-infection resistance mechanisms. Apoplastic water potential is known to be a critical factor for the growth of phytopathogenic bacteria in plant tissue.

2.4.5 Induced Resistance

Salicylic acid has been recognized as a central molecule in the signal transduction pathway that leads to systemic acquired resistance (SAR) to pathogens (Sticher et al., 1997). Additionally, exogenous SA induces the expression of PR genes and decreases disease symptoms (White, 1979; Uknes et al., 1992, 1993). JA, a ubiquitous wound signal known to activate various defenses against herbivores, is also involved in the activation of plant-disease resistance and of various defense-associated responses (Pieterse et al., 1998; Kenton et al., 1999; van Wees et al., 2000). SA-, JA-, and ET-dependent defense pathways can affect each other's signaling; some of these interactions occur at the level of phytohormone production (Lund et al., 1998; Reymond and Farmer, 1998; Pieterse and Van Loon, 1999; van Wees et al., 2000; Pieterse et al., 2001; O'Donnell et al., 2003). Following HR or cell death caused by necrogenic pathogens, tissues distal to the infection sites can develop resistance to secondary infection by the same or different pathogens.

The antimicrobial compounds that are likely responsible for SAR resistance include a number of PR proteins that accumulate locally in plant tissues undergoing HR and systemically in tissues where SAR is detected (van Loon, 1997). Pathogenesis-related proteins with chitinase and β-1,3-glucanase activities may have a role in SAR to bacterial diseases. Another form of induced resistance is ISR (induced systemic resistance), which is potentiated by plant growth-promoting rhizobacteria and is also effective against bacterial diseases (Vallad and Goodman, 2004). Unlike SAR, ISR does not involve the accumulation of PRs or SA and relies on pathways regulated by JA and ET (Pieterse et al., 1996).

2.5 PATHOGENICITY OF PHYTOPLASMA PATHOGENS

Phytoplasmas, inducing yellowing, little leaf, and virescence of floral organs, are less-studied microorganisms because of the difficulty in culturing them in artificial media. DNA hybridization, PCR, and RFLP have been employed to detect, identify, and differentiate phytoplasmas. Very little information is available on the different stages of phytoplasma-induced pathogenesis. All of the phytoplasmas must be introduced into the susceptible host-plant cells as the first step in the process of infection, via specific insect vector(s) (Narayanasamy, 2008). The requirement of adhesion of mycoplasma to host cells for successful colonization of host tissues has been reported. This step is mediated by mycoplasma surface proteins; among them, adhesions are thought to play an important role in invasion and pathogenicty (Rottem, 2002). An adhesion-like gene in *Spiroplasma kunkelii*, the causal agent of corn stunt disease, has been identified (Wei et al., 2006).

Plant pathogenic phytoplasmas are endocellular. They lack cell walls that allow secreted phytoplasmal proteins to directly interact with host plant and insect vector cells, which suggests possible roles for these proteins. Immunodominant membrane proteins of phytoplasmas constitute a major portion of the total cellular membrane proteins in most phytoplasmas. The presence of genes that encode immune-dominant membrane proteins has been detected in phytoplasmas such as aster yellows, clover phyllody (Barbara et al., 2002), pear decline, peach yellow leaf roll, and European stone fruit yellows (Morton et al., 2003). Variations in amino acid sequence and antigenicity in these proteins have been observed (Barbara et al., 2002). The genes of onion yellows that encode the SecA and SecY proteins, which form essential components of the Sec protein translocation system, were identified; this indicates the existence of a Sec system in phytoplasmas (Kakizawa et al., 2001).

Because at present it is impossible to culture phytoplasmas, information on their protein-secretion systems is scant. Future research in this area may allow better understanding of the phenomenon of phytoplasma of pathogenesis (Kakizawa et al., 2004). Some studies meant to promote the understanding of the molecular basis of the interaction between *Spiroplasma citri* and the leafhopper vector species have been launched. The propagative nature of the relationship between *S. citri* and *Circulifer haemotoceps* involves multiplication of the pathogen in the leafhopper cells. In *S. citri*, an adhesion-related protein, P89, was shown to be directly involved in the spiroplasma–insect cell interaction (Yu et al., 2000). Cellular recognition mediated by carbohydrates and lectins has also been suggested, because the basal lamina of insect organs is highly glycosylated. Therefore, carbohydrates are likely to be the targets for *S. citri* interaction (Altmann, 1996). These results indicate a key role for spiralin in the transmission of *S. citri* by mediating spiroplasma adherence to epithelial cells of insect vector gut or salivary gland (Killiny et al., 2005).

2.6 A CASE SAMPLE: INFECTION PROCESS TO *RH. RADIOBACTER* TO HOSTS

Rhizobium radiobacter (*Agrobacterium tumefaciens*), the causal agent of crown gall in many fruit and forest trees, vegetables, and herbaceous dicotyledonous plants from 90 families, is able to genetically transform healthy host cells in tumoral cells by inserting T-DNA, contained on the Ti plasmid, into the plant genome. T-DNA carries genes involved in the synthesis of plant growth hormones (auxins and cytokinins) and synthesis and secretion of amino compounds called opines; these are tumor-specific compounds exclusively assimilated by the pathogen as major carbon and nitrogen sources. Chromosomal virulence (*chv*) genes are involved in bacterial chemotaxis toward and attachment to the wounded plant-cell wall. The importance of motility in the infection process is illustrated by the fact that mutations in genes that encode flagellin abolish motility and reduce tumorigenesis.

Binding of *Rh. radiobacter* to plant surfaces, an essential stage for establishing long-term interaction with the host, is dependent on several plant factors. The first may involve a variety of bacterial polysaccharides (i.e., cyclic glucans), which bind host polysaccharides. Mutations in the *chv* genes reduce the binding of the bacterium to cultured plant cells and abolish tumorigenesis. The second binding step requires the synthesis of bacterial cellulose. The expression of *vir* genes, another crucial event for tumorigenesis, is stimulated by plant-released signals; these include specific phenolic compounds and monosaccharides in combination with acidic pH (5.2–5.7) and temperatures below 30°C. *Vir*-inducing conditions mainly occur during wounding, an event that is generally thought to be required for tumorigenesis.

At wound sites, phenolics accumulate as precursors of lignin biosynthesis during the wound-healing process; monosaccharides originate from mechanical and enzymatic degradation of plant cell-wall polysaccharides; and the pH level tends to be acidic, given the presence of acidic compounds in the wound sap. Conditions that induce *vir* genes are perceived by the VirA-VirG two-component system and the ChvE sugar-binding protein. After detecting plant-released phenolics, monosaccharides, and acidity, VirA (a transmembrane histidine kinase) phosphorylates VirG, which is the DNA-binding response regulator. Moreover, monosaccharides seem to be indirectly detected by VirA through ChvE, a periplasmic sugar-binding protein that is required for chemotaxis toward and uptake of monosaccharides and interacts directly with the VirA periplasmic domain. Phosphorylated VirG binds to the *vir* box, a conserved sequence in the promoter regions of *vir* genes, thereby activating the transcription of these genes. The successive cellular process involved in *Rh. radiobacter* tumorigenesis is the cleavage of T-DNA from the Ti plasmid by the action of VirD1 and VirD2. In addition, the VirD2 molecule covalently attaches to the 5′ end of the T-strand and forms the immature T-complex. Along with several other Vir proteins, the immature T-complex is exported into the host cell by a VirB/D4 type IV secretion system, which is encoded by the *virB* operon and the *virD4* gene.

The *virB* gene products, termed the mating pair formation (Mpf) proteins, elaborate a cell envelope-spanning structure required for substrate transfer, as well as an extracellular appendage (T pilus) that mediates attachment to recipient cells. Once inside the host-cell cytoplasm, the T-DNA is thought to exist as a mature T-complex (T-complex) in which the entire length of the T-strand molecule is coated with numerous VirE2 molecules; the latter confer to the T-DNA the structure and protection needed for its journey to the host-cell nucleus (Sheng and Citovsky, 1996; Chesnokova et al., 1997; Matthysse and Kijne, 1998; Christie, 2004; Brencic and Winans, 2005; Li et al., 2005; Tzfira and Citovsky, 2006).

2.7 CONCLUSION

Bacteria colonize the apoplast; that is, a plant's intercellular spaces or xylem vessels. The fitness of a bacterial pathogen can be quantified by measuring its reproductive rate, rate of multiplication, efficiency of infection, and amount of disease. New biotypes of bacteria arise by means of at least three sexual-like processes: conjugation, transformation, and transduction. A QS process has been shown to be involved in the regulation of many important physiological functions of bacteria such as symbiosis, conjugation, and virulence. Pathogenicity factors encoded by *pat* and *dsp* genes are involved in steps that are crucial for the establishment of disease. There are four requirements of pathogenicity: entry to the host tissues through these surfaces, multiplication in the environment *in vivo*, interference with host defense mechanisms, and damage to the host. Some plant cell wall-degrading enzymes, toxins, hormones, polysaccharides, proteinases, and siderophores are produced by bacterial pathogens in pathogen–plant interactions. Many plant bacterial pathogens require type III secretion systems to mount successful invasions of susceptible host-plant species.

In many host–pathogen interactions, plants react to attack by pathogens with enhanced production of SA, JA, and ET while a distinct set of gene-to-gene resistance defense-related genes is activated and attempts to block the infection. The antimicrobial compounds likely responsible for SAR resistance include a number of PR proteins that accumulate locally in plant tissues undergoing HR and systemically in tissues where SAR is detected. ISR does not involve the accumulation of PRs or SA and relies on pathways regulated by JA and ET. The severity of disease symptoms caused by bacteria in plants is determined by the production of a number of virulence factors, including phytotoxins, plant cell wall-degrading enzymes, EPSs, and phytohormones. Nonhost resistance has been overcome by individual phytopathogenic strains of a given bacterial species through the acquisition of virulence factors, which enable them to either evade or suppress plant defense mechanisms.

In general, cultivar-specific resistance conforms to the gene-for-gene hypothesis and is genetically determined by complementary pairs of bacterium-encoded *avr* genes and plant *R* genes. The overproduction of ROS is an early event that characterizes the HR induced by bacteria, which may generate unfavorable conditions for bacterial multiplication in the apoplast. Phenolics have been proposed to modify cell-wall polysaccharides to resist the action of lytic enzymes. Lignin accumulation and changes in apoplastic water potential are additional possible post-infection resistance mechanisms to bacterial infection.

REFERENCES

Abramovitch, R.B., Kim, Y.J., Chen, S., Dickman, M.B., Martin, G.B. 2003. *Pseudomonas* type III effector AvrPtoB induces plant disease susceptibility by inhibition of host programmed cell death. *EMBO Journal* 22: 60–69.

Agrios, G.N. 1997. *Plant Pathology*, 4th edn. Amsterdam, The Netherlands: Elsevier/Academic.

Agrios, G.N. 2005. *Plant Pathology*, 5th edn. Amsterdam, The Netherlands: Elsevier/Academic, 948p.

Alfano, J.R., Collmer, A. 1997. The type III (Hrp) secretion pathway of plant pathogenic bacteria: Trafficking harpins, Avr proteins, and death. *Journal of Bacteriology* 179: 5655–5662.

Alfano, J.R., Collmer, A. 2004. Type III secretion system effector proteins: Double agents in bacterial disease and plant defense. *Annual Reviews of Phyopathology* 42: 385–414.

Altmann, F. 1996. *N*-glycosylation in insects revisited. *Trends in Glycoscience and Glycotechnology* 8: 101–114.

Anderson, A.J. 1982. Preformed resistance mechanisms. In *Phytopathogenic Prokaryotes*, Vol. 2. Mount, M.S., Lacy, G.H. (eds.). New York: Academic Press, pp. 119–136.

Arnold, D.L., Pitman, A., Jackson, R.W. 2003. Pathogenicity and other genomic islands in plant pathogenic bacteria. *Molecular Plant Pathology* 4: 407–420.

Arrebola, E., Cazorla, F.M., Romero, D., Pérez-García, A., de Vicente, A. 2007. A nonribosomal peptide synthetase gene (*mg*A) of *Pseudomonas syringae* pv. *syringae* is involved in mangotoxin biosynthesis and is required for full virulence. *Molecular Plant-Microbe Interactions* 20: 500–509.

Asai, T., Tena, G., Plotnikova, J., Willmann, M.R., Chiu, W.L., Gomez-Gomez, L., Boller, T. et al. 2002. MAP kinase signaling cascade in *Arabidopsis* innate immunity. *Nature* 415: 977–983.

Ausubel, F.M. 2005. Are innate immune signaling pathways in plants and animals conserved? *Nature Reviews Immunology* 6: 973–979.

Axtell, M.J., Staskawicz, B.J. 2003. Initiation of RPS2-specified disease resistance in *Arabidopsis* is coupled to the AvrRpt2 directed elimination of RIN4. *Cell* 112: 369–377.

Baker, C.J., Orlandi, E.W. 1995. Active oxygen in plant pathogenesis. *Annual Review of Phytopathology* 33: 299–321.

Banar, D.W., Bogdanove, A.J., Beer, S.V., Collmer, A. 1994. *Erwinia chrysanthemi hrp* genes and their involvement in soft rot pathogenesis and elicitation of the hypersensitive response. *Molecular Plant-Microbe Interactions* 7: 573–581.

Barbara, D.J., Morton, A., Clark, M.F., Davies, D.L. 2002. Immuno-dominant membrane proteins from two phytoplasmas in the aster yellows clade (chloroante aster yellows and clover phyllody) are highly divergent in the major hydrophilic region. *Microbiology* 148: 157–167.

Barber, C.E., Tang, J.L., Feng, J.X., Pan, M.Q., Wilson, T.J.G., Slater, H. 1997. A novel regulatory system required for pathogenicity of *Xanthomonas campestris* is mediated by a small diffusible signal molecule. *Molecular Microbiology* 24: 555–566.

Barny, M.A., Gaudriault, S., Brisset, M.N., Paulin, J.P. 1999. Hrp-secreted proteins: Their role in pathogenicity and HR-elicitation. *Acta Horticulturae* 489: 353–358.

Barta, T.M., Kinscherf, T.G., Willis, D.K. 1992. Regulation of tabtoxin production by the *lemA* gene in *Pseudomonas syringae*. *Journal of Bacteriology* 174: 3021–3029.

Bauer, D. W., Bogdanove, A. J., Beer, S. V., Collmer, A. 1994. *Erwinia chrysanthemi hrp* genes and their involvement in soft rot pathogenesis and elicitation of the hypersensitive response. *Molecular Plant-Microbe Interactions* **7**: 573–581.

Beaulieu, C., Boccara, M., Vangijsegem, F. 1993. Pathogenic behaviour of pectinase-defective *Erwinia chrysanthemi* mutants on different plants. *Molecular Plant-Microbe Interactions* 6: 197–202.

Beier, D., Gross, R. 2006. Regulation of bacterial virulence by two component systems. *Current Opinion in Microbiology* 9, 143–152.

Bellemann, P., Bereswill, S., Berger, S., Geider, K. 1994. Visualization of capsule formation by *Erwinia amylovora* and assays to determine amylovoran synthesis. *International Journal of Biological Macromolecules* 16: 290–296.

Bellemann, P., Geider, K. 1992. Localization of transposon insertions in pathogenicity mutants of *Erwinia amylovora* and their biochemical-characterization. *Journal of General Microbiology* 138: 931–940.

Bender, C.L., Alarcón-Chaidez, F., Gross, D.C. 1999. *Pseudomonas syringae* phytotoxins: Mode of action, regulation, and biosynthesis by peptide and polyketide synthetases. *Microbiology and Molecular Biology Reviews* 63: 266–292.

Berks, B.C., Palmer, T., Sargent, F. 2003. The Tat protein translocation pathway and its role in microbial physiology. *Advances in Microbial Physiology* 47: 187–254.

Berks, B.C., Sargent, F., Palmer, T. 2000. The Tat protein export pathway. *Molecular Microbiology.* 35: 260–274.

Bermpohl, A., Dreier, J., Bahro, R., Eichenlaub, R. 1996. Exopolysaccharides in the pathogenic interaction of *Clavibacter michiganensis* subsp *michiganensis* with tomato plants. *Microbiology Research* 151: 391–399.

Bernhard, F., Coplin, D.L., Geider, K. 1993. A gene cluster for amylovoran synthesis in *Erwinia amylovora*: Characterization and relationship to *cps* genes in *Erwinia stewartii. Molecular General Genetics* 239: 158–168.

Bestwick, C.S., Brown, I.R., Bennett, M., Mansfield, J.W. 1997. Localization of hydrogen peroxide accumulation during the hypersensitive reaction of lettuce cells to *Pseudomonas syringae* pv *phaseolicola. Plant Cell* 9: 209–221.

Bins, A.N., Costantino, P. 1998. The *Agrobacterium* oncogenes. In *The Rhizobiaceae (Spainik)*. Kondorsi, H.P., Hooykaas, P.J.J. (eds.). Dordrecht, The Netherlands: Kluwer Academic Publishers, 252–256.

Bleecker, A.B., Kende, H. 2000. Ethylene: A gaseous signal molecule in plants. *Annual Review of Cell and Developmental Biology* 16: 1–18.

Bogdanove, A.J., Kim, J.F., Wei, Z.M., Kolchinsky, P., Charkowski, A.O., Collmer, A., Beer, S.V. 1998. Homology and functional similarity of an hrp-linked pathogenicity locus *dspEF* of *Erwinia amylovora*. *Proceedings of the National Academy of Science of the United States of America* 95: 1325–1330.

Bonas, U., Van den Ackerveken, G. 1999. Gene-for-gene interactions: Bacterial avirulence proteins specify plant disease resistance. *Current Opinion in Microbiology* 2: 94–98.

Brandl, M.T., Lindow, S.E. 1996. Cloning and characterization of a locus encoding an indolepyruvate decarboxylase involved in indole-3-acetic acid synthesis in *Erwinia herbicola*. *Applied Environmental Microbiology* 62: 4121–4128.

Brencic, A., Winans, S.C. 2005. Detection and response to signals involved in host–microbe interactions by plant associated bacteria. *Microbiology and Molecular Biology Reviews* 69: 155–194.

Bronstein, P.A., Marrichi, M., Cartinhour, S., Schneider, D.J., DeLisa, M.P. 2005. Identification of a twin-arginine translocation system in *Pseudomonas syringae* pv. *tomato* DC 3000 and its contribution to pathogenicity and fitness. *Journal of Bacteriology* 187: 8350–8461.

Buonaurio, R. 2008. Infection and plant defense responses during plant–bacterial interaction. In *Plant-Microbe Interactions* Barka, E.A., Clément, C. (eds.). Trivandrum, Kerala: India Research Signpost, pp. 169–197.

Cao, T., Duncan, R.A., McKenry, M.V., Shackel, K.A., DeJong, T.M., Kirkpatrick, B.C. 2005. Interaction between nitrogen-fertilized peach trees and expression of *syrB*, a gene involved in syringomycin production in *Pseudomonas syringae* pv. *syringae*. *Phytopathology* 95: 581–586.

Caponero, A., Contesini, A.M., Iacobellis, N.S. 1995. Population diversity of *Pseudomonas syringae* subsp. *savastanoi* on olive and oleander. *Plant Pathol* 44: 848–855.

Chen, C.Y., Lu, Y.Y., Chung, J.C. 2003. Induced host resistance against Botrytis leaf blight. İn *Advances in Plant Disease Management*. Huang, H.C., Acharya, S.N. (eds.). Research Signpost: Trivandrum, Kerala, India, pp. 259–267.

Chesnokova, O., Coutinho, J.B., Khan, I.H., Mikhail, M.S., Kado, C.I. 1997. Characterization of flagella genes of *Agrobacterium tumefaciens*, and the effect of a bald strain on virulence. *Molecular Microbiology* 23: 579–590.

Cho, Y.S., Wilcoxson, R.D., Frosheiser, F.I. 1973. Differences in anatomy, plant-extracts, and movement of bacteria in plants of bacterial wilt resistant and susceptible varieties of alfalfa. *Phytopathology* 63: 769–765.

Christie, P.J. 2004. Type IV secretion: The *Agrobacterium* VirB/D4 and related conjugation systems. *Biochimica et Biophysica Acta-Molecular Cell Research* 1694: 219–234.

Ciardi, J., Klee, H. 2001. Regulation of ethylene-mediated responses at the level of the receptor. *Annals of Botany* 88: 813–822.

Cohn, J., Martin, G.B. 2003. Pathogen recognition and signal transduction in plant immunity. In *Infection, Disease: Innate Immunity*. Ezekowitch, R.A.B.H., Hoffman, J.A. (eds.).. Totowa, NJ: Humana Press Inc., pp. 3–27.

Collmer, A., Badel, J.L., Charkowski, A.O., Deng, W.L., Fouts, D.W., Ramos, A.R., Rehm, A.H. et al. 2000. *Pseudomonas syringae* pv. *syringae* Hrp type III secretion system and effector proteins. *Proceedings of the National Academy of Science of the United States of America* 97: 8770–8777.

Collmer, A., Lindeberg, M., Petnicki-Ocwieja, T., Schneider, D.J., Alfano, R. 2002. Genomic mining type III secretion system effectors in *Pseudomonas syringae* yields new picks for all TTSS prospectors. *Trends in Microbiology* 10: 462–469.

Condemine, G., Castillo, A., Passeri, F., Enard, C. 1999. The PecT repressor coregulates synthesis of exopolysaccharides and virulence factors in *Erwinia chrysanthemi*. *Molecular Plant-Microbe Interactions* 12: 45–52.

Cristobal, S.J., deGier, J.W., Nielsen, H., von Heijne, G. 1999. Competition between Sec- and TAT-dependent protein translocation in *Escherichia coli*. *EMBO Journal* 18: 2981–2990.

Da Cunha, L., McFall, A.J., Mackey, D. 2006. Innate immunity in plants: A continuum of layered defenses. *Microbes Infection* 8: 1372–1381.

Dellagi, A., Brisset, M.-N., Paulin, J.-P., Expert, D. 1998. Dual role of desferrioxamine in *Erwinia amylovora* pathogenicity. *Molecular Plant-Microbe Interactions* 11: 734–742.

Denny, T.P., Schell, M.A. 1994. Virulence and pathogenicity of *Pseudomonas solanacearum*: Genetic and biochemical perspectives. In *Bacterial Pathogenesis and Disease Resistance*. Bills, D., Kung, S.D. (eds.). River Edge, NJ: World Science.

Denny, T.P. 1995. Involvement of bacterial polysaccharides in plant pathogenesis. *Annual Review of Phytopathology* 33: 173–197.

Denny, T.P. 1999. Auto regulator-dependent control of extracellular polysaccharide production in phytopathogenic bacteria. *European Journal of Plant Pathology* 105: 417–430.

Dong, X.N. 1998. SA, JA, ethylene, and disease resistance in plants. *Current Opinion in Plant Biology* 1: 316–323.

Dorman, C.J., Deighan, P. 2003. Regulation of gene expression by histone-like proteins in bacteria. *Current Opinion in Genetics and Development* 13: 179–184.

Dorman, C.J. 2004. H-NS: A universal regulator for a dynamic genome. *Nature Reviews Microbiology* 2: 391–400.

Dow, J.M., Crossman, L., Findlay, K., He, Y.Q., Feng, J.X., Tang, J.L. 2003. Biofilm dispersal in *Xanthomonas campestris* is controlled by cell–cell signaling and is required for full virulence to plants. *Proceedings of the National Academy of Science of the United States of America* 100: 10995–11000.

Dow, J.M., Newman, M.A., von Roepenack, E. 2000. The induction and modulation of plant defense responses by bacterial lipopolysaccharides. *Annual Review of Phytopathology* 38: 241–261.

Dreier, J., Meletzus, D., Eichenlaub, R. 1997. Characterization of the plasmid encoded virulence region *pat-1* of phytopathogenic *Clavibacter michiganensis* subsp. *michiganensis*. *Molecular Plant-Microbe Interactions* 10: 195–206.

Duan, Y.P., Castaneda, A., Zhao, G., Erdos, G., Gabriel, D.W. 1999. Expression of a single, host-specific bacterial pathogenicity gene in plant cells elicits divisions, enlargement and cell death. *Molecular Plant-Microbe Interactions* 12: 556–560.

Eriksson, R.R., Anderson, R.A., Pirhonen, M., Palva, E.T. 1998. Two component regulators involved in the global control of virulence of *Erwinia carotovora* subsp. *carotovora*. *Molecular Plant-Microbe Interactions* 11: 743–752.

Feldman, M.F., Cornelis, G.R. 2003. The multitalented type III chaperones, all you can do with 15-kDa. *FEMS Microbiology Letters* 219: 151–158.

Felix, G., Boller, T. 2003. Molecular sensing of bacteria in plants. The highly conserved RNA binding motif RNP-1 of bacterial cold shock proteins is recognized as an elicitor signal in tobacco. *Journal of Biological Chemistry* 278: 6201–6208.

Fellbrich, G., Romanski, A., Varet, A., Blume, B., Brunner, F., Engelhardt, S., Felix, G. et al. 2002. NPP1, a *Phytophthora*-associated trigger of plant defense in parsley and *Arabidopsis*. *Plant Journal* 32: 375–390.

Fry, S.M., Milholland, R.D. 1990. Response of resistant, tolerant, and susceptible grapevine tissues to invasion by the Pierces disease bacterium, *Xylella fastidiosa*. *Phytopathology* 80: 66–69.

Fuchs, T.M. 1998. Molecular mechanisms of bacterial pathogenicity. *Naturwissenschaften* 85: 99–108.

Geider, K., Zhang, Y., Ullrich, H., Langlotz, C. 1999. Expression of virulence factors secreted by *Erwinia amylovora*. *Acta Horticulturae* 489: 347–351.

Geier, G., Geider, K. 1993. Characterization and influence on virulence of the levansucrase gene from the fireblight pathogen *Erwinia amylovora*. *Physiological and Molecular Plant Pathology* 42: 387–404.

Gomez-Gomez. L., Boller, T. 2002. Flagellin perception: A paradigm for innate immunity. *Trends in Plant Science* 7: 251–256.

Graham, L.S., Sticklen, M.B. 1994. Plant chitinase. *Canadian Journal of Botany* 72: 1057–1083.

Gross, R. 1993. Signal transduction in human and animal pathogens. *FEMS Microbiol Rev* 104: 301–326.

Hao, G., Zhang, H., Zheng, D., Burr, T.J. 2005. *luxR* homolog *avhR* in *Agrobacterium vitis* affects the development of a grape-specific necrosis and a tobacco hypersensitive response. *Journal of Bacteriology* 187(1): 185–192.

Hauck, P., Thilmony, R., He, S.Y. 2003. A *Pseudomonas syringae* type III effector suppresses cell wall-based extracellular defense in susceptible *Arabidopsis* plants. *Proceedings of the National Academy of Science of the United States of America* 100: 8577–8582.

He, S.Y. 1998. Type III protein secretion systems in plant and animal pathogenic bacteria. *Annual Review of Phytopathology* 36: 363–392.

He, Y.W., Xu, M., Lin, K., Ng, Y.J.A., Wen, C.M., Wang, L.H., Liu, Z.D. et al. 2006. Genome scale analysis of diffusible signal factor regulon in *Xanthomonas campestris* pv. *campestris*: Identification of novel cell-cell communication-dependent genes and functions. *Molecular Microbiology* 59: 610–622.

Heurlier, K., Denervaud, V., Pessi, G., Reimmann, C., Haas, D. 2003. Negative control of quorum sensing by RpoN (_54) in *Pseudomonas aeruginosa* PAO1. *Journal of Bacteriology* 185: 2227–2235.

Hildebrand, M., Aldridge, P., Geider, K. 2006. Characterization of *hns* genes from *Erwinia amylovora*. *Molecular Genetics and Genomics* 275: 310–319.

Hoch, J.A., Silhavy, T.J. 1995. *Two-Component Signal Transduction*. Washington, DC: ASM Press.

Hong, J.K., H.W. Jung, Y.J. Kim, Hwang, B.K. 2000. Pepper gene encoding a basic class II chitinase is inducible by pathogen and ethephon. *Plant Science* 159: 39–49.

Hooykaas, P.J.J., den Dult-Ras, H., Schilperont, R.A. 1988. The *Agrobacterium tumefaciens* T-DNA gene *6b* is an *onc* gene. *Plant Molecular Biology* 11: 791–794.

Horino, O. 1984. Ultrastructure of water pores in *Leersia japonica* Makino and *Oryza sativa* L.: Its correlation with the resistance to hydathodal invasion of *Xanthomonas campestris* pv. *oryzae*. *Annals of the Phytopathological Society of Japan* 50: 72–76.

Hossain, M.M., Shibata, S., Aizawa, S.I., Tsuyumu, S. 2005. Motility is an important determinant for pathogenesis of *Erwinia carotovora* subsp. *carotovora*. *Physiological and Molecular Plant Pathology* 66: 134–143.

Hossain, M.M., Tsuyumu, S. 2006. Flagella-mediated motility is required for biofilm formation by *Erwinia carotovora* subsp. *carotovora*. *Journal of General Plant Pathology* 72: 34–39.

Huang, J., Schmelz, E.A., Alborn, H., Engelberth, J., Tumlinson, J.H. 2005. Phytohormones mediate volatile emissions during the interaction of compatible and incompatible pathogens: The role of ethylene in *Pseudomonas Syringae* infected tobacco. *Journal of Chemical Ecology* 31(3): 439–459.

Hugouvieux-Cotte-Pattat, N., Condemine, G., Nasser, W., Reverchon, S. 1996. Regulation of pectinolysis in *Erwinia chrysanthemi*. *Annual Review of Microbiology* 50: 213–257.

Jackson, R.W., Athanassopoulos, E., Tsiamis, G., Mansfield, J.W., Sesma, A., Arnold, D.L., Gibbon, M.J. et al. 1999. Identification of a pathogenicity island which contains genes for virulence and avirulence on a large native plasmid in the bean pathogen *Pseudomonas syringae* pv. *phaseolicola*. *Proceedings of the National Academy of Science of the United States of America* 96: 10875–10880.

Jahr, H., Bahro R., Eichenlaub, R. 1999. Genetics of phytopathology: Phytopathogenic bacteria. In *Progress in Botany*, Vol. 60. Esser, K. (ed.). Berlin, Heidelberg: Springer-Verlag, pp. 119–138.

Jamir, Y., Guo, M., Oh, H.S., Petnicki-Ocwieja, T., Chen, S., Tang, X., Dickman, M.B. et al. 2004. Identification of *Pseudomonas syringae* type III effectors that can suppress programmed cell death in plants and yeast. *Plant Journal* 37: 554–565.

Jha, G., Rajeshwari, R., Sonti, R.V. 2005. Bacterial type two secretion system secreted proteins: Double-edged swords for plant pathogens. *Molecular Plant-Microbe Interactions* 18: 891–898.

Jones, D.A., Takemoto, D. 2004. Plant innate immunity-direct and indirect recognition of general and specific pathogen-associated molecules. *Current in Opinion in Immunology* 16: 48–62.

Kakizawa, S., Oshima, K., Kuboyama, T. 2001. Cloning and expression analysis of phytoplasma translocation genes. *Molecular Plant-Microbe Interactions* 14: 1043–1050.

Kakizawa, S., Oshima, K., Nishigawa, H., Jung, H.Y., Wei, W., Suzuki, S., Tanaka, M. et al. 2004. Secretion of immunodominant membrane protein from onion yellows phytoplasma through Sec protein-translocation system in *Escherichia coli*. *Microbiology* 150: 135–142.

Kenton, P., Mur, L.A.J., Atzorn, R., Wasternack, C., Draper, J. 1999. Jasmonic acid accumulation in tobacco hypersensitive response lesions. *Molecular Plant-Microbe Interactations* 12: 74–78.

Killiny, N., Catroviejo, M., Saillard, C. 2005. *Spiroplasma citri* spiralin acts *in vitro* as a lectin binding to glycoproteins from its insect vector *Circulifer haematoceps*. *Phytopathology* 95: 541–548.

Kim, H.S., Desveaux, D., Singer, A.U., Patel, P., Sondek, J., Dangl, J.L. 2005. The *Pseudomonas syringae* effector AvrRpt 2 cleaves its C-terminally acylated target, RIN4, from *Arabidopsis* membranes to block RPM1 activation. *Proceedings of the National Academy of Science of the United States of America* 102: 6496–6501.

Kim, J.F., Alfano, J.R. 2002. Pathogenicity islands and virulence plasmids of bacterial plant pathogens. *Current Topics in Microbiology and Immunology* 264: 127–147.

Kim, J.G., Jeon, E., Oh, J., Moon, J.S., Hwang, I. 2004. Mutational analysis of *Xanthomonas* harpin HpaG identifies a key functional region that elicits the hypersensitive response in nonhost plants. *Journal of Bacteriology* 186: 6239–6247.

Kloek, A.P., Brooks, D.M., Kunkel, B.N. 2000. A *dsbA* mutant of *Pseudomonas syringae* exhibits reduced virulence and partial impairment of type III secretion. *Molecular Plant Pathology* 1: 139–150.

Knoester, M., Linthorst, H.J.M., Bol, J.F., Van Loon, L.C. 2001. Involvement of ethylene in lesion development and systemic acquired resistance in tobacco during hypersensitive reaction to tobacco mosaic virus. *Physiology and Molecular Plant Pathology* 59: 45–57.

Koutsoudis, M.D., Tsaltas, D., Minogue, T.D., von Bodman, S.B. 2006. Quorum-sensing regulation governs bacterial adhesion, biofilm development and host colonization in *Pantoea stewartii* subsp *stewartii*. *Proceedings of the National Academy of Science of the United States of America* 103: 5983–5988.

Krivanek, A.F., Stevenson, J.F., Walker, M.A. 2005. Development and comparison of symptom indices for quantifying grapevine resistance to Pierce's disease. *Phytopathology* 95: 36–43.

Kunkel, B.N., Z. Chen. 2006. Virulence strategies of plant pathogenic bacteria. In *The Prokaryotes*. Dworkin, M., Falkow, S., Rosenberg, E., Schleifer, K.H., Stackebrandt, E. (eds.). New York: Springer.

Lavin, J.L., Kiil, K., Resano, O., Ussery, D.W., Oguiza, J.A. 2007. Comparative genomic analysis of two-component regulatory proteins in *Pseudomonas syringae*. *BMC Genomics* 8: 397.

Lee, S., Sharma, Y., Lee, T.K., Chang, M., Davis, K.R. 2001. Lignification induced by pseudomonads harboring avirulent genes on *Arabidopsis*. *Molecules and Cells* 12: 25–31.

Leister, R.T., Katagiri, F. 2000. A resistance gene product of the nucleotide binding site/leucine rich repeats class can form a complex with bacterial avirulence proteins in vivo. *Plant Journal* 22: 345–354.

Li, J.X., Wolf, S.G., Elbaum, M., Tzfira, T. 2005a. Exploring cargo transport mechanics in the type IV secretion systems. *Trends in Microbiology* 13: 295–298.

Li, X., Lin, H., Zhang, W., Zou, Y., Zhang, J., Tang, X., Zhou, J.M. 2005b. Flagellin induces innate immunity in nonhost interaction that is suppressed by *Pseudomonas syringae* effectors. *Proceedings of the National Academy of Science of the United States of America* 102: 12990–12995.

Lin, N.C., Martin, G.B. 2005. An avrPto/avrPtoB mutant of *Pseudomonas syringae* pv. *tomato* DC3000 does not elicit Pto-mediated resistance and is less virulent on tomato. *Molecular Plant-Microbe Interaction* 18: 43–51.

Lindberg, M., Collmer, A. 1992. Analysis of eight *out* genes in a cluster required for pectic enzyme secretion by *Erwinia chrysanthemi*: Sequence comparison with secretion genes from other gram negative bacteria. *Journal of Bacteriology* 174: 7385–7397.

Lindgren, P.B. 1997. The role of *hrp* genes during plant–bacterial interactions. *Annual Review of Phytopathology* 35: 129–152.

Lojkowska, E., Masclaux, C., Boccara, M., Robertbaudouy, J., Hugouvieux-Cotte-Pattat, N. 1995. Characterization of the *PelL* gene encoding a novel pectate lyase of *Erwinia chrysanthemi*-3937. *Molecular Microbiology* 16: 1183–1195.

Lu, S.E., Wang, N., Wang, J., Chen, Z.J., Gross, D.C. 2005. Oligonucleotide microarray analysis of the Sal A regulon controlling phytotoxin production by *Pseudomonas syringae* pv. *syringae. Molecular Plant-Microbe Interactions* 18: 324–333.

Lund, S.T., Stall, R.E., Klee, H.J. 1998. Ethylene regulates the susceptible response to pathogen infection in tomato. *Plant Cell* 10: 371–382.

Mackey, D., Belkhadir, Y., Alonso, J.M., Ecker, J.R., Dangl, J.L. 2003. *Arabidopsis* RIN4 is a target of the type III virulence effector Avr Rpt2 and modulates *RPS2* mediated resistance. *Cell* 112: 379–389.

Makino, S., Sugio, A.,White, F., Bogdanove, A.J. 2006. Inhibition of resistance gene mediated defense in rice by *Xanthomonas oryzae* pv. *oryzicola. Molecular Plant-Microbe Interactions* 19: 240–249.

Manulis, S., Haviv-Chesner, A., Brandl, M.T., Lindow, S.E., Barash, I. 1998. Differential involvement of indole-3-acetic acid biosynthetic pathways in pathogenicity and epiphytic fitness of *Erwinia herbicola* pv, *gypsophilae. Molecular Plant-Microbe Interactions* 11: 634–642.

Marklund, B.I., Tennent, J.M., Garcia, E., Hamers, A., Baga, M., Lindberg, F., Gaastra, W., Normark, S. 1992. Horizontal gene transfer of the *Escherichia coli pap* and *prs* pili operons as a mechanism for the development of tissue-specific adhesive properties. *Molecular Microbiology* 6: 2225–2242.

Martin, G.B., Bogdanove, A.J., Sessa, G. 2003. Understanding the functions of plant disease resistance proteins. *Annual Reviews of Plant Biology* 54: 23–61.

Matern, U., Grimmig, B., Kneusel, R.E. 1995. Plant-cell wall reinforcement in the disease resistance response—Molecular composition and regulation. *Canadian Journal of Botany* 73: S511–S517.

Matthysse, A.G., Kijne, J.W. 1998. Attachment of Rhizobiaceae to plant cells. In *The Rhizobiaceae*. Spaink, H.P., Kondorosi, A., Hooykaas, P.J.J. (eds.). Dordrecht, The Netherlands: Kluwer Academic Publishers, pp. 235–249.

McKenny, D., Brown, K.E., Allison, D.G. 1995. Influence of *Pseudomonas aeruginosa* exoproducts on virulence of *Burkholderia cepacia*: Evidence for interspecies communication. *Journal of Bacteriology* 177: 6989–6992.

Medzhitov, R., and Janeway, Jr., C.A. 2002. Decoding the patterns of self and nonself by the innate immune system. *Science* 296: 298–300.

Melotto, M., Underwood, W., Koczan, J., Nomura, K., He, S.Y. 2006. Plant stomata function in innate immunity against bacterial invasion. *Cell* 126: 969–980.

Meng, X., Bonasera, J.M., Kim, J.F., Nissimen, R.M., Beer, S.V. 2006. Apple proteins that interact with Dsp A/E, a pathogenicity effector of *Erwinia amylovora*, the fire blight pathogen. *Molecular Plant-Microbe Interactions* 19: 53–61.

Miller, M.B., Bassler, B.L. 2001. Quorum sensing in bacteria. *Annual Review of Microbiology* 55: 165–199.

Mole, B.M., Baltrus, D.A., Dangl, J.L., Grant, S.R. 2007. Global virulence regulation networks in phytopathogenic bacteria. *Trends in Microbiology* 15: 363–371.

Morales, C.Q., Posada, J., Macneale, E., Franklin, D., Rivas, I., Bravo, M., Minsavage, J. et al. 2005. Functional analysis of the early chlorosis factor gene. *Molecular Plant-Microbe Interactions* 18: 477–486.

Morton, A., Davies, D.L., Blomquist, C.L., Barbara, D.J. 2003. Characterization of homologues of the apple proliferation immunodominant membrane protein gene from three related phytoplasmas. *Molecular Plant Pathology* 4: 109–114.

Mudgett, M.B. 2005. New insights to the function of phytopathogenic bacterial type III effectors in plants. *Annual Review of Plant Biology* 56: 509–531.

Mysore, K.S., Ryu, C.M. 2004. Nonhost resistance: How much do we know? *Trends in Plant Science* 9: 97–104.

Narayanasamy, P. 2002. *Microbial Plant Pathogens and Crop Disease Management*. Enfield, CT: Science Publishers.

Narayanasamy, P. 2008. *Molecular Biology of Plant Disease Development*, Vol. 2. The Netherlands: Springer, 257p.

Nasser, W., Reverchon, S., Vedel, R., Boccara, M. 2005. PecS and PecT coregulate the synthesis of HrpN and pectate lyases, two virulence determinants in *E. chrysanthemi* 3937. *Molecular Plant-Microbe Interaction* 18: 1205–1214.

Navarro, L., Zipfel, C., Rowland, O., Keller, I., Robatzek, S., Boller, T., Jones, J.D. 2004. The transcriptional innate immune response to flg22: Interplay and overlap with Avr gene-dependent defense responses and bacterial pathogenesis. *Plant Physiology* 135:1113–1128.

Nissinen, R.M., Ytterberg, A.J., Bogadanove, A.J., van Wijk, K.J., Beer, S.V. 2007. Analysis of the secretomes of *Erwinia amylovora* and selected *hrp* mutants reveal novel type III secreted proteins and an effect of HrpJ on extracellular harpin levels. *Molecular Plant Pathology* 8: 55–67.

Normura, K., Nasser, W., Kawagishi, H., Tsuyumu, S. 1998. The *pir* gene of *Erwinia chrysanthemi* EC16 regulates hyperinduction of pectate lyase virulence genes in response to plant signals. *Proceedings of the National Academy of Science of the United States of America* 95: 14034–14039.

Noueiry, A.O., Lucas, W.J., Gilbertson, R.L., Nomura, K., Nasser, W., Kawagishi, H., Tsuyumu, S. 1999. The *pir* gene of *Erwinia chrysanthemi* EC16 regulates hyperinduction of pectate lyase virulence genes in response to plant signals. *Proceedings of the National Academy of Science of the United States of America* 95: 14034–14039.

Nurnberger, T., Brunner, F. 2002. Innate immunity in plants and animals: Emerging parallels between the recognition of general elicitors and pathogen-associated molecular patterns. *Current Opinion in Plant Biology* 5: 318–324.

Nürnberger, T., Brunner, F., Kemmerling, B., Piater, L. 2004. Innate immunity in plants and animals: Striking similarities and obvious differences. *Immunology Review* 198: 249–266.

O'Donnell, P.J., Schmelz, E. A., Moussatche, P., Lund, S.T., Jones, J.B., Klee, H.J. 2003. Susceptible to intolerance: A range of hormonal actions in a susceptible *Arabidopsis* pathogen response. *Plant Journal* 33: 245–257.

Ochiai, H., Inoue, Y., Takeya, M., Sasaki, A., Kaku, H. 2005. Genome sequence of *Xanthomonas oryzae* pv. *oryzae* suggests contribution of large number of effector genes and insertion sequences to its race diversity. *Journal of Agricultural Research Quarterly* 39: 275–287.

Oh, C.S., Beer, S.V. 2005. Molecular genetics of *Erwinia amylovora* involved in the development of fire blight. *FEMS Microbiology Letters* 253: 185–192.

Oh, C.S., Kim, J.F., Beer, S.V. 2005. The Hrp pathogenicity island of *Erwinia amylovora* and identification of three novel genes required for systemic infection. *Molecular Plant Pathology* 6: 125–138.

Pautot, V., Holzer, F.M., Chaufaux, J., Walling, L.L. 2001. The induction of tomato leucine aminopeptidase genes (LapA) after *Pseudomonas syringae* pv. *tomato* infection is primarily a wound response triggered by coronatine. *Molecular Plant-Microbe Interaction* 14: 214–224.

Peng, Q., Yang, S., Charkowski, A.O., Yap, M.N., Steeber, D.A., Keen, N.T., Yang, C.H. 2006. Population behavior analysis of *dspE* and *pelD* regulation in *Erwinia chrysanthemi* 3937. *Molecular Plant-Microbe Interactions* 19: 451–457.

Perino, C., Gaudriceult, S., Vian, B., Barny, M.A. 1999. Visualizaiton of harpin secretion in planta during infection of apple seedlings by *Erwinia amylovora. Cell Microbiology* 1: 131–141.

Pierson, L.S., Wood, D.W., Pierson, E.A. 1997. Homoserine lactone-mediated gene regulation in plant-associated bacteria. *Annual Review of Phytopathology* 36: 207– 225.

Pieterse, C.M.J., Ton, J., Van Loon, L.C. 2001. Cross-talk between plant defence signalling pathways: Boost or burden? *AgBiotech Net* 3: 1–8.

Pieterse, C.M.J., Van Loon, L.C. 1999. Salicylic acid-independent plant defense pathways. *Trends in Plant Science* 4: 52–58.

Pieterse, C.M.J., Van Wees, S.C.M., Hoffland, E., Vanpelt, J.A., Vanloon, L.C. 1996. Systemic resistance in *Arabidopsis* induced by biocontrol bacteria is independent of salicylic acid accumulation and pathogenesis-related gene expression. *Plant Cell* 8: 1225–1237.

Pieterse, C.M., Van Wees, S.C.M., van Pelt, J.A., Knoester, M., Lann, R., Gerrits, H., Weisbeek PJ, van Loon, L.C. 1998. A novel signaling pathway controlling induced systemic resistance in *Arabidopsis. Plant Cell* 10: 1571–1580.

Pissavin, C., Robert-Baudouy, J., Hugouvieux-Cotte-Pattat, N. 1996. Regulation of *pelZ*, a gene of the pelB-pelC cluster encoding a new pectate lyase of *Erwinia chrysanthemi* 3937. *Journal of Bacteriology* 178: 7187–7196.

Poplawsky, A.R., Chun,W. 1998. *Xanthomonas campestris* pv. *campestris* requires a functional *pigB* for epiphytic survival and host infection. *Molecular Plant-Microbe Interactions* 11: 466–475.

Poplawsky, A.R., Chun, W., Slater, H., Daniels, M.J., Dow, J.M. 1998. Synthesis of polysaccharide extracellular enzymes and xanthomonad in *Xanthomonas campestris*: Evidence for the involvement of two intercellular regulatory signals. *Molecular Plant-Microbe Interactions* 11: 68–70.

Preston, G.M., Studholme, D.J., Caldelari, I. 2005. Profiling the secretomes of plant pathogenic Proteobacteria. *FEMS Microbiology Review* 29: 331–360.

Preston J.F., Rice J.D., Ingram L.O., Keen N.T. 1992. Differential depolymerization mechanisms of pectate lyases secreted by Erwinia chrysanthemi EC16. *J. Bacteriol.* 174: 2039–2042.

Pugsley, A.P. 1993. The complete general secretory pathway. *Microbiology Review* 57: 50–108.

Qian, W., Han, Z., He, C. 2008. Two-component signal transduction systems of *Xanthomonas* spp.: A lesson from genomics. *Molecular Plant-Microbe Interactions* 21: 151–161.

Rajeswari, R., Gopaljee, J., Sonti, R.V. 2005. Role of in planta-expressed xylanase of *Xanthomonas oryzae* pv. *oryzae* in promoting virulence on rice. *Molecular Plant-Microbe Interactions* 18: 830–837.

Ramonell, K., Berrocal-Lobo, M., Koh, S., Wan, J., Edwards, H., Stacey, G., Somerville, S. 2005. Loss of function mutation in chitin responsive genes show increased susceptibility to the powdery mildew pathogen *Erysiphe cichoracearum. Plant Physiology* 138: 1027–1036.

Ramos, L.J., Narayanan, K.R., McMillan, R.T. 1992. Association of stomatal frequency and morphology in *Lycopersicon* species with resistance to *Xanthomonas campestris* pv. *vesicatoria. Plant Pathology* 41: 157–164.

Ray, S.K., Rajeswari, R., Sharma, Y., Sonti, R.V. 2002. A high-molecular weight outer membrane protein of *Xanthomonas oryzae* pv. *oryzae* exhibits similarity to nonfibrial adhesions of animal pathogenic bacteria and is required for optimum virulence. *Molecular Microbiology* 46: 637–647.

Reeves, P.J., Whitecombe, D., Wharam, S., Gibson, M., Allison, G. 1993. Molecular cloning and characterization of 13 *out* genes from *Erwinia carotovora* subsp. *carotovora*: Genes encoding members of a general secretion pathway (GSP) widespread in Gram-negative bacteria. *Molecular Microbiology* 38: 443–456.

Reimers, P.J., Leach, J.E. 1991. *Race-specific resistance to Xanthomonas oryzae* pv. *oryzae* conferred by bacterial blight resistance gene Xa-10 in rice (*Oryza sativa*) involves accumulation of a lignin-like substance in host tissues. *Physiology Molecular Plant Pathology* 38: 39–55.

Reymond, P., Farmer, E.E. 1998. Jasmonate and salicylate as global signals for defense gene expression. *Current Opinion in Plant Biology* 1: 404–411.

Ried, J.L., Collmer, A. 1988. Construction and characterization of an *Erwinia chrysanthemi* mutant with directed deletions in all of the pectate lyase structural genes. *Molecular Plant-Microbe Interactions* 1: 32–38.

Roper, M.C., Greve, L.C.,Warren, J.G., Labavitch, J.M., Kirkpatrick, B.C. 2007. *Xylella fastidiosa* requires polygalacturonase for colonization and pathogenicity in *Vitis vinifera* grapevines. *Molecular Plant-Microbe Interactions* 20: 411–419.

Rossier, O., Van den Ackerveken, G., Bonas, U. 2000. HrpB2 and HrpF from *Xanthomonas* are type III-secreted proteins and essential for pathogenicity and recognition by the host plant. *Molecular Microbiology* 38: 828–838.

Rottem, S. 2002. Invasion of mycoplasmas into and fusion with host cells. In *Molecular Biology and Pathogenicity of Mycoplasmas*. Razin, S., Hermann, R. (eds.). New York: Kluwer Academic Plenum Publishers, pp. 391–401.

Sandkvist, M. 2001. Type II secretion and pathogenesis. *Infection and Immunity* 69: 3523–3535.

Sauvage, C., Expert, D. 1994. Differential regulation by iron of *Erwinia chrysanthemi* pectate lyases: Pathogenicity of iron transport regulatory (*cbr*) mutants. *Molecular Plant-Microbe Interactions* 7: 71–77.

Schroth, M., Hildebrand, D. 1965. β-Glucosidase in *Erwinia amylovora* and *Pseudomonas syringae. Phytopathology* 55: 31–33.

Sheng, J.S., Citovsky, V. 1996. *Agrobacterium* plant cell DNA transport: Have virulence proteins, will travel. *Plant Cell* 8: 1699–1710.

Shevchik, V.E., Robert Baudouy, J., Hugouvieux-Cotte-Pattat, N. 1997. Pectate lyase *PelI* of *Erwinia chrysanthemi* 3937 belongs to a new family. *Journal of Bacteriology* 179: 7321–7330.

Slater, H., Alvarez-Morales, A., Barber, C.E., Daniels, M.J., Dow, J.M. 2000. A two-component system involving an HD-GYP domain protein links cell–cell signaling to pathogenicity gene expression in *Xanthomonas campestris*. *Molecular Microbiology* 38: 986–1003.

Soylu, S., Brown, I., Mansfield, J.W. 2005. Cellular reactions in *Arabidopsis* following challenge by strains of *Pseudomonas syringae:* From basal resistance to compatibility. *Physiology Molecular Plant Pathology* 66: 232–243.

Sticher, L., Mauchmani, B., Metraux, J.P. 1997. Systemic acquired resistance. *Annual Review of Phytopathology* 35: 235–270.

Stock, A.M., Robinson, V.L., Goudreau, P.N. 2000. Two-component signal transduction. *Annual Review of Biochemistry* 69: 183–215.

Sugio, A., Yang, B., White, F.F. 2005. Characterization of the *hrpF* pathogenicity peninsula of *Xanthomonas oryzae* pv. *oryzae*. *Molecular Plant-Microbe Interactions* 18: 546–554.

Suzuki, F., Sawada, H., Azegami, K., Tsuchiya, K., 2004. Molecular characterization of the operon involved in toxoflavin biosynthesis of *Burkholderia glumae*. *Journal of General Plant Pathology* 70: 97–107.

Takamiya, K.I., Tsuchiya, T., Ohta, H. 2000. Degradation pathway(s) of chlorophyll: What has gene cloning revealed? *Trends in Plant Science* 5: 426–431.

Toth, I. K. Bell, K.S., Holeva, M.C., Birch, P.R.J. 2003. Soft rot erwiniae: From genes to genomes. *Molecular Plant Pathology* 4: 17–30.

Trans-Kersten, J., Guan, Y., Allen, C. 2001. *Ralstonia solanacearum* needs motility for invasive virulence on tomato. *Journal of Bacteriology* 183: 3597–3605.

Tsiamis, G., Mansfield, J.W., Hockenhull, R., Jackson, R.W., Sesma, A., Athanassopoulos, E., Bennett, M.A. et al. 2000. Cultivar-specific avirulence and virulence functions assigned to avrPphF in *Pseudomonas syringae* pv. *phaseolicola*, the cause of bean halo-blight disease. *EMBO Journal* 19: 3204–3214.

Tzfira, T., Citovsky, V. 2006. *Agrobacterium*-mediated genetic transformation of plants: Biology and biotechnology. *Current Opinion in Biotechnology* 17: 147–154.

Uknes, S., Mauch-Mani, B., Moyer, M., Potter, S., Williams, S., Dincher, S., Chandler, D. et al. 1992. Acquired resistance in *Arabidopsis*. *Plant Cell* 4: 645–656.

Uknes, S., Winter, G., Delaney, A.M.T., Vernooij, B., Morse, A., Friedrich, L., Nye, G. et al. 1993. Biological induction of systemic acquired resistance in *Arabidopsis*. *Molecular Plant-Microbe Interactions* 6: 692–698.

Vallad, G.E., Goodman, R.M. 2004. Systemic acquired resistance and induced systemic resistance in conventional agriculture. *Crop Science* 44: 1920–1934.

Van Gijsegem, F., Genin, S., Boucher, C. (1995a). *hrp* and *avr* genes-key determinants controlling the interactions between plants and Gram-negative phytopathogenic bacteria. In *Pathogens and Host Specificity in Plant Diseases: Histochemical, Biochemical Genetic and Molecular Bases*, Vol I. Singh, U.S., Singh, R.P. (eds.). UK: Pergamon Press, pp. 273–292.

Van Gijsegem, F., Gough, C., Zischek, C., Niqueux, E., Arlat, M., Genin, S., Berberis, P. et al. 1995b. The *hrp* gene locus of *Pseudomonas solanacearum* which controls the production of a type III secretion system, encodes eight proteins related to components of the bacterial flagellar biogenesis complex. *Molecular Microbiology* 15: 1095–1114.

Van Loon, L.C. 1997. Induced resistance in plants and the role of pathogenesis-related proteins. *European Journal of Plant Pathology* 1103: 753–765.

Van Wees, S.C.M., de Swart, E.A.M., van Pelt, J.A., van Loon, L.C., Pieterse, C.M.J. 2000. Enhancement of induced disease resistance by simultaneous activation of salicylate and jasmonate-dependent defense pathways in *Arabidopsis thaliana*. *Proceedings of the National Academy of Science of the United States of America* 97: 8711–8716.

Vojnov, A.A., Slater, H., Daniels, M.J., Dow, J.M. 2001 Expression of the *gum* operon directing xanthan biosynthesis in *Xanthomonas campestris* and its regulation in planta. *Molecular Plant-Microbe Interaction* 14: 768–774.

Von Bodman, S.B., Bauer, W.D., Coplin, D.L. 2003. Quorum sensing in plant-pathogenic bacteria. *Annual Review of Phytopathology* 41: 455–82.

Wang, N., Lu, S.E., Yang, Q., Sze, S.H., Gross, D.C. 2006. Identification of the *syr-syp* box in the promoter regions of genes dedicated to syringomycin and syringopeptin production by *Pseudomonas syringae* pv. *syringae* E301D. *Journal of Bacteriology* 188:160–168.

Weber, E., Koebnik, R. 2005. Domain structure of HrpE, the Hrp pilus subunit of *Xanthomonas campestris* pv. *vesicatoria*. *Journal of Bacteriology* 187: 6175–6186.

Weber, E., Ojanen-Reuhs, T., Huguet, E., Hause, G., Romantschuk, M., Korhonen, T.K., Bonas, U., Koebnik, R. 2005. The type III dependent Hrp pilus is required for productive interaction of *Xanthomonas campestris* pv. *vesicatoria* with pepper host plants. *Journal of Bacteriology* 187: 2458–2468.

Wei, W., Opgenorth, D.C., Davis, R.E., Chang, C.J., Summers, C.G., Zho, Y. 2006. Characterization of a novel adhesin-like gene and design of real-time PCR for rapid sensitive and specific detection of *Spiroplasma kunkelii*. *Plant Disease* 90: 1233–1238.

Weingart, H., Stubner S., Ullrich, M.S. 2003. Temperature-regulated biosynthesis of coronatine by *Pseudomonas syringae in vitro* and *in planta*. In *Pseudomonas syringae Pathovars and Related Pathogens, Biology and Genetic*. Iacobellis, N.S., Collmer, A., Hutcheson, S.W., Mansfield, J.W., Morris, C.E., Murillo, J., Schaad, N.W. et al. (eds.). Dordrecht, The Netherlands: Kluwer Academic Publishers, pp. 199–205.

Weingart, H., Völksch, B. 1997. Ethylene production by *Pseudomonas syringae* pathovars *in vitro* and *in planta*. *Applied Environmental Microbiology* 63: 156–161.

Weingart, H., Völksch, B., Ullrich, M.S. 1999. Comparison of ethylene production by *Pseudomonas syringae* and *Ralstonia solanacearum*. *Phytopathology* 89: 360–365.

Weller, D.M., Saettler, A.W. 1980. Colonization and distribution of *Xanthomonas phaseoli* and *Xanthomonas phaseoli* var *fuscans* in field-grown navy beans. *Phytopathology* 70: 500–506.

Whistler, C.A., Pierson, L.S. 2003. Repression of phenazine antibiotic production in *Pseudomonas aureofaciens* strain 30-84 by RpeA. *Journal of Bacteriology* 185: 3718–3725.

White, R.F. 1979. Acetyl salicylic acid (aspirin) induces resistance to tobacco mosaic virus in tobacco. *Virology* 99: 410–412.

Whitehead, N.A., Barnard, A.M.L., Slater, H., Simpson, N.J.L., Salmond, G.P.C. 2001. Quorum-sensing in gram-negative bacteria. *FEMS Microbiology Review* 25: 365–404.

Wisniewski-Dye, F., Downie, J.A. 2002. Quorum-sensing in *Rhizobium*. *Antonie van Leeuwenhoek* 81: 397–407.

Wolanin, P., Thomason, P.A., Stock, J.B. 2002. Histidine protein kinases: Key signal transducers outside the animal kingdom. *Genome Biology* 3: 3013.

Wright, C.A., Beattie, G.A. 2004. *Pseudomonas syringae* pv. *tomato* cells encounter inhibitory levels of water stress during the hypersensitive response of *Arabidopsis thaliana*. *Proceedings of the National Academy of Science of the United States of America* 101: 3269–3274.

Yang, B., White, F.F. 2004. Diverse members of the AvrBs3/PthA family of type III effectors are major virulence determinants in bacterial blight disease of rice. *Molecular Plant-Microbe Interactions* 17: 1192–1200.

Yang, B., Zhu, W., Johnson, L.B., White, F.F. 2000. The virulence factor AvrXa7 of *Xanthomonas oryzae* pv. *oryzae* is a type III secretion pathway-dependent nuclear-localized double-stranded DNA-binding protein. *Proceedings of the National Academy of Science of the United States of America* 97: 9807–9812.

Yao, J., Allen, C. 2006. Chemotaxis is required for virulence and competitive fitness of the bacterial wilt pathogen *Ralstonia solanacearum*. *Journal of Bacteriology* 188: 3697–3708.

Yoneyama, K., Kono, Y., Yamaguchi, I., Horikoshi, M., Hirooka, T. 1998. Toxoflavin is an essential factor for virulence of *Burkholderia glumae* causing rice seedling rot disease. *Annals of Phytopathological Society of Japan* 64: 91–96.

Yu, J., Wayadande, A.C., Fletcher, J. 2000. *Spiroplasma citri* surface protein P89 implicated in adhesion to cells of the vector *Circulifer tenallus*. *Phytopathology* 90: 716–722.

Yun, M.H., Torres, P.S., El Oirdi, M., Rigano, L.A., Gonzalez-Lamothe, R., Marano, M.R., Castagnaro, A.P. et al. 2006a. Xanthan induces plant susceptibility by suppressing callose deposition. *Plant Physiology* 141: 178–187.

Yun, M.H., Torres, P.S., Oirdi, M.E., Rigano, L.A., Gonzalez-Lamothe, R., Marano, M.R., Castagnaro, A.P. et al. 2006b. Xanthan induces plant susceptibility by suppressing callose deposition. *Plant Physiology* 141:178–187.

Zhang, L., Xu, J., Birch, R.G. 1999. Engineered detoxification confers resistance against a pathogenic bacterium. *Nature Biotechnology* 17: 1021–1024.

Zhao, Y.F., Sundin, G.W., Wang, D.P. 2009. Construction and analysis of pathogenicity island deletion mutants of *Erwinia amylovora*. *Canadian Journal of Microbiology* 55: 457–464.

Zinsou, V., Wydra, K., Ahohuendo, B., Schreiber, L. 2006. Leaf waxes of cassava (*Manihot esculenta* Crantz) in relation to ecozone and resistance to *Xanthomonas* blight. *Euphytica* 149: 189–198.

3 Epidemiology and Forecasting Systems of Plant Pathogenic Bacteria

Titus Alfred Makudali Msagati, Bhekie Brilliance Mamba, Venkataraman Sivasankar, Rathinagiri Sakthivel, and Mylsamy Prabhakaran

CONTENTS

ABSTRACT Plant productivity and corresponding yields are experiencing higher levels of devastation due to pathogenic bacteria. To address the catastrophy of plant immune systems, this chapter discusses various bacterial pathogens that causes diseases in plants. The epidemiology and forecasting of such diseases are also discussed according to several significant factors. Mathematical models such as multiple regression, non-linear, and growth models have enabled the forecasting of various plant diseases by parametric variables.

KEYWORDS: bacterial pathogens, plant diseases, epidemiology, forecasting, mathematical models

3.1 INTRODUCTION

3.1.1 Overview

Bacteria, an extremely diverse group of organisms from a metabolic standpoint, are found in vast numbers almost everywhere on Earth. Beneficial bacteria are involved in diverse processes such as digestion in animals, nitrogen fixation in the roots of certain legumes, decomposition of animal and plant remains, and sewage disposal. Conversely, pathogenic bacteria cause severe and often fatal diseases in humans, animals, and plants. The bacterial disease caused by *Bacillus anthracis*, anthrax, was first reported in 1876; the next, fire blight, was first discovered and reported between 1877 and 1885. The discovery of fastidious vascular bacteria in 1967 enabled bacteriologists to identify the causes of many diseases that previously had only been believed to be caused by viruses. These bacterial pathogens, which are difficult to culture in the laboratory, grow in either xylem or phloem tissues and interfere with the transport of water and nutrients in plants. As evidenced by corn stunt, many of these bacteria are vectored by sucking insects such as leafhoppers, plant hoppers, and psyllids.

3.1.2 Bacterial Plant Diseases

Bacteria are microscopic, single-celled organisms that multiply rapidly by simple cell division and absorb nutrients from their immediate environment. Bacteria are classified into two main groups based on their cell-wall structures, which can be determined by a simple staining procedure called the Gram stain. Gram-negative bacteria stain red or pink; Gram-positive bacteria stain purple. This color difference is directly related to the chemical composition and structure of their cell walls. The cells can be rod-shaped, spherical, spiral-shaped, or filamentous. Only a few of the latter are known to cause diseases in plants. Most bacteria are motile and have whip-like flagella that propel them through films of water. The most obvious reason for bacteria to seek interaction with plants is nutrition. The introduction of bacterial pathogens to new sites is often facilitated by contaminated seeds or transplants. The spread of bacteria is made feasible by splashing rain, water runoff, wind-driven rain or mists (aerosols), equipment, insects, and people. The latter include botanists, horticulturalists, gardeners, farmers, herbalists, and arbiculturists. Bacteria persist in and around tomato plantings in weeds, volunteer plants, and infested crop debris as symptomless colonizers of plant surfaces (Jackson et al., 2011; Fisher et al., 2012).

3.1.3 Epidemiology

The development of plant pathogenic bacteria (heterotrophic organisms) on host plants can occur in the form of parasites. The host plants of particular pathogens can be many or few; for example, several hundred plant species can host *Grobacterium tumefaciens* but only strawberry can host *Xanthomonas fragariae.* A pathovar, which is a bacterial strain or set of strains with the same or similar characteristics, is differentiated on the basis of different pathogenicities to one or more plant hosts. For example, the bacterium that causes citruscanker, *Xanthomonas axonopodis*, has several pathovars with different host ranges (Billing, 1987; Klement et al., 1990; Agrios, 1997; Sobiczewski and Schollenberger, 2002).

3.1.3.1 Measures of Association

Measures of association are relative clarifications of the magnitude of an association between exposure to pathogens by plants and the disease that is normally expressed. The relative effects of the association between disease cases and exposures can be described as ratios of rates or risk measures. A rate ratio (RR) is the ratio of a rate of pathogenic infection in one population of plants compared to the rate caused by the same pathogen in another plant population.

3.1.3.2 Measures of Covariates (Risk Factors)

In order to measure the effect of risk factors on a particular disease case, an investigation is normally conducted by subdividing a plant population according to the appropriate categories (e.g., species in the same genus, plants in the same locality, etc.).

3.1.4 Bacterial Infection and Symptoms

Infection does not occur directly upon plant tissues. Instead, it occurs through natural openings (e.g., stomata, hydatodes, nectaroides, and injuries of various origins). Bacterial multiplicity in plants begins after entry and adaptation into a new environment that consists of intercellular spaces and continues with dissemination into various tissues. The appearance and intensity of disease symptoms (e.g., leaf scars) occur because of conditional factors such as increased bacterial population at optimum temperature and high relative humidity. Symptoms of bacterioses include leaf spots, necroses, blights, cankers, scabs, wilting, tumors, galls and overgrowths, and specks and soft rots of roots, storage organs, and fruits (Table 3.1).

3.1.4.1 Above-Ground Pathogen Infections

Bacterial organisms such as epiphytes and saprophytes can infect above-ground plant organs. In such cases, the death of infected tissues will result initially in the decrease and ultimately in the decline of the bacterial population. In this case, the dissemination of bacteria is feasible by wind, rain, insects, birds, and infected plants as well but the infection source may drastically be reduced.

3.1.4.2 Pathogens That Infect Root Systems

The media that infect root systems are soil and soil fauna, aided by anthropogenic activities and management practices. Pathogens of this type with long-term survival in soil are *Agrobacterium tumefaciensor* and *Ralstonia solanacearumare*; their main development, however, takes place in host plants. These soil-borne pathogens build up their populations in host plants and are released into the soil where they can survive to act as primary inoculums for the next season (Agrios, 1997; Lucas, 1998; Trigiano et al., 2004). Mild, short winters are expected to have little effect on soil-borne pathogens. The survival period of pathogens depends on factors that can be abiotic (temperature, humidity, and nutrients) or biotic (antagonistic bacteria and fungi).

Plant disease epidemiology deals with the study of plant diseases that occur due to spread of pathogens such as bacteria, viruses, fungi, nematodes, phytoplasmas, protozoa, and parasites

TABLE 3.1
Diagnostic Symptoms of Bacterial Infections

Sl. No.	Bacteria	Target(s)	Diagnostic Symptoms and Physical Appearance
1	*Pseudomonas* and *Xanthomonas*	Leaves, blossoms, fruits, and stems	Leaf spots or blights Dicotyledonous plant leaves appear water-soaked and angular; presence of bacterial ooze; coalescing spots lead to large areas of necrotic tissue that have rotten or fishy odor
2	*Pseudomonas* and *Xanthomonas*	Stone fruit and pome fruit trees; citrus trunks, stems, twigs, and branches	Monocotyledonous plant leaves appear with streaks or stripes. Development of cankers and gum exudation (gummosis); slightly sunken, dark brown and longer cankers; cortical tissue can be orange brown to dark brown; cankers can also have a sour odor and be soft, sunken and moist; leaf stress caused by girdle trunks and branches.
3	*Agrobacterium* (*A. tumefaciens*, *A. rubi*, and *A. vitis*) and certain species of *Arthrobacter*, *Pseudomonas*, *Rhizobacter*, and *Rhodococcus*	More than 390 plant genera: Plant crown or main root Herbaceous plants such as vegetables, field crops, ornamentals, and other tropical plants	Bacterial galls Crown gall, crown knot, root knot, and root gall Crown gall: Appears as small, whitish, soft round over growths; appears as orange-brown gall with surface enlargement and dark brown gall when convoluted. Bacterial vascular wilts Wilting and death of above-ground parts of the plant; cracks on the surface of infected leaves or seepage of bacterial ooze
4	*Erwinia* spp. *Bacillus* spp. *Clostridium* spp.	Fleshy plant structures both above and below ground Underground parts of plants such as potatoes	Bacterial soft rots Rotting tissue becomes watery and soft; slimy foul-smelling ooze is exuded by infected tissue Bacterial scabs Localized scabby lesions on the outer surface of the tuber

(George N. Agrios 2005) and their interaction with the environment/ecology with respect to space and time. This area of study, which is often multidisciplinary, may incorporate biology, statistics, agronomy, and ecology in such a way that enables the life cycle of a pathogen and its adverse effect on a crop, including the influence of agronomic practices on specific diseases and the function of native species as reservoirs, to be expressed by statistical models (Arneson, 2001; Madden, 2006). The primary objective of plant disease epidemiology is to analyze the underlying mechanisms of pathogen dynamics, pathogenic distribution, and the spread of plant diseases to the heterogeneous environment (Table 3.2).

Spatial statistics is concerned investigating the distribution of a plant disease within a given field or area of a field according to environmental influence or the degree of human intervention. The incidence of plant diseases based on spatial patterns varies between high levels of aggregation and randomness. An aggregated spatial pattern is a relative measurement of pathogen density in unknown locations compared to the density in spatially referenced locations. Such measures may be unattainable, however, because of the lack of influence of one location on another.

3.2 EPIDEMICS AND THEIR ELEMENTS

Plant disease epidemics can cause huge losses in crop yields and also threaten to destroy entire plant species. The vital elements of such epidemics, which are usually represented in triangular form, are a susceptible host, a pathogen, and a conducive environment (Figure 3.1). Epidemic continuity

TABLE 3.2
Pathogens, Their Hosts, and Associated Diseases

Pathogen(s)	Host(s)	Disease
Xanthomonas euvesicatoria and *Xanthomonas perforans* = [*Xanthomonas axonopodis* (syn. *campestris*.) pv. *vesicatoria*], *Xanthomonas vesicatoria* and *Xanthomonas gardneri*	Economically important hosts include pepper (*Capsicum* spp.) and tomato (*Solanum lycopersicum*)	Bacterial Spot pepper and tomato Bacterial fruit blotch of cucurbits
Acidovoraxavenae subsp. *citrulli* (=*Pseudomonas pseudoalcaligenes* subsp. *citrulli*, *Pseudomonas avenae* subsp. *citrulli*)	Watermelon, melon (cantaloupe, honeydew), cucumber, pumpkin, squash, gourds, citron melon, and other cucurbits	Bacterial leaf scorch (BLS) of shade trees
Xylella fastidiosa *Erwinia carotovora* subsp. *atroseptica* (Synonym: *Pectobacterium atrosepticum*) *Xanthomonas axonopodis* pv. *citri* and *Xanthomonas axonopodis* pv. *aurantifolii*	Causes bacterial leaf scorch (BLS) that affects many shade tree species (Table 3.1) including American elm, red maple, sweet gum, sycamore, and London plane, as well as a number of species of oak (Figure 3.1). Identified in urban forest (landscapes, street plantings, small woodlots) throughout the Eastern United States and as far west as Texas Potato (*Solanum tuberosum*)	Blackleg of potato Citrus canker Crown gall
Agrobacterium tumefaciens *Erwinia amylovora*	Numerous species, cultivars, and hybrids of citrus and citrus relatives including orange, grapefruit, pummelo, mandarin, lemon, lime, tangerine, tangelo, sour orange, rough lemon, calamondin, trifoliate orange, and kumquat	Fire blight of apple and pear Lethal yellowing of palm Stewart's wilt (Syn. Stewart's disease Stewart's bacterial wilt)
Candidatus Phytoplasma palmae *Erwinia stewartii* (Syn. *Pantoea stewartii*) *Belonolaimus longicaudatus*, *Belonolaimus* spp.	Members of 93 families of plants including apple, pear, several rosaceous ornamentals Palms (species in the family Arecaceae); *Pandanus utilis* (Pandanaceae corn, *Zea mays*); artificial infection by inoculation in other secondary host plants Many hosts, including common agronomic crops, vegetables, grasses, grains, and trees	Sting nematode

describes the function of these elements over time. An increase in temperature leads to environmental changes that are not conducive for the pathogen to cause disease especially in temperature-resisting adult plants (Figure 3.2). In an environment that is conducive but also pathogen-free (Figure 3.3), susceptibility of the host is feasible but disease development is absent. The ability of a plant to resist disease, which is called ontogenic resistance, depends on changes in the plant as it matures.

3.2.1 Types of Epidemics

Monocyclic epidemics, which are typical of soil-borne diseases, can have only one infection cycle per season. Polycyclic epidemics, which typically result from airborne diseases, can have several infection cycles per season. In polyetic epidemics, which can be caused by both monocyclic and polycyclic pathogens, disease occurrence over several growing seasons is due to the prolonged travel of once-produced inoculums.

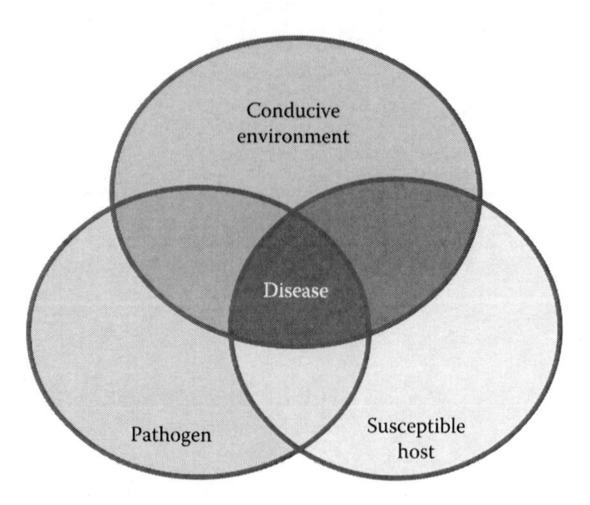

FIGURE 3.1 Epidemics and their associated elements.

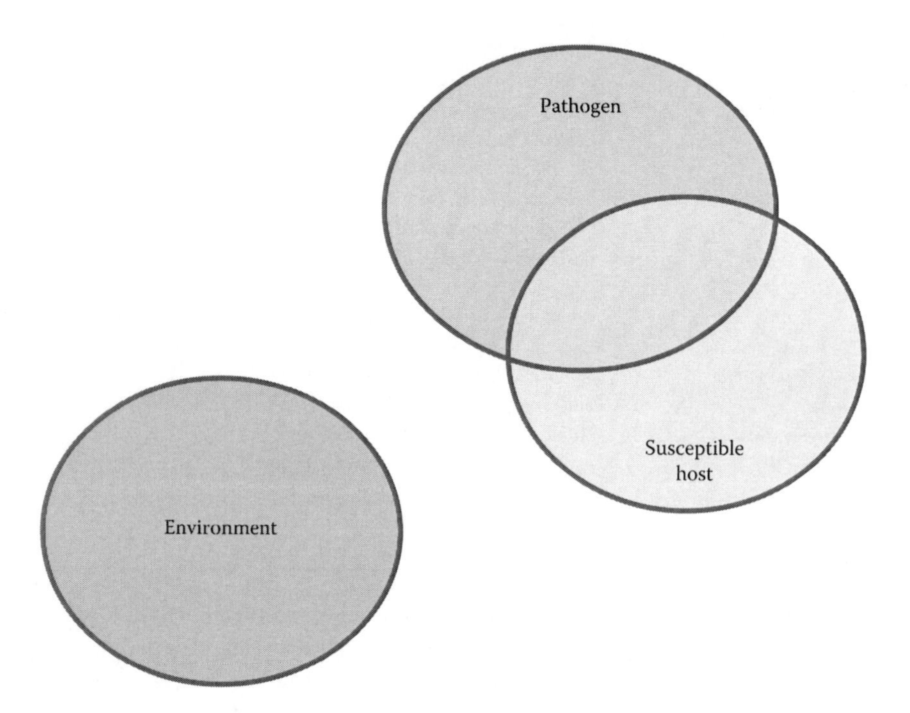

FIGURE 3.2 Disease triangle for temperature-resistant adult plants.

3.3 PLANT DISEASE FORECASTING SYSTEMS

These management systems, which are also called plant disease warning systems, accurately predict the occurrence of or the change in the severity of plant diseases, as well as subsequent economic losses. These systems help growers make economic decisions about disease management (Agrios, 2004) that help to avoid initial inoculums or to slow down the rate of an epidemic. Forecasting systems can also differentiate monocyclic diseases from polycyclic ones. Although some plant disease forecasts focus on avoiding initial inoculums as well as on reducing seasonal epidemic rates,

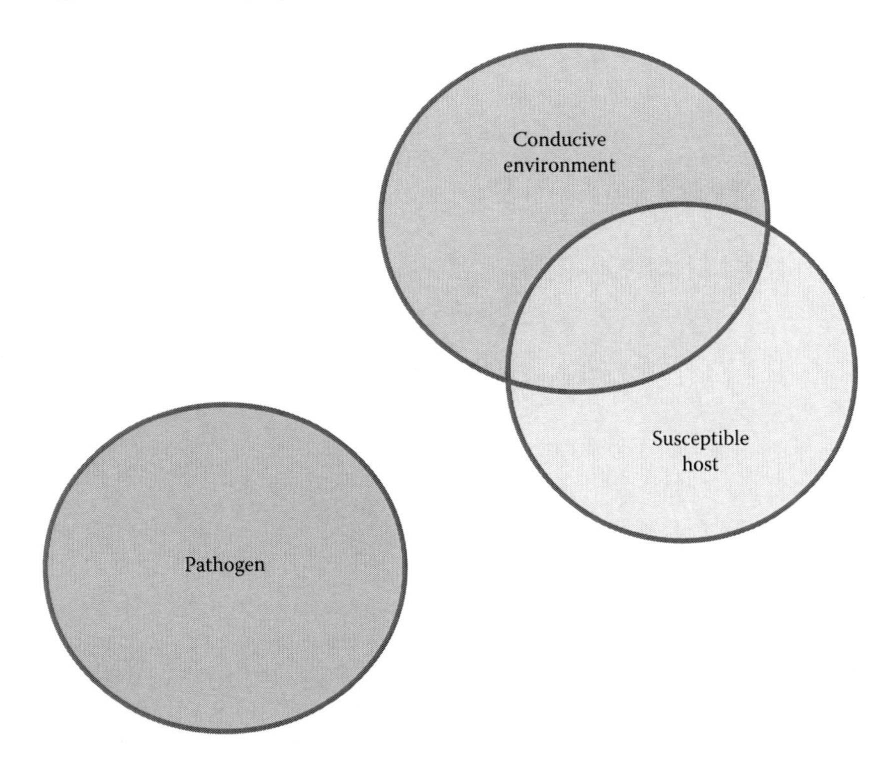

FIGURE 3.3 Disease triangle for a pathogen-free conducive environment.

systems usually emphasize plant diseases based on the following principles (Campbell and Madden, 1990; Agrio, 2004):

- Measures of initial inoculums or disease (e.g., Stewart's disease in corn)
- Favorable weather conditions for the development of secondary inoculums (e.g., late blight in potato)
- Both initial and secondary inoculums (e.g., apple scab)

A well-known example of a multiple disease/pest forecasting system is the EPIdemiology, PREdiction and PREvention (EPIPRE) system, developed in the Netherlands for winter wheat, that focuses on multiple pathogens (Reinink, 1986). When properly validated, a developed model results is successfully developed and reliable. Plant disease modelers and researchers may focus upon improving producers' profitability through validations based on cost quantifications of a model via false positive and/or negative predictions. False predictions by a plant disease forecasting system based on economic validation may occur because: (i) The forecast was made for a disease that cannot occur in a particular location; and (ii) The forecast was negative but the disease was found. These two types of false predictions may have different economic effects for producers (Madden, 2006).

3.3.1 Attributes

Campbell and Madden (1990) listed several criteria for a successful forecasting system (Figure 3.4). Essentially, it should be more useful for multipurpose applications (i.e., as a monitoring and decision-making tool for several diseases) based upon the available components of the disease triangle; and it should be simple to implement (i.e., user friendly), reliable with accurate data, and cost-effective relative to available disease-management tactics.

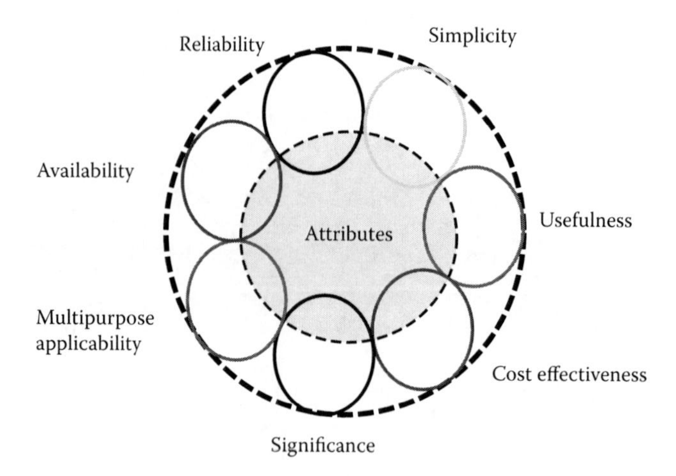

FIGURE 3.4 Attributes of a forecasting system.

The forecasting system, which is based on assumptions derived from the disease triangle, pertains to the interaction of a pathogen with the host crop and environment. The ability and nature of both host (resistant or susceptible) and the pathogen can often be ascertained from crop history or survey data. Nevertheless, environment is the controlling factor in disease dynamics because it affects the ability of the pathogen to cause disease. In such cases, the forecasting system helps to analyze a favorable environment for disease development. Systems are usually designed as optional for diseases that are irregular enough to necessitate a prediction system (rather than diseases that occur every season/year). A forecasting system that helps growers to make economic decisions is often based on a questionnaire that elicits details about host-crop susceptibility and current weather conditions. In addition, forecasts of future weather conditions can be used to develop recommendations about treatments for plant diseases.

3.4 DISEASE FORECASTING BASED ON MATHEMATICAL CONCEPTS

3.4.1 Multiple Regression Model

Multiple linear regressions generalize a methodology that allows multiple predictor variables such as mean temperature and mean relative humidity. Its goal may be an accurate prediction, and the regression coefficients ratings may be of direct interest (Maindonald and Braun, 2003).

The notation for a model including m predictors is $y = \beta_0 + \beta_1 x_1 + \beta_2 x_2 + \ldots + \beta_m x_m + \varepsilon$.

One common multiple linear regression model is the polynomial model: $y = \beta_0 + \beta_1 x_1 + \beta_2 x_1^2 + \beta_3 x_2 + \beta_4 x_2^2 + \beta_5 x_1 x_2 + \varepsilon$. The estimate of the parameter can be calculated by redefining variables: $w_1 = x_1$, $w_2 = x_1^2$, $w_3 = x_2$, $w_4 = x_2^2$, $w_5 = x_1 x_2$ and performing a multiple linear regression model using $y = \beta_0 + \beta_1 w_1 + \beta_2 w_2 + \beta_3 w_3 + \beta_4 w_4 + \beta_5 w_5 + \varepsilon$.

The above is now an ordinary multiple linear regression model that uses w as predictor variables. Multicollinearity or correlation among predictor variables (Figure 3.5) is significant in the use of multiple linear regressions. However, problems that arise from multicollinearity will tend to raise the standard errors and thus render the estimates illogical. The variance inflation factor (VIF) can be calculated for each independent variable; when its value is five or more, multicollinearity is indicated. Thus, the VIF values for temperature (22.9), relative humidity (21.3), and rainfall (2.4) are indicative of possible multicollinearity. It is well supported by the scatter plot that temperature and relative humidity are strongly correlated, which could explain the correspondingly large VIF value. Problems in this multicollinearity method can be eliminated by doing variable selection to identify and select a subset of predictor variables that backward elimination indicates are not

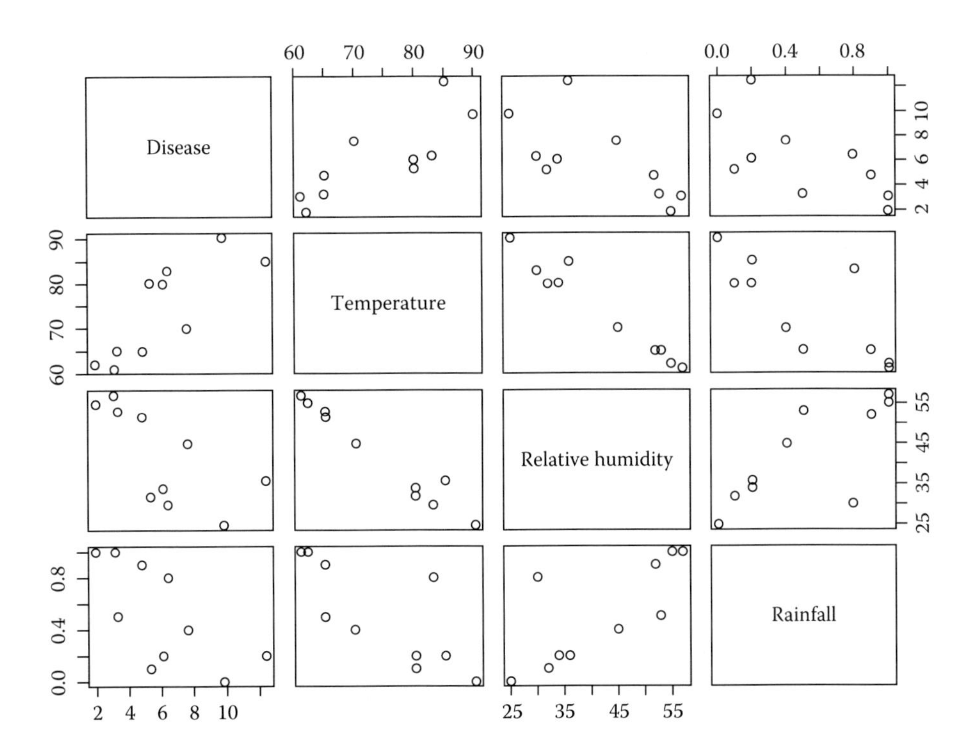

FIGURE 3.5 Multicollinearity among predictor variables. (Adapted from Esker, P.D. et al. 2008. *The Plant Health Instructor*. DOI: 10.1094/PHI-A-2008-0129-01.)

strongly correlated with each other (Neter et al., 1996). In the backward elimination model, all of the predictor variables are included and particular variables are successively removed to search iteratively for the best fit. The Akaike information criterion (AIC) can be used to select an optimal combination of parsimony among models, which leads to a perfect fit with the smallest number of parameters.

3.4.2 Non-Linear Model

A simple forecasting system can also be used with precipitation and temperature as the main predictors of disease severity. In this type of system, the potential predictors tend to be fairly constant and to stay within the disease-conducive range. The relationship among precipitation, temperature, and disease severity may take many forms. Usually, the best option is to model the rate of change in disease severity or pathogen populations as a function of the weather variables. The logistic growth is reviewed before accounting the environmental effects.

Logistic growth at a constant growth rate: When the discrete logistic curve gives knowledge of disease progress with respect to time, the rate of increase of the parameter R is

$$D_{t+1} = D_t + R * D_t * [(1 - D_t)/100]$$

where D_t = percentage infection at time t and R = growth rate.

The logistic growth for a plant disease using R programming is as follows.

```
## Input the data that include the variables time,
#   plant ID, and severity
```

```
time <-c(seq(0,10),seq(0,10),seq(0,10))
plant <-c(rep(1,11),rep(2,11),rep(3,11))
## Severity represents the number of
## lesions on the leaf surface, standardized
## as a proportion of the maximum
severity<-c(
        42,51,59,64,76,93,106,125,149,171,199,
        40,49,58,72,84,103,122,138,162,187,209,
        41,49,57,71,89,112,146,174,218,250,288)/288
data1 <-data.frame(
cbind(
time,
plant,
severity
)
)
## Plot severity versus time
## to see the relationship between
## the two variables for each plant
plot(
        data1$time,
        data1$severity,
xlab = "Time",
ylab = "Severity",
type = "n"
)
text(
        data1$time,
        data1$severity,
        data1$plant
)
title(main = "Graph of severity vs time")
```

From the graph shown in Figure 3.6, the possible fit of the data can be made with the logistic curve in the following equation using the initial values of the parameters. The initial values can be obtained from the R functions get Initial and SSlogis.

$$Y\ (t) = \frac{\alpha}{1 + e^{\beta - rt}}$$

The scatter plot (Figure 3.6) enables us to get some ideas regarding the initial parameters that support the formation of the shape of the curve. The parameter α represents the asymptotic limit of severity in the logistic curve. With reference to the inflection point at (β/λ, $\alpha/2$) from the scatter plot, the estimate for α might be 300 and λ is about 7. The respective estimated values of α and β are 2 and 15.

The initial value of the parameters using get Initial and SSlogis can thereby be obtained. SSlogis uses a different parameterization for the logistic model that uses SSlogis.

The first step is the parameterization from SSlogis, which is carried out as follows:

```
getInitial(
severity ~ SSlogis(time, alpha, xmid, scale),
data = data1
)
```

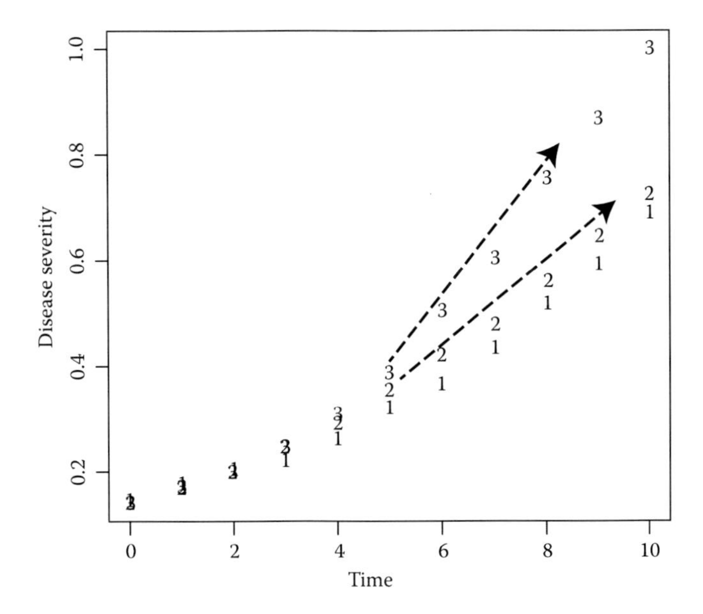

FIGURE 3.6 Scatter plot of disease severity versus time. (Adapted from Sparks, A.H. et al. 2008. *The Plant Instructor*. DOI: 10.1094/PHI-A-2008-0129-02.)

The output is as follows:

```
alphaxmid scale
2.212 12.507 4.572
```

Here, fitting the logistic curve to the data is executed for converting to our parameterization.

```
## Using the initial parameters above,
## fit the data with a logistic curve.
para0.st <- c(
alpha = 2.212,
        beta = 12.507/4.572, # beta in our model is xmid/scale
gama = 1/4.572 # gamma (or r) is 1/scale
)
fit0 <- nls(
severity ~ alpha/(1 + exp(beta-gamma*time)),
        data1,
start = para0.st,
trace = T
)
```

The output is

```
0.1621433 : 2.2120000 2.7355643 0.2187227
0.1621427 : 2.2124095 2.7352979 0.2187056
```

Using the scatter plot, the above data is plotted for its fitting with the logistic curve

```
## Plot to see how the model fits the data; plot the
## logistic curve on a scatter plot
plot(
     data1$time,
```

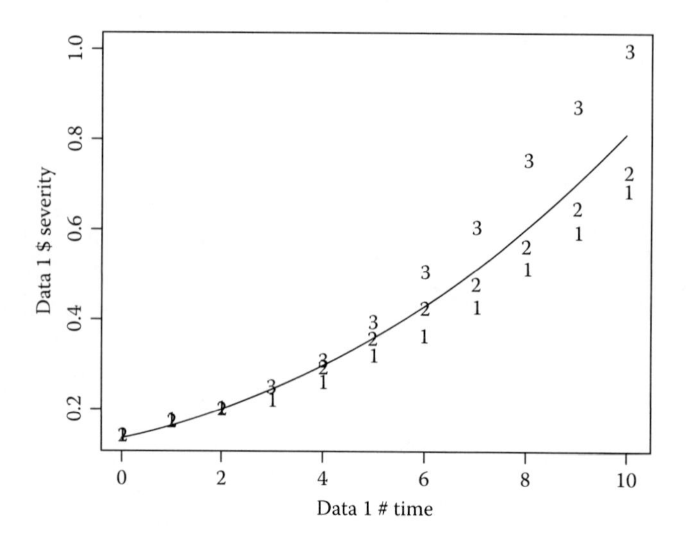

FIGURE 3.7 Scatter plot of disease severity versus time. (Adapted from Sparks, A.H. et al. 2008. *The Plant Instructor*. DOI: 10.1094/PHI-A-2008-0129-02.)

```
     data1$severity,
type = "n"
)
text(
        data1$time,
        data1$severity,
        data1$plant
)
title(main = "Graph of severity vs time")
curve(
        2.21/(1 + exp(2.74-0.22*x)),
from = time[1],
to = time[11],
add = TRUE
)
```

As is well illustrated in Figure 3.7, the logistic curve fits the data nicely. The disease-progress data is well analyzed with many nonlinear models using scatter plots of the data over time.

3.5 GROWTH MODELS

The study of disease progress over time is of frequent interest where time (t) is modeled as a continuous variable rather than a discrete variable. Growth models (Figure 3.8) for modeling disease progress curves have been reported by researchers (Gilligan, 1990; Madden et al., 2007).

The five common growth curve models are discussed below, along with certain assumptions.

3.5.1 Exponential Model

The model is applied with the equation of linear form: $\log y = \log y_0 + rt$. In this model, the absolute rate of disease is assumed to increase (dy/dt) proportionally to the disease intensity (y). The function plotexp, created by *R* code, shows the exponential relationship between disease incidence and

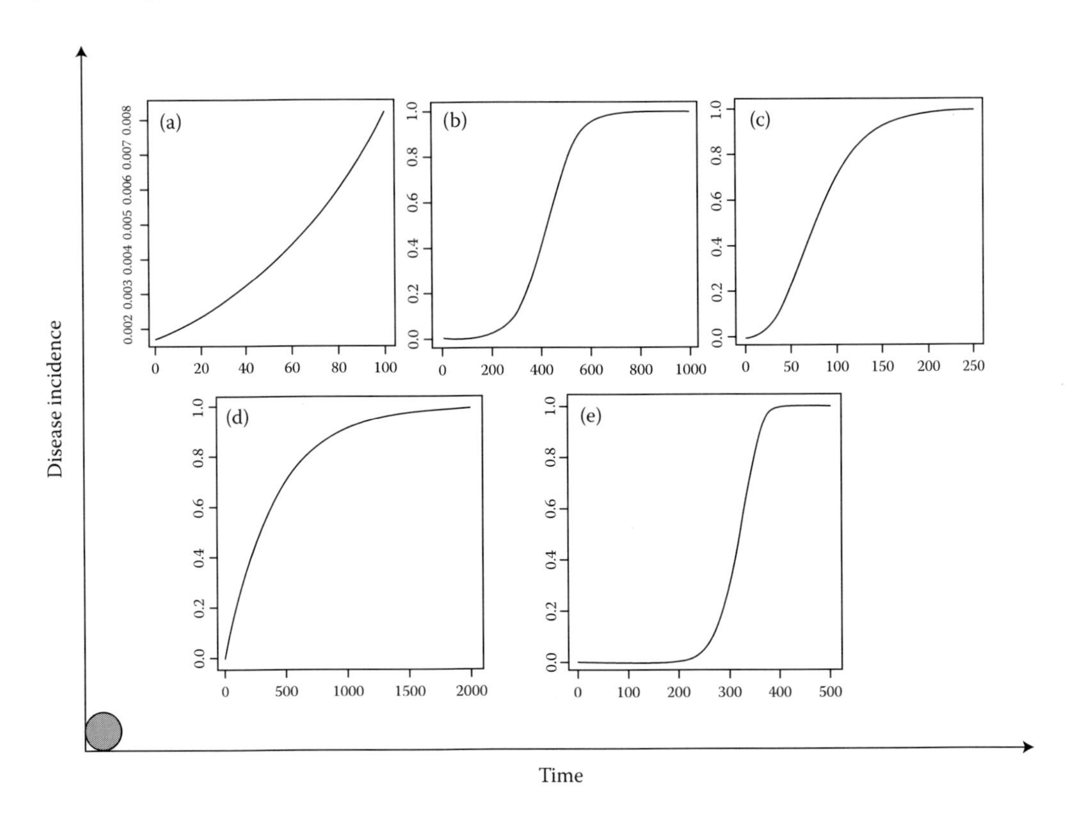

FIGURE 3.8 Growth models: (a) Exponential, (b) Logistic, (c) Gompertz, (d) Monomolecular, and (e) Weibull. (Adapted from Sparks, A.H. et al. 2008. *The Plant Instructor.* DOI: 10.1094/PHI-A-2008-0129-02.)

time. This function helps to explain the influence upon the shape of the curve of the parameters applied by us. In the illustration, parameters y_0 and r are respectively set to 0.0017 and 0.01579 with a maximum time of 100.

```
## Exponential Model Example
plotexp<- function(y0,r,maxt){
curve(
        y0*exp(r*x),
from=0,
to=maxt,
xlab='Time',
ylab='Disease Incidence',
col='mediumblue')
}
plotexp(0.0017, 0.01579, 100)
```

3.5.2 Mono Molecular Model

The linear form of the model is $\ln\{1/(1-y)\} = \ln\{1/(1-y_0)\} + rt$.

The assumption made by this model is that the carrying capacity or maximum level of the disease allows proportional measurement of disease severity or incidence. The diseased tissue of a plant falls between 0 (healthy) and 1 (complete disease) and assumes that the absolute rate of change is proportional to the healthy tissue $(1-y)$. The function, plotmono, is used or replaced with the set of parametric values that ascertains the change in the relationship through the shape of the graphical curve.

```
## Monomolecular Model Example
plotmono <- function(y0,r,maxt){
curve(
        1-(1-y0)*exp(-r*x),
from = 0,
to = maxt,
xlab = 'Time',
ylab = 'Disease Incidence',
col = 'mediumblue')
}
plotmono(0.0017, 0.00242, 2000)
```

3.5.3 Logistic Model

The linear form of the logistic model is

$$\ln(y/(1-y) = \ln\{y_0/(1-y_0) + rt\}$$

where the absolute rate of change in disease level is assumed and the model depends on both the healthy (y) and diseased tissues ($1 - y$) at a particular time. The symmetry of the curve is perfect with an inflection point at $t = 1/r \ln y_o (1 - y_o)$ when $y = 1/2$. The value of dy/dt increases until y is halved and then decreases. The parametric values are applied using the created plotlog function, followed by validation of the model.

```
## Logistic Model Example
plotlog <- function(y0,r,maxt){
curve(
        1/(1+ (1-y0)/y0*exp(-r*x)),
from = 0,
to = maxt,
xlab = 'Time',
ylab = 'Disease Incidence',
col = 'mediumblue'
        )
}
plotlog(0.001, 0.01636, 1000)
```

3.5.4 Gompertz Model

The linear form of the equation of the model is as follows:

$$-\ln(-\ln y) = -\ln(-\ln y_0) + rt$$

The present model is very similar to the logistic model which assumes that the absolute rate of change is a dependent of y and $(1/y)$. It is more asymmetric, with an inflection point at 0.37 ($1/e$). The plotgomp function is created and various parametric data are used to see the effect on the shape of the graphic curve.

```
## Gompertz Model Example
plotgomp<- function(y0,r,maxt){
curve(
exp(log(y0)*exp(-r*x)),
from=0, to=maxt, xlab='Time',
ylab='Disease Incidence',
col='mediumblue'39
      )
}
plotgomp(0.0017,0.02922, 250)
```

3.5.5 Weibull Model

The model involves the following linear equation:

$$\ln\left[\ln\{1/(1-y)\}\right] = -\text{clnb} + \ln(t-a)$$

This model contains a large number of parameters and can describe more complicated curves. The essential parameters are:

a. Units of time (indicating the time of disease onset)
b. The scale parameter (inversely related to the rate of disease increase)
c. Unit-less shape parameter (controlling the skewness of the curve)

If $c = 1$, this model is identical to the monomolecular model with the rate parameter $r = 1/b$ and the initial disease level $y_0 = 1 - \exp(a/b)$. With suitable values of parameters 1, 2, and 3, other models can be approximated by the Weibull model.

```
## Weibull Model Example
plotweib<- function(a,b,c,maxt){
curve(
                  1-exp(-((x-a)/b)^c),
from=0,
to=maxt,
xlab='Time',
ylab='Disease Incidence',
col='mediumblue'
      )
}
plotweib(1, 331.10, 10.04, 500)
```

3.6 DISEASE CYCLE AND EPIDEMIOLOGY OF BACTERIAL PATHOGENS

3.6.1 Bacterial Spot of Tomato and Pepper

Bacterial spot is caused by *Xanthomonas campestris pv. vesicatoria* (Minsavage et al., 1990; Sahin and Miller, 1996; Swords et al., 1996; Bouzar et al., 1999). It is periodically a severe disease of tomatoes and sweet peppers that produces similar symptoms in each that mainly consist of bacterial spot and speck. Both spot and speck, which occur on stems and petioles, are indistinguishable and therefore are often misdiagnosed. Infected leaves can appear small and irregular, with dark lesions that can coalesce to develop leaf yellowing. Infections in flowers and pedicels cause bacterial diseases that lead to bacterial spot (Jones, 1991; Jones et al., 1998 and 2004) and early blossom drop. These diseases are distinguishable based on fruit symptoms. For example, small water-soaked bacterial spots on green fruits can enlarge to 1/8–1/4 inch in diameter. The center becomes irregular, brown, and slightly sunken along with a rough and scabby surface.

Bacterial spot (Jones and Pernezny, 2003) is a disease that spreads under warm and humid regions. In addition, it can be developed under arid regions through practices such as rain or sprinkler irrigation and during pesticide application with high-pressure sprayers. In leaves, bacterial entry through the stomata is facilitated through leaf and fruit wounds caused by wind-driven sand-particle abrasion. The dynamics of disease development are associated with the relative humidity.

The potential sources of inoculum of these bacteria (Jones, 1991; Balogh et al., 2003) are volunteer tomato plants. Survival duration is very limited in the soil (e.g., from days to weeks) and is always associated with cultural remains of diseased or infected plants. These organisms are reported to persist in association with roots of wheat; however, with weeds, their survival is considered less important. Rare cases of survival can be observed in colder regions because of the death of the plant. In these regions, the reintroduction of the pathogen occurs primarily from contaminated or infected seedlings, in which its survival is possible both externally and internally in association with that of the seed. An externally infested seed can have infected cotyledons because of bacterial contact with the seed coat, which results in the formation of lesions as soon as it emerges from the soil. Rapid distribution of pathogens (e.g., within 24 h) results in the rapid multiplication and production of millions of cells. The disease thereby threatens seedling production, particularly because seedling trays favor the development of bacterial spot due to the irrigation and thickening of plants. This type of risk can be reduced by treatment with sodium salts of hypochlorite or phosphate. Treatment of external parts is achievable, but treatment of internal parts (by chemical or thermal methods) is more complicated because it damages and kills the seed.

3.6.2 Bacterial Fruit Blotch of Cucurbits

In bacterial fruit blotch (BFB) disease (Latin and Hopkins, 1995), the bacteria enter through leaf stomata and wounds and remain in the apoplast (intercellular spaces) of infected tissues. In the case of fruit infection, bacteria are deposited on the rinds during the 2–3 weeks after anthesis (flowering) and swim through open stomata to cause infections. After this period, entry via stomata becomes blocked by the deposition of waxes on the surface of the fruit. Mature fruit thus become inaccessible to natural infection by *A. avenae* subsp. *citrulli* (Hopkins and Thompson, 2002) unless wounding occurs. Bacterial fruit-blotch symptoms at early stages of fruit development are inconspicuous (or absent) but appear suddenly at the stage of harvest maturity. As a consequence of rapid expansion in surface lesions, the fruit may rot in the field. In decaying infested rind tissues or in infested seeds *A. avenae* subsp. *citrullican*, overwintering and production of volunteer plants in successive crops occur.

The epidemic development of BFB derives from the most important primary inoculum source, seed-borne (Webb and Goth, 1965) inoculums. Although all seeds of cucurbit can be infested and capable of transmitting BFB, knowledge of the survival of pathogens (on or under the seed coat) and the biology of seed infestation by pathogens remain both unclear and inadequate. The contaminated tissues in an infected fruit are able to infest seeds due to contact with each other. Because the seedling transmission of BFB through direct planting heavily depends on high relative humidity and high temperature, the risk of disease outbreak can be reduced. During transmission, bacterial movement between seed and seedling tissues causes multiplicity in intercellular spaces and thus initiates water-soaked lesions. Rapid epidemic development and high populations of infected seedlings can therefore be deemed results of secondary infections from splash-dispersed bacteria. Foliar lesions and epiphytic population, which can also contribute to secondary infections, serve as reservoirs for *A. avenae* subsp. *citrulli* inoculums. This inoculum, which is disseminated by wind-driven rain and irrigation water, causes foliar lesions and blight symptoms.

3.6.3 Bacterial Leaf Scorch of Shade Trees

Bacterial leaf scorch (BLS) is caused by the spread of *Xylella fastidiosa* (Figure 3.9), which is carried by leafhopper insects called sharpshooters (sub-family: Cicadellidae) and spittle bugs (family: Cercopidae). The carrying capacity of pathogens varies between insects and bugs. Insects at adult and nymph (immature) stages have piercing/sucking mouthparts and subsist in xylem fluid to acquire bacteria as they feed on the succulent tissues of infected hosts. As a result, the xylem fluid is drawn into the insect, which responds by attaching bacterial cells to its cibarial pump (suction-pump mechanism) and esophageal lining. Ultimately, by extracting nutrients from xylem fluid with a pumping mechanism, the proliferating bacteria form an encased and protected biofilm (Figure 3.9) in a bacterial glycocalyx.

After the early stages of feeding, bacterial cell dislodging followed by direct pumping into xylem occurs where there is systemic movement within the host. Transmission of bacterium to the new host begins within 1–2 h from an adult, and continues for the remainder of its life, whereas nymphs (which shed the foregut during molting) can only transmit until their next molt. Unlike Pierce's disease in grapes, ambiguity exists in the threshold population of bacteria before transmission by insect vectors to cause BLS in shade trees (Hewitt et al., 1946; Davis et al., 1978).

Leafhoppers (Purcell et al., 1979), which are xylem-feeding insects, can be polyphagous, act as hosts of *X. fastidiosa* as a food source, and overwinter as adults on alternative plants. Such alternative hosts may be the source for deriving considerable amounts of inocula for transmission by vectors to economically significant crops (e.g., grapes and peaches). The respective insects that vector some economically important diseases such as Pierce's disease of grapevine and phony peach disease are *Graphocephala* and *Homalodisca/Oncometopia* (Brlansky et al., 1983). The involvement of insects that vector BLS in shade trees and alternative hosts is still under investigation. Research studies in oak and other shade-tree hosts has ascertained that several known vectors of other diseases caused by *X. fastidiosa* are present in shade trees (Hammerschlag et al., 1986) during the growing season.

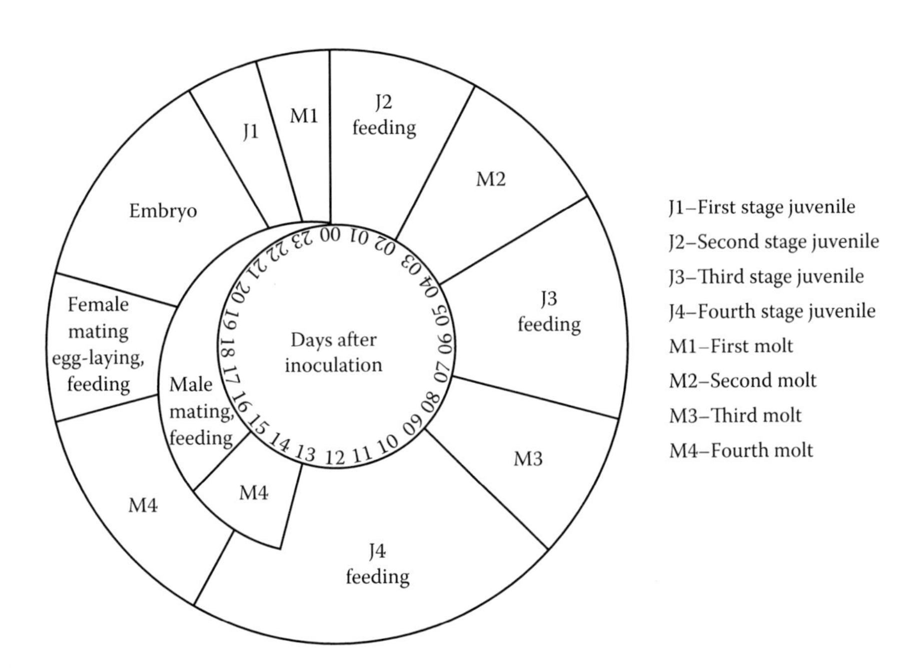

FIGURE 3.9 Schematic representation of the life cycle of *B. longicaudatus* on excited roots of *Zea mays*. (Huang, X., De. Bever, A., Becker, O. The *in-vitro* life cycle of *Belonolaimus longicaudatus*. *J Nematology*. 1997. Copyright Wiley-VCH Verlag GmbH & Co. KGaA.)

3.6.3.1 Disease Development Dynamics and Mechanisms

In xylem tissue, *X. fastidiosa* lives and multiplies within the tracheary elements, tracheids, vessels, and intercellular spaces. It also fluctuates seasonally within xylem tissues. The xylem fluid consists of amino, organic, and inorganic acids that do not have nutritional value. Bacterial and insect vectors use amino acids (i.e., glutamine and asparagines) for their growth. The quality and composition of xylem fluid (both between hosts and within a host) vary with respect to season, time, and the health and age of the host plant. The variation of fluid affects the feeding behavior of vectors. As the insects search for suitable nutrient sources, they dock *X. fastidiosa* to transmit the pathogen to new hosts. As a consequence, a biofilm is produced from the attachment of *X. fastidiosa* cells to xylem vessels as well as to the foreguts of insect vectors. The aggregated and encased bacterial cells in a self-produced matrix of polysaccharide are well protected and are deemed to enhance pathogenicity.

Participation of terminal fimbriae in biofilm formation (Marques et al., 2002) and twitching motility (which enables bacterial cells to move against the xylem stream; Meng et al., 2005) is of great significance. The results of water stress due to high bacterial populations in xylem tissue and the overproduction of defense compounds in host plants (pectins and tyloses) combine to reduce resistance to infection. The embolisms that eventually form help to plug xylem vessels, which reduce the function of xylem and water stress. In turn, this prolonged reduction in the function of xylem vessels and water stress leads to photosynthesis reduction and starch-reserve depletion that ultimately results in leaf scorching and premature senescence. Owing to the congregation of bacteria in the roots, stunted growth is the primary symptom of phony peach and alfalfa dwarf diseases. Nonetheless, these pathogens also reside above the ground; heavy populations accumulate in the veins and petioles of symptomatic leaves, leaving a primary symptom of leaf scorch in the BLS of shade trees. In shade trees, most strains of *X. fastidiosa* are sensitive to cold and are presumed to overwinter in parts such as trunks or roots. During the growing season, with an increased population there can be an acropetal movement of pathogens to distal portions of the canopy. In addition to vector movement, the geographic distribution of *Xylella*-associated diseases in cold conditions can contribute to the hardiness of the pathogen.

3.6.4 Blackleg of Potato

Blackleg disease (Perombelon, 1992) originates from contaminated seed pieces and is often seen in plants that are well grown and flowering. When the disease becomes severe, entire seed pieces and developing sprouts may rot before they emerge from the ground. Leaves of diseased plants tend to roll upward at the margins and often die. The tissues in potato tubers are very soft and watery, with granular consistency. The diseased tissue appears cream-to-tan colored within a black border that separates healthy from infected areas. Although soft rot decay is odorless at its initial stage, due to the invasion of secondary bacteria a stringy or slimy decay later develops a foul odor. These soft-rot organisms (Perombelon and Kelman, 1980) consume internal tuber tissues and sometimes leave only a shell of skin in the soil.

Blackleg bacterium survival is poor in soil; it does not survive outside of association with host-plant tissue. When the seed tuber (an important source of inoculum) becomes infected, it provides three routes that ultimately form the disease. Blackleg disease is due to: (1) the movement of blackleg bacteria into the growing plant through vascular bundles. If the tuber contamination is confined to lenticels, initially the seed tuber occurs, decays, and the growing stem (i.e., a moist, cool environment) is invaded by a high population of bacteria; (2) Favorable conditions for plant growth nullify the disease even in the presence of blackleg bacteria; and (3) Decay of a seed piece (Perombelon, 2000) prior to plant development.

The most frequent condition under which no blackleg disease occurs is the decay of contaminated seed pieces after the plant is established. In this case, the seeping blackleg bacteria from the decaying seed piece have the ability to contaminate the entire root zone, which includes the development of progeny tubers (Toth et al., 2003). The bacterially contaminated progeny tubers develop hard rot symptoms (Perombelon and Kelman, 1980) or remain symptomless during storage. When the symptomless but contaminated tubers are used for planting, the disease cycle repeats. The contamination of potato tubers is exacerbated during harvesting and storage. As a consequence of damage during storage, the tubers are vulnerable to decay by the *Pectolyticerwinia* (Elphinstone and Perombelon, 1986) and the environment is more conducive due to moisture. The formation of a film of water (due to soil flooding or improper drying of washed tubers) around tubers causes an anaerobic condition that inhibits their metabolic activity and impedes their normal resistance.

3.6.5 Citrus Canker

Citrus canker is a disease which affects citrus species caused by the bacterium called *Xanthomonas axonopodis*. This disease significantly affects the vitality of trees by causing leaves and fruit to drop prematurely but the infected fruit is safe to eat. These bacterial organisms propagate in lesions on the leaves, stems, and fruit. Owing to free moisture on lesions, the bacteria ooze out and get dispersed to infect a new growth. The main dispersal agent is the wind driven rain (Bock et al., 2005), however, the penetration is facilitated through stomatal pores, wounds (made by thorns or insects), and blowing sand. The spread of canker by wind and rain can either be within a tree or between trees at short distances. Canker is developed more severely on the side of the tree exposed to wind driven rain.

Pruning, hedging, and spray equipment have been involved in the spread of disease within and among plantings. Pruning can cause severe wounding that leads to infection. The expansion of lesions or the multiplication of bacterial organisms per lesion are related to general host susceptibility. Bacteria can survive (Gottwald et al., 2002) in the margins of lesions in leaves and fruit until the bacteria are abscised and fall on the ground. The survival of bacteria differs with respect to biotic and abiotic factors (Gottwald and Timmer, 1994); they expire upon exposure to rapid drying and direct sunlight. In soil, survival is limited to a few days but may extend for months if the soil is incorporated with plant refuse. Bacteria can survive for years in infected plant tissues (Graham et al., 1992) that have been kept dry and free from soil.

Although mature leaves, stems, and fruits are resistant to infection, they become infected when they are wounded. Once the growth is initiated, almost all infections occur on leaves and stems within the first 6 weeks; for fruit-rind infection, the most critical period is the initial 13 weeks after petal drop. Infection after this critical period, which is small, is observed by inconspicuous pustules because the supceptibility period for fruits (Verniere et al., 2002) is longer than for leaves. If there is more than one dispersal event, lesions of different ages may appear on the same fruit. Estimating the lesion age on a fruit helps in determining the time of infection and can also indicate when meteorological events such as storms occurred. A recent report revealed that 99% of infections that occur within 30 d under normal weather conditions are located within about 2000 ft of previously infected trees.

Canker is spread over distances of up to several miles by meteorological events such as tropical storms, hurricanes, and tornadoes. In 2004, the landfall of three different hurricanes in Florida had a near-fatal effect upon the commercial citrus industry. This long disease spread also results in affected propagating material (e.g., bud wood, rootstock seedlings, budded trees), commercial shipments of diseased fruit, and wooden harvesting boxes that contain diseased fruits and leaves.

3.6.6 Crown Gall

The neoplastic diseases known as crown gall and hairy root were described initially at the beginning of the twentieth century (Stewart et al., 1900; Smith and Townsend, 1907). *Agrobacterium tumefaciens* (*Rhizobium radiobacter*) (Kado, 1976) has been identified as the common soil-dwelling agent (Smith et al., 1911) of crown gall disease (the formation of tumors) in more than 140 species of Eudicot; it resides in the rhizoplane of woody and herbaceous weeds. *Agrobacterium tumefaciens* is a serious pathogen of walnuts, grapevines, stone fruits, nut trees, sugar beets, horseradish, and rhubarb. This Gram-negative soil bacterium causes disruption in the plant's hormone balance that results in hyperplasia without killing the host (Hooykaas and Beijersbergen, 1994). As a result, the bacterium causes symptoms by the insertion of a small segment of DNA from the plasmid into the plant cell; in turn, this DNA is incorporated at a semirandom location into the plant genome. Many researchers have tried to identify the consecutive steps that are involved in the formation of crown gall (Gelvin, 2000, 2003; Tzfira et al., 2004; Anand and Mysore, 2006; McCullen and Binns, 2006; Citovsky et al., 2007; Gelvin, 2010).

The bacterium infects the plant through its Ti plasmid, which integrates T-DNA into the chromosomal DNA (Zaenen et al., 1974) of its host plant cells. This bacterium either disseminates as infected plant material or spreads from the broken or sloughed galls of the infected plants during cultivation practices. Nursery stock from uncertified sources, which may include infected and infested plants, are primary spreaders of the disease. Secondary spread of the disease occurs through pruning and cultivation equipment during the manual removal of galls. Tilling equipment may be contaminated through galls at or near the bases of the trunks of infected trees. Roguing an infected tree and replanting a new one in the same place would lead to an abundant population of *A. Tumefaciens* because of the sloughed off galls that serve as sources for the bacterium.

Although a single-cell transformation into a gall is feasible, the size of the initial tumor depends upon the amount of inoculums. The severity of disease and the amount of gall production at the infection site depend upon how many *A. tumefaciens* cells enter a plant wound. It is often observed that a systematically infected plant will harbor the bacterium for extended periods of time in the absence of overt symptoms.

An accidental wound during cultivation practices or frost can also cause tissue injuries. These injuries can elicit new infections that may appear as galls along the vascular system of the host. Important horticultural crops that are affected by *Agrobacterium* are stone fruits such as cherry, peach, apricot, and plum (Ali et al., 2010); other fruits such as apple, pear, and grape (Ali et al., 2010; Rouhrazi and Rahimian, 2012); nuts such as almond, pecan, and walnut (Pulawska, 2010); vegetables such as tomato and sweet pepper (Cubero et al., 2006); and ornamentals such as rose, chrysanthemum, and aster (Aysan and Sahin, 2003; Cubero et al., 2006).

3.6.7 Fire Blight of Apple and Pear

Fire blight is a contagious systemic disease that affects apples, pears, and their associated family members (Baker, 1971; Beer, 1990). The disease-affected areas appear blackened, shrunken, and cracked, as if scorched by fire. The causal pathogen (Vanneste, 2000), *Erwinia amylovora*, is a Gram-negative bacterium that can destroy an entire orchard in a single growing season. Primary infections are established in tender new shoots and leaves during spring, when blossoms are open. Abiotic factors (e.g., rain and wind) and biotic factors (e.g., honeybees, other insects, and birds) can transmit the bacterium to susceptible tissue (preferably of pears). Punctures and tears caused by plant-sucking or -biting insects can develop injured tissues that make the plant susceptible to infection.

Open stomata are the pathways for the bacterium during primary infection. The entered bacterium causes blackened, necrotic lesions that lead to the production of viscous exudate. This laden exudate, which causes secondary infections, travels to other parts of the same plant or to susceptible areas of different plants by biotic and abiotic factors. The spread of the disease is active with respect to hot, wet weather conditions but dormant in winter. The spread of the pathogen in a tree occurs through the vascular system from the infected point; ultimately, it reaches the roots and/or graft junction. The release of free bacterial cells on the bark surface (as visible ooze) also attracts insects and makes dissemination by rain or insects from the canker to blossoms feasible. Available bacteria of an infected plant are dormant in winter, but can produce exudate after the arrival of warm weather in spring. This exudate then becomes the source for primary infections in the ensuing seasons. Overwintering sites, in which the pathogen survives in a small percentage of the annual cankers, are called "holdover cankers." Over-pruning and over-fertilization cause water-sprout and other midsummer growths that leave the tree more susceptible to infection and disease.

3.6.7.1 Floral Infections

The principal site for epiphytic colonization of *E. amylovora* is stigma. The ultimate population size of the pathogen is attained at the floral epiphytic phase by the influence of temperature and number of blossoms. Temperature facilitates the regulation of the pathogen's generation time, and the blossoms are the sites where the pathogens (~10^6 cells under ideal conditions) are established by pollinating insects.

In flowers, primary infection is associated with blossom-blight initiation due to external washing of *E. amylovora* cells from stigma to hypanthium (floral cup). On the hypanthium, the pathogen gains entry to the plant through cell secretors (nectarthodes) located on the surface. Blossom blight expands further according to the tendency of a species to produce secondary (rattail) blossoms during late spring or early summer, long after the period of primary bloom. Blossom blight is sporadic from season to season because it requires warm temperatures to drive the development of large epiphytic populations.

Several epidemiological models (e.g., COUGARBLIGHT, MARYBLYT) work by identifying the periods conducive to epiphytic growth of *E. amylovora* on blossoms, and therefore are widely used to aid decisions about the need for and timing of chemical applications. Blossom-blight risk models accumulate degree units above a threshold temperature of 15.5°C (60°F) or 18°C (64°F). Data on rain or blossom wetness during warm weather are used for more precise indications of the timing and likelihood of blossom infection. Other temperature-based models predict incubation-period length (i.e., the time to symptom expression after an infection event) based on heat-unit sums. These models are used to time orchard inspections and/or pruning activities.

Infection at blossom, which is apparent from diseased tissues, is carried to shoot, fruit, and root stock as a consequence of inoculums during the secondary phase. Wounds are created by insects such as plant bugs, psylla (during feeding), and strong winds, rain, and hail. Infection symptoms induced by severe weather are sometimes called "trauma blight." Both primary and secondary infections expand throughout summer and eventually result in severe infection according to the host species, the cultivar, the environment, and the age/nutritional status of host tissues. The susceptibility of fruit blight is rapid for young trees or those with vigorous tissues. Conversely, the rate of susceptibility is slow for trees that are older or have slow-growing tissues. The rate of susceptibility can also be correlated with the intake of nitrogen fertilizer or nutrient regime, because the growth of a tree depends on its nutrient intake. The rate of canker expansion is also proportional to certain abiotic factors such as water and temperature. Canker expansion in later summer was slow and the growth rate declined, but expansion was enhanced in poorly drained soils.

3.6.8 Lethal Yellowing of Palm

Lethal yellowing is a phytoplasma disease that attacks many species of palm, including coconut and date palms (Eden-Green, 1997; Harrison et al., 1999). Mycoplasma-like particles are found in the sieve-tube elements in the phloem (Broschat et al., 2002) of coconut and other palms that exhibit characteristic symptoms (Beakbane et al., 1972; Plavsic-Banjac et al., 1972). Marked differences in susceptibility to the phytoplasma are reported in coconut palms with respect to their heights; these include less resistant, as in Jamaica Tall (except Panama Tall), and notably resistant, as in Malayan Dwarf. Howard (1980) reported that the population density of *Haplaxius crudus* (*M. crudusare*) was up to 40 times higher in affected areas. Gradual introduction of *H. crudus* from an infected area onto caged susceptible palms resulted in the development of symptoms (Howard et al., 1983 and 1984).

The early symptom is the drying up of developing inflorescences, followed in coconut palms by the discoloration and blackening of spathes (which enclose flowers) and tips. The water-soaked streaks of the youngest leaves spread until the growing point rots. The progressive discoloration of older leaves proceeds rapidly on the younger leaves and can lead to foliage-color changes from light yellow to orange-yellow. The ultimate death of *Cocos nucifera* occurs four months after the initial symptoms.

This bacterium is spread by Florida's plant-hopper *H. crudus*, an insect with piercing and sucking mouthparts that feeds on the contents of the plant host's vascular system. The spread of phytoplasma is facilitated from palm to palm in a random pattern around an active focus of disease during the feeding activity of a plant-hopper; however, its survival outside the plant or insect hosts is not known. Purcell (1985) noted that *H. crudus* was an inefficient vector of lethal yellowing but that its abundant presence transmitted the disease at a very low rate. The continual reproduction of tropical *M. crudus* (Howard et al., 1983) throughout the year is limited by lack of winter die-off to areas with sufficiently mild winters.

Beyond this primary focus, further spread may occur in leaps of 100 km or more that establish a disease in new areas called foci. The disease is also characterized by a "jump-spread" pattern that indicates the involvement of an airborne vector in dissemination. Differing rates of lethal yellowing with respect to geographic locations have been reported. For example, the spread from the cities of Miami to Palm Beach (~130 km) occurred within three years in Florida, United States, whereas the spread over about 240 km from the west end to the east end of Jamaica took almost 60 years.

Inoculation, which is initiated through a susceptible and infected plant, is prolonged over the latent (incubation) phase of 112–262 d. The growth of infection, which is stimulated 80 d prior to symptom appearance, is followed by a period of gradual decline; growth ceases about 30 d before the end of the incubation phase. The lifetime of infected plants normally ends in three to six months. Prevention is the only effective remedy. Preventive measures include diverting the attraction of the plant-hopper toward resistant varieties of coconut palm. However, it was reported that heavy-turn grasses and similar green covers will induce plant-hoppers to lay eggs in locations that result in the development of nymphs at the roots. This situation threatens the crop in coconut-growing countries that import grass seed from the United States for golf courses and lawns. No noticeable evidence was detected for the lethal yellowing of palm groves when two important food palms were grown in traditional ways (without grasses) in plantations and along shorelines. The incidence of infection influenced by cutting or trimming instrument of palm was also found to be unjustifiable with regard to the spread of disease.

3.6.9 Stewart's Wilt of Corn

The first report on Stewart's bacterial wilt of corn was made in the United States in 1897. The disease is caused by the bacterium *Erwinia stewartii* (syn. *Pantoea stewartii*), which can systematically spread throughout the plant and ultimately kill it. The bacterium has two hosts, corn and cornflea beetles (*Chaetocnema pulicaria*), and cannot spread from plant to plant without a vector. About 10%–30% of the emerging beetles carry the bacteria and the prevalence of the disease in the previous season determines the percentage of emerging beetles that will carry it. The most probable mode of infection appears to be from the fecal contamination of feeding wounds. Experimental reports from Charlotte Elliott and Poos (1940) revealed that 28,500 insect specimens representing 94 species and 76 genera for *E. stewartii* affirmed the significance of *C. pulicaria* in harboring the bacterium and spreading of disease.

The bacterium can infect the plant at any age, but the greatest damage to corn occurs when plants are infected (Dodd, 1980) as seedlings, prior to the five-leaf stage, and when the bacterium is transmitted by the overwintering generation of flea beetles. Damage remission because of the plant's maturity is possible. In the seedling-wilt phase, the bacterium enters via feeding wounds on leaves created by corn flea beetles and then enters the developing stalk to kill the growing point (indicated by dead tassels). During the disease stage, leaves appear from pale green to yellow-streaked. The leaf-blight phase, which results from the infections of older plants, is characterized by long whitish to chlorotic streaks (with irregular wavy margins that run parallel to the veins) along the leaves. Vascular discoloration in the stem and at plant nodes can occur in highly susceptible varieties. Subsequent generations of flea beetles, which function as the overwintering populations, acquire the bacterium from infected plants.

Although the bacterium is seed-borne, seed transmission plays an insignificant role and is virtually meaningless in the epidemiology of Stewart's wilt (Pepper, 1967) in North America and other world regions. The infection of seeds with *E. stewartii* (Lamka, 1991) is due to the systematic infection of the seed-parent plant during the disease's seedling-wilt phase. Conversely, seed-parent plants are not infected systematically even during the leaf-blight phase of Stewart's wilt; as a result, seed infection is extremely unlikely. Recent research reports suggested transmission rates of the bacterium in seed are about 1 in 50,000 for seed produced on systematically infected and susceptible seed-parent plants, and 1 in 20,000,000 for seed produced on resistant plants that show symptoms of the leaf-blight phase of Stewart's wilt.

The bacterium has never become established through vectors outside North America but its spread has been reported across the world. More than 60 countries have placed quarantine restrictions on corn seeds in order to prevent the entry of *E. stewartii*. Stewart's wilt is endemic in the mid-Atlantic States, the Ohio River Valley, and the south portion of the Corn Belt. Its occurrence through corn flea beetles was also observed in other eastern and midwestern states and in portions of Canada. The bacterial dynamics that cause Stewart's wilt disease are as follows:

The stay of *Erwinia stewartii* in the alimentary tract or "gut" in winter

Beetles feeding causes the spread of the bacterium from alimentary tract to sweet corn plants during spring

Deposition of bacteria at the feeding sites and colonization of leaf tissues

Entry of bacteria into the vascular system

Spread of bacteria throughout the plant

The severity of the disease depends upon winter temperatures prior to planting (Stevens, 1934; Boewe, 1949), amount of Stewart's wilt in the previous season, and the susceptibility of the hybrid to the disease. In the early 1930s, a forecasting system was developed for Stewart's wilt disease by comparing maps of winter temperatures and occurrences of Stewart's wilt. Later, N. E. Stevens developed a prediction system based on a winter-temperature index. In 1945, G. H. Boewe modified Stevens's forecasting system to predict the disease of leaf blight phase more accurately in late summer. With the Stevens–Boewe Index, risk is calculated by summing the average monthly temperatures for December, January, and February. A sum below 80°F indicates a slight risk, 80–90°F is considered low-to-moderate risk, and greater than 90°F is considered high risk. Forecasting with the Stevens–Boewe Index shows that southern Iowa has a high risk of the late leaf-blight phase of Stewart's disease.

3.6.10 Sting Nematode

Belonolaimus, a genus of nematodes commonly known as sting nematodes, can cause severe damage to plants that may lead to complete crop losses. *Belonolaimus longicaudatus* was recognized as an important agricultural pest in the southeastern United States and is considered to be the worst pest species there. Some of the largest plant-parasitic nematodes can reach 3 mm in length and feed by inserting their stylets into roots to suck the contents of root cells. They can be found on fruits, vegetables, and other crops such as cotton and soybeans, turf grasses, and forestry trees (Abu-Gharbieh and Perry, 1970). The life cycle (Huang and Becker, 1999) of *B. longicaudatus* was studied on roots of corn (*Zea mays*) in axenic root culture. The life cycle of the sting nematode is comprised of the egg stage, first-stage juvenile (J_1), second-stage juvenile (J_2), third-stage juvenile (J_3), fourth-stage juvenile (J_4), and adult stage.

The process of embryogenesis of *B. longicaudatus* extends from one-celled stage to J_1, during which it molts inside the egg into J_2 and becomes ready to hatch. After hatching, J_2 moves through soil to the root system of the host plant, where it congregates around root hairs and feeds until it molts into J_3, which moves immediately to the meristems of lateral roots to further feeding. After a second molting, J_3 becomes J_4, which molts again to become an adult. Sting nematodes are amphimictic. Male and female nematodes are abundant and attracted to each other for reproduction. Because the female sting nematodes have a spermatheca (a sperm-storage organ), only a single mating is required for the fertilization of many eggs over time. These eggs are produced separately from each of two paired ovaries. An approximate count of 9–10 eggs was laid by a female nematode over a 10–15 h period (Figure 3.9). Recent findings that are pertinent to various plant diseases caused by bacterial pathogens in many countries are summarized in Table 3.3.

3.7 CONCLUDING REMARKS

The causes of bacterial infections by pathogens can be studied above and beneath the ground. Certain mathematical models (multiple regression and non-linear models) can forecast diseases based on parameters such as temperature, relative humidity, and precipitation.

The progress of a disease over time and its progress curve are modeled and well illustrated by five growth models. Various bacterial organisms that frequently and seasonally affect plants and trees, along with abiotic and biotic factors, have been elaborated upon in this chapter, which has been intended to convey that scientific opportunities for biotechnological applications in plant pathology are numerous and remarkably worth exploring.

3.8 FUTURE PERSPECTIVES

The observations and modes of action of effectors from pathogens (with diverse life haustorial) will help us to define a comprehensive set of host targets and the evolutionary pressures that act upon both hosts and pathogens. The comprehension of haustorical interface and its differentiation can be accomplished by rewiring host and microbe vesicle traffic. Knowledge of genomic sequences can result in comparative genomic analyses that will enable identification of novel virulence factors through targeted investigation of genes that are unique to specific pathogens.

An in-depth probe of the use of natural plant compounds as antibacterial agents characterized by diverse chemical structures and mechanisms is the need of the hour. Such a probe would pave the way for discovering bioactive products that could be established as spellbound therapeutic tools. The biggest challenge for the future lies in augmenting and applying our knowledge of plant immune systems to develop more disease-resistant crops and to develop crops that display durable resistance across time and space.

ACKNOWLEDGMENTS

Venkataraman Sivasankar thanks the principal and management of Pachaiyappa's College, Shenoy Nagar, Chennai 600 030. I am also greatly indebted to Dr. R. Prabhaakaran, former president and present trustee of Pachaiyappa's Board Trust, Chennai 600 030, for his constant support and encouragement.

TABLE 3.3
Summarized Research Reports on Various Bacterial Pathogens That Cause Diseases and Associated Symptoms

Sl. No.	Disease and Symptoms	Bacterial Pathogen	Location	Reference
01.	Necrosis of cotyledons, leaves, and stems in sweet basil (*Ocimum basilicum* L.) seedlings and plants Dark grey, water-soaked, circular-to-irregular lesions mainly on the leaf and cotyledon margins	*Pseudomonas viridiflava*	Liguria region, Northern Italy	Minuto et al. (2008)
02.	Wilt disease in Bottle gourd (*Lagenaria siceraria*) Wilting, yellowing of leaves extending to the stem in bottle gourd	*Ralstonia solanacearum*	China	Gao et al. (2007)
	Bacterial wilt of potato, tubers with typical brown rot symptoms		Uruguay	Siri et al. (2011)
03.	Blight of golden cane palms (*Dypsis lutescens*) Angular, water-soaked, and chlorotic haloed lesions typical of blight symptoms	*Burkholderia androogonis*	Northeastern Queensland, Australia	Young et al. (2007)
04.	Leaf spot in sorghum plants (*Sorghum bicolor*)	*Pantoea ananatis*	Brazil	Cota et al. 2010
	Red spots on the leaf surface followed by progression of necrotic lesions		Mexico	Perez-y-Terron et al. (2009)
	Leaf spot on maize		Argentina Brazil	Alippi and Lopez (2010) Paccola-Meirelles et al. (2001)
	Infection of onion leaves (leaf blight) that leads to bulb infection		Ithaca, New York	Carr et al. (2013)
05.	Sheath browning and grain discoloration of rice (*Oryza sativa*) Symptoms on glumes range from necrotic spotting to complete discoloration; leaf-sheath lesions are light brown to black and are generally irregular in shape, leading to mottling effect.	Complex rice pathogen Non-phytopathogenic endophytes	East Timor (Timor Leste), Australia	Adorada et al. (2013)
06.	Little leaf disease in moss-rose purslane (*Portulaca grandiflora*) Typical bud proliferation, downward curling and diminishing size of leaves followed by stunted growth	*Candidatus phytoplasma*	Ornamental gardens and pots in Central Institute of Medicinal and Aromatic Plants (CIMAP), Lucknow, India	Ajaykumar et al. (2007)
07.	Disease of hellebore (*Helleborus orientalis* and *Helleborus hybridus*) Circulur black spots, black stem lesions, and dry, gray-to-brown lesions with distinct margins on the flower petals	*Pseudomonas viridiflava*	Tauranga, New Zealand	Taylor et al. (2011)

(Continued)

TABLE 3.3 (*Continued*)
Summarized Research Reports on Various Bacterial Pathogens That Cause Diseases and Associated Symptoms

Sl. No.	Disease and Symptoms	Bacterial Pathogen	Location	Reference
08.	Rice orange leaf disease of phytoplasma in rice (*Oryza sativa*) Stunting; proliferating auxiliary shoots that result in sterile and deformed flowers; virescence; and phyllody	*Candidatus phytoplasma* asteris (16SrI) group	Cauvery Delta, Lower Bhavani Project Delta, Parambikulam Aliyar Delta, Periyar Vaigai Project Delta and Thamiraparani Delta zones in South India	Valarmathi et al. (2013)
09.	Huanglongbing (HLB) or Citrus greening in citrus (*Diaphorina communis*) Leaf defoliation, mottling, twig and tree dieback, poor fruit production and lopsided fruit *Murrayapaniculata* associated with *Diaphorina citri*	*Candidatusliberibacter* asiaticus (an α-proteobacterium)	Punakha, wangduephodrang and Tsirang Districts, Bhutan North–West Frontier Province of Pakistan Jammu, Kashmir, Punjab and Rajasthan Punjab and Peshawar Saint Luice County and Palm Beach County, Florida	Donovan et al. (2012) Chohan et al. (2007) Ahlawat and Raychaudhuri (1988) Akhtar and Ahmad (1999) Walter et al. (2012)
10.	Soft rot disease of potato (*Solanum tuberosum*) Brown-colored, rotten, and oozing tubers	*Pseudomonas marginalis* pv. *marginalis*	Zhangye District of Central Gansu, China	Li et al. (2007)
11.	Soft root on lettuce (*Lactuca sativa* var. *romana*) Chlorotic or necrotic leaves with a yellow color that exhibit extensive water-soaked lesions	*Pectobacterium carotovorum*	Malaysia	Nazerian et al. (2013)
12.	Necrosis of mango (*Mangifera indica*) Blight and necrosis of panicle, reduced flowering, and reduced-to-absent fruit set Discoloration in vasculature in traces down the petioles and infected stems Observation of young lesions in an alternating arrangement down either side of the leaf mid-rib, which may coalesce and result in complete death Also called black blight (in Israel) and bacterial apical necrosis (in Spain and Portugal)	*Pseudomonas syringae* pv. *syringae*	Australia	Young (2008) [a]Pinkas et al. (1996) [b]Cazorla et al. (1998)

(*Continued*)

TABLE 3.3 (*Continued*)
Summarized Research Reports on Various Bacterial Pathogens That Cause Diseases and Associated Symptoms

Sl. No.	Disease and Symptoms	Bacterial Pathogen	Location	Reference
13.	Bacterial canker of kiwi fruits (*Actinidia chinensis* Pl. *cultivars*) Brown discoloration of buds; dark brown spots surrounded by yellow haloes on leaves; cankers with reddish exudates on twigs, leaders and trunks; and collapse of fruits	*Pseudomonas syringae* pv. *actinidiae*	Central and Northern Italy	Balestra et al. (2009)
14.	Wilt disease of banana plants (*Musa paradisiaca*) Yellowing and leaf death	16SrVIII phytoplasmas	Papua New Guinea (PNG)	Davis et al. (2012)
15.	Incidence of crown gall in walnut root stock 'Paradox' (*Juglans hindisii* × *J. regia*) in nursery and orchards	*Agrobacterium tumefaciens biovar*1	California	Yakabe et al. (2014) Yakabe et al. (2012)
16.	Yellow bud disease of Italian rye grass (*Lolium multiflorum*) and curly dock (*Rumex crispus*)	*Pseudomonas* sp.	Georgia, USA	Dutta et al. (2014)
17.	Bacterial soft rot disease in banana (*Musa paradisiaca*) Leaf wilting, collapse of pseudo stems and unusual odor	*Dickeyazeae*	China	Zhang et al. (2014)
18.	Bacterial leaf scorch disease in high-bush blueberry stem and root sections Drought-like symptoms and eventual plant death	*Xylella fastidiosa*	Southeastern United States	Holland et al. (2014)
	Almond leaf scorch disease (ALSD)		California	Krugner et al. (2012)
19.	Lethal wilt disease of oil palm (*Elaeis guineensis*)	*Candidatus photoplasmaasteris*	Colombia, South America	Alvarez et al. (2013)
20.	Bacterial spot of rose (*Rosa* spp.)	*Xanthomonas* sp.	Florida and Texas	Huang et al. (2013)
21.	Bacterial canker of kiwi (*Cyanococcus vaccinium*) fruit Angular leaf spots with or without a yellow halo and wilting of young shoots	*Pseudomonas syringae* pv. *actinidiae*	New Zealand	Vanneste et al. (2013)
22.	Citrus variegated chlorosis (CVC) disease Variegated chlorosis on leaves located at the top of a single branch	*Xylella fastidiosa*	Brazil	Gracia et al. (2012)
23.	Bacterial spot on tomato (*Solanum lycopersicum*) and pepper (*Capsicum* spp.)	*X. perforans, X. gardneri,* and *X. vesicatoria* strains	Southwest Indian Ocean region	Hamza et al. (2010)
24.	Black rot of cabbage (*Brassica* spp.) V-shaped chlorotic lesions with black veins found on leaves Also exhibited blackening of the vascular tissue when cut in half	*X. campestris* pv. *Campestris* strains	Eastern, Central, and Western Nepal	Jensen et al. (2010)

REFERENCES

Abu-Gharbieh, W.I., Perry, V.G. 1970. Host differences among populations of *Belonolaimus longicaudatus* Rau. *J Nematol* 2: 209–216.

Adorada, D.L., Stodart, B.J., Tpoi, R.P., Costa, S.S., Ash, G.V. 2013. Bacteria associated with sheath browning and grain discoloration of rice in East Timor and implications for Australia's biosecurity. *Australazian Plant Dis* 8: 43–47.

Agrios G.N. 1997. *Plant Pathology*. Academic Press, London, 635.

Agrios, G.N. 2004. *Plant Pathology*. 5th edition. Elsevier, Academic Press, San Diego, pp. 922.

Agrios G.N. 2005. *Plant Pathology*, Elsevier, Academic Press, London, pp. 922.

Ahlawat, Y.S., Raychaudhuri, S.P. 1988. Status of citrus triseza and dieback diseases in India and their detection. In *Proceedings of the Sixth International Citrus Congress*. Tel Aviv, 871–879.

Ajayakumar, P.V., Samad, A., Shasany, A.K., Gupta, M.K., Alam, M., Rastogi, S. 2007. First record of a *Candidatus phytoplasma* associated with little leaf disease of *Portulaca grandiflora*. *Australasian Plant Dis* 2: 67–69.

Akhtar, M.A., Ahmad, I. 1999. Incidence of citrus greening in Pakistan. *Pak J Phytopathol* 11: 1–5.

Ali, H., Ahmed, K., Hussain, A., Imran, A. 2010. Incidence and severity of crown gall disease of cherry, apple and apricot plants caused by *Agrobacterium tumefaciens* in Nagar Valley of Gilgit-Baltistan, Pakistan. *Pak J Nut* 9: 577–581.

Alippi, A.M., Lopez, A.C. 2010. First report of leaf spot disease of maize caused by *Pantoea ananatis* in Argentina. *Plant Dis* 94(4): 487.

Alvarez, E., Mejia, J.F., Contaldo, N., Paltrinieri, S., Duduk, B. 2013. *Candidatus Phytoplasma asteris* strains associated with oil palm lethal wilt in Colombia. *Plant Dis* 97(12): 1524–1528.

Anand, A., Mysore, K.S. 2006. Agrobacterium biology and crown gall disease. In *Plant Associated Bacteria*. Edited by S.S. Gnanamanickam. Dordrecht: Springer, 359–384.

Arneson, P.A. 2001. Plant disease epidemiology: Temporal aspects. *Plant Health Instructor* (American Phytopathological Society). Doi:10.1094/PHI-A-2001-0524-01.

Aysan, Y., Sahin, F. 2003. An outbreak of crown gall disease on rose caused by *Agrobacterium tumefaciens* in Turkey. *Plant Pathol* 52: 780.

Baker, K.F. 1971. Fire Blight of pome fruits: The genesis of the concept that bacteria can be pathogenic to plants. *Hilgardia* 40: 603–633.

Balestra, G.M., Mazzaglia, A., Quattrucci, A., Renzi, M., Rossetti, A. 2009. Current status of bacterial canker spread on kiwifruit in Italy. *Australasian Plant Dis* 4: 34–36.

Balogh, B., Jones, J.B., Momol, M.T., Olson, S.M., Obradovic, A., King, P., Jackson, L.E. 2003. Improved efficacy of newly Formulated bacteriophages for management of bacterial spot on tomato. *Plant Disease* 87: 949–954.

Beakbane, A.B., Slater, C.H.W., Posnette, A.F. 1972. Mycoplasmas in the phloem of coconut, *Cocos nucifera* L. with lethal yellowing disease. *J. Horticultural Sci* 47: 256.

Beer, S.V. 1990. Fire Blight. In *Compendium of Apple and Pear Diseases* Edited by A.L. Jones, H.S. Aldwinckle in APS Press, St. Paul, MN. 61–63.

Billing E. 1987. Bacteria as plant pathogens. *Aspects of Microbiology*, 14th Edition. Wokingham (UK): Van Nostrand Reinhold, 79.

Bock, C.H., Parker, P.E., Gottwald, T.R. 2005. The effect of simulated wind-driven rain on duration and distance of dispersal of *Xanthomonas axonopodis* pv. *citri* from canker infected citrus trees. *Plant Dis* 89: 71–80.

Boewe, G.H. 1949. Late season incidence of Stewart's disease on sweet corn and winter temperatures in Illinois, 1944–1948. *Plant Dis Reporter* 33: 192–194.

Bouzar, H., Jones, J.B., Stall, R.E., Louws, F.J., Schneider, M., Rademaker, J.L.W., de Bruijn, F.J., Jackson, L.E. 1999. Multiphasic analysis of xanthomonads causing bacterial spot disease on tomato and pepper in the Caribbean and Central America: Evidence for common lineages within and between countries. *Phytopathology* 89: 328–335.

Broschat, T.K., Harrison, N.A., Donselman, H. 2002. Losses to lethal yellowing cast doubt on coconut cultivar resistance. *Palms* 46: 185–189.

Brlansky, R.H., Timmer, L.W., French, W.J., McCoy, R.E. 1983. Colonization of the sharpshooter vectors, *On cometopianigricans* and *Homalodiscaco agulata*, by xylem limited bacteria. *Phytopathology* 73: 530–535.

Campbell, C.L., Madden, L.V. 1990. *Introduction to Plant Disease Epidemiology*. New York: John Wiley and Sons, USA.

Carr, E.A., Zaid, A.M., Bonasera, J.M., Lorbeer, J.W., Beer, S.V. 2013. Infection of onion leaves by *Pantoea ananatis* leads to bulb infection. *Plant Dis* 97(12): 1524–1528.

Cazorla, F.M., Tores, J.A., Olalla, L., Perez-Garcia, A., Farre, J.M., de Vincente, A. 1998. Bacterial apical necrosis of mango in Southern Spain: A disease caused by *Pseudomonas syringae* pv. *syringae*. *Phytopathology* 88: 614–620.

Chohan, S.N., Qamar, R., Sadiq, I., Azam, M., Holford, P., Beattie, A. 2007. Molecular evidence for the presence of Huanglongbing in Pakistan. *Australasian Plant Dis* 2: 37–38.

Citovsky, V., Kozlovsky, S.V., Lacroix, B., Zaltsman, A., Dafny-Yelin, M., Vyas, S., Andriy, T., Tzvi, T. 2007. Biological systems of the host cell involved in *Agrobacterium* infection. *Cell Microbiol* 9: 9–20.

Cota, L.V., Costa, R.V., Silva, D.D., Parreira, D.F., Lana, U.G.P., Casela, C.R. 2010. First report of pathogenicity of *Pantoea ananatis* in sorghum (*Sorghum bicolor*) in Brazil. *Australasian Plant Dis* 5: 120–122.

Cubero J., Lastra, B., Salcedo, C.I., Piquer, J., López, M.M. 2006. Systemic movement of *Agrobacterium tumefaciens* in several plant species. *J Appl Microbiol* 101: 412–421.

Davis, M.J., Purcell, A.H., Thomson, S.V. 1978. Pierce's disease of grapevines: Isolation of the causal bacterium. *Science* 199: 75–77.

Davis, R.I., Kokoa, P., Jones, L.M., Mackie, J., Constable, F.E., Rodoni, B.C., Gunua, T.G., Rossel, J.B. 2012. A new wilt disease of banana plants associated with phytoplasmas in Papua New Guinea (PNG). *Australasian Plant Dis* 7: 91–97.

Dodd, J.L. 1980. The role of plant stress in the development of corn stalk rots. *Plant Dis* 64: 533–537.

Donovan, N.J., Beattie, G.A.C., Chambers, G.A., Holford, P., Englezou, A., Hardy, S., Wangdi, D.P., Om, T.N. 2012. First report of '*Candidatus liberibacterasiaticus*' in *Diaphorina communis*. *Australasian Plant Dis* 7: 1–4.

Dutta, B., Gitaitis, R.D., Webster, T.M., Sanders, H., Smith, S., Langston, D.B., Jr. 2014. Distribution and Survival of *Pseudomonas* sp. on Italian Ryegrass and Curly Dock in Georgia. *Plant Dis* 98(5): 660–666.

Eden-Green, S.J. 1997. History, distribution and present status of lethal yellowing-like disease of palms. *Proceedings of an International Workshop on Lethal Yellowing-Like Diseases of Coconut*, Elmina Ghana, November 1995. Eden-Green, S. J., Ofori, F. (eds.). UK: Natural Resources Institute, 9–25.

Elliott, C., Poos, F.W. 1940. Seasonal development, insect vectors and host range of bacterial wilt of sweet corn. *J Agric Res* 10: 645–686.

Elphinstone, J.G., Perombelon, M.C.M. 1986. Contamination of potatoes by *Erwinia carotovora* during grading. *Plant Pathol* 35: 25–33.

Esker, P.D., Sparks, A.H., Campbell, L., Guo, Z., Rouse, M., Silwal, S.D., Tools, S., Van Allen, B., Garrett, K.A. 2008. Ecology and epidemiology in R: Disease forecasting. *The Plant Health Instructor* DOI: 10.1094/PHI-A-2008-0129-01.

Fisher, M.C., Henk, D.A., Briggs, C.J., Brownstein, J.S., Madoff, L.C., McCraw, S.L., Gurr, S.J. 2012. Emerging fungal threats to animal, plant and ecosystem health. *Nature* 484: 186–194.

Gao, G., Jin, L.P., Xie, K.Y., Yan, G.Q., Qu, D.Y. 2007. Bottle gourd: A new host of *Ralstonia solanacearum* in China. *Australasian Plant Dis* 2: 151–152.

Garcia, A.L., Torres, S.C.Z., Heredia, M., Lopes, S.A. 2012. Citrus responses to *Xylella fastidiosa* infection. *Plant Dis* 96:1245–1249.

Gelvin, S.B. 2000. Agrobacterium and plant genes involved in T-DNA transfer and integration. *Annu Rev Plant Physiol Plant Mol Biol* 51: 223–256.

Gelvin, S.B. 2003. Agrobacterium-mediated plant transformation: The biology behind the "gene-jockeying" tool. *Microbiol Mol Biol Rev* 67: 16–37.

Gelvin, S.B. 2010. Plant proteins involved in *Agrobacterium*-mediated genetic transformation. *Annu Rev Phytopathol*. 48: 45–68.

Gilligan, C.A. 1990. Mathematical modeling and analysis of soilborne pathogens. In *Epidemics of Plant Diseases*, 2nd Kranz, J. (ed.). Berlin: Springer-Verlag, 96–142.

Gottwald, T.R., Graham, J.H., Schubert, T.S. 2002. Citrus canker: The pathogen and its impact. *Plant Health Progress* DOI: 10.1094/PHP-2002-0812-01-RV.

Gottwald T.R., Timmer, L.W. 1994. The efficacy of windbreaks in reducing the spread of citrus canker caused by *Xanthomonas campestris* pv. *citri Trop Agric* 72: 194–201.

Graham, J.H., Gottwald, T.R., Riley, T.D., Bruce, M.A. 1992. Susceptibility of citrus fruit to bacterial spot and citrus canker. *Phytopathology* 82: 452–457.

Hammerschlag, R., Sherald, J., Kostka, S. 1986. Shade tree leaf scorch. *J Arboriculture* 12: 38–43.

Hamza, A.A., Robène-Soustrade, I., Jouen, E., Gagnevin, L., Lefeuvre, P., Chiroleu, F., Pruvost, O. 2010. Genetic and pathological diversity among *Xanthomonas* strains responsible for bacterial spot on tomato and pepper in the southwest Indian Ocean region. *Plant Dis* 94: 993–999.

Harrison, N.A., Cordova, I., Richardson, P., DiBonito, R. 1999. Detection and diagnosis of lethal yellowing. In *Current Advances in Coconut Biotechnology.* Oropeza, C., Verdeil, J.L., Ashburner, G.R., Cardeña, R., Santamaría J.M. (eds.). Dordrecht, The Netherlands: Kluwer Academic Publications, 183–196.

Hewitt, W.B., Houston, B.R., Frazier, N.W., Freitag, J.H. 1946. Leafhopper transmission of the virus causing Pierce's disease of grape and dwarf of alfalfa. *Phytopathology* 36: 117–128.

Holland, R.M., Christiano, R.S.C., Gamlie-Atinsky, E., Scherm, H. 2014. Distribution of *Xylella fastidiosa* in blueberry stem and root sections in relation to disease severity in the field. *Plant Dis* 98(4): 443–447.

Hooykaas, P.J.J., Beijersbergen, A.G.M. 1994. The virulence system of *Agrobacterium tumefaciens. Annu Rev Phytopathol* 32: 157–179.

Hopkins, D.L., Thompson, C.M. 2002. Seed transmission of *Acidovorax avenae* subsp *citrulli* in cucurbits. *Hortscience* 37: 924–926.

Howard, F.W. 1980. Population densities of *Myndus crudus* Van Duzee (Homoptera: Cixiidae) in relation to coconut lethal yellowing distribution in Florida. *Principes* 24: 174–178.

Howard, F.W. et al. 1983. World distribution and possible geographical origin of palm lethal yellowing disease and its vectors. *FAO Plant Protection Bulletin* 31: 101–113.

Howard, F.W., Williams, D.S., Norris, R.C. 1984. Insect transmission of lethal yellowing to young palms. *Int J Entomol* 26: 331–338.

Huang, X., De Bever, A., Becker, O. 1997. The *in-vitro* life cycle of *Belonolaimus longicaudatus. J Nematology* New York: John Wiley and Sons.

Huang, X., Becker, J.O. 1999. Lifecycle and mating behavior of *Belonolaimus longicaudatus* in gnotobiotic culture. *Journal of Nematology* 31: 70–74.

Huang, C.H., Vallad, G.E., Adkison, H., Summers, C., Margenthaler, E., Schneider, C. 2013. A novel *Xanthomonas* sp. causes bacterial spot of rose (*Rosa* spp.). *Plant Dis* 97(10): 1301–1307.

Jackson, R.W., Johnson, L.J., Clarke, S.R., Arnold, D.L. 2011. Bacterial pathogen evolution: Breaking news. *Trends Genet* 27: 32–40.

Jensen, B.D., Vicente, J.G., Manandhar, H.K., Roberts, S.J. 2010. Occurrence and diversity of *Xanthomonas campestris pv. Campestris* in vegetable Brassica fields in Nepal. *Plant Dis* 94: 298–305.

Jones, J.B. 1991. Bacterial spot. In: *Compendium of Tomato Diseases*. Jones, J.B., Jones, J.P., Stall, R.E., Zitter, T.A. (eds.). MN: American Phytopathological Society, St. Paul, 27.

Jones, J.B., Stall, R.E., Bouzar, H. 1998. Diversity among xanthomonads pathogenic on pepper and tomato. *Annual Review of Phytopathology* 36: 41–58.

Jones, J.B., Pernezny, K. 2003. Bacterial spot. In: *Compendium of Pepper Diseases*. Pernezny, K., Roberts, P.D., Murphy, J.F., Goldberg, N.P. (eds.). MN: American Phytopathological Society, St. Paul, 6–7.

Jones, J.B., Lacy, G.H., Bouzar, H., Stall, R.E., Schaad, N.W. 2004. Reclassification of the xanthomonads associated with bacterial spot disease of tomato and pepper. *Systematic and Applied Microbiology* 27: 755–762.

Kado, C.I. 1976. The tumor-inducing substance of *Agrobacterium tumefaciens*. *Annu Rev Phytopathol* 14: 265–308.

Klement, Z., Rudolph, K., Sands, D.C. 1990. Methods in phytobacteriology, Akademiai Kiado, Budapest. 568.

Krugner, R., Ledbetter, C.A., Chen, J., Shrestha, A. 2012. Phenology of *Xylella fastidiosa* and its vector around California almond nurseries: An assessment of plant vulnerability to almond leaf scorch disease. *Plant Dis* 96: 1488–1494.

Lamka, G.L., Hill, J.H., McGee, D.C., Braun, E.J. 1991. Development of an immunosorbent assay for seed borne *Erwinia stewartii* in corn seeds. *Phytopathology* 81: 839–846.

Latin, R.X., Hopkins, D.L. 1995. Bacterial fruit blotch of watermelon: The hypothetical exam question becomes reality. *Plant Dis* 79: 761–765.

Li, J., Chai, Z., Yang, H., Li, G., Wang, D. 2007. First report of *Pseudomonas marginalis* pv. *marginalis* as a cause of soft rot of potato in China. *Australasian Plant Dis* 2: 71–73.

Lucas, J.A. 1998. *Plant Pathology and Plant Pathogens*. 3rd edition, Blackwell Science, United Kingdom. 274.

Madden L.V. 2006. Botanical epidemiology: Some key advances and its continuing role in disease management. *Eur J Plant Pathol* 115: 3–23.

Madden, L.V., Hughes, G., Van den Bosch, F. 2007. *Study of Plant Disease Epidemics*. St Paul, Minnesota, USA: The American Phytopathology Press.

Madden, L., Gareth, H., Frank Van Den Bosch. 2007. *Study of Plant Disease Epidemics*. American Phytopathological Society. ISBN 978-0-89054-354-2.

Maindonald, J.H., Braun, W.J. 2003. *Data Analysis and Graphics using R—An Example Based Approach.* Cambridge University Press, UK.

Marques, L.L., Ceri, R.H., Manfio, G.P., Reid, D.M., Olson, M.E. 2002. Characterization of biofilm formation by *Xylella fastidiosa in vitro*. *Plant Disease* 86: 633–638.

McCullen, C.A., Binns, A.N. 2006. *Agrobacterium tumefaciens* and plant cell interactions and activities required for inter kingdom macromolecular transfer. *Annu Rev Cell Dev Biol* 22: 101–127.

Meng, Y., Li, Y., Galvani, C.D., Hao, G., Turner, J.N., Burr, T.J., Hoch, H.C. 2005. Upstream migration of *Xylella fastidiosa* via pilus-driven twitching motility. *J Bacteriol* 187: 5560–5567.

Minsavage, G.V., Dahlbeck, D., Whalen, M., Kearney, B., Bonas, U., Staskawicz, B.J., Stall, R.E. 1990. Gene-for-gene relationship specifying disease resistance in *Xanthomonas campestris* pv. *vesicatoria*-pepper interactions. *Molecular Plant-Microbe Interactions* 3: 41–47.

Minuto, A., Minuto, G., Martini, P., Odasso, M., Biodi, E., Mucini, S., Scortichini, M. 2008. First report of *Pseudomonas viridiflava* in basil seedlings and plants in soilless crop in Italy. *Australasian Plant Dis* 3: 165.

Nazerian, E., Sijam, K., Ahmad, Z.A.M., Vadamalai, G. 2013. Characterization of *Pectobacterium carotovorum* subsp. *carotovorum* as a new disease on lettuce in Malaysia. *Australasian Plant Dis* 8: 105–107.

Neter, J., Kutner, M.H., Nachtsheim, C.J., Wasserman, W. 1996. *Applied Linear Statistical Models*, 4th Edn. WCB McGraw-Hill, Boston.

Paccola-Meirelles, L.D., Ferreira, A.S., Meirelles, W.F., Marriel, I.E., Casela, C.R. 2001. Detection of a bacterium associated with a leaf spot disease of maize in Brazil. *J. Phytopathol* 149: 275–279.

Pepper, E.H. 1967. Stewart's bacterial wilt of corn. *Monograph 4*. American Phytopathological Society. St. Paul, MN.

Perez-y-Terron, R., Villegas, M.C., Cuellar, A., Munoz-Rojas, J., Castaneda-Lucio, M., Hernandez-Lucas, I., Bustillos-Cristales, R. et al. 2009. Detection of *Pantoea ananatis*, causal agent of leaf spot disease of maize in Mexico. *Australasian Plant Dis*. 4: 96–99.

Perombelon, M.C.M., Kelman, A. 1980. Ecology of the soft rot erwinias. *Annu. Rev. Phytopathol* 18: 361–387.

Perombelon, M.C.M. 1992. Potato blackleg: Epidemiology, host-pathogen interaction and control. *Neth J Plant Pathol* 98: 135–146.

Perombelon, M.C.M. 2000. Blackleg risk potential of seed potatoes determined by quantification of tuber contamination by the causal agent of *Erwinia rotovora* subsp. *atroseptica*: A critical review. *EPPO Bull* 30: 413–420.

Pinkas, Y., Maymon, M., Smolewich, Y. 1996. Bacterial black blight of mango. *Alon Hanotea* 50: 475.

Plavsic-Banjac, B., Hunt, P., Maramorosch, K. 1972. Mycoplasma-like bodies associated with lethal yellowing disease of coconut palms. *Phytopathology* 62: 298–299.

Puławska, J. 2010. Crown gall of stone fruits and nuts, economic significance and diversity of its causal agents: Tumorigenic *Agrobacterium* spp. *J Plant Pathol* 92: S1.87–98.

Purcell, A.H. 1985. The ecology of plant diseases spread by leafhoppers and plant hoppers. In *The Leafhoppers and Plant Hoppers*. Nault, L.R., Rodriguez J.G. (eds.). New York, USA: Wiley publications, 351–380.

Purcell, A.H., Finlay, A.H., McLean, D.L. 1979. Pierce's disease bacterium: Mechanism of transmission by leafhopper vectors. *Science* 206: 839–841.

Reinink, K. 1986. Experimental verification and development of EPIPRE, a supervised disease and pest management for wheat. *Neth. J. Pl. Path*. 92: 3–14.

Rouhrazi, K., Rahimian, H. 2012. Genetic diversity of Iranian *Agrobacterium* strains from grapevine. *Ann Microbiol*. 62: 1661–1667.

Sahin, F., Miller, S.A. 1996. Characterization of Ohio strains of *Xanthomonas campestris* pv. *vesicatoria* of causal agent of bacterial spot of pepper. *Plant Disease* 80: 773–778.

Siri, M.I., Sanabria, A., Pianzzola, M.J. 2011. Genetic diversity and aggressiveness of *Ralstonia solanacearum* strains causing bacterial wilt of potato in Uruguay. *Plant Dis* 95: 1292–1301.

Sobiczewski, P., Schollenberger, M. 2002. Bakteryjnechorobyroślinogrodniczych.Podręcznikdlastudentów. (Bacterial diseases of horticultural plants–manual for students, in Polish).–PWRiL. *Warszawa* 156–187.

Smith, E.F., Townsend, C.O. 1907. A plant-tumor of bacterial origin. *Science* 25: 671–673.

Smith, E.F., Brown, N.A., Townsend, C.O. 1911. Crown-gall of plants: Its cause and remedy (U.S. Department of Agricultural Bureau-Plant Industry, Bulletin 213). Washington, DC: Government Printing Office Washington.

Sparks, A.H., Esker, P.D., Bates, M., Dall Acqua, W., Guo, Z., Segovia, V., Silwal, S.D., Tolos, S., Garrett, K.A. 2008. Ecology and epidemiology in R: Disease progress over time. *The Plant Instructor*. DOI: 10.1094/PHI-A-2008-0129-02.

Stevens, N.E. 1934. Stewart's disease in relation to winter temperatures. *Plant Dis. Reporter* 18: 141–149.

Stewart, F.C., Rolfs, F.M., Hall, F.H. 1900. A fruit-disease survey of western New York in 1900. *New York Agr Exp Stat Bull.* 191: 291–331.

Swords, K.M.M., Dahlbeck, D., Kearney, B., Roy, M., Staskawicz, B.J. 1996. Spontaneous and induced mutations in a single open reading frame to both virulence and a virulence in *Xanthomonas campestris* pv. *vesicatoria avrBs2*. *Journal of Bacteriology* 178: 4661–4669.

Taylor, T.K., Romberg, M.K., Alexander, B.J.R. 2011. A bacterial disease of hellebore caused by *Pseudomonas viridiflava* in New Zealand. *Australasian Plant Dis.* 6: 28–29.

Toth, I.K., Sullivan, L., Brierley, J.L., Avrova, A.O., Hyman, L.J., Holeva, M., Broadfoot, L., Perombelon, M.C.M., McNicol, J. 2003. Relationship between potato seed tuber contamination by *Erwinia carotovora* ssp. *atroseptica*, blackleg disease development and progeny tuber contamination. *Plant Pathol* 52: 119–126.

Trigiano, R.N., Windhum, M.T., Windham, A.S. 2004. *Plant Pathology, Concept and Laboratory Exercise*. CRC Press, New York.

Tzfira, T., Li, J., Lacroix, B., Citovsky, V. 2004. Agrobacterium T-DNA integration: Molecules and models. *Trends Genet.* 20: 375–383.

Valarmathi, P., Rabindran, R., Velazhahan, R., Suresh, S., Robin, S. 2013. First report of Rice orange leaf disease phytoplasma (16SrI) in rice (*Oryza sativa*) in India. *Australasian Plant Dis* 8: 141–143.

Vanneste, J.L. 2000. Fire Blight: The disease and its causative agent, *Erwinia amylovora*. CABI Publishing, Wallingford, UK.

Vanneste, J.L., Yu, J., Cornish, D.A., Tanner, D.J., Windner, R., Chapman, J.R., Taylor, R.K., Mackay, J.F., Dowlut, S. 2013. Identification, virulence, and distribution of two biovars of *Pseudomonas syringae pv. actinidiae* in New Zealand. *Plant Dis* 97: 708–719.

Verniere, C.J., Gottwald, T.R., Pruvost, O. 2002. Disease development and symptom expression of *Xanthomonas campestris* pv. *citri* in various citrus plant tissues. *Phytopathology* 93: 832–843.

Walter, A.J., Hall, D.G., Duan, Y.P. 2012. Low incidence of '*Candidatus liberibacterasiaticus*' in *Murraya paniculata* and associated *Diaphorina citri*. *Plant Dis* 96: 827–832.

Webb, R.E., Goth, R.W. 1965. A seed borne bacterium isolated from watermelon. *Plant Dis.Rep.* 49: 818–821.

Yakabe, L.E., Parker, S.R., Kluepfel, D.A. 2012. Role of systemic *Agrobacterium tumefaciens* populations in crown gall incidence on the walnut hybrid rootstock 'Paradox'. *Plant Dis* 96: 1415–1421.

Yakabe, L.E., Parker, S.R., Kluepfel, D.A. 2014. Incidence of *Agrobacterium tumefaciens* biovar 1 in and on 'Paradox' (*Juglans hindsii* × *Juglans regia*) Walnut seed collected from commercial nurseries. *Plant Dis* 98(6): 766–770.

Young, A.J., Grice, K.R.E., Trevorrow, P.R., Vawdrey, L.L. 2007. *Burkholderia andropogonis* blight of golden cane palms in North Queensland. *Australian Plant Dis* 2: 131–132.

Young, A. 2008. Notes on *Pseudomonas syringae pv. syringae* bacterial necrosis of mango (*Mangifera indica*) in Australia. *Australasian Plant Dis* 3: 138–140.

Zaenen, I., N., H. Van Larebeke, M. Teuchy, Van Montagu, J. Schell. 1974. Super coiled circular DNA in crown-gall inducing. *Agrobacterium* strains. *J Mol Biol* 86: 109–127.

Zhang, J., Shen, H., Pu, X., Lin, B. Identification of *Dickeyazeae* as a causal agent of bacterial soft rot in banana in China. *Plant Dis* 98(4): 436–442.

4 Controlling Strategies (Diagnosis/Quarantine/Eradication) of Plant Pathogenic Bacteria

Kubilay Kurtulus Bastas and Velu Rajesh Kannan

CONTENTS

ABSTRACT Epiphytic and endophytic bacteria can cause severe losses in plants at each stage of growth. Plant pathogenic bacteria induce characteristic leaf spots, chlorosis, blights, scabs, wilting, cankers, and tumors that may be produced in the infected leaves, fruits, roots, or stems. The traditional detection protocol, which is based on cultural, morphological, physiological, and biochemical properties, requires skilled taxonomical expertise to confirm the identity of the causal bacterium. New detection tools will be used not only for rapid, sensitive, and specific diagnosis but also to help to understand plant–pathogen relationships and the structure and function of pathogens and their communities. Molecular techniques are widely recognized as powerful plant pathogen-detection techniques. The DNA sequences from which the primers are designed for bacteria come from three main origins: pathogenicity/virulence genes, ribosomal genes, and plasmid genes. The selection of appropriate diagnostic methods is mainly focused on eradication, certification of mother plants, sanitation, quarantine programs or large surveys to evaluate incidence, and screening tests for surveillance of the spreading of a disease, especially for quarantine pathogens or in critical cases of export-import. Propagating material such as seeds, cuttings, root-stocks, buds, tubers, rhizomes, and bulbs can carry pathogens to new areas, either in or on plant tissues. Certification schemes ensure that seed and vegetative propagation material are free from particular diseases. Close association with seeds facilitates the long-term survival, introduction into new areas, and widespread dissemination of bacterial pathogens. Cultural practices are intended to make atmospheric, edaphic, or biological surroundings favorable to the crop plant and unfavorable to its parasites. Some bacterial pathogens that survive in soil can often be controlled by crop rotations with unsusceptible species. Sanitation is a basic requirement for preventing the spread of disease to new areas or, on a smaller scale, to new plants or plant products. Eradication of the pathogen and disease, as required for quarantine pathogens, often fails due to incomplete pathogen eradication, natural reinvasion, and reintroduction through short- or long-distance movement of infected material from contaminated areas.

KEYWORDS: plant pathogenic bacteria, controlling strategies, diagnosis, quarantine, eradication

4.1 INTRODUCTION

Plant bacterial pathogens cause substantial loss to the productivity of major crop plants. Bacterial pathogens cannot be contained effectively through chemical methods. Early detection in seeds and planting materials, or by ensuring disease-free planting materials through rapid diagnostics, are likely the effective means of reducing bacterial disease incidence. The traditional detection protocol, which is based on cultural, morphological, and biochemical properties, requires skilled taxonomical expertise to confirm the identity of the causal bacterium (Mondal and Shanmugam, 2013).

4.2 DIAGNOSIS

4.2.1 Morphology

Because the morphological characteristics of bacterial cells are less variable, biological, biochemical, physiological, immunological, and genomic characteristics must be determined to obtain reliable identification and meaningful classification of bacterial pathogens. Phytoplasmas are cell wall-less, nonculturable bacteria belonging to Mollicutes. Lack of cultural characteristics has made it obligatory to study the morphological characteristics of phytoplasmas in the phloem cells of infected plant hosts by electron microscopy. Because most of the phytoplasmas look alike in the ultrathin sections, their morphological characters have no diagnostic value (Narayanasamy, 2011).

The most important bacterial plant pathogens are very small, single-celled organisms, about 0.7–5.0 × 0.4–1.5 μm. They can be readily seen with a light microscope using an oil-immersion objective. The bacterial cell contains nuclear material that is not separated from the cytoplasm by nuclear membrane, as in eukaryotes, nor is there mitotic mechanism as in higher life forms. The genetic material is contained in a single chromosome with double-stranded DNA in a closed circular form. The bacterial cell may also contain plasmids that are capable of replicating independent of a chromosome. They have extrachromosomal DNA that governs characteristics such as pathogenicity, resistance to chemicals and antibiotics, and tumor formation. In addition to the rod-shaped plasmids, club-, Y-, or V-shaped and branched forms may also be observed. Pairs of cells or short chains of cells may also be enveloped by a thin or thick slime layer made of viscous gummy materials. The slime layer (capsule) may be found as a larger mass around the cells. Most of the plant pathogenic bacteria are motile. The flagella, which are organs of locomotion, may be present either singly or in groups at one or both ends of the bacterial cell or distributed over the entire cell surface. Bacterial colonies show variation in shape, size, color, elevation, edge forms, and so on and these characteristics may be useful in the identification of certain bacterial genera. The size of colonies may vary from 1 mm to several cm in diameter, and they may be circular, oval, or irregular with smooth, wavy, or angular edges. The elevation of the colonies may be flat, raised dome-like, or gray. No plant pathogenic bacterial species other than *Bacillus* spp. is known to produce endospores. Therefore, they are sensitive to desiccation (Narayanasamy, 2001, 2011).

Several genera of bacteria have been shown to cause plant disease. The first step in distinguishing them is to determine whether they are Gram-positive or Gram-negative by using the Gram stain technique, which indicates a major difference in cell-wall structure. Gram-negative bacteria have two cell membranes, an inner plasma membrane and an outer membrane. Their cell walls contain small amounts (usually less than 10%) of the structural polymer peptidoglycan. By contrast, Gram-positive bacteria have only one membrane, the plasma membrane, and peptidoglycan is a major component of the cell walls (usually more than 30%). In plant bacterial pathogens, flagella may be attached at the cell ends (polar), over the whole cell surface (peritrichous), or as a ring around the center of the cell. Flagellation is useful in distinguishing the genera of Gram-negative plant pathogenic bacteria, because morphological characters are of little value in the identification of bacteria; they are identified by what they do, rather than by what they look like.

For preliminary observations, a high-power or even a low-power objective can be used. Small pieces of diseased tissue should be removed, mounted in a drop of sterile water on a slide, and gently teased out. In nearly all cases where disease is caused by bacteria, masses of bacteria will ooze from infected tissue into water, which usually becomes milky in appearance. If present, bacteria will be seen as innumerable rod-shaped cells that are frequently motile, especially toward the edge of the coverslip where oxygen is abundant. Once bacteria have been shown to be present in abundance, they should be isolated and grown in pure culture. Some plant pathogenic bacteria are difficult to grow in culture and special media must be used for them. After streaking, agar plates should be incubated at 25–28°C. Depending on the organism, colonies will appear after 2–7 d, or even later in some cases.

4.2.2 Symptomatology

At the initial stages of infection, plant pathogenic bacteria induce characteristic water-soaked lesions in the infected tissues; later, these lesions may turn necrotic (brown). Formation of encrustations or bacterial ooze from infected tissue is another distinguishing feature of bacterial disease. If an infection is localized, leaf spots, blights, scabs, cankers, and tumors may be produced in the infected leaves, fruits, roots, or stems. If infection becomes systemic as the bacterial pathogen moves to other plant tissues and from the original site of infection, symptoms such as chlorosis, leaf spot, blight, rotting, and wilting of the whole may be observed (Figure 4.1).

Phytopathogenic bacteria induce hypersensitive reaction (HR) in leaf mesophyll tissues, but saprophytic bacteria cannot induce the response even when inoculated on plants. Production of visible necrotic lesions at the site of inoculation of a bacterial species thus indicates not only the presence of the target bacteria, but also its pathogenicity on the plant species. Most Gram-negative bacteria produce HR in tobacco (Paret et al., 2009). Generally, semi-selective/selective media may be required for the isolation of bacterial pathogens from soil samples, but nonselective media may be sufficient for the isolation of bacteria from plant tissues that can be surface-disinfected. A pathogenicity test must be performed for the ultimate verification of the presumptive pathogen once it has been isolated from infected plant tissues or other substrates such as soil, water, or air. Host-plant species that can rapidly produce symptoms of infection by bacterial pathogens may be used as diagnostic hosts. A simple, rapid, and reliable bioassay technique was developed for the detection of *Rhizobium radiobacter*, which causes tumors in a large number of plant species. Detached leaves of the highly susceptible *Kalanchoe tubiflora* were used as biological bait for trapping tumorigenic cells of *R. radiobacter* from soil as well as from tumor tissues of infected plants (Romeiro et al., 1999).

Recent taxonomic studies of plant pathogenic Gram-negative bacteria have been grouped to 25 genera: *Acetobacter*, *Acidovorax*, *Arthrobacter*, *Brenneria*, *Burkholderia*, *Clostridium*, *Dickeya*, *Enterobacter*, *Erwinia*, *Gluconobacter*, *Herbaspirillum*, *Pantoea*, *Pectobacterium*, *Pseudomonas*, *Ralstonia*, *Rhizobacter*, *Rhizobium*, *Rhizomonas*, *Samsonia*, *Serratia*, *Sphingomonas*, *Spiroplasma*, *Xanthomonas*, *Xylella*, and *Xylophilus*. Gram-positive bacteria have been allocated to nine genera: *Arthrobacter*, *Bacillus*, *Clavibacter*, *Curtobacterium*, *Leifsonia*, *Nocardia*, *Rhodococcus*,

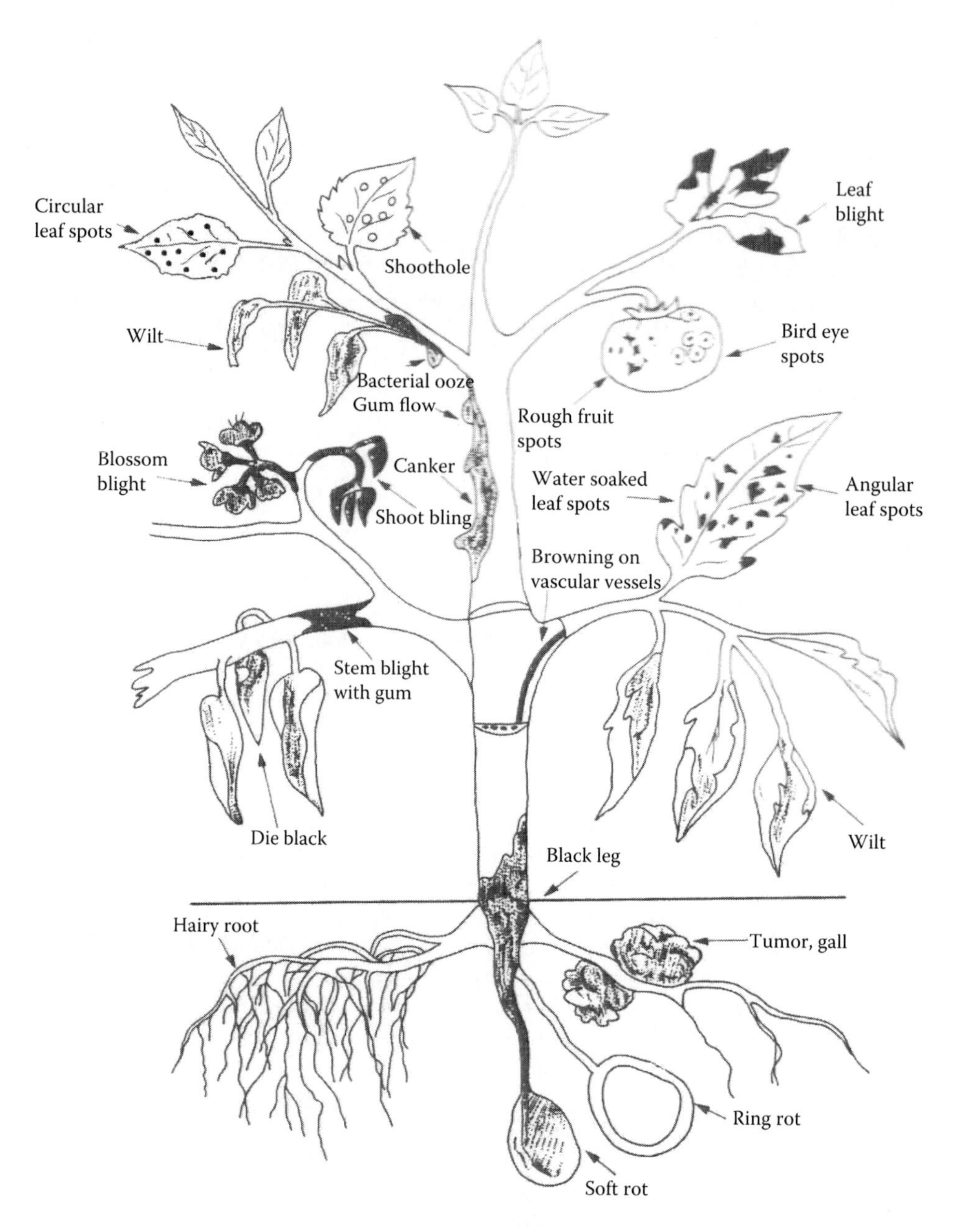

FIGURE 4.1 Symptoms of plant bacterial diseases. (Modified from Saygılı, H., Sahin, F., Aysan, Y. 2006. *Fitobakteriyoloji*, Meta Basım, İzmir, 630p.)

Rathayibacter, and *Streptomyces*. The cause of greening disease of citrus cannot be grown in culture but is known to be a member of the *Proteobacteria*. No genus has been designated although *Liberobacter* has been suggested.

Traditionally, the available detection and diagnostic techniques for plant pathogenic bacteria have been microscopical observation, isolation, biochemical characterization, serology (mainly through immunofluorescence and enzyme-linked immunosorbent assay or ELISA that uses polyclonal and/or monoclonal antibodies), bioassays, and pathogenicity tests. Standard protocols for detection of plant bacteria based on isolation and further identification are time consuming and not always sensitive and specific enough. Consequently, they are obviously not suited for routine analysis of a large number of samples. Other handicaps are the low reproducibility of identification by phenotypic traits, frequent lack of phylogenetic significance, and false negatives. The latter may be due to stressed or injured bacteria, or to those in the viable but nonculturable state (VBNC) that escapes isolation. The VBNC state is a survival strategy in which bacterial cells do not form visible colonies on nonselective solid medium, but instead remain viable according to culture-independent methods (Oliver, 2005). Detection of cells in particular physiological states is important, especially for quarantine organisms, because they can retain pathogenicity and thereby threaten plant health.

4.2.3 Biological Assays

Biological assays, which were developed first, are still in widespread use because they are simple, require minimal knowledge of the pathogen, and are polyvalent. Furthermore, biological indexing is still the only method of choice to detect uncharacterized but graft-transmissible agents. Its sensitivity is considered to be very high due to the viral multiplication that occurs in the host plant used as the indicator. However, despite their multiplication in the host, some isolates can induce no symptoms; thus, they escape detection. The major limitations of using biological assays are the high economic cost, the long time (weeks to months) required to obtain results, and the impossibility of large-scale use. Bacteriophages that infect phytopathogenic bacteria have been isolated from infected plant tissues, soil, and irrigation water. *Xanthomonas oryzae* pv. *oryzae* is susceptible to several phages (Kuo et al., 1967). Four different kinds of bacteriophages (øRSL, øRSA, øRDM, and øRSS) were isolated from the soilborne pathogen *Ralstonia solanacearum*. Results of this investigation suggest that the bacteriophages may be useful for both detection and control of bacterial pathogens that infect economically important crops (Yamada et al., 2007).

Detection and diagnostic tests may be interpreted as a function of several parameters that increase information about the sanitary status of a plant and strengthen or lessen the probability of infection. In general, methods of detection and diagnosis used to classify plants depend upon the presence or absence of one or several specific pathogens. The results of the analyses enable one or more conclusions to be drawn and facilitate effective decision making. Sensitivity is defined as the proportion of the true positives of infected plants identified by a specific technique or method. Because they accurately indicate pathogen-free status, the methods that afford the highest sensitivity must be used to discard the presence of a pathogen and to affirm the good health of plants.

Selection of appropriate diagnostic methods should involve critical appraisals that focus on the stated objective: (i) eradication, certification of mother plants, sanitation, or quarantine programs; (ii) large surveys to evaluate incidence, or screening tests for surveillance of the spread of a disease. For (i), the need to use the most sensitive method should be emphasized and the risk of obtaining false positives should be accepted. Evaluation of the sensitivity and specificity of techniques in order to select the most sensitive is therefore the main requirement. Use of the most sensitive method available would enable the presence of the pathogen to be discarded most effectively because the most sensitive method will afford the most accurate diagnosis of healthy plants with high confidence when the target pathogen is not detected. However, in the case of large-scale surveys or screening tests for surveillance, the selection of one, two, or several methods should be based on an evidence-based approach, evaluation of cost per analysis, calculation of post-test probability of disease, and consideration of different scenarios with different prevalences. For this reason, in order to achieve adequate risk assessment of the methods, the application of likelihood ratios to evaluate diagnostic tests is a must in present as well as future diagnoses. When plotting pre- and post-test probabilities, coupling the likelihood ratios of methods will aid decision making.

Post-test probability will support the evaluation of results as well as the risk management associated with the use of the methods. In addition, an interlaboratory evaluation that applies a kappa index will enable detailed and reliable protocols to be developed for routine testing. A transfer of these concepts to plant pathogen diagnosis in order to achieve better risk management of the techniques is one of the main challenges for the near future. Knowledge of how a molecular method performs in routine analysis will permit its adequate integration into diagnostic schemes, the correct interpretation of results, and the design of optimal risk management strategies. All of these elements will facilitate decision making.

4.2.4 Molecular Detection

Depending on the material to be analyzed, the extraction methods can be quite simple or more complex. The use of commercial kits, either general or specifically designed for one type of plant material, has gained acceptance for extraction, given the ease of use and avoidance of toxic reagents during the purification process. For instance, DNeasy and RNeasy Plant System (Qiagen), Easy-DNA-Extraction kit (Invitrogen), Wizard Genomic DNA and SV Total RNA Isolation System (Promega), Extract-N-Amp Plant PCR kit (Sigma), Powersoil DNA kit (MoBio), and RNA/DNA/Protein Purification kit (Norgen) are successfully applied in different models. In addition, several commercial automated systems allow the extraction and analysis of nucleic acids from plant and microorganisms and equipment that can perform automatic separation has even been developed. The latest systems employ miniaturized devices to achieve DNA extraction in a microchip via different approaches such as laser irradiation (Liu et al., 2007).

No standardized protocols are currently recommended for the detection of plant pathogenic bacteria in soil samples, because the protocols can be very complex. In addition, DNA yields could be variable, which would affect the diagnostic efficiency. Direct extraction methods of total microbial DNA from soils and sediments have been reviewed (Johnston et al., 1996; van Elsas and Smalla, 1999) and can be applied for detection of soil-borne plant pathogenic bacteria such as *R. radiobacter* and *R. solanacearum*. Methods to extract DNA from fresh water and sediments generally share the common feature of cell concentration on micropore membrane and removal of biological particles from water by prefiltering (Bej, 1999; Pickup et al., 1999). DNA extraction from sources such as insects that act as vectors of important bacterial pathogens (*Xylella fastidiosa*, sharp-shooters; *Erwinia amylovora* and *Pantoea stewartii*, bees; *Candidatus Liberibacter* spp., psyllids), can require specific protocols (Bextine et al., 2005; Meyer et al., 2007). Several genomes from causal agents of plant bacterial diseases have been completely sequenced and more are underway (www.integratedgenomics.com). Based on genome analysis, new specific sequences could be used to design detection probes for different pathogens (Van Sluys et al., 2002). The sequences of complete genomes in GenBank are available through NCBI (www.ncbi.nlm.nih.gov) and other databases.

Technological advances in polymerase chain reaction (PCR)-based methods enable the fast and accurate detection, quantification, and characterization of plant pathogens and are now being applied to solve practical problems. For example, the use of molecular techniques in bacterial taxonomy allows different taxa of etiologically significant bacteria to be separated (De Boer et al., 2007). Therefore, molecular diagnostics can provide the degree of discrimination needed to detect and monitor plant diseases, which is not always accomplished by other types of analysis. Despite the fact that nucleic acid technology is the only choice when cultures of bacteria or phytoplasma are not up to date, DNA-based methods have not yet completely replaced traditional culture and phenotypic tests in the most common plant pathogens detection because the information from different methods can be contradictory. For this reason, the current trend in the European Union (EU) and European and Mediterranean Plant Protection Organization (EPPO) protocols for the detection of plant pathogens is to combine conventional, serological, and molecular techniques in integrated approaches. The use of polyphasic or integrated approaches for detection is advised, especially when the targets are plant quarantine bacteria (López et al., 2003, 2005; Alvarez, 2004; Janse, 2005).

As an example, the most recently published official EU protocols for *Clavibacter michiganensis* ssp. *sepedonicus* and *R. solanacearum* (Anonymous, 2006) incorporate PCR as a screening test in an integrated protocol that includes serological techniques, isolation, and bioassays for higher-accuracy detection of these quarantine pathogens. This approach not only increases our ability to detect plant pathogens but can also provide new insights into their ecology and epidemiology (Martin et al., 2000). For quarantine bacteria, isolation and proof of pathogenicity is required in both the EU and EPPO current protocols. This could be substituted, after appropriate validation, with real-time PCR (RT-PCR) based on the detection of m-RNA of selected target genes, which correlates with cell viability and pathogenicity. Methodology for selecting and validating a test for routine diagnosis has also been discussed (Janse, 2005).

The DNA sequences from which the primers are designed for bacteria come from three main origins: pathogenicity/virulence genes, ribosomal genes, and plasmid genes. The pathogenicity genes used as targets can be involved in any of the several steps leading to symptom development and can be related to virulence factors (Coletta-Filho et al., 2006), virulence or avirulence genes (Haas et al., 1995; Kousik and Ritchie, 1998), toxin products (Lydon and Patterson, 2001), and other factors (Schonfeld et al., 2003; Tan-Bian et al., 2003). These genes can be situated in plasmids, as primers from Ti plasmid of *R. radiobacter* (Haas et al., 1995; Sawada et al., 1995), or from plasmids of *C. michiganensis* ssp. *sepedonicus* (Schneider et al., 1993), *Xanthomonas axonopodis* pv. *citri* (Hartung et al., 1993), or *E. amylovora* (Bereswill et al., 1992; McManus and Jones, 1995; De Bellis et al., 2007). They can also be located in the chromosome and be specific to a pathogen, or to a group of pathogens, such as the *pel* gene of soft-rot diseases caused by pectolytic subspecies of *Pectobacterium* (Darrasse et al., 1994; Louws et al., 1999), or belong to a cluster of genes that are involved in the virulence system of different bacterial families (*hrp*, *pth*, and *vir* genes). The ribosomal operon has also been employed as a source of primers in many models.

The advantage of this target is the universality of the ribosomes in all bacteria, and a size (1600–2300 bp) that enables the rapid sequencing of the whole operon and the selection of suitable primers. Genus-specific rDNA sequences of phytobacteria are now available, and many primers based on those sequences have been developed to detect many plant pathogens (Li and De Boer, 1995; Louws et al., 1999; Walcott and Gitaitis, 2000; Botha et al., 2001; Wang et al., 2006). The drawback of choosing this region is the relatively low complexity of the sequences, which mainly occur in bacteria belonging to the same genus or species. This drawback can be resolved by using sequences from the internal transcribed spacer (ITS), which is more variable in its nucleotide composition (Jensen et al., 1993). Despite this universality, several primer sets have been developed for the specific detection of some plant pathogenic bacteria from the DNA sequence comparison of the ITS region; these include *C. michiganensis* ssp. *sepedonicus* (Li and De Boer, 1995), *E. amylovora* (Maes et al., 1996), and *R. solanacearum* (Fegan et al., 1998).

Plasmid DNA is also widely employed in the design of primers for important bacterial diseases. The amplified plasmid genes may be associated to pathogenicity, as indicated above, or have functions unknown. One problem to be addressed is the stability of the target plasmids, except when the chosen plasmid provides special fitness or pathogenicity traits; in such cases they probably can be more stable (Louws et al., 1999). In some cases, primers that target plasmid genes have not been found to be useful for the universal detection of a pathogen, due to the lack of the plasmid in some strains. For example, primers have been used from pEA29 plasmid sequences for sensitive and specific detection of *E. amylovora* (Bereswill et al., 1992; McManus and Jones, 1995).

The advantages of nucleic acid-based technology in terms of sensitive, specific, and rapid diagnostics are provided by conventional PCR and its variants. Developing a detection method is both an art and a never-ending story. Similarly, the concept of accurate detection of plant pathogenic bacteria is moving from conventional methods to molecular techniques, and is included in integrated approaches (Alvarez, 2004; López et al., 2005).

For quarantine pathogens or in critical export-import situations, experience indicates the use of multiple techniques based on different biological principles to avoid the risk of false positives and false negatives. In practice, the success of this method will depend on the pathogen diversity, the functional significance of the selected genes in the disease progression, and their expression at the different steps in the host/pathogen interactions. The new qualitative and quantitative detection data generated should provide a more complete picture of the life cycle of each plant pathogen.

Consequently, more-appropriate sampling methodologies and systems will be set up for the efficient detection of latent infections and pathogen reservoirs (López et al., 2003). The accuracy of new detection protocols based on molecular methods will influence the availability, in the near future, of plants that are free of a wide range of pathogens.

Molecular techniques have recently been used to distinguish groups of bacteria. Nucleic acid-based techniques are widely recognized and powerful plant pathogen detection techniques. The target region mostly exploited for bacterial diagnostics is ribosomal DNA, which is present in all bacteria at high copy numbers per genome with highly conserved regions; these conditions allow for very sensitive detection (Mondal et al., 2004). Because the specificity of DNA-based techniques relies exclusively upon primer and probe sequences, such assays are easy to develop and can be transposed into virtually every pathosystem. BIO-PCR, nested PCR, multiplex PCR, multiplex nested PCR, co-operational PCR, RT-PCR, repetitive sequence-based PCR (Rep-PCR), molecular markers such as restriction fragment length polymorphism (RFLP), random amplified polymorphic DNA (RAPD), amplified fragment-length polymorphism (AFLP), fluorescence *in situ* hybridization (FISH), and loop-mediated isothermal amplification (LAMP), as well as *hrp* genes and *pth* gene-based markers, have been extensively employed to detect and identify pathogenic bacteria that affect different crop plants (Williams et al., 1990; Minsavage et al., 1994; Opio et al., 1996; Kerkoud et al., 2002; Berg et al., 2005). Biomarkers, the important housekeeping marker genes, were also used for multi-locus sequence typing (MLST) of bacterial pathogens.

The PCR protocols have been developed for the most important plant pathogenic bacteria. Sensitivity and specificity problems associated with conventional PCR and RT-PCR can be reduced by using nested PCR-based methods in two consecutive rounds of amplification (Porter-Jordan et al., 1990; Simmonds et al., 1990). Sensitivity increases by two orders of magnitude to about 10^2 bacterial cells per mL of extract. Some authors have proposed single-tube-nested PCR protocols for the bacteria *E. amylovora* (Llop et al., 2000) and for *Pseudomonas savastanoi* pv. *savastanoi* (Bertolini et al., 2003a). A new PCR concept is co-operational PCR, which is based on the simultaneous action of three or four primers (Olmos et al., 2002). The technique was first developed and used successfully for the detection of the bacterium *R. solanacearum* in water (Caruso et al., 2003), and includes protocols for phytoplasmas such as *Ca Phytoplasma mali*, *Ca Phytoplasma prunorum*, and *Ca Phytoplasma pyri* (Bertolini et al., 2007). The simultaneous detection of two or more DNA and/or RNA targets can be afforded by multiplex PCR in a single reaction when several specific primers are included in the PCR cocktail.

The design of a multiplex RT-PCR is based on the use of compatible primers specific to different targets, which must be evaluated theoretically *in silico* and empirically tested *in vitro*. The use of general and common primers to amplify different targets, such as those based on an 16SrRNA gene sequence, is not appropriate because the targets are competing and the reaction will be displaced to the most abundant target, which makes detection of the less abundant ones more difficult (López et al., 2006). Nested PCR has been demonstrated to be useful in the ultrasensitive detection of various plant pathogenic bacteria, including *C. michiganensis* ssp. *sepedonicus* (Lee et al., 1997), *E. amylovora* (Llop et al., 2000), *P. savastanoi* pv. *savastanoi* (Bertolini, 2003b), and *Pectobacterium carotovorum* ssp. *carotovorum* (Mahmoudi et al., 2007). The multiplex nested RT-PCR method in a single tube, which combines the advantages of the multiplex PCR with the sensitivity and reliability of the nested PCR, saves time and reagent costs because two reactions are sequentially performed using a single reaction cocktail. In addition, it enables simultaneous detection of RNA and DNA targets.

Multiplex nested PCR has been used for detection of phytoplasmas (Clair et al., 2003; Stukenbrock and Rosendahl, 2005), and the bacterium *P. savastanoi* pv. *savastanoi* in olive plant material using 20 compatible primers in both detection. The sensitivity achieved for the bacterium *P. savastanoi* pv. *savastanoi* by multiplex nested RT-PCR (1 cell mL^{-1}) was similar to the sensitivity achieved by applying the monospecific nested PCR, which was demonstrated to be a hundredfold more sensitive than conventional PCR (Bertolini et al., 2003a). The sensitivity of a PCR-based detection assay can be improved by adding a short culture-enrichment step before PCR amplification; this technique is referred to as BIO-PCR (Schaad et al., 1995). Increased sensitivity of the detection assay that includes a culture- enrichment step prior to PCR was observed during the detection of many bacterial plant pathogens, including *Pseudomonas syringae* pv. *phaseolicola* in bean seed extracts and *Acidovorax avenae* ssp. *avenae* in rice seeds (Schaad et al., 1995; López et al., 2003; Song et al., 2004). An optimized BIO-PCR procedure was developed for the detection of *C. michiganensis* ssp. *michiganensis* in tomato transplants during the early latent stage of infection (Narayanasamy, 2011).

In real-time PCR, the most advanced version of PCR, the amplification of target sequence can be quantified after each PCR cycle and amplification is pictographically displayed through an attached monitor (Espy et al., 2006). This and other advances in the chemistries of primers and probes indicate the establishment of new approaches that use real-time protocols but have different characteristics, depending on the target and assay requirements (e.g., quantification, discrimination between closely related subspecies, SNPs). The primer design and probe type must therefore first be evaluated in terms of the features required for the assay, so that the one(s) that best fit specific requirements may be chosen. RT-PCR technique was successfully applied for the detection of *Leifsonia xyli* ssp. *xyli*. Real-time technology is also being used in multiplex format for the detection and characterization of several bacteria (Weller et al., 2000; Berg et al., 2006; Abriouel et al., 2007).

RFLP analysis has been extensively used in the detection and identification of plant bacterial pathogens (Mondal et al., 2004). A small DNA segment from a known bacterium, which is pathogenic to the host plant in question, is used as a probe. DNA from both the known (as positive control) and suspected bacterial pathogens (isolated from infected plant samples) are digested with the same restriction enzyme(s). RAPD markers have been demonstrated to be useful for determining polymorphisms among phytopathogenic bacteria (Rafalski et al., 1994; Chen et al., 2003; Grover et al., 2006; Mondal et al., 2008). The utility of random primers for differentiating bacterial blight pathogens in beans *X. axonopodis* pv. *phaseoli* and *X. axonopodis* pv. *phaseoli* var. *fuscans* has been well documented (Birch et al., 1997).

AFLP involves amplification of a specific region of genomic DNA through PCR (using a single primer) followed by the cleaving of the amplified fragments using restriction endonuclease. The technique is thus a combination of RFLP and PCR techniques and is extremely useful in the detection of polymorphism between closely related bacterial pathogens. AFLP is preferred and very often employed in the detection and differentiation of several bacterial plant pathogens, including *X. axonopodis* pv. *phaseoli*, *X. axonopodis* pv. *phaseoli* var. *fuscans*, and *Ps. syringae* pv. *tomato*, using primers specific to ribosomal genes or rRNA operons (Manceau and Horvais, 1997; Mahuku et al., 2006). Diagnosis of bacterial plant pathogens using primers that correspond to specific repetitive sequences, such as enterobacterial repetitive intergenic consensus (ERIC), repetitive extragenic palindromic (Rep), and repetitive BOX elements (BOX), which are dispersed throughout the bacterial genome, has well been documented (Martin et al., 1992; Louws et al., 1994, 1999; Barak and Gilbertson, 2003). The distribution patterns of these repetitive sequences, which vary from one bacterium to another, comprises the basis of differentiation in a bacterial population. ERIC-PCR has been demonstrated to be an effective method in determining the genetic diversity among populations of many bacterial plant pathogenic genera, including *Xanthomonas* and *Pseudomonas* (Weingrat and Volksch, 1997; Mondal and Mani, 2009). Recently, the use of BOX-PCR in determining genetic variability among the bacterial flora associated with pomegranate leaf was documented (Mondal, 2011).

The LAMP technique has been shown to be a good approach for amplifying nucleic acid with high specificity, efficiency, and rapidity without the need for a thermal cycler (Kubota et al., 2007). Because this method has a sensitivity of approximately 10^5–10^6 CFU mL^{-1}, it is recommended for use only on symptomatic plant tissues (Braun-Kiewnick et al., 2011). The LAMP-based technique, which facilitates its use in the field and in less well-resourced settings, is recommended in the EPPO protocol (OEPP/EPPO, 2004a). The LAMP method has been successfully used for the detection of many plant pathogenic bacteria, including epiphytic *E. amylovora* in pear and apple (Temple et al.,

2008). More sensitive diagnostic tools for detecting latent infections of *E. amylovora* would be most welcome for surveillance and monitoring applications. *Erwinia amylovora* is detected at a sensitivity of 10^2–10^3 CFU mL^{-1} per apple or pear flower (Temple and Johnson, 2011).

FISH, which combines microscopical observation of bacteria and the specificity of hybridization (Wullings et al., 1998; Volkhard et al., 2000), is dependent on the hybridization of DNA probes to species-specific regions of bacterial ribosomes. Although FISH can detect single cells, the detection level in practice is near 10^3 CFU mL^{-1} of plant extract. This technique has been included in official diagnostic protocols for *R. solanacearum* and is recommended in the EPPO protocol for the same pathogen (OEPP/EPPO, 2004b). LAMP is another type of isothermal amplification whose increasing use in the diagnostic field provides sensitivity and is low cost (Notomi et al., 2000).

Four different fluorescence-detection techniques are currently used to detect amplicons: SYBR green-dye-based detection, TaqMan probes, fluorescent resonance energy transfer (FRET) probes, and molecular beacons (Mackay et al., 2002; Schaad and Frederick, 2002). The use of fluorescent dyes such as SYBR Green I (SG) in RT-PCR has become more important for the diagnostic applications of plant pathogens (Kiltie and Ryan, 1997; Vitzthum and Bernhagen, 2002). A real-time SYBR Green I assay was developed for the detection of *X. arboricola* pv. *pruni*, the causal agent of bacterial spot disease of stone fruit (Palacio-Bielsa et al., 2011). The TaqMan assay is based on the 5′–3′ exonuclease activity of *Taq* polymerase (Weller et al., 2000; Varma-Basil et al., 2004). These assays have been proposed for the detection of *C. michiganensis* ssp. *sepedonicus* (Schaad et al., 1999; van Beckhoven et al., 2002), *R. solanacearum* (Weller et al., 2000; Ozakman and Schaad, 2003), *E. amylovora* (Salm and Geider, 2004), *Candidatus Liberibacter asiaticus* (Liao et al., 2004; Li et al., 2006), *P. stewartii* ssp. *stewartii* (Qi et al., 2003), *X. fastidiosa* (Schaad and Frederick, 2002; Bextine et al., 2005), and *X. fragariae* (Weller et al., 2007). Molecular beacons are short fluorescent oligonucleotide probes that are designed to form stem-loop folding (Didenko, 2001). The probes contain a fluorescent chromophore at the 5′ end and a quencher molecule at the 3′ end (Cockerill and Smith, 2002).

Genome-wide comparison between pathogenic and nonpathogenic strains within a species is a useful strategy for identifying candidate genes that are important for virulence. DNA microarray-based genome composition analysis, which is a good alternative to full genome sequencing, has been used in comparative studies to analyze various bacterial pathogens including *Pseudomonas aeruginosa* (Wolfgang et al., 2003) and *X. fastidiosa* (Koide et al., 2004). The gene distribution among strains of *R. solanacearum*, a highly polymorphic plant pathogenic bacterium, has been a priority area in which to study the status of known or candidate pathogenicity genes. For bacterial detection, the material spotted to date has almost universally been oligonucleotides targeting the 16S and 23S rDNA genes (Crocetti et al., 2000; Loy et al., 2002, 2005; Fessehaie et al., 2003; Peplies et al., 2003; Franke-Whittle et al., 2005).

A particular feature of this system is that biotinylated immobilized molecules can be either oligo-capture probes or amplified PCR samples. Hybridization is detected and analyzed by fluorescent oligo probes. By regulating the electric-field strength, hybridization stringency can be adjusted for homologous interactions. Nanochips have shown high specificity and accuracy in the diagnosis bacterial pathogens that affect potato, due to their ability to discriminate single-nucleotide changes (Ruiz-García et al., 2004). A detection system comprising of amplification of 16S rDNA by isothermal and chimeric primer-mediated amplification of nucleic acids (ICAN) and detection of amplified products with cycling probe (cycleave) technology was developed for the detection of *Ca. Liberibacter asiaticus*.

Multilocus sequence typing (MLST), a useful way to detect and identify strains of a bacterial species, is based on the differences in the nucleotide sequences of a small number of genes. In this system, each allele of a gene is allotted a number and each characterized strain (for *n* loci) is represented by a set of *n* numbers that defines the alleles at each locus. A single-nucleotide difference always produces a new allele in an MLST dataset; this feature distinguishes this method from others such as multilocus enzyme electrophoresis and pulsed field gel electrophoresis (PFGE), both of which require greater substitution numbers for discrimination (Peacock et al., 2002). Strains of *X. fastidiosa* from grapevine, oleander, oak, almond, and peach were detected and differentiated with an initial set of sequences from 10 loci (9.3 kb). The MLST approach, which is characterized by its simplicity, offers a distinct advantage over phylogenetic investigations by providing information required for rapid recognition of the incidence of an unusual or new isolate of *X. fastidiosa* in a given location; this information can then be used to initiate damage control processes. In addition, MLST is also useful for efficiently cataloguing the genetic diversity within a bacterial species (Scally et al., 2005).

Flow cytometry is a technique for rapid identification of cells or other particles as they pass individually through a sensor in a liquid stream. Bacterial cells are identified by fluorescent dyes that are conjugated to specific antibodies and electronically detected with a fluorescence-activated cell sorter that measures several cellular parameters based on light scattering and fluorescence. Multiparameter analysis includes cell sizing, fluorescence imaging, and gating out (elimination of unwanted background associated with dead cells and debris) (Davey and Kell, 1996; Alvarez, 2001). Flow cytometry has excellent potential as a research tool and possibly for routine use in seed-health testing and other fields. The several parameters that can be analyzed simultaneously include total particle count, distinction between living and dead cells, and differentiation of target and non-target bacterial populations associated with seeds or other plant material.

The detection of bacterial plant pathogens with antisera is still the method of choice for many plant pathologists because of the relatively low costs and the presence of technical infrastructure based on automated ELISA. Development of mono- and polyclonal antibodies is required for specific immunological diagnostic assays. In bacterial plant pathogens, the preferred antigens are lipopolysaccharides, exopolysaccharide, cell-wall protein, flagellar protein, and outer membrane proteins. Through this technique, bacterial pathogens can be detected directly in infected plant materials without culturing the bacteria. The sensitivity of ELISA assay ranges between 10^5 and 10^6 CFU mL^{-1}. Several reports describe the employment of monoclonal antibodies to detect bacterial plant pathogens, including *Xanthomonas campestris* pv. *begoniae*, *X. campestris* pv. *pelargonii* (Benedict et al., 1990), *X. oryzae* pv. *oryzae* and *X. oryzae* pv. *oryzicola* (Benedict et al., 1989), *C. michiganensis* ssp. *sepedonicum* (De Boer and Wieczorek, 1984), and *E. amylovora* (Gorris et al., 1996). Shanmugam et al. (2004) used a polyclonal antibody developed against whole bacterial cells of *R. solanacearum* infecting potato to detect the survival of *R. solanacearum* in ginger rhizomes stored at different temperatures. Lateral flow devices are now available for the rapid detection of bacterial pathogens, including *R. solanacearum* (Danks and Barker, 2000), *C. michiganensis* ssp. *Michiganensis*, and *X. hortorum* pv. *pelargonii* (Alvarez, 2004). Dot-immunobinding assay (DIBA) is quite similar to the ELISA test in principle. Detection of bacterial pathogens in asymptomatic plants is essential in order to restrict the incidence and spread of disease(s) by eliminating infected plants/plant materials, as in the case of citrus canker disease. The pathogen *X. axonopodis* pv. *citri* could be detected in 38.4% of asymptomatic citrus plants. The detection threshold was 10^4 CFU mL^{-1} (de Lima et al., 1998).

Phytopathogenic bacteria can be identified and differentiated by thin-layer chromatography (TLC) profiles. Comparisons have been made of TLC of aminolipids extracted from *C. michiganensis*, *R. solanacearum*, *P. carotovorum*, *Pectobacterium chrysanthemi*, *Ps. syringae*, *X. campestris*, and *X. oryzae*. The TLC profiles of aminolipids can be prepared easily and used for the presumptive identification and differentiation of bacterial plant pathogens (Matsuyama et al., 2009). The fatty acid composition of phytopathogenic bacteria has been demonstrated as a basis for their identification. A dendrogram based on fatty acid composition showed that all of the pathovars of *Ps. syringae*, *Pseudomonas viridiflava*, and pear and radish strains were closely related (Khan et al., 1999).

A technique based on the differential utilization of 147 carbon sources was developed by API Systems. Results, which depend on the visualization of the growth of target bacterium, may require a long time. For example, with *Xanthomonas* spp., results took 1 week to obtain. This technique was employed to identify *Ps. syringae* pathovars (Gardan et al., 1984), to distinguish *X. campestris* pv. pathotypes (Verniere et al., 1991), and to identify strains of *Brenneria quercina* causing bark canker and drippy bud and nut disease in *Quercus ilex* and *Quercus pyrenaica* (Biosca et al., 2003). The BIOLOG-automated identification system is based on the differential utilization of 95 carbon sources by different bacterial species/isolates/strains (Klinger et al., 1992).

One must choose the best or the best combination of options, depending upon need. For example, when multiple pathogens are to be detected in a minimum time-frame, multiplex PCR would be the best option. Although the pathogen detection limit in a sample is at zero tolerance level, nested PCR and BIO-PCR should be carried out to detect even lower numbers of bacterial cells in the tested samples; this approach would also help to differentiate between viable and nonviable cells. In addition, for routine diagnosis of bacterial pathogens, integrated approaches including 16S rDNA sequencing, MLST, BIOLOG phenotyping, fatty acid methyl esterase (FAME) profiling, and pathogenicity assay (as in Koch's postulate) are to be preferred.

New detection tools will be used not only for rapid, sensitive, and specific diagnosis but also to help to understand plant–pathogen relationships and the structure and function of pathogens and their communities. Only when new technologies become fully integrated with other conventional tools, which they should complement not substitute, will they provide information that is useful for the understanding and prevention of plant diseases (López et al., 2003).

The possibility of next-generation sequencing methods for identifying microorganisms present in the microbiomes of diseased plant tissue, as is being done for other microbially complex milieus such as endophyte communities in plant roots (Manter et al., 2010), is becoming more realistic. Much research is being devoted to the development of on-the-spot diagnostic devices that use DNA-amplification strategies (e.g., the isothermal LAMP method) and multiplexing strategies (e.g., macro- and microarray technologies). New technologies that involve genomic barcoding, next-generation sequencing, biosensors, and mass spectrometry are also being explored for their applications as diagnostic and pathogen-identification tools (De Boer and López, 2012; Mondal and Shanmugam, 2013).

4.3 SEED-BORNE BACTERIAL PATHOGENS

Seed-borne pathogens present a serious threat to seedling establishment. Close association with seeds facilitates the long-term survival, introduction into new areas, and widespread dissemination of bacterial pathogens. Under greenhouse conditions, the risks of significant economic losses due to diseases are great because factors including high populations of susceptible plants, high relative humidity, high temperatures, and overhead irrigation promote explosive plant disease development. Under these conditions, the most effective disease-management strategy is exclusion, which is accomplished by using seed-detection assays to screen and eliminate infested seed lots before planting.

Unlike infected vegetative-plant tissues, infested seeds can be asymptomatic, which makes visual detection impossible. Seed-health assays should be sensitive, specific, rapid, robust, inexpensive, and simple to implement and interpret. A direct method of testing seeds is to allow pathogens to grow from them onto appropriate artificial media. This can be done by directly plating surface-sterilized seed samples or seed-wash liquid onto artificial media, followed by incubation under adequate conditions. Once a pathogen is isolated, it can be identified by its cultural or biochemical characteristics; for example, the production of a bluish-green fluorescent pigment on King's B medium (King et al., 1956) in the case of fluorescent *Pseudomonas* spp. or the production of dark. Unfortunately, seeds may be contaminated by saprophytic microorganisms (nonpathogens) that grow as well as or better than target organisms on nutrient-rich, artificial media.

Serology-based seed tests have several formats, including the widely applied ELISA (McLaughlin and Chen, 1990) and immunofluorescence microscopy (Franken, 1992). Because serological assays do not require pure isolations of the pathogen, they are applicable to biotrophic and necrotrophic seed-borne pathogens. The seedling grow-out assay is a direct measure of the seed lot's ability to transmit a disease. To conduct this assay, seed-lot samples are planted under greenhouse conditions conducive to disease development and seedlings are observed for the development of symptoms.

Seedling grow-out is one of the most applicable and widely used seed-detection assays (Capoor et al., 1986; Lee et al., 1990; Yang et al., 1997); however, for successful implementation, infected seedlings must display obvious and characteristic symptoms. Another drawback of the seedling grow-out assay is in diseases such as bacterial fruit blotch (*A. avenae* ssp. *citrulli*) of watermelon, large seed samples (10,000–50,000 seeds) must be tested to statistically ensure that one infested seed can be detected. In addition to losses associated with the destructive testing of expensive seeds, assaying this quantity of seed requires large areas of greenhouse space and adequate labor for assay set-up and evaluation. The seedling grow-out assay is also time-consuming; it requires up to 3 weeks for seedling germination and symptom development. Finally, seed-test evaluators must be familiar with the symptoms associated with each disease.

Many seed types contain compounds (e.g., tannins and phenolic compounds) that inhibit DNA amplification and therefore result in false-negative results when PCR is attempted directly on seed extracts. Because DNA from nonviable cells or tissues can yield false-positive results in seed assays, it is necessary to confirm positive results by recovering the target organism. The BIO-PCR has been developed for the detection of bacterial fruit blotch (*A. avenae* ssp. *citrulli*) of watermelon, halo blight (*Ps. syringae* ssp. *phaseolicola*) of beans, and bacterial ring rot (*C. michiganensis* ssp. *sepidonicum*) of potato (Schaad et al., 1995, 1999; Pryor and Gilbertson, 2001). This technique has been reported to significantly improve the sensitivity and the ease of implementation of PCR, displaying detection limits of 2–3 CFU mL^{-1} (Schaad et al., 1999).

Immunomagnetic separation refers to the use of microscopic magnetic beads (IMBs) coated with antibodies that have been produced against a specific microorganism to selectively sequester target cells from suspensions that contain heterogenous cell mixtures (Olsvik et al., 1994; Safarik and Safarikova, 1999). The IMS-PCR, which was developed for the detection of *A. avenae* ssp. *citrulli* in watermelon seeds (Walcott and Gitaitis, 2000), has significantly better detection efficiency and sensitivity than conventional PCR (Walcott and Gitaitis, 2000). The IMS consistently recovered viable target colonies from suspensions containing 10 target CFU mL^{-1}. The IMS-PCR proved to be more sensitive and reliable than hexacetyl dimethyl ethyl ammonium bromide (CTAB)-DNA extraction (Ausubel et al., 1987) followed by direct PCR and ELISA detection of *A. avenae* ssp. *citrulli* in watermelon seed lots. The IMS-PCR also facilitated the detection of *A. avenae* ssp. *citrulli* in seed lots with 0%, 1%, 5%, and 10% infection (Walcott and Gitaitis, 2000). In addition, an IMS-PCR-based seed assay has been reported for center rot of onion caused by *Pantoea ananatis* (Walcott et al., 2002).

Rapid-cycle RT-PCR promises to eliminate many of these barriers and thereby to make PCR more accessible for seed detection. With RT-PCR, DNA amplification is coupled with the production of a fluorescent signal that increases proportionally with the numbers of amplicons produced (Kurian et al., 1999; Cockerill and Smith, 2002). Detection of amplified DNA can be accomplished by staining with SYBR Green I, which binds double-stranded DNA indiscriminately, or with specific reporter probes such as TaqMan (Taylor et al., 2002). Other detection systems, including FRET and molecular-beacon probes, are also employed for RT-PCR (Cockerill and Smith, 2002). Compared to conventional PCR, RT-PCR has several key advantages that potentially make it more acceptable for use in routine seed testing. These include rapid cycling, cross-contamination risk reduction, no need for post-PCR electrophoresis, ability to detect multiple pathogens in the same reaction (Wittwer et al., 2001), and accurate determination of seed-infestation levels.

DNA chips or microarrays are other DNA-based detection assays that may be applied to test seeds for pathogens (Lemieux et al., 1998; Vernet, 2002). With DNA-chip technology, oligonucleotide probes are attached to small glass or silica-based surfaces (chips). These oligonucleotides can be complementary to DNA sequences that are unique to certain microorganisms and thus can be used to detect pathogens in seed samples. If the DNA from the pathogen of interest is present in the seed sample, then the oligonucleotide probe at the position on the chip that corresponds to that pathogen will display fluorescence. DNA chips are being used in many different fields for diagnosis (Lemieux et al., 1998; Anthony et al., 2000).

4.4 PREVENTIVE MEASURES FOR PLANT PATHOGENIC BACTERIA

Efficient control of bacterial plant diseases has been based on Koch's control system (Janse, 1996). These rules state that the main responsibility for controlling a disease lies with the country in which infections occur; that statistically meaningful surveys should be performed to assess the presence, absence, and eventual distribution of a disease; that the country involved should report the disease as quickly as possible; and that rapid and sure means of detection and diagnosis are vital. The rules also state that countries should try to contain and control a disease as soon as possible by imposing restrictions on the movements of infected crops/lots (when appropriate), by implementing appropriate measures on contaminated fields and/or premises (e.g., imposing hygienic protocols), by tracing the origins of an infection (clonal relationships of infected crops, trade lines, contaminated surface water), by performing surveys on the original host and all other potential hosts to determine pathogen distribution and spread, by establishing epidemiological risk factors and taking action on them (e.g., water or use of safe sources of water), and by imposing safe disposal procedures for any plant waste that may be contaminated with the pathogen in question. In addition, inspections should be done by importing countries and production and use of healthy planting material should be mandatory (e.g., indexing, and for latent infections, development/use resistant varieties, chemo- or thermotherapy of basic planting material) (Janse and Wenneker, 2002).

Cultural practices usually influence disease development in plants by affecting the environment. Such practices are intended to make atmospheric, edaphic, or biological surroundings favorable to a crop plant and unfavorable to its parasites. Cultural practices that lead to disease control have little effect on the climate of a region but can exert significant influence on the microclimate of the crop plants in a field. Three stages of a parasite's life-cycle (i.e., survival between crops, production of inoculum for the primary cycle, and inoculation) can be controlled with preventive measures. Some bacterial pathogens that survive in soil can often be controlled by crop rotations with unsusceptible species. Plant diseases caused by organisms that survive as parasites within perennial hosts or within the seed of annual plants may be controlled therapeutically. Therapeutic treatments of heat and surgery are applicable here; those involving the use of chemicals will be discussed later. Removal of cankered limbs (surgery) helps control fire blight of pears, and hot-water treatment of cabbage seed controls the bacterial disease known as black rot.

Environmental factors (particularly temperature, water, and organic and inorganic nutrients) significantly affect inoculum production. Warm temperature usually breaks the dormancy of overseasoning structures; rain may leach growth inhibitors from soil and permit germination of resting spores; and special nutrients may stimulate the growth of overseasoning structures that produce inoculum. Cultural practices that exemplify avoidance are sometimes used to prevent effective dissemination.

One of the preventive measures to control plant disease is sample inspection, which involves laboratory evaluation of a representative sample (drawn by a certification agency) for germination, moisture content, weed-seed content, admixture, purity, and seedborne pathogens. Propagating material includes seeds, cuttings, root-stocks, buds, tubers, rhizomes, bulbs, and corms. Because all of these can carry pathogens, it is essential to introduce only disease-free propagating material to new areas. This can be achieved by producing the propagating material in disease-free areas, inspecting the material carefully before it enters a disease-free area, and treating or destroying any diseased material that is found. Material can be inspected at any number of stages between production and planting, and again after planting to check that no disease has developed. Potentially diseased material can be successfully treated with heat for some pathogens, as long as the host's heat-tolerance level is higher than the pathogen's. This is a particularly useful method for deep-tissue infections caused by viruses and bacteria, which cannot be reached with chemical treatment. Alternatively, seeds can be treated with chemicals that kill pathogens on their surfaces. Some certification schemes ensure that seed and vegetative propagating material are free from particular diseases.

4.4.1 Quarantine

Because political boundaries seldom represent distinct ecological regions, pests introduced into one country may constitute a threat to many adjacent ones. Therefore, FAO was instrumental in suggesting and encouraging the establishment of regional plant-protection organizations. Their purpose is to standardize safeguards against pest introductions and to provide mutual assistance to that end.

As international trade in agriculture and plant protection organizations recognize their phytosanitary practices, it becomes ever more important for the assessment of risks presented by trading in plants and plant products to be based on sound, structured biological principles. International recognition that the best way to manage such risks is through the process of pest-risk analysis is also important. The first stage of disease-risk analysis is risk identification, which involves determining whether the organism in question causes effects that qualify as a quarantinable disease according to an international plant-protection organization's definition. The second stage, risk assessment, is the process of establishing an estimate of the probability of introduction of a target disease into a country. Pathogens carried on air currents or by insects or birds are not amenable to control by quarantine. However, pathogens spread on the host, on host parts, and on inert material such as packaging or containers; all of these can be controlled by quarantine. In this context, computer simulation programs can be used to decide where climatic conditions are conducive to the establishment of pathogens. Risk management, the third stage, involves establishing options to assist quarantine in reducing the risk of introduction and establishment of the target disease. Disease-risk analysis ends when it is decided whether or not a proposed import should be permitted. Accepting some risk reflects the requirement that quarantine must be flexible enough to meet changing demands, new technologies, and changing levels of resources as well as to avoid establishing unjustified trade barriers and to provide a level of security against the entry of unwanted diseases. Quarantine must also be deemed cost effective and scientifically justifiable.

The term "quarantine" means the establishment by public authorities of protective barriers against the dissemination of injurious pathogens. The goal of plant quarantine is to protect agriculture and the environment from avoidable damage by hazardous organisms that have been introduced by human activities. This goal is achieved by government-established restrictions on the movement of plants, plant products, soil, cultures of living organisms, packing materials, and commodities, as well as their containers and means of transport. These restrictions are designed to prevent the establishment of plant pathogens in areas where they do not occur. It was gradually recognized that an effective quarantine system prevents the entry of dangerous, exotic pathogens but does not block the movement of plant products and germ plasm. Since the late twentieth century, quarantine services have been using an acceptable-risk policy based on a process of risk assessment. Acceptable risk is the point at which the decision maker determines that the potential risks posed to a nation by a proposed import are small enough to be manageable (Ogle, 1997).

A number of regional organizations such as EPPO, the Inter-African Phytosanitary Organization, the North American Forestry Commission, and the Plant Protection Commission for Southeast Asia and the Pacific Region are now in operation. Good progress in providing for more uniform safeguards for larger geographical and ecological areas has been made. The World Health Organization has led the way in encouraging worldwide recognition of the risks involved in the international transmission of infectious human and animal diseases; perhaps plant-protection organizations throughout the world might learn from and adopt some of their techniques.

The European Plant Protection Organization aims to help its member countries prevent the entry or spread of dangerous pests (plant quarantine). The organization therefore identifies pests that may present a risk, makes proposals on relevant phytosanitary measures, and prepares quarantine lists. In recent years the identification of risk has been formalized, because transparent justifications of phytosanitary measures are required and such measures must be commensurate with the risk. The EPPO recommends that its member countries regulate the bacteria and phytoplasmas listed below (version 2014–09) as quarantine pests (A1; pests are absent from the EPPO region, A2; pests

TABLE 4.1
EPPO A1 and A2 Lists of Bacteria and Phytoplasmas Recommended for Regulation as Quarantine Pests (version 2014–09)

A1 List	A2 List
Bacteria and Phytoplasmas	
Acidovorax citrulli	*Burkholderia caryophylli*
Citrus huanglongbing (citrus greening)	*Clavibacter michiganensis* ssp. *insidiosus*
Liberibacter africanum	*C. michiganensis* ssp. *michiganensis*
Liberibacter asiaticum	*C. michiganensis* ssp. *sepedonicus*
Liberibacter solanacearum	*Curtobacterium flaccumfaciens* pv. *flaccumfaciens*
'Ca. Phytoplasma ulmi' (Elm phloem necrosis)	*Dickeya dianthicola* (*Erwinia chrysanthemi* pv. *dianthicola*)
Palm lethal yellowing phytoplasma	*Erwinia amylovora*
Peach rosette phytoplasma	'Ca. *Phytoplasma mali*' (Apple proliferation)
Peach yellows phytoplasma	'Ca. *Phytoplasma pyri*' (Pear decline)
Potato purple-top wilt phytoplasma	'Ca. *Phytoplasma solani*' (Stolbur)
Western X-disease phytoplasma	'Ca. *Phytoplasma vitis*' (Grapevine flavescence)
Xanthomonas axonopodis pv. *allii*	*Pantoea stewartii*
X. axonopodis pv. *citri*	*Pseudomonas syringae* pv. *actinidiae*
X. oryzae pv. *oryzae*	*Pseudomonas syringae* pv. *persicae*
X. oryzae pv. *oryzicola*	*Ralstonia solanacearum*
Xylella fastidiosa	*R. solanacearum* race 2
	R. solanacearum race 3
	Xanthomonas arboricola pv. *corylina*
	X. arboricola pv. *pruni*
	X. axonopodis pv. *dieffenbachiae*
	X. axonopodis pv. *phaseoli*
	X. axonopodis pv. *poinsettiicola*
	X. axonopodis pv. *vesicatoria* and *X. vesicatoria*
	X. fragariae
	X. translucens pv. *translucens*
	Xylophilus ampelinus

Source: Adapted from OEPP/EPPO, 2014. EPPO A1 and A2 lists of quarantine pests (version 2014-09).

are locally present in the EPPO region; Table 4.1). The EPPO lists are reviewed every year by the Working Party on Phytosanitary Regulations and approved by Council consisting of expert scientists and official members of the EU countries related with the subjects (OEPP/EPPO, 2014).

Propagating material such as seeds, cuttings, root stocks, buds, tubers, rhizomes, and bulbs can carry pathogens to new areas, either in or on plant tissues. It is essential that propagation material being introduced to new areas be free from disease. This can be achieved by producing planting material in areas free from a pathogen or in areas not suitable for a pathogen or its vector(s). It can also be achieved by carefully inspecting plant materials for the presence of pathogens. If pathogens are found, the materials can either be destroyed or treated to kill the harmful organisms. Inspection may involve visual appraisal of disease symptoms, classification procedures, and techniques for the isolation and identification of pathogens. Propagating material can be inspected during growth, at harvesting, in storage, at market, and before planting. Classification procedures include separating healthy from diseased or suspect propagation material by hand, sieving, putting seed through fanning mills (which involves sieves and a strong air current), or immersion of seed samples in liquids to facilitate density separation.

Many seed-borne pathogens and infections cannot be directly observed; consequently, various diagnostic tests have been developed. Seed-borne bacterial infections require more sophisticated methods of isolation and identification. A number of latent bacterial infections in plants are difficult to detect and may often be overlooked in the selection of propagation materials. The success of any heat treatment applied to propagation materials depends on the differential temperature tolerances of the host and the pathogen (i.e., the host tissues must be able to withstand higher temperatures than the pathogen during the treatment period). Hot-water treatments have been effective in controlling the ratoon stunt bacterium (*Clavibacter xyli*) in sugar cane planting material (setts). Seeds may also be treated with chemicals to kill pathogens on their surfaces. Tomato seed can be treated in acetic or hydrochloric acid to remove the pulp surrounding the seed and to eliminate bacterial canker (*C. michiganensis* ssp. *michiganensis*). Treating cotton seed with sulfuric acid to remove the fibers that remain after ginning reduces populations of the bacterial blight bacterium *X. axonopodis* pv. *malvacearum* on the seed.

Certification schemes ensure that seed and vegetative propagation material are free from particular diseases. Crops are grown by registered farmers and inspected at prescribed intervals by specialist staff from government agencies. The aim is to produce planting material that is guaranteed free from certain diseases and conforms to cultivar characteristics. The French bean is well known for the variety of diseases that can be transmitted by sowing contaminated seed. Some countries have instituted schemes for the production of certified bean seed that attempt to exclude halo blight (*P. savastanoi* pv. *phaseolicola*) and common blight (*X. axonopodis* pv. *phaseoli*). Crop rotation with nonsusceptible crops literally starves out bacteria, fungi, and nematodes that have a restricted host range. Some pathogens can survive only as long as the host residue persists, usually no more than a year or two.

4.4.2 Sanitation

Sanitation is a basic requirement for preventing the spread of disease to new areas or, on a smaller scale, to new plants or plant products. It involves all procedures that prevent such spread. It can also be used effectively, under appropriate conditions, to reduce the amount of inoculum available at the beginning of a new growing season in areas where a pathogen is already established. Removing and disposing of infected leaves, flowers, branches, and any other plant material reduces the amount of inoculum available to begin a new epidemic. In the past, ploughing under or burning of plant debris remaining in the field after a crop was harvested was a popular practice that did reduce inoculum levels at the commencement of the next season. However, with the current trend toward minimum tillage and stubble retention to prevent soil erosion and maintain soil fertility, ploughing under and burning are no longer practiced to the same extent. As a result, pathogens that survive between seasons on plant debris and some that were previously not economically important have become important (Ogle, 1997).

Prevention is essential to avoid pathogen dissemination through a variety of vehicles, including contaminated propagative plant material, vectors, irrigation water, soil, etc. (Martín et al., 2000; Janse and Wenneker, 2002; López et al., 2003; Alvarez, 2004; De Boer et al., 2007). Prevention measures demand pathogen detection methods of high sensitivity, specificity, and reliability because many phytopathogenic bacteria can remain latent in low numbers, and/or in some special physiological states in propagative plant material and other reservoirs (Helias et al., 2000; Grey and Steck, 2001; Janse et al., 2004; Biosca et al., 2006; Ordax et al., 2006). Accurate detection of phytopathogenic organisms is crucial for virtually all aspects of plant pathology, from basic research on the biology of pathogens to control of the diseases they cause. The need for rapid and highly accurate techniques is especially necessary for quarantine pathogens because the risk of disease and the spread of inoculum must be reduced to nearly zero (López et al., 2003).

Soil may contain pathogens that will damage plants grown in them. It can also introduce pathogens to new areas if infested material is moved from one place to another. Steam, steam/air mixtures, or other forms of heat can be used to pasteurize soil. In such processes, the soil is brought up to 60°C over a period of 30–45 min and held at that temperature for 30 min. This treatment kills pathogens but beneficial organisms that form resistant spores survive to help prevent recontamination. Levels of pathogen inoculum in soil should also be checked to determine whether it is necessary to fumigate before planting a particular crop or in a particular season. In the longer term, it is anticipated that disease management will be achieved by using a full range of strategies including sanitation, disease-free planting material, alternative forms of soil treatment, containerized planting or hydroponics, and resistant cultivars. Solarization uses heat from the sun to raise the temperature of soil to levels that are high enough to kill pests and diseases.

The presence of plant pathogens limit the increased uptake of composted organic waste by potential end-users in the horticultural and agricultural sectors (Noble and Roberts, 2004). The success of composting in eliminating pathogens is not solely a result of the heating process; it also depends upon the many and complex microbial interactions that may occur, as well as other compost parameters such as moisture content (Bollen, 1985). However, heat generated during the thermophilic high-temperature phase of aerobic composting appears to be the most important factor for the elimination of plant pathogens (Bollen and Volker, 1996). Standards for compost sanitization have been developed in the U.S. by the Composting Council of the United States (Leege and Thompson, 1997), in the U.K. jointly by the Waste and Resources Action Programme and the Composting Association (Anonymous, 2002), and in several European countries as well (Stentiford, 1996). These standards specify minimum compost temperatures of 55–65°C for periods of 3–14 d, depending upon the composting system (e.g., turned windrow, in-vessel, static aerated piles). Christensen et al. (2002) recommended even more stringent sanitary requirements: 70°C for 2 d or 65°C for 4 d, with at least five turnings for windrow systems.

For filtration to effectively remove propagules of pathogens from water, filters with pore sizes of 0.1–0.2 pm must be used. This means that a large amount of pump energy is required to push water through the filter. It also means that filter membranes must be cleaned frequently, unless the water supply is very clean. Unclean water can be pasteurized in the same way as soil and potting media. Heat from a steam sterilizer used to treat soil or potting media is often also used to treat water supplies. Chlorination is perhaps the cheapest and most reliable means of disinfesting water supplies. Three common sources of chlorine––calcium hypochlorite (chloride of lime), sodium hypochlorite, and chlorine gas––are probably the most economic and convenient. Chlorine reacts with water to produce hypochlorous acid, which is the active material. As such it kills propagules of many pathogens as well as algae and oxidizes organic matter and substances such as iron and manganese into easily removable compounds, thereby preventing them from precipitating and blocking the system. For effective disinfestation, a minimum concentration of 5 ppm free chlorine for 20 min is recommended. Bromination can also be used to remove contaminating microorganisms from water supplies. Bromine hydrolyses in water, in a similar way to chlorine, gives hypobromous acid, which effectively disinfests water at much higher pHs than chlorine. Ozone, an unstable gas with a characteristic pungent odor, can also be used to disinfest water in a process known as ozonation. The oxidation potential of ozone is twice that of chlorine. It also acts more rapidly and is less affected than chlorine by pH and temperature.

Epiphytic and endophytic bacteria can cause severe losses to micropropagated plants at each stage of growth (Leifen et al., 1991). Preventing or avoiding microbial contamination of plant tissue cultures is critical to successful micropropagation.

4.4.3 Eradication

Eradication, the process of removing all infected plant material in a specific area, usually involves the destruction of large numbers of infested plants. Eradication of both pathogen and disease is required for quarantine pathogens but often fails through (i) incomplete pathogen eradication (hidden foci), (ii) natural reinvasion, and (iii) reintroduction through short- or long-distance movement of infected material from contaminated areas (Janse and Wenneker, 2002). Although some diseases have been effectively eradicated after outbreaks, the time between introduction and identification of a disease in a new area can allow inoculum to accumulate to the point where eradication or even containment are impossible. Effective quarantine practices are thus especially important. Typical measures include destruction of the diseased plants, elimination of alternate host plants, pruning, disinfection, and heat treatments (Ogle, 1997).

The domestic quarantine policies of many countries have achieved notable successes in eradicating serious pathogens that affect important crop industries; for example, Australia has successfully eradicated bacterial citrus canker. The disease is a generic term that includes a number of diseases of citrus caused by *X. axonopodis* pv. *citri*. A small outbreak of citrus canker on Thursday Island in the Torres Strait in 1984 was eliminated by cutting down and incinerating trees with canker-like symptoms (5 of 200 trees inspected), burning all leaf trash, and igniting vegetation beneath infected trees with a pneumatic flame gun. The major expenditure of the US 40,000 eradication program was for survey work in remote locations (Ogle, 1997).

Successful eradication of *R. solanacearum* was achieved in 1972 in Sweden, where infections had been found in fields irrigated with river water later found to be contaminated by *R. solanacearum*; latent infection occurred on *Solanum dulcamara* (bittersweet) plants growing along the river with their roots and parts of their stems in the water. The source of contamination/infection appeared to be two potato-processing plants that used potatoes obtained from known brown-rot-contaminated areas and then dumped unprocessed waste and wastewater into the river. After six years of measures (e.g., eradication of bittersweet, prohibition of irrigation with surface water, taking contaminated fields out of potato production for a number of years, disinfection of premises and machines), the disease was eradicated (Olsson, 1976a,b; Persson, 1998).

Another pathogen that appears to have been successfully eradicated from Australia is the Moko disease bacterium. Moko disease or bacterial wilt of banana is caused by *R. solanacearum* biovar 1 race 2. The bacterium was introduced into Australia in 1989 from Hawaii with plants of Heliconia, a close relative of banana, but was not detected during the required three-month post-entry quarantine. When disease symptoms developed later, in the field, rapid quarantine action was taken to contain and eradicate the bacterium. All Heliconia plants in the consignment and banana plants within a 1 km radius of the contaminated site were destroyed. Close monitoring of affected sites has failed to detect the bacterium since then (Ogle, 1997).

4.5 AN OVERVIEW OF QUARANTINE AND ERADICATION OF FIRE BLIGHT DISEASE

Fire blight disease, which is caused by *E. amylovora*, is a very destructive bacterial pathogen to the Rosaceae family (e.g., apple, pear, quince). In 1991, related economic losses in the U.S. were estimated to be US $3.8 million in the state of Michigan alone (van der Zwet ve Beer, 1995). Other domestic losses included US $68 million in Washington and Oregon states in 1998 (Johnson ve Stockwell, 1998; Bonn and Van der Zwet, 2000). The annual loss suffered by the fruit industry due to the disease in the U.S. was determined to be US $100 million (Norelli et al., 2003). Losses in other countries include NZ$10 million in New Zealand (Vanneste, 2000), US$9 million in Sweden between 1997 and 2000 (Hasler et al., 2002), US$1.6 million in Constance lake locality, Germany in 2007 (Scheer, 2009), and US$148 million in Canada (Anonymous, 2010). Because fire blight, 5000 pome fruit trees died in Italy (Calzolari et al., 2000).

Erwinia amylovora is a quarantine pest whose introduction is so widely prohibited that all countries, even those where the disease exists, have imposed restrictions and ask for phytosanitary certificates for the introduction of susceptible host plants. All plant organs except seeds are considered to be potential sources of dissemination, but it is widely accepted that fruits present an insignificant risk in practice. There is no adequate chemical or other treatment for the elimination of the pathogen from plant material that does not destroy the plant tissues.

The disease is a major threat for the EPPO region, and *E. amylovora* is one of the most important pests on the EPPO A2 list (OEPP/EPPO, 2014). It is also considered to be a quarantine pest by COSAVE and IAPSC, and by numerous uninfested countries around the world (e.g., Australia, Japan). It presents a risk to the pear and apple industries as well as to the nursery trade, because many ornamental species are susceptible hosts. The presence of *E. amylovora* fire blight in a country is a major constraint for export trade of plants that host fire blight. For the Mediterranean region, the risks are more serious because of the favorable climatic conditions for disease development and the existence of self-rooted wild hosts. The disease has inflicted very severe damage in the Mediterranean countries where it has occurred (Psallidas, 1990); its damage to Mediterranean ecosystems is unpredictable.

The only sure method for preventing or postponing the spread of *E. amylovora* into uninfested areas is to impose strict phytosanitary measures on imported host-plant material and to maintain vigilance in orchards and nurseries. The EPPO (OEPP/EPPO, 1990) recommends countries at high risk to prohibit importation of host plants for planting. However, an exception can be made for importation during the winter months, in which case consignments should come from an area where *E. amylovora* does not occur, or from an area found by the EPPO to be free from the pest during the last growing season and where an official control campaign has minimized spread. To reduce the risk of spread in international trade, it is recommended that countries (even those where *E. amylovora* occurs) require area freedom or growing-season inspection.

Effective management of fire blight is multifaceted and largely preventative. For control of the disease, an integrated program of chemical control combined with sanitation, pruning, eradication, tree nutrition, and use of resistant or tolerant cultivars is recommended. Warning systems based mainly on climatic data have been developed for successful and economic control of the disease (Thomson et al., 1982; Lightner and Steiner, 1990; Billing, 2000). Sufficient evidence seems to indicate that fire blight can be spread into an area where the disease has not previously been found. As a modified form of quarantine in some areas, a common practice is to avoid planting both apple and pear trees in the same orchard. In the early 1920s, laws were passed in America to prevent pear trees from being planted within 1.5–2.5 km of an apple orchard. Where this was done, less fire blight occurred in the apple orchards. Similarly, because the proximity of hawthorn to pear and apple has increased the incidence and severity of fire blight, removal of hawthorn hedges in or near orchards is suggested. The most effective quarantine regulation is strict enforcement to prohibit importation of all fruit, seed, budwood, and other plant parts of all rosaceous plants from any country with fire blight. Soon after the introduction of fire blight into New Zealand, in about 1919, the Commonwealth of Australia passed a law prohibiting the importation of any plant parts in the family Rosaceae from any country where fire blight was present. This enforcement continues and, together with the country's isolated location, undoubtedly has contributed to the absence of fire blight in Australia today. Since the introduction of fire blight into Western Europe, strict quarantine regulations have been adopted by Norway, Sweden, and Switzerland (Van der Zwet and Keil, 1979).

Roberts and Reymond (1989) reported that citrate buffer, benzalkonium chloride, and sodium hypochlorite gave significant reductions in surviving *E. amylovora* cells on apple fruits. Van der Zwet et al. (1990) reported that apple fruits collected from apparently healthy trees or harvested a minimum of 100 cm from visible blight symptoms were free from *E. amylovora*, and thus incapable of disseminating the disease to areas or countries without fire blight. Some bacterial plant pathogens can eradicate at a temperature of 60°C, despite some apparent contradictions. Keck et al. (1995) found that *E. amylovora* was eradicated in apple budwood subjected to dry heat for 3 h at 45°C, but not when subjected to moist heat at 50°C for the same period. This result is odd because bacteria are usually killed more effectively by moist heat than dry heat (Turner, 2002).

In summer, established infections are principally controlled by pruning. Effective control through pruning requires cuts of 30–35 cm (12–14 inches; Bastas, 2011) to be made below the visible end of the expanding canker and for the pruning tools to be disinfected between cuts with a bleach or alcohol solution to prevent cut-to-cut transmission. Repeated trips through an orchard are necessary, as some infections are invariably missed and others become visible at later times. Prunings that harbor the pathogen are usually destroyed by burning. Application of these simple steps has indicated that fire blight can be controlled by pruning even under severe disease conditions. Although it is impractical to undertake such procedures in large orchards, pruning can help control the disease. However, summer pruning of apples in early June markedly enhanced infection of fire blight, and ringing the limbs of certain apple cultivars to induce early bearing resulted in considerable blight infection of the knife-cut girdle wounds. In some fruit areas, pruning blighted branches is particularly helpful when the practice supplements a chemical control schedule. Pruning tools should be dipped in disinfectant between cuts to prevent the spread of the bacteria. Several chemicals, including solutions of the corrosive sublimate mercury cyanide, various denatured alcohols, and the common household sodium hypochlorite, have been used successfully to decontaminate pruning tools. Sodium hypochlorite has been widely used in recent years because it is economical, easily accessible to the average household, and kills the bacteria when the tool is dipped in a 10% solution for about 5 s (Van der Zwet and Keil, 1979).

Once the disease has become established in orchards or on wild hosts, eradication measures have proved ineffective to prevent its spread in a given area. In newly infested areas, eradication is generally attempted despite past experience, simply because of the great importance of the disease. In a few cases, isolated imported nursery plants have been found to be infected and have been destroyed in enough time to prevent establishment. An extreme example occurred in 1906, when a nursery block of 10,000 pear trees was completely destroyed by fire blight. Holdover blight cankers present in the stock trees and pruning tools had apparently become contaminated and subsequently transmitted the disease to nearly every tree (Van der Zwet and Keil, 1979).

In addition to blighted twigs and branches, serious body blight in the form of cankers in large limbs and the trunk sometimes develops. Such blight infection often causes loss of the tree unless the bacteria in these cankers are eliminated or inactivated before the blossoms open. For eradication, cankers are scraped or scarified with a sharp-edged tool to allow them to dry out. They are usually also painted with poisonous solutions that kill the blight bacteria. Sometimes careful surgery is practiced whereby the diseased bark is removed, followed by proper disinfection. With this method, the cut should be made several centimeters into the apparently unaffected wood of the canker margin. In some countries, disinfectants are painted over the surface of the canker without any surgery. This method requires the least time and has proved effective in some orchards. Of the many chemicals tried in canker paint formulas, only a few have proved to be effective and relatively noninjurious to healthy wood. Combinations prepared by dissolving zinc chloride in denatured alcohol, water, and hydrochloric acid have been used extensively. Experimental studies by Gardiner (1957) using paints containing 550 ppm of streptomycin indicated that weekly applications on cankers in the spring and early summer may stop further development of the cankers and keep them inactive until they finally dry out. Coyier (1969) reported limited control of fire blight infection in cankers with 2,4-xylenol plus *m* cresol (Bacticin) developed for treatment of crown gall in various fruit and ornamental trees.

Eradication measures should also include the elimination of all susceptible wild host plants that are adjacent to apple and pear orchards. Any susceptible escaped wild plants that harbor the fire blight organism may serve as sources of inoculum for orchard trees.

4.6 CONCLUSION

The available detection and diagnostic techniques for plant pathogenic bacteria have been microscopic observation, isolation; biochemical, physiological and molecular characterization; serology, bioassays, and pathogenicity tests. Early detection in seeds and planting materials, and ensuring disease-free planting materials through rapid diagnostics are likely the effective means of reducing bacterial disease incidence. Phytopathogenic bacteria induce characteristic leaf spots, blights, scabs, cankers, and tumors, as well as rotting and wilting of whole plants. The selection of appropriate diagnostic methods is critical with regard to eradication, certification, and sanitation or quarantine programs, and to evaluate the incidence or spread of a disease. Technological advances in PCR-based methods, which enable rapid, accurate detection, quantification, and characterization of plant pathogens, are now being applied to solve practical problems. The current trend, EPPO protocols for the detection of plant pathogens, is to combine conventional, serological, and molecular techniques in integrated approaches. Because seedborne pathogens present a serious threat to seedling establishment, seed-health assays should be sensitive, specific, rapid, robust, inexpensive, and simple to implement and interpret. The efficient control of bacterial plant diseases has been based on determinations of pathogen distribution and spread, establishment of epidemiological risk factors, and taking action by monitoring water, using safe sources of water, and imposing safe disposal of any plant waste that may be contaminated with the pathogen. In addition, inspections should be made by importing countries and production and use of healthy planting material should be ensured by indexing and, for latent infections, developing/using resistant varieties as well as chemo- or thermotherapy of basic planting material.

Quarantine is the establishment by public authorities of protective barriers against the dissemination of injurious pathogens. Acceptable risk is the point at which the decision maker determines that the potential risks posed to a nation by a proposed import are small enough to be manageable. Certification schemes ensure that seed- and vegetative-propagating material are free from particular diseases. Sanitation is a basic requirement for preventing the spread of disease to new areas or, on a smaller scale, to new plants or plant products. Eradication, which is the process of removing all infected plant material in an area, usually involves the destruction of large numbers of plants in infested areas. The domestic quarantine policies of many countries have achieved notable successes in eradicating serious pathogens that affect important crop industries. When an integrated program is used, effective management of bacterial diseases is multifaceted and largely preventative.

REFERENCES

Abriouel, H., Ben Omar, N., Lucas López, R., Martínez Canamero, M., Ortega, E., Galvez, A. 2007. Differentiation and characterization by molecular techniques of *Bacillus cereus* group isolates from poto poto and degue, two traditional cereal-based fermented foods of Burkina Faso and Republic of Congo. *Journal of Food Protection* 70, 1165–1173.

Agrios, G.N. 2005. *Plant Pathology*, 5th ed.. Elsevier Academic Press, 903p.

Alvarez, A.M. 2001. Differentiation of bacterial populations in seed extracts by flow cytometry. In *Plant Pathogenic Bacteria*, De Boer, S.H. (ed.). Heidelberg: Springer Netherlands, pp. 393–396.

Alvarez, A.M. 2004. Integrated approaches for detection of plant pathogenic bacteria and diagnosis of bacterial diseases. *Annual Review of Phytopathology* 42: 339–366.

Anonymous. 2002. *Specification for Composted Materials PAS 100*. London: British Standards Institution.

Anonymous. 2006. Commission Directive 2006/63/EC of July 14, 2006 amending annexes II to VII to Council Directive 98/57/EC on the control of *Ralstonia solanacearum* (Smith) Yabuuchi et al. *Official Journal of European Union* L206: 36–106.

Anonymous. 2010. http://www.statcan.gc.ca/pub/22-003-x/2011001/t029-eng.pdf. 2010.

Anthony, R.M., Brown, T.J., French, G.L. 2000. Rapid diagnosis of bacteremia by universal amplification of 23S ribosomal DNA followed by hybridization to an oligonucleotide array. *Journal of Clinical Microbiology* 38: 781–788.

Ausubel, F.M., Bent, R., Kingston, R.E., Moore, D.J., Smith, J.A., Silverman, G., Struhl, K. 1987. *Current Protocols in Molecular Biology*. New York: John Wiley.

Barak, J.D., Gilbertson, R.L. 2003. Genetic diversity of *Xanthomonas campestris* pv. *vitians*, the causal agent of bacterial leafspot of lettuce. *Phytopathology* 93: 596–603.

Bastas, K.K. 2011. An integrated management program for fire blight disease on apples. *HortScience* 46(9) (suppl.). In *ASHS Annual Conference*, Waikoloa, Hawaii, S129.

Bej, A.K. 1999. Detection of microbial nucleic acids by polymerase chain reaction in aquatic samples. In *Molecular Microbial Ecology Manual*. Akkermans, A.D.L., Van Elsas, J.D., Bruijn, F.J. (eds.). Dordrecht, The Netherlands: Kluwer Academic Publishers, pp. 1–49.

Benedict, A.A., Alvarez, A.M., Berestecky, J., Imanaka, W., Mizumoto, C.Y., Pollard, L.W., Mew, T.W., Gonzalez, C.F. 1989. Pathovar-specific monoclonal antibodies for *Xanthomonas campestris* pv. *oryzae* and for *Xanthomonas campestris* pv. *oryzicola*. *Phytopathology* 79: 322–328.

Benedict, A.A., Alvarez, A.M., Pollard, L.W., 1990. Pathovar specific antigens of *Xanthomonas campestris* pv. *begoniae and X. campestris* pv. *pelargonii* detected with monoclonal antibodies. *Applied and Environmental Microbiology* 56, 572–574.

Bereswill, S., Pahl, A., Bellemann, P., Berger, F., Zeller, W., Geider, K. 1992. Sensitive and species-specific detection of *Erwinia amylovora* by polymerase chain reaction analysis. *Applied and Environment Microbiology* 58: 3522–3526.

Berg, T., Tesoriero, L., Hailstone, D.L. 2005. PCR-based detection of *Xanthomonas campestris* pathovers in Brassica seed. *Plant Pathology* 54: 416–427.

Berg, T., Tesoriero, L., Hailstones, D.L. 2006. A multiplex real-time PCR assay for detection of *Xanthomonas campestris* from brassicas. *Letters in Applied Microbiology* 42: 624–630.

Bertolini, E., Olmos, A., López, M.M., Cambra, M. 2003a. Multiplex nested reverse-transcription polymerase chain reaction in a single tube for sensitive and simultaneous detection of four RNA viruses and *Pseudomonas savastanoi* pv. *savastanoi* in olive trees. *Phytopathology* 93: 286–292.

Bertolini, E., Penyalver, R., García, A., Olmos, A., Quesada, J.M., Cambra, M., López, M.M. 2003b. Highly sensitive detection of *Pseudomonas savastanoi* pv. *savastanoi* in asymptomatic olive plants by nested PCR in a single closed tube. *Journal of Microbiological Methods* 52: 261–266.

Bertolini, E., Torres, E., Olmos, A., Martín, M.P., Bertaccini, A., Cambra, M. 2007. Cooperational PCR coupled with dot blot hybridization for detection and 16SrX grouping of phytoplasmas. *Plant Pathology* 56: 677–682.

Bextine, B., Blua, M., Harshman, D., Miller, T.A. 2005. A SYBR green-based real-time polymerase chain reaction protocol and novel DNA extraction technique to detect *Xylella fastidiosa* in *Homalodisca coagulata*. *Journal of Economic Entomology 3*: 667–672.

Billing, E. 2000. Fire Blight risk assessment systems and models. In *Fire Blight: The Disease and its Causative Agent, Erwinia Amylovora.* Vanneste, J.L. (ed.). Oxon and New York: CABI Publishing, pp. 293–318.

Biosca, E.G., Gonzalez, R., López-López, M.J., Soria, S., Monton, C., Perez-Laorga, E., López, M.M. 2003. Isolation and characterization of *Brenneria quercina*, causal agent of bark canker and drippy nut of *Quercus* spp. in Spain. *Phytopathology* 93: 485–492.

Biosca, E.G., Marco-Noales, E., Ordax, M., López, M.M. 2006. Long-term starvation-survival of *Erwinia amylovora* in sterile irrigation water. *Acta Horticulturae* 704: 107–112.

Birch, P.R.J., Hyman, L.J., Taylor, R., Opio, A.F., Bragardm, C., Toth, I.K. 1997. The RAPD PCR-based differentiation of *Xanthomonas campestris* pv. *phaseoli* and *Xanthomonas campestris* pv. *phaseoli* var. fuscans. *European Journal Plant Pathology* 103: 809–814.

Bollen, G.J. 1985. The fate of plant pathogens during composting of crop residues. In *Composting of Agricultural and Other Wastes*, Gasser, J.K.R. (ed.). London: Elsevier Applied Science, pp. 282–290.

Bollen, G.J., Volker, D. 1996. Phytogenic aspects of composting. In *The Science of Composting*. de Bertoldi, M., Sequi, P., Lemmes, B., Papi, T. (eds.). Glasgow, UK: Blackie Academic and Professional, pp. 233–246.

Bonn, W.G., Van der Zwet, T. 2000. Distribution and economic importance of fire blight. In *Fire Blight, the Disease and Its Causative Agent, Erwinia amylovora*. Vanneste, J.L. (ed.). Wallingford, UK: CAB International, pp. 37–54.

Botha, W.J., Serfontein, S., Greyling, M.M., Berger, D.K. 2001. Detection of *Xylophilus ampelinus* in grapevine cuttings using a nested polymerase chain reaction. *Plant Pathology* 50: 515–526.

Braun-Kiewnick, A., Altenbach, D., Oberhansli, T., Bitterlin, W., Duffy, B. 2011. A rapid lateral-flow immunoassay for phytosanitary detection or *Erwinia amylovora* and on-site fire blight diagnosis. *Journal of Microbiological Methods* 87: 1–9.

Calzolari, A., Finelli, F., Bazzi, C., Mazzucchi, U. 2000. Fireblight: Prevention and control in the Emilia-Romagna region. *Rivista di Frutticoltura e di Ortofloricoltora* 62: 23–28.

Capoor, S.P., Rao, D.G., Sawant, D.M. 1986. Seed transmission of French bean mosaic virus. *Indian Phytopathology* 39: 343–345.

Caruso, P., Bertolini, E., Cambra, M., López, M.M. 2003. A new and co-operational polymerase chain reaction (Co-PCR) for rapid detection of *Ralstonia solanacearum* in water. *Journal of Microbiological Methods* 55(1): 257–272.

Chen, Y.F., He, L.Y., Xu, J. 2003. RAPD analysis and group division of *Ralstonia solanacearum* strains in China. *Acta Phytopathologica Sinica* 33: 503–508.

Christensen, K.K., Carlsbaek, M., Kron, E. 2002. Strategies for evaluating the sanitary quality of composting. *Journal of Applied Microbiology* 92: 1143–1158.

Clair, D., Larrue, J., Aubert, G., Gillet, J., Cloquemin, G., Boudon-Padieu, E. 2003. A multiplex nested-PCR assay for sensitive and simultaneous detection and direct identification of phytoplasma in the Elm yellows group and Stolbur group and its use in survey of grapevine yellows in France. *Vitis* 42: 151–157.

Cockerill, F.R., Smith, T.F. 2002. Rapid-cycle real-time PCR: A revolution for clinical microbiology. *American Society For Microbiology News* 68: 77–83.

Coletta-Filho, H.D., Takita, M.A., de Souza, A.A., Neto, J.R., Destefano, S.A.L., Hartung, J.S., Machado, M.A. 2006. Primers based on the *rpf* gene region provide improved detection of *Xanthomonas axonopodis* pv. *citri* in naturally and artificially infected citrus plants. *Journal of Applied Microbiology* 100: 279–285.

Coyier, D.L. 1969. Inoculatıon of apple and pear roots with *E. amylovora*: control of fire blight with bacticin. First Workshop on Fire Blight Research, University of Missouri, Columbia, pp. 57–58.

Crocetti, G.R., Hugenholtz, P., Bond, P.L., Schuler, A., Keller, J., Jenkins, D., Blackall, L.L. 2000. Identification of polyphosphate-accumulating organisms and design of 16S rRNA-directed probes for their detection and quantitation. *Applied and Environmental Microbiology* 66: 1175–1182.

Danks, C., Barker, I. 2000. On-site detection of plant pathogens using lateral flow devices. *Bulletin of the OEPP/EPPO* 30: 421–426.

Darrasse, A., Kotoujansky, A., Bertheau, Y. 1994. Isolation by genomic subtraction of DNA probes specific for *Erwinia carotovora* subsp. *atroseptica. Applied and Environmental Microbiology* 60: 298–306.

Davey, H.M., Kell, D.B. 1996. Flow cytometry and cell sorting of heterogeneous microbial populations: The importance of single cell analyses. *Microbiological Review* 60: 641–696.

De Bellis, P., Schena, L., Cariddi, C. 2007. Realtime Scorpion-PCR detection and quantification of *Erwinia amylovora* on pear leaves and flowers. *European Journal of Plant Pathology* 118: 11–22.

De Boer, S.H., Elphinstone, J.G., Saddler G. 2007. Molecular detection strategies for phytopathogenic bacteria. In *Biotechnology and Plant Disease Management*. Punja, Z.K., De Boer, S.H., Sanfançon, H. (eds.). Oxfordshire, UK: CAB International, pp. 165–194.

De Boer, S.H., López, M.M. 2012. New grower-friendly methods for plant pathogen monitoring. *Annu. Rev. Phytopathol* 50:197–218.

De Boer, S.H., Wieczorek, A. 1984. Production of monoclonal antibodies to *Corynebacterium sepedonicum. Phtopathology* 74: 1431–1434.

de Lima, J.E.O., Miranda, V.S., Hartung, J.S., Brlansky, R.H., Coutinho, A., Roberto, S.R., Carlos, E.F. 1998. Coffee leaf scorch bacterium: Axenic culture, pathogenicity and comparison with *Xylella fastidiosa* in citrus. *Plant Disease* 82: 94–97.

Didenko, V.D. 2001. DNA probes using fluorescence resonance energy transfer (FRET): Designs and applications. *Biotechniques* 31: 1106–1120.

EPPO. 2004a. Diagnostic protocol for regulated pests. *Citrus tristeza closterovirus. EPPO Bulletin* 34:239–246.

EPPO. 2004b. Diagnostics protocols for regulated pests. *Ralstonia solanacearum. Bulletin of the OEPP/EPPO* 34:173–178.

Espy, M.J., Uhl, J.R., Sloan, L.M., Buck-Walter, S.P., Jones, M.F., Vetter, E.A., Yao, J.D.C. et al. 2006. Real-time PCR in clinical microbiology application testing. *Clinical Microbiology Reviews* 19:165–256.

Fegan, M., Croft, B.J., Teakle, D.S., Hayward, A.C., Smith, G.R. 1998. Sensitive and specific detection of *Clavibacter xyli* subsp. *xyli*, causal agent of ratoon stunting disease of sugarcane, with a polymerase chain reaction-based assay. *Plant Pathology* 47: 495–504.

Fessehaie, A.S.H., Boer, D., Levesque, C.A. 2003. An oligonucleotide array for the identification and differentiation of bacteria pathogenic on potato. *Phytopathology* 93: 262–269.

Franke-Whittle, I.H., Klammer, S.H., Insam, H. 2005. Design and application of an oligonucleotide microarray for the investigation of compost microbial communities. *Journal of Microbiological Methods* 62: 37–56.

Franken, A.A.J.M. 1992. Comparison of immunofluorescence microscopy and dilution plating for the detection of *Xanthomonas campestris* pv. *campestris* in crucifer seeds. *Netherlands Journal of Plant Pathology* 98: 169–178.

Gardan, L., Moreno, B., Guinebretiere, J.P. 1984. The use of a micromethod of auxanogram, and data analysis for identification of *Pseudomonas*. Proceedings of the 2nd Working Group, Athens Greece, pp. 37–39.

Gardiner, J.C. 1957. Control of fire blight on apple and pear. Canada Dept. Agr. Ann. Rpt. 1951–52: 31; 1953–54: 34; 1954–55: 32–33; 1955–56: 36; (1956–57): 40–41.

Gorris, M.T., Cambra, M., Llop, P., López, M.M., Lecomte, P., Chartier, R., Paulin, J.P. 1996. A sensitive and specific detection of *Erwinia amylovora* based on the ELISA DASI enrichment method with monoclonal antibodies. *Acta Horticulturae* 411: 41–45.

Grey, B.E., Steck, T.R. 2001. The viable but nonculturable state of *Ralstonia solanacearum* may be involved in long-term survival and plant infection. *Applied and Environmental Microbiology* 67: 3866–3872.

Grover, A, Azmi, W., Gadewar, A.V., Pattanayak, D., Naik, P.S., Shekhawat, G.S., Chakrabarti, S.K. 2006. Genotypic diversity in a localized population of *Ralstonia solanacearum* as revealed by random amplified polymorphic DNA markers. *Journal of Applied Microbiology* 101: 798–806.

Haas, J.H., Moore, L.W., Ream, W., Manulis, S. 1995. Universal PCR primers for detection of phytopathogenic *Agrobacterium* strains. *Applied and Environmental Microbiology* 61: 2879–2884.

Hartung, J.S., Daniel, J.F., Pruvost, O.P. 1993. Detection of *Xanthomonas campestris* pv. *citri* by the polymerase chain reaction. *Applied and Environmental Microbiology* 59: 1143–1148.

Hasler, T., Schaerer, H.J., Holliger, E., Vogelsanger, J., Vignutelli, A., Schoch, B. 2002. Fire blight situation in Switzerland. *Acta Horticulturae (ISHS)* 590, 73–79.

Helias, V., Andrivon, D., Jouan, B. 2000. Internal colonization pathways of potato plants by *Erwinia carotovora* ssp. *atroseptica*. *Plant Pathology* 49: 33–42.

Janse, J.D. 1996. Potato brown rot in Western Europe—History, present occurrence and some remarks on possible origin, epidemiology and control strategies. *Bulletin of the OEPP/EPPO* 26: 679–685.

Janse, J.D. 2005. Standardization, validation and approval of test methods for quarantine bacteria: Examples of harmonization in plant health laboratories in Europe. *Phytopathologica Polonica* 35: 19–27.

Janse, J., van den Beld, H.E., Elphinstone, J., Simpkins, S., Tjou-Tam-Sin, N.N.A., Van Vaerenbergh, J. 2004. Introduction to Europe of *Ralstonia solanacearum* biovar 2, race 3 in *Pelargonium zonale* cuttings. *Journal of Plant Pathology* 86: 147–155.

Janse, J.D., Wenneker, M. 2002. Possibilities of avoidance and control of bacterial plant diseases when using pathogen-tested (certified) or -treated planting material. *Plant Pathology* 51: 523–536.

Jensen, M.A., Webster, J.A., Straus, N. 1993. Rapid identification of bacteria on the basis of polymerase chain reaction-amplified ribosomal DNA spacer polymorphisms. *Applied and Environmental Microbiology* 59: 945–952.

Johnson, K.B., Stockwell, V.O. 1998. Management of fire blight: A case study in microbial ecology. *Annual Review of Phytopathology* 36: 227–248.

Johnston, W.H., Stapleton, R., Sayler, G.S. 1996. Direct extraction of microbial DNA from soils and sediments. In *Molecular Microbial Ecology Manual*. Akkermans A.D.L., van Elsas J.D., de Bruijn F.J. (eds). Dordrecht, the Netherlands: Kluwer Academic Publishers, vol. 1.3.2, pp. 1–9.

Keck, M., Chartier, R., Zislavsky, W., Lecomte, P., Paulin, J.P. 1995. Heat treatment of plant propagation material for the control of fire blight. *Plant Pathology* 44: 124–129.

Kerkoud, M., Manceau, C., Paulin, J.P. 2002. Rapid diagnosis of *Pseudomonas syringae* pv. *papulans*, the causal agent of blister spot of apple, by polymerase chain reaction using specifically designed *hrpL* gene primers. *Phytopathology* 92: 1077–1083.

Khan, A.A., Furuya, N., Matsumoto, M., Matsuyama, N. 1999. Trial for rapid identification of pathogens from blasted pear blossoms and rotted radish leaves by direct TLC and whole cellular fatty acid analysis. *Journal of the Faculty of Agriculture Kyushu University* 43: 327–335.

Kiltie, A.E. Ryan, A.J. 1997. SYBR Green I staining of pulsed field agarose gels is a sensitive and inexpensive way of quantitating DNA double-strand breaks in mammalian cells. *Nucleic Acids Research* 25: 2945–2946.

King, E.O., Ward, M.K., Raney, D.E. 1956. Two simple media for the demonstration of pyocyanin and fluorescin. *Journal of Laboratory Clinical Medicine* 44: 301–307.

Klinger, J.M., Stowe, R.P., Obenhuber, D.C., Groves, T.O., Mishra, S.K., Pierson, D.L. 1992. Evaluation of the Biolog automated microbial identification system. *Applied and Environmental Microbiology* 58:2089–2092.

Koide, T., Paulo, A.Z., Leandro, M.M., Ricardo, Z.N., Vencio, A.Y., Matsukuma, A.M.D., Diva, C.T. et al. 2004. DNA microarray-based genome comparison of a pathogenic and a nonpathogenic strain of *Xylella fastidiosa* delineates genes important for bacterial virulence. *Journal of Bacteriology* 186: 5442–5449.

Kousik, C.S., Ritchie, D.F. 1998. Response of bell pepper cultivars to bacterial spot pathogen races that individually overcome major resistance genes. *Plant Disease* 82: 181–186.

Kubota, R., Alvarez, A.M., Vine, B.G., Jenkins, D.M. 2007. Development of a loop mediated isothermal amplification method (LAMP) for detection of the bacterial wilt pathogen. *Ralstonia solanacearum. (Abstract) Phytopathology* 97: S60.

Kuo, T.T., Huang, T.C., Wu, R.Y., Yang, C.M. 1967. Characterization of three bacteriophages of *Xanthomonas oryzae* (Uyeda et Ishiyama) Dowson. *Botanical Bulletin Academia Sinica* 8:246–254.

Kurian, K.M., Watson, C.J., Wyllie, A.H. 1999. DNA chip technology. *Journal of Pathology* 187: 267–271.

Lee, I.M., Bartoszyk, I.M., Gundersen, D.E., Mogen, B., Davis, R.E. 1997. Nested PCR for ultrasensitive detection of the potato ring rot bacteria *Clavibacter michiganensis* subsp. *sepedonicus*. *Applied and Environmental Microbiology* 63: 2625–2630.

Lee, K.W., Lee, B.C., Park, H.C., Lee, Y.S. 1990. Occurrence of green mottle mosaic virus disease of watermelon in Korea. *Korean Journal of Plant Pathology* 6: 250–255.

Leege, P.B., Thompson, W. H. 1997. *Test Methods for the Examination of Composting and Compost*. Bethesda, MD: The US Composting Council.

Leifen, C., Camota, H., Wright, S.M., Waites, B., Cheyne, V.A., Waites, W.M. 1991. Elimination of *Laclobacillus planlarum, Corynebacterium* spp. *Staphylococcus saprophyticus* and *Pseudomonas paucimobilis* from micropropagate. Hemerocallis, Choisya and Delphinium cultures using antibiotics. *Journal of Applied Bacteriology* 71:307–330.

Lemieux, B., Aharoni, A., Schena, M. 1998. Overview of DNA chip technology. *Molecular Breeding* 4: 277–289.

Li, W., Hartung, J.S., Levy, L. 2006. Quantitative real-time PCR for detection and identification of *Candidatus Liberibacter* species associated with citrus huanglongbing. *Journal of Microbiological Methods* 66: 104–115.

Li, X., De Boer, S. 1995. Selection of polymerase chain reaction primers from an RNA intergenic spacer region for specific detection of *Clavibacter michiganensis* subsp. *sepedonicus*. *Phytopathology* 85: 837–842.

Liao, X., Zhu, S., Zhao, W., Luo, K., Qi, Y., Chen, H., He, K., Zhu, X. 2004. Cloning and sequencing of citrus Huanglongbing pathogen 16S rDNA and its detection by real-time fluorescent PCR. *Journal of Agricultural Biotechnology* 12: 80–85.

Lightner, G.W., Steiner, P.W. 1990. Computerization of blossom blight prediction model. *Acta Horticulturae* 273: 171–184.

Liu, Y., Cady, N.C., Batt, C.A. 2007. A plastic microchip for nucleic acid purification. *Biomedical Microdevices* 9: 769–776.

Llop, P., Bonaterra, A., Penalver, J., López, M.M. 2000. Development of a highly sensitive nested-PCR procedure using a single closed tube for detection of *Erwinia amylovora* in asymptomatic plant material. *Applied and Environmental Microbiology* 66: 2071–2078.

López, M.M., Bertolini, E., Caruso, P., Penyalver, R., Marco-Noales, E., Gorris, M.T., Morente, C. et al. 2005. Advantages of an integrated approach for diagnosis of quarantine pathogenic bacteria in plant material. *Phytopathol. Polonica* 35: 49–56.

López, M.M., Bertolini, E., Olmos, A., Caruso, P., Gorris, M.T., Llop, P., Penyalver, R., Cambra, M. 2003. Innovative tools for detection of plant pathogenic viruses and bacteria. *International Microbiology* 6: 233–243.

Lopez, R., Asensio, C., Guzman, M.M., Boonham, N. 2006. Development of real-time and conventional RT-PCR assays for the detection of Potato yellow vein virus (PYVV). *Journal of Virological Methods* 136: 24–9.

Louws, F.J., Fulbright, D.W., Stephens, C.T., de Bruijn, F.J. 1994. Specific genomic fingerprints of phytopathogenic *Xanthomonas* and *Pseudomonas* pathovars and strains generated with repetitive sequences and PCR. *Applied and Environmental Microbiology* 60:2286–2295.

Louws, F.J., Rademaker, J.L.W., Brujin, F.J. 1999. The three Ds of PCR-based genomic analysis of phytobacteria: Diversity, detection, and disease diagnosis. *Annual Review of Phytopathology* 37: 81–125.

Loy, A., Lehner, A., Lee, N., Adamczyk, J., Meier, H., Ernst, J., Schleifer, K.H., Wagner, M. 2002. Oligonucleotide microarray for 16S rRNA gene based detection of all recognized lineages of sulfate-reducing prokaryotes in the environment. *Applied and Environmental Microbiology* 68: 5064–5081.

Loy, A., Schulz, C., Lucker, S., Schopfer-Wendels, A., Stoecker, K., Baranyi, C., Lehner, A., Wagner, M. 2005. 16S rRNA gene-based oligonucleotide microarray for environmental monitoring of the betaproteobacterial order "Rhodocyclales." *Applied and Environmental Microbiology* 71: 1373–1386.

Lydon, J., Patterson, C.D. 2001. Detection of tabtoxin-producing strains of *Pseudomonas syringae* by PCR. *Letters in Applied Microbiology* 32: 166–170.

Maes, B.G., Niu, J.X., Morley-Bunker, M., Pan, L.Z., Maes, M., Garbeva, P., Crepel, C. 1996. Identification and sensitive endophytic detection of the fireblight pathogen *Erwinia amylovora* with 23S ribosomal DNA sequences and the polymerase chain reaction. *Plant Pathology* 45: 1139–1149.

Mackay, L.M., Arden, K.E., Nitsche, A. 2002. Survey and summary of real time PCR in virology. *Nucleic Acids Research* 30:1292–1305.

Mahmoudi, E., Soleimani, M.J., Taghavi, M. 2007. Detection of bacterial soft rot of crown imperial caused by *Pectobacterium carotovorum* subsp. *carotovorum*, using specific primers. *Phytopathologia Mediterranea* 46:168–176.

Mahuku, G.S., Jara, C., Henriquez, M.A., Castellanos, G., Cuasquer, J. 2006. Genotypic characterization of the common bean bacterial blight pathogens, *Xanthomonas axonopodis* pv. *phaseoli* and *Xanthomonas axonopodis* pv. *phaseoli* var. *fuscans* by rep-PCR and PCR-RFLP of the ribosomal genes. *Journal of Phytopathology* 154: 1–35.

Manceau, C., Horvais, A. 1997. Assessment of genetic diversity among strains of *Pseudomonas syringae* by PCR-restriction fragment length polymorphism analysis of rRNA operons with special emphasis on *P. syringae* pv. *tomato*. *Applied and Environmental Micriobiology* 63: 498–505.

Manter, D.K., Delgado, J.A., Holm, D.G., Stong, R.A. 2010. Pyrosequencing reveals a highly diverse and cultivar-specific bacterial endophyte community in potato roots. *Microbial Ecology* 60:157–166.

Martin, B., Humbert, O., Camara, M., Guenzi, E., Walker, J., Mitchell, T., Andrew, P. et al. 1992. A highly conserved repeated DNA element located in the chromosome of *Streptococcus pneumoniae*. *Nucleic Acids Research* 20: 3479–3483.

Martin, R.R., James, D., Levesque, C.A. 2000. Impacts of molecular diagnostic technologies on plant disease management. *Annual Review of Phytopathology* 38: 207–239.

Matsuyama, N., Daikohara, M., Yoshimura, K., Manabe, K., Khan, M.A.A., de Melo, M.S., Negishi, H. et al. 2009. Presumptive differentiation of phytopathogenic and non-pathogenic bacteria by improved rapid-extraction TLC method. *Journal of the Faculty of Agriculture, Kyushu University* 54: 1–11.

McLaughlin, R.J., Chen, T.A. 1990. ELISA methods for plant pathogenic prokaryotes. In *Serological Methods for Detection and Identification of Viral and Bacterial Plant Pathogens,* Hampton, R., Ball, E., De Boer, S.H. (eds.) St. Paul, MN: APS Press, pp. 197–204.

McManus, P.S., Jones, A.L. 1995. Detection of *Erwinia amylovora* by nested PCR and PCR-dot-blot and reverse blot hybridisations. *Phytopathology* 85: 618–623.

Meyer, J., Hoy, M., Singh, R. 2007. Low incidence of *Candidatus Liberibacter asiaticus* in *Diaphorina citri* (hemiptera: Psyllidae) populations between Nov 2005 and Jan 2006: Relevance to management of citrus greening disease in Florida. *Florida Entomologist* 90: 394–397.

Minsavage, G.V., Thompson, C.M., Hopkins, D.L., Leite, R.M.V., Stall, R.E. 1994. Development of a polymerase chain reaction protocol for detection of *Xylella fastidiosa* in plant-tissue. *Phytopathology* 84: 456–461.

Mondal, K.K. 2011. *Plant Bacteriology*. New Delhi: Kalyani Publishers, ISBN: 978-93-272-1631-8, p. 190.

Mondal, K.K., Bhattacharya, R.C., Kaundal, K.R. 2004. Biotechnological strategies in the detection, characterization and management of fungal diseases in plant. *Botanica* 54: 1–20.

Mondal, K.K., Mani, C. 2009. ERIC-PCR generated genomic fingerprints and their correlation with pathogenic variability of *Xanthomonas campestris* pv. *punicae*, the incitant of bacterial blight of pomegranate. *Current Microbiology* 59: 616–620.

Mondal, K.K., V. Shanmugam. 2013. Advancements in the diagnosis of bacterial plant pathogens: An overview. *Biotechnology and Molecular Biology Review* 8 (1): 1–11.

Mondal, K.K., Singh, D., Prasad, R., Mani, C. 2008. Differential characterization of three bacterial blights in legumes based on symptomatological, cultural, pigmentation and molecular profiles. In: *Proceedings of National Symposium on Plant Disease Scenario on Organic Agriculture for Eco-friendly Sustainability, Indian Phytopathological Society*, January 10–12, 2008, Mahabaleshwar, Satara, 69 pp.

Narayanasamy, P. 2001. *Plant Pathogen Detection and Disease Diagnosis*, 2nd edn. New York: Marcel Dekker.

Narayanasamy, P. 2011. Microbial plant pathogens-detection and disease diagnosis. In: *Bacterial and Phytoplasmal Pathogens*. The Netherlands: Springer. 2: 5–169.

Noble, R., S.J. Roberts. 2004. Eradication of plant pathogens and nematodes during composting: A review. *Plant Pathology* 53: 548–568.

Norelli, J.L., Jones, A.L., Aldwinckle, H.S. 2003. Fire blight management in the twenty-first century. *Plant Disease* 87:756–765.

Notomi, T., Okayama, H., Masubuchi, H., Yonekawa, T., Watanabe, K., Amino, N., Hase, T. 2000. Loop-mediated isothermal amplification of DNA. *Nucleic Acids Research* 28: E63.

OEPP/EPPO. 1990. Quarantine procedure 26: *Pseudomonas solanacearum*. *Bulletin of the OEPP/EPPO* 20:255–262.

OEPP/EPPO. 2004a. Diagnostic protocol for regulated pests. *Citrus tristeza closterovirus*. *OEPP/EPPO Bulletin* 34: 239–246.

OEPP/EPPO. 2004b. Diagnostics protocols for regulated pests. *Ralstonia solanacearum*. *Bulletin of the OEPP/EPPO* 34: 173–178.

OEPP/EPPO. 2014. EPPO A1 and A2 lists of quarantine pests (version 2014-09).

Ogle H.J. 1997. Disease management: Exclusion, eradication and elimination. In *Plant Pathogens and Plant Diseases*. Brown, J.I., Ogle, H. (eds). Australia: Rockvale Publications, pp. 358–372.

Oliver, J.D. 2005. The viable but nonculturable state in bacteria. *Journal of Microbiology* 43:93–100.

Olmos, A., Bertolini, E., Cambra, M. 2002. Simultaneous and cooperational amplification (Co-PCR): A new concept for detection of plant viruses. *Journal of Virological Methods* 106: 51–59.

Olsson, K. 1976a. Experience of brown rot caused by *Pseudomonas solanacearum* (Smith) Smith in Sweden. *Bulletin of the OEPP/EPPO* 6: 199–207.

Olsson, K. 1976b. Overwintering of *Pseudomonas solanacearum* in Sweden. In: *Proceedings of the First International Planning Conference and Workshop on the Ecology and Control of Bacterial Wilt Caused by Pseudomonas solanacearum*. Sequeira, L, Kelman, A, (eds.). Raleigh, NC: North Carolina State University, pp. 105–109.

Olsvik, O., Popovic, T., Skjerve, E., Cudjoe, K.S., Hornes, E., Ugelstad, J., Uhlen, M. 1994. Magnetic separation techniques in diagnostic microbiology. *Clinical Microbiological Review* 7: 43–54.

Opio, A.E., Allen, D.J., Teri, J.M. 1996. Pathogen variation in *Xanthomonas campestris* pv. *phaseoli*, the causal agent of common bacterial blight in *Phaseolus* beans. *Plant Pathology* 45: 1126–1133.

Ordax, M., Marco-Noales, E., López, M.M., Biosca, E.G. 2006. Survival strategy of *Erwinia amylovora* against copper: Induction of the viable-but-nonculturable state. *Applied and Environmental Microbiology* 72: 3482–3488.

Ozakman, M., Schaad, N.W. 2003. A real-time BIO-PCR assay for detection of *Ralstonia solanacearum* race 3, biovar 2, in asymptomatic potato tubers. *Canadian Journal of Plant Pathology* 25: 232–239.

Palacio-Bielsa, A., Cubero, J., Cambra, M.A., Collados, R., Berruete, I.M., López, M.M. 2011. Development of an efficient real-time quantitative PCR protocol for detection of *Xanthomonas arboricola* pv. *pruni* in Prunus species. *Applied and Environmental Microbiology* 7: 89–97.

Paret, M.L., de Silva, A.S, Alvarez, A.M. 2009. Bioindicators for *Ralstonia solanacearum* race 4: Plants in the Zingiberaceae and Costaceae families. *Australian Plant Pathology* 38:6–12.

Peacock, S.J., de Silva, G.D., Justice, A., Cowland, C.E., Moore, C.E., Winerals, C.G., Day, N.P. 2002. Comparison of multilocus sequence typing and pulsed-field gel electrophoresis as tools for typing *Staphylococcus aureus* isolates in a microepidemiological setting. *Journal of Clinical Microbiology* 40:3764–3770.

Peplies, J., Glockner, F.O., Amann, R. 2003. Optimization strategies for DNA microarray-based detection of bacteria with 16S rRNA-targeting oligonucleotide probes. *Applied and Environmental Microbiology* 69: 1397–1407.

Persson, P. 1998. Successful eradication of *Ralstonia solanacearum* from Sweden. *Bulletin of the OEPP/EPPO* 28: 113–119.

Pickup, R.W., Rhodes, G., Saunders, J.R. 1999. Extraction of microbial DNA from aquatic sources: Freshwater. In *Molecular Microbial Ecology Manual*. Akkermans, A.D.L., Van Elsas, J.D., Bruijn, F.J. (eds.). Dordrecht, the Netherlands: Kluwer Academic Publishers, pp. 1–11.

Porter-Jordan, K., Rosenberg, E.I., Keiser, J.F., Gross, J.D., Ross, A.M., Nasim, S., Garrett, C.T. 1990. Nested polymerase chain reaction assay for the detection of cytomegalovirus overcomes false positives caused by contamination with fragmented DNA. *Journal of Medicinal Virology* 30: 85–91.

Pryor, B.M., Gilbertson, R.L. 2001. A PCR-based assay for detection of *Alternaria radicina* on carrot seed. *Plant Disease* 85: 18–23.

Psallidas, P.G. 1990. Fire blight of pomaceous trees in Greece—Evolution of the disease and characteristics of the pathogen *Erwinia amylovora*. *Acta Horticulturae* 273: 25–32.

Qi, Y., Xiao, Q., Zhu, S. 2003. 16S rDNA gene clone and detection of *Pantoea stewartii* subsp. *stewartii* by real-time fluorescent PCR. *Journal of Hunan Agricultural University* 29: 183–187.

Rafalski, A., Tingey, S., Williams, J.G.K. 1994. Random amplified polymorphic DNA (RAPD) markers. In *Plant Molecular Biology*. Gelvin, S.B., Schilperoot, R.A. (eds.). Belgium: Kluwer, H4: 1–8.

Roberts, R.G., Reymond, S.T. 1989. Evaluation of post harvest treatments for eradication of *Erwinia amylovora* from apple fruit. *Crop Protection* 8: 283–288.

Romeiro, R.S., Silva, H.S.A., Beriam, L.O.S., Rodrigues, Neto, J., de Carvalho, M.G. 1999. A bioassay for the detection of *Agrobacterium tumefaciens* in soil and plant material. *Summa Phytopathology* 25: 359–362.

Ruiz-García, A.B., Olmos, A., Arahal, D.R., Antunez, O., Llop, P., Perez-Ortin, J.E., López, M.M., Cambra, M. 2004. Biochip electronico para la deteccion y caracterizacion simultánea de los principales virus y bacterias patógenos de la patata. XII Congreso de la Sociedad Española de Fitopatología. Lloret de Mar. p. 12.

Safarik, I., M. Safarikova. 1999. Use of magnetic techniques for the isolation of cells. *Journal of Chromatography B* 722: 33–53.

Salm, H., Geider, K. 2004. Real-time PCR for detection and quantification of *Erwinia amylovora,* the causal agent of fireblight. *Plant Pathology* 53: 602–610.
Sawada, H., Ieki, H., Matsuda, I. 1995. PCR detection of Ti and Ri plasmids from phytopathogenic *Agrobacterium* strains. *Applied and Environmental Microbiology* 61: 828–831.
Saygılı, H., Sahin, F., Aysan, Y. 2006. *Fitobakteriyoloji*, Meta Basım, İzmir, 630p.
Scally, M., Schuenzel, E.L., Stouthamer, R.A., Nunney, L. 2005. Multilocus sequence type system for the plant pathogen *Xylella fastidiosa* and relative contributions of recombination and point mutation of clonal diversity. *Applied and Environmental Microbiology* 71: 8491–8499.
Schaad, N.W., Berthier-Schaad, Y., Sechler, A., Knorr, D. 1999. Detection of *Clavibacter michiganensis* subsp. *sepedonicus* in potato tubers by BIO-PCR and an automated real-time fluorescence detection system. *Plant Disease* 83: 1095–1100.
Schaad, N.W., Cheong, S.S., Tamaki, S., Hatziloukas, E., Panopoulos, N.J. 1995. A combined biological and enzymatic amplification (BIO-PCR) technique to detect *Pseudomonas syringae* pv. *phaseolicola* in bean seed extracts. *Phytopathology* 85: 243–248.
Schaad, N.W., Frederick, R. 2002. Real-time PCR and its application for rapid plant disease diagnostics. *Canadian Journal of Plant Pathology* 24: 250–258.
Scheer, C. 2009. Feuerbrandsituation im Bodenseeraum und Ergebnisse der Feuerbrandversuche des KOB 2008. *Obstbau* 3:168–172.
Schneider, B.J., Zhao, J., Orser, C. 1993. Detection of *Clavibacter michiganensis* subsp. *sepedonicus* by DNA amplification. *FEMS Microbiological Letters* 109: 207–212.
Schonfeld, J., Heuer, H., van Elsas, J.D., Smalla, K. 2003. Specific and sensitive detection of *Ralstonia solanacearum* in soil on the basis of PCR amplification of fliC fragments. *Applied and Environmental Microbiology* 69: 7248–7256.
Shanmugam, V., Kumar, A., Sarma, Y.R. 2004. Survival of *Ralstonia solanacearum* in ginger rhizomes stored at different temperatures. *Plant Disease Research* 19:40–43.
Simmonds, P., Balfe, P., Peutherer, J.F., Ludlam, C.A., Bishop, J.O., Brown, A.J. 1990. Human immunodeficiency virus-infected individuals contain provirus in small numbers of peripheral mononuclear cells and at low copy numbers. *Journal of Virology* 64: 864–872.
Song, W.Y., Kim, H.M., Hwang, C.Y., Schaad, N.W. 2004. Detection of *Acidovorax avenae* ssp. *avenae* in rice seeds using BIO-PCR. *Journal of Phytopathology* 152: 667–676.
Stentiford, E.I. 1996. Composting control: Principles and practice. In *The Science of Composting.* de Bertoldi, M., Sequi, P., Lemmes, B., Papi, T. (eds.). London: Blackie Academic and Professional, pp. 49–59.
Stukenbrock, E.H., Rosendahl, S. 2005. Development and amplification of multiple codominant genetic markers from single spores of arbuscular mycorrhizal fungi by nested multiplex PCR. *Fungal Genetics and Biol* 42: 73–80.
Suárez-Estrella, F., Vargas-García, M.C., Elorrieta, M.A., López, M.J., Moreno, J. 2003. Temperature effect on *Fusarium oxysporum* f.sp. *melonis* survival during horticultural waste composting. *Journal of Applied Microbiology* 94: 475–482.
Tan-Bian, S., Yabuki, J., Matsumoto, S., Kageyama, K., Fukui, H. 2003. PCR primers for identification of opine types of *Agrobacterium tumefaciens* in Japan. *Journal of General Plant Pathology* 69: 258–266.
Taylor, E., Bates, J., Kenyon, D., Maccaferri, M., Thomas, J. 2002. Modern molecular methods for characterisation and diagnosis of seed-borne fungal pathogens. *Journal of Plant Pathology* 83: 75–81.
Temple, T.N., Johnson, K.B. 2011. Evaluation of loop-mediated isothermal amplification for rapid detection of *Erwinia amylovora* on pear and apple fruit flowers. *Plant Disease* 95: 423–430.
Temple, T.N., Stockwell, V.O., Johnson, K.B. 2008. Development of a rapid detection method for *Erwinia amylovora* by loop-mediated isothermal amplification (LAMP). *Acta Horticul* 793: 497–503.
Thomson, S.V., Schroth, M.N., Moller, W.J. 1982. A forecasting model for fire blight of pear. *Plant Disease* 66: 576–579.
Turner, C. 2002. The thermal inactivation of *E. coli* in straw and pig manure. *Bioresource Technology* 84: 57–61.
Van Beckhoven, J.R.C.M., Stead, D.E., van der Wolf, J.M. 2002. Detection of *Clavibacter michiganensis* subsp. *sepedonicus* by AmpliDet RNA, a new technology based on real time monitoring of NASBA amplicons with a molecular beacon. *Journal of Applied Microbiology* 93: 840–849.
Van der Zwet, T., Beer, S.V. 1995. Fire Blight—Its Nature, Preventation, Control. Agriculturae Information Bulletin. No: 631.
Van der Zwet, T., Keil, H.L. 1979. Fire blight bacterial disease of Rosaceous plants. United States Department of Agriculture, Agriculture Handbook Number 510, 205p.
Van der Zwet, T., Thomson, S.V., Covey, R.P., Bonn, W.G. 1990. Population of *Erwinia amylovora* on external and internal apple fruit tissues. *Plant Disease* 74: 711–716.

Van Elsas, J.D., Smalla, K. 1999. Extraction of microbial community DNA from soils. In: *Molecular Microbial Ecology Manual. Akkermans*. Akkermans, A.D.L., Van Elsas, J.D., Bruijn, F.J. (eds.). Dordrecht, the Netherlands: Kluwer Academic Publishers, pp. 1–11.

Van Sluys, M.A., Monteiro-Vitorello, C.B., Camargo, L.E.A., Miyaki, C.Y., Furlan, L.R., Camargo, L.E.A., Da Silva, A.C.R. et al. 2002. Comparative genomic analysis of plant-associated bacteria. *Annual Review of Phytopathology* 40: 169–190.

Vanneste, J.L. 2000. *Fire Blight: The Disease and its Causative Agent, Erwinia amylovora*. Wallingford, UK: CAB International.

Varma-Basil, M., El-Hajj, H., Marras, S.A., Hazbon, M.H., Mann, J.M., Connell, N.D., Kramer, F.R., Alland, D. 2004. Molecular beacons for multiplex detection of four bacterial bioterrorism agents. *Clin. Chem* 50: 1060–1062.

Vernet, G. 2002. DNA-Chip technology and infectious diseases. *Virus Research* 82: 65–71.

Verniere, C., Devaux, M., Pruvost, O., Couteau, A., Luisetti, J. 1991. Studies on the biochemical and physiological variations among strains of *Xanthomonas campestris* pv. *citri*, the causal agent of citrus bacterial canker disease. *Fruits* 46:162–170.

Vitzthum, F., Bernhagen, J. 2002. SYBR Green I: An ultrasensitive fluorescent dye for double-standed DNA quantification in solution and other applications. *Recent Research Developments Analytical Biochemistry* 2: 65–93.

Volkhard, A., Kempf, J., Trebesius, K., Autenrieth, I.B. 2000. Fluorescent *in situ* hybridization allows rapid identification of microorganisms in blood cultures. *Journal of Clinical Microbiology* 38: 830–838.

Walcott, R.R., Gitaitis, R.D. 2000. Detection of *Acidovorax avenae* subsp. *citrulli* in watermelon seed using immunomagnetic separation and the polymerase chain reaction. *Plant Disease* 84: 470–474.

Walcott, R.R., Gitaitis, R.D., Castro, A.C., Sanders, F.H., Diaz-Perez, J.C. 2002. Natural infestation of onion seed by *Pantoea ananatis*, causal agent of center rot. *Plant Disease* 86: 106–111.

Wang, Z., Yin, Y., Hu, H., Yuan, Q., Peng, G., Xia, Y. 2006. Development and application of molecular-based diagnosis for "*Candidatus* Liberibacter asiaticus," the causal pathogen of citrus huanglongbing. *Plant Pathology* 55: 630–638.

Weingrat, H., Volksch, B. 1997. Genetic fingerprinting of *Pseudomonas syringae* pv using ERIC, REP and 1S50 PCR. *Journal of Phytopathology* 145: 339–345.

Weller, S.A., Beresford-Jones, N.J., Hall, J., Thwaites, R., Parkinson, N., Elphinstone, J.G. 2007. Detection of *Xanthomonas fragariae* and presumptive detection of *Xanthomonas arboricola* pv. *fragariae*, from strawberry leaves, by real-time PCR. *Journal of Microbiologial Methods* 70: 379–383.

Weller, S.A., Elphinstone, J.G., Smith, N.C., Boonham, N., Stead, D.E. 2000. Detection of *Ralstonia solanacearum* strains with a quantitative, multiplex, real-time, fluorogenic PCR (TaqMan) assay. *Applied and Environmental Microbiology* 66: 2853–2858.

Williams, J.G.K., Kubelik, A.R., Livak, K.J., Rafalski, J.A., Tingey, S.V. 1990. DNA polymorphisms amplified by arbitrary primers are useful as genetic markers. *Nucleic Acids Research* 18: 6531–6535.

Wittwer, C.T., Herrmann, M.G., Gundry, C.N., Elenitoba-Johnson, K.S.J. 2001. Real-time multiplex PCR assays. *Methods* 25:430–442.

Wolfgang, M.C., Kulasekara, B.R., Liang, X.Y., Boyd, D., Wu, K., Yang, Q., Miyada, C.G., Lory, S. 2003. Conservation of genome content and virulence determinants among clinical and environmental isolates of *Pseudomonas aeruginosa*. *Proceedings of National Academy of Science. USA* 100: 8484–8489.

Wullings, B.A., Beuningen, A.R., van Janse, J.D., Akkermans, A.D.L., Van Beuningen, A.R. 1998. Detection of *Ralstonia solanacearum*, which causes brown rot of potato, by fluorescent *in situ* hybridization with 23S rRNA-targeted probes. *Applied and Environmental Microbiol* 64: 4546–4554.

Yamada, T., Kawasaki, T., Nagata, S., Fujiwara, A., Usami, S., Fujie, M. 2007. New bacteriophages that infect the phytopathogen *Ralstonia solanacearum*. *Microbiology* 153: 2630–2639.

Yang, Y., Kim, K., Anderson, E.J. 1997. Seed transmission of cucumber mosaic virus in spinach. *Phytopathology* 87: 924–931.

5 Agro-Traditional Practices of Plant Pathogens Control

Kanthaiah Kannan, Duraisamy Nivas, Velu Rajesh Kannan, and Kubilay Kurtulus Bastas

CONTENTS

ABSTRACT Despite enormous progress in agro science technology, considerable limitations remain due to the biological systems in question, the disadvantages of managing balanced activities with other organisms, and the inevitability of compromising quality or quantity. Implementations of new technologies such as agrochemicals have led to reductions in soil fertility and in native microflora and -fauna (including non-target organisms). Agro-traditional practices, which provide precious methods for controlling the diseases spread by plant pathogenesis, enhancing the advantages, and enriching soil fertility. The methods implemented directly or indirectly by traditional farmers are effective and reliable in controlling plant pathogens in storage places and in increasing crop yields. This chapter examines traditional plant pathogenic control strategies implemented by the ancients and presently adopted in ongoing practices.

KEYWORDS: agro-traditional practices, ancient periods, agrochemicals, vrikshayurvedha, panchagavya

5.1 TRADITIONAL AGRICULTURE

Human beings commenced cultivating land by utilizing their then-existing implements to produce agricultural edibles for survival during the Stone Age and the Hunter-Gatherer Age. Although they ate meat, from both birds and animals including fishes, these did not fulfill their needs for sustenance. They were therefore prompted to learn methods of agro cultivation, using nearby or available river water, and to harvest selected crops and grains. Ancient people developed sustainable agriculture practices that allowed them to produce food and fiber for thousands of years with few outside inputs; however, some of their traditional strategies were not so successful. In any case, even their successful practices have been forgotten or abandoned except in some developing countries. Considerable evidence shows the traditional farmers' routine practices, most of which were developed empirically through millennia of trial and error, natural selection, and keen observation (Thurston, 1992).

Traditional agriculture uses strategies of cultivation that were acquired through practices that began with the beginning of crop production 10,000 years ago. Most of the activities of the farmers in traditional agriculture are culture based and are exhibited in tasks such as seed sowing, planting, transplanting, growth, and harvest. Although little information is available about traditional methods of cultivation, the available information is easy to understand and administer. Still, recent research in the field of agriculture exhibits limitations that arise from traditional methods.

5.2 ROLE OF TRADITIONAL AGRICULTURE

Rapid changes have occurred in traditional agriculture due to the impact of settlement, commerce, and Western technology. These changes increased food production and some have been aimed to resolve the global food demand (Marten, 1986). The changes that have been accepted in traditional agriculture have been modernized in scientific agriculture, which has helped to improve food production and to reduce world starvation caused by human overpopulation. In situations where changes are necessary, traditional agriculture practices must be thoroughly understood and compared with modernized agriculture systems prior to implementation; in addition, environmental degradation must be taken into consideration when adapting modernized agriculture practices (Thurston, 1990).

Conversely, traditional agriculture practices have provided effective and sustainable means of disease control. The present agriculture system evolved from ancient traditional agriculture techniques; the latter have had a profound effect on modern agriculture. The gradual conversion to modern agriculture systems has resulted in the disappearance of traditional agriculture practices among farmers. This chapter reassesses beneficial traditional methods which, although largely forgotten, have had efficacious impact on contemporary agricultural disease control technology. Although these forgotten methods continue, as in current pest/disease management, agro-scientists find it difficult to comprehend these cultural practices in this essential area. We therefore need to reaffirm both the efficacy and limitations of traditional agro disease management (Marten, 1986).

5.3 TRADITIONAL AGRICULTURE PRACTICE: MYTH OR TRUTH?

In general, although today's traditional agriculture farmers are concerned with achieving stable and reliable profit, they are not always interested in acquiring the highest yields. These practices of indigenous people, which have provided rich storehouses of traditional beliefs, folklore, rituals, and rites, continue to hold value in the light of empirical sciences; therefore, it is essential to identify some of these beliefs as sound agricultural practices (Thurston, 1992). Based on these beliefs, seeds are collected and thrashed on the day of the new moon (Amawasia) for sowing in the next season. This and other customs prevent diseases spread by pests and pathogens, and the plant diseases caused by forming halo around the sun (Chhetry, 2008).

5.4 PATHOGEN LIFE CYCLES AND THE SPREAD OF PLANT DISEASES

Generally, plants are affected both directly and indirectly by numerous pests such as flies, insects, birds, and microbial pathogens. The indirect mode of microbial pathogen transfer occurs by various carriers or vectors when the microbes come into contact with plants (Figure 5.1). A detailed study

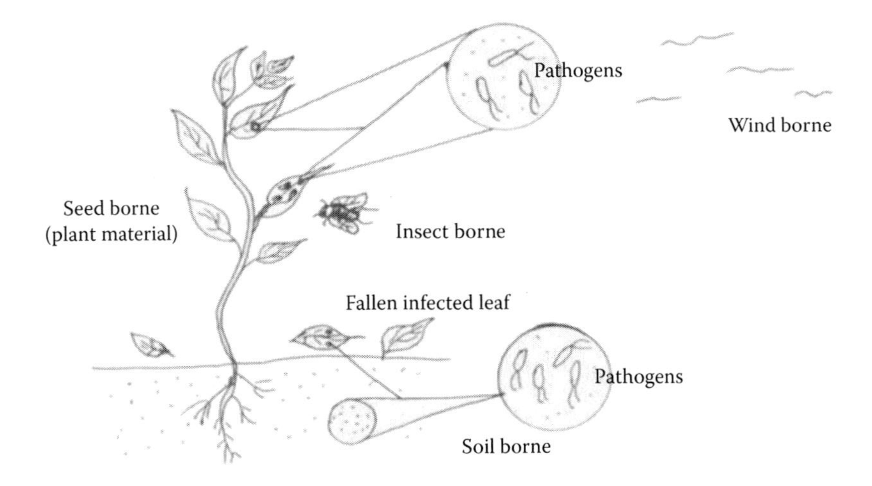

FIGURE 5.1 Primary mode of infection.

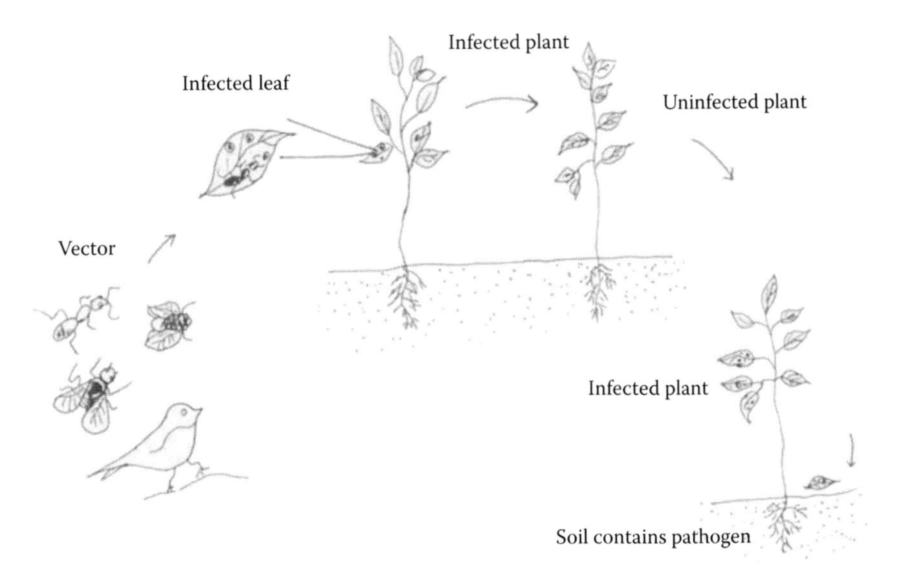

FIGURE 5.2 Life cycle and spread of diseases.

of the life cycles of different pathogens is helpful in controlling them through agricultural practices. They survive under favorable and unfavorable conditions; all parasitic conditions, including viral pathogen/diseases, are transmissible. Because the parasites or issues are infectious to suitable host plants they have the innate potential to spread from one host to other from one area to another. These phenomena are called dissemination or dispersal of the pathogen. The dissemination of plant diseases is recognized in relation to different phases of diseases as primary infection (initial infection in a suitable host) and secondary spread (rapid multiplication) (Niranjan, 2014). As stated above, the bacterial pathogens spread and survive due to carrier interaction with its host plant. By using traditional agricultural practices, plant bacterial pathogens can be controlled both directly and indirectly (Figure 5.2).

5.5 TRADITIONAL CULTURAL PRACTICES FOR DISEASE CONTROL

Cultural practices aim to improve plant growth by providing proper nutrition and moisture, and by creating conditions that are unfavorable to pathogens, thereby reducing disease development.

Ogle and Brown (1997) have stated that in greenhouses there is proper spacing between the plants and provision of good ventilation, which reduces humidity and inhibits infection by pathogens such as *Botrytis* spp. More sophisticated greenhouses frequently feature low humidity, especially at night, to dry the surfaces of plants and to encourage the spread of beneficial pathogens such as *Botrytis* spp. and *Diplocarpon rosae.*

5.5.1 Crop Rotation

Crop rotation, one of the most commonly used ancient agricultural practices, improves soil fertility, moisture, and texture. Successful crop rotation, which assists in weed and pathogen control, creates a time interval between susceptible crops that is longer than the known survival periods of some pathogens. Crop rotations are effective in controlling plant pathogens such as *Gaeumannomyces graminis*, *Pyrenophora tritici-repentis*, and various *Colletotrichum* and *Phoma* spp. Some pathogens only survive in the presence of a specific host. Perhaps crop rotation or some other approach is effective in controlling damping-off and root-rot fungi such as *Pythium*, *Aphanomyces*, and *Fusarium* spp. including the vascular-wilt pathogens *Sclerotinia* spp., *Macrophomina phaseoli*, and *Plasmodiophora brassicae*, which can survive for long periods in soil as saprophytes.

5.5.1.1 Crop Rotation in Winter Cereals

More than 85% of winter cereals are affected by *Fusarium* wilt diseases such as *F. framinearum*, *F. culmorum*, *F. moiliforms*, *F. arenaceum*, and *F. spotrichiela*. Wilt and rot diseases also occur in wheat and barley due to continuous and repeated cropping practices.

5.5.2 Sanitation Practices

Sanitation procedures are applied to eliminate the sources of inoculum in many nurseries, greenhouses, and the cultivation fields of high-value crops. *Mycosphaerella fragariae* is a leaf-spotting fungi that affects strawberry in early season. Infected plant parts are removed in cases of fire blight of pome fruit (*Erwinia amylovora*), apple canker (*Nectria galligena*), powdery mildew of apples (*Podosphaera Leucotricha*), and silver leaf of plums (*Chondrostereum purpureum*). Mummified fruits and diseased twigs of pome and stone fruits, the major inocula of *Sclerotinia* spp., are removed to reduce the disease incidence in plants (Ogle and Brown, 1997). Pathogens can utilize plants as alternative hosts for their survival between seasons, a practice that also provides a source of inoculum in the new growing season. For example, grasses such as *Hordeum leportnum* are hosts of the wheat take-all fungus *Gaeumannomyces graminis* and certain leguminous green manure crops maintain the halo blight organism *Pseudomonas syringe* pv. *phaseolicola*. However, the relationship between weeds and pathogens are very complicated; many weeds carry various viruses without any symptoms (Ogle and Brown, 1997).

5.5.3 Tillage

Tillage, one of the predominant practices in traditional agriculture systems, indirectly suppresses and prevents the spread of plant pathogens. Tillage could also reduce the pathogen populations of weeds and volunteer plants that harbor pathogens between crops. During the tillage practices, pathogenic organisms are transferred from topsoil and buried in the bottom layer of soil. In addition, physiological changes that influence disease incidence are created during tillage. These include moisture, bulk density, aeration, and temperature which in turn influence the disease incidence (Figure 5.3).

Preparation of seed beds can greatly alter soil texture, aeration, temperature, moisture levels, and density. Tillage can also influence nutrient release within soil and can generally benefit crops. Farming practices have been moved away from regular tillage, which has reduced damage to roots

FIGURE 5.3 Tillage practice.

and the spread of pathogens caused by tilling machinery. However, minimum tillage can also encourage some pathogens, such as those that feed on crop residues left on the surface of the soil. Healthy plants may be more resistant to some pathogens but the more humid microclimate within the crop can be conducive to the spread of other pathogens. Successive tillage operations can reduce inoculum levels of some pathogens through exposure of the inoculum to desiccation by the sun.

5.5.4 Raised Fields and Beds, Ridges, and Mounds

For millennia, the most widely used practices of traditional farmers were raised fields and beds, ridges, and mounds. These were used in the geographically separate areas of tropical America, Asia, and Africa, where the raised-bed system of agriculture was predominant. In raised-field cultivation, drainage, fertilization, frost control, and irrigation are important factors in crop establishment (Erickson, 1968). Planting in soil raised above the surrounding area also has significant implications for disease management especially for soil-borne pathogens (Figure 5.4).

FIGURE 5.4 Raised fields and beds.

In specialized traditional raised-bed systems, disease-spreading pathogens are controlled at least until the germinated seeds become saplings that are able to resist infection. Field flooding, which destroys many soil-borne pathogens in rice, can be used in the cultivation of other crops. In Asian countries, most vegetables are cultivated with the use of mounds, ridges, and raised beds; for root and tuber crops, these are the only approaches used because they reduce root rot. Waddell's (1972) discussion of the mound builders of New Guinea included a photograph of traditional farmers cultivating sweet potatoes (*Ipomoea batatas*) on mounds that produced high yields for long periods of time with no apparent disease problems.

Agroforestry systems are characterized by high levels of on-site nutrient conservation. Growing of trees such as *Azadirachta indica*, *Lannea coromandalica*, *Leucaena leucocephala*, and *Thespesia populnea* on field bunds is a regular practice in North east coastal region of Tamilnadu. This practice helps to optimize the supply of green leaf manure and thereby improves the nutrient availability, provides stress resistance against soil moisture, and reduces the intensity of soil-borne diseases, particularly wilt caused by *Fusarium oxysporum* (Schlecht. Emend. Synd. & Hans) in vegetables and root rot caused by *Marophomina phaseolina* (Tassi) Goid. In rice, fallow pulses and improvement in rhizosphere environment are favorable for plant growth (Immanuel et al., 2010).

5.5.5 Mulching

In 1990, Thurston analyzed the tropical region (the Gunanacaste region of northern Costa Rica) and identified the web blight of common bean (*Phaseolus vulgaris*) caused by the fungus *Thanatephorus cucumeris* (Frank) Donk (anamorph: *Rhizotonia solani* Kuhn). Web blight, which is probably a single destructive disease of beans, causes rapid defoliation and sometimes complete crop failure. An epidemic of web blight in the Gunanacaste region of northern Costa Rica resulted in 90% reduction in bean yields because beans were being cleanly cultivated. To prevent this web blight disease, which is caused by *T. cucumeris*, traditional farmers adopted a system called *frijol tapado* [covered beans] (Galindo et al., 1983) in which bean seeds are mulched with chopped weeds that cannot regrow but do conserve soil moisture.

Narayanasamy (2002) stated that the infection of apple roots by *Sclerotium rolfsii* in Israel could be markedly reduced by application of mulches, which reduced soil temperature enough to kill the pathogen on the root surfaces. Black spot disease of citrus caused by *Guignaradia bidwelli* could be reduced by mulching with *Panicum maximum* and fungicidal sprays; in one case, the percentage of disease-free fruit was increased to 13%–16%. In nurseries, the damping-off disease caused by *Pythium debaryanum*, *Rhizoctonia solani*, *Phytopthora* spp., and *Fusarium* spp. are reduced by mulching with sand and covering with sawdust. The application of inert materials to reduce the incidence of soil-borne diseases was also investigated in sugar beet and cabbage nurseries.

Organic mulches may introduce both beneficial and harmful microbes into soil. For example, actinomycetes isolated from an avocado plantation have an inhibitory effect on *Phytopthora cinnamomi,* apart from the isolates that would inhibit plant growth. Different types of mulches are used in traditional plant disease control. For example, cellulose-rich wood mulches are used for controlling *Phytophthora*, root rot caused by *P. cinnamomi*. In wood mulches, cellulose plays a major role in pathogen suppression by reducing the sporangial production of the pathogen of *P. cinnamomi*. These results are indicated by Narayanasamy (2002), who stated that the cellulose activity in mulch was sufficient to impair sporangial production of *P. cinnamomi* but not always sufficient to reduce the vegetative biomass of the pathogen.

5.6 CONTROL OF SEED-BORNE DISEASES

Seeds are the major sources of disease transmission that includes large numbers of pathogenic bacteria and fungi. Seed-borne diseases severely affect seed germination and seedling vigor. The pathogen-affected seeds transmit primarily sorghum grain smut (*Sphacelotheca sorgi*), rice bakanae

(*Fusarium moniliforme*), cabbage black leg (*Phoma lingam*), cabbage black rot (*Xanthomonas campestris* pv *campestris*), and common bean bacterial wilt (*Curtobacterium flaccumfaciens*). The use of non-affected seeds is a very vital strategy to minimize the spread of these diseases in fields or other places of cultivation.

5.7 DISEASE CONTROL THROUGH PHYSIOLOGICAL METHODS

Physiological methods of agro disease control are not only ecofriendly but also easy ways to manage soil-borne plant pathogens. Of these methods, solarization is one of the most important. It has been practiced worldwide in protected agriculture. Today, transparent polyethylene sheets are used to mulch moist, infested soil during periods of high ambient temperatures. These coverings increase the elevated temperature to levels that are lethal to the resting structures of soil-borne pathogens (Figure 5.5). This method is especially effective in temperate regions such as India and in arid countries. It is feasible for the repeated cultivation of high-value crops, mass production of AM fungal inoculum, floriculture, seed production, and post-plant effect (Lodha, 2011).

The main factors that influence the success of solarization are soil type and the type of pathogen that is present. Because it employs a combination of physical, chemical, and biological mechanisms, solarization is compatible with many other disinfestation methods that are used for integrated pest management. Additionally, solarization controls many important soil-borne fungal and bacterial plant pathogens, including those that cause *Verticillium* and *Fusarium* wilt, *Phytophthora* root rot, southern blight, damping-off, crown gall disease, tomato canker, potato scab, and others (Table 5.1). However, a few heat-tolerant fungi and bacteria are more difficult to control with solarization (Stapleton, 2000).

Evidently, soil solarization controls many important soil-borne fungal and bacterial plant pathogens. This technology is particularly useful for fungal pathogens that cause root rot diseases in the field, such as *Fusarium oxysporum* that causes wilt disease in cotton and tomato plants, and the pathogenic fungi *Rhizoctonia solani* and *Phytophthora cinnamoni*. Other harmful bacteria such as *Pseudomonas solanacearum* (causes bacterial wilt), *Agrobacterium tumefaciens* (causes crown gall disease), *Clavibacter michiganensis* (causes tomato canker), and *Streptomyces scabies* (causes scab disease in potato plants) and the diseases they produce can be controlled by soil solarization technology (Kapoor, 2013).

FIGURE 5.5 Solarization.

TABLE 5.1
Pathogens Controlled by Solarization

Pathogenic Microorganisms	Disease	Crops/Plants
Macrophomina phaseolina	Dry root rot	Legumes and oil seeds
Fusarium solani	Root rot	Jojoba, gayule, eucalyptus
Cylindrocarpon lichenicola	Dry root rot	Jojoba
Fusarium oxysporum f. sp. lycopersici	Wilt	Tomato
Fusarium f. sp. varsinfecum	Wilt	Cotton
Phytophthora cinnamomi	Phytophthora root rot	Many crops
Pythium ultimum	Seed rot	Many crops
Sclerotium rolfii	Southern blight	Many crops
Verticillium dahtiae	Wilt	Many crops

Source: Adapted from Lodha, S. 2011. *Desert Environment News Letter.* 13:1.

5.8 INDIRECT DISEASE CONTROL PRACTICES

Several mechanical control methods are practiced by Javanese farmers in West Java. These include scarecrows, labor systems, Brongsong system, smoke, manual removal systems, and washing systems for spores. Scarecrows are also used by the traditional farmers of Tamilnadu in India, who call their man made, life-sized straw dolls attired in human clothing *solakaattu pommai* (Figure 5.6). These devices are used to frighten birds away from crop fields in order to prevent the spread of pathogenic microorganisms that are present on plant leafs and other parts (Marten, 1986). Traditional farmers who use labor systems burn entire affected plants to prevent the spread of pathogenic microorganisms. In the Brongsong system, fruits are often wrapped in cloth or covered by inverted baskets to keep birds, bats, and caterpillars away until harvest. Smoke and manual removal systems are used by traditional farmers to eliminate pathogen survival and spread.

Ayurvedic methods can reduce the spread of pathogenic microorganisms as well as other plant diseases. *Vrkshayurveda* (the science of the life of plants) is one of the major and widely used

FIGURE 5.6 Scarecrow.

ayurvedic methods to control pests. There is an enormous literature on *vrkshayurveda* in Sanskrit as well as other regional Indian languages. During the Vedic, prehistoric, and historic periods, farmers used plant pathogen control methods such as the storage of grains in cylindrical pits, in granaries or containers made of rope and plastered with mud, and in well-baked clay pots. They also scared birds away by slinging balls, used mixed cropping techniques, and implemented controlled field-irrigation systems. Plant materials, cow dung, and urine (animal and human) are also frequently used to control plant pathogenic bacteria (Narayanasamy, 2002).

Only one ancient copy of *vrkshayurveda*, known as Surapala's Vrikshayurveda (~1000 AD) exists. Written on palm leaves, it is preserved at the Bodleian Library at Oxford. In 1994, the author obtained a microfiche from the library and created a printout. A bulletin of the Sanskrit text, its English translation, and commentaries by researchers was published in 1996 by AAHF (Asian Agri-History Foundation). Since 2000, the term "vrikshayurveda" has become widely known among agriculturists in other countries as well as India (Nene, 2012).

In general, *vrikshayurveda* deals with the detection of underground water, spacing between trees, methods of propagation, the fruits and seeds of Jaiur-blai (*Zanthoxylum oxyphyllum* Egdew) (Kalia et al., 1999; Firake et al., 2013) It contains methods of traditional farming practices from Tamilnadu and the northeastern Himalayas. Farmers in Kothapalli village of Andhra Pradesh in India have followed these biointensive pest-management practices in order to get high yields and simultaneously reduce their use of pesticides. They used five different formulations: (1) and (2) Two botanicals, a biological method, (3) vermicomposting method, (4) a cow urine solution and a mix of curd, jaggery (concentrated sugarcane juice) and bread yeast, (5) in addition, a mixture of three different bacterial strains also used (Ranga Rao et al., 2007), including the use of two plants, *Azadirachta indica* and *Gliricidia sepium*; a vermicomposting method of cow's urine mixed with curd, bread yeast, and jaggery; curd preparation; and a biological method. Cow urine serves as a repellent and curd mixture added after 50% flowering stage. The fifth method involves a mixture of three different bacterial strains such as *Pseudomonas fluorescens* (promotes plant growth, makes more phosphorus available to growing plants, and suppresses soil-borne fungi), *Azotobacter vinelandii* strain HT54 (a nitrogen-fixing bacterium), and *Bacillus licheniformis* (promotes plant growth); these were applied to the soil during the sowing pigeon pea, chickpea, other vegetables, and cotton (Ranga Rao et al., 2007).

Panchgavya, another important method, refers to combinations of five substances that have high antimicrobial abilities: cow dung, urine, milk, curd, and ghee. The use of cow urine is practiced in several places. *Panchgavya*, which includes many groups of beneficial microorganisms such as fungi, actinomycetes, and bacteria, comprises almost all of the macro- and micronutrients and growth hormones (IAA, GA) that are required for plant growth.

Kunapajala, a fermented liquid obtained from animal wastes containing flesh, dung, urine, marrow, and skin, contains basic constituents such as amino acids, sugars, fatty acids, keratins, and macro- and micronutrients in available form. Plants naturally respond very well to the nourishment provided by *kunapajala* and flourish with excellent growth, flowering, and fruiting (Nene, 2012). Some of the elements suggested for plant disease management in addition to *kunapajala* are mustard, honey, milk, neem bark, vidanga, and hair/nails/horns. The key properties of mustards include insecticidal activity, whereas honey is antimicrobial, antibacterial, and contains plant growth hormone as well as properties that promote wound healing and drought resistance. Lactoferrin, which is present in bovine milk, has antifungal, antibacterial, antiviral, and antinematode properties. Neem bark has antibacterial and insecticidal properties. The fruit of vidanga/bidanga (*Embelia ribes*) is an antihelmintic material. Embelin (2, 5-dihydroxy-3-undecyl-p-benzoquinone) is found to be the active principle in *Embelia ribes* and is reported to possess a wide spectrum of biological activities including antibacterial and insecticidal properties. Hair, nails, and horns contain keratin, which has large amounts of the sulfur-containing amino acid cysteine; when burnt, keratin emits a strongly sulfurous odor. Smoke from nails releases sulfur, which controls diseases and pests. Panchamula consists of a powdered mixture from the dried roots of five plants (Table 5.2).

TABLE 5.2
Plant Species Used in Preparing Panchamula and Their Relevant Properties

Plant Species	Properties
Aegle marmelos	Antifungal, nematicidal, insect antifeedant
Clerodendrum phlomides	Antifungal, antiviral, antibacterial, insect antifeedant
Gmelina arborea	Antiviral
Oroxylum indicum	Antimicrobial
Stereospermum suaveolens	Antifungal, antibacterial

Source: Adapted from Nene, Y.L. 2012. *Asian Agriculture History.* 16(1): 45–54.

5.9 OTHER TRADITIONAL METHODS OF PLANT DISEASE CONTROL

Other traditional methods for indirectly controlling plant pathogens involve the reduction or elimination of carriers or vectors of pathogens such as birds, rats, flies, and insects. In maize fields, farmers bend the ears of maize downward in order to protect them from rain; in fact, the grain dries better on the plant in the sun than in dark storage. Such maize showed only 1% of grain damage by fungi (David Thurston, 1990). Canker and lesions caused during winter seasons by fire blight on pear and apple trees are prevented from causing further pathogen infection with fungicide applications. Tree surgery is another method that has been used to eradicate infected plant materials. Bacterial canker caused by *Pseudomonas syringae* pv *syringae*, which has occurred in stone fruits since the late 1880s, can be a major component of peach tree short-life complex in southeastern U.S. peach orchards. This disease is sometimes referred to as sour sap, blast, die-back, or gummosis. However, in the southeastern United States, gummosis typically refers to peach fungal gummosis caused by *Botryisphaeria* spp. and also to a bacterial canker that causes serious disease in tomato by *Clavibacter michiganensis* subsp. *michiganensis.*

5.10 CONCLUSION

Agricultural research since 2000 has provided adequate knowledge of the consequences of agricultural development as well as the advantages of adopting traditional practices. The enormous amounts of useful traditional agricultural practices handed down by our ancestors are very simple, feasible, and efficient, and also easily adaptable with low costs. The traditional practices in our agriculture system maintain soil fertility without damaging the environment and minimize the use of agrochemicals. Traditional farmers follow standard cropping patterns and rotations in order to eliminate the occurrence of plant pathogenic organisms; by doing so, they simultaneously increase production with antagonistic plants and their parts. Farmers also have knowledge of using animal excretions to eliminate pathogens and their vectors. Methods such as solarization are efficient in controlling pathogens. However, the use of polyethylene films in arid regions such as India, while very easy for the farmers, include issues of polyethylene disposal, which is of prime importance in some areas. Cultural and mechanical methods such as bending maize ears and producing ethylene protect plants from rain and eliminate pathogenic germination of seed without contamination. It is evident that traditional agricultural practices, if implemented judiciously, have enormous potential to restore soil fertility.

ACKNOWLEDGMENT

The authors are grateful to Professor Dr. M. Murugesan for his constructive ideas about links between ancient and modern practices.

REFERENCES

Chhetry, G.K.N., Belbahri, L. 2008. Indigenous pest and disease management practices in traditional farming systems in north east India. A review. *Journal of Plant Breeding and Crop Science*. 1(3):28–38.

Erickson, C.L. 1968. Raised filed agriculture in the lake of Titicaca basin. *Expedition*. 30:3.

Firake, D.M., Lytan, D., Thubru, D.P., Behere, G.T., Firale, P.D., Thakur, N.S.A. 2013. Traditional pest management practices and beliefs of different ethnic tribes of Meghalaya, North Eastern Himalaya. *Indian Journal of Hill Farming*. 26(1):58–61.

Galindo, J.J., Abawi, G.S., Thurston, H.D., Galvez, G. 1983. Source of inoculum and development of bean web blight in Costa Rica. *Plant Disease*. 67:1016–1021.

Immanuel, R.R., Imayavaramban, V., Elizabeth, L.L., Kannan, T., Murugan, G. 2010. Traditional farming knowledge on agro ecosystem conservation in northeast coastal Tamilnadu. *Indian Journal of Traditional Knowledge*. 9(2):366–374.

Kalia, N.K., Singh, B., Sood, R.P. 1999. A new amide from *Zanthoxylum armatum*. *Journal of Natural Products* 62:311–312.

Kapoor, R.T. 2013. *Soil Solarization: Eco-Friendly Technology for Farmers in Agriculture for Pest Management. 2nd International Conference on Advances in Biological and Pharmaceutical Sciences (ICABPS'2013)*, Hong Kong, Sept 17–18.

Lodha, S. 2011. Soil Solarization: An Eco friendly Approach to manage soil borne plant pathogens. *Desert Environment News Letter*. 13:1.

Marten, G.G. 1986. *Traditional Agriculture in Southeast Asia: A Human Ecology Perspective*, Boulder, Colorado: Westview Press. 244–266.

Narayanasamy, P. 2002. Traditional pest control: A retrospection. *Indian Journal of Traditional Knowledge*. 1(1):40–50.

Nene, Y.L. 2012. Potential of some methods described in Vrikshayurvedas in crop yield increase and disease management. *Asian Agriculture History*. 16(1):45–54.

Niranjan, 2014. http://www.indiaagronet.com/indiaagronet/Disease_management/DiseaseMan.htm

Ogle, H.J., Brown, J.F. 1997. *Plant Pathogens and Plant Diseases. Ogle H and Dale M. Disease Management: Cultural Practices*. Australia: Rock vale publications. 390–404.

Ranga Rao, G.V., Rupela, O.P., Wani1, S.P., Rahman, S.J., Jyothsna, J.S., Rameshwar Rao, V., Humayun, P. 2007. Integrated pest management. *Pesticides News*. 76(3):16–17.

Stapleton, J.J. 2000. Soil solarization in various agricultural production systems. *Crop Protection* 19:837–841.

Thurston, H.D. 1990. Plant disease management practices of traditional farmers. *Plant Disease*. 74(2):96–102.

Thurston, H.D. 1992. *Sustainable Practices for Plant Disease Management in Traditional Farming Systems*. Westview, Boulder, CO, New Delhi. 279.

Waddell, E. 1972. *The Mound Builders: Agricultural Practices, Environment and Society in the Central Highlands of New Guinea*. Seattle: University of Washington Press.

6 Antimicrobial Polypeptides in the Control of Plant Pathogenic Bacteria

Vahap Eldem, Sezer Okay, Yakup Bakir, and Turgay Unver

CONTENTS

ABSTRACT Antimicrobial peptides (AMPs) are synthesized by all kingdoms in small-defense molecules. These peptides are short in length, generally contain fewer than 50 amino acids, have broad antimicrobial spectra, and can be categorized according to their structures and amino acid motifs. As sessile organisms, plants experience several types of biotic stress factors including viruses, bacteria, protozoa, fungi, nematodes, and insects. Therefore, plants can be expected to have resistance mechanisms to overcome the yield losses caused by phytopathogens. AMPs are also isolated by a wide variety of plant species to develop protection against phytopathogenic agents that cause decreases in quality and agricultural losses. To date, several types of plant AMPs such as thionins, defensins, lipid-transfer proteins, cyclotides, and snakins have been isolated from diverse plant species. Their structures, mechanisms, and activities for the regulation of plant disease resistance have been discovered. Protection against phytopathogens has been developed on the basis of AMPs by expressing the AMP-encoding genes from diverse types of organisms including animals, plants, fungi, and bacteria in transgenic plants. In addition, new synthetic peptides have been designed and constructed for the enhancement of plant disease resistance. The growing databases for AMPs resources, such as PhtyAMP, CyBase, and APD2, include valuable sequences, structures, and literature that presents subjects that are useful for further discoveries.

KEYWORDS: hypersensitive response, molecular plant–pathogen interaction, peptides, synthetic peptides, transgenic, type III secretion system

6.1 INTRODUCTION

The global distribution and extremely rich biodiversity of vascular plants are considered to be the most obvious indicators of successful adaptation. As sessile organisms, plants have developed a variety of adaptation mechanisms and resistance strategies to avoid or alleviate harmful effects of biotic and abiotic stresses (Mittler, 2006; Pieterse and Dicke, 2007). Besides abiotic stresses, because the blow-ground and aerial parts of plants are primary food sources for a large number of organisms, these parts are routinely infected by a variety of pathogenic organisms including viruses, bacteria, protozoa, fungi, nematodes, and insects (Lucas, 1998; Walters, 2011). Among pathogenic microbes, bacteria are more ubiquitous than those such as fungi and protists in natural environment and plants are presumably more exposed to bacteria because of their minute size, abundance, and species richness. Plant–bacteria interactions (mutualism, parasitism, and symbiosis) have been known for a long time (Sigee, 2005).

Once attacked by a group of bacterial pathogens, plants use both preexisting (constitutive) and induced (triggered upon attack) defense mechanisms to protect themselves. Physical barriers (e.g., cuticle, waxes, epidermis, stomata) and chemical compounds comprise the main arsenal of plant defense mechanisms. Whereas the organization and structures of antimicrobial barriers remain stable in the plants, chemical antimicrobial compounds (e.g., phenolics, terpenoids, and nitrogen-containing compounds) and small peptides can be secreted to infected areas at sufficient rates. These antimicrobial compounds can be either the products of secondary plant metabolism or directly produced by biosynthesis pathway (Castro and Fontes, 2005; Walters, 2011).

Vascular plants have considerable potential for producing a wide variety of the antimicrobial compounds and antimicrobial peptides (AMPs) that are involved in plant defense. As endogenous defense molecules, AMPs belong to the class of short (<100 amino acid residues), cationic, and amphipathic (but >30% highly hydrophobic) peptides, which exhibit microbicidal activity against most bacteria. The AMPs are largely considered to be primary antibacterial defense mechanisms; this function is probably the reason they are widely distributed in virtually all multicellular organisms. Although AMPs are thought to have developed from ancient molecules, they show extreme variability in both size and structure. No clear evolutionary conservation pattern has been observed in their sequences or (to some extent) structures, however (Broekaert et al., 1997, Pelegrini et al., 2011; Stotz et al., 2013). Plants can produce a wide range of structurally and functionally different AMPs (Table 6.1). According to the PhtyAMP database (http://phytamp.pfba-lab-tun.org/main.php), a total of 271 plant AMPs from nine families (Violaceae, Brassicaceae, Triticeae, Spermacoceae, Vicieae, Oryzeae, Santalaceae, Andropogoneae, and Amaranthaceae) have been described to date (Table 6.1) (Hammami et al., 2009). Plant AMPs have been classified into groups such as thionins, defensins, lipid-transfer proteins (LTPs), cyclotides, snakins, impatiens, shepherins, MBP-1 and MiAMP1 peptides, and hevein-, vicilin-, and knottin-like peptides (Table 6.1, Figure 6.1). These classifications are based mostly on the tertiary structures of these small peptides. Upon bacterial infection, these AMPs from different classes act either individually or synergistically (when produced by the same tissue) to minimize the harmful effects of infection.

6.1.1 THIONINS

The first evidence for the existence of small AMPs in plants was obtained by isolation of crystallized thionins from the endosperm of wheat (Balls et al., 1942). A variety number of thionins that regulate plant responses to bacteria have subsequently been identified by experimental and computational methods (Benko-Iseppon et al., 2010; Pestana-Calsa and Calsa, 2011). Basically, thionins are small (generally, 45–47 amino acid residues), low molecular-weight, sulfur-rich (cysteine, arginine, and lysine) cationic peptides. Although amino acid sequences of plant thionins are highly variable, multiple sequence alignments showed that arrangements of six cycteine residues at positions 3, 4, 16, 27, 33, and 41 and that arginine at position 10 is conserved (Broekaert et al., 1997). X-ray

TABLE 6.1
Summary of the Main Characteristic Features of Plant AMPs

Plant Antimicrobial Peptide Families	Peptide Length (avg.)	Molecular Masses (Dalton)	Average Polar Residues (%)	Average Hydrophobic Residues (%)	Average Basic Residues (%)	Average Acidic Residues (%)	Average Hydrophobicity	Target Organism
Thionins	37–47 (45)	3845.11–5319.94 (4884.27)	22.97	9.02	16.49	4.4	−0.36	Antibacterial, Antifungal, Antiyeast
Defensins	45–57 (48)	4761.1–6549.62 (5461.43)	22.34	9.54	18	7.48	−0.56	Antibacterial, Antifungal
Lipid transfer proteins (LTP)	67–94 (91)	7007.16–10864.32 (9175.30)	40.2	29.82	10.87	3.12	0.09	Antibacterial, Antifungal, Antiyeast
Cyclotides	28–37 (30)	2920.79–3947.06 (3164.18)	17.05	6.9	6.82	5.18	0.29	Antibacterial, Antifungal, Antiviral
Snakins	63–67 (66)	6924.83–7696.18 (7257.08)	31.55	9.95	19.27	4.33	−0.63	Antibacterial, Antifungal
Hevein	29–45 (38)	3025–5038.34 (4030.09)	23.5	4.85	10.86	3.87	−0.46	Antibacterial, Antifungal, Antiyeast
Vicilin-like	35–67 (47)	4660.64–8883.69 (6154.80)	9.5	1.16	26.16	22.28	−2.53	Antibacterial, Antifungal
Knottins	36–38 (37)	3912–4306.59 (4048)	20.5	6	13.37	2.68	−0.61	Antibacterial, Antifungal

Note: This information was compiled from the PhytAMP database. Numbers in parentheses (as in columns 2 and 3) represent average values.

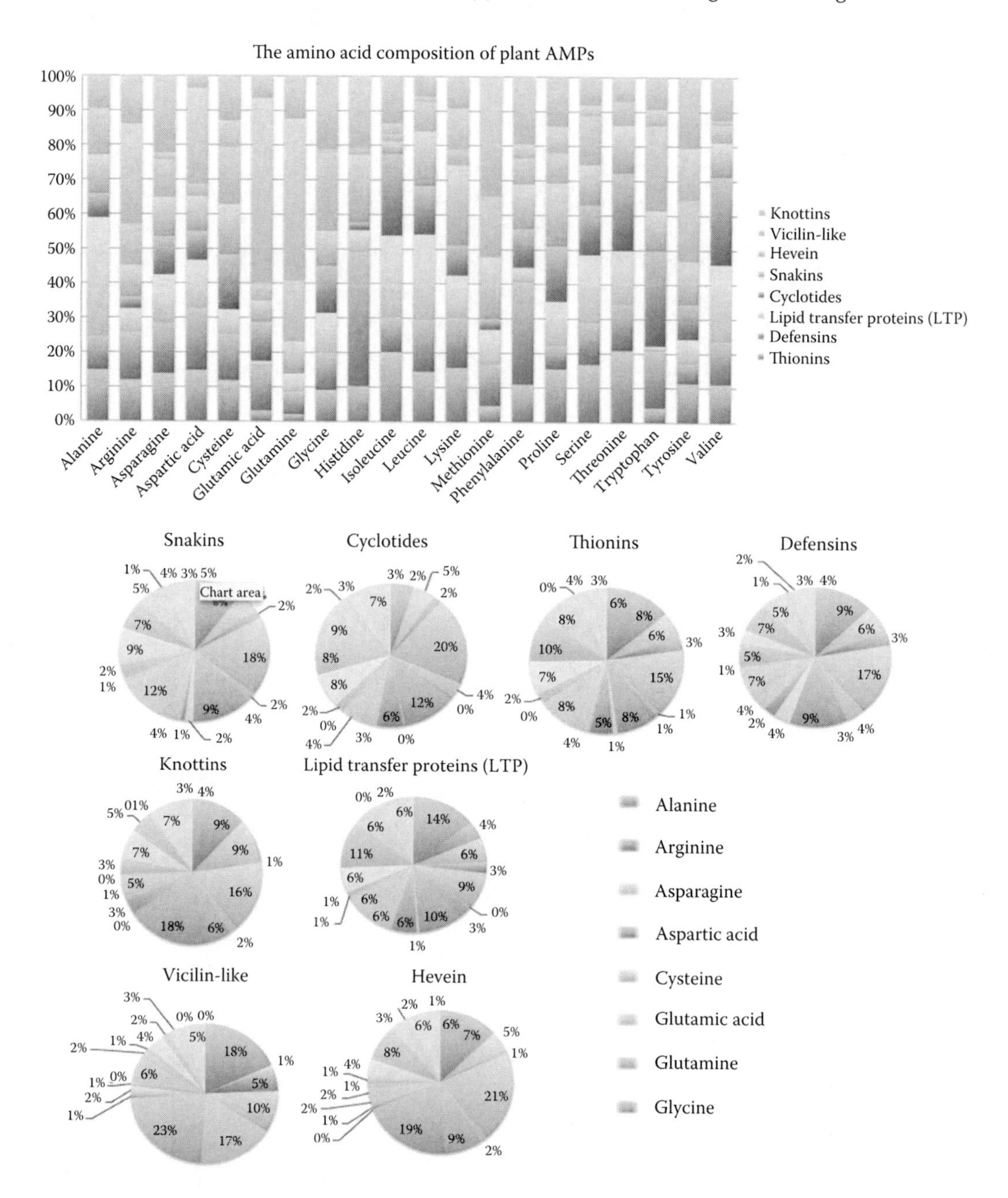

FIGURE 6.1 The average amino acid composition of the main plant AMP families (all information on the AMP was obtained from the PhytAM database (http://phytamp.pfba-lab-tun.org/about.php).

crystallography and NMR spectroscopy showed that thionins share common structural features that are characterized by a stem consisting of two antiparallel α-helixes (helix-turn-helix motif) and an arm consisting of antiparallel and double-stranded β sheets (Stec, 2006; Kaas et al., 2010; Padovan et al., 2010) (Figure 6.2a). The unique structure of plant thionins may be associated with their amphipathic nature; the hydrophilic residues are generally located on the inner surface of the arm, whereas the hydrophobic residues are located on the outer surface of the arm.

Plant thionins were once classified as either subgroup I thionins with four disulfide bridges among eight cysteine residues or subgroup II thionins with four disulfide bridges among six cysteine

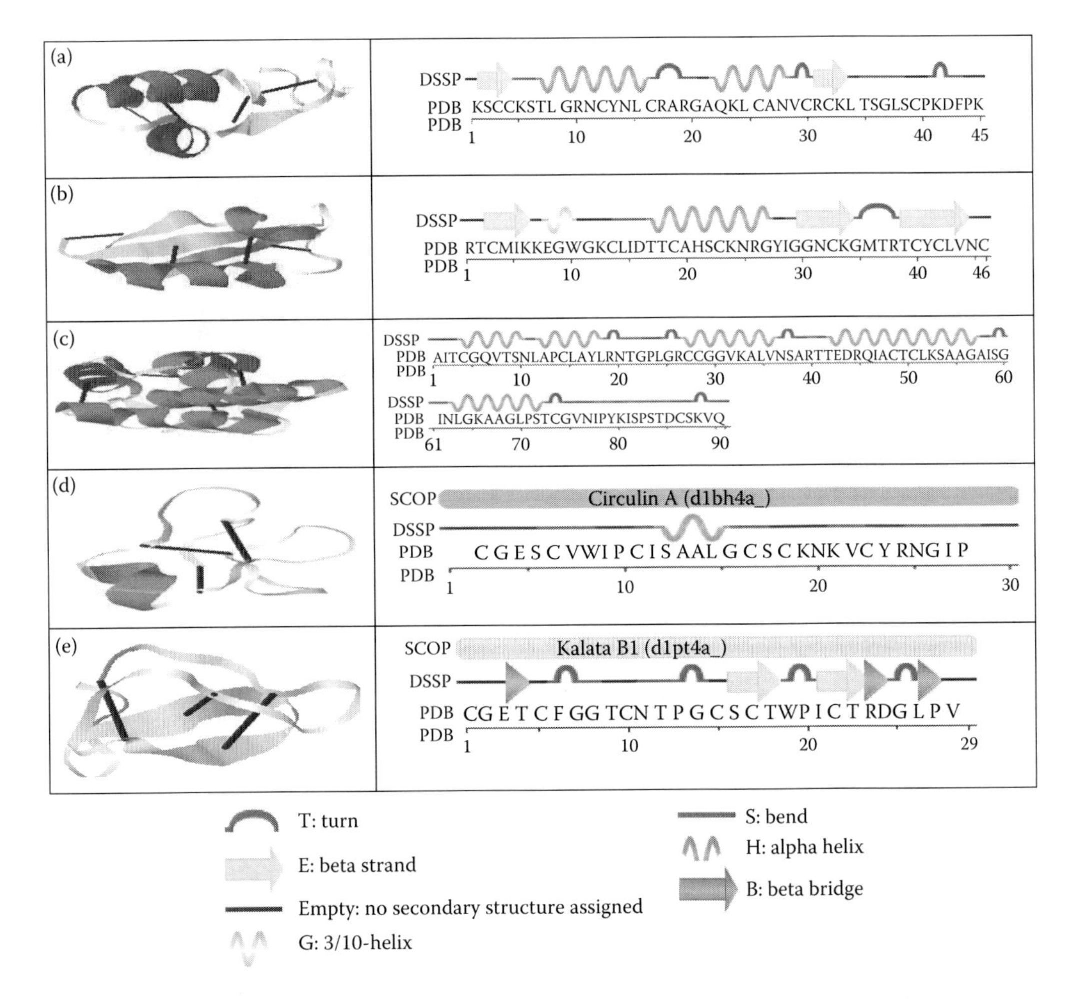

FIGURE 6.2 Schematic diagram of the 3D structure of plant antimicrobial peptides and their structural annotations. The secondary structures of peptides were extracted from the plant antimicrobial peptide database PhytAMP. (a) Thionin (alfa 1, purothionin), (b) defensin-2, (c) lipid transfer protein (LTP-1), (d) Circulin-A braclets, and (e) Katala-B1. The graphic representation of plant antimicrobial peptide sequences was obtained from the datasets of the Protein Data Bank (PDB). The broad black bars represent disulfide bridges in peptides.

residues (Broekaert et al., 1997). Plant thionins have been recently reclassified based on their disulfide bonds as well as other criteria. These include the number of basic residues and the total charges of amino acid residues. Accordingly, plant thionins fall into five classes: type I (four disulfide bridges, positively charged [+10], and highly alkaline), type II (four disulfide bridges, positively charged [+7], and moderately alkaline), type III (three disulfide bridges, positively charged [+7]), type IV (three disulfide bridges, neutral), and type V (various, modified version of other thionins) (Stec, 2006; Kaas et al., 2010).

The thionins produced and secreted by plant tissue have inhibitory effects against bacteria and fungi. These thionins are ubiquitously or selectively expressed in organs or tissues such as seed (endosperm), leaves, stems, flower buds, roots, and rhizomes (De Caleya et al., 1972; Reimann-Philipp et al., 1989; Franco et al., 2006; Lee et al., 2011). A survey in the PhtyAMP database revealed that a total of 43 thionins have been detected in several plant families. Additionally, 780 thionins (59.2% of the total) belonging to nine botanical families have been predicted using expressed sequence-tag-based data mining (Pestana-Calsa and Calsa, 2011). Although a significant number of plant-derived thionins in the database have antifungal activity, some have been found to have antibacterial effects. For instance, it was demonstrated that a mixture of α- and β-purothionins isolated from *Triticum aestivum* showed highly toxic effects against *Pseudomonas solanacearum*, *Xanthomonas phaseoli*, and *Corynebacterium michiganense*. However, α-purothionin in particular appeared to have greater activity against *X. phaseoli* than *Ps. solanacearum* (De Caleya et al., 1972; Castro and Fontes, 2005).

Empirical evidence for the antibacterial effects of thionins was also obtained from two transgenic studies. In the first study, when the expression of genes encoding α-thionin from barley endosperm were increased in tobacco plants, this overexpression generated resistance to *Pseudomonas syringae* in tobacco (Carmona et al., 1993). In the second study, Iwai et al. (2002) showed that when one of the thinning genes from oat was overexpressed in rice coleoptile, it made rice more resistant to two major seed-transmitted phytopathogenic bacteria, *Burkholderia plantarii* and *Burkholderia glumae*. Moreover, two types of thionin (Pp-AMP1 and Pp-AMP2) found in moss bamboo (*Phyllostachys pubescent*) had some toxic effect against *Er. carotovora* (IC50 = 22 μg/mL), *Agrobacterium radiobacter* (IC50 = 15 μg/mL), *Agrobacterium rhizogenes* (IC50 = 15 μg/mL), *Clavibacter michiganensis* (IC50 = 14 μg/mL), and *Curtobacterium flaccumfaciens* (IC50 = 25 μg/mL) (Fujimura et al., 2005). It was also shown that two novel AMPs (Tu-AMP1 and Tu-AMP2) belonging to the thioin family in tulip bulbs (*Tulipa gesneriana* L.) exhibited roughly similar toxic effects against these bacteria (Fujimura et al., 2004). Furthermore, a thionin (Pp-thionin) from *Pyrularia pubera* had significant inhibitory effects on *Rhizobium meliloti*, *Xanthomonas campestris*, *Micrococcus luteus*, and *Cl. michiganensis* at an IC_{50} < 50 μg/mL (Vernon et al., 1985). Two thionin-like AMPs (Fractions 1 and 3) have been recently identified in *Capsicum annuum*; in addition, it was shown that a thionin-like peptide (Fraction 1) caused growth reduction to the bacteria species *Escherichia coli* and *Pseudomonas aeruginosa* (Taveira et al., 2013).

Thionins exert their major toxic effects by changing the permeability of membranes or the formation of membrane-ion channels (Caaveiro et al., 1997; Hughes et al., 2000; Stec et al., 2004; Oard, 2011). In particular, thionins have been proposed to interact with head-groups of phospholipids in membranes and artificial liposomes (Vernon and Rogers, 1992).

This interaction between thionin and phospholipids was partly explained by Stec et al. (2004), who showed that Lys1 and Arg10 residues in thionin contribute to phosphate binding, whereas Ser2 and Tyr13 form the glycerol-binding site. Also, it was observed that membrane permeabilization triggered by thionins was dependent on the membrane phospholipid composition of Gram-negative and Gram-positive bacteria (Caaveiro et al., 1997). In other words, the membrane phospholipid composition of bacteria can alter the response to thionins secreted by a plant. For instance, a mutagenic effect of Pp-TH (a thionin from *Py. pubera*) against Gram-negative and Gram-positive bacteria was most likely related to differences in membrane phospholipid composition (Vila-Perelló et al., 2003).

Thionins can also cause the formation of ion channels in lipid bilayers when they interact with negatively charged phospholipids such as phosphatidylserin. The experiments conducted by Richard et al. (2002, 2005) showed that the functional interaction of β-purothionin with an artificial membrane bilayer caused the formation of protein channels rather than membrane disruption. The release of the vacuolar contents, which include high concentrations of AMPs, will rapidly inactivate infectious bacteria and give plants an additional protection system from bacterial phytopathogens (Romero et al., 1997). Like other members of AMPs (except cyclopeptide), mature thionins are initially synthesized as preproproteins that include a mature thionin domain located between an N-terminal signal peptide domain and a C-terminal pro-domain. These preproproteins are intracellularly cleaved into proproteins that are further processed to generate mature thionins of approximately 3.8–5.4 kDa (Ponz et al., 1986; Bohlmann et al., 1988; Schrader-Fischer and Apel, 1994; Epple et al., 1995; Broekaert et al., 1997).

Upon pathogen infection, the expression of genes involved in thionin synthesis is likely induced by a jasmonate family of signaling compounds such as methyl jasmonate, which have a central role in plant defense. In *Arabidopsis thaliana*, the expression patterns of the *Thi2.1* and *Thi2.2* genes that encode thionins differ from each other (Epple et al., 1995). Although *Thi2.1* is expressed at high constitutive levels in flowers in response to methyl jasmonate via inoculation with the fungal pathogen *Fusarium oxysporum*, *Thi2.2* appears to be expressed in seedlings but not inducible by jasmonate (JA) (Bohlmann et al., 1998; Stotz et al., 2013). It was also reported that the overexpression of the *Arabidopsis* thionin-related gene (*AtTHI2.1*) in tomato substantially reduced the number of Gram-negative bacteria *Ralstonia solanacearum* in stem tissue and that the overexpression of *AtTHI2.1* induced high resistance to wilt bacteria in tomato (Chan et al., 2005).

6.1.2 Defensins

Like other plant AMPs, plant defensins are cationic and short, basic AMPs that play an important role in plant innate immunity against invasive phytopathogenic organisms (Kragh et al., 1995; Neumann et al., 1996; Houlne et al., 1998; Odintsova et al., 2007; Carvalho and Gomes, 2009). Currently, a total of 55 plant defensins belonging to 23 species have been reported; experimental research on interactions between plant defensins and pathogens is still in progress. The presence of plant defensins has been detected in main vegetative and generative tissues, including leaves, stems, tubers, flowers, and seeds (Mendez et al., 1990; Moreno et al., 1994; Kragh et al., 1995). The molecular weight distribution of cysteine-rich plant defensins ranged from 5 to 7 kDa; and mature domains within pre-defensins are 45–50 amino acids in length. Plant defensins were previously considered to be a subgroup of the thionin family and were termed γ-thionins due to their positively charged residues, cysteine content, and size (Colilla et al., 1990; Mendez et al., 1990).

They are differentiated from other thionins (α- and β-thionins), however, in terms of their structures and their numbers of disulfide bonds among their cysteine residues. Whereas plant defensins consist of a triple-stranded antiparallel β-sheet and an α-helix located at an adjacent but antiparallel β-sheet, thionins contain two antiparallel α-helixes and one double-stranded antiparallel β-sheet (Figure 6.2b). Another difference between plant defensins and thionins concerns a structural motif; two cysteines in the Cys-XXX-Cys segment of the α helical structure are connected by two disulfide bridges to cysteine residues of β-sheet in defensin structures (Broekaert et al., 1997). Therefore, plant defensins or γ-thionins are now considered to be a new superfamily of AMPs. Previous studies have also revealed that although animal and fungi defensin structures generally contain three, four, or five disulfide bonds, plant defensins have four or five disulfide bonds. The defensins that exhibit toxicity against bacteria are commonly animal or fungi defensins, because they contain three disulfide bonds (Cândido et al., 2011).

Studies of plant defensins have mostly concentrated on antimicrobial effects against fungi rather than bacteria (Pelegrini and Franco, 2005; Van der Weerden et al., 2010). The cellular toxicity of plant defensins, which has been particularly well studied in fungi, includes work on protein synthesis inhibition (Mendez et al., 1990), membrane permeabilization (Van der Weerden et al., 2010), cell death and granulation of cytoplasm (Van der Weerden et al., 2008), inhibition of growth (Carvalho and Gomes, 2009), ROS accumulation and the induction of apoptosis (Aerts et al., 2011), and inhibition of α-amylases and proteases (Colilla et al., 1990; Thomma et al., 2002).

More-recent defensin studies of plants suggested that defensins may be crucial to protection against bacterial infection, despite the fact that the detailed mechanisms of plant defensins against bacteria have not been precisely resolved. For example, a peptide derived from tomato defensin (SolyC07g007760) has shown strong antimicrobial activity against Gram-negative bacteria *Salmonella enterica* and *Helicobacter pylori* (Avitabile et al., 2013). Another endogenous defensin peptide isolated from *Medicago sativa* seeds exhibited a strong antimicrobial activity against Gram-negative bacteria such as methicillin-resistant *Staphylococcus aureus* and strains of vancomycin-resistant *Enterococcus faecium*, but this peptide possesses lower antimicrobial potency for a Gram-negative strain of *E. coli* (Aliahmadi et al., 2012). It is also possible that the transgenic expression of defensin genes in plants can increase their antibacterial effects.

For instance, the wasabi defensin gene was overexpressed in *Phalaenopsis* orchid that used *Agrobacterium tumefaciens* carrying a plasmid that contained a wasabi defensin gene selectable marker *npt*II and *hpt* genes. After transformation, most of the transgenic *Phalaenopsis* orchids displayed striking resistance to *Er. carotovora*, which causes soft rot disease in the control plant (Sjahril et al., 2006). When fabatine (which, like the defensins, comprises small peptides with antimicrobial properties) was extracted from the broad bean *Vicia faba*, it exhibited strong bactericidal activity against the Gram-negative bacterium *Ps. aeruginosa* but less antibacterial activity against *E. coli* (Zhang and Lewis, 1997). Cp-thionin II (a small peptide within a defensin isolated from cowpea) had a strong lethal effect on *St. aureus* and *E. coli* (Franco et al., 2006). Both butterfly-pea defensin (*Clitoria ternatea*, Ct-AMP1) and a potato tuber-deposited defensin displayed strong antimicrobial activity against, respectively, *Bacillus subtilis* and *Ps. solanacearum* and *Cl. michiganensis* (Moreno et al., 1994; Osborn et al., 1995).

Silverstein et al. (2005) showed that more than 300 genes encoding defensin-like (DEFL) peptides were found in *A. thaliana* genome. Along with their antibacterial activity, some of these peptides have strong mutual relationships with bacteria. For example, the nodule cysteine-rich (NCR) group of defensin-like (*DEFL*) genes is expressed in the nodules of the model legume *Medicago truncatula*. Researchers employed an Affymetrix microarray chip to monitor defensin-like (*DEFL*) gene expression changes in *M. truncatula* after inoculation with the nitrogen-fixing bacterium *Sinorhizobium meliloti*. Approximately 82% of *Medicago* DEFLs (566 of 684 genes) detected that expression in response to inoculation with *Si. meliloti*. To date, most plant defensins that have been studied have been synthesized with signal peptide sequences that are not in propeptide form, although tomato and potato flower-specific defensins are synthesized as preproproteins with a propeptide located in the C-terminal region (Chiang and Hadwiger, 1991; Gu et al., 1992; Karunanandaa et al., 1994; Milligan and Gasser, 1995; Terras et al., 1995). The differences between biosynthetic pathways of plant defensins are probably related to their modes of action.

6.1.3 Lipid-Transfer Proteins

Specific and non-specific LTPs (nsLTPs) belong to a family of AMPs that has the longest peptide sequence (>90 amino acids) among plant AMPs. The first experimental evidence of the presence and function of LTPs in plants was obtained from an *in vitro* study which showed that the transfer of radiolabeled phosphatidylcholine from artificial liposomes to chloroplast membranes was directed by an LTP from spinach leaves (Miquel et al., 1988). To date, a total of 45 LTPs belonging to 17 species with an average length of 91 amino acids have been identified in the current version of PhytAMP. However, according to Pestana-Calsa and Calsa (2011), plant LTPs also constitute 16% of all AMP sequences deposited in various databases, which a makes plant LTPs the most abundant group of AMPs. This value is probably exaggerated due to the multifunctionality of LTP genes.

Based on their molecular mass values, plant AMPs generally fall into two subfamilies (LTP1 and LTB2). The peptides in the LTP1 family are 90–95 amino acids long, with a mass of approximately 10 kDa. LTP2 has some similar features to LTB1, but its members are typically 70 amino acids in length, with a mass of 7 kDa (Broekaert et al., 1997; Castro and Fontes, 2005; Pelegrini et al., 2011; Stotz et al., 2013). Several experimental tools such as NMR, X-ray, and Raman crystallography have been used for structural determination of LTPs (Kader, 1996, 1997; Yeats and Rose, 2008). The first tertiary structure of a plant LTP from *Zea mays* was determined by X-ray crystallography (Shin et al., 1995). The structures of LTPs from wheat (Gincel et al., 1994) and barley (Heinemann et al., 1996) were resolved by NMR. These findings showed that plant LTPs form a region whose structural motif consists of four a-helix bundles (nearly half of the whole structure) connected by disulfide bridges (Figure 6.2c). There also appears to be an internal cavity that is located centrally and generally bounded by hydrophobic residues. This internal hydrophobic cavity serves as an interaction motif with the aliphatic chain of lipids.

Plant LTPs can have the capacity to bind and transfer polar lipids such as phosphatidylcholine, phosphatidylethanolamine, and phosphatidylinositol, as well as galactolipids (Castro and Fontes, 2005). When plant LTPs interacted with those of lipids, subtle conformational changes occurred in their C-terminal regions (Lerche et al., 1997; Lerche and Poulsen, 1998). The expression of plant LTP genes has been observed in various tissues, including roots (Blilou et al., 2000; Pii et al., 2009), leaves (Molina et al., 1993; Maldonado et al., 2002; Sarowar et al., 2009; Edstam et al., 2013), stems and flower buds (Pyee et al., 1994; Wang et al., 2012), trichomes (Choi et al., 2012), and cell walls mainly in epidermal cells (Thoma et al., 1994; Nieuwland et al., 2005; DeBono et al., 2009). The wide distribution of plant LTPs in virtually all plant tissue is possibly related to their other biological functions, in addition to their antimicrobial effects. Basically, the main function of plant LTPs is the transference of lipids from donor to acceptor membranes. However, recent experimental studies about the function of plant LTPs revealed that the antifungal activity of wheat LTPs was not correlated with their lipid-binding function (Sun et al., 2008). Moreover, a mutation in some rice LTPs showed that the lipid-binding activity of its LTP was unrelated to anti-fungal activity (Ge et al., 2003).

The antimicrobial activities of LTPs obtained from various plant species have been investigated by microbiological assay methods. Four homogeneous proteins (Cw_{18}, Cw_{20}, Cw_{21}, Cw_{22}) and one homologous protein (Cw_{41}) corresponding to LTPs were isolated and purified from barley and maize leaves, respectively. These purified LTPs showed a growth-inhibition effect against pathogen *Cl. michiganensis* ssp. *sepedonicus* and *Ps. solanacearum* (Molina et al., 1993). Transgenic tobacco that expressed LTP2 from barley reduced the symptoms of wildfire disease caused by *Ps. syringae* pv. *tabaci*. Additionally, pathogen growth and lesion areas in transgenic tobacco were reduced compared to control tobacco (Molina and Garcia-Olmedo, 1997). A nsLTP displaying features similar to those found in wheat and maize seeds was extracted from seeds of mung bean (*Phaseolus mungo*). This compound exhibited antibacterial activity toward *St. aureus* but not *Salmonella typhimurium* (Wang et al., 2004).

The expression level of CALP cDNA clones (CALTPI, CALTPII, and CALTPIII) that encoded LTPs in pepper (*Ca. annuum*) was induced in response to *X. campestris* pv. *vesicatoria*, *Phytophthora capsici*, and *Colletotrichum gloeosporioides* infections in leaf, stem, and fruit tissues. Using *in situ* hybridization, the localization of CALTPI mRNA was detected in phloem cells of the vascular tissues of pepper leaf, stem, and fruit tissues upon infection by these pathogens. Although the expression levels of CALTPI and CALTPIII were found to be high in various tissues upon pathogen infection in pepper, the expression of CALTPII was induced by bacterial infection (*X. campestris* pv. *vesicatoria)* and not by fungal infection. This selective toxic effect of pepper LTP (CALTPII) might be related to its mechanism of action against prokaryotic and eukaryotic organisms (Jung et al., 2003).

The antifungal and antibacterial properties of LTP110 from rice were also tested against two rice pathogens, *Pyricularia oryzae* and *Xanthomonas oryzae*, which cause blast and blight disease of rice. LTP110 inhibited the germination of fungal (*Pyricularia oryzae*) spores, but its growth-inhibitory effects against *Xanthomonas oryzae* spores appeared to be slightly lower than its antifungal effects (Ge et al., 2003). Overexpressions of barley LTP gene in *A. thaliana* and *Nicotiana tabacum* have been shown to enhance the bacterial resistance of these transgenic plants to *Ps. syringae* (Molina and Garcia-Olmedo, 1997). A non-specific LTP-like antimicrobial protein that was isolated and characterized from a medicinal plant (*Leonurus japonicus*), the heat-stable antimicrobial (LJAMP2) peptide, was active against *Ba. subtilis*. The half-maximal inhibitory concentration (IC 50) of LJAMP2 for *Ba. subtilis* is about 15 microM (Yang et al., 2006).

6.1.4 Cyclotides

The existence of cyclotides has been observed in some dicotyledons and monocotyledons; they are particularly abundant within species belonging to Violaceae, Rubiaceae, Apocynaceae,

Curcubitaceae, Poaceae, Euphorbiaceaes and Fabaceae (Mulvenna et al., 2006; Tan and Zhou, 2006; Gruber et al., 2008; Poth et al., 2011; Nguyen et al., 2013). Members of this novel class of plant-derived, disulphide-rich small peptides (27 to 37 amino acids in length) are characterized by their unique structure, which comprises a head-to-tail cyclic peptide backbone and a cystine knot formed by three conserved disulfide bonds (Simonsen et al., 2005; Ireland et al., 2006; Padovan et al., 2010). The N- and C-terminal amino acids of plant cyclotides are connected via a peptide bond to form a stable circular structure. This unique structural pattern provides high levels of resistance to thermal, chemical (chaotropic agents), and enzymatic (proteases) degradations (Kaas et al., 2010). Unlike other plant AMPs, cyclotides show no cationic character; however, they do contain a group of solvent-exposed hydrophobic residues (Stotz et al., 2013). Plant cyclotides fall into two main groups based on their 3D structure and residue composition (the latter is indicated by the presence or absence of cis-proline residue in loop 5). The first group contains bracelets whose tertiary structure is composed of α-helices and strands, whereas the second group are Möbius peptides composed of β-sheets and turns (Figures 6.2d and 6.2e). The main difference between the groups is the existence of a cis-Pro bond in loop 5 of the Möbius cyclotides, which creates a 180° twist in the loop of the peptide backbone; this is not seen in the bracelet cyclotides (Craik et al., 2006; Pinto et al., 2012). It was also reported that bracelet cyclotides appeared to be more hydrophobic than Möbius cyclotides, due to the positions of the hydrophobic residues within the molecule (Ireland et al., 2006).

A brief survey of plant antimicrobial databases showed that a total of 76 entries belonging to 11 species and 431 entries (natural and engineered) belonging to 42 species have been described just in PhytAMP and CyBase (http://www.cybase.org.au) (Wang et al., 2008; Hammami et al., 2009). In addition to anticancer (Gerlach et al., 2010), anti-HIV (Henriques et al., 2011) and antifungal (Prema and Pruthvi, 2012) activities, plant cyclotides have antibacterial activity. The hydrophobic patches of the plant cyclotides or their partly amphipathic nature designate them as appropriate candidates for antibacterial potentiality (Tam et al., 1999; Kaas et al., 2010). For instance, four macrocyclic cystine-knot peptides (kalata, CirA, CirB, and cyclopsychotride) were isolated from coffee plant (*Coffea arabica*) and then obtained by solid phase peptide synthesis for further research. It was found that Kalata and CirA showed antibacterial activities toward *St. aureus* but not toward Gram-negative bacteria such as *E. coli* and *Ps. aeruginosa*. CirB showed particularly antibacterial effect against *E. coli*, however (Tam et al., 1999), and Kalata B1 was also active against *E. coli* (Gran et al., 2008). The inconsistency between these two studies was probably from the result of their methodologies and the concentration levels of Kalata B1.

The antibacterial activity of bracelet cyclotide cycloviolacin O2 (cyO2) isolated from *Viola odorata* was tested against bacteria such as *Sa. enterica* svar. *typhimurium* (LT2), *E. coli*, and *St. aureus*. CyO2 was found to exert strong inhibition on the growth of *Sa. enterica* svar. *typhimurium* LT2 and *E. coli*; only *St. aureus* was resistant to its bactericidal effects. A time-kill assay showed that cyO2 efficiently killed the Gram-negative bacteria species *Klebsiella pneumoniae* and *Ps. aeruginosa*. Therefore, it was concluded that cy02 has high antibacterial activity against Gram-negative bacteria in particular (Pränting et al., 2010). Using SPE-C18 column chromatography, another cyclotide was isolated and semi-purified from Iranian *Vi. odorata*. Such semi-purified isolates (minimal inhibitory concentration or MIC = $1 \cdot 6$ mg mL^{-1}) contain cyclotides, as shown by their displays of inhibitory activity against Gram-positive *St. aureus* (Zarrabi et al., 2013). One possible explanation for the modes of action of cyclic peptides that include cyclotides is the selective interaction of these cyclic peptides with the lipid membranes of bacteria. The unique structures and hydrophobic nature of cyclotides play an important role in their antibacterial activity (Barbosa et al., 2011).

6.1.5 Snakins

Snakins, a new family of plant AMPs, are small cationic AMPs that are present in various plants but mainly in the Solanaceae family. Their expression is tissue-specific. Although the primary sequence of plant snakins is 63–66 amino acids in length, with an average of ~65 residues, a newly identified snaking, Snakin-Z derived from *Zizyphus jujuba*, is 50 amino acids in length (Segura et al., 1999; Berrocal-Lobo et al., 2002; Daneshmand et al., 2013). Generally, cysteines appear to be more-abundant residues of snaking peptides (they constitute approximately 18% of the total peptides). Snakin-1 (SN1) and snakin-2 (SN2) isolated from potato tubers were conventionally named snakins due to their sequence similarity to disintegrin-like proteins from various snake venoms. Sequence-similarity analysis also revealed that snakins share homology with gibberellin-stimulated transcripts GAST and GASA, which are involved in different plant developmental process. Snakin-2 is synthesized as preproproteins with an N-terminal peptide, which is proteolytically processed to a mature active form, whereas snakin-1 is preceded by a signal peptide sequence (Stotz et al., 2013). Northern blot analysis detected expression of a *Snakin-1* gene in some tomato organs (tubers, stems, axillary buds, and various flower parts) but not in tomato roots, leaves, and stolons (Segura et al., 1999). The expression level of the *Snakin-1* gene in tomato petals and carpels was found to be significantly higher than in sepals and stamens. Expression of the *Snakin-2* gene was detected in tubers, stems, shoot apex, and leaves, but not in roots and stolons (Berrocal-Lobo et al., 2002). The function of snaking genes in seed tissue was shown for the first time in avocado seed (Daneshmand et al., 2013).

Snakins, which are known to have both antibacterial and antifungal activities, present different mechanisms of action than other families of AMPs. Generally, snakins bind to lipid membranes to cause a rapid aggregation of liposomes rather than degradation of the membranes themselves (Caaveiro et al., 1997). A recent study discovered that a snakin gene (*Snakin-Z*) isolated from jujube fruit (*Z. jujuba*) had antibacterial and antifungal activities against *St. aureus* with the MIC value of 28.8 μg/mL, and against *Phomopsis azadirachtae* with the MIC value of 7.65 μg/mL (Daneshmand et al., 2013). The snakin gene from avocado seeds (*Persea americana* var. *drymifolia*) has been expressed in a bovine endothelial cell line (BVE-E6E7). Conditioned media from the transfected cells exhibited antibacterial activities against *E. coli* and *St. aureus* (Guzmán-Rodríguez et al., 2013). In order to evaluate the synergistic effects of snakin-1 (*SN1*) and defensin-1 (*PTH1*) genes from *Solanum tuberosum*, a fusion gene (*SAP*) was generated and co-expressed in *E. coli*. Synergistic expression of *SN1* and *PTH1* displayed higher antibacterial activity than individual expression against *Ps. syringae* pv. *syringae*, *Ps. syringae* pv. *tabaci*, and *Colletotrichum coccoides*.

The *SAP* expression showed the highest antimicrobial activity against *Cl. michiganensis* ssp. *sepedonicus*, causal agent of bacterial ring rot disease of potato (Kovalskaya et al., 2011). Transgenic tomato plants that express snakin-2 (*SN2*) or extensin-like protein (*ELP*) were produced using an *Ag. tumefaciens*-mediated gene transfer. Overexpression of both *SN2* and *ELP* mRNAs was confirmed by Northern blot analysis. A reduction in the viable *Cl. michiganensis* ssp. *michiganensis* population was detected in transgenic tomato plants that overexpressed snakin-2 (*SN2*) or extensin-like protein (*ELP*) genes. Transgenic tomato plants were delayed in wilt-symptom development caused by *Cl. michiganensis* ssp. *michiganensis* (Balaji and Smart, 2012). In transgenic potato plants expressing *SN1* gene from other *Solanum* species, showed an up-regulation of the expression level of *SN1* in response to *Botritys cinerea* infection; however, this expression was decreased following infection with the bacteria *Ra. solanacearum* and *Erwinia chrysanthemi*. Transgenic potato did show significant protection against the important potato pathogen *Er. carotovora* (Almasia et al., 2008).

6.2 MECHANISMS OF ACTION OF PLANT AMPs ON BACTERIA

The unique structures, amphipathic qualities, and cationic acceptor abilities of plant AMPs make them attractive as antimicrobial agents. The mechanism of action of plant AMPs is predominantly governed by electrostatic interactions between charged peptide residues and charged surfaces, which results in membrane collapse (Brogden, 2005). The positively charged residues of cationic AMPs can bind electrostatically to anionic phospholipids (phosphatidylglycerol and cardiolipin) and lipopolysaccharides of Gram-negative bacteria, as well as to anionic teichoic and lipoteichoic acids located in cell walls of Gram-positive bacteria. The interactions among AMPs and membrane phospholipids can stimulate various signal transduction pathways, leading to cell death.

Sufficient concentrations of AMPs on the surfaces of bacterial membranes is critical for their subsequent actions. It was suggested that determining an optimum threshold concentration for AMPs is affected by both intrinsic (the peptide's ability to induce self-assembly) and extrinsic factors (phospholipid membrane composition, liquidity of membrane). Membrane permeation and the disruption of bacterial membrane integrity by pore formation appear to be the main mechanisms of AMP actions on bacterial membranes. Three main models have been proposed to describe the pore-formation processes: (i) carpet model, (ii) barrel-stave model, and (iii) toroidal pore or wormhole model.

In the carpet model, which describes an electrostatically driven interaction, the hydrophobic surfaces of cationic AMPs interact with negatively charged head-groups of phospholipids in bacterial membranes. The peptides then align themselves parallel to the membrane surface and cover ("carpet") it. Once the critical concentration of peptides is reached, the peptide monomers on the membrane surface orient preferentially toward the membrane's hydrophobic core, thereby causing membrane disintegration and permeabilization by pore formation. These secondarily formed pores permit the passage of low-molecular-weight molecules and ions into cells (Stec, 2006; Wimley, 2010; Pelegrini et al., 2011).

In the barrel-stave model, cationic AMPs with α-helical structures interact with each other on the outer surfaces of the bacterial lipid bilayer to form a complex structure. The hydrophilia of the peptides plays a central role in the formation of this complex structure. When the peptides have accumulated to sufficient concentrations to be effective, the peptides in the structure are oriented perpendicular to the plane of the membrane. This type of orientation permits the interaction of hydrophobic surfaces of peptides with the hydrophobic region of the phospholipids bilayer. The hydrophilic peptides orient longitudinally to surround a hydrophilic pore that extends along the width of the cell membrane. Briefly, a barrel stave-like protein complex formed by AMP monomers acts as a hydrophilic channel in bacterial membranes, where it causes cell lysis due to membrane disintegration (López-Meza et al., 2011; Pelegrini et al., 2011; Rahnamaeian, 2011; Park and Hahm, 2005).

The other primary model of plant AMP interactions with polar and nonpolar residues of bacterial membranes, the toroidal pore model, occurs in two stages. It is reported that the concentration levels of AMPs on bacterial membranes is a determining factor in the active mode of this model (Meincken et al., 2005). In the first step of toroidal pore formation, AMPs are aligned parallel to bacterial membrane surfaces at low peptide concentrations (inactive state) and their polar regions interact with the lipid head-groups of membranes. In the second step, the peptide concentration exceeds the critical level and thus induces perpendicular reorientation of AMPs to hydrophobic regions of phospholipid bilayers (active state) to form a hydrophilic transmembrane pore. In a toroidal pore model, cationic peptides with α-helical structures are stabilized by lipid head-groups in pores.

6.3 RESISTANCE MECHANISMS OF PATHOGENIC BACTERIA AGAINST ANTIMICROBIAL DEFENSES IN PLANTS

Plant pathogenic bacteria possess strategies to overcome the detrimental effects of the antimicrobial defenses in plants. Pathogenic bacteria may respond to plant defenses with two-component systems (TCSs); these involve a sensor kinase (His protein kinase, HPK) and a response-regulator protein (RR). In response to an environmental stimulus, HPK regulates the activity of its cognate RR through phosphorylation (Chang and Stewart, 1998).

Pleiotropic TCS PhoP/PhoQ is involved in the AMP resistance in *Xanthomonas oryzae* pv. *oryzae* (Lee et al., 2008), *Er. chrysanthemi* (Llama-Palacios et al., 2003), and *Erwinia amylovora* (Nakka et al., 2010). Rice lines carrying XA21 protein are resistant to *X. oryzae* pv. *oryzae* strains that have AvrXA21 activity, but not the strains that lack AvrXA21 activity. PhoPQ TCS is required for AvrXA21 to induce a pathogenic state in response to changing extracellular conditions (Lee et al., 2008). By contrast, PhoPQ confers resistance in *Er. chrysanthemi* and *Er. amylovora* against AMPs (Llama-Palacios et al., 2003; Nakka et al., 2010). PhoPQ TCS also provides resistance to AMPs in *Salmonella typhymurium* (Gunn and Miller, 1996; Shi et al., 2004), which is hosted by plants as well (Schikora et al., 2012). PhoP-activated *pagB* and *ugtL* are responsible for resistance to the AMPs magainin, polymyxin, azurocidin, bactericidal/permeability-increasing proteins (BPI or CAP57), protamine, and polylysine (Gunn and Miller, 1996; Shi et al., 2004). Another TCS, BfdR, takes a role in biofilm formation in *X. axonopodis* pv. *citri* (Huang et al., 2013). Biofilms form a diffusion barrier to positively charged cationic AMPs (CAMPs) (Gooderham et al., 2008). Additionally, the lipid lysyl-phosphatidyl glycerol in membranes of *Rhizobium tropici* was shown to be protective against polymyxin B under acidic growth conditions (Sohlenkamp et al., 2007).

A type III secretion system (TTSS) encoded by *hrp* (hypersensitive reaction and pathogenicity) genes is required for the transfer of virulence proteins to plant cells in several plant pathogenic bacteria (Alfano and Collmer, 1997; Abramovitch and Martin, 2004; Nomura et al., 2005). After the assembly of a basal body into the inner and outer bacterial membranes, early and middle substrates are used for the respective formations of needle and tip. When the tip comes into contact with a host cell, a pore is formed on the host-cell membrane and the effector proteins are secreted into the host-cell cytoplasm (Deane et al., 2010). *Pseudomonas syringae* pathovars resist pathogen-associated molecular pattern-triggered immunity (PTI) and/or effector-triggered immunity (ETI) responses in plants by using a number of TTSS effectors (Guo et al., 2009). Production of antimicrobial phytoalexins was inhibited in bean after infiltration with *Ps. syringae* pv. *phaseolicola* (Jakobek et al., 1993).

Hou et al. (2011) reported that the *Ps. syringae* pv. *phaseolicola* effector HopF1 inhibits PTI in common bean. The TTSS effector protein AvrPtoB in *Ps. syringae* pv. *tomato* induces plant susceptibility to bacterial infection (Abramovitch et al., 2003). *Pseudomonas syringae* TTSS also inhibits the production of defense proteins in *Arabidopsis* (Hauck et al., 2003). A TTSS effector, HopAM1 from *Ps. Syringae*, manipulates abscisic acid responses to suppress pathogen defenses in *Arabidopsis* (Goel et al., 2008). Many TTSS effector proteins, including nodulation outer protein L (NopL) in *Rhizobium*, also suppress production of pathogenesis-related defense proteins in plants such as tobacco and lotus (Bartsev et al., 2004). A TTSS effector, cyclic β-(1,2)-glucan, in bacterial phytopathogen *X. campestris* pv. *campestris* locally and systemically suppresses particular host defenses (Rigano et al., 2007).

Another resistance mechanism of phytopathogenic bacteria during the course of infection in plants is the production of extracellular metabolites such as phytotoxin coronatine (COR) that mimics methyl jasmonate (MeJA); the latter triggers plant hormonal-defense signaling (García Véscovi et al., 1996; Lee et al., 2013). Cor gene mutant *Arabidopsis* (Kloek et al., 2001) and tomato (Zhao et al., 2003) plants were identified as resistant against *Ps. syringae*. Extracellular metabolites of the *Ps. syringae* pv. *tabaci* and *Ps. syringae* pv. *tomato* suppress early defense responses, stomal closure, and hypersensitive response (HR) in their respective host plants, *Nicotiana benthamiana* and tomato (Lee et al., 2013).

6.4 TRANSGENIC PLANTS THAT EXPRESS AMPs

Phytopathogens cause important agricultural production decrements all around the world. Because of environmental concerns such as water pollution, microorganism resistance to chemicals, and economic issues, utilization of pesticides as chemical substances to combat viral, bacterial, and fungal pathogens is not preferred. Therefore, advances in plant biotechnology and genetic engineering provide novel approaches to overcome phytopathogen-caused yield losses (Sarika et al., 2012). One application involves the construction of genetically modified plants that are resistant to plant pathogens. Studies on the expression of the genes that encode AMPs in plants began in the late 1990s. For example, *Thi2.1* (a gene-encoding thionin) overexpressed in *Arabidopsis* was used to enhance resistance against *F. oxysporum* f.sp. *matthiolae* (Epple et al., 1997).

The structural and functional properties of the AMPs employed for the production of transgenic plants that are resistant to phytopathogens are divided into five categories. Category i AMPs are directly toxic to pathogens or reduce their growth; category ii AMPs destroy or neutralize some phytopathogen components; category iii AMPs enhance a plant's structural defenses; category iv AMPs regulate signaling pathways that can activate plant defenses; and category v AMPs are capable of interacting with avirulence (Avr) factors and are involved in HR (Punja, 2001).

AMPs detected from humans and other vertebrates were found to be protective against phytopathogens. For example, a human cathelicidin AMP, LL-37 (hCAP18), was stably expressed in homozygous Chinese cabbage lines as resistant against a bacteria, *Pectobacterium carotovorum* ssp. *carotovorum* (*Pcc*), and fungal pathogens *F. oxysporum*, *Colletotrichum higginsianum*, and *Rhizoctonia solani* (Jung et al., 2012). The same peptide, LL-37, also leads to enhanced protection against *Pcc* and *X. campestris* pv. *vesicatoria* (*Xcv*) in tomato (Jung, 2013). MSI-99, an analog of magainin 2 that is secreted as a defense peptide from the skin of African clawed frog (*Xenopus laevis*), was expressed in tobacco to enhance resistance against *Ps. syringae* pv. *tabaci* and the fungi *Aspergillus flavus*, *Fusarium moniliforme*, and *Verticillium dahliae* (DeGray et al., 2001), as well as *Sclerotinia sclerotiorum*, *Alternaria alternata*, and *Bo. cinerea*. In addition, the MSI-99 was expressed in banana to confer protection against *F. oxysporum* f.sp. *cubense* and *Mycosphaerella musicola* (Chakrabarti et al., 2003).

An N-terminal-modified AMP temporin A (MsrA3), secreted from skin of the European red frog (*Rana temporaria*), was expressed in potato and offered resistance against *Er. carotovora* as well as the fungi *Phytophthora infestans* and *Phytophthora erythroseptica* (Osusky et al., 2004). Esculentin-1, a 46-residue AMP of the green frog (*Rana esculenta*) skin secretion, was expressed in tobacco and provided enhanced resistance against *Ps. syringae* pv. *tabaci* and *Ps. aeruginosa* (Ponti et al., 2003). Transgenic potato that expressed dermaseptin from the waxy monkey leaf frog (*Phyllomedusa sauvagii*) and the chicken (*Gallus gallus*) lysozyme was found to be highly resistant against *Er. carotovora* and *Streptomyces scabies* as well as the fungi *Ph. infestans*, *Rh. solani*, and *Fusarium solani* (Rivero et al., 2012).

Antimicrobial peptides from diverse invertebrates also have the ability to protect plants against phytopathogens. For example, when the cationic lytic peptide cecropin B from the giant silk moth (*Hyalophora cecropia*) was expressed in transgenic tomato, the plants were shown to be resistant against bacterial pathogens *Ra. solanacearum* and *X. campestris* pv. *vesicatoria* (Jan et al., 2010). An AMP from shrimp, Penaeidin 4-1, provides enhanced resistance against fungal disease agents *Sclerotinia homoecarpa* and *Rh. solani* in creeping bentgrass (*Agrostis stolonifera*) (Zhou et al., 2011).

Antimicrobial peptides from diverse microorganisms also confer enhanced protection against phytopathogens in plants. When expressed in rice, the *Aspergillus giganteus* antifungal protein (AFP) gave enhanced resistance against *Magnaporthe grisea* (Coca et al., 2004). The HrpZ gene of *Ps. syringae* pv. *phaseolicola* (HrpZ_{Psph}) was expressed in tobacco. The fusion protein was constructed as an *N*-terminal HrpZ and PR1 signal peptide (SP/HrpZ). The transgenic tobacco plants presented enhanced resistance against rhizomania, which causes beet necrotic yellow vein virus (Pavli et al., 2011).

Antimicrobial peptide genes taken from different plant species as transgenes also provide resistance against phytopathogens. For example, β-purothionin from wheat expressed in *Arabidopsis* conferred enhanced resistance against *Ps. syringae* pv. *tomato* and *F. oxysporum* (Oard and Enright, 2006). In another case, transgenic rice plants constitutively expressing the puroindoline genes *pinA* and/or *pinB* from wheat exhibited increased resistance against *Ma. grisea* and *Rh. solani* (Krishnamurthy, 2001). Defensin-type AMPs are widely used to develop resistance against phytopathogens via transgenic approach in tobacco, tomato, oilseed rape, rice, and papaya (Stotz et al., 2009). Transgenic tobacco plants that express a mustard defensin (*BjD*) gene are resistant to the fungal pathogens *F. moniliforme* and *Phytophytora parasitica* pv. *nicotianae*; in addition, the *BjD* provides enhanced resistance against the *Pheaoisariopsis personata* and *Cercospora arachidicola* pathogens in peanut (Swathi Anuradha et al., 2008). Alfalfa antifungal defensin peptide (alfAFP) isolated from *M. sativa* seeds offered protection against *Ve. dahliae*, an agronomically important fungal pathogen (Gao et al., 2000). When it was overexpressed in rice (*Oryza sativa* cv. Pusa basmati 1), an *Rs-AFP2* defensin gene from radish (*Raphanus sativus*) led to the suppression of *Magnaporthe oryzae* and *Rh. solani* growth (Jha and Chattoo, 2010). Construction of a hybrid peptide by snakin-1 (SN1) and defensin-1 (PTH1) AMPs from potato had protective effects against *Co. coccoides* in tobacco and *Cl. michiganensis* ssp. *sepedonicus* in potato (Kovalskaya et al., 2011).

6.5 SYNTHETIC AMPs

Although AMPs are naturally biosynthesized by several organisms, new molecules with improved activity, defined specificity, biodegradability, and enhanced toxicity can be designed with combinatorial chemistry that uses resources stored in databases (Montesinos and Bardají, 2008). For example, APD2 (http://aps.unmc.edu/AP/main.php), an AMP database, is frequently used to screen the amino acid residues and motifs in AMPs. Ala and Gly are the most abundant residues in bacterial peptides; Gly and Cys in plant peptides; Ala, Gly, and Lys in insect peptides; and Leu, Ala, Gly, and Lys in frog peptides. The abundance of Cys residue in AMPs results in disulfide bonds (which are related to β-sheets) that confer greater stability to the structure. Likewise, the increased frequency of Leu, Ala, Gly, and Lys residues in frog peptides shows a strong tendency to adopt an amphipathic helix (Wang et al., 2009). When combinatorial optimization was used for the derivation of 12 new synthetic peptides from the previously described lead compound PAF19, the new peptides PAF26, PAF32, and PAF34 were reported to have stronger activity than PAF19 against *Penicillium digitatum*, *Penicillium italicum*, and *Bo. cinerea* (López-García et al., 2002). Because of the resistance to degradation, D-enantiomers of amino acids are commonly used for designing new peptides (Blondelle et al., 2003; Marcos et al., 2008).

Because the production of stable transgenic plants is difficult and time consuming, screening promising AMPs for their ability to control plant pathogens is not straightforward. Therefore, prior to the introduction of an efficient AMP into a plant, an *in vitro* prescreening method can be performed (Visser et al., 2012). In a study, four synthetic cationic peptides (pep6, pep7, pep11, and pep20) were tested alone and in combinations *in vitro* against phytopathogens. All four peptides were found to be effective against *Ph. infestans* and *Alternaria solani*, whereas *Er. carotovora* ssp. *carotovora* and *Er. carotovora* ssp. *atroseptica* were inhibited only by pep11 and pep20 (Ali and Reddy, 2000). Additionally, the antimicrobial activity of pep3 and 22 analogues were effective against plant pathogenic bacteria *Er. amylovora*, *Ps. syringae*, and *X. vesicatoria* (Ferre et al., 2006).

Chimeric peptides can be used to control phytopathogens. N terminus-modified, cecropin-melittin cationic peptide chimera (MsrA1) was expressed in potato and conferred protection against *Phytophthora cactorum*, *Fusarium solani*, and *Er. carotovora* (Osusky et al., 2000). Another family of synthetic peptides, which was generated by linking four amino acids to fatty acids, was reported to have antimicrobial activity against *Bo. cinerea*, *Ps. syringae* pv. *lachrimans*, and *Ps. syringae* pv. *tomato* (Brotman et al., 2009).

Antimicrobial activity of a synthetic peptide, D4E1 (FKLRAKIKVRLRAKIKL), was assayed *in vitro* against many pathogens. Fungal pathogens *Thielaviopsis basicola*, *Ve. dahliae*, *F. moniliforme*, *Phytophthora cinnamomi*, and *Ph. parasitica* were found to be highly sensitive to D4E1, whereas the sensitivities of *Al. alternata*, *Colletotrichum destructivum*, and *Rh. solani* to D4E1 were lower. Two bacterial pathogens, *Ps. syringae* pv. *tabaci* and *X. campestris* pv. *Malvacearum*, were reported to be most sensitive to D4E1 (Rajasekaran et al., 2001). *In planta* antimicrobial activity of D4E1 in tobacco was shown against *C. destructivum* (Cary et al., 2000). D4E1 synthetic peptide expressed in poplar supplied protection against *Ag. tumefaciens* and *Xanthomonas populi* pv. *populi* (Mentag et al., 2003). D4E1 was also expressed in transgenic cotton but did not confer protection *in vitro* against pregerminated spores of *As. flavus*, whereas the transgenic cotton seeds inhibited extensive colonization and spread by the fungus in cotyledons and seed coats *in planta* (Rajasekaran et al., 2005).

According to the nature of cellular response and environment, *in vitro* and *in planta* activities of AMPs might not generate the same results. To resolve this issue, Visser et al. (2012) developed a prescreening assay of AMP efficiency in the plant environment. In their study, the efficacy of D4E1 synthetic peptide was tested *in planta* against *Xylophilus ampelinus* and *Agrobacterium vitis* in grapevine leaves. A comparison was made with transient D4E1-expressing or non-expressing (empty binary vector) groups to determine the difference in pathogen titers via qPCR. When the obtained values were analyzed, the average pathogen concentration in the treatment group was found to be lower than the average concentration in the control group. This work showed a reduction in pathogen concentrations as a result of transient AMP expression in plant tissue.

Another synthetic peptide, BP100, which is a strongly cationic α-helical undecapeptide with high and specific antibacterial activity, was expressed in rice; this expression provided increased resistance to the pathogens *Dickeya chrysanthemi* and *Fusarium verticillioides* as well as tolerance to oxidative stress (Nadal et al., 2012).

6.6 DISTRIBUTION OF AMPs IN DIFFERENT LIFE FORMS

All forms of life, from multicellular organisms to bacterial cells, produce a variety of ribosomally synthesized natural AMPs (Table 6.2) (Koczulla and Bals, 2003). Bactocins, the AMPs produced by bacteria, confer certain ecological niches to their producers by killing their competitors, whether these occur in the same species (narrow spectrum) or in different genera (broad spectrum) (Hassan et al., 2012). There are four classes of bacteriocins: (i) lantibiotics, which are small membrane-active peptides (<5 kDa) that contain lanthionin, an unusual amino acid, (ii) small heat-stable peptides, (iii) large molecules sensitive to heat, and (iv) a group of complex proteins that are associated with other lipid or carbohydrate moieties (Rajaram et al., 2010). For instance, lactic acid bacteria produce class ii AMPs that are effective on food spoilage bacteria; this ability makes them usable as food preservatives (Nes and Holo, 2000). AMPs have also been identified from fungi. The saprophytic ascomycete *Pseudoplectania nigrella* produces an AMP called plectasin, which is active against *Streptococcus pneumoniae* (Mygind et al., 2005). Multicellular organisms synthesize distinct types of AMPs (Zasloff, 2002). A high content of cysteine repeats connected by disulfide bridges in plant AMPs confers stability to the peptides (Broekaert et al., 1997). Members of the plant kingdom produce AMPs that are classified into distinct families such as thionins, plant defensins, LTP, cyclotide, and sinakin-type AMPs, according to their homologies at the primary-structure level (Broekaert et al., 1997).

Insect AMPs are low-molecular-weight, positively charged peptides that are classified into three families based on their sequences and structural features: (i) linear peptides forming α-helices and deprived of cysteine residues, (ii) cyclic peptides containing cysteine residues, and (iii) peptides with overrepresentations in proline and/or glycine residues (Bulet et al., 1999). AMPs in *Drosophila* are grouped according to their main biological targets, defensins against Gram-positive bacteria; cecropins, drosocin, attacins, diptericin, and MPAC against Gram-negative bacteria; and drosomycin and metchnikowin against fungi (Imler and Bulet, 2005). AMPs in other invertebrates have also been reported, such as callinectin in the crab *Callinectes sapidus* (Khoo et al., 1999), penaeidin in the shrimp *Penaeus vannamei* (Destoumieux et al., 2000), and mytilins, myticins, and mytimycins in the mussel *Mytilus galloprovincialis* (Mitta et al., 2000).

Vertebrates produce very distinct types of AMPs. Even some tissues and organs are the sources of specific active compounds. Amphibian skin contains various AMPs such as magainins/peptidylglycine-scrine (PGS) peptides, peptidylglycine-leucine carboxyarnide (PGLa), and a multitude of fragments derived from the precursors of caerulein, xenopsin, and laevitide (Barra and Simmaco, 1995). Mammalian blood is rich in AMPs because they are an important part of the immune system. Porcine leukocytes contain protegrins that combine the features of corticostatic defensins and tachyplesins (Kokryakov et al., 1993). Seven thrombin-releasable AMPs were identified from human platelets (Tang et al., 2002). Human AMPs are classified into three groups, according to

TABLE 6.2
Classification of AMPs among Some Life Forms

Organisms	Classification Base	I	II	III	IV	Reference
Bacteria	Structure	Lantibiotics	Small heat-stable peptides	Large molecules sensitive to heat	Complex proteins, with lipid or carbohydrate moieties	Rajaram et al. (2010)
Insects	Structure	Linear peptides without cysteine residues	Cyclic peptides containing cysteine residues	Peptides rich in proline and/or glycine residues	—	Bulet et al. (1999)
Drosophila	Target	Against gr(-): defensins	Against gr(+): Cecropins, drosocin, attacins, diptericin, MPAC	Against fungi: drosomycin, metchnikowin	—	Imler and Bulet (2005)
Plant	Structure	Plant defensins	Thionins	Lipid transfer proteins	hevein- and knottin-type AMPs	Broekaert et al. (1997)
Human	Structure	Defensins	Histatins	Cathelicidin	—	De Smet and Contreras (2005); Niyonsaba et al. (2010)

their structure: (i) defensins, which are cationic, non-glycosylated peptides containing six cysteine residues that form three intramolecular disulfide bridges; (ii) histatins, which are small, cationic, histidine-rich peptides present in human saliva; and (iii) cathelicidin, which is derived proteolytically from the C-terminal end of the human CAP18 protein (De Smet and Contreras, 2005; Niyonsaba et al., 2010).

6.7 CONCLUDING REMARKS

Advances in biotechnology and molecular biology can aid the development of novel approaches to combat plant diseases. Discovery of new AMPs and effectors and their detailed characterization lead to the use of AMPs against a broad spectrum of plant diseases. Studying the nature of phytopathogens and their infection mechanisms at the molecular level will provide better understanding of plant–pathogen interactions. The genomic, transcriptomic, proteomic, and metabolomic abilities of plants to deal with disease-causing pathogens will be better analyzed by novel next-generation tools. These have already generated high-throughput data that allows the development of novel strategies about AMP-based plant protection. Understanding the structural basis of AMPs will help to generate new synthetic versions and fusion proteins to regulate plant disease resistance. By using genomic data obtained from pathogen and plant species, new transgenic lines with enhanced resistance against broad-spectrum phytopathogens will be developed in the near future.

REFERENCES

Abramovitch, R.B., Khim, Y.J., Chen, S., Dickman, M.B., Martin, G.B. 2003. *Pseudomonas* type III effector AvrPtoB induces plant disease susceptibility by inhibition of host programmed cell death. *EMBO Journal* 22(1): 60–69.

Abramovitch, R.B., Martin, G.B. 2004. Strategies used by bacterial pathogens to suppress plant defenses. *Current Opinion in Plant Biology* 7: 356–364.

Aerts, A.M., Bammens, L., Govaert, G., Carmona-Gutierrez, D., Madeo, F., Cammue, B.P.A., Thevissen, K. 2011. The antifungal plant defensin HsAFP1 from *Heuchera sanguinea* induces apoptosis in *Candida albicans*. *Frontiers in Microbiology* 2: 47.

Alfano, J.R., Collmer, A. 1997. The type III (Hrp) secretion pathway of plant pathogenic bacteria: Trafficking harpins, Avr proteins, and death. *Journal of Bacteriology* 179(18): 5655–5662.

Ali, G.S., Reddy, A.S. 2000. Inhibition of fungal and bacterial plant pathogens by synthetic peptides: *In vitro* growth inhibition, interaction between peptides and inhibition of disease progression. *Molecular Plant-Microbe Interactions* 13(8): 847–859.

Aliahmadi, A., Roghanian, R., Emtiazi, G., Mirzajani, F., Ghassempour, A. 2012. Identification and primary characterization of a plant antimicrobial peptide with remarkable inhibitory effects against antibiotic resistant bacteria. *African Journal of Biotechnology* 11(40): 9672–9676.

Almasia, N.I., Bazzini, A.A., Hopp, H.E., Vazquez-Rovere, C. 2008. Overexpression of snakin-1 gene enhances resistance to *Rhizoctonia solani* and *Erwinia carotovora* in transgenic potato plants. *Molecular Plant Pathology* 9(3): 329–338.

Avitabile, C., Capparelli, R., Rigano, M.M., Fulgione, A., Barone, A. 2013. Antimicrobial peptides from plants: Stabilization of the γ core of a tomato defensin by intramolecular disulfide bond. *Journal of Peptide Science* 13(4): 240–245.

Balaji, V., Smart, C.D. 2012. Over-expression of snakin-2 and extensin-like protein genes restricts pathogen invasiveness and enhances tolerance to *Clavibacter michiganensis* subsp. *michiganensis* in transgenic tomato (*Solanum lycopersicum*). *Transgenic Research* 21(1): 23–37.

Balls, A.K., Hale, W.S., Harris, T.H. 1942. A crystalline protein obtained from a lipoprotein of wheat flour. *Cereal Chemistry* 19: 279–288.

Barbosa Pelegrini, P., Del Sarto, R.P., Silva, O.N., Franco, O.L., Grossi-de-Sa, M.F. 2011. Antibacterial peptides from plants: What they are and how they probably work. *Biochemistry Research International* 3: 250–349.

Barra, D., Simmaco, M. 1995. Amphibian skin: A promising resource for antimicrobial peptides. *Trends in Biotechnology* 13(6): 205–209.

Bartsev, A.V., Deakin, W.J., Boukli, N.M., McAlvin, C.B., Stacey, G., Malnoē, P., Broughton, W.J., Staehelin, C. 2004. NopL, an effector protein of *Rhizobium* sp. NGR234, thwarts activation of plant defense reactions. *Plant Physiology* 134(2): 871–879.

Benko-Iseppon, A.M., Galdino, S.L., Calsa, J., Kido, E.A., Tossi, A., Belarmino, L.C., Crovella, S. 2010. Overview on plant antimicrobial peptides. *Current Protein and Peptide Science* 11(3): 181–188.

Berrocal-Lobo, M., Segura, A., Moreno, M., López, G., García-Olmedo, F., Molina, A. 2002. Snakin-2, an antimicrobial peptide from potato whose gene is locally induced by wounding and responds to pathogen infection. *Plant Physiology* 128(3): 951–961.

Blilou, I., Ocampo, J.A., García-Garrido, J.M. 2000. Induction of Ltp (lipid transfer protein) and Pal (phenylalanine ammonia-lyase) gene expression in rice roots colonized by the arbuscular mycorrhizal fungus *Glomus mosseae*. *Journal of Experimental Botany* 51(353): 1969–1977.

Blondelle, S.E., Pinilla, C., Boggiano, C. 2003. Synthetic combinatorial libraries as an alternative strategy for the development of novel treatments for infectious diseases. *Methods in Enzymology* 369: 322–344.

Bohlmann, H., Clausen, S., Behnke, S., Giese, H., Hiller, C., Reimann-Philipp, U., Schrader, G., Barkholt, V., Apel, K. 1988. Leaf-specific thionins of barley—A novel class of cell wall proteins toxic to plant-pathogenic fungi and possibly involved in the defence mechanism of plants. *EMBO Journal* 7(6): 1559–1565.

Bohlmann, H., Vignutelli, A., Hilpert, B., Miersch, O., Wasternack, C., Apel, K. 1998. Wounding and chemicals induce expression of the *Arabidopsis thaliana* gene Thi2. 1, encoding a fungal defense thionin, via the octadecanoid pathway. *FEBS Letters* 437(3): 281–286.

Broekaert, W.F., Cammue, B.P.A., De Bolle, M.F.C., Thevissen, K., De Samblanx, G.W., Osborn, R.W. 1997. Antimicrobial peptides from plants. *Critical Reviews in Plant Science* 16(3): 297–323.

Brogden, K.A. 2005. Antimicrobial peptides: Pore formers or metabolic inhibitors in bacteria? *Nature Reviews Microbiology* 3: 238–250.

Brogden, K.A. 2005. Antimicrobial peptides: Pore formers or metabolic inhibitors in bacteria? *Nature Reviews Microbiology* 3(3): 238–250.

Brotman, Y., Makovitzki, A., Shai, Y., Chet, I., Viterbo, A. 2009. Synthetic ultrashort cationic lipopeptides induce systemic plant defense responses against bacterial and fungal pathogens. *Applied and Environmental Microbiology* 75(16): 5373–53739.

Bulet, P., Hetru, C., Dimarcq, J.L., Hoffmann, D. 1999. Antimicrobial peptides in insects; structure and function. *Developmental and Comparative Immunology* 23(4–5): 329–44.

Caaveiro, J.M.M., Molina, A., González-Mañas, J.M., Rodríguez-Palenzuela, P., García-Olmedo, F., Goi, F.M. 1997. Differential effects of five types of antipathogenic plant peptides on model membranes. *FEBS Letters* 410(2): 338–342.

Cândido, E.S., Porto, W.F., Amaro, D.S., Viana, J.C., Dias, S.C., Franco, O.L. 2011. Structural and functional insights into plant bactericidal peptides. In: Mendez-Vilas A (ed.), *Science against Microbial Pathogens: Communicating Current Research and Technological Advances*, vol. 2. Spain: Formatex Research Center, pp. 951–960.

Carmona, M.J., Molina, A., Fernández, J.A., López-Fando, J.J., García-Olmedo, F. 1993. Expression of the α-thionin gene from barley in tobacco confers enhanced resistance to bacterial pathogens. *The Plant Journal* 3(3): 457–462.

Carvalho, A.D.O., Gomes, V.M. 2009. Plant defensins—Prospects for the biological functions and biotechnological properties. *Peptides* 30(5): 1007–1020.

Cary, J.W., Rajasekaran, K., Jaynes, J.M., Cleveland, T.E. 2000. Transgenic expression of a gene encoding a synthetic antimicrobial peptide results in inhibition of fungal growth *in vitro* and in planta. *Plant Science* 154(2): 171–181.

Castro, M.S., Fontes, W. 2005. Plant defense and antimicrobial peptides. *Protein and Peptide Letters* 12(1): 11–16.

Chakrabarti, A., Ganapathi, T.R., Mukherjee, P.K., Bapat, V.A. 2003. MSI-99, a magainin analogue, imparts enhanced disease resistance in transgenic tobacco and banana. *Planta* 216(4): 587–596.

Chan, Y.L., Prasad, V., Chen, K.H., Liu, P.C., Chan, M.T., Cheng, C.P. 2005. Transgenic tomato plants expressing an *Arabidopsis* thionin (Thi2. 1) driven by fruit-inactive promoter battle against phytopathogenic attack. *Planta* 221(3): 386–393.

Chang, C., Stewart, R.C. 1998. The two-component system, regulation of diverse signaling pathways in prokaryotes and eukaryotes. *Plant Physiology* 117: 723–731.

Chiang, C.C., Hadwiger, L.A. 1991. The *Fusarium* solani-induced expression of a pea gene family encoding high cysteine content proteins. *Molecular Plant-Microbe Interactions*. 4: 324–331.

Choi, Y.E., Lim, S., Kim, H.J., Han, J.Y., Lee, M.H., Yang, Y., Kim, J.-A. et al. 2012. Tobacco NtLTP1, a glandular-specific lipid transfer protein, is required for lipid secretion from glandular trichomes. *The Plant Journal* 70(3): 480–491.

Coca, M., Bortolotti, C., Rufat, M., Peñas, G., Eritja, R., Tharreau, D., Martinez del Pozo, A. et al. 2004. Transgenic rice plants expressing the antifungal AFP protein from *Aspergillus giganteus* show enhanced resistance to the rice blast fungus *Magnaporthe grisea. Plant Molecular Biology* 54(2): 245–259.

Colilla, F.J., Rocher, A., Mendez, E. 1990. γ-Purothionins: Amino acid sequence of two polypeptides of a new family of thionins from wheat endosperm. *FEBS Letters* 270(1): 191–194.

Craik, D.J., Cemazar, M., Wang, C.K., Daly, N.L. 2006. The cyclotide family of circular miniproteins: Nature' s combinatorial peptide template. *Biopolymers* 84: 250–266.

Daneshmand, F., Zare-Zardini, H., Ebrahimi, L. 2013. Investigation of the antimicrobial activities of Snakin-Z, a new cationic peptide derived from Zizyphus jujuba fruits. *Natural Product Research*, 27(24): 2292–2296.

Deane, J.E., Abrusci, P., Johnson, S., Lea, S.M. 2010. Timing is everything: The regulation of type III secretion. *Cellular and Molecular Life Science* 67(7): 1065–1075.

DeBono, A., Yeats, T.H., Rose, J.K., Bird, D., Jetter, R., Kunst, L., Samuels, L. 2009. *Arabidopsis* LTPG is a glycosylphosphatidylinositol-anchored lipid transfer protein required for export of lipids to the plant surface. *The Plant Cell Online* 21(4): 1230–1238.

De Caleya, R.F., Gonzalez-Pascual, B., García-Olmedo, F., Carbonero, P. 1972. Susceptibility of phytopathogenic bacteria to wheat purothionins *in vitro. Applied Microbiology* 23(5): 998–1000.

DeGray, G., Rajasekaran, K., Smith, F., Sanford, J., Daniell, H. 2001. Expression of an antimicrobial peptide via the chloroplast genome to control phytopathogenic bacteria and fungi. *Plant Physiology* 127(3): 852–862.

De Smet, K., Contreras, R. 2005. Human antimicrobial peptides: Defensins, cathelicidins and histatins. *Biotechnology Letters* 27(18): 1337–1347.

Destoumieux, D., Munoz, M., Bulet, P., Bachère, E. 2000. Penaeidins, a family of antimicrobial peptides from penaeid shrimp (Crustacea, Decapoda). *Cellular and Molecular Life Science* 57(8–9): 1260–1271.

Edstam, M.M., Blomqvist, K., Eklöf, A., Wennergren, U., Edqvist, J. 2013. Coexpression patterns indicate that GPI-anchored non-specific lipid transfer proteins are involved in accumulation of cuticular wax, suberin and sporopollenin. *Plant Molecular Biology*, 83: 625–649.

Epple, P., Apel, K., Bohlmann, H. 1995. An *Arabidopsis thaliana* thionin gene is inducible via a signal transduction pathway different from that for pathogenesis-related proteins. *Plant Physiology* 109(3): 813–820.

Epple, P., Apel, K., Bohlmann, H. 1997. Overexpression of an endogenous thionin enhances resistance of *Arabidopsis* against *Fusarium oxysporum. Plant Cell* 9(4): 509–520.

Ferre, R., Badosa, E., Feliu, L., Planas, M., Montesinos, E., Bardají, E. 2006. Inhibition of plant-pathogenic bacteria by short synthetic cecropin A-melittin hybrid peptides. *Applied and Environmental Microbiology* 72(5): 3302–3308.

Franco, O.L., Murad, A.M., Leite, J.R., Mendes, P.A.M., Prates, M.V., Bloch, Jr., C. 2006. Identification of a cowpea γ-thionin with bactericidal activity. *FEBS Journal* 273: 3489–3497.

Fujimura, M., Ideguchi, M., Minami, Y., Watanabe, K., Tadera, K. 2004. Purification, characterization, and sequencing of novel antimicrobial peptides, Tu-AMP 1 and Tu-AMP 2, from bulbs of tulip (*Tulipa gesneriana* L.). *Bioscience, Biotechnology, and Biochemistry* 68(3): 571–577.

Fujimura, M., Ideguchi, M., Minami, Y., Watanabe, K., Tadera, K. 2005. Amino acid sequence and antimicrobial activity of chitin-binding peptides, Pp-AMP 1 and Pp-AMP 2, from Japanese bamboo shoots (*Phyllostachys pubescens*). *Bioscience, Biotechnology, and Biochemistry* 69(3): 642–645.

Gao, A.G., Hakimi, S.M., Mittanck, C.A., Wu, Y., Woerner, B.M., Stark, D.M., Shah, D.M., Liang, J., Rommens, C.M.T. 2000. Fungal pathogen protection in potato by expression of a plant defensin peptide. *Nature Biotechnology* 18(12): 1307–1310.

García Véscovi, E., Soncini, F.C., Groisman, E.A. 1996. Mg^{2+} as an extracellular signal: Environmental regulation of *Salmonella* virulence. *Cell* 84(1): 165–174.

Ge, X., Chen, J., Sun, C., Cao, K. 2003. Preliminary study on the structural basis of the antifungal activity of a rice lipid transfer protein. *Protein Engineering* 16: 387–390.

Gerlach, S.L., Rathinakumar, R., Chakravarty, G., Göransson, U., Wimley, W.C. et al. 2010. Anticancer and chemosensitizing abilities of cycloviolacin O2 from *Viola odorata* and psyle cyclotides from *Psychotria leptothyrsa. Peptide Science* 94(5): 617–625.

Gincel, E., Simorre, J.P., Caille, A., Marion, D., Ptak, M., Vovelle, F. 1994. Three-dimensional structure in solution of a wheat lipid-transfer protein from multidimensional 1H-NMR data. *European Journal of Biochemistry* 226(2): 413–422.

Goel, A.K., Lundberg, D., Torres, M.A., Matthews, R., Akimoto-Tomiyama, C., Farmer, L. 2008. The *Pseudomonas syringae* type III effector HopAM1 enhances virulence on water-stressed plants. *Molecular Plant-Microbe Interactions* 21(3): 361–370.

Gooderham, W.J., Bains, M., McPhee, J.B., Wiegand, I., Hancock, R.E. 2008. Induction by cationic antimicrobial peptides and involvement in intrinsic polymyxin and antimicrobial peptide resistance, biofilm formation, and swarming motility of PsrA in *Pseudomonas aeruginosa*. *Journal of Bacteriology* 190(16): 5624–5634.

Gran, L., Sletten, K., Skjeldal, L. 2008. Cyclic peptides from *Oldenlandia affinis* DC. Molecular and biological properties. *Chemistry and Biodiversity* 5: 2014–2022.

Gruber, C.W., Elliott, A.G., Ireland, D.C., Delprete, P.G., Dessein, S., Göransson, U., Trabi, M. et al. 2008. Distribution and evolution of circular miniproteins in flowering plants. *The Plant Cell Online* 20(9): 2471–2483.

Gu, Q., Kawata, E.E., Morse, M.J., Wu, H.M., Cheung, A.Y. 1992. A flower-specific cDNA encoding a novel thionin in tobacco. *Molecular General Genetics* 234: 89–96.

Gunn, J.S., Miller, S.I. 1996. PhoP-PhoQ activates transcription of *pmrAB*, encoding a two-component regulatory system involved in *Salmonella typhimurium* antimicrobial peptide resistance. *Journal of Bacteriology* 178(23): 6857–6864.

Guo, M., Tian, F., Wamboldt, Y., Alfano, J.R. 2009. The majority of the type III effector inventory of *Pseudomonas syringae* pv. *tomato* DC3000 can suppress plant immunity. *Molecular Plant-Microbe Interactions* 22(9): 1069–1080.

Guzmán-Rodríguez, J.J., Ibarra-Laclette, E., Herrera-Estrella, L., Ochoa-Zarzosa, A., Suárez-Rodríguez, L.M., Rodríguez-Zapata, L.C., López-Gómez, R. 2013. Analysis of expressed sequence tags (ESTs) from avocado seed (*Persea americana* var. *drymifolia*) reveals abundant expression of the gene encoding the antimicrobial peptide snakin. *Plant Physiology and Biochemistry* 70: 318–324.

Hammami, R., Hamida, J.B., Vergoten, G., Fliss, I. 2009. PhytAMP: A database dedicated to antimicrobial plant peptides. *Nucleic Acids Research* 37(1): D963–D968.

Hassan, M., Kjos, M., Nes, I.F., Diep, D.B., Lotfipour, F. 2012. Natural antimicrobial peptides from bacteria: Characteristics and potential applications to fight against antibiotic resistance. *Journal of Applied Microbiology* 113(4): 723–736.

Hauck, P., Thilmony, R., He, S.Y. 2003. A *Pseudomonas syringae* type III effector suppresses cell wall-based extracellular defense in susceptible *Arabidopsis* plants. *Proceedings of the National Academy of Science of the United States of America* 100(14): 8577–8582.

Heinemann, B., Andersen, K.V., Nielsen, P.R., Bech, L.M., Poulsen, F.M. 1996. Structure in solution of a four-helix lipid binding protein. *Protein Science* 5(1): 13–23.

Henriques, S.T., Huang, Y.H., Rosengren, K.J., Franquelim, H.G., Carvalho, F.A., Johnson, A., Sonza, S. et al. 2011. Decoding the membrane activity of the cyclotide Kalata B1—The importance of phosphatidylethanolamine phospholipids and lipid organization on hemolytic and anti-HIV activities. *Journal of Biological Chemistry* 286(27): 24231–24241.

Hou, S., Mu, R., Ma, G., Xu, X., Zhang, C., Yang, Y., Wu, D. 2011. *Pseudomonas syringae* pv. *phaseolicola* effector HopF1 inhibits pathogen-associated molecular pattern-triggered immunity in a RIN4-independent manner in common bean (*Phaseolus vulgaris*). *FEMS Microbiology Letters* 323(1): 35–43.

Houlne, G., Meyer, B., Schantz, R. 1998. Alteration of the expression of a plant defensin gene by exon shuffling in bell pepper (*Capsicum annuum* L.). *Molecular and General Genetics* 259(5): 504–510.

Huang, T.P., Lu, K.M., Chen, Y.H. 2013. A novel two-component response regulator links rpf with biofilm formation and virulence of *Xanthomonas axonopodis* pv. *citri*. *PLoS One* 8(4): e62824.

Hughes, P., Dennis, E., Whitecross, M., Llewellyn, D., Gage, P. 2000. The cytotoxic plant protein, beta-purothionin, forms ion channels in lipid membranes. *Journal of Biological Chemistry* 275(2): 823–827.

Imler, J.L., Bulet, P. 2005. Antimicrobial peptides in *Drosophila*: Structures, activities and gene regulation. *Chemical Immunology and Allergy* 86: 1–21.

Ireland, D., Colgrave, M., Craik, D. 2006. A novel suite of cyclotides from *Viola odorata*: Sequence variation and the implications for structure, function and stability. *Biochemistry Journal* 400: 1–12.

Iwai, T., Kaku, H., Honkura, R., Nakamura, S., Ochiai, H., Sasaki, T., Ohashi, Y. 2002. Enhanced resistance to seed-transmitted bacterial diseases in transgenic rice plants overproducing an oat cell-wall-bound thionin. *Molecular Plant-Microbe Interactions* 15(6): 515–521.

Jakobek, J.L., Smith, J.A., Lindgren, P.B. 1993. Suppression of bean defense responses by *Pseudomonas syringae*. *Plant Cell* 5(1): 57–63.

Jan, P.S., Huang, H.Y., Chen, H.M. 2010. Expression of a synthesized gene encoding cationic peptide cecropin B in transgenic tomato plants protects against bacterial diseases. *Applied and Environmental Microbiology* 76(3): 769–775.

Jha, S., Chattoo, B.B. 2010. Expression of a plant defensin in rice confers resistance to fungal phytopathogens. *Transgenic Research* 19(3): 373–384.

Jung, H.W., Kim, W., Hwang, B.K. 2003. Three pathogen-inducible genes encoding lipid transfer protein from pepper are differentially activated by pathogens, abiotic, and environmental stresses. *Plant, Cell and Environment* 26(6): 915–928.

Jung, Y.J. 2013. Enhanced resistance to bacterial pathogen in transgenic tomato plants expressing cathelicidin antimicrobial peptide. *Biotechnology and Bioprocess Engineering* 18: 615–624.

Jung, Y.J., Lee, S.Y., Moon, Y.S., Kang, K.K. 2012. Enhanced resistance to bacterial and fungal pathogens by overexpression of a human cathelicidin antimicrobial peptide (hCAP18/LL-37) in Chinese cabbage. *Plant Biotechnology Report* 6(1): 39–46.

Kaas, Q., Westermann, I.C., Henriques, S.T., Craik, D.J. 2010. Antimicrobial peptides in plants. In *Antimicrobial Peptides: Discovery, Design and Novel Therapeutic Strategies*. Wang, G. (ed.), Oxfordshire, UK: CABI, pp. 40–71, ISBN: 13:978-1-84593-657-0.

Kader, J.C. 1996. Lipid-transfer proteins in plants. *Annual Review of Plant Biology* 47(1): 627–654.

Kader, J.C. 1997. Lipid-transfer proteins: A puzzling family of plant proteins. *Trends in Plant Science* 2(2): 66–70.

Karunanandaa, B., Singh, A., Kao, T.H. 1994. Characterization of a predominantly pistil-expressed gene encoding a gamma-thionin-like protein of *Petunia inflata. Plant Molecular Biology* 26: 459–464.

Khoo, L., Robinette, D.W., Noga, E.J. 1999. Callinectin, an antibacterial peptide from blue crab, *Callinectes sapidus* hemocytes. *Marine Biotechnology* 1: 44–51.

Kloek, A.P., Verbsky, M.L., Sharma, S.B., Schoelz, J.E., Vogel, J., Klessig, D.F., Kunkel, B.N. 2001. Resistance to *Pseudomonas syringae* conferred by an *Arabidopsis thaliana* coronatine-insensitive (*coi1*) mutation occurs through two distinct mechanisms. *Plant Journal* 26(5): 509–522.

Koczulla, A.R., Bals, R. 2003. Antimicrobial peptides: Current status and therapeutic potential. *Drugs* 63(4): 389–406.

Kokryakov, V.N., Harwig, S.S., Panyutich, E.A., Shevchenko, A.A., Aleshina, G.M., Shamova, O.V., Korneva, H.A., 1993. Protegrins: Leukocyte antimicrobial peptides that combine features of corticostatic defensins and tachyplesins. *FEBS Letters* 327(2): 231–236.

Kovalskaya, N., Zhao, Y., Hammond, R.W. 2011. Antibacterial and antifungal activity of a snakin-defensin hybrid protein expressed in tobacco and potato plants. *Open Plant Science Journal* 5: 29–42.

Kragh, K.M., Nielsen, J.E., Nielsen, K.K., Dreboldt, S., Mikkelsen, J.D. 1995. Characterization and localization of new antifungal cysteine-rich proteins from *Beta vulgaris. MPMI-Molecular Plant Microbe Interactions* 8(3): 424–434.

Krishnamurthy, K., Balconi, C., Sherwood, J.E., Giroux, M.J. 2001. Wheat puroindolines enhance fungal disease resistance in transgenic rice. *Molecular Plant-Microbe Interactions* 14(10): 1255–1260.

Lee, O.R., Kim, Y.J., Devi Balusamy, S.R., Kim, M.K., Sathiyamoorthy, S., Yang, D.C. 2011. Ginseng γ-thionin is localized to cell wall-bound extracellular spaces and responsive to biotic and abiotic stresses. *Physiological and Molecular Plant Pathology* 76(2): 82–89.

Lee, S., Yang, D.S., Uppalapati, S.R., Sumner, L.W., Mysore, K.S. 2013. Suppression of plant defense responses by extracellular metabolites from *Pseudomonas syringae* pv. *tabaci* in *Nicotiana benthamiana. BMC Plant Biology* 13: 65.

Lee, S.W., Jeong, K.S., Han, S.W., Lee, S.E., Phee, B.K., Hahn,T.-R., Ronald, P. 2008. The *Xanthomonas oryzae* pv. *oryzae* PhoPQ two-component system is required for AvrXA21 activity, *hrpG* expression, and virulence. *Journal of Bacteriology* 190(6): 2183–2197.

Lerche, M.H., Kragelund, B.B., Bech, L.M., Poulsen, F.M. 1997. Barley lipid-transfer protein complexed with palmitoyl CoA: The structure reveals a hydrophobic binding site that can expand to fit both large and small lipid-like ligands. *Structure* 5(2): 291–306.

Lerche, M.H., Poulsen, F.M. 1998. Solution structure of barley lipid transfer protein complexed with palmitate. Two different binding modes of palmitate in the homologous maize and barley nonspecific lipid transfer proteins. *Protein Science* 7(12): 2490–2498.

Llama-Palacios, A., López-Solanilla, E., Poza-Carrión, C., García-Olmedo, F., Rodríguez-Palenzuela, P. 2003. The *Erwinia chrysanthemi phoP-phoQ* operon plays an important role in growth at low pH, virulence and bacterial survival in plant tissue. *Molecular Microbiology* 49(2): 347–357.

López-García, B., Pérez-Payá, E., Marcos, J.F. 2002. Identification of novel hexapeptides bioactive against phytopathogenic fungi through screening of a synthetic peptide combinatorial library. *Applied and Environmental Microbiology* 68(5): 2453–2460.

López-Meza, J.E., Ochoa-Zarzosa, A., Aguilar, J.A., Loeza-Lara, P.D. 2011. Antimicrobial peptides: Diversity and perspectives for their biomedical application. In: M. A. Komorowska and S.Olsztynska-Janus (eds.), *Biomedical Engineering,Trends, Research and Technologies*, vol. 1. Croatia: Intech, pp. 275–304.

Lucas, J. A. 1998. *Plant Pathology and Plant Pathogens*. 3rd edn. Oxford: Blackwell Science Ltd. ISBN 0-632-03046-1.

Maldonado, A.M., Doerner, P., Dixon, R.A., Lamb, C.J., Cameron, R.K. 2002. A putative lipid transfer protein involved in systemic resistance signalling in *Arabidopsis*. *Nature* 419(6905): 399–403.

Marcos, J.F., Muñoz, A., Pérez-Payá, E., Misra, S., López-García, B. 2008. Identification and rational design of novel antimicrobial peptides for plant protection. *Annual Review of Phytopathology* 46: 273–301.

Meincken, M., Holroyd, D.L., Rautenbach, M. 2005. Atomic force microscopy study of the effect of antimicrobial peptides on the cell envelope of *Escherichia coli*. *Antimicrobial Agents and Chemotherapy* 49(10): 4085–4092.

Mendez, E., Moreno, A., Colilla, F., Pelaez, F., Limas, G.G., Mendez, R., Soriano, F. et al. 1990. Primary structure and inhibition of protein synthesis in eukaryotic cell-free system of a novel thionin, y-hordothionin, from barley endosperm. *European Journal of Biochemistry* 194: 533–539.

Mentag, R., Luckevich, M., Morency, M.J., Séguin, A. 2003. Bacterial disease resistance of transgenic hybrid poplar expressing the synthetic antimicrobial peptide D4E1. *Tree Physiology* 23(6): 405–411.

Milligan, S.B., Gasser, C.S. 1995. Nature and regulation of pistil-expressed genes in tomato. *Plant Molecular Biology* 28: 691–711.

Miquel, M., Block, M.A., Joyard, J., Dorne, A.J., Dubacq, J.P., Kader, J.-C., Douce, R. 1988. Protein-mediated transfer of phosphatidylcholine from liposomes to spinach chloroplast envelope membranes. *Biochimica et Biophysica Acta* 937: 219–228.

Mitta, G., Vandenbulcke, F., Roch, P. 2000. Original involvement of antimicrobial peptides in mussel innate immunity. *FEBS Letters* 486(3): 185–190.

Mittler, R. 2006. Abiotic stress, the field environment and stress combination. *Trends in Plant Science* 11(1): 15–19.

Molina, A., Garcia-Olmedo, F. 1997. Enhanced tolerance to bacterial pathogens caused by the transgenic expression of barley lipid transfer protein LTP2. *Plant Journal* 12: 669–675.

Molina, A., Segura, A., García-Olmedo, F. 1993. Lipid transfer proteins (nsLTPs) from barley and maize leaves are potent inhibitors of bacterial and fungal plant pathogens. *FEBS Letters* 316(2): 119–122.

Montesinos, E., Bardají, E. 2008. Synthetic antimicrobial peptides as agricultural pesticides for plant-disease control. *Chemistry and Biodiversity* 5(7): 1225–1237.

Moreno, M., Segura, A., García-Olmedo, F. 1994. Pseudothionin-St1, a potato peptide active against potato pathogens. *European Journal of Biochemistry* 223: 135–139.

Mulvenna, J.P., Mylne, J.S., Bharathi, R., Burton, R.A., Shirley, N.J., Fincher, G.B., Anderson, M.A. et al. 2006. Discovery of cyclotide-like protein sequences in graminaceous crop plants: Ancestral precursors of circular proteins? *The Plant Cell Online* 18(9): 2134–2144.

Mygind, P.H., Fischer, R.L., Schnorr, K.M., Hansen, M.T., Sönksen, C.P., Ludvigsen, S., Raventos, D. et al. 2005. Plectasin is a peptide antibiotic with therapeutic potential from a saprophytic fungus. *Nature* 4377061: 975–980.

Nadal, A., Montero, M., Company, N., Badosa, E., Messeguer, J., Montesinos, L., Montesinos, E. et al. 2012. Constitutive expression of transgenes encoding derivatives of the synthetic antimicrobial peptide BP100: Impact on rice host plant fitness. *BMC Plant Biology* 12: 159.

Nakka, S., Qi, M., Zhao, Y. 2010. The *Erwinia amylovora* PhoPQ system is involved in resistance to antimicrobial peptide and suppresses gene expression of two novel type III secretion systems. *Microbiology Research* 165(8): 665–673.

Nes, I.F., Holo, H. 2000. Class II antimicrobial peptides from lactic acid bacteria. *Biopolymers* 55(1): 50–61.

Neumann, G.M., Condron, R., Polya, G.M. 1996. Purification and mass spectrometry-based sequencing of yellow mustard (*Sinapis alba* L.) 6 kDa proteins identification as antifungal proteins. *International Journal of Peptide and Protein Research* 47(6): 437–446.

Nguyen, G.K.T., Lian, Y., Pang, E.W.H., Nguyen, P.Q.T., Tran, T.D., Tam, J.P. 2013. Discovery of linear cyclotides in monocot plant panicum laxum of Poaceae family provides new insights into evolution and distribution of cyclotides in plants. *Journal of Biological Chemistry* 288(5): 3370–3380.

Nieuwland, J., Feron, R., Huisman, B.A., Fasolino, A., Hilbers, C.W., Derksen, J., Mariani, C. 2005. Lipid transfer proteins enhance cell wall extension in tobacco. *The Plant Cell Online* 17(7): 2009–2019.

Niyonsaba, F., Ushio, H., Hara, M., Yokoi, H., Tominaga, M., Takamori, K., Kajiwara, N. et al. 2010. Antimicrobial peptides human beta-defensins and cathelicidin LL-37 induce the secretion of a pruritogenic cytokine IL-31 by human mast cells. *Journal of Immunology* 184(7): 3526–3534.

Nomura, K., Melotto, M., He, S.Y. 2005. Suppression of host defense in compatible plant-*Pseudomonas syringae* interactions. *Current Opinion in Plant Biology* 8(4): 361–368.

Oard, S.V. 2011. Deciphering a mechanism of membrane permeabilization by α-hordothionin peptide. *Biochimica et Biophysica Acta (BBA)-Biomembranes* 1808(6): 1737–1745.

Oard, S.V., Enright, F.M. 2006. Expression of the antimicrobial peptides in plants to control phytopathogenic bacteria and fungi. *Plant Cell Report* 25(6): 561–572.

Odintsova, T.I., Egorov, T.A., Musolyamov, A.K., Odintsova, M.S., Pukhalsky, V.A., Grishin, E.V. 2007. Seed defensins from *i* T. kiharae and related species: Genome localization of defensin-encoding genes. *Biochimie* 89(5): 605–612.

Osborn, R.W., De Samblanx, G.W., Thevissen, K., Goderis, I., Torrekens, S., Van Leuven, F., Attenborough, S. et al. 1995. Isolation and characterisation of plant defensins from seeds of Asteraceae, Fabaceae, Hippocastanaceae and Saxifragaceae. *FEBS Letters* 368(2): 257–262.

Osusky, M., Osuska, L., Hancock, R.E., Kay, W.W., Misra, S. 2004. Transgenic potatoes expressing a novel cationic peptide are resistant to late blight and pink rot. *Transgenic Research* 13(2): 181–190.

Osusky, M., Zhou, G., Osuska, L., Hancock, R.E., Kay, W.W., Misra, S. 2000. Transgenic plants expressing cationic peptide chimeras exhibit broad-spectrum resistance to phytopathogens. *Nature Biotechnology* 18(11): 1162–1166.

Padovan, L., Scocchi, M., Tossi, A. 2010. Structural aspects of plant antimicrobial peptides. *Current Protein and Peptide Science* 11(3): 210–219.

Park, Y., Hahm, K.S. 2005. Antimicrobial peptides (AMPs): Peptide structure and mode of action. *Journal of Biochemistry and Molecular Biology* 38(5): 507–516.

Pavli, O.I., Kelaidi, G.I., Tampakaki, A.P., Skaracis, G.N. 2011. The *hrpZ* gene of *Pseudomonas syringae* pv. *phaseolicola* enhances resistance to rhizomania disease in transgenic *Nicotiana benthamiana* and sugar beet. *PLoS One* 6(3):e17306.

Pelegrini, P.B., Del Sarto, R.P., Silva, O.N., Franco, O.L., Grossi-de-Sa, M.F. 2011. Antibacterial peptides from plants: What they are and how they probably work. *Biochemistry Research International*, Article ID 250349, p. 9.

Pelegrini, P.B., Franco, O.L. 2005. Plant γ-thionins: Novel insights on the mechanism of action of a multi-functional class of defense proteins. *The International Journal of Biochemistry and Cell Biology* 37(11): 2239–2253.

Pestana-Calsa, M.C., Calsa, Jr., T. 2011. *In silico* identification of plant-derived antimicrobial peptides. In: Ning-Sun Yang (ed.), *Systems and Computational Biology–Molecular and Cellular Experimental Systems*, Intech, pp. 275–304.

Pieterse, C.M., Dicke, M. 2007. Plant interactions with microbes and insects: From molecular mechanisms to ecology. *Trends in Plant Science* 12(12): 564–569.

Pii, Y., Astegno, A., Peroni, E., Zaccardelli, M., Pandolfini, T., Crimi, M. 2009. The *Medicago truncatula* N5 gene encoding a root-specific lipid transfer protein is required for the symbiotic interaction with *Sinorhizobium meliloti*. *Molecular Plant-Microbe Interactions* 22(12): 1577–1587.

Pinto, M.F., Almeida, R.G., Porto, W.F., Fensterseifer, I.C., Lima, L.A., Dias, S.C., Franco, O.L. 2012. Cyclotides from gene structure to promiscuous multifunctionality. *Journal of Evidence-Based Complementary and Alternative Medicine* 17(1): 40–53.

Ponti, D., Mangoni, M.L., Mignogna, G., Simmaco, M., Barra, D. 2003. An amphibian antimicrobial peptide variant expressed in *Nicotiana tabacum* confers resistance to phytopathogens. *Biochemistry Journal* 370(Pt 1): 121–127.

Ponz, F., Paz-Ares, J., Hernandez-Lucas, C., Garcia-Olmedo, F., Carbonero, P. 1986. Cloning and nucleotide sequence of a cDNA encoding the precursor of the barley toxin α-hordothionin. *European Journal of Biochemistry* 156(1): 131–135.

Poth, A.G., Colgrave, M.L., Philip, R., Kerenga, B., Daly, N.L. et al. 2011. Discovery of cyclotides in the Fabaceae plant family provides new insights into the cyclization, evolution, and distribution of circular proteins. *ACS Chemical Biology* 6(4): 345–355.

Pränting, M., Lööv, C., Burman, R., Göransson, U., Andersson, D.I. 2010. The cyclotide cycloviolacin O2 from *Viola odorata* has potent bactericidal activity against Gram-negative bacteria. *Journal of Antimicrobial Chemotherapy* 65(9): 1964–1971.

Prema, G.U., Pruthvi, T.P.M. 2012. Antifungal plant defensins. *Current Biotica* 6(2): 254–270.

Punja, Z.K. 2001. Genetic engineering of plants to enhance resistance to fungal pathogens—A review of progress and future prospects. *Canadian Journal of Plant Pathology* 23: 216–235.

Pyee, J., Yu, H.S., Kolattukudy, P.E. 1994. Identification of a lipid transfer protein as the major protein in the surface wax of broccoli (*Brassica oleracea*) leaves. *Archives of Biochemistry and Biophysics* 311(2): 460–468.

Rahnamaeian, M. 2011. Antimicrobial peptides: Modes of mechanism, modulation of defense responses. *Plant Signaling and Behavior* 6(9): 1325–1332.

Rajaram, G., Manivasagan, P., Thilagavathi, B., Saravanakumar, A. 2010. Purification and characterization of a bacteriocin produced by *Lactobacillus lactis* isolated from marine environment. *Advanced Journal of Food Science Technology* 2(2): 138–144.

Rajasekaran, K., Cary, J.W., Jaynes, J.M., Cleveland, T.E. 2005. Disease resistance conferred by the expression of a gene encoding a synthetic peptide in transgenic cotton (*Gossypium hirsutum* L.) plants. *Plant Biotechnology Journal* 3(6): 545–554.

Rajasekaran, K., Stromberg, K.D., Cary, J.W., Cleveland, T.E. 2001. Broad-spectrum antimicrobial activity *in vitro* of the synthetic peptide D4E1. *Journal of Agriculture and Food Chemistry* 49(6): 2799–2803.

Reimann-Philipp, U., Schrader, G., Martinoia, E., Barkholt, V., Apel, K. 1989. Intracellular thionins of barley. A second group of leaf thionins closely related to but distinct from cell wall-bound thionins. *Journal of Biological Chemistry* 264(15): 8978–8984.

Richard, J.A., Kelly, I., Marion, D., Auger, M., Pézolet, M. 2005. Structure of beta-purothionin in membranes: A two-dimensional infrared correlation spectroscopy study. *Biochemistry* 44: 52–61.

Richard, J.A., Kelly, I., Marion, D., Pezolet, M., Auger, M. 2002. Interaction between β-purothionin and dimyristoylphosphatidylglycerol: A ^{31}P-NMR and infrared spectroscopic study. *Biophysics Journal* 83: 2074–2083.

Rigano, L.A., Payette, C., Brouillard, G., Marano, M.R., Abramowicz, L., Torres, P.S., Yun, M. et al. 2007. Bacterial cyclic beta-(1,2)-glucan acts in systemic suppression of plant immune responses. *Plant Cell* 19(6): 2077–2089.

Rivero, M., Furman, N., Mencacci, N., Picca, P., Toum, L., Lentz, E., Bravo-Almonacid, F. et al. 2012. Stacking of antimicrobial genes in potato transgenic plants confers increased resistance to bacterial and fungal pathogens. *Journal of Biotechnology* 157(2): 334–343.

Romero, A., Alamillo, J.M., Garcia-Olmedo, F. 1997. Processing of thionin precursors in barley leaves by a vacuolar proteinase. *Europear Journal of Biochemistry* 243: 202–208.

Sarika, Iquebal, M.A., Rai, A. 2012. Biotic stress resistance in agriculture through antimicrobial peptides. *Peptides* 36(2): 322–330.

Sarowar, S., Kim, Y.J., Kim, K.D., Hwang, B.K., Ok, S.H., Shin, J.S. 2009. Overexpression of lipid transfer protein (LTP) genes enhances resistance to plant pathogens and LTP functions in long-distance systemic signaling in tobacco. *Plant Cell Reports* 28(3): 419–427.

Schikora, A., Garcia, A.V., Hirt, H. 2012. Plants as alternative hosts for *Salmonella*. *Trends in Plant Science* 17(5): 245–249.

Schrader-Fischer, G., Apel, K. 1994. Organ-specific expression of highly divergent thionin variants that are distinct from the seed-specific crambin in the crucifer *Crambe abyssinica*. *Molecular and General Genetics*, 245(3): 380–389.

Segura, A., Moreno, M., Madueño, F., Molina, A., García-Olmedo, F. 1999. Snakin-1, a peptide from potato that is active against plant pathogens. *Molecular Plant-Microbe Interactions* 12(1): 16–23.

Shi, Y., Latifi, T., Cromie, M.J., Groisman, E.A. 2004. Transcriptional control of the antimicrobial peptide resistance *ugtL* gene by the *Salmonella* PhoP and SlyA regulatory proteins. *Journal of Biological Chemistry* 279(37): 38,618–28,625.

Shin, D.H., Lee, J.Y., Hwang, K.Y., Kyu Kim, K., Suh, S.W. 1995. High-resolution crystal structure of the nonspecific lipid-transfer protein from maize seedlings. *Structure* 3(2): 189–199.

Sigee, D.C. 2005. *Bacterial Plant Pathology: Cell and Molecular Aspects*. Cambridge: Cambridge University Press. ISBN: 0-521-35064-6.

Silverstein, K.A., Graham, M.A., Paape, T.D., VandenBosch, K.A. 2005. Genome organization of more than 300 defensin-like genes in *Arabidopsis*. *Plant Physiology* 138(2): 600–610.

Simonsen, S.M., Sando, L., Ireland, D.C., Colgrave, M.L., Bharathi, R., Göransson, U., Craik, D.J. 2005. A continent of plant defense peptide diversity: Cyclotides in Australian Hybanthus (Violaceae). *The Plant Cell Online* 17(11): 3176–3189.

Sjahril, R., Chin, D.P., Khan, R.S., Yamamura, S., Nakamura, I., Amemiya, Y., Mii, M. 2006. Transgenic Phalaenopsis plants with resistance to *Erwinia carotovora* produced by introducing wasabi defensin gene using *Agrobacterium* method. *Plant Biotechnology* 23(2): 191–194.

Sohlenkamp, C., Galindo-Lagunas, K.A., Guan, Z., Vinuesa, P., Robinson, S., Thomas-Oates, J., Raetz, C.R.H. et al. 2007. The lipid lysyl-phosphatidylglycerol is present in membranes of *Rhizobium tropici* CIAT899 and confers increased resistance to polymyxin B under acidic growth conditions. *Molecular Plant-Microbe Interactions* 20(11): 1421–1430.

Stec, B. 2006. Plant thionins—The structural perspective. *Cellular and Molecular Life Sciences* 63: 1370–1385.

Stec, B., Markman, O., Rao, U., Heffron, G., Henderson, S., Vernon, L.P., Brumfeld, V. et al. 2004. Proposal for molecular mechanism of thionins deduced from physico-chemical studies of plant toxins. *The Journal of Peptide Research* 64(6): 210–224.

Stotz, H.U., Thomson, J.G., Wang, Y. 2009. Plant defensins: Defense, development and application. *Plant Signaling and Behavior* 4(11): 1010–1012.

Stotz, H.U., Waller, F., Wang, K. 2013. Innate immunity in plants: The role of antimicrobial peptides. In: S. Hiemstra and S. A. J. Zaat, (eds.), *Antimicrobial Peptides and Innate Immunity.* Springer, Basel: Springer Science & Business Media, pp. 29–51, ISBN: 3034805403.

Sun, J.Y., Gaudet, D.A., Lu, Z.X., Frick, M., Puchalski, B., Laroche, A. 2008. Characterization and antifungal properties of wheat nonspecific lipid transfer proteins. *Molecular Plant-Microbe Interactions* 21: 346–360.

Swathi Anuradha, T., Divya, K., Jami, S.K., Kirti, P.B. 2008. Transgenic tobacco and peanut plants expressing a mustard defensin show resistance to fungal pathogens. *Plant Cell Report* 27(11): 1777–1786.

Tam, J.P., Lu, Y.A., Yang, J.L., Chiu, K.W. 1999. An unusual structural motif of antimicrobial peptides containing end-to-end macrocycle and cystine-knot disulfides. *Proceedings of the National Academy of Sciences of the United States of America* 96: 8913–8918.

Tan, N.H., Zhou, J. 2006. Plant cyclopeptides. *Chemical Reviews* 106(3): 840–895.

Tang, Y.Q., Yeaman, M.R., Selsted, M.E. 2002. Antimicrobial peptides from human platelets. *Infection and Immunity* 70(12): 6524–6533.

Taveira, G.B., Mathias, L.S., Motta, O.V., Machado, O.L., Rodrigues, R., Carvalho, A.O., Teixeira-Ferreira, A., et al. 2013. Thionin-like peptides from *Capsicum annuum* fruits with high activity against human pathogenic bacteria and yeasts. *Peptide Science* 102: 30–39.

Terras, F.R.G., Eggermont, K., Kovaleva, V., Raikhel, N.V., Osborn, R.W., Kester, A., Rees, S.B., et al. 1995. Small cysteine-rich antifungal proteins from radish: Their role in host defence. *Plant Cell* 7: 573–588.

Thoma, S., Hecht, U., Kippers, A., Botella, J., De Vries, S., Somerville, C. 1994. Tissue-specific expression of a gene encoding a cell wall-localized lipid transfer protein from *Arabidopsis. Plant Physiology* 105(1): 35–45.

Thomma, B.P., Cammue, B.P., Thevissen, K. 2002. Plant defensins. *Planta* 216(2): 193–202.

Van der Weerden, N.L., Hancock, R.E., Anderson, M.A. 2010. Permeabilization of fungal hyphae by the plant defensin NaD1 occurs through a cell wall-dependent process. *Journal of Biological Chemistry* 285(48): 37513–37520.

Van der Weerden, N.L., Lay, F.T., Anderson, M.A. 2008. The plant defensin, NaD1, enters the cytoplasm of *Fusarium oxysporum* hyphae. *Journal of Biological Chemistry* 283(21): 14,445–14,452.

Vernon, L.P., Evett, G.E., Zeikus, R.D., Gray, W.R. 1985. A toxic thionin from *Pyrularia pubera*: Purification, properties, and amino acid sequence. *Archives of Biochemistry and Biophysics* 238(1): 18–29.

Vernon, L.P., Rogers, A. 1992. Binding properties of *Pyrularia* thionin and *Naja naja* kaouthia cardiotoxin to human and animal erythrocytes and to murine P388 cells. *Toxicon* 30(7): 711–721.

Vila-Perelló, M., Sánchez-Vallet, A., García-Olmedo, F., Molina, A., Andreu, D. 2003. Synthetic and structural studies on *Pyrularia pubera* thionin: A single-residue mutation enhances activity against Gram-negative bacteria. *FEBS Letters* 536(1): 215–219.

Visser, M., Stephan, D., Jaynes, J.M., Burger, J.T. 2012. A transient expression assay for the in planta efficacy screening of an antimicrobial peptide against grapevine bacterial pathogens. *Letters in Applied Microbiology* 54(6): 543–551.

Walters D.R. 2011. *Plant Defense: Warding off Attack by Pathogens, Herbivores, and Parasitic Plants.* 1st edn. Wiley-Blackwell, UK, ISBN: 978-1-4051-7589-0.

Wang, C.K., Kaas, Q., Chiche, L., Craik, D.J. 2008. CyBase: A database of cyclic protein sequences and structures, with applications in protein discovery and engineering. *Nucleic Acids Research* 36: D206–D210.

Wang, G., Li, X., Wang, Z. 2009. APD2: The updated antimicrobial peptide database and its application in peptide design. *Nucleic Acids Research* 37(database issue): D933–D937.

Wang, H.W., Hwang, S.G., Karuppanapandian, T., Liu, A., Kim, W., Jang, C.S. 2012. Insight into the molecular evolution of non-specific lipid transfer proteins via comparative analysis between rice and sorghum. *DNA Research* 19(2): 179–194.

Wang, S.Y., Wu, J.H., Ng, T.B., Ye, X.Y., Rao, P.F. 2004. A non-specific lipid transfer protein with antifungal and antibacterial activities from the mung bean. *Peptides* 25(8): 1235–1242.

Wimley, W.C. 2010. Describing the mechanism of antimicrobial peptide action with the interfacial activity model. *ACS Chemical Biology* 5(10): 905–917.

Yang, X., Li, J., Li, X., She, R., Pei, Y. 2006. Isolation and characterization of a non-specific lipid transfer protein-like antimicrobial protein from motherwort (*Leonurus japonicus* Houtt) seeds. *Peptides* 27(12): 3122–3128.

Yeats, T.H., Rose J.K. 2008. The biochemistry and biology of extracellular plant lipid-transfer proteins (LTPs). *Protein Science* 17(2): 191–198.

Zarrabi, M., Dalirfardouei, R., Sepehrizade, Z., Kermanshahi, R.K. 2013. Comparison of the antimicrobial effects of semipurified cyclotides from Iranian *Viola odorata* against some of plant and human pathogenic bacteria. *Journal of Applied Microbiology* 115: 367–375.

Zasloff, M. 2002. Antimicrobial peptides of multicellular organisms. *Nature* 4156870: 389–395.

Zhang, Y., Lewis, K. 1997. Fabatins: New antimicrobial plant peptides. *FEMS Microbiology Letters* 149(1): 59–64.

Zhao, Y., Thilmony, R., Bender, C.L., Schaller, A., He, S.Y., Howe, G.A. 2003. Virulence systems of *Pseudomonas syringae* pv. *tomato* promote bacterial speck disease in tomato by targeting the jasmonate signaling pathway. *Plant Journal* 36(4): 485–499.

Zhou, M., Hu, Q., Li, Z., Li, D., Chen, C.F., Luo, H. 2011. Expression of a novel antimicrobial peptide Penaeidin 4-1 in creeping bentgrass (*Agrostis stolonifera* L.) enhances plant fungal disease resistance. *PLoS One* 6(9): e24677.

7 Nutrient Supplements for Plant Pathogenic Bacteria

Their Role in Microbial Growth and Pathogenicity

Penumatsa Kishore Varma, Uppala Naga Mangala, Koothala Jyothirmai Madhavi, and Kotamraju Vijay Krishna Kumar

CONTENTS

ABSTRACT Plant pathogenic bacteria (PPB) cause significant economic losses globally in a wide variety of crops. In order to devise robust management strategies against these bacterial diseases, comprehensive understanding of varied aspects of their growth, multiplication, virulence, and modes of action is mandatory. Specifically, the factors responsible for the flare-ups of these bacterial diseases in economically important plants require much attention. This endeavor must include a thorough understanding of the growth habits and preferences of PPB. Herein, we highlight the most economically significant bacterial diseases worldwide and critically discuss the different major and minor nutrient sources commonly used for *in vitro* culturing of PPB. The preferences by PPB to these nutrient supplements for growth and multiplication are also reviewed, and the roles of these nutrient supplements in promoting virulence or its suppression under *in vitro* conditions are compiled. Aspects related to exopolysaccharide (EPS) production, hypersensitive response and pathogenicity (*Hrp*) genes, and their induction in different PPB are reviewed as well. Last, the types of nutrients that trigger hypersensitive response (HR) in resistant plants and non-hosts is considered.

KEYWORDS: plant pathogenic bacteria, exopolysaccharides, nutrients, virulence, *Hrp* genes

7.1 INTRODUCTION

Plant diseases incited by bacterial pathogens have assumed greater significance because of the huge economical losses they cause globally. For the majority of these diseases, comprehensive management practices are available; however, the devastating losses they cause annually in economically important plant species remain of huge concern. Multiple reasons are attributed to these losses despite the availability of concrete control methods; in this context, the variability of plant pathogenic bacteria (PPB) is of paramount significance. When adapted to various microclimates of a specific crop, a bacterial pathogen exhibits greater virulence and thus greater pathogenicity. Another important reason for a bacterial disease outbreak could be the breakdown of resistance in crop germplasm. Breeders often come up with biotechnological approaches to tackle such resistance breakdowns, which are now a topic within resistance breeding (Maruthasalam et al., 2007). However, variability studies of microbial plant diseases in general and of bacterial diseases in particular have not yet gained momentum. The majority of the research is directed toward identifying suitable control measures that cover chemical, biological, and host-plant resistance, as well as the use of culturally based control methods. However, field research often requires a thorough backdrop of laboratory studies that, in turn, rely upon ample knowledge of pathogen growth, multiplication, nutrient preference, ambient conditions for disease development, and other related virulence factors.

Plant pathogenic bacteria (PPB) belong to the genera *Erwinia*, *Xanthomonas*, *Pseudomonas*, *Agrobacterium*, *Pantoea*, *Ralstonia*, *Bacillus*, *Serratia*, *Rhizobacter*, *Rhizomonas*, *Candidatus liberobacter*, *Xylophilus*, *Sphingomonas*, *Clavibacter*, *Burkholderia*, *Acidovorax*, *Xylella*, *Spiroplasma*, and *Phytoplasma* (Schaad et al., 2001). Cultivable genera among these pathogens exhibit specific nutrient preferences and therefore also have specific tissue and crop preferences for initiating disease. Although carbon, nitrogen, and water are the essential requirements for these pathogens (Masao Goto, 1992), their specific types and sources dictate their multiplication. This aspect is more relevant to *in vitro* studies, in which the types of nutrient supplements obtained are often significant. Supplement type often dictates virulence factors of PPB such as production of toxins (Gallarato et al., 2012), exopolysaccharides (Ashok Kumar et al., 2003), pathogen-related enzymes and proteins (PR) (Smith, 1958), and their effect on type III secretion systems (TTSS) (Tang et al., 2006). Cell-to-cell communication in PPB (Quorum Sensing or QR) is the key factor in regulation of the production of virulence factors such as extracellular polysaccharides, cell wall-degrading enzymes, antibiotic production (for competitive saprophytic ability), iron-chelating agents (siderophores), pigments, and the *hrp* gene expression that regulates disease development in a susceptible host (Arlat et al., 1991; Salmond, 1994; Sunish Kumar and Sakthivel, 2001; Dellagi et al., 2009; Raaijmakers and Mark Mazzola, 2012).

Phytotoxin production is also dictated by nutrient availability and preferences in PPB. For example, coronatine production by *Pseudomonas syringae* pv. *glycinea* on soybean is influenced by several nutritional factors. Coronatine production under *in vitro* conditions is enhanced by supplementing millimolar levels of KNO_3 or micro-molar levels of $FeCl_3$. However, no significant enhancement in this phytotoxin production was noticed through supplementation of plant extracts, plant-derived secondary metabolites, or zinc (Zn). Specific carbon and nitrogen sources such as glucose and NH_4Cl have significant positive influences on pathogen growth and its coronatine production (Palmer and Bender, 1993). In the case of syringomycin by *P. syringae* pv. *syringae*, supplementing the syringomycin minimal medium with arbutin and D-fructose has significantly enhanced its production (Mo and Gross, 1991).

Among the other pathogenic factors in PPB, enzyme-production levels vary with different nutrient supplements and their sources. For example, pectinolytic enzymes produced by soft rot group *Erwinia caratovora* is influenced by carbon sources. Shevchik et al. (1992) reported that production of polygalacturonase (PG), pectin methyl esterase (PME) and pectate lyase (PL), and isozymes of PL by *Erwinia* were regulated in the presence of diverse carbon sources. They also demonstrated the scope of using specially constructed media for the production of these pectinolytic enzymes. The nutrient factors that lead to host-plant colonization through biofilm formation by PPB are also important for understanding the mechanism of action by PPB. For example, the important steps in biofilm formation by PPB, such as the initial attachment to the plant surface and the final detachment, are directly dependent upon nutrient availability. Environmental signals that have been found to influence the initial attachment of bacteria include osmolarity, pH, iron availability, oxygen tension, and temperature (Fletcher and Loeb, 1979; Pratt and Kolter, 1998; Davey and O'Toole, 2000). Inorganic phosphates also influence biofilm formation by *Pseudomonads* (Laville et al., 1992).

Thorough understanding of PPB virulence mechanisms and their nutrient preferences in activating these mechanisms is essential for devising robust management practices, especially for dreadful bacterial diseases. It is precisely at this juncture that a critical review of these aspects is essential. This chapter covers key components of nutritional requirements for important PPB; preferred sources for their *in vitro* culturing; nutritional factors that mediate their genetic expressions for virulence, enzyme, and toxin production; host preferences; and their tissue colonization as influenced by the availabilities of various nutrients *in vivo.*

7.2 MAJOR PLANT PATHOGENIC BACTERIA

Based on their economic and scientific importance as rated by molecular plant pathologists in 2012, the major bacterial plant pathogens are: (1) *Pseudomonas syringae* pathovars; (2) *Ralstonia solanacearum*; (3) *Agrobacterium tumefaciens*; (4) *Xanthomonas oryzae* pv. *oryzae*; (5) *Xanthomonas campestris* pathovars; (6) *Xanthomonas axonopodis* pathovars; (7) *Erwinia amylovora*; (8) *Xylella fastidiosa*; (9) *Dickeya* (*dadantii* and *solani*); and (10) *Pectobacterium caratovorum* (and *P. atrosepticum*). Other major bacteria include *Clavibacter michiganensis* (*michiganensis* and *sepedonicus*), *Pseudomonas savastanoi*, and *Candidatus liberobacter asiaticus* (Mansfield et al., 2012). Some of the major plant diseases associated with these bacterial pathogens are shown in Table 7.1.

7.3 NUTRIENT SUPPLEMENTS FOR *IN VITRO* CULTURING OF PLANT PATHOGENIC BACTERIA

Nutrient requirements for a PPB can be broadly categorized based on the biosynthetic pathways they possess and also on the pathways they lack. However, common nutrients *in vitro* must be supplemented for their growth and multiplication. These nutrients include carbon, nitrogen, and oxygen,

TABLE 7.1
Some Examples of Plant Diseases Induced by Most Economically Significant Bacterial Pathogens

Bacterial Pathogen[a]	Crop/Disease	References
Pseudomonas syringae pv. tomato	Tomato (Bacterial speck)	Uppalapati et al. (2011)
Ralstonia solanacearum	Ginger (Wilts)	Kumar et al. (2004)
Agrobacterium tumefaciens	Rose (Crown gall)	Aysan and Sahin (2003)
Xanthomonas oryzae pv. oryzae	Rice (Bacterial blight)	Ansari and Sridhar (2001)
Xanthomonas campestris pv. campestris	Crucifers (Black rot)	Vicente (2012)
Xanthomonas axonopodis pv. punicae	Pomegranate (Bacterial blight)	Kumar et al. (2009)
Erwinia amylovora	Pear (Fire blight)	Miller and Schroth (1972)
Xylella fastidiosa	Grapevine (Pierce disease)	Hopkins and Purcell (2002)
Dickeya solani	Potato (Soft rot)	Garlant et al. (2013)
Pectobacterium caratovorum	Potato (Black leg)	Haan et al. (2008)
Clavibacter michiganensis	Tomato (Bacterial canker)	Kasselaki et al. (2011)
Pseudomonas savastanoi pv. savatsanoi	Olive (Knot disease)	Ramos et al. (2012)
Candidatus liberibacter asiaticus	Citrus (Greening/Huanglongbing)	Graca (1991)

[a] Pathogen ranking derived from Mansfield, J. et al. 2012. *Molecular Plant Pathology*. 13(6):614–629.

along with micronutrients such as zinc, copper, manganese, selenium, tungsten, and molybdenum. Trace elements such as magnesium, iron, and manganese may also be required for optimal growth (Magasanik, 1957). PPB differ from one another owing to their diversified nutrient requirements. This specificity in supplementing the right carbon and other nutrient sources assumes significance for the acquisition and maintenance of pure cultures of PPB. Despite varied requirements in terms of nutrients, optimal growth is attained for a particular organism only by optimal supplementation of the right combination and dosages of these nutrients under *in vitro* conditions. This specificity is particularly useful for laboratory studies wherein the optimal growth of PPB is mandatory. Herein, we discuss the requirements of major elements such as carbon and nitrogen in detail.

7.3.1 Carbon Sources and Preferences

Carbon, an essential element, is a component of nucleic acids, proteins, cell walls, and other cellular materials. Various carbon sources in both inorganic and organic forms are generally supplemented *in vitro* for their growth. The commonly used carbon supplements in the cultivation of PPB include glucose, sucrose, galactose, fructose, mannose, arabinose, trehalose, cellobiose, maltose, xylose, rhamnose, raffinose, inulin, mannitose, sorbitol, quinose dextrin, citric acid, and others (Table 7.2). Each of these carbon sources is specifically required for the cultivation of several PPB and the dosage of a particular type of carbon source was found to influence optimal growth.

For example, glucose is an important carbon source for the growth of *Xanthomonas campestris* (Jackson et al., 1998). Tanaka (1964) reported glucose as the best carbon source for the growth of selective PPB. For the growth of other PPB such as *X. arboricola* pv. *corylina* (Xac), the causal agent of bacterial blight of hazelnut, glucose fortification into the media is recommended. Glucose favors production of mucoid polysaccharides (Xanthan gum) (Schaad et al., 2001; Scortichini, 2002).

Optimum doses of glucose in the medium also influence the growth of PPB. For example, excess doses of glucose at 3%–4% levels in nutrient broth medium were found to inhibit the growth of PPB such as *Xanthomonas oryzae* pv. *oryzae*. Similarly, 1%–2% glucose levels inhibited the same pathogen in potato medium (Fang et al., 1957). Oxidative metabolism of glucose by isolates of *X. campestris* pv. *manihotis* (the causal agent of cassava blight); however, acid production by *X. c. manihotis* and other carbon sources such as maltose, xylose, and galactose were found to be more suitable than glucose (Ogunjobi et al., 2010).

Sucrose is another important carbon source that is suitable for the growth and multiplication of PPB. Sucrose as an ideal carbon source (other than glucose) for *X. campestris* was reported by Jackson et al. (1998). Fang et al. (1956) reported that sucrose is an ideal carbon source compared with glucose for the growth of *X. oryzae*. Similar reports on the growth of PPB with sucrose as an ideal carbon were reported by Tanaka (1964). Sucrose levels of up to 1% were found to be an ideal component for the cultivation of B-strain of *X. campestris* pv. *citri*, the causal agent of Cancrosis B of citrus in Argentina and Uruguay. A dose of 5 g/L sucrose in a semi-selective medium was found to be optional for isolating and culturing *X. campestris* pv. *malvacearum* and for its detection in cotton seed (Dezordi et al., 2009). For acid production of *X. c.* pv. *manihotis* (cassava blight), sucrose was found to exhibit varied responses in terms of stimulating acid production by *X. c. malvacearum* (Ogunjobi et al., 2010).

Other carbon sources have also been commonly used as nutrient supplements for PPB growth. For optimum acid production of *X. c.* pv. *manihotis*, maltose, xylose, and galactose were suitable (Ogunjobi et al., 2010). However, varied reports have been made on the influence of other carbon sources such as lactose, rhamnose, and raffinose for acid production by *X. c.* pv. *manihotis*. Oxidative metabolyses of other carbon sources such as mannose, tetrahalose, cellobiose, and fructose by cassava blight pathogen were also reported (Ogunjobi et al., 2010). However, carbon sources such as inulin, mannitol, and sorbitol could not be oxidatively metabolized by cassava blight pathogen. Among other important carbon sources, galactose is a favorable supplement (as are mannose and maltose) for the growth of *X. o.* pv. *oryzae*.

TABLE 7.2
Preferred Carbon Supplements for Different Plant Pathogenic Bacteria

Pathogen	Carbon source	Reference
X. campestris	Sucrose, glucose, DL-alanine, L-glutamate, and L-proline	Jackson et al. (1998); Patel and Kulkarni (1949)
Xanthomonas oryzae pv. oryzae	Sucrose, glucose	Yuan (1990)
Xanthomonas campestris pv. manihotis	Glucose, mannose, arabinose, trehalose, cellobiose, and fructose	Ogunjobi et al. (2010)
X. campestris pv. citri	Sucrose	Canteros de Echenique et al. (1985)
Xanthomonas axonopodis pv. malvacearum	Sucrose	Dezordi et al. (2009)
Xanthomonas phaseoli var. fuscans	Arabinose, cellobiose, dextrose, fructose, galactose, glycogen, lactose, maltose, mannose, sucrose, trehalose, xylose, melibiose, and dextrin	Basu and Wallen (1996)
Agrobacterium tumefaciens	Sucrose, D-mannitol, D-sorbitol, indol, inositol, melibioze, D-galactose, L arabinose, rhamnose, amygdalin, lactose, and glucose	Setti and Bencheikh (2013)
Erwinia amylovora	Ribose, trehalose, citrate, formate, and lactate	Vantomme et al. (1986)
Pseudomonas solanacearum	Glycerol, acetate, citrate, dextrose, sucrose, mannose, galactose, levulose, malonate, tartrate, L-phenyl Alanine	Granada and Sequiera (1975)
Xylella fastidiosa	Chitin, citrate, and succinate	Killiny et al. (2010); Rodrigo et al. (2004)
Dickeya solani	D-galactonic acid γ-lactone, m-tartaric acid, citric acid	Pedron et al. (2014)
Pectobacterium caratovorum	Glucose, sucrose, trehalose, D-galactose, gentiobiose, D-gluconic acid, m-inositol, D-fructose	Ni et al. (2010)
Clavibacter michiganensis	Sodium acetate, succinate	Kaneshiro et al. (2006)
Pseudomonas savastanoi pv. savastanoi	Betaine, DL-Glycerate, innositol, mannitol, L-tartrate	Hall et al. (2004)

Detrimental effects on the growth of PPB with the use of specific carbon sources were also reported. For example, fructose, dextrin, and citric acid were found to be detrimental to the growth of *X. o.* pv. *oryzae* under *in vitro* conditions (Tanaka, 1964). Some specific sources of carbon, such as glucose, are also used for differentiating xanthomonads from other saprophytes based on color differences. For example, for *X. arboricola* pv. *corylina* on glucose-containing media, the pathogen will be differentiated based on excess production of mucoid polysaccharides (Schaad et al., 2001; Scortichini, 2002). Similarly, when quinate is used as a carbon source in succinate-quinate (SQ) medium, differentiation of *Xanthomonas arboricola* on hazelnut could be done based on color development. The pathogen produces a deep green color that diffuses around the bacterial streak, whereas other xanthomonads show no color (Lee et al., 1999). Development of diffusible pigments to various degrees was also reported by some PPB in the presence of various carbon sources. For example, brown diffusible pigment is produced by *Xanthomonas phaseoli* var. *fuscans*, the causal organism of fuscous blight of bean seeds to various degrees by different carbon sources. These carbon sources can be grouped as: (1) Sources showing good growth, lowered pH, and no browning of media (i.e., arabinose, cellobiose, dextrose, fructose, galactose, glycogen, lactose, maltose, mannose, sucrose, trehalsoe, xylose, melibiose, and dextrin); (2) Sources showing fair growth, little change of pH, and no browning of media (i.e., raffinose, ribose, and salicin); and (3) Sources showing poor growth, increased pH, and browning of media (i.e., adonitol, dulcitol, inulin, methyl-a-D-glucoside, rhamnose, sorbitol, and mannitol). However, the production of a brown diffusible pigment by *Xanthomonas phaseoli* var. *fuscans* was lowered in the presence of dextrose in the medium (Basu and Wallen, 1996).

Physiological variations based on carbon-source utilization also occur among PPB (Picard et al., 2008). For example, in *Xanthomonas axonopodis* pv. *allii*, the causal agent of bacterial blight of *Allium* species, carbon-source utilization varied among strains. In a study from the Mascarene Archipelago, Brazil, and Japan, all strains produced pale yellow colonies and were identified as an AFLP group A utilizing L-fructose, D-mellibiose, and citric acid. The other group is called AFLP group B, in contrast to the first group (Picard et al., 2008).

Strain differentiation among *Xanthomonas* species such as *X. euvesicatoria*, *X. vesicatoria*, *X. perforans*, and *X. gardneri* based on carbon-utilization patterns was also reported by Jones et al. (2000). The carbon sources screened included dextrin, glycogen, N-acetyl-D-glucosamine, D-galactose, gentibiose, α-D-lactose lactulose, acetic acid, Cis-aconitic acid, malonic acid, propionic acid, D-alanine, Glycyl-L-aspartic acid, and L-threonine. Among these strains, *X. gardneri* did not utilize any of the carbon sources tested. However, all *X. euvesicatoria* strains utilized dextrin, glycogen, N-acetyl-D-glucosamine, D-galactose, gentibiose, cis-acotinic acid, and malonic acid. For *X. vesicatoria*, dextrin was utilized by all the strains under study. In the case of *X. perforans*, except for glycogen and Cis-acotinic acid, uniform carbon utilization from other sources was noticed in all of the strains.

7.3.2 Nitrogen Sources and Preferences

The majority of PPB utilize nitrogen in the form of inorganic compounds. However, some PPB prefer both inorganic and organic nitrogen supplements for their growth (Starr and Mandel, 1950). Inorganic nitrogen sources include ammonium salts and nitrates, and others. Commonly used organic nitrogen sources include amino acids, peptone, beef extract, and yeast extracts. The nitrogen requirement of a PPB varies during the growth of the pathogen. The compositions of intra- and extracellular amino acids of a PPB change considerably during its growth. For example, in a developing culture of *Xanthomonas citri*, 14 amino acids were detected in the intracellular amino compound pool. However, prominent among these were alanine, glutamic acid, and aspartic acid, which are constituted in the cell walls of *X. citri* (Prasad, 1979). Similarly, 12 amino acids were detected in the developing culture of *Xanthomonas campestris* pv. *malvacearum*. However, of these, only six amino acids were present in the bacterial filtrate. Only glutamic acid and alanine were present, at a 2:1 ratio in the cell-wall component of the bacterium (Lal et al., 1988).

Amino acid preferences vary among PPB. For example, in the case of *X. oryzae* pv. *oryzae*, glutamic acid, aspartic acid, methionine, cystine, and aspargine are good nitrogen sources (Tanaka, 1963). Of these, L-glutamic acid and L-aspartic acid were found to be the best (Hsu, 1966). Pathovars of *Xanthomonas campestris* could be differentiated according to their preferences toward different amino acids. In a study of 16 pathovars of *X. campestris*, it was found that all of the pathovars would accept DL-alanine, L-glutamate, L-proline, DL-Methionine, DL-treonine, DL-aspartate, L-aspargine, L-hydroxyproline, and L-hystidine as sole nitrogen sources. None of the pathovars preferred DL-serine, DL-norleucine, or L-tyrosine. However, the utilization of glycine, DL-valine, L-tryptophan, L-leucine, DL-isoleucine, L-arginine, DL-Lysine, and L-cystine varied with the pathovars (Kotasthane et al., 1965).

The role of amino acids in enhancing the growth of PPB is well established. In another study on the nutrient requirements of four strains of *X. oryzae* pv. *oryzae* (*Xoo*) in a modified Watanabe medium, it was observed that methionine enhanced the pathogen growth. Other organic nitrogen sources in the medium, such as tryptone and peptone, also enhanced the growth of *Xoo*.

The role of other organic nitrogen supplements in enhancing the growth of PPB is also well established. For example, peptone at the rate of 0.5% in medium supported the growth of Cancrosis bacterium, which causes cancrosis B of citrus in Argentina and Uruguay (Canteros de Echenique et al., 1985). Another example of PPB with good growth in a semi-selective culture medium that contains peptone (@5g/L of distilled water) is *X. axonopodis* pv. *malvacearum*.

This technique is also used to detect the bacterium in cotton seeds. Similarly beef extract (@3g/L) is another organic nitrogen source that favors the growth of *X. axanopodis* pv. *malvacearum* in

TABLE 7.3
Nitrogen Sources Utilized by Different Plant Pathogenic Bacteria

Pathogen	Nitrogen source	Reference
Pseudomonas solanacearum	Phenylalanine and phenylacetic acid	Agarwal et al. (1997)
X. campestris pv. citri	Peptone, dipotassium phosphate	Canteros de Echenique et al. (1985)
Xanthomonas axonopodis pv. malvacearum	Peptone, beef extract, potassium nitrate	Dezordi et al. (2009); Patel and Kulkarni (1949)
Xanthomonas oryzae pv. oryzae	Potassium dihydrogen phosphate, ammonium chloride, methionine	Yuan (1990)
Xanthomonas campestris	Yeast extract and synthetic aminoacid mixtures	Jackson et al. (1998)
Xanthomonas oryzae pv. oryzae	Ammonium dihydrogen phosphate, L-glutamic acid, cystine, methionine, L-aspartic acid, aspargine	Fang et al. (1957); Tanaka (1963); Hsu (1966)
Xanthomonas campestris	DL-alanine, L-glutamate, L-proline, DL-Methionine, DL-treonine, DL-aspartate, L-aspargine, L-hydroxyproline, and L-hystidine	Kotasthane et al. (1965)

the same selective medium (Dezordi et al., 2009). The growth of another xanthomonad, *Xoo*, was found to be enhanced by peptone in a modified Watanabe medium (Yuan, 1990). Similar reports on the growth enhancement of the PPB *Erwinia herbicola* through the use of peptone were reported (Yuan, 1990). In another study on the role of organic nitrogen supplements, it was reported that the growth of *X. campestris* was favored by the combination of organic nitrogen sources such as yeast extract and synthetic amino acid mixtures. These organic nitrogen sources resulted in optimum growth and cell yield (Jackson et al., 1998).

Inorganic nitrogen compounds are the commonly used nitrogen sources for PPB cultivation. For example, ammonium sulfate and ammonium dihydrogen phosphate are used in growth studies of *Xoo*. In one study, when the two inorganic supplements were present in a medium, ammonium sulfate was used to a slight extent by the PPB, whereas the other nitrogen source was not utilized (Fang et al., 1957). In the presence of ammonium dihydrogen phosphate, glucose was also found to be a favorable carbon source (Fang et al., 1957). Details of the different nitrogen preferences take both organic and inorganic forms for PPB (Table 7.3).

7.4 ROLE OF NUTRIENT SUPPLEMENTS IN ENHANCING PATHOGENESIS

7.4.1 Extracellular Polysaccharides (EPS)

Nutrient supplements to PPB enhance the factors that are responsible for pathogenesis in host plants. One of the important virulence factors of PPB is extracellular polysaccharide (EPS), which plays a major role in pathogenesis (Coplin and Cook, 1990). Several mechanisms are proposed for the pathogenesis by these EPS. For example, EPS enhances the susceptibility of host plants by suppressing defense responses through the production of callose (Yun et al., 2006). Other important mechanisms include masking the bacterium from recognition by the host, which enables PPB to colonize the host surface (Alvarez, 2000), and biofilm formation, which leads to bacterial resistance to host-defense mechanisms (Dow et al., 2003). EPS produced by PPB also interferes with the water-transport system of host plants by plugging the xylem vessels, thereby causing wilt (Kao et al., 1994) as in the case of *Ralstonia solanacearum*. Extracellular polysaccharides also chelate heavy metals and increase the tolerance of PPB to toxic substances (Kao et al., 1992). The role of EPS as a major virulence factor in the bacterial wilt of several agricultural crops was established by Hayward (1991).

Nutrient supplements play a major role in EPS production and thus in the virulence of a PPB. Several reports on the role of carbon sources in EPS production specific PPB are available. A study on *Xanthomonas campestris pv. campestris* (*Xcc*) and its carbohydrate metabolism indicated that the pathogen metabolizes sugars (e.g., glucose and sucrose). These two carbon sources were also found to be the best carbon sources for EPS production in *Xcc* (Garcia-Ochoa et al., 2000). In addition, it was demonstrated that sucrose utilization is mandatory for the full pathogenicity of *Xcc* (Blanvillain et al., 2007). The sucrose metabolism includes transportation of extracellular sucrose through the outer membrane of *Xcc* via a TonB-dependent receptor (SuxA), as well as through the inner membrane via a sugar transporter (SuxC). The intracellular sucrose is later hydrolyzed by sucrose hydrolase (SUH) to yield glucose and fructose (Blanvillain et al., 2007).

Xanthan gum is an important polysaccharide produced by *Xanthomonas*. Its production is attributed to a shift from balanced to unbalanced growth by the *Xanthomonas*. The virulence of *Xcc* in crucifer black rot is associated with the production of this polymer in the vascular system of a host plant; the viscosity of the polymer blocks nutrient supplies (Sutton and Williams, 1970). Research has indicated that the virulence of *Xanthomonas* is influenced by the quality of xanthan gum produced (Ramirez et al., 1988). However, contradictory reports have indicated that this polymer has no role in the pathogenicity of *Xcc* (Shaw et al., 1988).

Production of xanthan gum is influenced by carbon sources. In one study, among different carbon sources tested, sucrose was found to produce maximum xanthan gum. After sucrose, galacatose was found to be a better carbon source for enhancing gum production (Kumaraswamy et al., 2012). In another study, by De Vuyst and Vermeire (1994), it was demonstrated that xanthan production is influenced by the types and initial concentrations of carbon as well as by nitrogen. The optimal sole carbon source was either 4% glucose, 4%–5% sucrose, or 2.8% Sirodex A (a glucose syrup with glucose content of 95%–96%). The PPB pseudomonads produce the EPS molecules levan and alginate (Fett et al., 1986; Conti et al., 1994). Several nutrient sources (e.g., sodium chloride and ethanol) were found to significantly enhance these polysaccharides (Singh et al., 1992). In another study, it was reported that sucrose and glucose are the primary carbon sources for production of levan and alginate by the soybean pathogen *P. syringae* pv. *glycinea* (Osman et al., 1986).

Nitrogen sources also influence the xanthan gum production by *Xanthomonas* (Kumaraswamy et al., 2012). For example, yeast extract is an efficient source of nitrogen for xanthan gum production. Xanthan production was observed up to 3.6 g/L in batch fermentation and 5.2 g/L in fed batch fermentation. In addition, polysaccharide production was enhanced with increases in yeast extract concentration; this result was attributed to increased nitrogen uptake (Palaniraj et al., 2011). In another study, it was observed that yeast extract was an ideal source of nitrogen over other sources such as beef extract, ammonium sulfate, peptone, and tryptone in a production medium for xanthan gum production by *X. campestris* (Kumaraswamy et al., 2012). Among other nitrogen sources that influence xanthan production, 2% corn-steep liquor (a combined source of nitrogen and phosphorus) and 4% peptone were found to be effective (Bruggeman et al., 1998). Reports on the combination of carbon and nitrogen for optimizing xanthan yields are also available. In a study on *X. campestris* using glucose (carbon source) and yeast extract (nitrogen source), it was reported that the xanthan yield and specific production rate increased with increases in glucose/yeast extract in the medium. However, the cell yield and specific growth rate decreased as the glucose/yeast extract increased. A two-stage batch fermentation with a glucose/yeast-extract shifted from the initially low level (2.5% glucose/0.3% yeast extract) to a high level (5.0% glucose/0.3% yeast extract) at the end of the exponential growth phase; this method was ideal for xanthan production (Yang-Ming Lo et al., 1997).

7.4.2 Nutrient Supplements and Phytotoxins

Most PPB produce phytotoxins that play a major role in their virulence. However, these are non-host-specific and are not involved in pathogenicity. Important phytotoxins produced by PPB include thaxtomin, phaseolotoxin, tabtoxin, coronatine, syringomycin, 3-methylthiopropionicacid (MTPA), and others. The production of these phytotoxins by PPB is influenced by several macro- and micronutrients. For example, MTPA production by *X. campestris* pv. *manihotis* in cassava is influenced by methionine and stands as a precursor for MTPA (Ewbank and Marait, 1990). The other important toxin produced by many *Streptomyces* such as *S. scabies*, *S. acidiscabies*, and *S. turgidiscabies* is thaxtomine. *In vitro* studies indicated that thaxtomine production is favored in an oat-based culture medium (Wach et al., 2007). Thaxtomin A (ThxA) production is also stimulated by oat-bran broth due to the xylans and glucans in the broth. Other supplements that stimulate ThxA include xylans from wheat and tamarind. However, ThxA production is not stimulated by starches and simple sugars. Similarly, in potato plants, ThxA production is not stimulated by the glycoalkaloids, solanine, or chaconine.

In the case of phaseolotoxin produced by *Pseudomonas syringae* pv. *phaseolicola*, the causal agent of yellow halo in bush beans, toxin secretion is influenced by oxygen concentration. An increase in toxin levels observed in a synthetic medium with an increase in oxygen concentration was highest at 12 mg O_2/I (air saturation). The toxin content degraded with the depletion of nitrogen sources in the medium (Lehmann-Danzinger et al., 1997). The tabtoxin (wildfire toxin) produced by *Pseudomonas syringae* pv. *tabaci* in tobacco is regulated by carbon sources such as sugars, amino acids, and organic acids. Significant quantities of this phytotoxin were obtained with some selective amino acids. Interestingly, fructose, mannose, lactose, and ribose as supplements hinder the synthesis of tabtoxin and simultaneously stimulate the production of inactive isotabtoxins. Certain amino acids are also found to enhance tabtoxin production. For example, with a supplement of 3 mM of aspartate, methionine, and serine, a threefold increase of tabtoxin was observed (Messaadia and Harzallah, 2011). The best carbon sources for tabtoxin production are sorbitol, xylose, and sucrose; with glucose, production was low (Dehbi et al., 2001). A study by Gallarato et al. (2012) showed a 150% increase in the production of tabtoxin by *P. syringae* (the halo blight pathogen of oats, rye, barley, wheat, and sorghum) when choline, betaine, or dimethylglycine were used as nitrogen sources.

Studies on syringomycin produced by *Pseudomonas syringae* pv. *syringae* (brown spot of beans) indicated that the phytotoxin production is modulated by both nutritional and plant signal molecules, and that iron exerts a positive effect on the toxin production. It was also reported that inorganic phosphate concentrations of 1 mM or higher repressed the toxin production (Mo and Gross, 1991). In another study on *P. syringae*, it was found that syringomycin production could be maximized when micromolar ferric ion concentrations were supplied to the growth medium (Gross, 1985).

Regarding the phytotoxin coronatine by *P. syringae* pv. *glycinea* PG4180, both the growth and quantity of the phytotoxin were significantly affected by carbon sources, nutrient levels (glucose, NH_4Cl, phosphate, Mg, and SO_4), amino acid supplements, and complex carbon and nitrogen sources. However, when the medium was fortified with plant extracts, plant-derived secondary metabolites, or zinc, coronatine production was not affected. Significant enhancement of the phytotoxin was noticed when millimolar levels of KNO_3 or micromolar levels of $FeCl_3$ were added (Palmer and Bender, 1993).

7.4.3 Nutrient Supplements and *Hrp* Genes

The *hrp* (hypersensitive reaction and pathogenicity) genes in PPB are responsible for initiating the pathogenitic interaction with a susceptible host and for inducing the hypersensitivity response in both resistant host and non-host plants (Bonas et al., 1991). The *hrp* gene expression in PPB is induced both *in planta* as well as during growth in minimal medium. Different PPB have different preferences for nutrients, as well as for their growth and *hrp* gene expression. In general, *hrp* gene expression is more prominent in a minimal medium than in a rich medium, and different nutrient sources will have either inductive or repressive effects on *hrp* gene expression. For example, pyruvate was shown to have an inductive effect on the *hrp* gene expression in *Xanthomonas campestris* pv. *vesicatoria* (Schulte and Bonas, 1992). However, the same nutrient was shown to have a repressor effect on the *hrp* gene expression in *Pseudomonas syringae* pv. *glycinea* (Huynh et al., 1989). The Hrp proteins are components of type III secretion systems, regulatory proteins, and elicitors for hypersensitive response in plants (Lindgren, 1997). The details of different nutrient supplements that have inductive or repressor effects on different PPB are given in Table 7.4.

TABLE 7.4
Examples of Different Plant Pathogenic Bacteria with Varied Influential Reactions on Bacterial Hrp Gene Expression

Pathogen	Nutrient Supplement	*Hrp* Gene Reaction			References
		Induction/ Activation	Repression	No Effect	
Ralstonia solanacearum	Pyruvate	√			Arlat et al. (1991)
Pseudomonas syringae pv. glycinea				√	Huynh et al. (1989)
Xanthomonas campestris pv. vesicatoria		√			Schulte and Bonas (1992)
Pseudomonas syringae pv. glycinea	Fructose, sorbitol, glycerol		√		
Ralstonia solanacearum				√ (Very little)	Arlat et al. (1991)
Erwinia amylovora	Mannitol	√			Wei et al. (1992)
Xanthomonas campestris pv. vesicatoria			√		
R. solanacearum, P. syringae pv. glycinea, E. amylovora, X. campestris pv. vesicatoria	Sucroe	√			Huynh et al. (1989)
E. amylovora	Nicotinic acid		√		Wei et al. (1992)
	Rich medium		√		Beer et al. (1991)
X. c. pv. vesicatoria (Bacterial spot of pepper and tomato)	Sulfur containing aminoacids	√			Schulte and Bonas (1992)
X. c. pv. vesicatoria X. c. pv. vesicatoria					
	Sucrose or fructose and low conc. of casamino acids	√			Schulte and Bonas (1992); Bonas et al. (1991)
	High concentrations of casamino acids		√		Bonas et al. (1991)
	High concentrations of phosphate or sodium chloride		√		Schulte and Bonas (1992); Wengelnik and Bonas (1996)

(*Continued*)

TABLE 7.4 (*Continued*)
Examples of Different Plant Pathogenic Bacteria with Varied Influential Reactions on Bacterial Hrp Gene Expression

		Hrp Gene Reaction			
Pathogen	**Nutrient Supplement**	**Induction/ Activation**	**Repression**	**No Effect**	**References**
P. solanacearum	Peptone or casamino acids		√		Arlat et al. (1990; 1992)
P. syringae pv. phaseolicola	Osmotic strength, low pH and carbon sources	√			Rahme et al. (1992)
	Rich medium		√		Rahme et al. (1991)
P. syringae pv. glycinea	Carbon sources		√		Huynh et al. (1989)
P. syringae pv. syringae	Nitrogen and osmolarity	√			Xiao et al. (1992)
P. syringae pv. tomato DC3000 (Bacterial speck in *Arabidopsis thaliana*)	Iron	√			Kim et al. (2009)
Xanthomonas campestris and *P. solanacearum, Erwinia amylovora*	Rich medium		√		Arlat et al. (1991; 1992); Wei et al. (1992)
R. solanacearum, Erwinia amylovora	Minimal medium	√			Genin et al. (1992); Jacobs et al. (2012); Arlat et al. (1992); Wei et al. (1992)
	Rich medium		√		Jacobs et al. (2012)

7.5 CONCLUSION

Nutrient supplements that show inductive, repressive, or neutral effects on growth, phytotoxin production, and *hrp* gene expression have been investigated thoroughly over many years. The research applications in the available literature will enable future investigators to understand the *in planta* effects of these nutrient supplements in promoting pathogenicity in plants through enhanced virulence. However, nutrient supplements that annul the deleterious effects of *hrp* genes in PPB, and thereby lead to recognition and activation of HR activity in resistant or non-host plants, should be thoroughly investigated. Such investigations should include alterations of the nutrient compositions of exudates in rhizoplane or phylloplane, which are the target sites of majority of PPB, and through genetic trials that modify surface-nutrient composition in ways that suit the recognition and regulation of the HR system rather than induct virulence in the pathogen. Other applications of current knowledge could be along the lines of devising management strategies against PPB that cause dreadful diseases by including dose-dependent and need-based identified nutrient supplements that simultaneously promote plant growth and suppress pathogen virulence through *hrp* genes that trigger HR in plants.

REFERENCES

Agarwal, P., Latha, S., Mahadevan, A. 1997. Utilization of phenylalanine and phenylacetic acid by *Pseudomonas solanacearum*. *Applied Biochemistry and Biotechnology*. 61:379–391.

Alvarez, A.M. 2000. Black rot of crucifers. In *Mechanisms of Resistance to Plant Diseases*, Slusarenko, A.J. Fraser, R.S.S. van Loon, L.C. (eds.). Dordrecht: Kluwer Academic Publications. pp. 21–52.

Ansari, M.M., Sridha, R. 2001. Iron nutrition and virulence in *Xanthomonas oryzae* pv. *oryzae*. *Indian Phytopathology*. 54 (3):279–283.

Arlat, M., Barberis, P., Trigalet, A., Boucher, C. 1990. Organization and expression of hrp genes in *Pseudomonas solanacearum*. In: *Proc. 7th Int. Conf. Plant Pathogenic bacteria* Klement, Z. (ed.). Budapest, Hungary: Akademiai Kiado. pp. 419–424.

Arlat, M., C. Gough, C.E. Barber, C. Boucher, Daniels, M. 1991. *Xanthomonas campestris* contains a cluster of *hrp* genes related to the larger *hrp* cluster of *Pseudomonas solanacearum*. *Mol. Plant-Microbe Interact.* 4:593–601.

Arlat, M., Gough, C.L., Zischek, C., Barberis, P.A., Trigalet, A., Boucher, C.A. 1992. Transcriptional organization and expression of the large *hrp* gene cluster of *Pseudomonas solanacearum*. *Mol. Plant-Microbe Interact.* 5:187–193.

Ashok Kumar, R., Sunish Kumar, N., Sakthivel. 2003. Compositional difference of the exopolysaccharides produced by the virulent and virulence-deficient strains of *Xanthomonas oryzae* pv. *Oryzae*. *Curr. microbiology*. 46:251–255.

Aysan, Y., Sahin, F. 2003. An outbreak of crown gall disease on rose caused by *Agrobacterium tumefaciens* in Turkey. *Plant Pathology*. 52(6):780.

Basu, P.K., Wallen, V.R. 1996. Influence of temperature on the viability, virulence, and physiologic characteristics of *Xanthomonas phaseoli* var. *fuscans in vivo* and *in vitro*. *Canadian Journal of Botany*. 44(10):1239–1245.

Beer, S.V., Bauer, D.W., Jiang, X.H., Lady, R.G., Sneath, B.J., Wei, Z.M., Wilcox, D.A., Zumoff, C.H. 1991. The hrp gene cluster of *Erwinia amylovora. In: Proceedings of 5th International Symposium on Molecular and Genetic Plant-Microbe Interactions* Hennecke, H. Verma, D.P. (eds.). Dordrecht: Academic publisher. pp. 53–60.

Blanvillain, S., Meyer, D., Boulanger, A., Lautier, M., Guynet, C., Denance, N., Vasse, J., Lauber, E., Arlat, M. 2007. Plant carbohydrate scavenging through TonB-dependent receptors: a feature shared by phytopathogenic and aquatic bacteria. *PLoS ONE 2*, 2(2):e224.

Bonas, U., Schulte, R., Fenselau, S., Minsavage, G.V., Staskawicz, B.J., Stall, R.E. 1991. Isolation of a gene cluster from *Xanthomonas campestris* pv. *vesicatoria* that determines Pathogenicity and the hypersensitive response on pepper and tomato. *Plant-Microbe Interact*. 4:81–88.

Bruggeman, G., Leclercq, A., Neves, A., Harnie, E., Smeets, D., Vandamme, E.J., Vervust, T., Akerkart, G. 1998. Improved Xanthan Production by *Xanthomonas campestris* using an optimized medium containing peptone PS as sole nitrogen source. 2nd Symposium of the Belgian Society for Microbiology, Apoptosis and Microorganisms, Leuven, Belgium, November 6, p.2.

Canteros de Echenique, B.I., Zagory, D., Stall, R.E. 1985. A medium for cultivation of the B-strain of *Xanthomonas campestris* pv. *citri*. cause of cancrosis B in Argentina and Uruguay. *Plant Disease*. 69: 122–123.

Conti, E., Flaibani, A., O'Regan, M., Sutherland, I.W. 1994. Alginate from *Pseudomonas fluorescens* and *P. putida*: Production and properties. *Microbiology*. 140:1125–1132.

Coplin D.L., Cook, D. 1990. Molecular genetics of extracellular polysaccharide biosynthesis in vascular phytopathogenic bacteria. *Molecular Plant-Microbe Interactions*. 3:271–279.

Davey, M.E., O'Toole, G.A. 2000. Microbial biofilms: From ecology to molecular genetics. *Microbiology and Molecular Biology Reviews*. 64(4):847.

De Vuyst, L., Vermeire, A. 1994. Use of industrial medium components for xanthan production by *Xanthomonas campestris* NRRL-B-1459. *Appl. Microbiol. Biotechnol.* 42:187–191.

Dehbi,F., Harzallah, D., Larous, L. 2001. Effects of nutritional factors on production of tabtoxin, a phytotoxin, by *Pseudomonas syringae* pv. *tabaci*. MededRijksunivGentFak LandbouwkdToegepBiolWet. 66(2): 241–247.

Dellagi, A., Segond, D., Rigault, M., Fagard, M., Simon, C., Saindrenan, P., Expert, D. 2009. Microbial siderophores exert a subtle role in arabidopsis during infection by manipulating the immune response and the iron status. *Plant Physiology*. 150:1687–1696.

Dezordi, C., Maringoni, A.C., Menten, J.O.M., Camara, R.C. 2009. Semi-selective culture medium for *Xanthomonas axonopodis* pv. *malvacearum* detection in cotton seeds. *Asian Journal of Plant Pathology*. 3(2):39–49.

Dow, J.M., Crossman, L., Findlay, K., He, Y.Q., Feng, J.X., Tang, J.L. 2003. Biofilm dispersal in *Xanthomonas campestris* is controlled by cell–cell signaling and is required for full virulence to plants. *Proc. Natl. Acad. Sci., U.S.A.* 100:10995–11000.

Ewbank, E., Marait, H. 1990. Conversion of methionine to phytotoxic 3-methylthiopropionic acid by *Xanthomonas campestris* pv. *manihotis*. *Journal of General Microbiology*. 136:185–189.

Fang, C., Liu, C., Chu, C. 1957. The inhibition of glucose to the growth of *Xanthomonas oryzae*. *Acta Phytopathologica sinica*. 3(2):125–136.

Fang, H.N., Lin, C. F., Chu, C. L. 1956. A preliminary study on the disease cycle of the bacterial leaf blight of rice. *Acta Phytopathologica Sinica*. 2:173–185.

Fett, W.F., Osman, S.F., Fishman, M.L., Siebles, T.S. 1986. Alginate production by plant-pathogenic pseudomonads. *Appl. Environ. Microbiol*. 52:466–473.

Fletcher, M., Loeb, G.I. 1979. Influence of substratum characteristics on the attachment of a marine pseudomonad to solid surfaces. *Applied and Environmental Biology*. 37:67–72.

Gallarato, L.A., Primo, E.D., Lisa, A.T., Garrido, M.N. 2012. Choline promotes growth and tabtoxin production in a *Pseudomonas syringae* strain. *Advances in Microbiology*. 2:327–331.

Garcia-Ochoa, F., Santos, V.E., Casas, J.A., Gomex, E. 2000. Xantham gum: Production, recovery and properties. *Biotechnology Advances*. 18:549–579.

Garlant, L., Koskinen, P, Rouhiainen, L., Laine, P., Paulin, L., Auvinen, P., Holm, L., Pirhonen, M. 2013. Genome sequence of *Dickeya solani*, a new soft rot pathogen of potato, suggests its emergence may be related to a novel combination of non-ribosomal peptide/polyketide synthetase Clusters. *Diversity*. 5:824–842.

Genin, S., Gough, C.L., Zischek, C., Boucher, C.A. 1992. Evidence that the *hrp B* gene encodes a positive regulator of pathogenicity genes from *Pseudomonas solanacearum*. *Mol. Microbiol*. 6:3065–3076.

Goto, M. 1992. *Fundamentals of bacterial plant Pathology*. California, USA: Academic press Inc. pp. 342.

Graca, J.V. 1991. Citrus greening disease. *Annual Review of Phytopathology*. 29:109–136.

Granada, G.A., Sequeira, L. 1975. Characteristics of Colombia isolates of *Pseudomonas solanacearum* from tobacco. *Phytopathology*. 65:1004–1009.

Gross, D.C. 1985. Regulation of syringomycin synthesis in *Pseudomonas syringae* pv. *syringae* and defined conditions for its production. *J. Appl. Bacteriol*. 58:167–174.

Haan, E.G., Dekker-Nooren, T.C. E.M., Bovencamp, G.W., Speksnijder, A.G.C.L., Zouwen, P. S., Wolf, J.M. 2008. *Pectobacterium carotovorum* subsp. *carotovorum* can cause potato blackleg in temperate climates. *European Journal of Plant Pathology*. 122 (4):561–569.

Hall, B.H., Cother, E.J., Whattam, M., Noble, D., Luck, J., Cartwright, D. 2004. First report of olive knot caused by *Pseudomonas savastanoi* pv. *savastanoi* on olives (*Olea europea*) in Australia. *Australian Plant Pathology*. 33:433–436.

Hayward A.C. 1991. Biology and epidemiology of bacterial wilt caused by *Pseudomonas solanacearum*. *Annu. Rev. Phytopathol*. 29:65–108.

Hopkins, D.L., Purcell, A.H. 2002. *Xylella fastidiosa*: Cause of Pierce's disease of grapevine and other emerging diseases. *Plant Disease*. 86(10):1056–1066.

Hsu, S.T. 1966. Nutritional requirements *in vitro* of *Xanthomonas oryzae* (Uyeda and Ishiyama) Dowson and its effect on host plant resistance. MS thesis, University of Phillippines, College of Agriculture, Quezon City, 58pp.

Huynh, T.V., Dahlbeck, D., Staskawicz, B.J. 1989. Bacterial blight of soybean: Regulation of a pathogen gene determining host cultivar specificity. *Science*. 245:1374–1377.

Jackson, M.A., Frymier, J.S., Wilkinson, B.J., Zomer, P., Evans, S. 1998. Growth requirements for production of stable cells of the bioherbicidal bacterium *Xanthomonas campestris*. *Journal of Industrial Microbiology and Biotechnology*. 21:237–241.

Jacobs, J.M., Babujee, L., Meng, F., Miling, A., Allen, C. 2012. The *In planta* transciptome of *Ralstonia solanacearum*: Conserved physiological and virulence strategies during bacterial wilt of tomato. *mBio* 3(4): doi:10.1128/mBio.00114-12.

Jones, J.B., Bouzar, H., Stall, R.E., Almira, E.C., Roberts, P., Bowen, B.W., Sudberry, J., Strickler, P., chun, J. 2000. Systematic analysis of xanthomonads (*Xanthomonas* spp.) associated with pepper and tomato lesions. *International Journal of Systematic Bacteriology*. 50:1211–1219.

Kaneshiro, W.S., Mizumoto, C.Y., Alvarez, A.M. 2006. Differentiation of *Clavibacter michiganensis* subsp. *michiganensis* from seed-borne saprophytes using ELISA, Biolog and 16S rDNA sequencing. *European Journal of Plant Pathology*. 116:45–56.

Kao, C.C., Barlow, E., Sequeira, L. 1992. Extracellular polysaccharide is required for wild-type virulence of *Pseudomonas solanacearum*. *J. Bacteriol*. 174:1068–1071.

Kao, C.C., Gosti, F., Huang, H., Sequeira, L. 1994. Characterization of a negative regulator of exopolysaccharide production by the plant-pathogenic bacterium *Pseudomonas solanacearum*. *Mol. Plant-Microbe Interact*. 7:121–130.

Kasselaki, A.M., Goumas, D., Tamm, L., Fuchs, J. Cooper, J., Leifert, C. 2011. Effect of alternative strategies for the disinfection of tomato seed infected with bacterial canker (*Clavibacter michiganensis* subsp. *michiganensis*). *NJAS—Wageningen Journal of Life Sciences*. 58:145–147.

Killiny, N., Prado, S.S., Almeida, R.P.P. 2010. Chitin utilization by the insect-transmitted bacterium *Xylella fastidiosa*. *Applied and Environmental Microbiology*. 76:6134–6140.

Kim, B.J., Park, J.H., Park, T.H., Bronstein, P.A., Schneider, D.J., Cartinhour, S.W., Shuler, M.L. 2009. Effect of iron concentration on the growth rate of *Pseudomonas syringae* and the expression of virulence factors in *hrp*-inducing minimal medium. *Appl. Environ. Microbiol*. 75(9):2720–2726.

Kotasthane, W.V., Padhya, A.C., Patel, M.K. 1965. Utilization of amino acids as sole source of carbon and nitrogen by some xanthomonads. *Indian Phytopathology*. 18:154–159.

Kumar, A., Sarma, Y.R., Anandaraj, M. 2004. Evaluation of genetic diversity of *Ralstonia solanacearum* causing bacterial wilt of ginger using REP–PCR and PCR–RFLP. *Current Science*. 87(11):1555–1561.

Kumar, R., Shamarao Jahagirdar, M.R., Yenjerappa, S.T., and Patil, H.B. 2009. Epidemiology and management of bacterial blight of pomegranate caused by *Xanthomonas axonopodis* pv. *punicae*. *Acta Hort*. (ISHS) 818:291–296.

Kumaraswamy, M., Khan Behlol, A., Rohit, K.C., Purushotham, B. 2012. Effect of carbon and nitrogen sources on the production of xanthan gum from *Xanthomonas campestris* isolated from soil. *Archives of Applied Science Research*. 4(6):2507–2512.

Lal, K., Lal, B.B., Prasad, M. 1988. The free amino acid composition during growth of the *Xanthomonas campestris* pv. *malvacearum* (Smith) Dowson. ZentralblattfürMikrobiologie. 143(8):591–594.

Laville, J., Voisard, C., Keel, C., Maurhofer, M., Defago, G., Haas, D. 1992. Global control in *Pseudomonas fluorescens* mediating antibiotic synthesis and suppression of black root rot of tobacco. *Proc Natl Acad Sci USA*. 89:1562–1566.

Lee Y.A., Lo Y.C., Yu P.P. 1999. A gene involved in quinate metabolism is specific to one DNA homology group of Xanthomonas campestris. *Journal of Applied Microbiology*. 87:649–658.

Lehmann-Danzinger, H., Jarchow-Redecker, K., Rudolph, K. 1997. Influence of Oxygen Concentration on Growth and Phaseolotoxin Secretion of *Pseudomonas syringae* pv. *phaseolicola Developmentsin Plant Pathology*. 9:255–260.

Lindgren, P.B. 1997. The role of *hrp* genes during plant bacterial interactions. *Annual Review of Phytopathology*. 35:129–152.

Magasanik, B. 1957. Nutrition of bacteria and fungi. *Annual Review of Microbiology*. 11:221–252.

Mansfield, J., Genin, S., Magori, S., Citovsky, V., Sriariyanum, M., Ronald, P., Dow, M. et al. 2012. Top 10 plant pathogenic bacteria in molecular plant pathology. *Molecular Plant Pathology*. 13(6):614–629.

Maruthasalam, S., Kalpana, K., Kumar, K.K., Loganathan, M., Poovannan, K., Raja, J.A., Kokiladevi, E., Samiyappan, R., Sudhakar, D., Balasubramanian, P. 2007. Pyramiding transgenic resistance in elite indica rice cultivars against the sheath blight and bacterial blight. *Plant Cell Reporter*. 26(6):791–804.

Messaadia N., Harzallah, D. 2011. *Regulatory Effects of Carbon Sources on Tabtoxin Production (A β-lactam Phytotoxin of Pseudomonas syringae pv. tabaci)* World Academy of Science, Engineering and Technology. 5:1–22.

Miller, T.D., Schroth, M.N. 1972. Monitoring the epiphytic population of *Erwinia amylovora* on pear with a selective medium. *Phytopathology*. 62:1175–1182.

Mo, Y.Y., Gross, D.C. 1991. Plant signal molecules activate the syrB gene, which is required for syringomycin production by *Pseudomonas syringae pv. syringae*. *Journal of Bacteriology*. 173(18):5784–5792.

Ni, L., Guo. L., Custers, J. B. M., Zhang, L. 2010. Characterization of calla lily soft rot caused by *Pectobacterium Carotovorum* subsp. *carotovorum* ZT0505: Bacterial growth and pectate lyase activity under different conditions. *Journal of Plant Pathology*. 92(2):421–428.

Ogunjobi, A.A., Fagade, O.E., Dixon, A.G.O. 2010. Physiological studies on *Xanthomonas axonopodis* pv *manihotis* (Xam) strains isolated in Nigeria. *European Journal of Biological Sciences*. 2(4):84–90.

Osman, S.F., Fett, W.F., Fishman, M.L. 1986. Exopolysaccharides of the phytopathogen *Pseudomonas syringae* pv. *glycinea*. J. Bacteriol. 166(1): 66–71.

Palaniraj, A., Jayaraman, V., Hariram, S.B. 2011. Influence of nitrogen sources and agitation in xanthan gum production by *Xanthomonas campestris*. *International Journal of Advanced Biotechnology and Research*. 2(3):305–309.

Palmer, D.A., Bender, C.L. 1993. Effects of environmental and nutritional factors on production of the polyketide phytotoxin coronatine by *Pseudomonas synringae* pv. *glycinea*. *Applied and Environmental Microbiology*. 59(5):1619–1626.

Patel, M.K., Kulkarni, Y.S. 1949. Nitrogen utilization by *Xanthomonas malvacearum* (Sm.) Dowson. Indian *Phytopathology*. 2:62–64.

Pedron, J., Mondy, S., Essarts, Y.R., Gijsegem, F.V., Faure, D. 2014. Genomic and metabolic comparison with *Dickeya dadantii* 3937 reveals the emerging *Dickeya solani* potato pathogen to display distinctive metabolic activities and T5SS/T6SS-related toxin repertoire. *BMC Genomics*. 15:283.

Picard, Y., Roumagnac, P., Legrand, D., Humeau, L., Robène-Soustrade, I., Chiroleu, F., Gagnevin, L., Pruvost, O. 2008. Polyphasic characterization of *Xanthomonas axonopodis* pv. *allii* associated with outbreaks of bacterial blight on three *Allium* species in the Mascarene archipelago. Phytopathology. 98:919–925.

Prasad, M. 1979. The free aminoacid pool composition during growth of the culture of *Xanthomonas citri* (Hasse) Dowson. Zentralbl. Bakteriol. Naturwiss. 134(8):692–696.

Pratt, L.A., Kolter, R. 1998. Genetic analysis of *Escherichia coli* biofilm formation: Roles of flagella, motility, chemotaxis and type I pili. *Molecular Microbiology*. 30(2):285–293.

Raaijmakers, J.S., Mark Mazzola. 2012. Diversity and natural functions of antibiotics produced by beneficial and plant pathogenic bacteria. *Annual Review of Phytopathology*. 50:403–424.

Rahme, L.G., Mindrions, M.N., Panopuolos, N.J. 1991. Genetic and transcriptional organization of the hrp cluster of Pseudomonas syringae pv. phaseolicola. *Journal of Bacteriology*. 173:575–586.

Rahme, L.G., Mindrinos, M.N., Panopoulos, N.J. 1992. Plant and environmental sensory signals control the expression of hrp genes in Pseudomonas syringae pv. phaseolicola. *Journal of Bacteriology*. 174:3499–3507.

Ramirez, M.E., Fucikousky, L., Garcia-Jimenez, F., Quintero, R., Galindo, E. 1988. Xanthan gum production by altered pathogenicity variants of *Xanthomonas campestris*. *Appl. Microbiol. Biotechnol*. 29:5–10.

Ramos,C., Matas, I.M., Bardaji, L., Aragón, I.M., Murillo, J. 2012. *Pseudomonas savastanoi pv. savastanoi*: Some like it knot. *Molecular Plant Pathology*. 13(9):998–1009.

Rodrigo, P., Almeida, P., Mann, R., Purcell, A.H. 2004. *Xylella fastidiosa* cultivation on a minimal solid defined medium. *Current Microbiology*. 48:368–372.

Salmond, G.P.C. 1994. Secretion of extracellular virulence factors by plant pathogenic bacteria. *Annual Review of Phytopathology*. 32:181–200.

Schaad, N.W., Jones, J.B., Chun, W. 2001. *Laboratory guide for identification of Plant pathogenic bacteria*. Third edition, Minnesota: APS press, The American Phytopathological society, St. Paul, pp. 373.

Schulte, R., Bonas, U. 1992. A *Xanthomonas* pathogenicity locus is induced by sucrose and sulfur-containing amino acids. *Plant Cell*. 4:79–86.

Scortichini, M. 2002. Bacterial canker and decline of Euro-pean hazelnut. *Plant Disease*. 86:704–709.

Setti, B., Bencheikh, M. 2013. Isolation and characterization of the *Agrobacterium tumefaciens* from almond nurseries in Chlef region in western Algeria. *European Scientific Journal*. 9:192–198.

Shaw, J.J., Settles, L.G., Kado, C.I. 1988. Transposon Tn4431 mutagenesis of *Xanthomonas campestris* pv. *campestris*: Characterization of a nonpathogenic mutant and cloning of a locus for pathogenicity. *Molecular Plant Microbe Interactions*. 1:39–45.

Shevchik, V.E., Evtushenkov, A.N., Babitskaya, H.V., Fomichev, Y.K. 1992. Production of pectolytic enzymes from *Erwinia* grown on different carbon sources. *World Journal of Microbiology and Biotechnology*. 8:115–120.

Singh, S., Koehler, B., Fett, W.F. 1992. Effect of osmolarity and dehydration on alginate production by fluorescent pseudomonads. *Curr. Microbiol*. 25:335–339.

Smith, W.K. 1958. A survey of the production of pectic enzymes by plant pathogenic and other bacteria. *Microbiology*. 18(1):33–41.

Starr, M.P., Mandel, M. 1950. The nutrition of Phytopathogenic bacteria. IV. Minimal nutritive requirements of the genus *Erwinia*. *Journal of Bacteriology*. 60(5):669.

Sunish Kumar, R., Sakthivel, N. 2001. Exopolysaccharides of Xanthomonas pathovar strains that infect rice and wheat crops. *Applied Microbiology and Biotechnology*. 55:782–786.

Sutton, J.C., Williams, P.H. 1970. Comparison of extracellular polysaccharide of *Xanthomonas campestris* from culture and from infected cabbage leaves. *Canadian Journal of Botany*. 48:645–651.

Tanaka, Y. 1963. *Studies on the Nutritional Physiology of Xanthomonas Oryzae (Uyeda et Ishiyama) Dowson. I. On the Nitrogen Sources*. Science bulletin of Faculty of Agriculture, Kyushu University, Japan. 20:151–155.

Tanaka, Y. 1964. *Studies on the Phage Resistant Strains of Xanthomonas oryzae (Uyeda et Ishiyama) Dowson. II. On the Carbon Sources and New Synthetic Medium*. Science bulletin of Faculty of Agriculture, Kyushu University, Japan. 21:149–153.

Tang, X., Xiao, Y., Zhou, J. 2006. Regulation of the Type III Secretion System in Phytopathogenic Bacteria. *Molecular Plant Microbe Interactions*. 19(11):1159–1166.

Uppalapati, S., Ishiga, Y., Ryu, C., Ishiga, T., Wang, K., Noe, L.D., Parker, J.E., Mysore, K.S. 2011. SGT1 contributes to coronatine signaling and *Pseudomonas syringae* pv. *tomato* disease symptom development in tomato and Arabidopsis. *New Phytologist*. 189:83–93.

Vantomme, R., Rijckaert, C., Swings, J., De Ley, J. 1986. Characterization of further Erwinia amylovora strains and the application of the API 20E system in diagnosis. *Journal of Phytopathology*. 117:34–42.

Vicente, J. G., Holub, E.B. 2012. *Xanthomonas campestris pv. campestris* (cause of black rot of crucifers) in the genomic era is still a worldwide threat to brassica crops. *Molecular Plant Pathology*. 14(1):2–18.

Wach, M.J., Krasno, S.B., Loria, R., Gibson, D.M. 2007. Effect of carbohydrates on the production of thaxtomin A by *Streptomyces acidiscabies*. *Arch Microbiol*. 188:81–88.

Wei, Z.M., Sneath, B.J., Beer, S.V. 1992. Expression of *Erwinia amylovora hrp* genes in response to environmental stimuli. *Journal of Bacteriology*. 174:1875–1882.

Wengelnik, K., Bonas, U. 1996. HrpXv, an AraC-type regulator, activates expression of five of the six loci in the hrp cluster of *Xanthomonas campestris* pv. vesicatoria. *Journal of Bacteriology*. 178:3462–3469.

Xiao, Y., Lu, Y., Heu, S., Hutcheson, S.W. 1992. Organization and environmental regulation of the *Pseudomonas syringae* pv. *syringae* 61 *hrp* cluster. *Journal of Bacteriology*. 174:1734–1741.

Yang-Ming Lo., Shang-Tian Yang., David B. Min. 1997. Effects of yeast extract and glucose on xanthan production and cell growth in batch culture of *Xanthomonas campestris*. *Appl Microbiol Biotechnol*. 47: 689–694.

Yuan, W. 1990. Culture medium for *Xanthomonas campestris* pv. *oryzae*. *Journal of Applied Bacteriology*. 69:798–805.

Yun, M.H., Torres, P.S., El Oirdi, M., Rigano, L.A., Gonzalez-Lamothe, R., Marano, M.R., Castagnaro, A.P., Dankert, M.A., Bouarab, K. Vojnov, A.A. 2006. Xanthan induces plant susceptibility by suppressing callose deposition. *Plant Physiol*. 141:178–187.

8 Biocontrol Mechanisms of Siderophores against Bacterial Plant Pathogens

Vellasamy Shanmugaiah, Karmegham Nithya, Hariharan Harikrishnan, Mani Jayaprakashvel, and Natesan Balasubramanian

CONTENTS

ABSTRACT Siderophore is an iron-healing compound that has an immense role in microbial interaction, especially in the rhizosphere. Siderophore is present in one of the major mechanisms of bacteria that is involved in the biological control of plant diseases. Both plant pathogenic fungi and bacteria are reported to be inhibited by siderophore-producing biocontrol agents. These siderophores are produced in iron-limited conditions to sequester the less-available iron from the environment and thereby deprive the pathogen of iron, which ultimately leads to inhibition. Three main groups of siderophores have been reported: hydroxymate siderophore, catecholate siderophore, and mixed siderophore. Siderophores are synthesized through two different biosynthetic pathways: NRPSs-dependent and non-NRPSs-dependent. The ABC-type transporter proteins are also involved in the delivery of iron-siderophore complex into the cytosol of the producing organism, where it can be utilized. Some pathogenic bacteria capable of producing siderophores are highly virulent. Siderophore interacts with the H_2O_2 and peroxidases in the affected tissue either to enhance oxidative stress induced by harpin, coded by an *hrpN* gene, or to protect bacterial cells by inhibiting the generation of reactive oxygen species. In this chapter we discuss control of bacterial plant pathogens with the production of siderophore by antagonistic rhizobacterium. Emphasis is given to the most economically important bacterial plant diseases such as bacterial soft rot of potato, tomato bacterial wilt, bacterial canker of tomato, rice bacterial blight, bacterial blight of cotton, bacterial leaf spot of mungbean, and fire blight disease in apple. An overview is included of the significance of siderophores in the context of the inhibition of plant pathogenic bacteria.

KEYWORDS: rhizosphere, fluorescent pseudomonads, siderophore, mechanisms, bacterial plant pathogens

8.1 IRON IMPORTANCE

Iron, the most abundant element on Earth, is an essential nutrient for almost all microorganisms. Ferrous iron (Fe^{+2}) is highly soluble up to a concentration of 100 mM at pH 7. However, the ferric form of iron (Fe^{+3}), which is needed for growth by organisms, is soluble at biological pH only to a concentration of 10^{-9} M, which makes the bioavailability of iron very low (Petsko, 1985). Fe^{+3} is not readily consumed by living organisms because of its extremely low solubility; this restriction means that iron bioavailability is a major limiting factor for living organisms. At neutral pH, iron bioavailability is much more limited and leads to competition among microoganisms for limited nutrients. This competition for iron nutrition is one of the mechanisms of biological control for both bacterial and fungal phytopathogens. Most microorganisms (e.g., bacteria and fungi) require at least 10^{-6} M iron for growth and development (McMorran et al., 2001).

8.2 SIDEROPHORE

In order to sequester and solubilize an available ferric form, many microorganisms synthesize siderophore, a low-molecular-weight molecule in the range of approximately 400–1500 Da that has a high affinity for ferric ion. When it is recognized by membrane receptor proteins within a microorganism, siderophore forms an Fe-Siderophore complex that is transported into cell periplasm, where it facilitates iron uptake in response to low-iron stress (Neilands, 1981; Hider and Kong, 2010). The term "siderophore," coined by Lankford in 1973, is derived from the Greek term meaning "iron carrier" (Ishimaru, 1993). To date, more than 500 siderophores have been identified. Among these, the chemical structures of 270 have been determined (Hider and Kong, 2010). According to the generally accepted definition, siderophores are "ferric specific microbial iron chelator compound[s] and [their] biosynthesis [is] regulated by the availability of iron in the surrounding environment" (Orsi, 2004). Kloepper et al. (1980b), who were the first to demonstrate the importance of siderophore production in the biocontrol mechanism of *Erwinia carotovora*, used pseudobactin, a type of siderophore (Table 8.1) produced by the biocontrol agent *Pseudomonas fluorescens*.

TABLE 8.1
Types of Siderophore Produced by Different Microorganisms

Organisms	Siderophore Produced	Types	Reference
Bacteria			
Escherichia coli	Enterobactin	Catecholate	Furrer et al. (2002)
Aeromonas hydrophila	Amonabactin	Catecholate	Barghouthi et al. (1989)
Pseudomonas aeruginosa	Pyoverdin and pyochelin	Salicylate-derived	Buysens et al. (1996)
Aerobacter aerogenes	Aerobacin	Hydroxamate	Buyer et al. (1991)
Salmonella	Salmochelin	Catechocolate	Müller et al. (2009)
Vibrio cholera	Vibriobactin	Catecholate	Griffiths et al. (1984)
Mycobacteriumtuberculosis	Mycobactin	Salicylate-derived	De Voss et al. (2000)
Yersinia pestis	Yersniabactin	Hydroxymate	Perry et al. (1999)
Fungi			
Ustilago sphaerogina	Ferrichrome	Hydroxymate	Jalal et al. (1998)
Ustilago maydis	Ferriaxamine B	Catecholate	Ardon et al. (1998)
Rhizopus microspores	Rhizoferin	Carboxylate	Drechsel et al. (1995)
Aspergillus nidulans	Triacetylfusarinine	Hydroxymate	Eisendle et al. (2004)
Actinomycetes			
Actinomadura madurae	Madurastatin	Catecholate	Harada et al. (2004)
Streptomyces griseus	Desferrioxamine B	Catecholate	Yamanaka et al. (2005)

8.2.1 Siderophores: Structures and Binding Sites

Several types of siderophores are produced by many soil microorganisms. Based on iron-coordinating functional groups, structural features, and types of ligands, siderophores are classified as hydroxymate, catecholate, carboxylates, or mixed (Crowley, 2006; Meneely, 2007). Hydroxymates are produced by both bacteria and fungi, catecholates by bacteria, and carboxylates by only a few bacteria (*Rhizobium* sp. and *Staphylococcus* sp.) and exclusively by fungi belonging to Mucorales. Mixed types are produced by fluorescent pseudomonads.

The three groups of siderophores engage two sites of the iron center, where they form a very stable and strainless five-member ring with the iron atom. The three binding sites always coordinate the iron with the lone pair of an oxygen atom. Because oxygen is a very hard donor ligand, it shows a high affinity to the hard Fe (III) cation. This interaction enhances the strength of the coordination as well as the selectivity of the siderophore ligands for iron (Albrecht-Gary and Crumbliss, 1998).

8.2.2 Hydroxymate Siderophore

As one of the major classes of siderophores, hydroxymates are well studied for their structural and functional properties. The iron chelation is afforded by a hydroxymate group (-CO-N(O^-)-) formed from aceylated or formylated hydroxylamines that are usually derived from lysine or ornithine (Miethke and Marahiel, 2007). Lysine derivatives are used for the synthesis of aerobactin (*E. coli*) and mycobactin (*Mycobacterium* spp.) (Crosa and Walsh, 2002). Ornithine derivatives are used in the synthesis of pyoverdin (*P. aeruginosa*), exochelin (*Mycobacterium* spp.), ornibactin (strains of pseudomonads), ferrichrome, fusarinine, and coprogens (variety of fungi) (Neilands, 1981). Histamine derivatives have also been found in anguibactin, which is the siderophore produced by *Vibrio anguillarum*, a fish pathogen (Crosa and Walsh, 2002). The hydroxymate group is assembled in a two-step process that begins with hydroxylation of the primary side-chain amine of ornithine or lysine by a flavinadenosine dinucleotide-dependent monooxygenase (Challis, 2005). The second step involves formylation by a methyl transferase for pyoverdin and ornibactin, or acetylation by an acetylase for all other hydroxymate siderophores (Miethke and Marahiel, 2007).

8.2.3 Catecholate Siderophores

The second most common siderophore class, the phenol-catecholates, contain a mono- or dihydroxybenzoic acid group to chelate the iron (Neilands, 1981; Meneely, 2007). This class of siderophores has been observed only in bacteria. The catecholate group is derived from salicylate or dihydroxybenzoic acid; its siderophores have iron-binding affinities that range from very tight in enterobactin from *E. coli* to fairly weak in pyochelin from *P. aeruginosa* (Cox and Graham, 1979). Pyochelin, an example of catecholate siderophores produced by *P. aeruginosa*, is a 324 Da peptide composed of salicylate and two cysteines that form a thiazoline and a thiozolidine heterocyclic ring (Ankenbauer et al., 1988). Pyochelin has been shown to bind several metals other than iron, including Zn(II), Mo(IV), Ni(II), and Co(II) (Perry and Brubaker, 1979; Namiranian et al., 1997). Unlike pyoverdin, which is highly divergent among strains of *P. aeruginosa*, the same pyochelin molecule is produced by a wide variety of *Pseudomonas* species and in some *Burkholderia cenocepacia* strains as well (Poole and McKay, 2003). The production of pyochelin by *P. aeruginosa* increases the lethality of virulent strains but not non-virulent strains, which indicates that although pyochelin is involved in the acquisition of iron, it is not a required virulence factor (Cox, 1982). It has been suggested that the mechanism for secretion of pyochelin into the environment is similar to that for pyoverdin in that both mechanisms use the RND efflux pump system.

8.2.4 Carboxylate (Complexion) Siderophore

Carboxylate siderophores contain hydroxyl or carboxyl that function as donor groups to iron. The first siderophore of this class, rhizobactin, is produced by *Rhizobium meliloti* strain DM4. It has an amino poly (carboxylic acid) with ethylenediaminedicarboxyl and hydroxycarboxyl moieties as iron-chelating groups (Smith et al., 1985). Staphyloferrin A, produced by *Staphylococcus hyicus* DSM20459, is another member of this class of complexion siderophores. Satphyloferrin A consists of one D-ornothine and two citric acid residues that are linked by two amide bonds (Meiwes et al., 1990).

8.2.5 Mixed Siderophores

Several other siderophores use multiple functional groups to chelate iron; these are known as mixed siderophores (Meneely, 2007). One example, the mycobactins produced by *Mycobacterium* spp., consist of hydroxymate and phenol-catecholate classes and are highly lipid-soluble. This class of siderophores is assumed to reside in the outer membrane of the producing bacterium *Mycobacterium* spp. and to work in conjunction with water-soluble chelating agents (exochelins) for iron sequestration. Structural studies using x-ray crystallography have revealed that iron binding in mycobactins is accomplished by two hydroxamates, a phenolate group, and oxazoline nitrogen. It is generally assumed that in oxazoline and thiazoline that contain donor-deficient siderophores, the imine-N will participate in ferric ion complexation (Raymond et al., 1984).

8.2.6 Pyoverdin

Pyoverdin is a water-soluble siderophore consisting of 6–12 amino acids, depending on the strain, with a dihydroxyquinoline fluorescent chromophore and a small dicarboxylic acid (Meyer, 2000; Poole and McKay, 2003). Pyoverdin has exceptional affinity for ferric iron with a dissociation constant of 10–30.8 M and has been demonstrated to chelate iron from transferrin (Wolz et al., 1994; Boukhalfa and Crumbliss, 2002). The virulence of *P. aeruginosa* is strongly influenced by the production of pyoverdin (Takase et al., 2000). The pyoverdin locus in the *Pseudomonas* genome, the most divergent region, results in highly variable pyoverdin molecules that differ in almost every strain of *Pseudomonas* (Smith et al., 2005).

Pyoverdins are classified into three main categories (Lamont et al., 2006):

- Type I: The pyoverdins with formyl-hydroxyornithines;
- Type II: The pyoverdins with one formyl-hydroxyornithine and a terminal cyclized hydroxyornithine; and
- Type III: The pyoverdins with two formyl-hydroxyornithines in a different arrangement from type I.

8.2.7 Siderophore Biosynthesis

Siderophores act as a virulence factor in many pathogenic microorganisms, including *E. coli*, *Salmonella typhimurium*, *Vibrio anguillarum*, *N. gonorrhoeae*, and *P. aeruginosa* (Crosa, 1980; Cox, 1982; Ankenbauer et al., 1985; Sokol, 1987). Many biocontrol agents produce siderophores as one of their biocontrol mechanisms to outcompete microbial pathogens in soil and plant systems. Therefore, siderophore biosynthetic pathways are always relevant for research and study. In general, the biosynthesis of siderophore includes two pathways. The first is by a large family of modular multi-enzymes called NRPSs and the other is NRPS-independent. The NRPSs assemble amino acids into peptides by forming peptide bonds that do not require an RNA template (Crosa and Walsh, 2002). The other pathway, the NRPS-independent pathway, involves linking dicarboxylic acid and diamine or amino alcohol building blocks with amide or ester bonds (Challis, 2005). The NRPS-independent pathway is used for the assembly of hydroxymate and carboxylate siderophores such as aerobactin, alcaligin, stephlobactin, and petrobactin (Challis, 2005; Miethke and Marahiel, 2007). The NRPS-dependent pathway is used for the biosynthesis of enterobactin, yersiniabactin, pyochelin, pyoverdin, vibriobactin, and mycobactin (Mossialos et al., 2002).

Proteins belonging to the NRPS family are a hallmark in the biosynthesis of peptide siderophores (Manuela Di Lorenzo et al., 2004). The NRPS synthesizes peptide siderophores or antibiotics in the absence of an RNA template via a multistep process (Keating and Walsh, 1999; Mootz et al., 2002). Diverse peptide structures are derived from catalytic domains of NRPSs. Sets of catalytic domains constitute a functional module that contains the information needed to complete an elongation step in peptide biosynthesis. Each module combines the catalytic functions for activation by ATP hydrolysis of a substrate amino acid in order to transfer the corresponding adenylate to the enzyme-bound 4′-phosphopantetheinyl cofactor, as well as for peptide bond formation. A classic module thus consists of an adenylation (A) domain, a peptidyl carrier protein (PCP) domain, and a condensation (C) domain; additional domains are also present that can, if required, lead to modification of the substrates in the synthesis of the peptide (Von Dohren et al., 1997).

Two anguibactin biosynthetic proteins, AngR and AngB, were characterized as NRPSs in *V. anguillarum* (Wertheimer et al., 1999; Welch et al., 2000). The biosynthetic step was catalyzed by AngM. The A domain of AngR activates the cysteine, which is then tethered to the PCP domain of AngM. The C domain of AngM catalyzes peptide-bond formation between the dihydroxyphenylthiazol (DHPT) group loaded on the PCP domain of AngM and the secondary amine group of hydroxyhistamine. AngE, AngB, AngN, AngU, and AngH are other NRPSs and tailoring enzymes that intervene in anguibactin biosynthesis.

8.2.8 Mechanism of Siderophore Export

The mechanism involved in the secretion or export of siderophores outside the cell is enacted by a transport protein or a pump. Three major types of proteins are involved in this process: the major facilitator super (MFS) family; the resistance, nodulation, and cell division (RND) superfamily; and the ABC superfamily. Siderophore-mediated iron uptake in microorganisms is both a receptor- and an energy-dependent process (Sigel and Sigel, 1998). Other components include two specific outer-membrane receptor proteins, Fec A and Fep A, a TonB-ExbB-ExbD protein complex in the inner

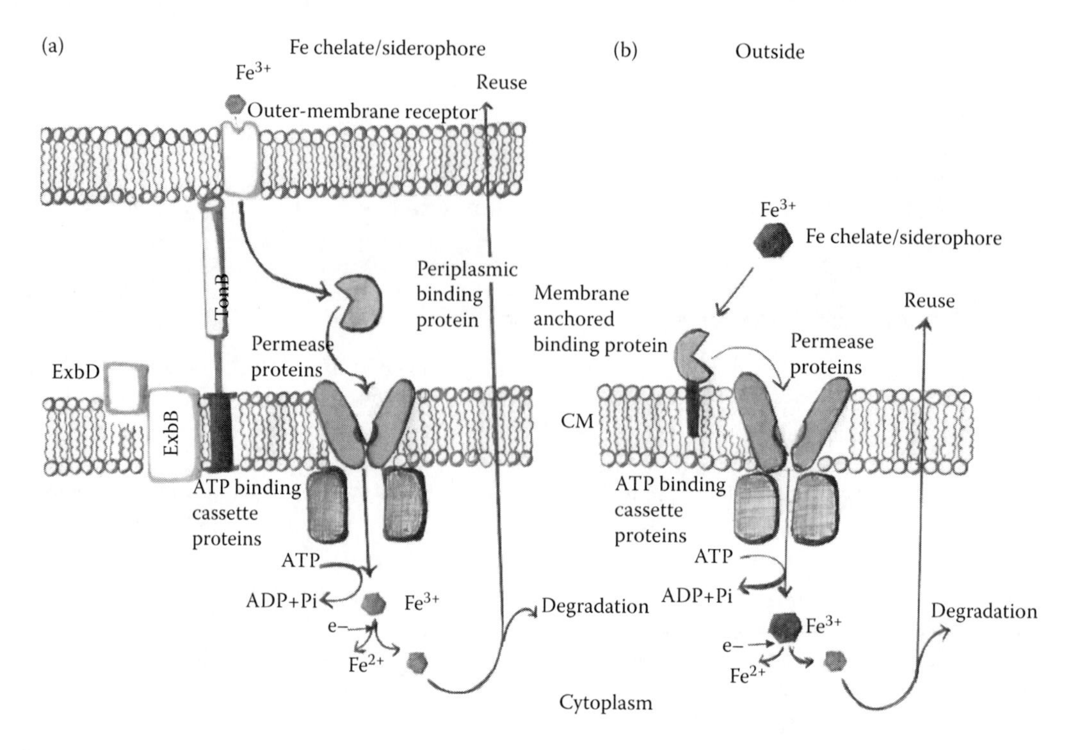

FIGURE 8.1 Schematic representation of iron uptake in (a) Gram-negative bacteria; (b) Gram-positive bacteria.

membrane, a periplasmic binding protein, and an inner-membrane ATP-dependent Fec CDE-Fep CDE protein (Figure 8.1). Under iron deficiency, bacteria synthesize siderophore and increase the number of receptor molecules. Once the siderophore is excreted outside the cell, it binds through the membrane receptor with an iron complex and transports the iron into the cell via the Fec A and Fep A outer membrane receptors, after it is transported to Fec C,D,E and Fep C,D,E, the so-called ABC transporter systems (from ATP-binding cassette) (Davidson and Nikaido, 1991; Boos and Eppler, 2001). The cassette is assembled of two proteins: one to span the membrane, which acts as a permease, and the other to hydrolyze ATP to provide the energy for transport.

In *E. coli*, enterobactin export is carried out by a MFS protein, EntS, which is encoded by the gene *ybda* (Lugtenberg et al., 2001). Recently, bacillibactin secretion in *B. subtilis* was found to be carried out by a similar MFS-type transporter YmfE (Mathesius et al., 2003).

In *P. aeruginosa*, the secretion of pyoverdine was thought to be carried out by a typical RND superfamily transport protein, the efflux system MexA-MexB-OprM (Mayak et al., 2004). Representatives of the ABC-type transporters that are involved in siderophore export are found in *S. aureus* (Poole et al., 1993), *Mycobacterium tuberculosis* (Rodriguez and Smith, 2006), and *M. smegmatis* (Farhana et al., 2008). Studies have shown that the ferric-pyoverdine complex in *P. aeruginosa* is dissociated in the periplasm; iron is released by reduction; and pyoverdine is then recycled from the periplasm to the external medium by the efflux pump PvdRT-OpmQ (Greenwald et al., 2007). The ferrichrome siderophore, by contrast, is transported across the outer membrane via the FiuA and FoxB receptors (Hannauer et al., 2010).

8.2.9 Transport of Iron-Siderophore Complex

The mechanism of release of iron from the Fe^{3+} siderophore complex varies among microbes. For example, in *E. coli* ferrichrome and ferrienterobactin pathways, iron is released in the cytoplasm, whereas in *P. aeruginosa* ferric-pyoverdine pathway, iron is released in the periplasm (Imperi et al., 2009). The uptake of Fe^{3+}-siderophore complex in *E. coli* is mediated through outer-membrane receptor proteins such as FepA (for enterobactin), FhuA (for ferrichrome), and FecA (for ferric citrate) (Ma et al., 2007) and by FpvA and FptA in *P. aeruginosa* (Nader et al., 2011). Studies of the structures and mechanisms of action of these receptors have suggested that a conformational change of particular domains of these receptors leads to the passage of substrate (Eisenhauer et al., 2005). The energy required for the transport is supplied by the TonB protein complex (Takase et al., 2000), which consists of two inner-membrane proteins, ExbD and ExbB (Pawelek et al., 2006). The role of TonB-ExbD-ExbB complex in ferric-siderophore transfer has been extensively studied in *P. aeruginosa* (Schalk et al., 2009) and *E. coli* (Ogierman and Braun, 2003).

Chatfield et al. (2011) reported a novel iron-uptake receptor, LbtU, in *Legionella pneumophila* independent of TonB. It has been reported that iron transport in freshwater bacterium *Aeromonas hydrophila* is found to occur by means of indiscriminant siderophore transport system composed of single multifunctional receptor. The ligand-exchange step occurs at the cell surface and has been reported in *E. coli* (Stintzi et al., 2000). The ABC-type transporter proteins are also involved in the delivery of iron-siderophore complex into the cytosol.

In Gram-negative bacteria, these are extracytoplasmic substrate-binding proteins located in the periplasm, whereas, in Gram-positive bacteria, these are present as lipoproteins attached to the external surface of the cell membrane (Speziali et al., 2006). Upon interaction with the extracytoplasmic substrate-binding unit of the ABC transporter, the Fe-siderophore complex is channeled through the membrane. The energy required for this process is provided by the cytoplasmic subunits of the ABC receptors, which undergo a dimerization or conformational change due to NTP binding or hydrolysis. Four subtypes of ABC transporters described in Gram-negative bacteria have been associated with Fe-siderophore uptake (Miethke and Marahiel, 2007).

8.2.10 Metabolism of Iron

Iron gets released from the complex by several mechanisms, including *fes* gene product (ferric enterobactin esterase Fes), which in *E. coli* helps in the hydrolytic release of iron from the siderophores (Brickman and Mcintosh, 1992). It was reported that in pathogenic strains of *E. coli* and *Salmonella*, five genes (*iroB*, *iroC*, *iroD*, *iroE*, *iroN*) encoded by the *iroA* locus were also responsible for the hydrolytic release of iron (Lin et al., 2005), In addition, an iron sulfur protein, FhuF, helps in mobilization of iron from hydroxamate siderophores by reductase activity (Matzanke et al., 2004); the general flavin reductases in prokaryotes reduces FMN, FAD, and riboflavin by utilizing NADH and NADPH; and the reduction of flavins leads to the reduction of iron-siderophore complexes (Pierre et al., 2002). The soluble Fe (II) can be readily removed from the siderophores and directly utilized in metabolic and physiological processes. All of these mechanisms involve chemical modification of the siderophores. However, studies have shown that in *P. aeruginosa*, iron is released from the ferric-pyoverdine complex simply by reduction and that released pyoverdine is recycled into the external medium without any chemical modifications (Schalk et al., 2011).

8.2.11 Iron Regulation in Bacteria

Iron regulation in Gram-negative bacteria is carried out by the ferric iron-uptake regulator (Fur) protein (Ollinger et al., 2006). Fur-like proteins is also found in many Gram-positive bacteria, such as *B. subtilis* (Bsat et al., 1998). The *E. coli* Fur protein has a molecular weight of 17 KDa and acts as a transcriptional repressor. It negatively regulates transcription of iron-transport genes by binding to the Fur-binding region located upstream of the iron-transport gene known as Fur box, and is made up of 19 bp-inverted repeat sequences. It is activated by divalent iron (corepressor), after which it represses the downstream iron-transport genes. Fur binds to DNA only under excess iron but under iron-deficient conditions, the metal dissociates from the protein and the genes are

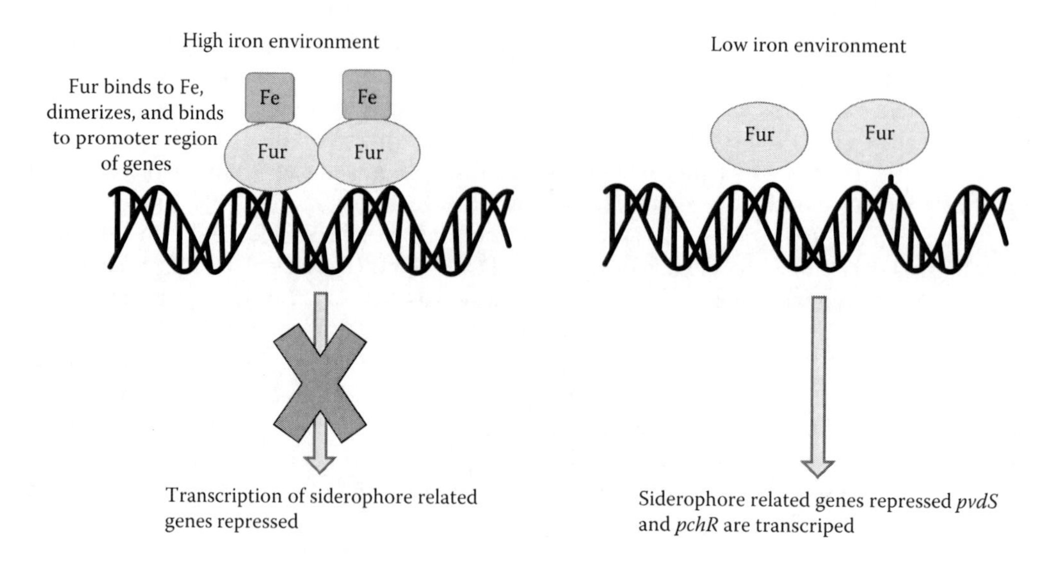

FIGURE 8.2 Schematic representation of Fur-mediated gene expression.

expressed (Mchugh et al., 2003) (Figure 8.2). In *E. coli*, the Fur protein controls about 90 genes at the transcriptional level (Escolar et al., 1999).

8.3 APPLICATIONS OF SIDEROPHORE

Siderophores have wide applications in various fields such as environmental sciences and medicine. In agriculture, they are used to improve soil fertility and biocontrol. In medicine, iron excesses as well as the primary iron-overload diseases (i.e., hemochromatosis, hemosiderosis, and accidental iron poisoning) require the removal of iron from the body and especially from the liver. Such diseases can be efficiently treated with siderophore-based drugs, for which siderophore acts as the principal model (Pietrangelo, 2002). Siderophore can be used for antibiotic delivery in antibiotic-resistant bacteria. This application uses the iron-transport abilities of siderophores to carry drugs into cells by preparation of conjugates between siderophores and antimicrobial agents, which is called the Trojan Horse strategy. Examples of siderophore-based antibiotics in nature include albomycins (Benz et al., 1982), ferrimycins (Bickel et al., 1966), and salimycins (Vértesy et al., 1995). For Fe^{3+} chelation, the albomycins use a part of the ferrichrome structure that is attached via a serine spacer to a toxic molecule.

Several microorganisms introduce albomycin through the ferrichrome-uptake system into the cell, where the toxic part is released enzymatically with detrimental effects to the cell. Siderophore has been used as an iron chelator in cancer drugs such as dexrazoxane, O-trensox, the desferriexochelins, desferrithiocin, and tachpyridine (Miethke and Marahiel, 2007). Siderophore is also used for the clearance of non-transferrin-bound iron in serum that occurs as a result of some chemotherapy (Chua et al., 2003). Siderophore produced by *Klebsiella pneumoniae* acts as an antimalarial agent (Gysin et al., 1991). Desferrioxamine B produced by *Streptomyces pilosus* is active against *P. falciparum* both *in vitro* and *in vivo*. Siderophore enters the *P. falciparum* cell and causes intracellular iron depletion. Therefore, siderophores used as environmental applications by naturally occurring ligands may affect actinide mobility in waste repositories and in the environment (Von Gunten and Benes, 1995; Ruggiero et al., 2000). Excessive accumulation of heavy metals is toxic to most plants and contaminates the soil, which results in decreased soil microbial activity as well as losses in soil fertility and decreased yields (McGrath et al., 1995). In this context, hydroxamate-type siderophore present in soil plays an important role in the immobilization of metals. Siderophores also have wide applications as biocontrol agents by suppressing the growth of many bacteria and pathogens and thereby decreasing yield losses of economically important crops.

8.4 MECHANISMS OF SIDEROPHORE IN BIOCONTROL: OVERVIEW

In soil, siderophore production activity plays a central role in determining the ability of different microorganisms to improve plant development. The mechanism illustrated in Figure 8.3 is for siderophore-mediated disease suppression by fluorescent pseudomonads. During this process, fluorescent siderophores (which have a very high affinity for ferric iron) are secreted during growth under low-iron conditions. The resulting ferric-siderophore complex is unavailable to other organisms, but the producing strain can utilize this complex via a very specific receptor in its outer-cell membrane (Buyer and Leong, 1986). In this way, fluorescent *Pseudomonas* strains may restrict the growth of deleterious bacteria and fungi at the plant root (Loper and Buyer, 1991). This efficient iron-uptake mechanism may also be a significant contributing factor to the ability of these strains to aggressively colonize plant roots, thus aiding the physical displacement of deleterious organisms.

Although iron competition among the rhizosphere population has attracted some research attention, it is not yet clear how this affects the iron requirements of plants. Iron deprivation in plants leads to a form of chlorosis (Julian et al., 1983). The influence of fluorescent pseudomonads upon siderophore production spans a wide variety of factors, including iron concentration (Kloepper et al., 1980), nature and concentration of carbon and nitrogen sources (Park et al., 1988), phosphate levels (Barbhaiya and Rao 1985), pH and light (Greppin and Gouda 1965), degree of aeration (Lenhoff, 1963), temperature (Weisbeek et al., 1986), and the presence of trace elements such as magnesium (Georgia and Poe 1931), zinc (Chakrabarty and Roy, 1964), and molybdenum (Lenhoff et al., 1956).

Siderophores have been isolated from different soils (Powell et al., 1980). Microbial siderophores enhance iron uptake by plants that are able to recognize the bacterial ferric-siderophore complex (Masalha et al., 2000; Katiyar and Goel, 2004; Dimkpa et al., 2009) and are also important in the iron uptake by plants in the presence of metals such as nickel and cadmium (Burd et al., 1998; Dimkpa et al., 2008). However, whether bacterial siderophore complexes can significantly contribute to the iron requirements of the plant remains unclear. Siderophore production confers competitive advantages to plant growth by promoting rhizobacteria (PGPR) that can colonize roots and exclude other microorganisms from this ecological niche (Haas and Défago, 2005). Under highly competitive conditions, the ability to acquire iron via siderophores may determine the outcome of competition for carbon sources that have been made available by root exudation or rhizodeposition (Crowley, 2006). Among the bacterial siderophores that have been studied, those produced by

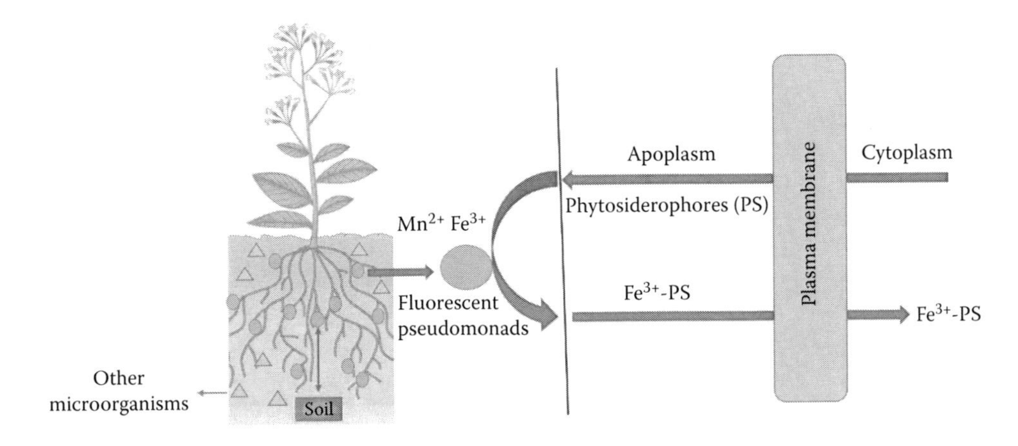

FIGURE 8.3 Iron acquisition and inhibition of root pathogens by siderophore produced from fluorescent pseudomonads.

pseudomonads are known for their high affinity to ferric ion. For example, the potent siderophore pyoverdin can inhibit the growth of bacteria and fungi that produce less potent siderophores in *in vitro* condition suppressing the growth of other pathogens (Kloepper et al., 1980a).

8.5 PHYTOPATHOLOGY

Application of epiphytic bacteria as control agents is considered to be a nonpolluting approach for alternative plant protection, and a number of potential antagonistic isolates have been described. However, only a few of these isolates have proven to be as effective under field conditions as they are in laboratory setups (Agrios, 2005). It has been proposed that several attributes contribute to biocontrol, including competition for nutrients, antibiosis, niche exclusion, and interference with cell-signaling systems (May et al., 1997; Duffy and Défago, 1999). Many potential antagonists have been selected from the fluorescent pseudomonad group because this group includes various nonpathogenic species that are adapted to plant colonization and are well known for their competitiveness (Haas and Défago, 2005). Increased thrust in biological control has recently intensified because of imminent bans on effective chemical controls such as methyl bromide, widespread development of fungicide resistance in pathogens, and a general need for more sustainable disease control strategies. Unfortunately, the seemingly inherent variable performance of most biocontrol strains between field locations and cropping seasons has hampered commercial development, and relatively few biological agents are registered for use in production agriculture (Cook, 1993). Much of this variability has been attributed to differences in the physical and chemical properties among the natural environments in which biocontrol agents are applied (Thomashow and Weller, 1996; Duffy et al., 1997). Understanding which environmental factors are important and how these influence disease suppression are widely recognized as key to improving the level and reliability of biocontrol.

Considerable progress has been made since the mid-1990s to elucidate the mechanisms by which fluorescent pseudomonads suppress disease. One of the mechanisms that underlies disease suppression by PGRB is the production of siderophores such as pyoverdine and pyochelin (Kloepper et al., 1980). The siderophores capture Fe in the vicinity of roots and thus limit the amount of iron required for the growth of pathogens such as *Fusarium oxysporum*, *Pythium ultimum*, and others that cause wilt and root rot disease in crops (Sahu and Sindhu, 2011). Some examples of siderophore-producing pseudomonads that have been proposed as biocontrol agents against soil-borne plant diseases include *P. fluorescens* CHA0 (Couillerot et al., 2009), *P. putida* WCS strains (Weller, 2007), and *P. syringae* pv. *syringae* strain 22d/93 (Wensing et al., 2010). Studies have shown that coinoculation of *Pseudomonas* strains with *Bradyrhizobium* and *Ralstonia solani* strains highly promoted legume growth and completely suppressed root rot disease under experimental conditions (Sahu and Sindhu, 2011).

Siderophores, including salicylic acid, pyochelin, and pyoverdine, which chelate iron and other metals, also contribute to disease suppression by conferring a competitive advantage to biocontrol agents for the limited supply of essential trace minerals in natural habitats (Höfte et al., 1992). Siderophores may indirectly stimulate the biosynthesis of other antimicrobial compounds by increasing the availability of these minerals to the bacteria. Antibiotics and siderophores may further function as stress factors or signals that induce local and systemic host resistance (Leeman et al., 1996). Biosynthesis of antibiotics and other antifungal compounds is regulated by a cascade of endogenous signals that includes sensor-kinase and response regulators encoded by lemA and gacA (Laville et al., 1992; Corbell and Loper, 1995), sigma factors encoded by rpoD (Schnider et al., 1995) and rpoS (Sarniguet et al., 1995), and quorum-sensing autoinducers such as N-acyl-homoserine lactones (Pierson et al., 1998). A pseudobactin siderophore produced by *P. putida* B10 was able to suppress *Fusarium oxysporum* in iron-deficient soil; however, this suppression was lost when the soil was replenished with iron, a condition that represses the production of iron chelators by microorganisms (Kloepper et al., 1980b).

Studies have demonstrated the suppression of soil-borne fungal pathogens through the release of iron-chelating siderophores by fluorescent pseudomonads, which render the iron unavailable to other organisms (Dwivedi and Johri, 2003). Maximum siderophore production of two fluorescent pseudomonads, *P. fluorescens* NCIM5096 and *P. putida* NCIM2847, produced respective maximum yields of hydroxamate-type siderophores of 87% and 83% (Sayyed et al., 2005). Reports have shown a number of mineral and/or carbon source amendments that stimulate siderophore production in *P. fluorescens* (Duffy and Défago, 1999). It has been reported that zinc stimulates the production of pyochelin and pyoverdin in the *Pseudomonas aeruginosa* biocontrol strain 7NSK2 (Höfte et al., 1994) and plant-associated *Azotobacter vinelandii* (Huyer and Page, 1989). From the study by Duffy and Défago (1999), it seems clear that environmental factors regulate the biosynthesis of antimicrobial compounds by disease-suppressive strains of *P. fluorescens*; this regulation is an essential step toward improving the level and reliability of their biocontrol activity.

8.6 BACTERIA–PATHOGEN INTERACTIONS

Bacterial pathogens are significant factors that reduce yields of agriculturally important plants worldwide.

8.6.1 Bacterial Soft Rot of Potato

Erwinia carotovora causes preemergence seed-piece decay, blackleg, soft stem rot, and soft rot of tubers. As the pathogen colonizes the potato roots and tubers, the population exceeds 10^6 colony-forming units per gram of soil under favorable environmental conditions; at the same time, fluorescent pseudomonads effectively inhibit the pathogen by colonizing the rhizosphere and increasing their population (Azad et al., 1985). It has been reported that, when applied to potato seeds, fluorescent pseudomonads respectively reduced the population of *E. carotovora* in the subsequent roots and tubers by 95%–100% and 27%–100%, compared with untreated plants (Sunaina et al., 1997). Strain B10 suppressed the growth of *E. carotovora* both *in vivo* and *in vitro* by producing pseudobactin. Soft rot *erwinias* can produce catechol and hydroxamate siderophores (Persmark et al., 1989; Ishimaru and Loper, 1992). Some attention has been focused on the role of iron-uptake systems (siderophores) as a virulence determinant of infecting *Saintpaulia* (Enard et al., 1988). A similar role for such systems has been described for several animal pathogenic bacteria (Payne, 1993). Soluble forms of iron, which are essential for all forms of life, are not readily available in plant or animal tissues.

Under iron-limiting conditions, growth of bacteria that can express high-affinity iron acquisition systems is favored; if these bacteria are also pathogenic, their virulence is apparently increased. However, siderophores probably influence the saprophytic free-living phase of these bacteria, for example at the iron-limiting oxygen levels and pH ranges of many soils (Bossier et al., 1988). It is relevant to note that siderophores are also produced by many rhizobacteria such as fluorescent *Pseudomonas* spp., and sometimes are used as biological control agents acting by competing with *erwinias* on potato (Kloepper, 1983). An additional role has been attributed to the *E. amylovora* siderophore, desferrioxamine, which has a more direct bearing on pathogenicity. Depending on its concentration, this siderophore interacts with the H_2O_2 and peroxidases in the affected tissue either to enhance oxidative stress induced by harpin, which is coded by an *hrpN* gene, or to protect bacterial cells by inhibiting the generation of reactive oxygen species (Dellagi et al., 1998). Potato tubers treated with an antibiotic-producing strain of *P. putida* M17 showed 6.8%–18.2% reduction in rotted tissue in the harvested tubers (Colyer and Mount, 1984).

8.6.2 Tomato Bacterial Wilt

Bacterial wilt caused by *Ralstonia solanacearum* (Smith) Yabuuchi is an important constraint in the production of eggplant and many other crops in the world's tropical, subtropical, and temperate regions (Buddenhagen and Kelman, 1964; Hayward, 1991). *Ralstonia solanacearum*, which limits the production of diverse crops such as potato, tomato, eggplant, pepper, banana, and peanut (Williamson et al., 2002; Ramesh et al., 2009), is a widespread and economically important bacterial plant pathogen (Horita and Tsuchiya, 2001). Its wide host range makes it most destructive and most difficult to control (Kelman et al., 1994). Different bacterial species, such as *Alcaligenes* spp. and *Kluyvera* spp. (Assis et al., 1998); *P. fluorescens*, *P. alcaligenes*, *P. putida*, *Flavobacterium* spp., and *Bacillus megaterium* (Reiter et al., 2002 and 2003); *B. pumilus* (Benhamou et al., 1998); and *Microbacterium* spp., *Clavibacter michiganensis*, *Curtobacterium* spp., and *B. subtilis* (Zinniel et al. 2002) have been reported as endophytes and as inhibitory to plant pathogens. Toyota and Kimura (2000) reported the suppressive effect of some antagonistic bacteria on *R. solanacearum*.

Ciampi-Panno et al. (1989) proved the use of antagonistic microbes in the control of *R. solanacearum* under field conditions. Anuratha and Gnanamanickam (1990) and Gamliel and Katan (1993) reported that utilization of antagonistic rhizosphere bacteria such as *Bacillus* spp., *P. fluorescens*, and *P. putida* significantly increased respective survival rates of potato, tomato, eggplant, and cotton by 60%–90%, 90%, and 84%–90% against bacterial wilt disease. *In vitro* results clearly confirm that plants treated with *P. aeruginosa* and *B. thuringiensis* significantly reduced disease compared with the infected control. Disease reduction by *P. aeruginosa* and *B. thuringiensis* in the colonization of plant roots may occur directly (i.e., through competition for space, nutrients, and ecological niches or production of antimicrobial substances) and indirectly (i.e., through induction of systemic resistance (ISR) (Kloepper and Beauchamp, 1992; Liu et al., 1995). *Pseudomonas aeruginosa* and *B. thuringiensis* may induce plant growth promotion by direct or indirect modes of action (Lazarovits and Nowak, 1997).

Direct mechanisms involve the production of plant growth regulators (auxins, cytokinins, gibberellins) and facilitation of the uptake of nutrients (nitrogen fixation, solubilization of phosphorus). The indirect mechanisms of *P. aeruginosa* and *B. thuringiensis* to prevent the deleterious effects of plant pathogens include the production of inhibitory substances (antibiotics, antifungal metabolites, iron-chelating siderophores, cell wall-degrading enzymes, and competition for sites on roots) and the increase of the host's natural resistance (induced systemic resistance). *Bacillus* spp. and *Pseudomonas* spp. increased the yield of tomato plants by reducing *R. solanacearum* wilt disease (Guo et al., 2004; Abeer et al. 2013). Seleim et al. (2011) recorded a 96% reduction of tomato bacterial wilt disease under greenhouse conditions using *Pseudomonas* spp.

8.6.3 Rice Bacterial Blight

Bacterial leaf blight (BLB) caused by *Xanthomonas oryzae* is one of the most devastating diseases of rice. The bacterial pathogen infects the host plant at the maximum tillering stage, resulting in 20%–40% yield reductions. The disease was first observed by Japanese farmers in 1884 (Tagami and Mizukami, 1962). It was reported that some strains of *Lysobacter* spp. (e.g., 3.1T8, SB-K88, and C3) effectively controlled both soil-borne and foliar diseases of rice (Folman et al., 2004; Islam et al., 2005; Jochum et al., 2006). The whole bacterial-broth culture of *L. antibioticus* strain 13-1 has been reported as the most effective form of control of rice BLB (Guang-Hai Ji et al., 2008). Further experiments in Ji et al.'s study showed that strain 13-1 and the pathogen *Xanthomonas oryzae* strain 53, applied by foliar spray, provided significant control of BB under field conditions (mean reduction in disease severity was approximately 70%). This reduction, which exceeded other bacterial biocontrol agents (e.g., *P. fluorescens*, *P. putida*, and *Bacillus* spp.), produced respective maximum reductions of ~64%, ~57%, and ~60% in comparison with previously published reports (Sivamani et al., 1987; Gnanamanickam et al., 1999; Vasudevan et al., 2002; Johri et al., 2003; Velusamy et al., 2006).

In *Xanthomonas oryzae* pv. *oryzae*, the causal agent of the serious bacterial blight disease of rice, a *fur* mutant exhibits a severe in planta growth deficiency that appears to be due to increased oxidative stress (Subramoni and Sonti, 2005). Alok Pandey (2010) characterized two different

X. oryzae pv. *oryzae* iron uptake systems. The *X. oryzae* pv. *oryzae feo-ABC* genes are homologues of well-characterized ferrous transporters in other bacteria wherein the *feoB* gene encodes the major bacterial ferrous ion transporter. The xss operon is homologous to siderophore biosynthetic genes in other bacteria and is essential for siderophore production by *X. oryzae* pv. *oryzae*. Further study revealed the regulation of these *X. oryzae* pv. *oryzae* iron-uptake systems, which are required for growth on low-iron media, and assessed their roles in virulence. This (Alok Pandey, 2010) is the first report on the role of specific iron-uptake systems in the virulence of any member of the important xanthomonad group of plant pathogens.

The *Xanthomonas oryzae* pv. *oryzae* mutants for *rpfF* (regulation of pathogenicity factor; a global regulator) are deficient for virulence and growth under low-iron conditions and are siderophore overproducers (Chatterjee and Sonti, 2002). Exogenous iron supplementation has been shown to promote the growth of *X. oryzae* pv. *oryzae rpfF* mutants in plants. It has been reported that FeoB protein is required for *X. oryzae* pv. *oryzae* to grow inside the rice plant; the study concluded that *X. oryzae* pv. *oryzae* FeoB can be used as a target for developing new molecules that can be used for application on rice fields against bacterial blight disease and possibly against diseases caused by other xanthomonads as well (Pandey and Sonti, 2010).

8.6.4 Bacterial Canker of Tomato

The *Clavibacter michiganensis* subsp. *michiganensis* (Cmm) causes bacterial canker of tomato (*Lycopersicon esculentum*) (Strider, 1969; Davis et al., 1984). This disease, which can be considered the most important bacterial disease of tomato, causes substantial economic losses worldwide. The plant-associated bacterium *C. michiganensis* produces high molecular weight and acidic exopolysaccharides (EPS) (Rai and Strobel, 1968a,b; Van den Bulk et al., 1989) that have multiple and important biological functions. Exopolysaccharides may protect bacteria against components of the plant defense system by blockage of agglutinins or lectins, detoxification of phytoalexins, or production of reactive oxygen species (Bradshaw-Rouse et al., 1981; Romeiro et al., 1981; Young and Sequeira, 1986; Kiraly et al., 1997). Cmm produces numerous extracellular enzymes other than EPS; in cultures containing supernatant endocellulase (Meletzus et al., 1993), polygalacturonase (Beimen et al., 1992), pectinmethylesterase (Strider, 1969) and xylanase (Beimen et al., 1992), activities have been detected. Tomato produces a number of phytoalexins; these include glycosylated alkaloid saponin and α-tomatine. The latter is known as an important component in the defense against fungi and possibly against bacterial pathogens as well (El-Shanshoury et al., 1995; Roldan-Arjona et al., 1999).

A survey of the Cmm genome data revealed at least 10 sigma factors, three in the subfamily of primary/secondary sigma factors and seven that are alternative sigma factors (Wösten, 1998). Inactivation, especially of alternative sigma factors by gene replacement, will show whether these sigma factors play a role in virulence. Of potential interest for biotechnological applications are NRPS and a polyketide synthethase (PKS) identified in the genome of Cmm. These enzymes are involved in the production of secondary metabolites (antibiotics, siderophores, and pigments). Elucidation of the chemical structure of the products formed by these enzymes and the accompanying modifying enzymes may lead to new compounds that can be used as antibiotics to inhibit the growth of *C. michiganensis* (Gartemann et al., 2003).

8.6.5 Bacterial Blight of Cotton

Bacterial blight of cotton (BBC) is caused by *Xanthomonas campestris* pv. *malvacearum* (Smith). Dye (Xcm) causes considerable yield loss in cotton-growing areas of India, which already ranks only fourth in production worldwide (Mishra and Krishna, 2001). Bioagents such as *Trichoderma harzianum*, *Pseudomonas fluorescens*, and *Bacillus subtilis* were isolated from cotton rhizosphere soil and tested individually for their effectiveness in controlling BBC.

Bioagents are reported to produce hydrolytic enzymes such as chitinase, β-1, 3-glucanase, proteases, and volatile and nonvolatile antibiotics (Elad et al., 1982). *Pseudomonas fluorescens* showed antibiotic activities through the production of secondary metabolites or siderophores or cyanide or pyrollonitrin, which are known to suppress the root-rot-causing fungus in cotton (Laha and Verma, 1998). The disease reduction of cotton bacterial blight by biocontrol agents might be attributed to the suppression of pathogen activity in the host and soil by the antagonists through overcolonization (Cook, 1991). Raghavendra et al. (2013) reported that bioagents showed differences in their degrees of protection against bacterial blight. Each bioagent activates different defense mechanisms within the reduced resistance pathway, resulting in differences in the reduction of disease. The concept of PGPR could protect plants against pathogens such as *Xanthomonas campestris* by inducing defense mechanisms by the iron-binding siderophore HCN and others (Raghavendra et al., 2013).

8.6.6 Bacterial Leaf Spot of Mungbean

Bacterial leaf spot of mungbean incited by *Xanthomonas axonopodis* pv. *vignaeradiatae* (Xav) (Sabet et al., 1969; Vauterin et al., 1995), one of the most important bacterial diseases of mungbean, has caused 5%–15% loss since the 1970s. Current defense protocols emphasize the use of both biocontrol agents and plant defense activators (Wei et al., 1991; Yamaguchi, 1998). Borah et al. (2000) observed that seed bacterization with phylloplane bacteria (isolated from mungbean phylloplane) increased the seed germinability of a mungbean crop. Because bacterial leaf spot (BLS) of mungbean is a seed-borne disease, the reduction of seed-borne primary inoculum is one of the most important management strategies. Dutta et al. (2005) found that seed treatments with $FeCl_3$ at 2 mM, and SA at 1 mM, reduced seedling infection by 50.6% and 49.4%, respectively.

Salicylic acid is a primary signal in systemic acquired resistance (Neuenschwander et al., 1996), and its application to plants induces resistance against a wide range of pathogens (Kessmann et al., 1994). *Pseudomonas aeruginosa* and *Bacillus subtilis* as seed dressing or soil treatment significantly prevented severe infection of mungbean by root-knot nematodes under both greenhouse and field conditions (Siddique et al., 2001). Fluorescent pseudomonads isolated from cotton, maize, soybean, and tomato rhizosphere significantly increased the seed health of cotton seedlings by producing siderophore and other secondary metabolites (Dutta et al., 2005).

8.6.7 Fire Blight Disease

Fire blight of rosaceous plants is an economically important disease, caused by *Erwinia amylovora*, that affects mainly apple and pear production but also several woody ornamentals. Several bacterial strains have been reported to be effective antagonists of *E. amylovora*; these include *Pseudomonas fluorescens* (Lindow et al., 1996), *Pantoea agglomerans* (syn. *Erwinia herbicola*) (Beer and Rundle, 1983), and *Bacillus subtilis* (Broggini et al., 2005). *Pseudomonas fluorescens* A506, *P. agglomerans* E325, and *Bacillus subtilis* QST713 are already registered as components in biological products for fire-blight control. Several mechanisms have been proposed to explain the inhibition of *E. amylovora* and the control of fire blight, depending on the strain of the antagonist, but most studies have focused on antibiosis (Stockwell et al., 2002). The antagonistic activity was lost by iron amendment due to siderophore production (Loper and Henkels, 1999). Siderophores in *Pseudomonas fluorescens* EPS62e were confirmed in Schwin-Neidlands medium (Jordi Cabrefiga et al., 2007). The role of *Pseudomonas*-produced siderophores in the control of some plant diseases has been described (Whipps, 2001). In some reports, siderophores have been shown to suppress several pathogen-induced diseases by conferring a competitive advantage upon biocontrol agent over pathogen under conditions in which natural habitats offer limited supplies of essential trace minerals (Duffy and Défago, 1999).

8.7 CONCLUSION

The production of siderophores is an efficient strategy used by bacteria, fungi, and plants to overcome a lack of iron. The stability of siderophore-iron complexes is an important factor for the siderophone efficiency. Siderophore application is widespread in various areas of medicine and environmental microbiology, but many scientific challenges remain for research on biocontrol pseudomonads. It will be important to exploit molecular techniques to study the genome expression of plant beneficial and plant pathogenic microorganisms *in situ*, and to obtain a fuller picture of the mechanisms behind the suppression by siderophore production of the growth of plant pathogens (phytopathogens). In the future, novel siderophores can be biosynthesized and used for the biocontrol of bacterial as well as fungal pathogens. It has also been reported that survival of specific proteins (e.g., FeoB) in some phytopathogens is required for *X. oryzae pv. oryzae* to grow inside the rice plant and induce pathogenicity. Therefore, when a specific protein necessary to produce pathogenicity, it can be used as a target for developing new molecules that can, in turn, be applied in fields against various diseases that affect economically valuable plants.

ACKNOWLEDGMENTS

The authors express their deep appreciation to the University Grant Commission, New Delhi, for financial support by the UGC-Major Research Project (MRP No: 39-214/2010). The central facilities of the UGC-CEGS, DST-PURSE Programme, the DBT-IPLS at MKU, and the coordinator at NRCBS, School of Biological Sciences, Madurai Kamaraj University, Madurai, Tamil Nadu, India are also gratefully acknowledged.

REFERENCES

Abeer, H., Makhlouf, I., Hamedo, H.A. 2013. Suppression of bacterial wilt disease of tomato plants using some bacterial strains. *Life Science* 10(3): 1732–1741.

Agrios, G.N. 2005. *Plant Pathology*, 5th Edn. San Diego, CA: Elsevier Academic Press.

Albrecht-Gary, A.M., Crumbliss, A.L. 1998. Coordination chemistry of siderophores: Thermodynamics and kinetics of iron chelation and release. *Metal Ions Biol. Sys* 35: 239–327.

Ankenbauer, R., Sriyosachati, S., Cox, C.D. 1985. Effects of siderophores on the growth of *Pseudomonas aeruginosa* in human serum and transferrin. *Infect Immun* 49: 132–140.

Ankenbauer, R.G., Toyokuni, T., Staley, A., Rinehart, K.L. Jr., Cox, C.D. 1988. Synthesis and biological activity of pyochelin, a siderophore of Pseudomonas aeruginosa. *J Bacteriol* 170: 5344–5351.

Anuratha, C.S., Gnanamanickam, S.S. 1990. Biological control of bacterial wilt caused by *Pseudomonas solanacearum* in India with antagonistic bacteria. *Plant and Soil* 124: 109–116.

Ardon, O., Nudelman, R., Caris, C., Libman, J., Shanzer, A., Chen, Y., Hadar, Y. 1998. Iron uptake in *Ustilago maydis*: Tracking the iron path. *J. Bacteriol* 180: 2021–2026.

Assis, S.M.P., Silveira, E.B., Mariano, R.L.R., Menezes, D. 1998. Bactérias endofi´ticas-Método deisolamento e potencial antagônico no controle da podridão negra do repolho. *Summa Phytopathol* 24: 216–220.

Azad, H.R., Davis, J.R., Schnathorst, W.C., Hado, C.I. 1985. Relationship between rhizoplane and rhizosphere bacteria and verticullium wilt resistance in potato. *Arch. Microbiol* 140: 347–351.

Barbhaiya, H.B., Rao, K.K. 1985. Production of pyoverdine, the fluorescent pigment of *Pseudomonas aeruginosa* PAO1. FEMS. *Microbiol. Lett* 27: 233–235.

Barghouthi, S., Young, R., Olson, M.O.J., Arceneaux, J.E.L., Clem, L.W., Byers, B.R. 1989. Amonabactin, a novel tryptophan-containing or phenylalanine-containing phenolate siderophore in *Aeromonas-hydrophila*. *J Bacteriol* 171: 1811–1816.

Beer, S.V., Rundle, J.R. 1983. Suppression of *Erwinia amylovora* by *Erwinia herbicola* in immature pear fruits. *Phytopathology* 73: 1346.

Beimen, A., Bermpohl, A., Meletzus, D., Eichenlaub, R., Barz, W. 1992. Accumulation of phenolic compounds in leaves of tomato plants after infection with *Clavibacter michiganense* subsp. *michiganense* strains differing in virulence. *Z. Naturforsch* 47c: 898–909.

Benhamou, N., Kloepper, J.W., Tuzun, S. 1998. Induction of resistance against Fusarium wilt of tomato by combination of chitosan with an endophytic bacterial strain: Ultrastructure and cytochemistry of the host response. *Planta* 204: 153–168.

Benz, G., Schroder, T., Kurz, J., Wunsche, C., Karl, W., Steffens, G., Pfitzner, J., Schmidt, D. 1982. Konstitution der Desferriform der Albomycine d1, d2 and e. *Angew. Chem* 94: 552–553.

Bickel, H., Mertens, P., Prelog, V., Seibl, J., Walser, A. 1966. Über die Konstitution von Ferrimycin A1. *Tetrahedron Suppl* 8/I: 171–179.

Boos, W., Eppler, T. 2001. Prokaryotic binding protein-dependent ABC transporters. In *Microbial Transport Systems*. Winkelmann, G. (ed.). Weinheim: Wiley-VCH, pp. 77–114.

Borah, P.K., Jindal, J.K., Verma, J.P. 2000. Biological management of bacterial leaf spot of mungbean caused by *Xanthomonas axonopodis* pv. *vignaeradiatae*. *Indian Phytopath* 53: 384–394.

Bossier, P., Höfte, M., Verstraete, W. 1988. Ecological significance of siderophores in soil. *Adv. Microb. Ecol.* 10: 385–414.

Boukhalfa, H., Crumbliss, A.L. 2002. Chemical aspects of siderophore mediated iron transport. *Biometals* 15: 325–339.

Bradshaw-Rouse, J.J., Whatley, M.H., Coplin, D.L., Woods, A., Sequeira, L., Kelman, A. 1981. Agglutination of *Erwinia stewartii* strains with a corn agglutinin: Correlation with extracellular polysaccharide production and pathogenicity. *Appl. Environ. Microbiol* 42: 344–350.

Brickman, T.J., Mcintosh, M.A. 1992. Overexpression and purification of ferric enterobactin esterase from *Escherichia coli*. Demonstration of enzymatic hydrolysis of enterobactin and its iron complex. *J. Biol. Chem* 267: 12350–12355.

Broggini, G., Duffy, B., Holliger, E., Schärer, H., Gessler, C., Patocchi, A. 2005. Detection of the fire blight biocontrol agent *Bacillus subtilis* BD170 (Biopro®) in a Swiss apple orchard. *Eur J Plant Pathol* 111: 93–100.

Bsat, N., Herbig, A., Casillas-Martinez, L., Setlow, P., Helmann, J.D. 1998. *Bacillus subtilis* contains multiple Fur homologues: Identification of the iron uptake (Fur) and peroxide regulon (PerR) repressors. *Mol. Microbiol* 29: 189–198.

Buddenhagen, I.W., Kelman, A. 1964. Biological and physiological aspects of bacterial wilt caused by *Pseudomonas solanacearum*. *Annu Rev Phytopathol* 2: 203–230.

Burd, G.I., Dixon, D.G., Glick B.R. 1998. A plant growth promoting bacterium that decreases nickel toxicity in seedlings. *Appl Environ Microbiol* 64: 3663–3668.

Buyer, J.S., De Lorenzo, V., Neilands, J.B. 1991. Production of the siderophore aerobactin by a halophilic pseudomonad. *Applied and environmental microbiology*, 57(8): 2246–2250.

Buyer, J.S., Leong, J. 1986. Iron transport-mediated antagonism between plant growth-promoting and plant-deleterious *Pseudomonas* strains. *J. Biol. Chem* 261: 791–794.

Buysens, S., Heungens, K., Poppe, J., Höfte, M. 1996. Involvement of Pyochelin and Pyoverdin in Suppression of *Pythium*-Induced Damping-Off of Tomato by *Pseudomonas aeruginosa 7NSK2*. *Appl. Environ. Microbiol* 62(3): 865.

Chakrabarty, A.M., Roy, S.C. 1964. Effects of trace elements on the production of pigments by a pseudomonad. *Biochem. J* 93: 228–231.

Challis, G.L. 2005. A widely distributed bacterial pathway for siderophore biosynthesis independent of nonribosomal peptide synthetases. *Chembiochem* 6: 601–611.

Chatfield, C.H., Mulhern, B.J., Burnside, D.M., Cianciotto, N.P. 2011. *Legionella pneumophila* LbtU acts as a novel, TonB-independent receptor for the legiobactin siderophore. *J. Bacteriol* 193: 1563–1575.

Chatterjee, S., Sonti, R.V. 2002. *RpfF* mutants of *Xanthomonas oryzae pv.oryzae* are deficient for virulence and growth under low iron conditions. *Mol. Plant Microbe Interact* 15: 463–471.

Chua, A.C., Ingram, H.A., Raymond, K.N., Baker, E. 2003. Multidentate pyridinones inhibit the metabolism of nontransferrin-bound iron by hepatocytes and hepatoma cells. *European Journal of Biochemistry* 270: 1689–1698.

Ciampi-Panno, L., Fernandez C, Bustamante, P., Andrade, N., Ojeda, S., Contreras A. 1989. Biological control of bacterial wilt of potatoes caused by pseudomonas solanacearum. *Am. Potato J* 66(5): 315–332.

Colyer, P.D., Mount, M.S. 1984. Bacterization of potatoes with *Pseudomonas putida* and its influence on post-harvest soft rot diseases. *Plant Dis* 68: 703–706.

Cook, R.J. 1991. Biological control of plant diseases: Broad concepts and applications. *Biological Control of Plant Diseases, FFTC Book series* 42: 1–29.

Cook, R.J. 1993. Making greater use of introduced microorganisms for biological control of plant pathogens. *Annu. Rev. Phytopathol* 31: 53–80.

Corbell, N., Loper, J.E. 1995. A global regulator of secondary metabolite production in *Pseudomonas fluorescens* Pf-5. *J. Bacteriol* 177: 6230–6236.

Couillerot, O., Prigent-Combaret, C., Caballero-Mellado, J., Moënne Loccoz, Y. 2009. Pseudomonas fluorescens and closely-related fluorescent pseudomonads as biocontrol agents of soil-borne phytopathogens. *Lett. Appl. Microbiol* 48: 505–512.

Cox, C.D. 1982. Effect of pyochelin on the virulence of *Pseudomonas aeruginosa. Infect Immun* 36: 17–23.

Cox, C.D., Graham, R. 1979. Isolation of an iron-binding compound from *Pseudomonas aeruginosa. J Bacteriol* 137: 357–364.

Crosa, J.H. 1980. A plasmid associated with virulence in the marine fish pathogen *Vibrio anguillarum* specifies an iron-sequestering system. *Nature* 284: 566–568.

Crosa, J.H., Walsh, C.T. 2002. Genetics and assembly line enzymology of siderophore biosynthesis in bacteria. *Microbiol Mol Biol Rev* 66: 223–249.

Crowley, D.E. 2006. Microbial siderophores in the plant rhizospheric. In *Iron Nutrition in Plants and Rhizospheric Microorganisms*. Dordrecht: Springer, pp. 169–198.

Davidson, A.L., Nikaido, H. 1991. Purification and characterization of the membrane-associated components of the maltose transport system from *Escherichia coli. J. Biol.Chem* 266: 8946–8951.

Davis, M.J., Gillaspie, A.G. Jr., Vidaver, A.K., Harris, R.W. 1984. *Clavibacter*: A new genus containing some phytopathogenic coryneform bacteria, including *Clavibacter xyli* subsp. *Xyli* sp. nov., subsp., nov. and *Clavibacter xyli* subsp. *Cynodontis* subsp. nov., pathogens that cause ratoon stunting disease of sugarcane and bermudagrass stunting disease. *Int. J. Syst.Bacteriol* 34: 107–117.

Dellagi, A., Brisset, M-N., Paulin, J-P., Expert, D. 1998. Dual role of desferrioxamine in *Erwinia amylovora* pathogenicity. *Molecular Plant–Microbe Interactions* 11: 734–742.

De Voss, J.J., Rutter, K., Schroeder, B.G., Su, H., Zhu, Y., Barry, III. 2000. The salicylate-derived mycobactin siderophores of *Mycobacterium tuberculosis* are essential for growth in macrophages. *Proc. Natl. Acad. Sci. USA* 97(3): 1252–1257.

Dimkpa, C., Svatos, A., Merten, D., Büchel, G., Kothe, E. 2008. Hydroxamate siderophores produced by Streptomyces acidiscabies E13 bind nickel and promote growth in cowpea (Vigna unguiculata L.) under nickel stress. *Can J Microbiol* 54: 163–172.

Dimkpa, C.O., Merten, D., Svatos, A., Büchel, G., Kothe, E. 2009. Siderophores mediate reduced and increased uptake of cadmium by *Streptomyces tendae* F4 and sunflower (*Helianthus annuus*), respectively. *J. Appl. Microbiol.* 107: 1687–1696.

Drechsel, H., Tschierske, M., Thieken, A., Jung, G., Zähner, H., Winkelmann, G. 1995.The carboxylate type siderophore rhizoferrin and its analogs produced by directed fermentation. *J Ind. Microbiol.* 114: 105–112.

Duffy, B.K., Défago, G. 1999. Environmental factors modulating antibiotic and siderophore biosynthesis by *Pseudomonas fluorescens* biocontrol strains. *Appl. Environ. Microbiol.* 65: 2429–2438.

Duffy, B.K., Ownley, B.H., Weller, D.M. 1997. Soil chemical and physical properties associated with suppression of take-all of wheat by *Trichoderma koningii*. *Phytopathology* 87: 1118–1124.

Dutta, S., Singh, R.P., Jindal, J.K. 2005. Effect of antagonistic bacteria and plant defence activators on management of bacterial leaf spot of mungbean. *Indian Phytopath* 58(3): 269–275.

Dwivedi, D., Johri, B.N. 2003. Antifungals from fluorescent pseudomonads: Biosynthesis and regulation. *Curr Sci* 12: 1693–1703.

Eisendle, M., Oberegger, H., Buttinger, R., Illmer, P., Haas, H. 2004. Biosynthesis and uptake of siderophores is controlled by the PacC-mediated ambient-pH Regulatory system in *Aspergillus nidulans. Eukaryot Cell* 3(2): 561–563.

Eisenhauer, H.A., Shames, S., Pawelek, P.D., Coulton, J.W. 2005. Siderophore transport through *Escherichia coli* outer membrane receptor FhuA with disulfide-tethered cork and barrel domains. *J. Biol. Chem* 280: 30574–30580.

Elad, Y., Chet, I., Henis, Y. 1982. Degradation of plant pathogenic fungi by *Trichoderma harzianum. Can. J. Microbiol* 28: 719–725.

El-Shanshoury, A.E.-R.R., El-Sououd, S.M.A., Awadalla, O.A., El-Bandy, N.B. 1995. Formation of tomatine in tomato plants infected with *Streptomyces* species and treated with herbicides, correlated with reduction of *Pseudomonas solanacearum* and *Fusarium oxysporum* f.sp. *lycopersici*. *Acta Microbiol. Polonica* 44: 255–266.

Enard, C., Diolez, A., Expert, D. 1988. Systemic virulence of *Erwinia chrysanthemi* 3987 requires a functional iron assimilation system. *J. Bacteriol.* 170: 2419–2426.

Escolar, L., Pérez-Martín, J., De Lorenzo, V. 1999. Opening the iron box: Transcriptional metalloregulation by the Fur protein. *J. Bacteriol.* 181: 6223–6229.

Farhana, A., Kumar, S., Rathore, S.S., Ghosh, P.C. 2008. Mechanistic insights into a novel exporter-importer system of *Mycobacterium tuberculosis* unravel its role in trafficking of iron. *PLoS One* 3: 2087.

Folman, L.B., De Klein, M.J.E.M., Postma, J., van Veen, J.A. 2004. Production of antifungal compounds by *Lysobacter enzymogenes* isolate 3.1 T8 under different conditions in relation to its efficacy as a biocontrol agent of *Pythium aphanidermatum* in cucumber. *Biological Control* 31: 145–154.

Furrer, J.L., D.N. Sanders, I.G. Hook-Barnard, M.A. McIntosh. 2002. Export of the siderophore enterobactin in *Escherichia coli*: Involvement of a 43 kDa membrane exporter. *Mol. Microbiol* 44:1225–1234.

Gamliel, A., Katan, J. 1993. Suppression of major and minor pathogen by *fluorescent pseudomonas* in solarized soils. *Phytopathology* 83: 68–75.

Gartemann, K.-H., Kirchner, O., Engemann, J., Gräfen, I., Eichenlaub, R., Burger, A. 2003. *Clavibacter michiganensis* subsp. *michiganensis*: First steps in the understanding of virulence of a Gram-positive phytopathogenic bacterium. *J. Biotechnol* 106: 179–191.

Georgia, F.R, Poe, C.P. 1931. Study of bacterial fluorescence in various media. 1. Inorganic substances necessary for bacterial fluorescence. *J. Bacteriol.* 22: 349–361.

Gnanamanickam, S.S., Brindha, P.V., Narayanan, N.N. 1999. An overview of bacterial blight disease of rice and strategies for its management. *Current Science* 77: 1435–1443.

Greenwald, J., Hoegy, F., Nader, M., Journet, L. 2007. Real time fluorescent resonance energy transfer visualization of ferric pyoverdine uptake in *Pseudomonas aeruginosa. J Biol Chem* 282: 2987–2995.

Greppin, H., Gouda, S. 1965. Action de la lumiere sur le pigment de *Pseudomonas fluorescens* Migula. *Arch. Sci* 18: 721–725.

Griffiths, G.L., Sigel, S.P., Payne, S.M., Neilands, J.B. 1984. Vibriobactin, a siderophore from *Vibrio chlorea. J. Biol. Chem* 259: 383–385.

Guo, J.H., Qi, H.Y., Guo, Y.H., Ge, H.L., Gong, L.Y., Zhang, L.X., Sun, P.H. 2004. Biocontrol of tomato wilt by plant growth-promoting rhizobacteria. *Biol. Control* 29: 66–72.

Gysin, J., Crenn, Y., Pereira da silva, L., Breton, C. 1991. Siderophores as antiparasitic agents. US patent. 5: 192–807.

Haas, D., Défago, G. 2005. Biological control of soil-borne pathogens by fluorescent pseudomonads. *Nat Rev Microbiol* 3: 307–319.

Hannauer, M., Barda, Y., Mislin, G.L.A., Shanzer, A., Schalk, I.J. 2010. The ferrichrome uptake pathway in *Pseudomonas aeruginosa* involves an iron release mechanism with acylation of the siderophore and recycling of the modified desferrichrome. *J. Bacteriol* 192: 1212–1220.

Harada, K.I., Tomika, K., Fujii, K., Masuda, K., Minkmi, Y., Yazawa., K., Komaki, H. 2004. Isolation and structural characterization of siderospores, Madura statins, produced by a pathogenic *Actinomadura madurae. J. Antibiot* 57(2): 125–135.

Hayward, A.C. 1991. Biology and epidemiology of bacterial wilt caused by *Pseudomonas solanacearum. Annu Rev Phytopathol* 29: 65–87.

Hider, R.C., Kong, X. 2010. Chemistry and biology of siderophores. *Natural Product Reports* 27: 637–657.

Höfte, M., Boelens, J., Verstraete, W. 1992. Survival and root colonization of mutants of plant growth-promoting pseudomonads affected in siderophore biosynthesis or regulation of siderophore production. *J. Plant Nutr* 15: 2253–2262.

Höfte, M., Dong, Q., Kourambas, S., Krishnapillai, V., Sherratt, D., Mergeay, M. 1994. The *sss* gene product, which affects pyoverdine production in *Pseudomonas aeruginosa* 7NSK2, is a site-specific recombinase. *Mol Microbiol* 14: 1011–1020.

Horita, M., Tsuchiya, K. 2001. Genetic diversity of Japanese strains of *Ralstonia solanacearum. Phytopathology* 91: 399–407.

Huyer, M., Page, W.J. 1989. Ferric reductase activity in *Azotobacter vinelandii* and its inhibition by Zn21. *J. Bacteriol* 171: 4031–4037.

Imperi, F., Tiburzi, F., Visca, P. 2009. Molecular basis of pyoverdine siderophore recycling in *Pseudomonas aeruginosa. Proc. Natl. Acad. Sci* 106: 20440–20445.

Ishimaru, C.A., Loper, J.E. 1992. High-affinity iron uptake systems present in *Erwinia carotovora* subsp. *carotovora* include the hydroxamate siderophore aerobactin. *Journal of Bacteriology* 174: 2993–3003.

Ishimaru, C.A., Loper, J.E. 1993. Biochemical and genetic analysis of siderophores produced by plant-associated Pseudomonas and Erwinia species. *In Iron Chelation in Plants and Soil Microorganisms*. Barton, L.B. and Hemming, B.C. (Eds.). Academic Press, Inc.

Islam, M.T., Hashidoko, Y., Deora, A., Ito, T., Tahara, S. 2005. Suppression of damping-off disease in host plants by the rhizoplane bacterium *Lysobacter* sp. strain SB-K88 is linked to plant colonization and antibiosis against soil borne Peronosporomycetes. *Applied and Environmental Microbiology* 71: 3786–3796.

Jalal, M.A.F., Love, S.K., Van der Helm, D. 1998. Nα Dimethylcoprogens. Three novel trihydroxamate siderophores from pathogenic fungi. *Bio. Met* 1: 4–8.

Ji, G.H., Wei, L.F., He, Y.Q., Wu, Y.P., Bai, X.H. 2008. Biological control of rice bacterial blight by *Lysobacter antibioticus* strain 13-1. *Biol. Control* 45(3): 288–296.

Jochum, C.C., Osborne, L.E., Yuen, G.Y. 2006. *Fusarium* head blight biological control with *Lysobacter enzymogenes* strain C3. *Biol. Control* 39: 336–344.

Johri, G.H., Sharma, A., Virdi, J.S. 2003. Rhizobacterial diversity in India and its influence on soil and plant health. *Adv. Biochem. Engg. Biotechnol.* 84: 49–89.

Jordi, C., Anna B., Emilio M. 2007. Mechanisms of antagonism of *Pseudomonas fluorescens* EPS62e against *Erwinia amylovora*, the causal agent of fire blight. *Int. Microbiol.* 10: 123–132.

Julian, G., Cameron, H.J., Olsen, R.A. 1983. Role of chelation by ortho dihydroxy phenols in iron absorption by plant roots. *J. Plant Nutr* 6: 163–175.

Katiyar V, Goel R. 2004. Siderophore-mediated plant growth promotion at low temperature by mutant of fluorescent pseudomonad. *Plant Growth Regul* 42: 239–244.

Keating, T.A., Walsh, C.T. 1999. Initiation, elongation, and termination strategies in polyketide and polypeptide antibiotic biosynthesis. *Curr. Opin. Chem. Biol* 3: 598–606.

Kelman,A.; Hartman, G.L.; Hayward, A.C. 1994. Introduction. In: Bacterial wilt: The disease and its causative agent. In *Pseudomonas Solanacearum*. Hayward, A.C., Hartman, G.L. (ed.). CAB International, Wallingford, UK, pp. 1–7.

Kessmann, H., Staub, T., Hofmann, C., Maetzke, T., Herzog, J. Ward, E., Uknes, S., Ryals, J. 1994. Introduction of systemic acquired resistance in plants by chemicals. *Annu. Rev. Phytopathol* 32: 439–459.

Kiraly, Z., El-Zahaby, H.M., Klement, Z. 1997. Role of extracellular polysaccharides (EPS) slime of plant pathogenic bacteria in protecting cells to reactive oxygen species. *J. Phytopathol* 145: 59–68.

Kloepper, J.W. 1983. Effect of seed piece inoculation with plant growth-promoting rhizobacteria on populations of *Erwinia carotovora* on potato roots and in daughter tubers. *Phytopathology* 73: 217–219.

Kloepper, J.W., Beauchamp, C.J. 1992. A review of issues related to measuring of plant roots by bacteria. *Can. J. Microbiol* 38: 1219–1232.

Kloepper, J.W., Leong, J., Teintze, M., Schroth, M.N. 1980a. Enhancing plant growth by siderophores produced by plant growth-promoting rhizobacteria. *Nature* 286: 885–886.

Kloepper, J.W., Leong, J., Teintze, M., Schroth, M.N. 1980b. *Pseudomonas* siderophores: A mechanism explaining disease suppressive soils. *Curr Microbiol* 4: 317–320.

Laha, G.S., Verma, J.P. 1998. Role of fluorescent *Pseudomonad's* in suppression of root-rot and damping off of cotton. *Ind. Phytopathol* 51(3): 275–278.

Lamont, I.L., Martin, L.W., Sims, T., Scott, A., Wallace, M. 2006. Characterization of a gene encoding an acetylase required for pyoverdine synthesis in *Pseudomonas aeruginosa*. *J. Bacteriol* 188: 3149–3152.

Lankford, C.E. 1973. Bacterial assimilation of iron. *Crit. Rev. Microbiol.* 2: 273–331.

Laville, J., Voisard, C., Keel, C., Maurhofer, M., Défago, G., Haas, D. 1992. Global control in *Pseudomonas fluorescens* mediating antibiotic synthesis and suppression of black root rot of tobacco. *Proc. Natl. Acad. Sci* 89: 1562–1566.

Lazarovits, G., Nowak, J. 1997. Rhizobacteria for improvement of plant growth and establishment. *Hortiscience* 32: 188–192.

Leeman, M., Den Ouden, F.M., Van Pelt, J.A., Dirkx, F.P.M, Steijl, H., Bakker, P.A.H.M., Schippers, B. 1996. Iron availability affects induction of systemic resistance to Fusarium wilt of radish in commercial greenhouse trials by seed treatment with *Pseudomonas fluorescens* WCS374. *Phytopathology* 85: 149–155.

Lenhoff, H.M. 1963. An inverse relationship of the effects of oxygen and iron on the production of fluorescin and cytochrome C by *Pseudomonas fluorescens*. *Nature* 199: 601–602.

Lenhoff, H.M., Nicholas, D.J.D. Kaplan, N.O. 1956. Effects of oxygen, iron and molybdenum on routes of electron transfer in *Pseudomonas fluorescens*. *J. Biol. Chem* 220: 983–995.

Lin, H., Fischbach, M.A., Liu, D.R., Walsh, C.T. 2005. *in vitro* characterization of salmochelin and enterobactin trilactone hydrolases IroD, IroE, and Fes. *J. Am. Chem. Soc* 127: 11075–11084.

Lindow, S.E., McGourty, G., Elkins, R. 1996. Interactions of antibiotics with *Pseudomonas fluorescens* strain A506 in the control of fire blight and frost injury to pear. *Phytopathology* 86: 841–848.

Liu, L., Kloepper, J.W., Tuzun, S. 1995. Induction of systemic resistance in cucumber against *Fusarium* wilt by plant growth promoting rhizobacteria. *Phytopathology* 85: 695–698.

Loper, J.E., Buyer, J.S. 1991. Siderophores in microbial interactions on plant surfaces. *Mol. Plant-Microbe Interact* 4: 5–13.

Loper, J.E., Henkels, M.D. 1999. Utilization of heterologous siderophores enhances levels of iron available to *Pseudomonas putida* in the rhizosphere. *Appl Environ Microbiol* 65: 5357–5363.
Lugtenberg, B.J., Dekkers, J.L., Bloemberg, G.V. 2001. Molecular determinants of rhizosphere colonization by *Pseudomonas. Annu. Rev. Phytopathol* 39: 461–490.
Ma, L., Kaserer, W., Annamalai, R., Scott, D.C. 2007. Evidence of ball-and-chain transport of ferric enterobactin through FepA. *J. Biol. Chem* 282: 397–406.
Manuela, D.L., Sophie, P., Michiel, S., Maho, N., Marcelo E. T., Jorge H. C. 2004. A Nonribosomal Peptide Synthetase with a Novel Domain Organization Is Essential for Siderophore Biosynthesis in *Vibrio anguillarum. J. Bacteriol.* 186: 7327–7336.
Masalha, J., Kosegarten, H., Elmaci, Ö., Mengel, K. 2000. The central role of microbial activity for iron acquisition in maize and sunflower. *Biol Fert Soils* 30: 433–439.
Mathesius U, Mulders S, Gao M, Teplitski M, CaetanoAnolles G, Rolfe B.G, Bauer W.D. 2003. Extensive and specific responses of a eukaryote to bacterial quorum-sensing signals. *Proc. Natl Acad. Sci. USA* 100: 1444–1449.
Matzanke, B.F., Anemüller, S., Schünemann, V., Trautwein, A.X., Hantke, K. 2004. FhuF, part of a siderophore reductase system. *Biochemistry* 43: 1386–1392.
May, R., Völksch, B., Kampmann, G. 1997. Antagonistic activities of epiphytic bacteria from soybean leaves against *Pseudomonas syringae* pv. *glycinea in vitro* and in planta. *Microb. Ecol* 34: 118–124.
Mayak, S., Tirosh, T., Glick, B.R. 2004. Plant growth-promoting bacteria confer resistance in tomato plants to salt stress. *Plant Physiol. Biochem* 42: 565–572.
McGrath, S.P., Chaudri, A.M., Giller, K.E. 1995. Long-term effects of metals in sewage sluge on soils, microorganisms and plants. *J. Ind. Microbiol* 14(2): 94–104.
Mchugh, J.P., Rodríguez-Quiñones, F., Abdul-Tehrani, H., Svistunenko, D.A. 2003. Global iron-dependent gene regulation in *Escherichia coli. J. Biol. Chem* 278: 29478–29486.
McMorran, B.J., Kumara, H.M., Sullivan, K., Lamont, I.L. 2001. Involvement of a transformylase enzyme in siderophore synthesis in *Pseudomonas aeruginosa. Microbiology* 147: 1517–1524.
Meiwes, J., Fiedler, H.P., Haag, H., Zähner, H., Konetschny-Rapp, S., Jung, G. 1990. Isolation and characterization of staphyloferrin A, a compound with siderophore activity from *Staphylococcus hyicus* DSM 20459. *FEMS Microbiol Lett* 55(1–2): 201–205.
Meletzus, D., Bermpohl, A., Dreier, J., Eichenlaub, R. 1993. Evidence for plasmid encoded virulence factors in the phytopathogenic bacterium *Clavibacter michiganensis* subsp. *michiganensis* NCPPB382. *J. Bacteriol* 175: 2131–2136.
Meneely, K.M. 2007. The biochemistry of siderophore biosynthesis. PhD., Thesis. University of Kansas, USA.
Meyer, J.M. 2000. Pyoverdines: Pigments, siderophores and potential taxonomic markers of fluorescent *Pseudomonas* species. *Arch Microbiol* 174:135–142.
Miethke, M., Marahiel, M.A. 2007. Siderophore-Based Iron Acquisition and Pathogen Control. *Microbiol Mol Biol Rev* 71: 413–445.
Mishra, S.P., Krishna. A. 2001. Assessment of yield losses due to bacterial blight of cotton. *J. Mycology and Pl. Pathol* 31: 232–233.
Mootz, H.D., Schwarzer, D., Marahiel, M.A. 2002. Ways of assembling complex natural products on modular nonribosomal peptide synthetases. *Chem. Biochem* 3: 490–504.
Mossialos, D., Ochsner, U., Baysse, C., Chablain, P., Pirnay, J.P., Koedam, N., Budzikiewicz, H. et al. 2002. Identification of new, conserved, non-ribosomal peptide synthetases from fluorescent pseudomonads involved in the biosynthesis of the siderophore pyoverdine. *Mol Microbiol* 45: 1673–1685.
Müller, S.I., Valdebenito, M., Hantke, K. 2009. Salmochelin, the long-overlooked catecholate siderophore of *Salmonella. Biometals* 22: 691–695.
Nader, M., Journet, L., Meksem, A., Guillon, L., Schalk, I.J. 2011. Mechanism of ferripyoverdine uptake by *Pseudomonas aeruginosa* outer membrane transporter FpvA: No diffusion channel formed at any time during ferri-siderophore uptake. *Biochemistry* 50: 2530–2540.
Namiranian, S., Richardson, D.J., Russell, D.A., Sodeau, J.R. 1997. Excited state properties of the siderophore pyochelin and its complex with zinc ions. *Photochem Photobiol* 65: 777–782.
Neilands, J.B. 1981. Microbial iron compounds. *Annu Rev Biochem* 50: 715–731.
Neuenschwander, U., Lawton, K., Ryals, J. 1996. Systemic acquired resistance, Vol. 1. In *Plant Microbe interactions*. Stacey, G., Keen, N.T. (eds.). New York: Chapman and Hall, 81–106.
Ogierman, M., Braun, V. 2003. Interactions between the outer membrane ferric citrate transporter FecA and TonB: Studies of the FecA TonB box. *J. Bacteriol* 185: 1870–1885.
Ollinger, J., Song, K-B., Antelmann, H., Hecker, M., Helmann, J.D. 2006. Role of the Fur regulon in iron transport in *Bacillus subtilis. J. Bacteriol* 188: 3664–3673.

Orsi, N. 2004. The antimicrobial activity of lactoferrin: Current status and perspectives. *Biometals* 17: 189–196.

Pandey, A., Sonti, R.V. 2010. Role of the FeoB protein and siderophore in promotingvirulence of Xanthomonas oryzae pv. oryzae on rice. *Journal of Bacteriology* 192(12): 3187–3203.

Park, C.S., Paulitz, T.C., Baker, R. 1988. Biocontrol of *Fusarium* wilt of cucumber resulting from interactions between *Pseudomonas putida* and non pathogenic isolates of *Fusarium oxysporum*. *Phytopathology* 78: 190–194.

Pawelek, P.D., Croteau, N., Ng-Thow-Hing, C., Khursigara, C.M. 2006. Structure of TonB in complex with FhuA, *E. coli* outer membrane receptor. *Science* 312: 1399–1402.

Payne, S.M. 1993. Iron acquisition in microbial pathogenesis. *Trends Microbiol.* 1: 66–69.

Perry, R.D., Balbo, P.B., Jones, H.A., Fetherston, J.D. and DeMoll, E. 1999. Yersiniabactin from *Yersinia pestis*: Biochemical characterization of the siderophore and its role in iron transport and regulation. *Microbiology* 145: 1181–1190.

Perry, R.D., Brubaker, R.R. 1979. Accumulation of iron by yersiniae. *J Bacteriol* 137: 1290–1298.

Persmark, M.D., Expert, D., Neilands, J.B. 1989. Isolation, characterization and synthesis of chrysobactin, a compound with siderophore activity from *Erwinia chrysanthemi*. *J. Biol. Chem.* 264: 3187–3193.

Petsko, G.A. 1985. Preparation of isomorphous heavy-atom derivatives. *Methods Enzymol* 114: 147–156.

Pierre, J.L., Fontecave, M., Crichton, R.R. 2002. Chemistry for an essential biological process: The reduction of ferric iron. *BioMetals* 15: 341–346.

Pierson, L.S., Wood, D.E., Pierson, E.A., Chancey, S.T. 1998. *N*-Acyl-homoserine lactone-mediated gene regulation in biological control by fluorescent pseudomonads: Current knowledge and future work. *Eur. J. Plant Pathol* 104: 1–9.

Pietrangelo, A. 2002. Mechanism of iron toxicity. In *Iron Chelation Theraphy*, Vol. 509. 1st Edn. Hershko, C. (ed.). New York: Kluwer Academic/Plenum Publishers, 19–43.

Poole, K., McKay, G.A. 2003. Iron acquisition and its control in *Pseudomonas aeruginosa*: Many roads lead to Rome. *Front Biosci* 8: 661–686.

Poole, K., Neshat, S., Krebes, K., Heinrichs, D.E. 1993. Cloning and nucleotide sequence analysis of the ferripyoverdine receptor gene fpvA of *Pseudomonas aeruginosa*. *J. Bacteriol* 175: 4597–4604.

Powell, P.E., Cline, G.R., Reid, C.P.P., Szaniszlo, P.J. 1980. Occurrence of hydroxamate siderophore iron chelators in soils. *Nature* 287: 833–834.

Raghavendra, V.B., Siddalingaiah, L., Sugunachar, N.K., Nayak, C., Ramachandrappa, N.S., 2013. Induction of systemic resistance by biocontrol agents against Bacterial blight of cotton caused by *Xanthomonas campestris Pv.Malvacearun*. *eSci. J. Plant Pathol* 2: 59–69.

Rai, P.V., Strobel, G.A. 1968a. Phytotoxic glycopeptides produced by *Corynebacterium michiganense*. II. Biological properties. *Phytopathology* 59: 53–57.

Rai, P.V., Strobel, G.A. 1968b. Phytotoxic glycopeptides produced by *Corynebacterium michiganense*. I. Methods of preparation, physical and chemical characterization. *Phytopathology* 59: 47–52.

Ramesh, R., Joshi, A.A., Ghanekar, M.P. 2009. Pseudomonads: Major antagonistic endophytic bacteria to suppress bacterial wilt pathogen, *Ralstonia solanacearum* in the eggplant (*Solanum melongena* L.). *World J Microbiol Biotechnol* 25: 47–55.

Raymond, K, Müller, G, Matzanke, B. 1984. Complexation of iron by siderophores a review of their solution and structural chemistry and biological function. *Top Curr Chem.* 123: 49–102.

Reiter, B., Pfeifer, U., Schwab, H., Sessitsch, A. 2002. Response of endophytic bacterial communities in potato plants to infection with *Erwinia carotovora* subsp. *atroseptica*. *Appl. Environ. Microbiol*. 68: 2261–2268.

Reiter, B., Wermbter, N., Gyamfi, S., Schwab, H., Sessitsch, A. 2003. Endophytic *Pseudomonas* spp. populations of pathogen-infected potato plants analyzed by 16S rDNA- and 16S rRNA-based denaturating gradient gel electrophoresis. *Plant Soil* 257: 397–405.

Rodriguez, G.M., Smith, I. 2006. Identification of an ABC transporter required for iron acquisition and virulence in *Mycobacterium tuberculosis*. *J. Bacteriol* 188: 424–430.

Roldan-Arjona, T., Perez-Espinosa, A., Ruiz-Rubio, M. 1999. Tomatinase from *Fusarium oxysporum* f. sp. *lycopersici* defines a new class of saponinases. *Mol. Plant-Microbe Interact* 12: 852–861.

Romeiro, R., Karr, A., Goodman, R. 1981. Isolation of a factor from apple that agglutinates *Erwinia amylovora*. *Plant Physiol* 68: 772–777.

Ruggiero, C.E., Neu, M.P., Matonic, J.H., Reilly, S.D. 2000. Interactions of Pu with desferrioxamine siderophores can affect bioavailability and mobility. *Actinide Res. Q.* 2000: 16–18.

Sabet, K.A., Farida, I., Khalil, O. 1969. Studies on the bacterial diseases of sudan crops. VII. New records. *Ann. Appl. Biol* 63: 357–369.

Sahu, G., Sindhu, S. 2011. Disease control and plant growth promotion of green gram by siderophore producing *Pseudomonas* sp. *Res. J. Microbiol* 6: 735–749.

Sarniguet, A., Kraus, J., Henkels, M.D., Muehlchen, A.M., Loper, J.E. 1995. The sigma factor ss affect antibiotic production and biological control activity of *Pseudomonas fluorescens* Pf-5. *Proc. Natl. Acad. Sci* 92: 12255–12259.

Sayyed, R.Z., Badgujar, M.D., Sonawane, H.M., Mhaske, M.M., Chincholkar, S.B. 2005. Production of microbial iron chelators (siderophores) by *Fluorescent pseudomonads. Indian J. Biotechnol.* 4: 484–490.

Schalk, I., Lamont, I., Cobessi, D. 2009. Structure, function relationships in the bifunctional ferrisiderophore FpvA receptor from *Pseudomonas aeruginosa. Bio Metals* 22: 671–678.

Schalk, I.J., Hannauer, M., Braud, A. 2011. New roles for bacterial siderophores in metal transport and tolerance. *Environ. Microbiol* 13: 2844–2854.

Schnider, U., Keel, C., Blumer, C., Troxler, J., Défago, G., Haas, D. 1995. Amplification of the housekeeping sigma factor in *Pseudomonas fluorescens* CHA0 enhances antibiotic production and improves biocontrol abilities. *J. Bacteriol* 177: 5387–5392.

Seleim, F.A., Saead, K.M.H., Abd-El-Moneem, Abo-ELyousr, K.A.M. 2011. Biological control of Bacterial Wilt of Tomato by Plant Growth Promoting Rhizobacteria. *Plant Pathology* 10: 146–153.

Siddique, I.A., Ehetshamul-Haque, S., Shaukat, S.S. 2001. Use of rhizobacteria in the control of root-rot-root-knot disease complex of mungbean. *J. Phytopathology* 149: 337–346.

Sigel, A., Sigel, G. 1998. *Iron Transport and Storage in Microorganism, Plants, and Animals and Metal Ions in Biological System*, Vol. 35. Basel: Marcel Dekker.

Sivamani, E., Anuratha, C.S., Gnanamanickam, S.S. 1987. Toxicity of Pseudomonas fluorescens towards bacterial pathogens of bananas (*Pseudomonas solanacearum*) and rice (*Xanthomonas campestris* pv. *oryzae*). *Curr. Sci.* 56: 547–548.

Smith, E.E., Sims, E.H., Spencer, D.H., Kaul, R., Olson, M.V. 2005. Evidence for diversifying selection at the pyoverdine locus of *Pseudomonas aeruginosa. J Bacteriol* 187: 2138–2147.

Smith, M.J., Shoolery, J.N., Schwyn, B., Holden, I., Neilands. J.B. 1985. Rhizobactin, a structurally novel siderophore from *Rhizobium meliloti. J. Am. Chem. Soc.* 107(6): 1739–1743.

Sokol, P.A. 1987. Surface expression of ferripyochelin-binding protein is required for virulence of *Pseudomonas aeruginosa. Infect Immun* 55: 2021–2025.

Speziali, C.D., Dale, S.E., Henderson, J.A., Vinés, E.D., Heinrichs, D.E. 2006. Requirement of *Staphylococcus aureus* ATP-binding cassette-ATPase FhuC for iron restricted growth and evidence that It functions with more than one iron transporter. *J. Bacteriol* 188: 2048–2055.

Stintzi, A., Barnes, C., Xu, J., Raymond, K.N. 2000. Microbial iron transport via a siderophore shuttle: A membrane ion transport paradigm. *Proc. Natl. Acad. Sci* 97: 10691–10696.

Stockwell, V.O., Johnson, K.B., Sugar, D., Loper, J.E. 2002. Antibiosis contributes to biological control of fire blight by *Pantoea agglomerans* strain Eh252 in orchards. *Phytopathology* 92: 1202–1209.

Strider, D.L. 1969. Bacterial canker of tomato caused by *Corynebacterium michiganense*: A literature review and bibliography. *N.C. Agric. Exp. Stn. Tech. Bull,* 193: 1–110.

Subramoni, S., Sonti, R.V. 2005. Growth deficiency of a *Xanthomonas oryzae* pv. Oryzae *fur* mutant in rice leaves is rescued by ascorbic acid supplementation. *Mol. Plant Microbe Interact* 18: 644–651.

Sunaina, V., Kishore, V., Shekhowat, G.S., Kumar, M. 1997. Control of bacterial wilt of potatoes in naturally infested soil by bacterial antagonist. *J. Plant Disease and Protection*, 104(4): 362–369.

Tagami, Y., Mizukami, T. 1962. Historical review of the researches on bacterial leaf blight of rice caused by *Xanthomonas oryzae* (Uyeda et Ishiyama) Dowson. Special report of the plant diseases and insect pests forecasting service No. 10. Plant protection Division, Ministry of Agriculture and Forestry, Tokyo, Japan. pp.112.

Takase, H., Nitanai, H., Hoshino, K., Otani, T. 2000. Impact of siderophore production on *Pseudomonas aeruginosa* infections in immune-suppressed mice. *Infect Immun* 68: 1834–1839.

Thomashow, L.S., Weller, D.M. 1996. Current concepts in the use of introduced bacteria for biological disease control: Mechanisms and antifungal metabolites. *Plant-Microbe Interactions* 1: 187–235.

Toyota, K., Kimura, M. 2000. Suppression of *Ralstonia solanacearum* in soil following colonization by other strains of *R. solanacearum. Soil Sci Plant Nutr* 46: 449–459.

Van den Bulk, R.W., Löffler, H.J.M., Dons, J.J.M. 1989. Effect of phytotoxic compounds produced by *Clavibacter michiganense* subsp. *michiganensis* on resistant and susceptible tomato plants. *Neth. J. Plant Pathol* 95: 107–117.

Vasudevan, P., Kavitha, S., Priyadarisini, V.B., Babujee, L., Gnanamanickam, S.S. 2002. Biological control of rice diseases. *Biological Control of Crop Diseases* 11–32.

Vauterin, L., Hoste, B., Kersters, K., Swings, J. 1995. Reclassification of *Xanthomonas. Inter. J. Sys. Bact* 45: 472–489.

Velusamy, P., Immanuel, J.E., Gnanamanickam, S.S. 2006. Biological control of rice bacterial blight by plant-associated bacteria producing 2,4-diacetylphloroglucinol. *Can. J. Microbiol.* 52: 56–64.

Vértesy, L., Aretz, W., Fehlhaber, H.W., Kogler, H. 1995. Salimycin A-D, Antibiotoka aus *Streptomyces violaveus*, DSM 8286, mit Siderophor- Aminoglycosid-Struktur. *Helv. Chim. Acta* 78: 46–60.

Von Dohren, H., Keller, U., Vater, J., Zocher, R. 1997. Multifunctional peptide synthetases. *Chem. Rev* 97: 2675–2706.

Von Gunten, H.R., Benes, P. 1995. Speciation of radionuclides in the environment. *Radiochim. Acta* 69: 1–29.

Wei, G., Kloepper, J.W., Tuzun, S. 1991. Induction of systemic resistance of cucumber to *Colletotrichum orbiculare* by select strains of plant growth promoting rhizobacteria. *Phytopathology* 81: 1508–1512.

Weisbeek, P.J., Van der Hofstad, G.A.J.M., Schippers, B., Marugg, J.D. 1986. Genetic analysis of the iron uptake system of two plant growth promoting *Pseudomonas* strains. *NATO ASI Ser. A* 117: 299–313.

Welch, T.J., Chai, S., Crosa, J.H. 2000. The overlapping *angB* and *angG* genes are encoded within the trans-acting factor region of the virulence plasmid in *Vibrio anguillarum*: Essential role in siderophore biosynthesis. *J. Bacteriol* 182: 6762–6773.

Weller, D.M. 2007. Pseudomonas biocontrol agents of soil borne pathogens: Looking back over 30 years. *Phytopathology* 97: 250–256.

Wensing, A., Braun, S.D., Büttner, P., Expert, D. 2010. Impact of siderophore production by *Pseudomonas syringae* pv. *syringae 22d/93* on epiphytic fitness and biocontrol activity against *Pseudomonas syringae* pv. glycinea 1a/96. *Appl. Environ. Microbiol* 76: 2704–2711.

Wertheimer, A.M., Verweij, W., Chen, Q., Crosa, L.M., Nagasawa, M., Tolmasky, M.E., Actis, L.A., Crosa, J.H. 1999. Characterization of the *angR* gene of *Vibrio anguillarum*: Essential role in virulence. *Infect. Immun* 67: 6496–6509.

Whipps, M.J. 2001. Microbial interactions and biocontrol in the rhizosphere. *J Exp Botany* 52: 487–511.

Williamson, L., Nakaho, K., Hudelson, B., Allen, C. 2002. *Ralstonia solanacearum* race 3, biovar 2 strains isolated from geranium are pathogenic on potato. *Plant Dis* 86: 987–991.

Wolz, C., Hohloch, K., Octaktan, A., Poole, K., Evans, R.W., Rochel, N., Albrecht-Gary, A.M., Abdallah, M.A., Döring, G 1994. Iron release from transferrin by pyoverdin and elastase from *Pseudomonas aeruginosa. Infect Immun* 62: 4021–4027.

Wösten, M.M.S.M. 1998. Eubacterial sigma-factors. *FEMS Microbiol. Rev* 22: 127–150.

Yamaguchi, S. 1998. Activators for systemic acquired resistance. In *Fungicidal Activity*. Hutson, D.H., Miyamoto, J. (eds.). Japan: John Wiley & Sons, 193–217.

Yamanaka, K., Oikawa, H., Ogawa, H.O., Hosono, K., Shinmachi, F., Takano, H., Sakuda, S., Beppu, T., Ueda, K. 2005. Desferrioxamine E produced by Streptomyces griseus stimulates growth and development of Streptomyce tanashiensis. *Microbiology* 151: 2899–2905.

Young, D.H., Sequeira, L. 1986. Binding of *Pseudomonas solanacearum* fimbriae to tobacco leaf cell walls and its inhibition by bacterial extracellular polysaccharides. *Physiol. Mol. Plant Pathol* 28: 393–402.

Zinniel, D.K., Lambrecht, P., Harris, N.B., Feng, Z., Kuczmarski, D., Higley, P. 2002. Isolation and characterization of endophytic colonizing bacteria from agronomic crops and prairie plants. *Appl Environ Microbiol* 68: 2198–2208.

9 Plant Metabolic Substances and Plant Pathogenic Bacterial Control

Sundaram Rajakumar, Subramanian Umadevi, and Pudukadu Munusamy Ayyasamy

CONTENTS

ABSTRACT Plants produce a vast array of natural products, many of which evolved to confer selective against pathogenic microbial attack. Among the microbes, the bacteria that are prominent invade plant cells and cause diseases that have major economic impact. In nature, other than the structural barriers, the so-called metabolic substances that are a characteristic feature of plants are important and can protect against a variety of bacteria. Elicited by bacterial infection, plants recognize such substances by pattern-recognition receptors that activate metabolic signaling cascades. This mechanism leads to long-lasting disease resistance that is effective against bacteria. Many different genetic strategies have been used to engineer plant resistance to bacterial diseases by enhancing natural plant defenses.

KEYWORDS: plant metabolites, microbial resistance, immune system, defense mechanism, bacterial control

9.1 INTRODUCTION

Because plants are rich sources of nutrients, microorganisms invade plant cells, where they may cause infectious diseases. Microorganisms can cause a large range of symptoms in most cultivated plants, which can be affected in different parts with various agronomic impacts (Lopez et al., 2003). Across large regions and many crop species, it is estimated that diseases typically reduce plant yields by 10% every year in more-developed nations, often exceeds 20% in less-developed nations, and are a negative factor in 15% of global crop production (Dangl et al., 2013).

9.2 PLANT PATHOGEN INTERACTIONS

Plant pathogens include fungi, bacteria, oomycetes, and viruses. Pathogens have devised different strategies to invade a plant, as well as to feed on and reproduce within it. The locations of pathogenic invasion, feeding, and reproduction are regarded as important in the classification of the attacking microorganisms as bacteria or fungi (Oliver and Ipcho, 2004). Biotrophic pathogens require living tissue for growth and reproduction; hemi-biotropic pathogens kill this tissue in the late stages of the infection. By contrast, necrotrophic pathogens kill the host tissue at the beginning of the infection and feed on the dead tissue. In general, viruses require living tissue for nutrition, whereas biotrophic as well as necrotrophic strategies can be found among bacteria and fungi. Similarities have been described between the pathways that are involved in plant defenses against biotrophic fungi and bacteria and the pathways that are involved in plant defenses against necrotrophic fungi and bacteria. The jasmonate/ethylene pathway is more important in defending against necrotrophic pathogens, whereas salicylic acid-dependent responses are more effective against biotrophic pathogens (Thomma et al., 2001).

9.3 PLANT PATHOGENIC BACTERIA

To successfully infect a plant, pathogens must overcome three layers of defense: performed physical barriers, a cell-surface-based surveillance system, and an intracellular surveillance system. Bacterial pathogens overcome the first layer either by invading through natural openings and wounds and/or by secreting hydrolytic enzymes that break down surface layers. Bacteria typically overcome the second layer by injecting effectors that interfere with defense signaling. Bacteria overcome the third layer by modifying or eliminating existing effectors, or by evolving new effectors that suppress defense activation (Ade and Innes, 2007). Bacterial pathogens have evolved sophisticated mechanisms to interact with their hosts. With most annual agricultural plants, pathogenic bacteria survive from season to season in seed or propagative parts found in debris from plants that have been diseased. Some also survive on or in living crop or weed plants; this type of survival is common in perennial woody species. The bacterial pathogen that incite yellows diseases survive within the vascular tissues of perennial plants (Leben, 1974). Bacterial pathogens access the plant milieu through mechanical openings such as wounds and pruning cuts, or through natural openings such as hydathodes (the termini of leaf veins located on the edges of leaves) and stomata (pores in the surfaces of leaves through which gas exchanges occur). Disease symptoms caused by bacterial pathogens include wilts, galls, specks, spots, cankers, and chlorosis (yellowing). For example, wilt-causing bacteria clog the vascular tissue, preventing movement of water and nutrients. The most-studied plant pathogenic bacteria belong to the genera *Pseudomonas*, *Xanthomonas*, *Erwinia*, *Ralstonia*, and *Agrobacterium* (Ade and Innes, 2007).

9.4 PLANT DISEASE RESISTANCE

Despite the enormous number and ubiquity of microorganisms that have pathogenic potential, the majority of plants remain healthy. To withstand the hostile environment of pathogenic microorganisms, plants have evolved several effective mechanisms of resistance. Although lacking immune systems comparable with those of animals, plants have developed a stunning array of structural, chemical, and protein-based defenses to detect invading organisms and stop them before they can cause extensive damage. Plant disease resistance protects plants from pathogens by histological defenses, accumulation of toxic substances, and hypersensitive defense reactions, as well as defenses through enzymes and modification of substrates to resist pathogens' enzymes and enhance detoxification and disease tolerance. Plants have two main forms of resistance: structural defenses that limit pathogen invasion and attachment, and induced immune systems that reduce pathogen growth (Freeman and Beattie, 2008).

9.4.1 Variations in Disease Resistance

The degree of resistance to a disease may vary under certain conditions. Several agronomical, climatic, and nutritional factors are known to influence plant resistance qualities. The application of successively higher doses of nitrogen increases the susceptibility of a variety of plant diseases. Phosphorus is believed to have a beneficial effect on disease resistance. Potassium is also beneficial because it promotes development of thicker outer-cell walls. Potash-starved potato crops are reported to succumb heavily to diseases.

Climatic conditions, particularly temperature, are known to have great influence on disease resistance in plants. Some wheat varieties that show resistance to rust infection at 10°C are susceptible at 21°C. Some sugarcane varieties in India that show high resistance to mosaic are highly susceptible to diseases in other countries. Even within a country, some varieties may be highly resistant in one tract but susceptible in others. Geographic variations and climatic factors, which significantly influence crop growth and its physiology, also influence resistance qualities. Primary influence factors include genetic changes in hosts, vitality of host plants, nutrition, environmental temperature, atmospheric humidity, light, CO_2 in air (with increasing CO_2 in the air, susceptibility diminishes in tomato to *Clasdosporium fulvum*), and physical characteristics of soil (Rangaswami and Bagyaraj, 2005).

9.5 PLANT IMMUNE SYSTEMS

Plants protect themselves by inducing immune systems that in turn produce high diversities of natural products or secondary metabolites that act against microbial pathogens on the basis of their toxicity. The plant immune system basically consists of two interconnected tiers of receptors, one outside and one inside the cell. Both systems sense the intruder, respond to the intrusion, and (optionally) signal the presence of the intruder to the rest of the plant and sometimes to neighboring plants as well. These two systems detect different types of pathogen molecules and classes of plant receptor proteins (Dangl et al., 2013). The first tier is primarily governed by pattern recognition receptors (PRRs) that are activated by the recognition of evolutionarily conserved pathogen or microbial-associated molecular patterns (PAMPs or MAMPs, here P/MAMP).

Activation of PRRs leads to intracellular signaling, transcriptional reprogramming, and biosynthesis of a complex output response that limits colonization. This system is known as PAMP-triggered immunity (PTI) (Jones and Dangl, 2006; Dodds and Rathjen, 2010). The second tier governs effector-triggered immunity (ETI) and consists of another set of LRRs, the nucleotide-binding LRRs (NLRs). These operate within the cell, encoded by R genes. The presence of specific pathogen effectors activates specific NLR proteins that limit pathogen proliferation.

Receptor responses include ion-channel gating, oxidative bursts, cellular redox changes, and protein kinase cascades that directly activate cellular changes (e.g., cell-wall reinforcement or antimicrobial production) or activate changes in gene expression that then elevate other defensive responses (Dangl et al., 2013). In addition to signaling by PTI and ETI, plant defenses can be activated by the sensing of damage-associated compounds (DAMP) such as the portions of the plant cell wall that are released during pathogenic infection. Many receptors for MAMPs, effectors, and DAMPs have been discovered. Effectors are often detected by NLRs, whereas MAMPs and DAMPs are often detected by transmembrane receptor-kinases that carry LRR or LysM extracellular domains (Dodds and Rathjen, 2010).

9.6 METABOLIC SUBSTANCES IN DEFENSE MECHANISMS

After pathogen recognition, ion channels open and form oxygen-reactive intermediates. The subsequent activation of ion fluxes results in the translocation of oxygen into the nucleus. Reactive oxygen intermediates are required for the following peroxidase-mediated cross-linking of cell-wall components and for the activation of the numerous genes that are involved in the synthesis of various defense compounds. Major components of the signal-transduction chain in cells, from elicitor perception to gene activation, are schematically represented in Figure 9.1 (Somssich and Hahlbrock, 1998). Numerous genes and/or proteins were identified that are involved in defense-signal transduction. Plant resistance (R) proteins recognize pathogen avirulence (Avr) determinants and in turn trigger signal-transduction cascades that lead to rapid defense mobilization. The major classes of R genes isolated in plant varieties are listed in Table 9.1 (Hannond-Kosack and Parker, 2003).

9.6.1 Secondary Metabolites in Defense Mechanisms

Secondary metabolites throughout the plant kingdom have very restricted distribution compared to primary metabolites; that is, they are often found only in one plant species or in a taxonomically related group of species. A high concentration of secondary metabolites might result in a more-resistant plant. It is believed (although it has not been conclusively proven) that most of the 1,00,000 known secondary metabolites are involved in plant chemical-defense systems, which in turn have been formed over the millions of years that plants have coexisted with their attackers (Wink, 1999). Such secondary metabolites have generally been viewed as waste products resulting from "mistakes" of primary metabolism, and therefore of little importance to plant metabolism and growth. It has become clear that such views are largely inaccurate and misguided, and that many secondary products are key components of active and potent defense mechanisms—in other words, soldiers in the age-long chemical warfare between plants and their pests and pathogens (Bennett and Wallsgrove, 1994).

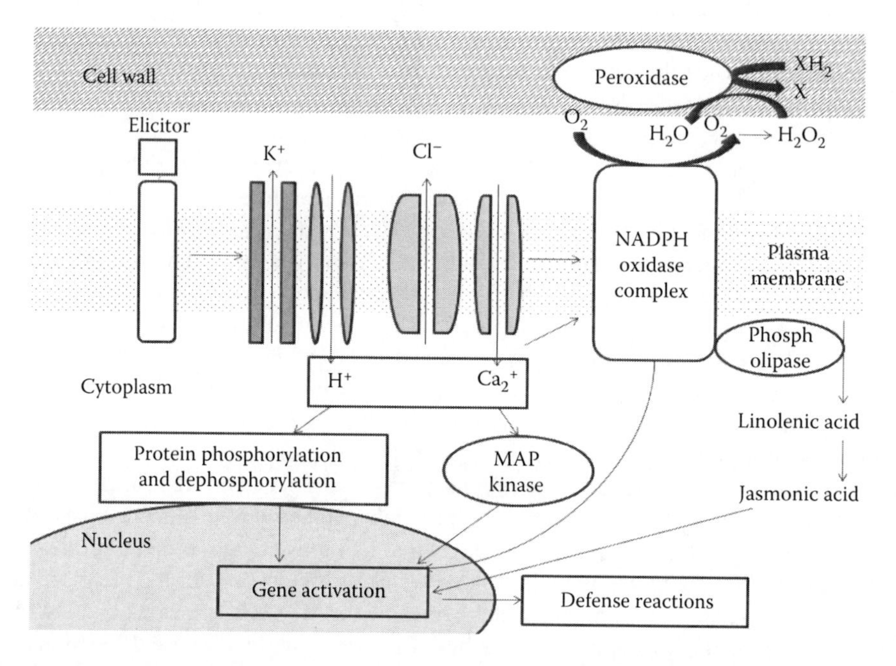

FIGURE 9.1 Major components of plant signal transduction. (Adapted from Somssich., I.E., Hahlbrock, K. 1998. *Trends Plant Sci.* 3(3): 86–90.)

TABLE 9.1
Plant Bacterial Resistance Genes

S. No	Gene	Plant	Pathogen
1	*Pto*	Tomato	*Pseudomonas syringae* p.v. tomato (*avrPto*)
2	*PSB1*	*Arabidopsis*	*Pseudomonas syringae* p.v. *phaseolicola* (*avrPphB*)
3	*RPS2*	*Arabidopsis*	*Pseudomonas syringae* p.v. *maculicola* (*avrRpt2*)
4	*Bs2*	Pepper	*Xanthomonas campestris* pv. *vesicatoria* (*avrBs2*)
5	*RRS-1*	*Arabidopsis*	*Ralstonia solanacearum* (race 1)
6	*Xa-21*	Rice	*Xanthomonas oryzae* p.v. *oryzae* (all races)
7	*FLS2*	*Arabidopsis*	Multiple bacteria (flagellin)

Source: Adapted from Hannond-Kosack, K.E., Parker, J.E. 2003. *Curr. Opin. Biotechnol.* 14(2): 177–193.

Secondary metabolites can be divided into three chemically distinct groups: terpenes, phenolics, and nitrogen (N)- and sulfur (S)-containing compounds. They defend plants against a variety of herbivores and pathogenic microorganisms as well as various kinds of abiotic stresses (Mazid et al., 2011). Terpenes composed of 5-C isopentanoid units are toxins and feeding deterrents to many herbivores. Phenolics, which are synthesized primarily from products of the shikimic acid pathway, have several important defensive roles in plants. Members of the N- and S-containing compounds are synthesized principally from common amino acids (Rosenthal and Berenbaum, 1992; Etten et al., 2001). In addition, an enormous variety of non-protein amino acids is found in plants; the legumes in particular contain a diverse range of compounds and high concentrations. The toxicity of many of these compounds and their roles in plant defense are well established. Detailed studies with some particular non-protein amino acids clearly demonstrated their deterrent effect on non-specialist herbivores, as well as the mechanisms whereby certain pest organisms can overcome this toxicity (Bennett and Wallsgrove, 1994).

9.6.2 Secondary Metabolites against Bacteria

By their nature, plants produce chemicals to protect themselves from bacterial attack. These mainly include lactone and aldehyde groups (eugenol, catechols, protoanemonin, helanins, hexanal, etc.) as essential oils or resins. Some of the plant varieties that have microbial resistance that especially control bacteria are allspice (*Pimenta dioica*), balsam pear (*Momordica charantia*), betel pepper (*Piper betel*), buttercup (*Ranunculus bulbosus*), castor bean (*Ricinus communis*), cockle (*Agrostemma githago*), coltsfoot (*Tussilago farfara*), echinacea (*Echinaceae angustifolia*), gamboge (*Garcinia hanburyi*), mace or nutmeg (*Myristica fragrans*), mesquite (*Prosopis juliflora*), mountain tobacco (*Arnica Montana*), olive (*Olea europaea*), poinsettia (*Euphorbia pulcherrima*), prostrate knotweed (*Polygonum aviculare*), smooth hydrangea or seven barks (*Hydrangea arborescens*), snake plant (*Rivea corymbosa*), lantana (*Lantana camara*), and Tua-Tua (*Jatropha gossyphiifolia*) (Cowan, 1999).

9.6.2.1 Terpenes

Terpenes, which constitute the largest class of secondary metabolites, occur in all plants. They are united by their common biosynthetic origin from acetyl-coA or glycolytic intermediates. Monoterpenes in plants are known to have mainly ecological roles as deterrents to feeding by herbivores, as antifungal defenses, and as attractants for pollinators (Langenheim, 1994). In mammals, terpenes are involved in stabilizing cell membranes, in metabolic pathways, and as regulators of enzymatic reactions. For example, cholesterol and related steroids are triterpenes that are derived from 6 isoprene units.

Herbs and higher plants containing terpenoids and their oxygenated derivatives have been used as fragrances and flavors for centuries. More than 22,000 individual terpenoids are presently known, which makes terpenoids the largest group of natural products. Terpenes have drawn increasing commercial attention because of better understanding of their roles in the prevention and therapy of several diseases, including cancer. They are natural insecticides and antimicrobial agents—properties that can be useful in storing agricultural produce (e.g., as a sprouting inhibitor in potatoes). They are also building blocks for the synthesis of many highly valuable compounds (Da Carvalho and Da Fonseca, 2006).

The biotransformation of terpenes is of interest because it allows the production of enantiomerically pure flavors and fragrances under mild reaction conditions. Products produced by biotransformation processes may be considered "natural." Industrial use of monoterpenes as substitutes of ozone-depleting chlorofluorocarbons is also flourishing (Kirchner, 1994). Terpenes may be used as substitutes for chlorinated solvents in applications such as cleaning of electronic components and cables, degreasing of metal, and cleaning of aircraft parts (Brown et al., 1992).

Essential oils, the common terpenes, are particularly found in basil (*Ocimum basilicum*), dill (*Anethum graveolens*), lemon verbena (*Aloysia triphylla*), buchu (*Barosma setulina*), rosemary (*Rosmarinus officinalis*), valerian (*Valeriana officinalis*), bay (*Laurus nobilis*), and tansy (*Tanacetum vulgare*). The terpene-class compounds that are found in plant varieties and have antibacterial properties are listed in Table 9.2. Terpene-mediated defenses against bacteria are found in ginseng (*Panax notoginseng*), grapefruit peel (*Citrus paradise*), horseradish (*Armoracia rusticana*), and lavender-cotton (*Santolina chamaecyparissus*); however, specific identification of their metabolites has not been made (Jones and Luchsinger, 1986; Apisariyakul et al., 1995; Cichewicz and Thorpe, 1996; Kadota et al., 1997).

9.6.2.2 Phenolic Compounds

Phenolic compounds are derived from aromatic amino acids and produce a wide variety of secondary products that contain a phenol group, a hydroxyl functional group, and an aromatic ring. Some secondary products and plant metabolites and their pathway relationships are shown in Figure 9.2 (Bennett and Wallsgrove, 1994). The flavonoids constitute a large group of secondary plant

TABLE 9.2
Significant Terpene Compounds in Plants

Plant Common Name	Scientific Name	Terpene Compound
Gotu kola	*Centella asiatica*	Asiatocoside
Chili peppers, paprika	*Capsicum annuum*	Capsaicin
Savory	*Satureja montana*	Carvacrol
Turmeric	*Curcuma longa*	Curcumin
Clove	*Syzygium aromaticum*	Eugenol
Peppermint	*Mentha piperita*	Menthol
Brazilian pepper tree	*Schinus terebinthifolius*	Terebinthone
(Japanese) herb	*Rabdosia trichocarpa*	Trichorabdal A

Source: Adapted from Cichewicz, R.H., Thorpe, P.A. 1996. *J. Ethnopharmacol* 52: 61–70; Apisariyakul, A., Vanittanakom, N., Buddhasukh, D. 1995. *J. Ethnopharmacol* 49: 163–169; Jones, S.B., Luchsinger, A.E. 1986. *Plant Systematics*. New York: McGraw-Hill Book Co.; Kadota, S. et al. 1997. *Zentbl. Bakteriol* 286: 63–67.

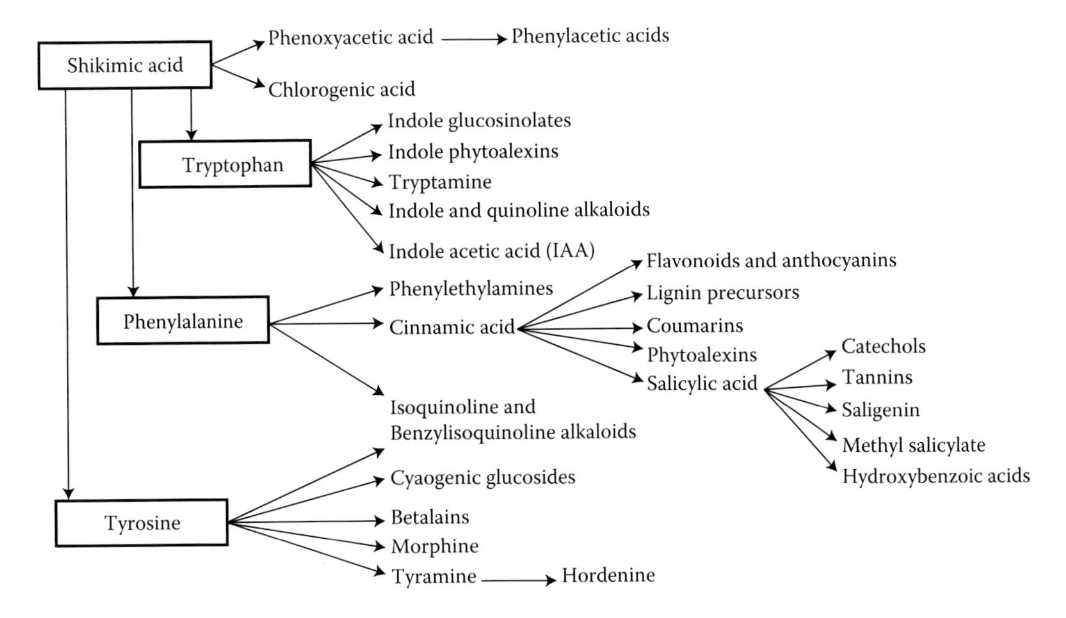

FIGURE 9.2 Phenolic compounds derived from shikimic acid. (Adapted from Bennett, R.N., Wallsgrove, R.M. 1994. *New Phytol.* 127: 617–633.)

metabolites that is ubiquitous among higher plants. They are polyphenolic compounds that generally occur as glycosylated derivatives. As dietary compounds, they are widely known antioxidants that inhibit the oxidation of low-density lipoproteins and reduce thrombotic tendencies (Hertog et al., 1993). Although attention has also been paid to their antimicrobial activity, no dramatic evidence of their effectiveness has been reported (Mori et al., 1987, Nishino et al., 1987, Barnabas and Nagarajan, 1988; Tsuchiya et al., 1996).

The phenolic metabolites that have antibacterial activity in plants are listed in Table 9.3. These compounds exhibit inhibitory effects against multiple bacteria. Several plants, such as burdock (*Arctium lappa*), eucalyptus (*Eucalyptus globulus*), and ceylon cinnamon (*Cinnamo mumverum*) have mixed terpene and phenolic compounds (Figure 9.3). Notably, thyme (*Thymus vulgaris*) has caffeic acid, thymol, and tannins that resist bacteria. Willow is effective against many pathogens because of the presence of salicin, tannins, and other essential oils (Bose, 1958; Berkada, 1978; Hamburger and Hostettmann, 1991; Kubo et al., 1992, 1993, 1994; Hufford et al., 1993; Hunter and Hull, 1993; Jones et al., 1994; Wild, 1994).

9.6.2.3 Sulfur-Containing Secondary Metabolites

Plants contain a variety of sulfate transporters that have specific functions in the uptake of sulfate by roots, its transport to shoots, and its subcellular distribution (Hawkesford and Smith, 1997; Hawkesford and Wray, 2000; Hawkesford, 2000 and 2003; Hawkesford et al., 2003). Roots contain all of the enzymes that are required to reduce sulfate to sulfide, although the chloroplast appears to be the primary site for this reduction and the subsequent incorporation of sulfide into cysteine (Brunold, 1990 and 1993; Davidian et al., 2000). Cysteine is the sulfur donor for most of the other organic sulfur compounds in plants. The predominant proportion of the sulfur is present in proteins as cysteine and methionine residues, and is highly significant in protein structure, conformation, and function (De Kok et al., 2002). Sulfur is also required for the synthesis of various other compounds, such as thiols (glutathione), sulfolipids, and secondary sulfur compounds (alliins, glucosinolates, phytochelatins), which play important roles in the physiology of plants and in the protection and adaptation of plants against stress and pests. Sulfur deficiency will result in the loss

TABLE 9.3
Significant Phenolic Compounds Present in Plants

Plant Common Name	Scientific Name	Phenolic Compound
Pasque flower	*Anemone pulsatilla*	Anemonins
Chamomile	*Matricaria chamomilla*	Anthemic acid
Green tea	*Camellia sinensis*	Catechin
Caraway	*Carum carvi*	Coumarins
Woodruff	*Galium odoratum*	Coumarin
Henna	*Lawsonia inermis*	Gallic acid, Lawsone
Licorice	*Glycyrrhiza glabra*	Glabrol
St. John's wort	*Hypericum perforatum*	Hypericin, others
Hops	*Humulus lupulus*	Lupulone, humulone
Chapparal	*Larrea tridentate*	Nordihydroguaiaretic acid
Purple prairie clover	*Petalostemum*	Petalostemumol
Apple	*Malus sylvestris*	Phloretin
Senna	*Cassia angustifolia*	Rhein
Cashew	*Anacardium pulsatilla*	Salicylic acids
Cascara sagrada	*Rhamnus purshiana*	Tannins
Sainfoin	*Onobrychis viciifolia*	Tannins
Oak	*Quercus rubra*	Tannins
Wintergreen	*Gaultheria procumbens*	Tannins
Tree bard	*Podocarpus nagi*	Totarol, Nagilactone
Ashwagandha	*Withania somniferum*	Withafarin A
Hemp	*Cannabis sativa*	β-Resercyclic acid

Source: Adapted from Hunter, M.D., Hull, L.A. 1993. *Phytochemistry* 34: 1251–1254; Berkada, B. 1978. *J. Irish Coll. Phys. Surg* 22: 56; Bose, P.K. 1958. *J. Indian Chem. Soc* 58: 367–375; Hamburger, H., Hostettmann, K. 1991. *Phytochemistry* 30: 3864–3874; Kubo, I., Muroi, H., Himejima, M. 1992. *J. Nat. Prod* 55: 1436–1440; Kubo, I., Muroi, H., Himejima, M. 1993. *J. Nat. Prod* 56: 220–226; Kubo, I., Muroi, H., Kubo, A. 1994. *J. Nat. Prod* 57: 9–17; Hufford, C.D. et al. 1993. *J. Nat. Prod* 56: 1878–1889; Jones, G.A. et al. 1994. *Appl. Environ. Microbiol* 60: 1374–1378; Wild, R. (Ed.). 1994. *The Complete Book of Natural and Medicinal Cures*. Emmaus, PA: Rodale Press, Inc.

of plant fitness, decreased resistance to environmental stress and pests, and decreased food quality and safety (Durenkamp and De Kok, 2004).

Plants assimilate sulfate and reduce it to sulfide, which is incorporated into cysteine and further converted to methionine. Sulfur-containing compounds play crucial roles in a number of cellular processes such as redox reactions, detoxification of heavy metals and xenobiotics, and metabolism of secondary products (Dubuis et al., 2005). Although sulfur-containing products of secondary metabolism are rather unusual plant constituents, they play important roles in the chemical defenses of *Brassicaceae*, *Alliaceae*, and *Astraceae*. In this group of metabolites, glucosinolates and allicins are involved in activated plant defense mechanisms (Burow et al., 2008).

The metabolite glucosides in mustard oil, allicin in onion (*Allium cepa*) and garlic (*Allium sativum*), and the additional compound ajoene in garlic, as well as the fabatin in fava bean (*Vivia faba*) are the volatile defensive substances that exhibit toxins and control the unfavorable effects of bacteria (Leustek, 2002). The elemental sulfur that is produced in cocoa (*Sterculiaceae* spp.), tomato, tobacco (*Solanaceae* spp.), cotton (*Malvaceae* spp.), and French bean (*Leguminosae* spp.) is created in response to xylem-invading bacterial pathogens (Naganawa et al., 1996; Cooper and Williams, 2004).

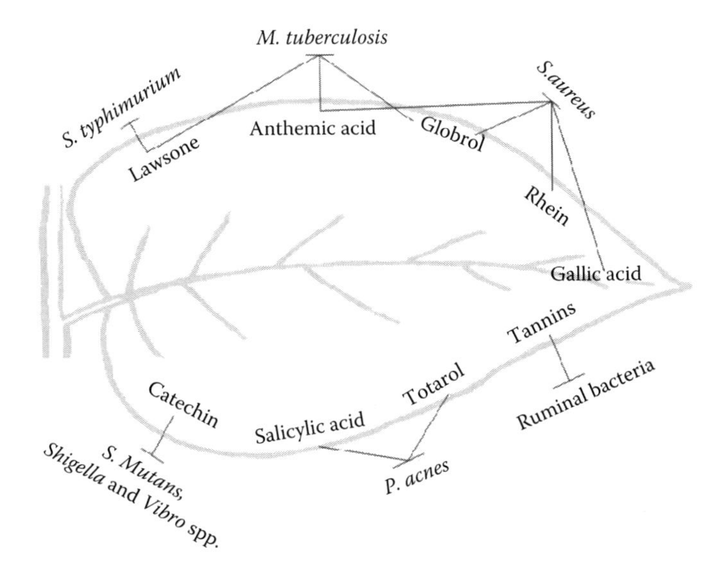

FIGURE 9.3 Phenolic compounds against bacteria.

9.6.2.4 Nitrogen-Containing Secondary Metabolites

Most of the members of this group, which includes alkaloids, cyanogenic glucosides, and non-protein amino acids, are biosynthesized from common amino acids. All are of considerable interest because of their roles in pathogen defense and their toxicity to humans (Mazid et al., 2011). Alkaloids, a large class of bitter-tasting nitrogenous compounds, are found in many vascular plants; they include caffeine, cocaine, morphine, and nicotine. They are derived from the amino acids aspartate, lysine, tyrosine, and tryptophan and may have powerful effects on animal physiology. Caffeine is an alkaloid found in plants such as coffee (*Coffea arabica*), tea (*Camellia sinensis*), and cocoa (*Theobroma cacao*). It is toxic to both insects and fungi. In fact, high levels of caffeine produced by coffee seedlings can even inhibit the germination of other seeds in the vicinity of the growing plants, a phenomenon called allelopathy. Allelopathy allows one plant species to defend itself against other plants that may compete for growing space and nutrient resources.

Members of the nightshade family (*Solanaceae*) produce many important alkaloid compounds. Nicotine is an alkaloid that is produced in the roots of tobacco plants (*Nicotiana tabacum*) and transported to leaves, where it is stored in vacuoles. It is released by nonhuman animals when herbivores graze on the leaves and break open the vacuoles. Atropine is a neurotoxin and cardiac stimulant produced by the deadly nightshade plant (*Atropa belladonna*). Although it is toxic in large quantities, it has been used medicinally by humans in small amounts as a pupil dilator, and as an antidote for some nerve-gas poisonings. Capsaicin and related capsaicinoids produced by members of the genus *Capsicum* are the active components of chili peppers; they produce the characteristic burning sensation caused by peppers in foods.

Cyanogenic glycosides are a particularly toxic class of nitrogenous compounds that break down to produce hydrogen cyanide (HCN), a lethal chemical that halts cellular respiration in aerobic organisms. Plants that produce cyanogenic glycosides also produce enzymes that convert these compounds into hydrogen cyanide, including glycosidases and hydroxynitrile lyases, but these are stored in separate compartments or tissues within the plant. When herbivores feed on these tissues, the enzymes and substrates mix and produce lethal hydrogen cyanide. Glucosinolates, also known as mustard oil glycosides, are sulfur-containing compounds synthesized by members of the mustard family (*Brassicaceae*) that produce cyanide gas when broken down by enzymes called thioglucosidases (Freeman and Beattie, 2008).

TABLE 9.4
Significant Alkaloid Compounds Present in Plants

Plant Common Name	Scientific Name	Phenolic Compound
Barberry	*Berberis vulgaris*	Berberine
Goldenseal	*Hydrastis canadensis*	Berberine, hydrastine
Coca	*Erythroxylum coca*	Cocaine
Glory lily	*Gloriosa superba*	Colchicine
Peyote	*Lophophora williamsii*	Mescaline
Poppy	*Papaver somniferum*	Opium
Black pepper	*Piper nigrum*	Piperine
Periwinkle	*Vinca minor*	Reserpine
Rauvolfia, chandra	*Rauvolfia serpentina*	Reserpine

Source: Adapted from Omulokoli, E., Khan, B., Chhabra, S.C. 1997. *J. Ethnopharmacol* 56: 133–137; Freiburghaus, F.R. et al. 1996. *J. Ethnopharmacol* 55: 1–11; McDevitt, J.T. et al. 1996. *Program and Abstracts of the 36th Interscience Conference on Antimicrobial Agents and Chemotherapy*, Louisiana, USA (September 15–18, 1996); Ghoshal, S., Krishna Prasad, B.N., Lakshmi, V. 1996. *J. Ethnopharmacol* 50: 167–170.

The alkaloid metabolites in plants with antibacterial activity are listed in Table 9.4. Among them, piperine acts against multiple bacteria (i.e., *Lactobacillus*, *Micrococcus*, *E. coli,* and *E. faecalis*). Some of the bactericide compounds are made of monosaccharides, latex, or mixed groups; the ones that do not belong to any of these groups are often found in papaya (*Carica papayaalfalfa*), (*Medicago sativa*), aloe (*Aloe barbadensis*), aveloz (*Euphorbia tirucalli*), blueberry (*Vaccinium* spp.), coriander or cilantro (*Coriandrum sativum*), and cranberry (*Vaccinium* spp.). Plants in which antibacterial activity was found but the compounds could not be categorically identified include dandelion (*Taraxacum officinale*), harmel or rue (*Peganum harmala*), marigold (*Calendula officinalis*), potato (*Solanum tuberosum*), sweet flag or calamus (*Acorus calamus*), and yellow dock (*Rumex crispus*) (Freiburghaus et al., 1996; Ghoshal et al., 1996; McDevitt et al., 1996; Omulokoli et al., 1997).

9.7 TRANSGENIC ENGINEERED SECONDARY METABOLITES

Improvements in plant-transformation techniques and progress in the understanding of plant–pathogen interactions have enabled the use of genetic engineering for the rational creation of disease-resistant plants. Such plants include antibacterial proteins of non-plant origin that inhibit bacterial pathogenicity or virulence factors, enhance natural plant defenses, and artificially induce programmed cell death at the site of infection (Mourgues et al., 1998). Recent *in vitro* experiments on genetically engineered bacterial resistance plants and the mechanisms behind them are listed in Table 9.5. Although higher

TABLE 9.5
Transgenic Bacterial Resistance Projects

Crop	Disease Resistance	Mechanism
Tomato	Bacterial spot	R gene from pepper
Apple	Fire blight	Antibacterial protein from moth
Rice	Bacterial blight and bacterial streak	Engineered E gene
Tomato	Multibacterial resistance	PRR from Arabidopsis
Banana	Xanthomonas wilt	Novel gene from pepper
Rice	Bacterial streak	R gene from maize

Source: Adapted from Siemens, D.H. et al. 2002. *Ecology* 83(2): 505–517.

concentrations of metabolites result in more-resistant plants, the production of secondary metabolites is costly and may reduce plant growth and reproduction (Siemens et al., 2002).

9.8 CONCLUSION AND FUTURE PERSPECTIVES

Plants have evolved multiple defense mechanisms against bacteria through R-protein-mediated cellular surveillance, which producies secondary metabolites. Terpenes are often found in many plant varieties; of these, the phenol groups have the largest variety of compounds. Advanced experiments are required to investigate the correct balance between crop growth and crop-resistance management. In the long term, it should be possible to produce valuable defensive secondary metabolites to improve plant resistance against bacteria as well as other pathogens.

ACKNOWLEDGMENT

The authors gratefully acknowledge Bharathidasan University for providing facilities.

REFERENCES

Ade, J., Innes, R.W. 2007. *Resistance to Bacterial Pathogens in Plants*. In eLS. Chichester: John Wiley & Sons Ltd,. http://www.els.net [doi: 10.1002/9780470015902.a0020091]

Apisariyakul, A., Vanittanakom, N., Buddhasukh, D. 1995. Antifungal activity of turmeric oil extracted from *Curcuma longa* (Zingiberaceae). *J. Ethnopharmacol* 49: 163–169.

Barnabas, C.G.G., Nagarajan, S. 1988. Antimicrobial activity of flavonoids of some medicinal plants. *Fitoterapia* 59(6): 508–510.

Bennett, R.N., Wallsgrove, R.M. 1994. Secondary metabolites in plant defence mechanisms. *New Phytol* 127: 617–633.

Berkada, B. 1978. Preliminary report on warfarin for the treatment of herpes simplex. *J. Irish Coll. Phys. Surg* 22: 56.

Bose, P.K. 1958. On some biochemical properties of natural coumarins. *J. Indian Chem. Soc* 58: 367–375.

Brown, L.M., Springer, J., Bower, M. 1992. Chemical substitution for 1,1,1-trichloroethane and methanol in an industrial cleaning operation. *J Hazard Mater* 29: 179–188.

Brunold, C. 1990. Reduction of sulphate to sulphide. In *Sulphur Nutrition and Sulphur Assimilation in Higher Plants Fundamental, Environmental and Agricultural Aspects*. Rennenberg, H., Brunold, C., De Kok, L.J., Stulen, I. (eds.). The Hague, The Netherlands: SPB Academic Publishing, 13–31.

Brunold, C. 1993. Regulatory interactions between sulphate and nitrate assimilation. In *Sulphur Nutrition and Assimilation in Higher Plants Regulatory, Agricultural and Environmental Aspects*. De Kok, L.J., Stulen, I., Rennenberg, H., Brunold, C., Rauser, W.E. (eds.). The Hague, The Netherlands: SPB Academic Publishing, 61–75.

Burow, M., Wittstock, U., Jonathan, G. 2008. Sulfur-containing secondary metabolites and their role in plant defense. *Adv. Photosynth Resp* 27: 201–222.

Cichewicz, R.H., Thorpe, P.A. 1996. The antimicrobial properties of chile peppers (Capsicum species) and their uses in Mayan medicine. *J. Ethnopharmacol* 52: 61–70.

Cooper, R.M., Williams, J.S. 2004. Elemental sulphur as an induced antifungal substance in plant defence. *J. Exp. Bot* 55(404): 1947–1953.

Cowan, M.M. 1999. Plant products as antimicrobial agents. *Clin Mircobiol Rev* 12(4): 564–582.

Da Carvalho, C.C.C.R., Da Fonseca, M.R.M. 2006. Biotransformation of terpenes. *Biotechnol. Adv* 24(2): 134–142.

Dangl, J.L., Horvath, D.M., Staskawicz, B.J. 2013. Pivoting the plant immune system from dissection to deployment. *Science* 3416147: 746.

Davidian, J.C., Hatzfeld, Y., Cathala, N., Tagmount, A., Vidmar, J.J. 2000. Sulphate uptake and transport in plants. In *Sulphur Nutrition and Sulphur Assimilation in Higher Plants: Molecular, Biochemical and Physiological Aspects*. Brunold, C., Rennenberg, H., De Kok, L.J., Stulen, I., Davidian, J.-C. (eds.). Bern, Switzerland: Paul Haupt Verlag, 19–40.

De Kok, L.J., Castro, A., Durenkamp, M., Stuiver, C.E.E., Westerman, S., Yang, L., Stulen, I. 2002. *Sulphur in Plant Physiology. Proceedings No. 500*. York, UK: International Fertiliser Society, 1–26.

Dodds, P.N., Rathjen, J.P. 2010. Plant immunity: Towards an integrated view of plant pathogen interactions. *Nat Rev Genetics* 11(8): 539.

Dubuis, P.H., Marazzi, C., Stadler, E., Mauch, F. 2005. Sulphur deficiency causes a reduction in antimicrobial potential and leads to increased disease susceptibility of oilseed rape. *J. Phytopathology* 153: 27–36.

Durenkamp, M., De Kok, L.J. 2004. Impact of pedospheric and atmospheric sulpher nutrition on sulpher metabolism of *Allium cepa* L., a species with a potential sink capacity for secondary sulpher compound. *J. Exp. Bot.* 55(404): 1821–1830.

Etten, V.H., Temporini, E., Wasmann, C. 2001. Phytoalexin (and phytoanticipin) tolerance as a virulence trait: Why is it not required by all pathogens? *Physiol. Mol. Plant Pathol.* 59: 83–93.

Freeman, B.C., Beattie, G.A. 2008. An overview of plant defenses against pathogens and herbivores. *The Plant Health Instructor*. DOI: 10.1094/PHI-I-2008-0226-01.

Freiburghaus, F.R., Kaminsky, M., Nkunya, H.H., Brun, R. 1996. Evaluation of African medicinal plants for their *in vitro* trypanocidal activity. *J. Ethnopharmacol* 55: 1–11.

Ghoshal, S., Krishna Prasad, B.N., Lakshmi, V. 1996. Antiamoebic activity of Piper longum fruits against Entamoeba histolytica *in vitro* and in vivo. *J. Ethnopharmacol* 50: 167–170.

Hamburger, H., Hostettmann, K. 1991. The link between phytochemistry and medicine. *Phytochemistry* 30: 3864–3874.

Hannond-Kosack, K.E., Parker, J.E. 2003. Deciphering plant-pathogen communication: Fresh perspectives for molecular resistance breeding. *Curr. Opin. Biotechnol.* 14(2): 177–193.

Hawkesford, M.J. 2000. Plant responses to sulphur deficiency and the genetic manipulation of sulphate transporters to improve S-utilization efficiency. *J. Exp. Bot.* 51: 131–138.

Hawkesford, M.J., Smith, F.W. 1997. Molecular biology of higher plant sulphate transporters. In *Sulphur Metabolism in Higher Plants—Molecular, Ecophysiological and Nutritional Aspects*. Cram, W.J., De Kok, L.J., Stulen. I., Brunold. C., Rennenberg. H., (eds.). Leiden, The Netherlands: Backhuys Publishers, pp. 13–25.

Hawkesford, M.J., Wray, J.L. 2000. Molecular genetics of sulphate assimilation. *Adv. Bot. Res.* 33: 159–223.

Hawkesford, M.J., Buchner, P., Hopkins, L., Howarth, J.R. 2003. Sulphate uptake and transport. In *Sulphur in Plants*. Abrol, Y.P., Ahmad. A. (eds.). Dordrecht, The Netherlands: Kluwer Academic Publishers, pp. 71–86.

Hertog, M.G.L., Feskens, E.J.M., Hollman, P.C., Katan, M.B., Kromhout, D. 1993. Dietary antioxidant flavonoids and risk of coronary heart disease: The Zutphen elderly study. *Lancet* 342: 1007–1011.

Hufford, C.D., Jia, Y., Croom, E.M., Muhammed, I., Okunade, A.L., Clark, A.M., Rogers, R.D. 1993. Antimicrobial compounds from *Petalostemum purpureum*. *J. Nat. Prod* 56: 1878–1889.

Hunter, M.D., Hull, L.A. 1993. Variation in concentrations of phloridzin and phloretin in apple foliage. *Phytochemistry* 34: 1251–1254.

Jones, G.A., McAllister, T.A., Muir, A.D., Cheng, K.J. 1994. Effects of sainfoin (Onobrychis viciifolia scop.) condensed tannins on growth and proteolysis by four strains of ruminal bacteria. *Appl. Environ. Microbiol* 60: 1374–1378.

Jones, J.D., Dangl, J.L. 2006. The plant immune system. *Nature* 444(7117): 323–329.

Jones, S.B., Luchsinger, A.E. 1986. *Plant Systematics*. New York: McGraw- Hill Book Co.

Kadota, S., Basnet, P., Ishii, E., Tamura, T., Namba, T. 1997. Antibacterial activity of trichorabdal from *Rabdosia trichocarpa* against *Helicobacter pylori*. *Zentbl. Bakteriol* 286: 63–67.

Kirchner, E.M. 1994. Environment, health concerns force shift in use of organic solvents. *Chem Eng News* 72: 13–20.

Kubo, I., Muroi, H., Himejima, M. 1992. Antibacterial activity of totarol and its potentiation. *J. Nat. Prod* 55: 1436–1440.

Kubo, I., Muroi, H., Himejima, M. 1993. Combination effects of antifungal nagilactones against Candida albicans and two other fungi with phenylpropanoids. *J. Nat. Prod* 56: 220–226.

Kubo, I., Muroi, H., Kubo, A. 1994. Naturally occurring anti-acne agents. *J. Nat. Prod* 57: 9–17.

Langenheim, J.H. 1994. Higher plant terpenoids: A phytocentric overview of their ecological roles. *J Chem Ecol* 20: 1223–1280.

Leben, C. 1974. Survival of plant pathogenic bacteria. *Ohio Agric Res Dev Cent. Spec. Cir* 100: 21.

Leustek, T. 2002. Sulfate metabolism. In *The Arabidopsis Book*. Somerville, C.R., Meyerowitz, E.M. (eds.). Rockville, MD: American Society of Plant Biologists.

Lopez, M.M., Bertolini, E., Olmos, A., Llop, P., Penyalver, R., Cambra, M. 2003. Innovative tools for the detection of plant pathogenic viruses and bacteria. *Int Microbiol* 6: 233–243.

Mazid, M., Khan, T.A., Mohammad, F. 2011. Role of secondary metabolites in defense mechanisms of plants. *Biol. Med.* 3(2): 232–249.

McDevitt, J.T., Schneider, D.M., Katiyar, S.K., Edlind, T.D. 1996. Berberine: A candidate for the treatment of diarrhea in AIDS patients, abstr. 175. In *Program and Abstracts of the 36th Interscience Conference on Antimicrobial Agents and Chemotherapy*, Louisiana, USA (September 15–18, 1996).

Mori, A., Nishino, C., Enoki, N., Tawata, S. 1987. Antibacterial activity and mode of action of plant flavonoids against *Proteus Vulgaris* and *Staphylococcus Aureus*. *Phytochemistry* 26(8): 2231–2234.

Mourgues, F., Brisset, M., Chevreau, E. 1998. Strategies to improve plant resistance to bacterial diseases through genetic engineering. *Trends Biotechnol.* 16(5): 203–210.

Naganawa, R., Iwata, N., Ishikawa, K., Fukuda, H., Fujino, T., Suzuki, A. 1996. Inhibition of microbial growth by ajoene, a sulfur-containing compound derived from garlic. *Appl. Environ. Microbiol* 62: 4238–4242.

Nishino, C., Enoki, N., Tawata, S., Mori, A., Kobayashi, K., Fukushima, M. 1987. Antibacterial activity of flavonoids against Staphylococcus epidermidis, a skin bacterium. *Agric. Biol. Chem* 51(1): 139–143.

Oliver, R.P., Ipcho, S.V.S. 2004. Arabidopsis pathology breathes new life into the necrotrophs-vs.-biotrophs classification of fungal pathogens. *Mol. Plant Pathol.* 5: 347–352.

Omulokoli, E., Khan, B., Chhabra, S.C. 1997. Antiplasmodial activity of four Kenyan medicinal plants. *J. Ethnopharmacol* 56: 133–137.

Rangaswami, G., Bagyaraj, D.J. 2005. *Plant Pathogenic Microorganisms. Agricultural Microbiology*, 2nd Edn. New Delhi: Prentice-Hall of India (Pvt.). pp. 272–273.

Rosenthal, G.A., Berenbaum, M.R. 1992. *Herbivores: Their Interaction with Secondary Plant Metabolites, Vol II Ecological and Evolutionary Processes*, 2nd Edn. San Diego: Academic Press.

Siemens, D.H., Garner, S.H., Mitchell-Olds, T., Callaway, R.M. 2002. Cost of defense in the context of plant competition: *Brassica rapa* may grow and defend. *Ecology* 83(2): 505–517.

Somssich., I.E., Hahlbrock, K. 1998. Pathogen defence in plants—A paradigm of biological complexity. *Trends Plant Sci.* 3(3): 86–90.

Thomma, B.P.H.J., Penninckx, I.A., Broekaert, W.F., Cammue, B.P. 2001. The complexity of disease signaling in Arabidopsis. *Curr. Opin. Immunol.* 13: 63–68.

Tsuchiya, H., M. Sato, T. Miyazaki, S. Fujiwara, S. Tanigaki, M. Ohyama, T. Tanaka, M. Iinuma. 1996. Comparative study on the antibacterial activity of phytochemical flavanones against methicillin-resistant *Staphylococcus aureus J. Ethnopharmacol* 50: 27–34.

Wild, R. (Ed.). 1994. *The Complete Book of Natural and Medicinal Cures*. Emmaus, PA: Rodale Press, Inc.

Wink M, 1999. Functions of plant secondary metabolites and their exploitation in biotechnology. In *Annu. Plant Rev. Vol. 3*. Boca Raton, FL: CRC Press.

10 Host Resistance
SAR and ISR to Plant Pathogenic Bacteria

Ömür Baysal

CONTENTS

ABSTRACT Interactions between plant and bacteria result in the activation of various plant mechanisms during the resistance phase. These responses to bacteria can be associated with cell signaling of bacteria in pathogenesis. Evidence shows that bacteria can be inhibited by two types of resistance: one that is constitutive and another that occurs for a limited period by chemical and biological stimulants. Constitutive systems involve the inhibition of bacteria in the pre-form phase of resisitance (i.e., during the invasion period) by producing toxic compounds when cells are injured, or by a combination of physiological factors. Known interactions, particularly between plants and pathogens, result in either successful infection (compatible response) or resistance (incompatible response). Advances in studies on bacterial pathogens have revealed the possibilities of using these systems to gain new insight into the nature of plant resistance to pathogens. Induced resistance systems include hypersensitive reaction (HR) as well as other protection responses which depend upon stimulants that mimick pathogen effectors. HR is elicited by most phytopathogenic bacteria (in particular, *Pseudomonads*, *Xanthomonads*, and especially *Erwinia amylovora*) when introduced into non-host plants, but not by most saprophytic bacteria. The second type of induced resistance is a protective reaction that can be induced by prior infiltration of living cells with avirulent or incompatible strains. Resistance is also induced by heat-killed cells of the bacteria. Responses to strains and chemical stimulants are generally local in plant organs, at the original site(s) of attack, or at distant parts that are systemically unaffected. The most prominent of these plant responses against pathogens are induced systemic resistance (ISR) and systemic acquired resistance (SAR). Excellent reviews on SAR and ISR have been published; therefore, this chapter presents an overview of dynamic findings related to SAR and ISR. These findings chiefly highlight ISR and SAR responses that have drawn the most attention from researchers in this context.

KEYWORDS: host–microbe interactions, induced resistance, systemic acquired resistance, induced systemic resistance, signaling cascades

10.1 INTRODUCTION

Many plants respond to pathogens with a *de novo* production of compounds in order to reduce or inhibit further attack. Infection that results from bacteria elicits a set of localized responses in and around the infected cells, which are indicators of incompatible interactions. Depending on the resistance stimulants, these responses include an oxidative burst and cell death (Lamb and Dixon, 1997; Baysal et al., 2005). When the pathogen is trapped in dead cells, further spread from the site of initial infection can be prevented. Local responses in cells surrounding the initial site of infection include changes in cell-wall composition that can inhibit penetration by the pathogen, and *de novo* synthesis of antimicrobial compounds such as phytoalexins (Kuc, 1995; Hammerschmidt, 1999a,b) and pathogenesis-related (PR) proteins (Van Loon, 1999).

10.2 WHAT IS HOST RESISTANCE TO BACTERIA?

Bacterial pathogenicity to plants has evolved through an arms race of attack and defense. The pathogens' key soldiers are bacterial effector proteins, which are delivered through type III secretion systems and have the ability to suppress basal defenses (Alfano and Collmer, 2004). In plants, varietal resistance to disease is based on recognition of effectors by the products of resistance (*R*) genes. When invasive pathogens are recognized, the effector (in this case, an avirulence or Avr protein) triggers a hypersensitive resistance reaction (HRR) in the plant. In turn, HRR generates antimicrobial compounds that have inhibitory properties. However, such gene-for-gene-based resistance commonly fails due to the emergence of virulent strains of the pathogen, which no longer trigger the HRR. In these cases, the plant offers non-host resistance. This a broad-spectrum defense that provides immunity to all members of a plant species against all isolates of a microorganism that is pathogenic to other plant species (Nomura et al., 2005). Upon contact with the surfaces of a non-host plant species, bacterial pathogens initially encounter preformed and, later, induced plant defenses.

One such initial defense response is pathogen-associated molecular pattern (PAMP)-triggered immunity (PTI). Plants have also evolved several other defense strategies against bacterial invasion. Unlike vertebrates, plants do not have an adaptive immune system; therefore, they must rely solely on an innate immune system for self-defense against invading pathogens (Jones and Dangl, 2006). Plant bacterial pathogens reveal themselves to the host immune system through pathogen-associated molecular patterns (PAMPs) such as flagellin or bacterial lipopolysaccharides (LPSs). Because nonpathogenic bacteria also contain these structures, PAMPs are also referred to as microbe-associated molecular patterns. Plants have evolved specialized cell-surface receptors to detect conserved features of PAMPs and activate defense responses. Nonhost plants, which also have mechanisms to detect nonhost–pathogen effectors, can trigger a defense response called effector-triggered immunity (ETI). This non-host resistance response often results in a hypersensitive response (HR) at the infection site (Hahn, 1996).

Phenolic compounds in plants, which have roles in several physiological mechanisms, are often involved in plant–pathogen interactions as well. The composition of the phenols in plant tissue is markedly influenced by endogenous factors as well as environmental conditions. However, although the antimicrobial potential of some plant phenolics have been demonstrated and results have been published, the variability of the phenolic compounds and their relationships to bacterial pathogen sensitivity are not yet sufficiently understood (Li and Steffens, 2002).

During initial contact between a plant and a bacterial pathogen, basal defense begins when the plant detects bacterial PAMPs. This perception activates signal-transduction cascades, which are the actual initiators of basal defenses. In turn, basal defense responses activate pathways. These include callose and silicone deposition to reinforce cell walls, and production of reactive oxygen species (ROS) and ethylene. Transcriptional induction of a large suite of defense genes, such as PR genes and post-transcriptional suppression of the auxin-signaling pathway, also has a role in defense (Navarro et al., 2006). These responses are triggered by pattern-recognition receptors (PRRs), which are plant extracellular receptors that specialize in the recognition of PAMPs. These basal defenses are generally sufficient to halt the growth of nonpathogenic microbes and prevent their establishment but are not always effective against pathogenic ones.

10.3 AN OVERVIEW OF INDUCED RESISTANCE AND BACTERIAL PATHOGENS

Systemic acquired resistance (SAR) activates resistance in uninfected plant parts when a necrotizing pathogen stimulates HR at post-local infection sites. The induction of SAR can be accompanied by local and systemic accumulations of endogenous levels of the plant hormone salicylic acid (SA), followed by PR-gene expression (Malamy et al., 1990; Métraux et al., 1990). Studies on this process have been initiated using transgenic NahG plants, which are able to express the bacterial salicylate hydroxylase *nahG* gene but fail to accumulate SA (Gaffney et al., 1994). Other studies have confirmed that NahG plants show no SAR response (Ryals et al., 1996). Exogenous application of SA or BTH (the latter is a functional analogue of SA) leads to the full expression of SAR on the bacterial pathogen *Erwinia amylovora* on apple seedlings (Baysal and Zeller, 2004). Other studies on SA production-deficient mutants sid1 (also called eds5) and sid2 (also called eds16) (Figure 10.1)

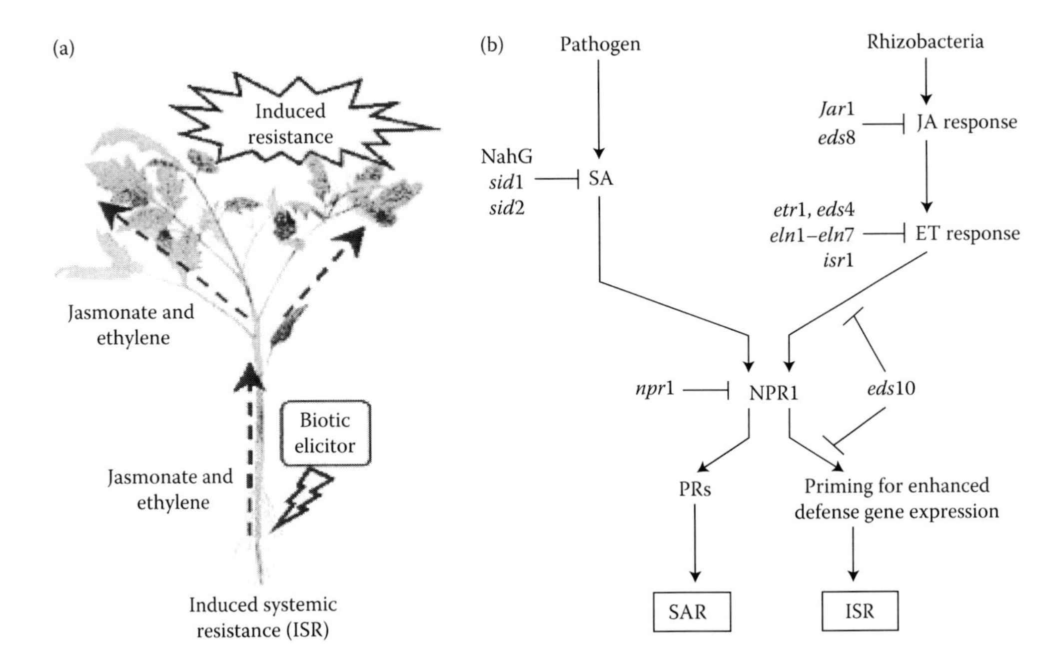

FIGURE 10.1 Induced systemic resistance. (a) Induced by the exposure of roots to specific strains of plant growth-promoting rhizobacteria, dependent upon the phytohormones ethylene and jasmonate (jasmonic acid) and independent of salicylate. (b) Pathogen-induced SAR and rhizobacteria-mediated ISR signal transduction pathways. (Adapted from Vallad, G.E., Goodman, R.M. 2004. *Crop Science* 44(6): 1920–1934.)

showed no SAR response after infection with a necrotizing pathogen (Wildermuth et al., 2001). These results indicate that SA is required for the induction of SAR. It has been shown that SA that originated from infected tobacco leaves accumulated at rates of up to 70% in noninfected leaves (Shulaev et al., 1995).

Ethylene (ET) perception has also been indicated in the generation of systemic signals; however, it is not required for the systemic signal responses that lead to SAR (Verberne et al., 2003). A putative apoplastic lipid transfer protein (DIR1) that interacts with lipid-derived molecules to promote long-distance signaling during SAR was characterized (Maldonado et al., 2002). The regulatory protein NPR1, which is required for transduction of the SA signal into PR-gene expression, has also been reported (Cao et al., 1994). Other studies on mutant npr1 plants showed an independent accumulation of SA after pathogen response, which led to PR-gene expression and SAR. Based on their impaired SAR expression, several mutants were studied before the clarifying *NPR1* gene was discovered (Delaney et al., 1995; Glazebrook et al., 1996).

The most recent findings on SAR systems indicated that, during the induction of SAR, SA triggers a change that leads to reduction of the disulfide bonds in a cellular redox potential. This reduction, in turn, results in the conversion of NPR1 into monomeric form. Monomers are thereafter translocated into the nucleus, where they interact with members of the TGA/OBF subclass of basic leucine-zipper (bZIP) transcription factors. These transcription factors are involved in the SA-dependent activation of PR genes (Zhang et al., 2003). An interaction between NPR1 and specific TGA transcription factors causes binding of the complex within the promoter of the PR genes.

In brief, SAR is a resistance response in the whole plant that occurs after localized exposure to a pathogen. There is evidence that SAR in plants and the innate immunity in these living organisms may be evolutionarily conserved. Plants use PRRs (elicitors) to recognize microbial signatures; in turn, this recognition triggers different cascades that are involved in immune response. In other words, SAR is important for plants to both resist and recover from disease. Generally, SAR is also associated with the induction of the so-called PR genes (Van Loon, 1989) and its activation requires the accumulation of endogenous SA. These cascades activate pathogen-induced SA signals through molecular signal-transduction pathways that are identified by a gene variously called NIM1, NPR1, or SAI1 in the model genetic system formed with *Arabidopsis thaliana*. Recent studies on the interactions between plants and pathogenic microorganisms indicate some uncoupling of disease symptom development from pathogen growth. Symptoms, as active host responses, are associated with disease when a pathogen is present (Baysal and Zeller, 2000).

These responses (i.e., symptoms) are frequently mediated by phytohormones. Possibly, such hormone-mediated responses function to induce systemic defense responses. It is not apparent why extensive tissue death is integral to a defense response if it does not have the effect of limiting pathogen proliferation. Ethylene and SA mediate symptom development but such interactions do not influence bacterial growth. A similar case has been observed between tomato (*Lycopersicon esculentum*) and virulent *Xanthomonas vesicatoria* (Xv) where virulent and avirulent Xv did not induce a systemic response, as evidenced by the expression of defense-associated PR genes in an ethylene- and SA-dependent manner (Ciardi et al., 2001). This systemic response reduced cell death but not bacterial growth during a subsequent challenge with virulent Xcv. These results can be ascribed to systemic acquired tolerance (SAT), which consists of reduced tissue damage in response to a secondary challenge by a virulent pathogen. SAT, which has also been associated with rapid ethylene and PR-gene induction upon challenge, can be induced by infection with *Pseudomonas syringae* pv. *tomato*. As the above results indicate, SAT resembles SAR without its inhibition of pathogen growth.

Multiple phytohormones are essential components of both local and systemic responses of a plant to pathogen invasion (Kumar and Klessig, 2003). Phytohormone synthesis or perception of released hormones on mutant individuals well demonstrate the roles of ethylene, SA, and jasmonates in disease-symptom development. For example, decreased symptom development in response to several virulent bacterial pathogens has been shown using the ethylene-insensitive ein2 mutant of Arabidopsis (*Arabidopsis thaliana*) (Bent et al., 1992). Correspondingly, in response to virulent *X. vesicatoria*, ethylene- and SA-deficient tomato showed decreased symptom development (O'Donnell et al., 2001). This reaction, known as tolerance, indicates phytohormone-mediated defenses; the latter are not associated with reduced symptoms or pathogen grown and therefore can be uncoupled from these activities.

Disease development in a host infected with virulent bacteria involves two phases, one primary and the other secondary.

- Primary disease development is the formation and localization of a lesion that is unaltered in tolerant lines.
- Secondary disease development, which requires the cooperative action of ethylene and SA, is reduced in tolerant lines. It also includes the formation of chlorosis and necrosis that spread from primary lesions (O'Donnell et al., 2003).

If host plants have survived infection by phytopathogenic viruses, bacteria, or fungi, they can be protected from further such attacks. It appears that the first infecting pathogen can immunize the plant with homologous pathogens, if even the plant does not carry a gene that determines cultivar-specific resistance. As stated above, the readiness of the plant to repel subsequent pathogen attacks and pathogen spread throughout the whole plant is called SAR. The development of SAR is often associated with various cellular defense responses, such as synthesis of PR proteins, phytoalexins, accumulation of AOS, rapid alterations in cell walls, and the enhanced activity of various defense-related enzymes (Baysal et al., 2003).

10.3.1 Avirulent/Virulent Bacterial Pathogens and Induction of Resistance

Once a bacterial pathogen is capable of inducing a systemic response in a host, SA and ethylene may participate in systemic signal generation. Inoculation with either virulent or avirulent strains can lead to an SA- and ethylene-dependent induction of defense genes and sensitize the plant to subsequent pathogen challenges. The term used to describe such tolerance to subsequent challenges by virulent strains, SAT, includes the reduction of host-tissue damage that occurs in response to virulent pathogen infection; however, as previously explained, SAT does not impact pathogen growth.

10.3.2 Host–Pathogen Interactions

Resistance in plant species is often divided into host- or nonhost-specific resistance. Host-specific resistance involves interactions between specific host and pathogen genotypes, which give a pathogen race-specific resistance. Non-host resistance, which is shown by a whole plant species against a specific parasite or pathogen, is the most common form of resistance in plants toward the majority of potential pathogens. The biochemical changes that occur during infection are very similar in host and non-host resistant plants. Disease spreads only in susceptible plants (through compatible interactions) that are unable to recognize the pathogen or respond too slowly.

HR is triggered by a plant when it recognizes a pathogen. Pathogen identification typically occurs when avirulence (*Avr*) gene products, secreted by the pathogen, bind or indirectly interact with the product (elicitors or molecules) of a plant resistance (*R*) gene; this interaction is known as the gene-for-gene model. When the *R* gene and corresponding *Avr* genes are both present, recognition occurs that leads to active resistance by the plant and the avirulence of the pathogen. If either the *Avr* gene in the pathogen or the *R* gene in the host is absent or mutated, no recognition occurs; therefore, the outcome will be compatible reaction and disease formation. When the pathogen's *Avr* gene and the plant's *R* gene interact, however, a signal transduction cascade is activated that leads to the activation of a variety of defense responses which are associated with the restriction of pathogen growth. Because resistance (*R*) gene products are highly polymorphic, many plants produce several different types of *R* gene products that can act as a receptor of the Avr proteins produced by many different pathogens.

10.3.3 Hypersensitive Response

Direct physiological contact between host and infecting parasite is necessary to initiate HR, which was first described as cell death in resistant wheat plants upon infection by rust fungi (Stakman, 1915). HR, which is a rapid defense reaction induced in an incompatible host by the effectors of a plant pathogen, causes the death of a limited number of host cells and a concomitant localization of the pathogen. Some investigators have observed reactions similar to HR during apoptosis (i.e., programmed cell death in many animal-cell types), which accounts for the expression of defense genes in addition to cell death. HR in plants, which is analogous to the immune responses of animals, provides resistance to biotrophic pathogens that obtain their energy from living cells (Morel and Dangl, 1997).

Almost 50 years after Stakman first described HR, Klement et al. (1964) discovered the ability of pathogenic bacteria to feed on living cells by inserting an avirulent bacterium (*Ps. syringae* pv. *syringae*) into the intercellular spaces of leaves of a non-host plant (tobacco). They observed rapid, localized, hypersensitive necrosis (i.e., the death of most of the plant cells in the infiltrated leaf tissue). A saprophytic bacterium (*Pseudomonas fluorescens*) did not elicit any HR, whereas a virulent bacterium (*Ps. syringae* pv. *tabaci*) caused a slowly spreading tissue necrosis (Klement et al., 1964).

In the early 1980s, a number of researchers started using transposon-mediated mutagenesis to reveal the bacterial genes that have important roles in various plant bacterial interactions. Clusters of bacterial genes, known as *hrp* (for HR and pathogenicity) genes, were identified in *Ps. syringae* pv. *phaseolicola* causing bacterial spot on bean (Lindgren et al., 1986). Transposon-induced mutations in *hrp* genes were found to abolish the ability of *Ps. syringae* to elicit HR in non-host plants or to cause disease in host plants.

Molecular genetic studies have provided clear explanations of *hrp* genes that were isolated through many plant pathogenic bacteria. These genes were characterized most extensively from *Ps. syringae* pv. *syringae*, *Ps. syringae* pv. *phaseolicola*, *Pseudomonas solanacearum* (which causes wilt in many solanaceous plants), *X. vesicatoria* (which causes bacterial spot on tomato and pepper), and *Erwinia amylovora* (which causes fire blight on rosaceous plants).

10.3.4 Generation of ROS

The first report of the rapid generation of ROS during plant–pathogen interactions was made by Doke (1983) on *Phytophthora infestans*–potato interaction. Studies of bacteria and cell suspensions have also been done on incompatible interactions. ROS such as O_2^-, OH^-, and H_2O_2, which are commonly produced by plants under stress conditions, are known to be strong oxidizing species that rapidly attack all types of biomolecules and cause damage. For protection from this oxidative damage, plant cells contain oxygen radical-detoxifying enzymes such as catalase, peroxidase, and superoxide dismutase, as well as nonenzymatic antioxidants such as ascorbate peroxidase and glutathione-*S*-transferase (Baysal et al., 2007). These enzymes are crucial to the protection of plant cells from oxidative damage. Earlier defense responses include the opening of specific ion channels across the plasma membranes, the rapid production of AOS such as H_2O_2 (i.e., the oxidative burst), and the phosphorylation or dephosphorylation of specific proteins (Baysal et al., 2007). These initial reactions are the prerequisites for initiation of the signaling networks that will trigger the overall defense responses within plant cells.

Reactive oxygen species are toxic intermediates that generate through the sequential one-electron reduction steps of molecular oxygen (Mehdy, 1994). Various enzyme systems have been proposed as the sources of ROS in plants. Because an NADPH oxidase system shows similarity to that of mammalian systems, two sources of oxidative burst can be suggested as a pH-dependent cell-wall peroxidase (Wojtaszek, 1997). NADPH oxidase activity, which is not a ROS-generating system, is usually rapidly dismutated to H_2O_2 via SOD. The first reduction of O_2 forms the superoxide anion and hydroperoxyl radical; the second step forms hydrogen peroxide (H_2O_2); and the third step produces an unstable hydroxyl radical. Uncharged H_2O_2 is more stable because it cannot migrate in solution and instead reacts locally, notably with molecular targets by physiologically modifying their structure and activity. H_2O_2 reacts also with polyunsaturated lipids in membranes to form lipid peroxides that promote biological membrane destruction (Grant and Loake, 2000). The cooperative function of these antioxidants is important in the scavenging of ROS and in maintaining the physiological redox status of organisms.

10.3.5 Signaling Associated with the Establishment of SAR

The onset of SAR in noninfected plant organs is triggered by the phloem mobile signal that is released after pathogen infection. The signal travels throughout the plant and is transduced in target tissues. Following signal transduction, resistance is maintained for several days or even weeks, likely because of *de novo* gene expression (Vernooij et al., 1994). The biochemical changes that occur during SAR can be divided into two phases: initiation and maintenance. Physiological changes during the initiation phase may be transient and short-lived but, depending upon the levels of pathogen invasion and virulence, the maintanance of the resistance status displays changes.

10.3.6 Elicitors and Their Mode of Action

Originally, the term "elicitor" was used for molecules that are capable of inducing phytoalexin production. Now, it is commonly used for compounds that stimulate any type of plant defense (Ebel and Cosio, 1994). The induction of defense responses may eventually lead to enhanced resistance. This broader definition of elicitors includes substances of pathogen origin (exogenous elicitors) and compounds released from plants by the action of the pathogen itself (endogenous elicitors) (Boller, 1995). Elicitors are classified as physical or chemical, biotic or abiotic, and complex or defined according to their origin and molecular structure. Elicitors may be divided into two groups, general and race-specific. General elicitors are able to trigger defenses both in host and non-host plants, whereas race-specific elicitors induce defense responses that lead to disease resistance only in specific host cultivars. A complementary pair of genes in a particular pathogen race and a host cultivar determines this cultivar-specific (gene-for-gene) resistance. Thus, a race-specific elicitor that is encoded or produced by the action of an avirulence gene in a particular race of a pathogen will elicit resistance in a host that carries the corresponding resistance gene, and the absence of either gene product will often result in disease (Cohn et al., 2001). In contrast, general elicitors perceive signals of the presence of potential pathogens in both host and non-host plants (Shibuya and Minami, 2001).

Studies have indicated remarkable similarities between the defense mechanisms triggered by general elicitors and the innate immunities of animals, because similar recognition mechanisms in the latter are related to the general elicitors that lead to plant innate immunity (Nürnberger and Brunner, 2002). An elicitor signal-transduction mechanism activates plant primary immune response (Figure 10.1). Ethylene production and SA involve in plant defense response to pathogens after their actions. These responses have failed to limit pathogen growth. In some cases even Ethylene and SA-mediated responses are observed and defense mechanisms are activated, some result remains unclear because pathogen growth is not inhibited by the extensive necrosis associated with secondary-disease development. One possible function of extensive tissue death is the induction of systemic responses.

Similarly, based on results from other organisms, the function of ethylene or SA may be related to systemic accrued resistance (SAR), which involves the sensitization of systemic defense responses that are initiated by infection with certain pathogens and leads to resistance to subsequent pathogen infections (Ryals et al., 1996). Many factors influence the induction of SAR, including host-cell death that is associated with an incompatible or compatible interaction (Hunt and Ryals, 1996). SAR results in the development of broad-spectrum, systemic resistance, but is not effective against or induced by all pathogens. For example, infection of *Arabidopsis* with *Botrytis cinerea* failed to induce SAR and inoculation with *Ps. syringae* did not affect a subsequent *B. cinerea* challenge (Govrin and Levine, 2002). Interestingly, although SAR can be induced in tomato by pathogens such as tobacco necrosis virus and *Ph. infestans*, there is no appearance of systemic SA accumulation either upon inoculation or subsequent challenge (Jeun et al., 2000). In *Arabidopsis*, the evidence supports a model in which ethylene and jasmonates coordinately mediate one defense response while SA mediates a distinct and antagonistic response (McDowell and Dangl, 2000). Interactions within jasmonates and ethylene by SA on the tomato variety that is susceptible to Xv can be classified, to some extent, as a species-specific function (O'Donnell et al., 2003). During the onset of SAR, a locally altered transcriptional response precedes the HR and a second wave of transcriptional reprogramming, which is not apparent in a virulent attack, marks the transition from basal to induced resistance (de Torres et al., 2006).

Plant immune responses refer to a plant's ability to ward off subsequent pathogen attacks. Activation of proteins that encode ETI induces SAR in leaves distal to the inoculation site, which in turn triggers broad-spectrum resistance to biotrophic pathogens. Systemic-accrued resistance relies on local accumulations of the plant hormone SA, which regulates small diffusible signals that originate from the point of infection. Intensive investigation on detecting pathogen-inducible mobile signals has identified several signals. Methyl salicylate, one of the mobile signal molecules, is essential in triggering SAR (Spoel and Dong, 2008). Several mobile signals have also been found in *Arabidopsis*. The novel nonprotein amino acid pipecolic acid has been shown to have a role in defense amplification and priming during SAR (Navarova et al., 2012).

Lipid-based signals derived from the apoplastic lipid-transfer protein are defective in induced resistance 1 (DIR1); the most recently reported SAR inducers are azelaic acid and glycerol-3-phosphate (Yun et al., 2012). DIR1 is required only for systemic resistance, in which it cooperatively functions with other mobile signals. For example, DIR1 activity is required for resistance induced by exogenously applied azelaic acid, and glycerol-3-phosphate requires DIR1 for mobile signal transduction (Chanda et al., 2011). One explanation for these findings is that different hormone- and lipid-based mobile SAR signals may function cooperatively, which would allow more-specialized responses.

The presence of SAR signals in distal tissues triggers the accumulation of SA, which results in defense-gene expression. SA-induced transcriptional reprogramming requires the transcription cofactor nonexpressor of PR gene 1 (NPR1) in distally activated tissues in order to regulate the differential expression of more than 2000 genes, including PR genes. A model of SA-induced transcriptional reprogramming suggests that NPR1 is a redox-sensitive protein that exists as an oligomer in the absence of SAR. In a reducing environment, the disulfide bonds in NPR1 are indeed reduced, which enables NPR1 monomers to enter the nucleus. The nonexpressor of PR gene 1 is also post-transcriptionally regulated by the proteasome. Upon immune activation, the phosphodegredon motif of NPR1 is phosphorylated, which results in ubiquitination through association with Cullin 3, a scaffold for the E3 ligase complex. Ubiquitinated NPR1 is subsequently degraded by the proteasome (Spoel et al., 2009). Based on these results, the current model suggests the interaction of NPR1 with promoters of target genes that are transcripting factors in the recruitment of transcriptional machinery. The NPR1 phosphorylation could occur after transcriptional initiation, which would result in degradation and enable fresh, active NPR1 to recruit another round of gene expression (Figure 10.1).

The link between NPR1 activation and SA accumulation remains unclear, although progress has been made in understanding NPR1-mediated transcriptional reprogramming via the identification of multiple SA receptors (Fu et al., 2012; Wu et al., 2012). Recombinant NPR3 and NPR4, paralogs of NPR1, strongly bind SA (Fu et al., 2012). Genetic and biochemical analyses provide evidence that NPR3 and NPR4 act as adapter proteins for Cullin 3 to mediate the degradation of NPR1.

Interestingly, in contrast to the nprl mutant's compromise in basal resistance, npr3/npr4 mutant has been shown exhibit enhanced disease resistance to virulent bacteria. However, the npr3/npr4 mutant was not able to elicit SAR in response to inoculation with the avirulent pathogen *Ps. syringae* pv. *maculicola* expressing AvrRpt2, and was also partly compromised in ETI (Fu et al., 2012). This finding suggests that while NPR3 and NPR4 are required for SA-defense pathways, other NPR1-mediated defense responses are functional in the npr3/npr4 mutant. Evidence that supports NPR1 as the SA receptor has also been published (Wu et al., 2012). Copper was required for NPR1 binding to SA (Wu et al., 2012). Side-by-side comparisons of binding affinities for all NPR proteins in the presence of a transition metal shed light upon their relative affinities for SA. Future research on how NPR1 and its paralogs dynamically associate in response to SA perception in planta will be important for understanding the mechanism of SAR induction. Plants also carry immune receptors that recognize highly variable pathogen effectors. These include the NBS-LRR class of proteins, which are mostly molecular-assisted selection markers and are used in breeding programs to cultivate resistant plant varieties.

10.4 INDUCED SYSTEMIC RESISTANCE

This phenomenon is caused by local responses in which a signal spreads through the plant and induces subtle changes in gene expression in its distal uninfected parts. The systemic response involves *de novo* production, phytoalexins, and PR proteins (Van Loon, 1997; Neuhaus, 1999; Van Loon and van Strien, 1999). Whereas phytoalexins are mainly characteristic of the local response, PR proteins occur both locally and systemically (Li et al., 2012).

Originally, PR proteins were thought to be initially absent in healthy plants but to accumulate in large amounts after infection; they have been found in more than 40 species belonging to at least 13 families (Van Loon, 1999). Two groups of PR proteins can be distinguished. The first, acidic PR proteins, are predominantly located in intercellular spaces. The second, basic PR proteins, are functionally similar to acidic proteins but have different molecular weights and amino acid sequences; they are mainly located intracellularly, in the vacuole (Legrand et al., 1987; Niki et al., 1998; Van Loon, 1999).

Some PR proteins exhibit chitinase (Legrand et al., 1987) or β-1,3-glucanase activity. Chitinases are a functionally and structurally diverse group of enzymes that can hydrolyse chitin; several are believed to contribute to plant defenses against certain fungal pathogens (Jackson and Taylor, 1996). Chitinases exhibit pronounced antifungal activity (Schlumbaum et al., 1986; Baysal et al., 2003) and plants that overexpress chitinase show decreased susceptibility to infection by fungi as indicated by chitin-containing cell walls (Broglie et al., 1991; Datta and Datta, 1999). The function of other PR proteins is still unknown (Van Loon and van Strien, 1999). In addition, many of them may be functionally active only when combined. Studies have demonstrated that the expression of typical defense-related genes such as PR-1 and β-glucanase 2 (which are often used as ISR markers) (Baysal et al., 2003) can be uncoupled from phenotypic pathogen resistance (Greenberg et al., 2000). These results indicate that these compounds are not absolutely necessary for an effective resistance phenotype.

Pathogenesis-related proteins, which are generally used as ISR markers, have no antiviral or antibacterial activity. Similarly, phytoalexins generally exhibit only *in vitro* antibacterial or antifungal effects; assumptions about their role in phenotypic plant resistance are mainly based on correlative evidence. Phenotypically, systemic resistance is manifested as a protection of the plant not only against the attacking pathogen, but also against other types of pathogens. Although some specificity has been described, this resistance seems to be nonspecific and long-lasting. Most of the research on this topic has been conducted on a restricted number of model species. Differences in biochemistry and efficacy, which have been found among various resistance forms, have yet to be investigated in detail. ISR is also generally considered to be a widespread and conserved trait that is effective against all three major types of pathogens (viruses, bacteria, and fungi).

10.4.1 Priming

Some of the compounds normally associated with ISR (i.e., PR proteins and phenolic compounds) are expressed in uninfected tissue in response to the initial stage of infection. Biochemical changes are characteristic properties of ISR-expressing plants that become obvious only in response to further pathogen invasion, and only in plant parts where effective resistance is required. This phenomenon is referred to as priming, conditioning, or sensitization (Sticher et al., 1997; Conrath et al., 2001). Priming effects can be elicited by chemical ISR inducers such as β-aminobutyric acid (Jakab et al., 2001; Baysal et al., 2007). After these effects have been initiated, responses such as phytoalexin synthesis or cell-wall lignification become more rapid and more strong than during the primary infection and thereby enable a more effective response to the new infection. The molecular mechanisms that underlie priming and its importance in the overall plant resistance have yet to be investigated.

10.4.2 Local and Systemic Signaling

Salicylic acid (Raskin, 1992) is an important component in the signaling pathway that leads to ISR (Mauch-Mani and Métraux, 1998; Cameron, 2000; Métraux, 2001). After infection, endogenous levels of SA increase locally and systemically; in addition, SA levels increase in the phloem before ISR occurs (Malamy et al., 1990; Métraux et al., 1990; Rasmussen et al., 1991). Because SA is synthesized in response to infection both locally and systemically, *de novo* production of SA in noninfected plant parts might contribute to the systemic expression of ISR (Meuwly et al., 1995). The level of resistance of plants that exhibit constitutive expression of SA is positively correlated with their SA levels. This is true for natural cultivars of rice (Silverman et al., 1995), for within-plant differences in SA levels in potato (Coquoz et al., 1995), and for arabidopsis plants that express a novel hybrid enzyme with salicylate synthase activity and thus have elevated SA levels (Mauch et al., 2001). Key experiments that have established a role for SA in certain forms of ISR utilized transgenic plants that expressed the bacterial *nahG* gene encoding for naphthalene hydroxylase G. Such plants cannot accumulate SA and their ISR response is blocked (Delaney et al., 1994; Gaffney et al., 1994).

Experiments that used reciprocal combinations of nahG and wild-type shoots grafted onto nahG and wild-type plants showed that ISR was elicited in the wild-type tissue even when the nahG-transformed part of the plant received the inducing infection; these results suggest that the signal emanating from the inducing tissue is not SA (Vernooij et al., 1994). Research has shown that nahG plants might suffer from further, as yet unknown, defects (Cameron, 2000). Rasmussen et al. (1991) reported that timelines of the induction and appearance of SA in phloem, when combined with leaf-removal experiments, were not consistent with SA as the primary systemic signal in the investigated system (cucumber). These and other experiments suggest that SA and other systemic signals are involved in ISR signaling (Sticher et al., 1997).

10.4.3 Elicitation

Plants can activate separate defense pathways according to the type of pathogen they encounter. Jasmonic acid (JA)- and ethylene-dependent responses seem to be initiated by necrotrophs, whereas an SA-dependent response is activated by biotrophic pathogens (Garcia et al., 2006). The mechanisms responsible for this differential recognition and response may involve crosstalk among these three different signal transduction pathways: JA, ethylene, and SA.

Salicylic acid is synthesized in response to mechanical damage, necrosis, and oxidative stress. Compounds that result from the degradation of cells or cell walls might be involved in eliciting the systemic signal, which means that ISR can be induced by different types of enemies. Correspondingly, JA can be induced in response to cell-wall degradation. Any factor leading to necrosis or activating some of these factors might thus elicit both IRH and ISR pathways (Figure 10.1).

Therefore, events at the elicitation level will mainly lead to the expression of a rather nonspecific cross-resistance.

Plants are challenged by a variety of biotic stresses such as fungal, bacterial, or viral infections, which lead to great losses in yield. Some resistance methods are the development of resistant cultivars and the uses of biological control, crop rotation, tillage, and chemical pesticides. Nearly all chemical pesticides or fungicides have a direct antibiotic principle and are carcinogenic. Therefore, considerable efforts have been made to devise environmentally friendly strategies for controlling plant disease.

Better understanding of plant signaling pathways has led to the discoveries of natural and synthetic compounds, called elicitors, that induce similar defense responses in plants; these are, in turn, induced by pathogen infection (Gómez et al., 2004). Different types of elicitors that have been characterized include carbohydrate polymers, lipids, glycopeptides, and glycoproteins. In plants, a complex array of defense responses is induced after microorganism detection via recognition of the elicitor molecules that are released during plant–pathogen interaction. Following elicitor perception, the activation of signal transduction pathways generally leads to the production of active oxygen species (AOS). In such cases, phytoalexin biosynthesis, reinforcement of plant cell walls associated with phenyl propanoid compounds, deposition of callose, synthesis of defense enzymes, and the accumulation of PR proteins may have antimicrobial properties (Van Loon and van Strien, 1999).

Active oxygen species lead to HR in plants. As previously discussed, HR is the localized or rapid death of one or more cells at the infection site to delimit pathogen growth. Following HR activation, uninfected distal parts of the plant may develop resistance to further infection by a phenomenon known as SAR, which is effective against diverse pathogens (viruses, bacteria, and fungi) (Heil and Bostock, 2002).

10.4.4 Signaling

Signaling is another step in the interaction phase. The different activities of various intermediates that have been reported for the octadecanoid cascade can lead a wide diversity of outcomes (Koch et al., 1999). Similar regulatory properties might characterize SA-dependent signaling. An inhibition of the JA pathway by SA has been described in different plant species. Induction by pathogens (although this probably elicits the early steps of the octadenanoid pathway) leads to the synthesis of high concentrations of SA and thus blocks later steps in octadecanoid signaling. Phenotypically, pathogen attack induces mainly (or only) ISR compounds (Figure 10.1). At the signaling level, apparent evolutions include the elicitation of ISR-typical compounds, formation of pathogen resistance by IR as well as by ISR elicitors, and inhibition of the JA pathway by compounds in the ISR pathway.

Production of defensive compounds can be limited by the supply of available precursors such as amino acids, ATP, and other biosynthetic cofactors, which do not depend only upon the outcomes of events at the signaling level. Niki et al. (1998) reported accumulated mRNA levels for the SA-responsive acidic types and the JA-responsive basic types of PR-1 genes in the presence of various JA and SA concentrations. Inductions of both SA-responsive and JA-responsive genes appeared to be triggering factors in the total amounts of defensive compounds (Heil, 2001).

Oligosaccharides (Bishop et al., 1981) and oligogalacturonides (Norman et al., 1999) released from damaged cell walls might have roles in the elicitation of general wound response; in addition, specific elicitors such as systemin have been reported (Pearce et al., 1991). Systemin is an 18-amino-acid polypeptide that is released upon wounding from a 200-amino-acid precursor (prosystemin). This release leads to the release of linolenic acid, which activates the octadecanoid signaling cascade (Ryan, 2000). Both JA (Zhang and Baldwin, 1997) and systemin (Ryan, 2000) can be transported in phloem and thus might act as systemic signals.

To date, systemin has been described for tomato only, and not even for other solanaceous plants such as tobacco (Ryan, 2000; León et al., 2001). The importance of cell-wall fragments in elicitation is supported by the finding that cellulysin can induce several JA-responsive volatiles in lima bean

(*Phaseolus lunatus*) (Piel et al., 1997). The action of cellulysin is followed by a rapid increase in endogenous JA (Koch et al., 1999). Studies on systemin are continuing.

10.4.5 Cross-Talk

Many studies have assumed the existence of at least two main signaling pathways: SA-dependent ISR, which is involved in resistance caused by and effective against pathogens, and JA-dependent IRH, which is effective against herbivores. The term "cross-talk" refers to resistance elicited by one group of enemies against another (e.g., resistance against pathogens that is induced by herbivores and resistance against herbivores that is induced by pathogens).

Cross-resistance has been found in different systems. A bacterial phytopathogen, *Ps. syringae* pv. *tomato* (Stout et al., 1998; Bostock et al., 2001), can induce PIs in tomato leaves; this reaction is characteristic of wound-response and induced herbivore resistance rather than ISR. Noctuid larvae feeding on leaves of *Pseudomonas*-induced plants performed significantly less well than on control leaves (Bostock et al., 2001).

10.5 COMMERCIALIZATION OF ISR AND SAR INDUCERS

Alternatives to chemicals that have fungitoxic properties in plant protection have arisen with the discovery of disease-resistance inducers of biotic and abiotic origins. The latter induce localized or systemic resistance in susceptible plants, which then become resistant to subsequent infections. Depending on their efficacy, these compounds can be used in fields either alone or in combination with fungicides. In some countries, many compounds (e.g., Bion, Actigard) have been commercially released as plant health promoters of annual crops (Oostedorp et al., 2001).

The SA-dependent defense pathway can be activated by the treatment of plants with chemical inducers such as benzo (1,2,3)-thiadiazole-7-carbothioic acid-S-methyl ester (acibenzolar-*S*-methyl, ASM or BTH, Bion). These have been developed as potent SAR activators which do not possess antimicrobial properties but instead increase crop resistance to diseases by activating SAR signal-transduction pathways in several species. Benzothiadiazole (BTH), a chemical analogue of SA, has been used successfully to induce resistance to a wide range of diseases in field crops. The nonprotein amino acid β-aminobutyric acid (BABA) protects numerous plants against various pathogens (Baysal et al., 2005, 2007). Several products have also been used as inducers of resistance in plants against pathogens; these include chitosan, SA analogues, living or processed fungal products, and seaweed extracts (Washington et al., 1999). Certain synthetic compounds with no direct antimicrobial effect, such as 2,6-dichloroisonicotinic acid (INA) and potassium salts, have been reported to induce SAR in plants (Oostendorp et al., 2001). This phenomenon was observed in the context of disease resistance.

Challenge inoculation after exposure to induced and noninduced plants revealed that the air coming from induced plants mainly primed resistance; specifically, the expression of PR protein 2 (PR-2) was significantly stronger in exposed than in nonexposed individuals in response to subsequent challenges with *Ps. syringae*. Among others, the plant-derived volatile nonanal was present in the headspace of BTH-treated plants and significantly enhanced PR-2 expression in the exposed plants, which resulted in reduced appearance of symptoms. Negative effects on the growth of BTH-treated plants, which usually occur as a consequence of the high costs of direct resistance induction, were not observed in plants that were exposed to volatile organic compounds. Volatile-mediated priming appears to be a highly attractive way to tailor SAR against plant pathogens. Three plant hormones that are central to the long-distance signaling that underlies this systemic response to local attack are JA, ethylene, and SA. In particular, SA and JA are transported, as themselves or in the form of derivatives within the plant, to elicit systemic responses (Heil and Ton, 2008).

Recent studies have revealed that long-distance signaling is not only caused by molecules that are transported in the vascular system; signals can also be volatile compounds that move in the headspace outside the plant (Heil and Ton, 2008). In particular, green-leaf volatiles and other herbivore-induced volatile organic compounds (VOCs) can mediate the systemic responses of plants to local herbivore damage (Karban et al., 2006; Frost et al., 2007; Heil and Silva Bueno, 2007). Because such VOCs move freely in the air, they may also affect neighboring plants and thus mediate the phenomenon of plant–plant communication. Such communication via VOCs thus appears to be a common phenomenon in herbivore resistance. In addition, similar volatile compounds can mediate the beneficial effects of plant growth-promoting rhizobacteria (Ryu et al., 2003, 2004).

In contrast, the SAR to biotrophic pathogens in many plant species is mediated by SA signaling, which increases the expression of phytoalexins and several PR proteins (Van Loon, 1997; Durrant and Dong, 2004). SA signaling is usually thought to act as an antagonist to JA signaling (Korneef and Pieterse, 2008). The volatile derivative of SA, methyl salicylate (MeSA), has been proposed as the most likely systemic signal (Park et al., 2007). In tobacco (*Nicotiana tabacum*), MeSA is converted back to SA, which then forms the active resistance-inducing compound (Forouhar et al., 2005). This mechanism might underlie the resistance induction in tobacco plants that were exposed to high MeSA concentrations (Shulaev et al., 1997). In a study on the role of MeSA as a mobile signal, Park et al. (2007) found evidence for the vascular transport of this compound.

A common phenomenon involved in disease resistance is priming, which prepares the plant to respond more rapidly and/or effectively to subsequent attack (Bruce et al., 2007; Goellner and Conrath, 2008) but incurs much lower costs than direct resistance induction (Walters and Heil, 2007). Plants were exposed to the VOCs emitted from neighbors that had been treated with the chemical SAR elicitor benzothiadiazole [BTH; benzo(1,2,3)thiadiazole-7-carbothioic acid *S*-methyl ester] or that had been induced biologically, and resulting changes in resistance were monitored at the phenotypic and gene-expression levels. In this way, we investigated whether VOCs also can prime resistance to pathogens by first exposing plants to VOCs coming from directly induced plants and then challenging them with *Ps. syringae* pv. *syringae*. Finally, VOCs released from the induced plants were analyzed, and the most likely candidates were evaluated for their effects on the expression of the resistance marker gene PR-2 in order to understand the chemical nature of the signal.

10.6 CONCLUSION

The whole genome sequences of several plant bacterial pathogens are now available. In addition, advanced analytical tools and techniques such as transcriptomics, proteomics, and metabolomics will continue to provide new insights into the impacts of these pathogens. Such tools are beneficial to our understanding of the mechanisms of the microbe-genes that are related to pathogenicity. Moreover, they can be efficiently used to reveal novel virulence factors through targeted investigations of genes that are unique to specific pathogens. These analyses will be a powerful tool for the identification of genes that are involved in host specificity and virulence. Identification of these targets is critical to our understanding of how bacterial pathogens cause disease in plants. In addition, to date we have no detailed information on the R proteins of plants. Finally, the biggest challenge for the future will be applying our increasing knowledge of plant immune systems to the development of disease-resistant crops that exhibit durable resistance across time and space. Further studies on host–microbe interaction may facilitate our understanding of the mystery of pathogenesis.

REFERENCES

Alfano, J.R., Collmer, A. 2004. Type III secretion system effector proteins: Double agents in bacterial disease and plant defense. *Annual Review of Phytopathology* 42: 385–414.

Baysal, Ö., Gürsoy, Y.Z., Duru, A., Örnek, H. 2005. Induction of oxidants in tomato leaves treated with DL-β-amino butyric acid (BABA) and infected with *Clavibacter michiganensis* ssp. michiganensis European. *Journal of Plant Pathology* 4: 361–369.

Baysal, Ö., Gürsoy, Y.Z., Örnek, H., Çetinel, B., Teixeira da Silva, J.A. 2007. Enhanced systemic resistance to bacterial speck disease caused by *Pseudomonas syringae* pv. *tomato* by DL-β-aminobutyric acid under salt stress. *Physiologia Plantarum* 129: 493–506.

Baysal, Ö., Soylu, S., Soylu, E.M. 2003. Induction of defence-related enzymes and resistance by the plant activator acibenzolar-*S*-methyl in tomato seedlings against bacterial canker caused by *Clavibacter michiganensis* ssp. *Michiganensis Plant Pathology* 52: 747–753.

Baysal, Ö., Zeller, W. 2000. Studies on control of fire blight (*Erwinia amylovora*) with the plant activator BION®. In *Proceeding of 10th International Conference Plant Pathogenic Bacteria in Canada*. Kluwer Academic Publishers, pp. 324–331.

Baysal, Ö., Zeller, W. 2004. Extract of Hedera helix induces resistance on apple rootstock M26 similar to acibenzolar-*S*-methyl against fire blight (*Erwinia amylovora*). *Physiological and Molecular Plant Pathology* (65)6: 305–315.

Bent A.F., Innes R.W., Ecker J.R., Staskawicz B.J. 1992. Disease development in ethylene-insensitive Arabidopsis thaliana infected with virulent and avirulent Pseudomonas and Xanthomonas pathogens. *Mol. Plant-Microbe Interact.* 5: 372–378.

Bishop, P.D., Makus, D.J., Pearce, G., Ryan, C.A. 1981. Proteinase inhibitor-inducing factor activity in tomato leaves resides in oligosaccharides enzymatically released from cell walls. *Proceedings of the National Academy of Sciences of the United States of America* 78: 3536–3540.

Boller, T. 1995. Chemoperception of microbial signals in plant cells. *Annual Review of Plant Physiology and Plant Molecular Biology* 46: 189–214.

Bostock, R.M., Karban, R., Thaler, J.S., Weyman, P.D., Gilchrist, D. 2001. Signal interactions in induced resistance to pathogens and insect herbivores. *European Journal of Plant Pathology* 107: 103–111.

Bruce, T.J.A., Matthes, M.C., Napier, J.A., Pickett, J.A. 2007. Stressful "memories" of plants: Evidence and possible mechanisms. *Plant Science* 173: 603–608.

Broglie, K., Chet, I., Holliday, M., Cressman, R., Riddle, P., Knowlton, S., Mauvais, C.J., Broglie, R. 1991. Transgenic plants with enhanced resistance to the fungal pathogen *Rhizoctonia solani*. *Science* 254: 1194–1197.

Cameron, R.K. 2000. Salicylic acid and its role in plant defense responses: What do we really know? *Physiological and Molecular Plant Pathology* 56: 91–93.

Cao, H., Bowling, S.A., Gordon, A.S., Dong, X. 1994. Characterization of an Arabidopsis mutant that is nonresponsive to inducers of systemic acquired resistance. *The Plant Cell Online*, 6(11): 1583–1592.

Chanda, B., Xia, Y., Mandal, M.K., Yu, K., Sekine, K.T., Gao, Q.-M., Selote D. et al. 2011. Glycerol-3-phosphate is a critical mobile inducer of systemic immunity in plants. *Nature genetics*, 43(5): 421–427.

Ciardi, J.A., Tieman, D,M,, Jones, J.B., Klee, H.J. 2001. Reduced expression of the tomato ethylene receptor gene LeETR4 enhances the hypersensitive response to *Xanthomonas campestris* pv. *vesicatoria*. *Molecular Plant-Microbe Interaction* 14: 487–495.

Cohn, J., Sessa, G., Martin, G.B. 2001. Innate immunity in plants. *Current Opinion in Immunology* 13(1): 55–62.

Conrath, U., Thulke, O., Katz, V., Schwindling, S., Kohler, A. 2001. Priming as mechanism in induced systemic resistance of plants. *European Journal of Plant Pathology* 107: 113–119.

Coquoz, J.L., Buchala, A.J., Meuwly, P., Métraux, J.P. 1995. Arachidonic acid treatment of potato plants induces local synthesis of salicylic acid and confers systemic resistance to Phytophthora infestans and Alternaria soltani. *Phythopathology* 85: 1219–1224.

Datta, K.N., Datta, S.K.1999. Expression and function of PR-protein genes in transgenic plants. In *Pathogenesis-Related Proteins in Plants*. Datta, S., Muthukrishnan, S., (eds.). Boca Raton: CRC Press, pp. 261–277.

Delaney, T.P., Friedrich, L., Ryals, J.A. 1995. Arabidopsis signal transduction mutant defective in chemically and biologically induced disease resistance. *Proceedings of the National Academy of Sciences,* 92(14): 6602–6606.

Delaney, T.P., Uknes, S., Vernooij, B., Friedrich, L., Weymann, K., Negrotto, D., Gaffney, T. et al. 1994. A central role of salicylic acid in plant disease resistance. *Science*. 266: 1247–1249.

Doke, N. 1983. Involvement of superoxide anion generation in the hypersensitive response of potato tuber tissues to infection with an incompatible race of *Phytophthora infestans* and to the hyphal wall components. *Physiological Plant Pathology* 23(3): 345–357.

Durrant, W.E., Dong, X. 2004. Systemic acquired resistance. *Annual Review of Phytopathology* 42: 185–209.

Ebel, J., Cosio, E.G. 1994. Elicitors of plant defense responses. *International Review of Cytology* 148: 1–36.

Farag, M.A., Fokar, M., Zhang, H.A., Allen, R.D., Paré, P.W. 2005. (Z)-3-Hexenol induces defense genes and downstream metabolites in maize. *Planta* 220: 900–909.

Forouhar, F., Yang, Y., Kumar, D., Chen, Y., Fridman, E., Park, S.W., Chiang, Y., Acton, T.B., Montelione, G.T., Pichersky, E. 2005. Structural and biochemical studies identify tobacco SABP2 as a methyl salicylate

esterase and implicate it in plant innate immunity. *Proceedings of National Academy of Sciences of the United States of America* 102: 1773–1778.
Frost, C., Appel, H., Carlson, J., De Moraes, C., Mescher, M., Schultz, J. 2007. Within-plant signalling by volatiles overcomes vascular constraints on systemic signalling and primes responses against herbivores. *Ecological Letters* 10: 490–498.
Fu, Z.Q., Yan, S., Saleh, A., Wang, W., Ruble, J., Oka, N., Dong, X. 2012. NPR3 and NPR4 are receptors for the immune signal salicylic acid in plants. *Nature*, 486(7402): 228–232.
Gaffney, T., Friedrich, L., Vernooij, B., Negrotto, D., Nye, G., Uknes, S., Ward, E., Kessmann, H., Ryals, J. 1994. Requirement of salicylic acid for the induction of systemic acquired resistance. *Science* 261: 754–756.
Garcia, A.,Brugger, Lamotte, O., Vandelle, E. 2006. Early signaling events induced by elicitors of plant defenses. *Molecular Plant-Microbe Interactions* 19(7): 711–724.
Glazebrook, J., Rogers, E.E., Ausubel, F.M. 1996. Isolation of Arabidopsis mutants with enhanced disease susceptibility by direct screening. *Genetics*, 143(2): 973–982.
Goellner, K., Conrath, U. 2008. Priming: it's all the world to induced disease resistance. *European Journal of Plant Pathology* 121: 233–242.
Gómez V., Day, R., Buschmann, H., Randles, S., Beeching, J.R., Coop, R.M. 2004. Phenylpropanoids, phenylalanine ammonia lyase and peroxidases in elicitor-challenged cassava (*Manihot esculenta*) suspension cells and leaves. *Annals of Botany* 94(1): 87–97.
Govrin, E.M., Levine, A. 2002. Infection of Arabidopsis with a necrotrophic pathogen, Botrytis cinerea, elicits various defense responses but does not induce systemic acquired resistance (SAR). *Plant molecular biology* 48(3): 267–276.
Grant, J.J., Loake, G.J. 2000. Role of reactive oxygen intermediates and cognate redox signaling in disease resistance. *Plant Physiology* 124(1): 21–30.
Greenberg, J.T., Silverman, F.P., Liang, H. 2000. Uncoupling salicylic acid-dependent cell death and defense-related responses from disease resistance in the *Arabidopsis mutant* acd5. *Genetics* 156: 341–350.
Hahn, M.G. 1996. Microbial elicitors and their receptors in plants. *Annual Review of Phytopathology* 34: 387–412.
Hammerschmidt, R. 1999a. Induced disease resistance: how do induced plants stop pathogens? *Physiological and Molecular Plant Pathology* 55: 77–84.
Hammerschmidt, R. 1999b. Phytoalexins: What have we learned after 60 years? *Annual Review of Phytopathology* 37: 285–306.
Hammerschmidt, R., Smith-Becker, J.A. 1999. The role of salicylic acid in disease resistance. In *Induced Plant Defenses against Pathogens and Herbivores: Biochemistry, Ecology, and Agriculture*. Agrawal, A.A., Tuzun, S., Bent, E. (eds.). St. Paul.: American Phytopathological Society Press, pp. 37–53.
Heil, M., 2001. The ecological concept of costs of induced systemic resistance (ISR). *European Journal of Plant Pathology* 107: 137–146.
Heil, M., Bostock, R.M. 2002. Induced systemic resistance (ISR) against pathogens in the context of induced plant defences. *Annals of Botany* 89(5): 503–512.
Heil, M., Ton, J. 2008. Long-distance signalling in plant defence. *Trends in Plant Science* 13: 264–272.
Hunt, M.D., Ryals, J.A. 1996. Systemic acquired resistance signal transduction. *Critical Reviews in Plant Sciences* 15: 583–606.
Jackson, A.O., Taylor, C.B. 1996. Plant–microbe interactions: Life and death at the interface. *The Plant Cell* 8: 1651–1668.
Jakab, G., Cottier, V., Toquin, V., Rigoli, G., Zimmerli, L., Métraux, J.-P, Mauch-Mani, B. 2001. β-Aminobutyric acid-induced resistance in plants. *European Journal of Plant Pathology* 107: 29–37.
Jeun Y.C., Siegrist J., Buchenauer H. 2000. Biochemical and cytological studies on mechanisms of systemic induced resistance in tomato plant against Phytophthora infestans. *Journal of Phytopathology.,* 148: 129–140.
Jones, J.D., Dangl, J.L. 2006. The plant immune system. *Nature* 444: 323–329.
Karban, R., Shiojiri, K., Huntzinger, M., McCall, A.C. 2006. Damage-induced resistance in sagebrush: Volatiles are key to intra-and interplant communication. *Ecology* 87(4): 922–930.
Klement, Z., Farkas, G.L., Lovrekovich, L. 1964. Hypersensitive reaction induced by phytopathogenic bacteria in the tobacco leaf. *Phytopathology* 54: 474–477.
Koch, T., Krumm, T., Jung, V., Engelberth, J., Boland, W. 1999. Differential induction of plant volatile biosynthesis in the lima bean by early and late intermediates of the octadecanoid-signaling pathway. *Plant Physiology* 121: 153–162.
Korneef, A., Pieterse, C.M.J. 2008. Cross talk in defense signaling. *Plant Physiology* 146: 839–844.

Kuc, J. 1995. Phytoalexins, stress metabolism, and disease resistance in plants. *Annual Review of Phytopathology* 33: 275–297.

Kumar, D., Klessig, D.F. 2003. High-affinity salicylic acid-binding protein 2 is required for plant innate immunity and has salicylic acid-stimulated lipase activity. *Proceedings of National Academy of Sciences of the United States of America* 100: 16101–16106.

Lamb, C., Dixon, R.A. 1997. The oxidative burst in plant disease resistance. *Annual Review of Plant Physiology and Plant Molecular Biology* 48: 251–275.

Legrand, M., Kauffmann, S., Pierrette, G., Fritig, B. 1987. Biological function of pathogenesis-related proteins: Four tobacco pathogenesis-related proteins are chitinases. *Proceedings of the National Academy of Sciences of the USA* 84: 6750–6754.

León, J., Rojo, E., Sánchez-Serrano, J.J. 2001. Wound signalling in plants. *Journal of Experimental Botany* 52: 1–9.

Li, L., Steffens, J.C. 2002. Overexpression of polyphenol oxidase in transgenic tomato plants results in enhanced bacterial disease resistance. *Planta* 215: 239–247.

Li, W., Shao, M., Zhong, W., Yang, J., Okada, K. 2012. Ectopic expression of Hrf1 enhances bacterial resistance via regulation of diterpene phytoalexins, silicon and reactive oxygen species burst in rice. *PLoS One* 7(9): e43914.

Lindgren, P.B., Peet, R.C., Panopoulos, N.J. 1986. A gene cluster of Pseudomonas syringae pv. phaseolicola controls pathogenicity on bean and hypersensitivity on non-host plants. *J. Bacteriol.* 168: 512–522.

Malamy, J., Carr, J.P., Klessig, D.F., Raskin, I. 1990. Salicylic acid: A likely endogenous signal in the resistance response of tobacco to viral infection. *Science* 250: 1002–1004.

Maldonado, A.M., Doerner, P., Dixon, R.A., Lamb, C.J., Cameron, R.K. 2002. A putative lipid transfer protein involved in systemic resistance signalling in Arabidopsis. *Nature* 419(6905): 399–403.

Mauch, F., Mauch-Mani, B., Gaille, C., Kull, B., Haas, D., Reimmann, C. 2001. Manipulation of salicylate content in *Arabidopsis thaliana* by the expression of an engineered bacterial salicylate synthase. *The Plant Journal* 25: 67–77.

Mauch-Mani, B., Métraux, J.-P. 1998. Salicylic acid and systemic acquired resistance to pathogen attack. *Annals of Botany* 82: 535–540.

McDowell, J.M., Dangl, J.L. 2000. Signal transduction in the plant immune response. *Trends in biochemical Sciences* 25(2): 79–82.

Mehdy, M.C. 1994. Active oxygen species in plant defense against pathogens. *Plant Physiology* 105(2): 467–472.

Meuwly, P., Mölders, W., Buchala, A., Métraux, J.P. 1995. Local and systemic biosynthesis of salicylic acid in infected cucumber plants. *Plant Physiology* 109: 1107–1114.

Métraux, J.P. 2001. Systemic acquired resistance and salicylic acid: Current state of knowledge. *European Journal of Plant Pathology* 107: 13–18.

Métraux, J.-P., Signer, H., Ryals, J., Ward, E., Wyss-Benz, M., Gaudin, J., Raschdorf, K., Schmid, E., Blum, W., Inverardi, B. 1990. Increase in salicylic acid at the onset of systemic acquired resistance. *Science* 250: 1004–1006.

Morel, J.B., Dangl, J.L. 1997. The hypersensitive response and the induction of cell death in plants. *Cell Death and Differentiation* 4(8): 671–683.

Návarová, H., Bernsdorff, F., Döring, A.C., Zeier, J. 2012. Pipecolic acid, an endogenous mediator of defense amplification and priming, is a critical regulator of inducible plant immunity. *The Plant Cell Online* 24(12): 5123–5141.

Navarro, L., Dunoyer, P., Jay, F. 2006. A plant miRNA contributes to antibacterial resistance by repressing auxin signaling. *Science* 312: 436–439.

Neuhaus, J.M. 1999. Plant chitinases (PR-3, PR-4, PR-8, PR-11). In *Pathogenesis-Related Proteins in Plants.* Datta, S.K., Muthukrishnan, S. (eds.). Boca Raton: CRC Press, pp. 77–105.

Niki, T., Mitsuhara, I., Seo, S., Ohtsubo, N., Ohashi, Y. 1998. Antagonistic effects of salicylic acid and jasmonic acid on the expression of pathogenesis-related (PR) protein genes in wounded mature tobacco leaves. *Plant Cell Physiology* 39: 500–507.

Nomura, K., Melotto, M., He, S.Y. 2005. Suppression of host defense in compatible plant *Pseudomonas syringae* interactions. *Current Opinion in Plant Biology* 8: 361–368.

Norman, C., Vidal, S., Palva, E.T. 1999. Oligogalacturonide-mediated induction of a gene involved in jasmonic acid synthesis in response to the cell-wall-degrading enzymes of the plant pathogen *Erwinia carotovora. Molecular Plant-Microbe Interactions* 12: 640–644.

Nürnberger, T. 1999. Signal perception in plant pathogen defense. *Cellular and Molecular Life Science* 55: 167–182.

Nürnberger, T., Brunner, F. 2002. Innate immunity in plants and animals: emerging parallels between the recognition of general elicitors and pathogen-associated molecular patterns. *Current Opinion in Plant Biology* 5(4): 318–324.
O'Donnell, P.J., Jones, J.B., Antoine, F.R., Ciardi, J., Klee, H.J. 2001. Ethylene-dependent salicylic acid regulates an expanded cell death response to a plant pathogen. *Plant Journak* 25: 315–323.
O'Donnell, P.J., Schmelz, E., Block, A., Miersch, O., Wasternack, C., Jones, J.B., Klee, H.J. 2003. Multiple hormones act sequentially to mediate a susceptible tomato pathogen defense response. *Plant Physiology* 133: 1181–1189.
Oostendorp, M., Kunz, W., Dietrich, B., Staub, T. 2001. Induced disease resistance in plants by chemicals. *European Journal of Plant Pathology* 107(1): 19–28.
Park, S.W., Kaimoyo, E., Kumar, D., Mosher, S., Klessig, D.F. 2007. Methyl salicylate is a critical mobile signal for plant systemic acquired resistance. *Science* 318: 113–116.
Pearce, G., Strydom, D., Johnson, S., Ryan, C.A. 1991. A polypeptide from tomato leaves induces wound-inducible proteinase inhibitor proteins. *Science* 253: 895–898.
Piel, J., Atzorn, R., Gäbler, R., Kühnemann, F., Boland, W. 1997. Cellulysin from the plant parastic fungus Trichoderma viride elicits volatile biosynthesis in higher plants via the octadecanoid signaling cascade. *FEBS Letters* 416: 143–148.
Raskin, I. 1992. Role of salicylic acid in plants. *Annual Review of Plant Physiologyand Plant Molecular Biology* 43: 439–463.
Rasmussen, J.B., Hammerschmidt, R., Zook, M.N. 1991. Systemic induction of salicylic acid accumulation in cucumber after inoculation with *Pseudomonas syringae* pv. *syringae*. *Plant Physiology* 97: 1342–1347.
Ryals, J.A., Neuenschwander, U.H., Willits, M.G., Molina, A., Steiner, H.Y., Hunt, M.D. 1996. Systemic acquired resistance. *The Plant Cell* 8(10): 1809–1819.
Ryan, C.A. 2000. The systemin signaling pathway: Differential activation of plant defensive genes. *Biochimica et Biophysica Acta* 1477: 112–121.
Ryu, C.M., Farag, M.A., Hu, C.H., Reddy, M.S., Kloepper, J.W., Paré, P.W. 2004. Bacterial volatiles induce systemic resistance in Arabidopsis. *Plant Physiology* 134(3): 1017–1026.
Ryu, C.M., Farag, M.A., Hu, C.H., Reddy, M.S., Wei, H.X., Pare, P.W., Kloepper, J.W. 2003. Bacterial volatiles promote growth in *Arabidopsis*. *Proceedings of National Academy of Sciences of the United States of America* 100: 4927–4932.
Schlumbaum, A., Mauch, F., Vogeli, U., Boller, T. 1986. Plant chitinases are potent inhibitors of fungal growth. *Nature* 324: 365–367.
Shibuya, N., Minami, E. 2001. Oligosaccharide signalling for defence responses in plant. *Physiological and Molecular Plant Pathology* 59(5): 223–233.
Shulaev, V., León, J., Raskin, I. 1995. Is salicylic acid a translocated signal of systemic acquired resistance in tobacco? *The Plant Cell Online* 7(10): 1691–1701.
Shulaev, V., Silverman, P., Raskin, I. 1997. Airborne signalling by methyl salicylate in plant pathogen resistance. *Nature* 385: 718–721.
Silverman P., Seskar. M., Kanter, D., Schweizer, P., Métraux, J. 1995. Salicylic acid in rice. *Plant Physiology* 108: 633–639.
Spoel, S.H. Dong, X. 2008. Making sense of hormone crosstalk during plant immune response. *Cell Host and Microbe* 3: 348–351.
Spoel, S.H., Mou, Z., Tada, Y., Spivey, N.W., Genschik, P., Dong, X. 2009. Proteasome-mediated turnover of the transcription coactivator NPR1 plays dual roles in regulating plant immunity. *Cell* 137(5): 860–872.
Stakman, E.C. 1915. Relation between Puccinia graminis and plants highly resistant to its attack. *Agricultural Research* 4: 193–299.
Sticher, L., Mauch-Mani, B., Métraux, J.P. 1997. Systemic acquired resistance. *Annual Review of Phytopathology* 35: 235–270.
Stout, M.J., Workman, K.V., Bostock, R.M., Duffey, S.S. 1998. Stimulation and attenuation of induced resistance by elicitors and inhibitors of chemical induction in tomato (*Lycopersicon esculentum*) foliage. *Entomologia Experimentalis et Applicata* 86: 267–279.
Torres, M.A., Jones, J.D., Dangl, J.L. 2006. Reactive oxygen species signaling in response to pathogens. *Plant physiology* 141(2): 373–378.
Vallad, G.E., Goodman, R.M. 2004. Systemic acquired resistance and induced systemic resistance in conventional agriculture. *Crop Science* 44(6): 1920–1934.

Van Loon, L.C. 1997. Induced resistance in plants and the role of pathogenesis-related proteins. *European Journal of Plant Pathology* 103: 753–765.

Van Loon, L.C. 1999. Occurrence and properties of plant pathogenesis-related proteins. In: Datta, S.K., Muthukrishnan, S. (eds.) *Pathogenesis-Related Proteins in Plants* Boca Raton: CRC Press, pp. 1–19.

Van Loon, L.C., van Strien, E.A. 1999a. The families of pathogenesis-related proteins, their activities, and comparative analysis of PR-1 type proteins. *Physiological and Molecular Plant Pathology* 55: 85–97.

Van Loon, L.C., Van Strien, E.A. 1999b. The families of pathogenesis-related proteins, their activities, and comparative analysis of PR-1 type proteins. *Physiological and Molecular Plant Pathology* 55(2): 85–97.

Verberne, M.C., Hoekstra, J., Bol, J.F., Linthorst, H.J. 2003. Signaling of systemic acquired resistance in tobacco depends on ethylene perception. *The Plant Journal* 35(1): 27–32.

Vernooij, B., Friedrich, L., Morse, A., Reist, R., Kolditz-Jawhar, R., Ward, E., Uknes, S., Kessmann, H., Ryals, J. 1994. Salicylic acid is not the translocated signal responsible for inducing systemic acquired resistance but is required in signal transduction. *Plant Cell* 6: 959–965.

Walters, D., Heil, M. 2007. Costs and trade-offs associated with induced resistance. *Physiological and Molecular Plant Pathology* 71: 3–17.

Washington, S., Engleitner, S., Boontjes, G., Shanmuganathan, N. 1999. Effect of fungicides, seaweed extracts, tea tree oil, and fungal agents on fruit rot and yield in strawberry. *Australian Journal of Experimental Agriculture* 39(4): 487–494.

Wildermuth, M.C., Dewdney, J., Wu, G., Ausubel, F.M. 2001. Isochorismate synthase is required to synthesize salicylic acid for plant defence. *Nature* 414(6863): 562–565.

Wojtaszek, P. 1997. Oxidative burst: An early plant response to pathogen infection. *Biochemical Journal* 322(3): 681–692.

Wu, Y., Zhang, D., Chu, J.Y., Boyle, P., Wang, Y., Brindle, I.D., Després, C. 2012. The Arabidopsis NPR1 protein is a receptor for the plant defense hormone salicylic acid. *Cell reports* 1(6): 639–647.

Yun, B.W., Spoel, S.H., Loake, G.J. 2012. Synthesis and signalling by small, redox active molecules in the plant immune response. *Biochimica Biophysica Acta* 1820: 770–776.

Zhang, Y., Goritschnig, S., Dong, X., Li, X. 2003. A gain-of-function mutation in a plant disease resistance gene leads to constitutive activation of downstream signal transduction pathways in suppressor of npr1-1, constitutive 1. *The Plant Cell Online* 15(11): 2636–2646.

Zhang, Z.-P., Baldwin, I.T. 1997. Transport of (2-14C) jasmonic acid from leaves to roots mimics wound-induced changes in endogenous jasmonic acid pools in *Nicotiana sylvestris*. *Planta* 203: 436–441.

11 Quorum Sensing in Plant Pathogenic and Plant-Associated Bacteria

Mani Jayaprakashvel and Vellasamy Shanmugaiah

CONTENTS

ABSTRACT Scientists used to think that microorganisms, especially unicellular bacteria, are simple entities that function individually. However, research in the 1970s found that unicellular microorganisms use small signaling molecules, which function analogously to language, to determine their local population densities and to exhibit certain phenotypes. Collectively, the processes involved in the production and recognition of these signals (called autoinducers) are known as quorum sensing (QS). Using this form of cell–cell communication allows unicellular organisms to function as multicellular systems. Studies of QS systems in bacteria have also revealed that these bacterial languages can be intercepted, modulated, or destroyed by other organisms (quorum quenching [QQ]). The basic chemistry, biosynthesis, and mechanisms of autoinducers and types of QS systems contain an ocean of scientific information about how plants combat microbial infections. The influence of QS systems in plant pathology, plant microbial interaction, and rhizosphere ecology has been well documented. QQ, also known as anti-QS, is a promising concept for plant disease-management programs. Several beneficial microorganisms, such as *Rhizobium*, have been found to operate QS systems that govern plant–microbe interactions and symbiotic associations. This chapter, which contains a comprehensive overview of QS systems in bacteria, considers their mechanisms, autoinducers, QQ, and mechanisms of QQ, as well as applications of QS in biological control from the perspective of plant pathology.

KEYWORDS: Quorum sensing, quorum quenching, plant pathogens, plant-associated bacteria, biocontrol of plant diseases

11.1 INTRODUCTION

Spoken language is considered to be one of the most important developments in the history of mankind. In an anthropological context, the histories of both spoken and written languages are very important indexes of human development. Microorganisms, especially bacteria, also have a way of communicating: through chemical language, in a process known as quorum sensing. Through QS, a cell–cell communication system, bacteria monitor the presence of other bacteria by producing and responding to signaling molecules known as autoinducers (AIs) (Taga and Bassler, 2003). Generally, QS is a population-density-dependent expression of certain phenotypes of bacteria (Bassler, 2002).

In bacteria, QS is enacted in many coordinated behaviors. Bacteria utilize QS for synchronous behavior, including responses to adverse or challenging environmental and physiological conditions. The discovery of QS led to the realization that bacteria are capable of coordinated activity based on intercellular communication; until this discovery, both explicit communication and coordinated activity were believed to be restricted to multicellular, higher-order organisms. The capacity to behave collectively confers obvious advantages, such as the ability to migrate to a more suitable environment/better nutrient supply and to adopt new modes of growth. One example of the latter, sporulation or biofilm formation, may afford protection from deleterious environments (de Kievit and Iglewski, 2000). Moreover, QS allows bacteria to share information about cell density, to adjust their gene expressions accordingly, and to express energetically expensive processes as a collective only when the impact of those processes on the environment or on a host will be maximized. Both Gram-positive and Gram-negative bacteria use QS machinery to control a diverse array of physiological activities such as symbiosis, virulence, competence, conjugation, antibiotic production, motility, sporulation, biofilm formation, and others (Miller and Bassler, 2001).

Despite several decades of mechanistic research, empirical studies have only recently addressed the benefits of QS and affirmations of its social nature and role in optimizing cell-density-dependent group behaviors (Schuster et al., 2013). Recently, several groups have demonstrated artificial intra- and inter-species communication through synthetic circuits that incorporate components of bacterial QS systems. Engineered QS-based circuits have a wide range of applications, for example in biochemical production, tissue engineering, and mixed-species fermentations. They are also highly useful in designing microbial biosensors to identify bacterial species in the environment and also within living organisms (Choudhary and Schmidt-Dannert, 2010). QS has applications in medicine (de Kievit and Iglewski, 2000), food (Skandamis and Nychas, 2012), industry (Davies, 1998), agriculture (Pirhonen et al., 1993), and other related fields (Choudhary and Schmidt-Dannert, 2010).

11.2 QS TYPES AND MECHANISMS

QS is operated in different classes of bacteria by various modes and mechanisms, but the molecules that are involved in the signaling processes are generally known as autoinducers (AIs). Although several QS systems are known, perhaps the two most thoroughly described systems are the acyl-homoserine lactone (acyl-HSL) systems of many Gram-negative bacterial species and the peptide-based signaling systems of some Gram-positive bacterial species. Taga and Bassler (2003) made a comprehensive overview of the different types of QS regulatory systems in bacteria. In both Gram-positive and Gram-negative bacteria, the population density of the producing bacterium influences AI concentration (Figure 11.1). At low-cell densities, in Gram-negative QS bacteria, the LuxI-type proteins catalyze the formation of a specific acyl-HSLAI that freely diffuses into and out of the cell and increases in concentration in proportion to cell-population density (Figure 11.2a). The LuxI and LuxR systems typically possess proteins that are homologous to the LuxI and LuxR proteins of *Vibrio fischeri*, the bacterium in which they were originally identified. At higher cell densities,

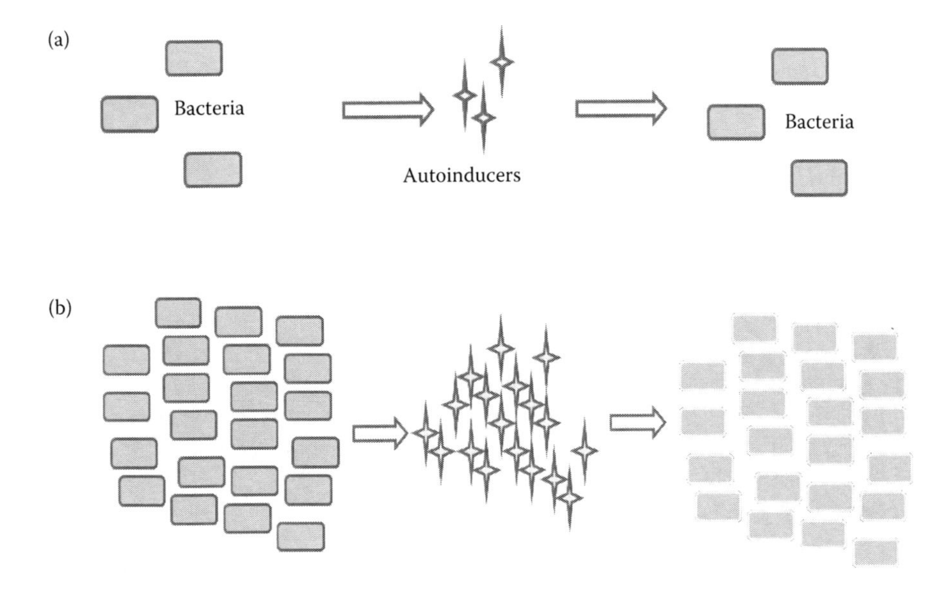

FIGURE 11.1 Common system of QS-regulated phenotypes in bacteria during different cell densities. (a) At low cell densities, there is no activation of target genes and hence no coordinated behavior expressed, (b) at high cell densities, bacteria produce more autoinducers; Autoinducer mediated QS dependent gene expression occurs, which results in the coordinated expression of phenotypes such as bioluminescence.

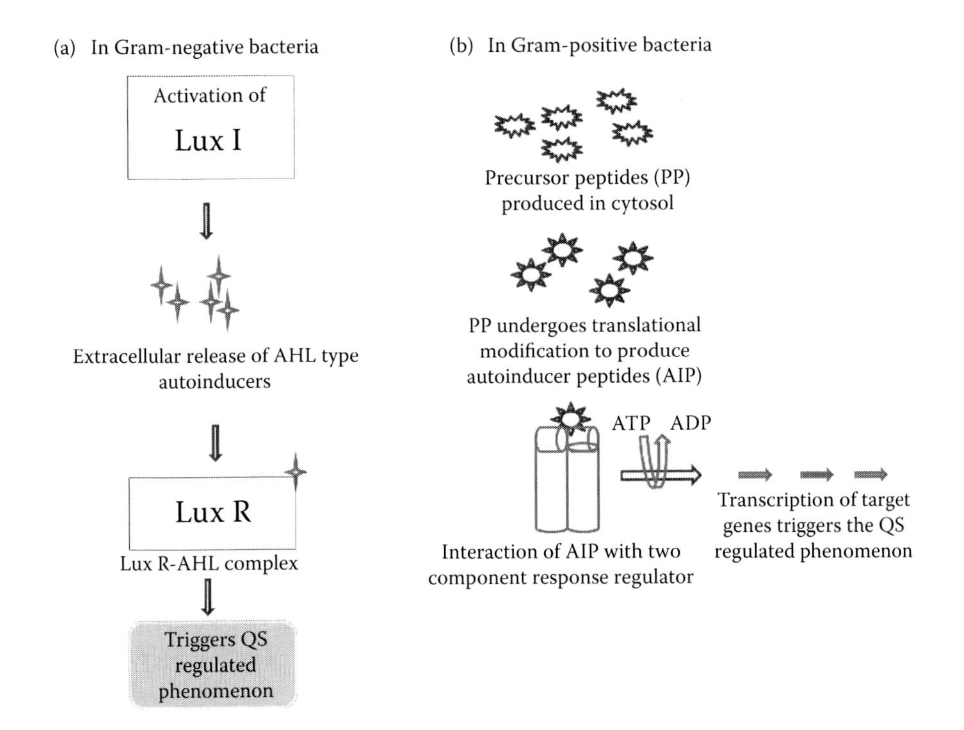

FIGURE 11.2 Series of events in (a) Gram-negative and (b) Gram-positive bacteria during QS regulation.

TABLE 11.1
Examples of Phenotypes Regulated by Quorum Sensing in Gram-Negative and Gram-Positive Bacteria

Sl. No.	Bacterium	Phenotype Regulated	References
	Gram-Negative Bacteria (LuxI-LuxR Type Regulatory System with Acyl-HSL Autoinducers)		
1	*Agrobacterium tumefaciens*	Plasmid conjugation	Piper et al. (1993)
2	*Erwinia carotovora*	Antibiotic production	Bainton et al. (1992)
3	*Sinorhizobium meliloti*	Symbiosis	Marketon et al. (2002)
4	*Pseudomonas aeruginosa*	Biofilm formation	Davies et al. (1998)
5	*Pseudomonas aeruginosa*	Bacterial virulence	Passador et al. (1993)
6	*Ralstonia solanacearum*	Virulence factors	Flavier et al. (1998)
7	*Xanthomonas oryzae* pv. *oryzicola*	Virulence factors; Exopolyaccarides	Zhao et al. (2011)
8	*Xanthomonas oryzae* pv. *oryzae*	Host–pathogen Interaction	Ferluga and Venturi (2008)
	Vibrio fisheri and *Vibrio harveyi*	Bioluminescence	Nealson and Hastings (1979)
	Gram-Positive Bacteria (Two-Component Regulatory System with AIP Autoinducers)		
9	*Bacillus subtilis*	Genetic competency and sporulation	Magnuson et al. (1994)
10	*Streptococcus pneumoniae*	Competence for DNA uptake during transformation	Håvarstein et al. (1999)
11	*Enterococcus faecalis*	Virulence factors	Qin et al. (2000)
12	*Staphylococcus aureus*	Virulence factors	Novick (1999)

the LuxR-type protein binds a specific acyl-HSL AI when the concentration of AI reaches a threshold level. The newly formed LuxR–acyl-HSL complex then triggers the target genes to initiate transcription by recognizing and binding specific DNA sequences in quorum-sensing-regulated promoters (Engebrecht et al., 1983; Whitehead et al., 2001; Taga and Bassler, 2003).

Although intercellular communication systems in Gram-negative bacteria often utilize homoserine lactones as signaling molecules, autoinducing peptides (AIPs) are involved in intercellular communication in Gram-positive bacteria (Figure 11.2b). Many of these peptides are exported by dedicated systems, post-translationally modified in various ways, and finally sensed by other cells via membrane-located receptors that belong to two-component regulatory systems (Sturme et al., 2002). AIPs, which are produced in cytoplasm as precursor peptides, are subsequently cleaved, modified, and exported. AIPs specifically interact with the external domains of membrane-bound two-component sensor kinase proteins. Interaction of the AI with its cognate sensor stimulates the kinase activity of the sensor kinase protein, which results in the phosphorylation of its partner response regulator protein. The phosphorylated response regulator protein binds DNA and alters the transcription of target genes (Sturme et al., 2002; Taga and Bassler, 2003). From the literature, which is extensive, it is evident that many QS bacteria utilize this regulatory mechanism for different effects on QS-controlled phenotypes (Table 11.1).

11.3 AUTOINDUCERS IN QS

QS, or the control of gene expression in response to cell density, is used by both Gram-negative and Gram-positive bacteria to regulate a variety of physiological functions. In all cases, QS involves the production and detection of autoinducers (AIs), which are extracellular signaling molecules (Bassler, 1999). The various functions controlled by QS reflect the need of a particular species of bacteria to inhabit a given niche. Three major QS circuits have been described: one used primarily by Gram-negative bacteria, one used primarily by Gram-positive bacteria, and one that has been proposed to be universal (Reading and Sperandio, 2006).

In general, Gram-negative bacteria use acyl homoserine lactones (acyl-HSLs) as AIs and Gram-positive bacteria use processed oligopeptides as AIs (Miller and Bassler, 2001). Acyl-HSLs are important intercellular signaling molecules in QS control of gene expression; their signals are synthesized by members of the LuxI family of proteins (Parsek et al., 1999). Research in acyl-HSL QS has been considerably aided by simple methods devised to detect acyl-HSLs using bacterial biosensors that phenotypically respond when exposed to exogenous acyl-HSLs. Such methods are extensively reviewed by Steindler and Venturi (2007).

Although all acyl-HSL molecules possess an HSL ring, the length of the acyl side chain and the substitutions on the side chain differ and are specificity determinants for different quorum sensors (Parsek et al., 1999). Acyl-HSL molecules contain 4–14 carbon acyl side chains and either an oxo, a hydroxy, or no substitution at the third carbon (de Kievit and Iglewski, 2000). The chains range from four to 14 carbon atoms, can contain double bonds, and often contain an oxo or hydroxyl group on the third carbon. The overwhelming majority of microbial acyl-HSLs identified so far have even numbers of carbons in their acyl chains. Many of the individual acyl-HSL species are synthesized by representatives of different bacterial genera. Likewise, many bacterial species can produce more than one type of acyl-HSL. Some of the more common microbial acyl-HSLs are *N*-butanoyl-*l*-homoserine lactone (BHL), *N*-(3-hydroxybutanoyl)-*l*-homoserine lactone (HBHL), *N*-hexanoyl-*l*-homoserine lactone (HHL), *N*-(3-oxohexanoyl)-*l*-homoserine lactone (OHHL), *N*-octanoyl-*l*-homoserine lactone (OHL), *N*-(3-oxooctanoyl)-*l*-homoserine lactone (OOHL), *N*-(3-hydroxy-7-*cis*-tetradecenoyl)-*l*-homoserine lactone (HtdeDHL), and *N*-(3-oxododecanoyl)-*l*-homoserine lactone (OdDHL) (Whitehead et al., 2001).

Gram-positive bacteria also utilize QS to regulate a variety of processes in response to population density. However, unlike Gram-negative bacteria, which use acyl-HSL AIs and LuxI/LuxR-type signaling systems, Gram-positive bacteria communicate using modified oligopeptides as AIs and two-component-type membrane-bound sensor histidine kinases as receptors. In general, the AIP is secreted outside the cell through a dedicated ATP-binding cassette transporter and increases in concentration as a function of cell density. In most cases, signal cleavage and modification are concomitant with signal release (Long, 2010).

In the processing of AIPs, Gram-positive bacteria secrete precursors known as pro-AIPs that are diverse in sequence and structure. Due to the impermeability of cell membranes to peptides, specialized transporters are also required to process pro-AIPs. The final processed AIPs range in size from 5 to 17 amino acids, undergo post-translational modification, and can be linear or cyclized (Rutherford and Bassler, 2012). Gram-positive bacteria contain three major classes of structurally diverse AIPs (Sturme et al., 2002): lantibiotic nisin A (found in *Lactococcus lactis*), double-glycine-type prepeptides (found in *Streptococcus pneumoniae*), and cyclic thiolactone and lactone peptides found in *Staphylococcus aureus* and *Enterococcus faecalis* respectively.

Hong et al. (2012) have provided a comprehensive list of AI molecules of diverse chemical classes. They range from low-molecular-weight molecules such as *N*-acyl-HSL, furanosyl borate diester (AI-2), 4,5-dihydroxy-2,3-pentanedione (DPD), 3-hydroxypalmitic acid methyl ester (3OH-PAME), *cis*-11-methyl-2-dodecenoic acid (diffusible signal factor, DSF), 2-isocapryloyl-3R-hydroxymethyl-γ-butyrolactone (A-factor), diketopiperazines (DKP), 2-heptyl 3-hydroxy-4-quinolone (*Pseudomonas* quinolone signal, PQS), and 4-hydroxy-2-heptyl-quinoline (HHQ), to high-molecular-weight molecules such as oligopeptide AIs.

11.4 QUORUM QUENCHING: SYSTEMS AND MOLECULES

QS is prevalent throughout the Eubacteria domain. It allows bacteria to regulate gene expression in a population-dependent manner in response to concentrations of diffusible chemical signals that are produced and released into the local environment by them or other bacteria, either of the same or different species. In short, QS allows synchronized bacterial behaviors to occur in coordination with each other. Interference with AI-dependent QS is known as quorum quenching (QQ).

In a polymicrobial community, while some bacteria are using QS to communicate with neighboring cells, others are using QQ to interrupt the communication. This conflict can be thought of as a constant arms race of intercellular communications (Hong et al., 2012). Numerous QQ phenomena have been observed, and QQ strategies have been tested with promising results.

In theory, any mechanism that can effectively interfere with any one of the key processes in QS could be potentially used for quenching QS (Dong et al., 2003, 2007). Five categories of effective QQ mechanisms have been found (Figure 11.3).

1. Inhibition of signal molecule biosynthesis, in which the precursor molecules involved in the synthesis of signal molecules are replaced by their structural analogues. For example, *S*-adenosylmethionine is the precursor for homoserine lactone moiety. In an attempt to block acyl-HSL biosynthesis, Parsek et al. (1999) found that analogues of *S*-adenosylmethionine (such as *S*-adenosylcysteine) inhibited QS activity that was meant to inhibit the activity of the *Pseudomonas aeruginosa*. *l*-Canavanine, an arginine analog found exclusively in the seeds of legumes, has the potential to affect the population biology of *Bacillus cereus* (Emmert et al., 1998).
2. Inhibition of signal processing by QS antagonists, in which molecules are produced that, although they can bind to QS response regulators, fail to activate them (Defoirdt et al., 2004). Because QS-mediated processes are often involved in interactions with plant and animal hosts, it might not be surprising that these higher organisms have developed QQ mechanisms of their own; the production of QS antagonists is one such mechanism. For example, the red marine alga *Delisea pulchra* produces halogenated furanones as antagonists of acyl-HSL-based QS to protect itself from extensive bacterial colonization (Givskov et al., 1996). Several synthetic acyl-HSL and furanone analogues with respect to QS were also developed and were demonstrated to quench the QS in different systems (Smith et al., 2003).
3. Chemical inactivation of QS signals, which takes place by alkaline hydrolysis and oxidization. Prospective agents include halogen antimicrobials, ozone, and strong alkalis. Some

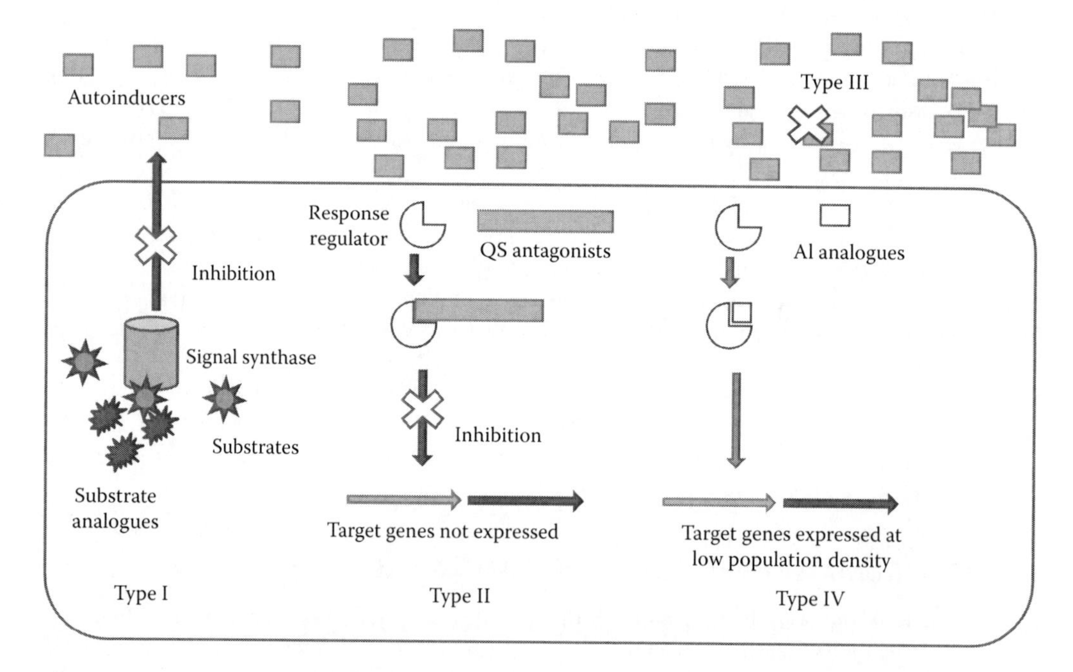

FIGURE 11.3 Types of QQ systems. Type I: inhibition of signal molecule biosynthesis, Type II: inhibition of signal processing by quorum sensing antagonists, Type III: chemical inactivation of QS signals, Type IV: expression of target genes at very low population density.

volatiles produced by rhizobacteria were also characterized to have QQ activity (Chernin et al., 2011).

4. Biodegradation of signal molecule by QQ enzymes. Enzymes that are able to inactivate and degrade QS signals are widespread in many organisms. Examples include bacterial acyl-HSL lactonases (Dong et al., 2000; Molina et al., 2003), bacterial acyl-HSL acylases (Leadbetter and Greenberg, 2000; Flagan et al., 2003), eukaryotic acylases (Xu et al., 2003) and dioxygenases (Pustelny et al., 2009).
5. Expression of target genes at very low population density, which occurs when analogues of AIs bind with receptors. This binding results in the expression of target genes even at low cell densities. In addition, regulatory RNAs (sRNAs) are involved specifically in repressing QS (Lenz et al., 2004).

11.4.1 QS and Drug Targeting

Naturally occurring QQ mechanisms, which appear to play important roles in microbe–microbe and pathogen–host interactions, have been used or have served as lead compounds in the development and formulation of a new generation of antimicrobials (Dong et al., 2007). Excessive and indiscriminate use of antibiotics to treat bacterial infections has resulted in the emergence of multiple-drug-resistant strains. Because most infectious diseases are caused by bacteria that proliferate within QS-mediated biofilms (Kalia, 2013), targeting QS systems is a potent defense strategy against bacterial infection. Recent developments in this area include the design of antivirulence deception strategies that can disrupt QS through signal-molecule inactivation, inhibition of signal-molecule biosynthesis, or the blockade of signal-transduction pathways (Rampioni et al., 2014).

Numerous anti-QS approaches have been documented. In terms of combating bacterial infection, anti-QS is a promising strategy because it does not impose any selection pressure and therefore is unlikely to produce multidrug-resistant pathogens. Plant-based natural products have been extensively studied in this context (Koh et al., 2013), as has the anti-QS activity of known medicinal plants (Adonizio et al., 2006). *Rhizophora* spp. was demonstrated to have antipathogenic potential against the QS-mediated virulence factors production in drug resistant *Ps. aeruginosa* (Annapoorani et al., 2013). Dietary phytochemicals, which are secondary metabolites in edible plants, are known to confer several health benefits, including antimicrobial activity. However, their ability to inhibit QS has never been studied. Vattem et al. (2007) used model bioassay test systems to investigate the effect of sublethal concentrations of bioactive dietary phytochemicals in common dietary fruit, herb, and spice extracts on modulating AI-mediated QS. This study's conclusion that all extracts significantly inhibited QS against *Escherichia coli* O157:H7 and *Ps. aeruginosa* PA-01 paved the way for a new antimicrobial chemotherapy strategy and led to the discovery of a new category of antibiotics that can overcome issues related to antimicrobial resistance. Wynendaele et al. (2012) suggested that QS peptide agonists and antagonists can be applied in oncology.

11.4.2 QS and Plant–Microbe Interaction

The rhizosphere is a densely populated area in which plant roots must compete not only with the invading root systems of neighboring plant species for space, water, and mineral nutrients, but also with other competitors for organic material. The latter include soilborne microorganisms such as bacteria and fungi, as well as insects (Ryan and Delhaize, 2001). Our literature survey indicates that, under QS, plants can produce structural analogues of AIs, AI mimics, and anti-QS compounds; manipulate QS for their own benefit; and also become affected by bacteria.

To prove that plants produce structural analogues of AIs, Teplitski et al. (2000) demonstrated that exudates from pea (*Pisum sativum*) seedlings contain several distinct activities that mimick acyl-HSL signals in well-characterized bacterial reporter strains; by doing so, they stimulate acyl-HSL-regulated behaviors in some strains and inhibit such behaviors in others. They also concluded that various other species of higher plants secrete acyl-HSL mimics. The acyl-HSL signal–mimic compounds could prove to be important in determining the outcome of interactions between higher plants and a diversity of pathogenic, symbiotic, and saprophytic bacteria. However, crude strawberry extracts were also demonstrated to inhibit the antibiotic production (an acyl-HSL response) of the *Es. coli* lawn adjacent to a well that contained fruit extract (Fray et al., 1999). This result proved that plants can produce anti-QS compounds. The ability of plants to produce substances that affect QS regulation may provide plants with important tools to manipulate gene expression and behavior in the bacteria they encounter. *Medicago truncatula*, a model leguminous plant, appears to produce at least 15–20 separable substances that are capable of specifically stimulating or inhibiting responses in QS reporter bacteria; primarily, these substances affect QS regulation that is dependent upon *N*-acyl-HSL signals (Gao et al., 2003).

Bacterial QS molecules such as *N*-acyl-HSLs modulate plant responses to contact with bacteria. Often, the AIs and QS responses are affected by environmental conditions such as the presence of other acyl-HSL-producing bacterial species. Plant-derived metabolites, including products that are direct results of bacterial infection, may also profoundly influence acyl-HSL-regulated behaviors. These plant products can interact both directly and indirectly with QS networks, and can profoundly affect QS behaviors (Newton and Fray, 2004). In short, plants can manipulate QS for their own benefit. For example, plant pathogenic *Ps. aeruginosa* strains PAO1 and PA14 are capable of infecting the roots of *Arabidopsis* and sweet basil. Upon *Ps. aeruginosa* infection, sweet basil roots secrete rosmarinic acid (RA), a multifunctional caffeic acid ester that exhibits *in vitro* antibacterial activity against planktonic cells of both *Ps. aeruginosa* strains. Moreover, exogenous supply and induction of RA in *Arabidopsis* plants inhibited the biofilm formation of *Ps. aeruginosa* and, in turn, the infection (Walker et al., 2004). Some plants enhance QS in bacteria. For example, nodulation-gene-inducing flavonoids increase the overall production of AIs and the expression of *N*-acyl-HSL-synthesis genes in *Sinorhizobium fredii* SMH12 and *Rhizobium etli* ISP42 (Pérez-Montaño et al., 2011).

The presence of acyl-HSL-mimic QS molecules in diverse *Oryza sativa* (rice) and *Phaseolus vulgaris* (bean) plant samples have been detected. It has also been demonstrated that bean and rice seed extracts contain molecules that lack the typical lactone ring of acyl-HSLs. Interestingly, these molecules specifically alter the QS-regulated biofilm formation of two plant-associated bacteria, *Si. fredii* SMH12 and *Pantoea ananatis* AMG501. This result suggests that plants are able to enhance or inhibit bacterial QS systems, depending upon the bacterial strain. Rice-root exudates were found to decrease the antibiotic production in a few biocontrol pseudomonads, which suggests that plants may also negatively influence QS (Jayaprakashvel, 2008).

11.4.3 QS in Plant Pathology

Plant pathogens coordinate multifaceted life histories and deploy stratified virulence determinants via complex, global regulation networks that belong to five types: centralized cell-to-cell communication systems, pervasive two-component signal-transduction systems, post-transcriptional regulation systems, AraC-like regulators, and sigma factors. Although these common regulatory systems control virulence, in diverse species each functions in different capacities and to differing ends. Therefore, the virulence regulation network of each species determines its survival and success in various life histories and niches (Mole et al., 2007). AI-mediated QS by bacteria regulates traits that are involved in symbiotic, pathogenic, and surface-associated relationships between microbial populations and their plant hosts. Microbes that inhabit plant surfaces also produce and respond to a diverse mixture of acyl-HSL signals, which suggests that both bacteria and plants utilize this method of bacterial communication as a key control point for influencing the outcomes of their interactions (Loh et al., 2002).

In an elaborate review, von Bodman et al. (2003) illustrated how plant pathogenic bacteria use QS signals to regulate genes for epiphytic fitness, such as motility in *Ralstonia solanacearum*, antibiosis in *Erwinia carotovora*, and UV-light resistance in *Xanthomonas campestris*. Such bacteria can also use QS signals to regulate genes for major pathogenicity factors, including EPS in *Pantoea stewartii*, *X. campestris*, and *R. solanacearum*; type III secretion systems in *Pa. stewartii* and *Er. carotovora*; and exoenzyme production in *Er. carotovora*, *X. campestris*, and *R. solanacearum*. Wevers et al. (2009) investigated the role of QS and specific QS-dependent properties in the colonization and spoilage of carrot slices by *Serratia plymuthica* RVH1, a strain that was isolated from a vegetable washing-and-cutting machine in an industrial kitchen.

It is now apparent that acyl-HSLs are used for regulating diverse behaviors in epiphytic, rhizosphere-inhabiting, and plant pathogenic bacteria and that some plants may produce metabolites that interfere with this signaling. The creation of transgenic plants that produce high levels of acyl-HSLs or can degrade bacterial-produced acyl-HSLs has dramatically altered susceptibilities to infection by pathogenic *Erwinia* species. In addition, such plants will be useful in determining the roles of acyl-HSL-regulated density-dependent behavior in the growth promotion and biological control of pathogenic plant-associated bacterial species (Fray, 2002). *Serratia liquefaciens* MG1, a C4- and C6-side chain acyl-HSL-producing bacterium, was found to influence specific induction of systemic resistance proteins after root inoculation. This wild-type *Se. liquefaciens* MG1 was also demonstrated to increase plant resistance to the foliar fungal pathogen *Alternaria alternata* in tomato, as compared to the acyl-HSL-negative mutant (Hartmann et al., 2004; Schuhegger et al., 2006). In knot disease of olive tree (*Olea europaea*), two bacterial species (i.e., *Pantoea agglomerans* and *Erwinia toletana*) that are not pathogenic and are olive plant epiphytes and endophytes were found to aggravate infection by the causative agent *Pseudomonas savastanoi* pv. *savastanoi* by producing the *N*-acyl-HSL family of QS signals (Hosni et al., 2011). Phytopathogenic bacteria depend upon the exchange of external QS signal molecules for normal infection and pathogenesis. Consequently, these signals present attractive targets for the control of bacterial diseases through the genetic engineering of plants (von Bodman et al., 2003).

11.4.4 QS and *Erwinia*

Members of the genus *Erwinia* are the most-studied plant pathogenic bacteria in the contexts of both QS and QQ. The background mechanisms of virulence factors, regulation of virulence gene expression, and virulence-related production of exoenzymes are particularly well-studied QS-related phenomena in the genus *Erwinia* (Jones et al., 1993; Mukherjee et al., 1997; Pirhonen et al., 2003). In *Erwinia*, *N*-acyl-HSL controls the expression of various traits, including extracellular enzyme/protein production and pathogenicity. *Erwinia* (soft rot pathogens belonging to the genus *Erwinia*, renamed *Pectobacterium*) utilize two types of QS-signaling molecules: *N*-acyl-HSLs and AI-2-type (Barnard and Salmond, 2007). *Erwinia carotovora* ssp. *atroseptica*, which is responsible for potato blackleg disease, produces acyl-HSLs for tissue maceration (an important step in the disease's development). However, the use of QQ strategies for biological control in *Er. carotovora* ssp. *atroseptica* could not prevent initial infection and multiplication of this pathogen (Smadja et al., 2004).

Venturi et al. (2004) first reported the production of acyl-HSLs by *Erwinia amylovora*, an important quarantine bacterial pathogen that causes fire blight in plants. The fact that *Er. amylovora* produces one *N*-acyl-HSL QS signal molecule both *in vitro* and in planta strongly implies that acyl-HSL-dependent QS could play an important role in the regulation of *Er. amylovora* virulence. Bainton et al. (1992) first reported that the production of carbapenem antibiotic by *Er. carotovora* is also regulated by identical self-produced acyl-HSLs. Acyl-HSLs were found to be responsible not only for antibiotic production but also for the induction of a glycine-rich protein (Harpin) that strongly elicits HR (Mukherjee et al., 1997). Even more importantly, acyl-HSLs provide global control of the production of exoenzymes such as pectin lyase, pectate lyase, polygalacturonase, cellulase, and protease, which are essential to soft rot *Erwinia* (Fray, 2002).

11.4.5 QS and *Xanthomonas*

Members of the genus *Xanthomonas* are known producers of exopolysaccharides. These exopolysaccharides have been proven to be one of the virulence factors in the genus *Xanthomonas* and to incite infections in various plants. For example, it has been demonstrated in *Xanthomonas oryzae* pv. *oryzicola* (*Xoc*), which causes bacterial leaf streak in rice, that a mutation in the *rpfF* gene that encodes DSF synthase caused it to lose the ability to produce DSF molecules; in addition, the bacterium exhibited a significant reduction of virulence in rice compared to the wild-type strain. Furthermore, the mutation of *rpfF* impaired EPS production and led to *Xoc* cell aggregation (Zhao et al., 2011). Similarly, a cell–cell signaling system encoded by genes within the *rpf* cluster was found to be necessary for the full virulence of *X. campestris* pv. *campestris,* the causal agent of black rot disease of cruciferous plants. This system has also been implicated in the regulation of the formation and dispersal of *Xanthomonas* biofilms (Crossman and Dow, 2004).

Xanthomonas oryzae pv. *oryzae* (*Xoo*), the second-most-important rice pathogen, causes bacterial leaf blight. The tissue necrosis and wilting caused by *Xoo* colonization and infection of plant vascular tissue result in significant yield losses worldwide (González et al., 2012). *Xoo* has been reported to contain a regulator, OryR, that is encoded in the genome and belongs to the *N*-acyl-HSL–dependent QS LuxR subfamily of proteins. OryR does not regulate production of the quorum-sensing DSF in the genus *Xanthomonas* but does serve as an important regulator in inter-kingdom communication between host and pathogen (Ferluga and Venturi, 2009). *Xanthomonas campestris* pv. *campestris*, a cabbage pathogen, produces a diffusible extracellular factor (DSF) that has yet to be chemically characterized but is not a acyl-HSL; this information confirms that a novel regulatory system (i.e., a system other than acyl-HSL) is required for the pathogenicity of *X. campestris* (Barber et al., 1997). The production of compounds with acyl-HSL activity seems to be very rare among *Xanthomonas* species. In addition, production of DSF appears to be limited to *Xanthomonas* species, largely to *X. campestris* strains (Barber et al., 1997; Whitehead et al., 2001).

11.4.6 QS in Plant Pathogenic *Pseudomonas*

A number of Gram-negative bacteria have QS systems and produce the *N*-acyl-HSL as a signal molecule. *Pseudomonas chlororaphis* ssp. *aurantiaca* StFRB508 produces one of the phenazine derivatives, phenazine-*l*-carboxylic acid (PCA). The phenazine antibiotic production and antifungal activity of *Ps. chlororaphis* ssp. *aurantiaca* StFRB508 are regulated by multiple quorum-sensing systems (Morohoshi et al., 2013). *Pseudomonas* sp. DF41 is able to suppress the fungal pathogen *Sclerotinia sclerotiorum* through production of a lipopeptide called sclerosin.

RfiA, a transcriptional activator in the QS system of *Pseudomonas* sp. DF41, indirectly controls the suppression of *Sclerotinia* by the bacterium (Berry et al., 2014). Two closely related phytopathogenic bacteria, *Pseudomonas corrugata* and *Pseudomonas mediterranea*, are both causal agents of tomato pith necrosis. *Pseudomonas corrugata* produces phytotoxic and antimicrobial cationic lipodepsipeptides (LDPs), which are thought to act as major virulence factors.

N-acyl-homoserine lactone QS in tomato phytopathogenic *Pseudomonas* spp. has been found to be involved in the regulation of LDP production, and thus to contribute to the pathogenicity of the bacteria (Licciardello et al., 2012). In most bacteria, Fur (ferric uptake regulator) is a crucial global regulator that operates not only in the regulation of iron homeostasis but also in a variety of other cellular processes. In the plant pathogenic bacterium *Pseudomonas syringae* pv. *tabaci* 11528, it has been demonstrated that the *fur* genes are responsible for the virulence of the bacterium by involving in siderophore production and swarming motility, which are controlled by QS (Cha et al., 2008). Cell-free lysate of endophytic bacteria, *Bacillus firmus* PT18 and *Enterobacter asburiae* PT39, exhibited potent acyl-HSL degrading ability resulted in the inhibition of biofilm formation in *Ps. aeruginosa* PAO1 due to the presence of QQ enzyme lactonase (Rajesh and Rai, 2014).

A two-component system comprising GacS and GacA affects a large number of traits in many Gram-negative bacteria, including the plant pathogen *Ps. syringae*. Cha et al. (2012) proved that the virulence traits related to the pathogenesis of *Ps. syringae* pv. *tabaci* 11528 are coordinately regulated by GacA and iron availability. They also concluded that several systems coordinately regulate *gacA* gene expression in response to iron concentration and bacterial cell density, and that GacA and iron together control the expression of several virulence genes in *Ps. syringae* pv. *tabaci*. *Pseudomonas syringae* pv. *phaseolicola* is the causal agent of halo blight disease of beans (*Ph. vulgaris* L.), which is characterized by water-soaked lesions surrounded by a chlorotic halo that results from the action of a non-host-specific toxin known as phaseolotoxin. Phaseolotoxin production, in turn, is controlled by several genes in the *Pht* cluster. It has been suggested that phaseolotoxin biosynthesis involves elements within and outside the Pht cluster and is controlled by a two-component signal transduction system that is involved in important pathogenicity and virulence mechanisms of *Ps. syringae* pv. *phaseolicola* (Torre-Zavala et al., 2011).

11.4.7 QS in *Ralstonia*

Expression of several virulence factors in the plant pathogen bacterium *R. solanacearum* is controlled by a complex regulatory network; at the center of this network is PhcA. Under the strong influence of QS and complex nitrogen sources, PhcA represses the expression of *hrp* genes that code for the type III protein secretion system, which is a major pathogenicity determinant in this bacterium (Genin et al., 2005). As the LysR-type regulator, PhcA is also central to the complex regulation of EPS and extracellular enzymes, and therefore to the pathogenicity of *R. solanacearum*. Consequently, a phcA mutant has decreased levels of EPS and extracellular enzymes and is almost avirulent (Brumbley et al., 1993). Endo- and exopolygalacturonases of *R. solanacearum* are inhibited by polygalacturonase-inhibiting protein (PGIP) activity in tomato stem extracts. PGIP activity was shown to be concentration-dependent, constitutively present, and not related to resistance nonsusceptibility of tomato recombinant inbred lines to *R. solanacearum* (Schacht et al., 2011).

Ralstonia solanacearum also contains a typical acyl-HSL-based QS system that employs two signals; one is probably OHL but the identity of the other is currently under debate. The acyl-HSL-dependent autoinduction system in *R. solanacearum* is part of a more complex auto regulatory hierarchy because the transcriptional activity of PhcA is controlled by a novel auto-regulatory system that responds to 3-hydroxypalmitic acid methyl ester (Flavier et al., 1998 and 1997). In summary, many of the key extra cytoplasmic virulence and pathogenicity factors are transcriptionally controlled in *Ralstonia* by a five-gene *Phc* system that regulates exopolysaccharide, cell-wall-degrading exoenzymes, and other factors in response to a self-produced signal molecule that monitors the pathogen's growth status and environment. Four additional environmentally responsive two-component systems work both independently and with the *Phc* system to fine-tune virulence gene expression (Schell, 2000).

11.4.8 QS and Biological Control

Many plant-associated bacteria employ QS for regulation of specific phenotypes as part of their pathogenic or symbiotic lifestyles. The ability to block or promote these QS systems may reveal new strategies for managing plant diseases and increasing crop productivity (de Kievit and Iglewski, 2000). The first application of QQ strategy in protection against plant disease was demonstrated by Dong et al. (2000), who transformed the *aiiA* gene into the phytopathogen *Er. carotovora* to attenuate its pathogenicity in Chinese cabbage. Later, *Rhodococcus erythropolis* W2 isolated from tobacco rhizosphere was demonstrated to degrade *N*-acyl-HSL and to markedly reduce the pathogenicity of *Pectobacterium carotovorum* ssp. *carotovorum* in potato tubers. In a classic experiment, this bacterium was proven to offer good control against soft rot in potato (Uroz et al., 2003).

In another study, plasmid pME6863 carrying the *aiiA* gene from the soil bacterium *Bacillus* sp. A24, which encodes a lactonase enzyme that is able to degrade acyl-HSLs, was introduced into the rhizosphere isolate *Pseudomonas fluorescens* P3. Although this strain is not an effective biological control agent against plant pathogens, the transformant *Ps. fluorescens* P3/pME6863 acquired the ability to degrade AHLs. In planta, *Ps. fluorescens* P3/pME6863 significantly reduced potato soft rot caused by *Er. carotovora* and crown gall of tomato caused by *Agrobacterium tumefaciens* to a similar level as *Bacillus* sp. A24 (Molina et al., 2003). Although Han et al. (2010) isolated a novel QQ bacterium, *Bacillus marcorestinctum* sp. nov., that strongly quenches the AHL QS signal, when this soil bacterium was applied to sliced potato tubers with *Pe. carotovorum*, the soft rot was effectively attenuated. This result suggests that the coculture of QQ microbes with QS-mediated pathogens could be used to prevent plant diseases. In another study, *Bacillus* sp. and *Arthrobacter* sp. capable of degrading both synthetic and natural N-AHL produced by *Pectobacterium atrosepticum* SM1 were not only significantly reduced by the pathogenicity of *Pe. atrosepticum* in potato tubers but were also prevented from causing any tissue maceration or blackleg symptoms in potato plants (Mahmoudi and Ahmadi, 2012).

In a recent study, among 1177 leaf-associated bacterial isolates, 168 strains (14%) were capable of interfering with AHL activity and 63% of 168 strains could enzymatically degrade AHL molecules using lactonases (Ma et al., 2013). This study clearly concluded that the naturally occurring wide diversity of bacterial quenchers might be exploited as effective biocontrol reagents for the suppression of plant pathogens *in situ*. Chernin et al. (2011) showed that the volatile organic compounds (VOCs) produced by rhizosphere strains of *Ps. fluorescens* B-4117 and *Se. plymuthica* IC1270 may act as inhibitors of a QS network mediated by *N*-acyl-HSL signal molecules produced by plant pathogenic bacteria such as *Agrobacterium*, *Chromobacterium*, and *Pectobacterium*. *Bacillus cereus* U92, isolated from tomato rhizosphere and capable of inactivating both short and long 3-oxo-substituted acyl-HSLs, was found to be efficient in alleviating QS-regulated crown gall incidence on tomato roots (up to 90%) as well as attenuating *Pectobacterium* soft rot on potato tubers (up to 60%) (Zamani et al., 2013).

Studies have also been done on genetic engineering biocontrol strains for better performance. When an *N*-acyl-HSL gene *aiiA* was transformed into *Lysobacter enzymogenes* OH11 to create strain *Lys. enzymogenes* OH11A, the new strain demonstrated the ability to degrade acyl-HSL molecules produced by *Ag. tumefaciens*, *Pe. carotovorum*, *Ps. syringae* pv. *tomato* DC3000, and *Acidovorax avenae* ssp. *citrulli*. Strains further expressed *in vivo* inhibitory activity against *Pectobacterium* virulence on Chinese cabbage, and on cactus. In antimicrobial activity assays, strains OH11A and OH11 showed similar antimicrobial activities against *Phytophthora capsici* and *Sclerotinia sclerotiorum* (Qian et al., 2010). Therefore, we may conclude that QS can be used as a strategy to develop genetically engineered biocontrol agents for superior performance. The ability to generate bacterial QS-signaling molecules in plants offers novel opportunities for disease control and for manipulating plant–microbe interactions.

Because QQ mechanisms can be engineered in plants, they might be used in the control of bacterial pathogens and the construction of proactive defense barriers (Zhang, 2003). Fray (2002) developed transgenic tobacco plants that can synthesize cognate acyl-HSL-signaling molecules (OHHL and HHL), which in turn complemented the disease-control ability of the biocontrol bacterium *Pseudomonas aureofaciens* 30-84. The non-target effect of QQ strategies in disease control, which has also been discussed, may be a double-edged sword in biological control of plant diseases (Zamani et al., 2013). However, in an important study by D'Angelo-Picard et al. (2011), it was found the root-associated bacterial populations in transgenic plants expressing the QQ lactonase AttM did not significantly alter.

11.4.9 QQ in Plant-Associated Bacteria

Microbes that are inhabiting plant surfaces also produce and respond to a diverse mixture of AIs. Both the production of AI mimics by plants and the identification of acyl-HSL–degradative pathways suggest that bacteria and plants utilize this method of bacterial communication as a key control point for influencing the outcomes of their interactions. QS and QQ regulatory systems are often studied in plant-associated bacteria in attempts to identify their roles in plant–microbe interactions and microbial physiology. The presence of acyl-HSL-mimic QS molecules in rice and bean were found to specifically alter the QS-regulated biofilm formation of two plant-associated bacteria, *Si. fredii* SMH12 and *Pa. ananatis* AMG501.

A more intensive analysis using biosensors that carried lactonase enzyme showed that bean and rice seed extracts contain molecules that lack the typical lactone ring of acyl-HSLs. This result suggests that plants are able to either enhance or inhibit bacterial QS systems, depending on the bacterial strain (Pérez-Montaño et al., 2013). When a QQ approach was exploited in order to identify functions regulated by QS in the plant-growth-promoting bacterium *Azospirillum lipoferum*, the AttM lactonase from *Ag. tumefaciens* was shown to enzymatically inactivate *N*-acyl-HSLs produced by two *Az. lipoferum* strains (Boyer et al., 2008).

11.5 CONCLUSION

QS, a fascinating field of science, is gaining momentum. Both QS and QQ have potential influence on plant pathogenic and other plant-associated bacteria. QQ can be a tool in controlling bacterial infections in plants. Genetically engineered plants have also been developed to increase plant virulence against selected bacterial infections. The understanding of QS mechanisms in plant-associated beneficial bacteria will increase the potential to enhance positive plant– microbial interactions.

ACKNOWLEDGMENTS

The authors thank the management and administration of AMET University and Madurai Kamaraj University for providing facilities and support. The first author thanks Ms S. Vinothini and Mr D. Anand of the Department of Marine Biotechnology, AMET University, Chennai, for their help in formatting the cited references.

REFERENCES

Adonizio, A.L., Downum, K., Bennett, B.C., Mathee, K. 2006. Anti-quorum sensing activity of medicinal plants in southern Florida. *Journal of Ethnopharmacology* 105(3): 427–435.

Annapoorani, A., Kalpana, B., Musthafa, K.S., Pandian, S.K., Ravi, A.V. 2013. Antipathogenic potential of *Rhizophora* spp. against the quorum sensing mediated virulence factors production in drug resistant *Pseudomonas aeruginosa*. *Phytomedicine* 20(11): 956–963.

Bainton, N.J., Stead, P., Chhabra, S.R., Bycroft, B.W., Salmond, G.P., Stewart, G.S., Williams, P. 1992. *N*-(3-oxohexanoyl)-*l*-homoserine lactone regulates carbapenem antibiotic production in *Erwinia carotovora*. *Biochemistry Journal* 288: 997–1004.

Barber, C.E., Tang, J.L., Feng, J.X., Pan, M.Q., Wilson, T.J., Slater, H., Dow, J.M., Williams, P., Daniels, M.J. 1997. A novel regulatory system required for pathogenicity of *Xanthomonas campestris* is mediated by a small diffusible signal molecule. *Molecular Microbiology* 24(3): 555–566.

Barnard, A.M., Salmond, G.P., 2007. Quorum sensing in Erwinia species. *Analytical and Bioanalytical Chemistry* 387: 415–423.

Bassler, B.L. 1999. How bacteria talk to each other: Regulation of gene expression by quorum sensing. *Current Opinion in Microbiology* 2(6): 582–587.

Bassler, B.L. 2002. Small talk. Cell-to-cell communication in bacteria. *Cell* 109: 421–424.

Berry, C.L., Nandi, M., Manuel, J., Brassinga, A.K.C., Fernando, W.G.D., Loewen, P.C., de Kievit, T.R. 2014. Characterization of the *Pseudomonas* sp. DF41 quorum sensing locus and its role in fungal antagonism. *Biological Control* 69: 82–89.

Boyer, M., Bally, R., Perrotto, S., Chaintreuil, C., Wisniewski-Dyé, F. 2008. A quorum-quenching approach to identify quorum-sensing-regulated functions in *Azospirillum lipoferum*. *Research in Microbiology* 159: 699–708.

Brumbley, S.M., Carney, B.F., Denny, T.P. 1993. Phenotype conversion in *Pseudomonas solanacearum* due to spontaneous inactivation of *PhcA*, a putative LysR transcriptional regulator. *Journal of Bacteriology* 175: 5477–5487.

Cha, J.Y., Lee, D.G., Lee, J.S., Oh, J.I., Baik, H.S. 2012. GacA directly regulates expression of several virulence genes in *Pseudomonas syringae* pv. *tabaci* 11528. *Biochemical and Biophysical Research Communications* 417(2): 665–672.

Cha, J.Y., Lee, J.S., Oh, J.I., Choi, J.W., Baik, H.S. 2008. Functional analysis of the role of Fur in the virulence of *Pseudomonas syringae* pv. *tabaci* 11528: Fur controls expression of genes involved in quorum-sensing. *Biochemical and Biophysical Research Communications* 366(2): 281–287.

Chernin, L., Toklikishvili, N., Ovadis, M., Ki, S., Julius, B., Inessa Khmel, K., Vainstein, A. 2011. Quorum-sensing quenching by rhizobacterial volatiles. *Environmental Microbiology Reports* 3(6): 698–704.

Choudhary, S., Schmidt-Dannert, C. 2010. Applications of quorum sensing in biotechnology. *Applied Microbiology and Biotechnology* 86(5): 1267–1279.

Crossman, L., Dow, J.M. 2004. Biofilm formation and dispersal in *Xanthomonas campestris*. *Microbes and Infection* 6(6): 623–629.

D'Angelo-Picard, C., Chapelle, E., Ratet, P., Faure, D., Dessaux, Y. 2011. Transgenic plants expressing the quorum quenching lactonase AttM do not significantly alter root-associated bacterial populations. *Research in Microbiology* 9: 951–958.

Davies, D.G., Parsek, M.R., Pearson, J.P., Iglewski, B.H., Costerton, J.W., Greenberg, E.P. 1998. The involvement of cell-to-cell signals in the development of a bacterial biofilm. *Science* 280(5361): 295–298.

de Kievit, T.R., Iglewski, B.H. 2000. Bacterial quorum sensing in pathogenic relationships. *Infection and Immunity* 68(9): 4839–4849.

Defoirdt, T., Boon, N., Bossier, P., Verstraete, W. 2004. Disruption of bacterial quorum sensing: An unexplored strategy to fight infections in aquaculture. *Aquaculture* 240(1–4): 69–88.

Dong, J., Chen, C.H., Chen, Z.X. 2003. Expression profiles of the Arabidopsis WRKY gene superfamily during plant defense response. *Plant Molecular Biology* 51: 21–37.

Dong, Y.-H.,Wang, L.-H., and Zhang, L.-H. 2007. Quorum-quenching microbial infections: Mechanisms and implications. *Philos Trans R Soc Lond B Biol Sci.* 362(1483): 1201–1211.

Dong, Y.H., Xu, J.L., Li, X.Z., Zhang, L.H. 2000. AiiA, an enzyme that inactivates the acylhomoserine lactone quorum-sensing signal and attenuates the virulence of *Erwinia carotovora*. *Proceedings of National Academic Science of the United States of America* 97(7): 3526–3531.

Emmert, EA.B., Milner, J.L., Lee, J.C., Pulvermacher, K.L., Olivares, H.A., Clardy, J., Handelsman, J. 1998. Effect of canavanine from alfalfa seeds on the population biology of *bacillus cereus*. *Applied Environmental Microbiology* 64(12): 4683–4688.

Engebrecht, J., Nealson, K., Silverman, M. 1983. Bacterial bioluminescence: Isolation and genetic analysis of functions from *Vibrio fischeri*. *Cell* 32(3): 773–781.

Ferluga, S., Venturi, V. 2009. *OryR* is a LuxR-family protein involved in interkingdom signaling between pathogenic *Xanthomonas oryzae* pv. *oryzae* and rice. *Journal of Bacteriology* 191(3): 890–897.

Flagan, S., Ching, W.K., Leadbetter, J.R. 2003. *Arthrobacter* sp. strain VAI-A utilizes acyl-homoserine lactone inactivation products and stimulates quorum signal biodegradation by *Variovorax paradoxus*. *Applied Environmental Microbiology* 69: 909–916.

Flavier, A.B., Ganova-Raeva, L.M., Schell, M.A., Denny, T.P. 1997. Hierarchical autoinduction in *Ralstonia solanacearum*: Control of acyl-homoserine lactone production by a novel autoregulatory system responsive to 3-hydroxypalmitic acid methyl ester. *Journal of Bacteriology* 179(22): 7089–7097.

Flavier, A.B., Schell, M.A., Denny, T.P. 1998. An RpoS (ss) homologue regulates acylhomoserine lactone-dependent autoinduction in *Ralstonia solanacearum*. *Molecular Microbiology* 28: 475–486.

Fray, R.G. 2002. Altering plant-microbe interaction through artificially manipulating bacterial quorum sensing. *Annals of Botany* 89(3): 245–253.

Fray, R.G., Throup, J.P., Wallace, A., Daykin, M., Williams, P., Stewart G.S.A.B., Grierson, D. 1999. Plants genetically modified to produce N-acylhomoserine lactones communicate with bacteria. *Nature Biotechnology* 17: 1017–1020.

Gao, M., Teplitski, M., Robinson, J.B., Bauer, W.D. 2003. Production of substance by *Medicago truncatula* that affect bacterial quorum sensing. *Molecular Plant-Microbe Interactactions* 16: 827–834.

Genin, B.B., Denny, T.P., Boucher, C. 2005. Control of the *Ralstonia solanacearum* Type III secretion system (Hrp) genes by the global virulence regulator PhcA. *FEBS Letters* 579(10, 11): 2077–2081.

Givskov, M., de Nys, R., Manefield, M., Gram, L., Maximilien, R., Eberl, L., Molin, S., Steinberg, P.D., Kjelleberg, S. 1996. Eukaryotic interference with homoserine lactone-mediated prokaryotic signalling. *Journal of Bacteriology* 178: 6618–6622.

González, J.F., Degrassi, G., Devescovi, G., Vleesschauwer, D.D., Höfte, M., Myers, M.P., Venturi, V. 2012. A proteomic study of *Xanthomonas oryzae* pv. *oryzae* in rice xylem sap. *Journal of Proteomics* 75(18): 5911–5919.

Han, Y., Chen, F., Li, N., Zhu, B., Li, X. 2010. *Bacillus marcorestinctum* sp. nov., a novel soil acylhomoserine lactone quorum-sensing signal quenching bacterium. *International Journal of Molecular Science* 11(2): 507–520.

Hartmann, A., Gantner, S., Schuhegger, R., Steidle, A., Dürr, C., Schmid, M., Dazzo, F.B., Eberl, L., Langenbartels, C. 2004. *N*-acyl-homoserine lactones of rhizosphere bacteria trigger systemic resistance in tomato plants. In *Biology of Molecular Plant-Microbe Interactions*, Vol. 4. Lugtenberg, B., Tikhonovich, I., Provorov, N. (eds.), *Biology of Molecular Plant-Microbe Interactions*, Vol. 4. St. Paul, MN: MPMI Press, pp. 554–556.

Håvarstein, L.S., Morrison, D.A. 1999. Quorum sensing and peptide pheromones in streptococcal competence for genetic transformation. In *Cell-Cell Signaling in Bacteria*. Dunny, G.M., Winans, S.C. (eds.). Washington, D.C: ASM Press, pp. 9–26.

Hong, K., Koh, C.L., Sam, C.K., Yin, W.F., Chan, K.G. 2012. Quorum quenching revisited from signal decays to signalling confusion. *Sensors* 12: 4661–4696.

Hosni, T., Moretti, C., Devescovi, G., Suarez-Moreno, Z.R., Fatmi, M.B., Guarnaccia, C., Pongor, S. et al. 2011. Sharing of quorum-sensing signals and role of interspecies communities in a bacterial plant disease. *ISME Journal* 5(12): 1857–1870.

Jayaprakashvel, M. 2008. Development of a synergistically performing bacterial consortium for sheath blight suppression in rice. PhD thesis, University of Madras, Chennai, India.

Jones, S., Yu, B., Bainton, N.J., Birdsall, M., Bycroft, B.W., Chhabra, S.R., Cox, A.J. et al. 1993. The Lux autoinducer regulates the production of exoenzyme virulence determination in *Erwinia carotovora* and *Pseudomonas aeruginosa*. *EMBO Journal* 12: 2477–2482.

Kalia, V.C. 2013. Quorum sensing inhibitors: An overview. *Biotechnology Advances* 31(2): 224–245.

Koh, C., Sam, C., Yin, W., Tan, L.Y., Krishnan, T., Chong, Y.M., Chan, K. 2013. Plant-derived natural products as sources of anti-quorum sensing compounds. *Sensors* 13(5): 6217–6228.

Leadbetter, J.R., Greenberg, E.P. 2000. Metabolism of acyl-homoserine lactone quorum-sensing signals by *Variovorax paradoxus*. *Journal of Bacteriology* 182: 6921–6926.

Lenz, D.H., Mok, K.C., Lilley, B.N., Kulkarni, R.V., Wingreen, N.S., Bassler, B.L. 2004. The small RNA chaperone Hfq and multiple small RNAs control quorum sensing in *Vibrio harveyi* and *Vibrio cholerae*. *Cell* 118(1): 69–82.

Licciardello, G., Strano, C.P., Bertani, I., Bella, P., Fiore, A., Fogliano, V., Venturi, V., Catara, V. 2012. N-acyl-homoserine-lactone quorum sensing in tomato phytopathogenic *Pseudomonas* spp. is involved in the regulation of lipodepsipeptide production. *Journal of Biotechnology* 159(4): 274–282.

Loh, J., Pierson, E.A., Pierson III, L.S., Stacey, G., Chatterjee, A. 2002. Quorum sensing in plant-associated bacteria. *Current Opinion in Plant Biology* 5(4): 285–290.

Long, T. 2010. Signal processing in bacterial quorum sensing. PhD thesis. Submitted to Princeton University, Princeton, USA.

Ma, A., Lv, D., Zhuang, X., Zhuang, G. 2013. Quorum quenching in culturable phyllosphere bacteria from tobacco. *International Journal of Molecular Sciences* 14607–14619.

Magnuson, R., Solomon, J., Grossman, A.D. 1994. Biochemical and genetic characterization of a competence pheromone from *B. subtilis*. *Cell* 77(2): 207–216.

Mahmoudi, E., Ahmadi, A. 2012. The effect of quorum sensing inhibitor bacteria on pathogenicity of *Pectobacterium atrosepticum* causal agent of potato blackleg. *Journal of Research in Agricultural Science* 8(2): 171–181.

Marketon, M.M., Gronquist, M.R., Eberhard, A., González, J.E. 2002. Characterization of the *Sinorhizobium meliloti sinR/sinI* locus and the production of novel N-acyl homoserine lactones. *Journal of Bacteriology* 184(20): 5686–5695.

Miller, M.B., Bassler, B.L. 2001. Quorum sensing in bacteria. *Annual Review of Microbiology* 55: 165–199.

Mole, B.M., Baltrus, D.A., Dangl, J.L., Grant, S.R. 2007. Global virulence regulation networks in phytopathogenic bacteria. *Trends in Microbiology* 15(8): 363–371.

Molina, L., Constantinescu, F., Michel, L., Reimmann, C., Duffy, B., Defago, G. 2003. Degradation of pathogen quorum-sensing molecules by soil bacteria: A preventive and curative biological control mechanism. *FEMS Microbiology Ecology* 45: 71–81.

Morohoshi, T., Wang, W.Z., Suto, T., Saito, Y., Ito, S., Someya, N., Ikeda, T. 2013. Phenazine antibiotic production and antifungal activity are regulated by multiple quorum-sensing systems in *Pseudomonas chlororaphis* subsp. *aurantiaca* StFRB508. *Journal of Bioscience and Bioengineering* 116(5): 580–584.

Mukherjee, A., Cui, Y., A.K., Chatterjee, L.Y. 1997. Molecular characterization and expression of the *Erwinia carotovora hrpNEcc* gene, which encodes an elicitor of the hypersensitive reaction. *Molecular Plant-Microbe Interactions* 10: 462–471.

Nealson, K.H., Hastings, J.W. 1979. Bacterial bioluminescence: Its control and ecological significance. *Microbiology Review* 43: 396–518.

Newton, J.A., Fray, R.G. 2004. Integration of environmental and host-derived signals with quorum sensing during plant-microbe interactions. *Cell Microbiology* 6(3): 213–224.

Novick, R.P. 1999. Cell–cell signaling in bacteria. In Dunney, G.M., Winans, S.C. (eds.). Washington, DC: *American Society of Microbiology*, 129–146.

Parsek, M.R., Val, D.L., Hanzelka, B.L., Cronan, J.R., Greenberg, E.P. 1999. Acyl homoserine-lactone quorum-sensing signal generation. *Proceedings of the National Academy of Science of the United States of America* 96: 4360–4365.

Passador, L., Cook, J.M., Gambello, M.J., Rust, L., Iglewski, B.H. 1993. Expression of *Pseudomonas aeruginosa* virulence genes requires cell-to-cell communication. *Science* 260: 1127–1130.

Pérez-Montaño, F., Guasch-Vidal, B., González-Barroso, S., López-Baena, F.J., Cubo, T., Ollero, F.J., Gil-Serrano, A.M. et al. 2011. Nodulation-gene-inducing flavonoids increase overall production of autoinducers and expression of *N*-acyl-homoserine lactone synthesis genes in *Rhizobia*. *Research in Microbiology* 162: 715–723.

Perez-Montano, F., Jimenez-Guerrero, I., Sanchez-Matamoros, R.C., Lopez-Baena, F.J., Ollero, F.J., Rodriguez-Carvajal, M.A., Bellogin, R.A., Espuny, M.R. 2013. Rice and bean AHL-mimic quorum-sensing signals specifically interfere with the capacity to form biofilms by plant-associated bacteria. *Research in Microbiology* 164(7): 749–60.

Piper, K.R., von Bodman, S.B., Farrand, S.K. 1993. Conjugation factor of *Agrobacterium tumefaciens* regulates Ti plasmid transfer by autoinduction. *Nature* 362(6419): 448–450.

Pirhonen, M., Flego, D., Heikinheimo, R., Palva, E. 1993. A small diffusible signal molecule is responsible for the global control of virulence and exoenzyme production in the plant pathogen *Erwinia carotovora*. *EMBO Journal* 12: 2467–2476.

Pustelny, C., Albers, A., Büldt-Karentzopoulos, K., Parschat, K., Chhabra, S.R., Cámara, M., Williams, P., Fetzner, S. 2009. Dioxygenase-mediated quenching of quinolone-dependent quorum sensing in *Pseudomonas aeruginosa*. *Chemistry and Biology* 16(12): 1259–1267.

Qian, G., Fan, J., Chen, D., Kang, Y., Han, B., Hu, B., Liu, F. 2010. Reducing *Pectobacterium virulence* by expression of an N-acyl homoserine lactonase gene Plpp-aiiA in *Lysobacter enzymogenes* strain OH11. *Biological Control* 52(1): 17–23.

Qin, X., Singh, K.V., Weinstock, G.M., Murray, B.E. 2000. Effects of *Enterococcus faecalis fsr* genes on production of gelatinase and a serine protease and virulence. *Infection and Immunity* 68(5): 2579–2586.

Rajesh, P.S., Ravishankar Rai, V. 2014. Quorum quenching activity in cell-free lysate of endophytic bacteria isolated from *Pterocarpus santalinus* Linn. and its effect on quorum sensing regulated biofilm in *Pseudomonas aeruginosa* PAO1. *Microbiological Research* 169(7–8): 561–569.

Rampioni, G., Leoni, L., Williams, P. 2014. The art of antibacterial warfare: Deception through interference with quorum sensing–mediated communication. *Bioorganic Chemistry* 55: 60–68.

Reading, N.C., Sperandio, V. 2006. Quorum sensing: The many languages of bacteria. *FEMS Microbiology Letters* 254(1): 1–11.

Rutherford, S.T., Bassler, B.L. 2012. Bacterial quorum sensing: Its role in virulence and possibilities for its control. *Cold Spring Harbor Perspectives Medicines* 2(11): 1–25.

Ryan, P.R., Delhaize, E. 2001. Function and mechanism of organic anion exudation from plant roots. *Annual Review of Plant Physiology and Molecular Biology* 52: 527–560.

Schacht, T., Unger, C., Pich, A., Wydra, K. 2011. Endo- and exopolygalacturonases of *Ralstonia solanacearum* are inhibited by polygalacturonase-inhibiting protein (PGIP) activity in tomato stem extracts. *Plant Physiology and Biochemistry* 49(4): 377–387.

Schell, M.A. 2000. Control of virulence and pathogenicity genes of *Ralstonia solanacearum* by an elaborate sensory network. *Annual Review of Phytopathology* 38: 263–292.

Schuhegger, R., Ihring, A., Gantner, S., Bahnweg, G., Knappe, C., Vogg, G., Hutzler, P. et al. 2006. Induction of systemic resistance in tomato plants by *N*-acyl-homoserine lactone-producing rhizosphere bacteria. *Plant, Cell and Environment* 29: 909–918.

Schuster, M., Sexton, D.I., Diggle, S.P., Greenberg, E.P. 2013. Acyl-homoserine lactone quorum sensing: From evolution to application. *Annual Review of Microbiology* 67: 43–63.

Skandamis, P.N., Nychas, G.E. 2012. Quorum sensing in the context of food microbiology. *Applied Environmental Microbiology* 78(16): 5473–5482.

Smadja, B., Latour, X., Faure, D., Chevalier, S., Dessaux, Y., Orange, N. 2004. Involvement of *N*-acylhomoserine lactones throughout the plant infection by *Erwinia carotovora* subsp. *Atroseptica* (*Pectobacterium atrosepticum*). *Molecular Plant-Microbe Interactions* 17: 1269–1278.

Smith, K.M., Bu, Y., Suga, H. 2003. Induction and inhibition of *Pseudomonas aeruginosa* quorum sensing by synthetic autoinducer analogs. *Chemistry and Biology* 10: 81–89.

Steindler, L., Venturi, V. 2007. Detection of quorum-sensing N-acyl homoserine lactone signal molecules by bacterial biosensors. *FEMS Microbiology Letters* 266(1): 1–9.

Sturme, M.H., Kleerebezem, M., Nakayama, J., Akkermans, A.D., Vaugha, E.E., de Vos, W.M. 2002. Cell to cell communication by autoinducing peptides in gram-positive bacteria. *Antonie Van Leeuwenhoek* 81(1–4): 233–243.

Taga, M.E., Bassler, B.L. 2003. Chemical communication among bacteria. *Proceeding of the National Academy of Sciences of the United States of America* 100: 14549–14554.

Teplitski, M., Robinson, J.B., Bauer, W.D. 2000. Plants secrete substances that mimic bacterial N-acyl homoserine lactone signal activities and affect population density-dependent behaviors in associated bacteria. *Molecular Plant-Microbe Interaction* 13: 637–648.

Torre-Zavala, S.D.L., Aguilera, S., Ibarra-Laclette, E., Hernandez-Flores, J.L., Hernández-Morales, A., Murillo, J., Alvarez-Morales, A. 2011. Gene expression of Pht cluster genes and a putative non-ribosomal peptide synthetase required for phaseolotoxin production is regulated by GacS/GacA in *Pseudomonas syringae* pv. *phaseolicola*. *Research in Microbiology* 5: 488–498.

Uroz, S., Angelo-Picard, C., Carlier, A., Elasri, M., Sicot, C., Petit, A., Oger, P., Faure, D., Dessaux, Y. 2003. Novel bacteria degrading N-acylhomoserine lactones and their use as quenchers of quorum-sensing-regulated functions of plant-pathogenic bacteria. *Microbiology* 149: 1981–1989.

Vattem, D.A., Mihalik, K., Crixell, S.H., McLean, R.J.C. 2007. Dietary phytochemicals as quorum sensing inhibitors. *Fitoterapia* 78(4): 302–310.

Venturi, V., Venuti, C., Devescovi, G., Lucchese, G., Friscina, A., Degrassi, G., Aguilar, C., Mazzucchi, U. 2004. The plant pathogen *Erwinia amylovora* produces acyl-homoserine lactone signal molecules *in vitro* and in planta. *FEMS Microbiology Letters* 241(2): 179–183.

von Bodman, S.B., Dietz Bauer, W., Coplin, D.L. 2003. Quorum sensing in plant-pathogenic bacteria. *Plant Science,* paper 16.

Walker, T.S., Bais, H.P., Deziel, E., Schweizer, H.P., Rahme, L.G., Fall, R., Vivanco, J.M. 2004. *Pseudomonas aeruginosa*-plant root interactions. Pathogenicity, biofilm formation, and root exudation. *Plant Physiology* 134: 320–331.

Wevers, W., Moons, P., Houdt, R.V., Lurquin, R., Aertsen, A., Michiels, C.W. 2009. Quorum sensing and butanediol fermentation affect colonization and spoilage of carrot slices by *Serratia plymuthica*. *International Journal of Food Microbiology* 134(1–2): 63–69.

Whitehead, NA, Barnard, A.M.L., Slater, H., Simpson, N.J.L., Salmond, G.P.C. 2001. Quorum-sensing in Gram-negative bacteria. *FEMS Microbiology Reviews* 25: 365–404.

Wynendaele, E., Pauwels, E., Van de Wiele, C., Burvenich, C., De Spiegeleer, B. 2012. The potential role of quorum-sensing peptides in oncology. *Medical Hypotheses* 78(6): 814–817.

Xu, F., Byun, T., Deussen, H.-J., Duke, K.R. 2003. Degradation of N-acylhomoserine lactones, the bacterial quorum-sensing molecules by acylase. *Journal of Biotechnology* 101: 89–96.

Zamani, M., Behboudi, K., Ahmadzadeh, M. 2013. Quorum sensing blockade in *Pseudomonas aeruginosa*: Biodegrading of N-acyl homoserine lactone by *Bacillus cereus* UT26 and its consequent effects on their interaction. *Journal of Paramedical Sciences* 4(1): 21–28.

Zhang, L.-H. 2003. Quorum quenching and proactive host defense. *Trends in Plant Science* 8, 238–244.

Zhao, Y., Qian, G., Yin, F., Fan, J., Zhai, Z., Liu, C., Hu, B., Liu, F. 2011. Proteomic analysis of the regulatory function of DSF-dependent quorum sensing in *Xanthomonas oryzae* pv. *oryzicola*. *Microbial Pathogenesis* 50(1): 48–55.

12 Cyanobacteria and Algae

Potential Sources of Biological Control Agents Used against Phytopathogenic Bacteria

Sumathy Shunmugam, Gangatharan Muralitharan, and Nooruddin Thajuddin

CONTENTS

ABSTRACT Cyanobacteria and algae have attracted much attention because of their potential biotechnological applications. These primary producers are ubiquitous in distribution, possess distinct features of evolutionary significance, and are the prolific sources of varied biologically active metabolites with potent biomedical and biotechnological applications. This chapter provides insight into the importance of cyanobacteria and algae as biological control agents against many phytopathogens. It is evident that many cyanobacteria and algae produce unique antibacterial and antifungal bioactive metabolites that are eco-friendly and may be suitably applied to the control of phytopathogens. Biological control through the use of microorganisms offers an alternative and attractive approach for disease control without the negative impact of chemicals. Biocontrol agents have become important for sustainable agriculture because they are easy to deliver, may activate plant resistance mechanisms such as systemic/induced resistance, and thereby can indirectly improve plant growth and yields. The comparative efficacy of cyanobacteria and algal extracts against phytopathogens shows that these organisms have future potential as potent biocontrol agents.

KEYWORDS: cyanobacteria, algae, biocontrol agents, biotechnological potential, phytopathogens

12.1 INTRODUCTION

Plants are often exposed to highly destructive microbial pathogens that can kill them (Subramanian et al., 2011). Such phytopathogens include fungi, bacteria, viruses, insects, pests, and nematodes. Some bacterial, fungal, and viral pathogens operate specifically by infecting single plant species, whereas other pathogens are opportunistic and infect a wide variety of host species (Jiménez et al., 2011). Combating invasions by microbial pathogens as well as controlling plant diseases are very important objectives because of the rising demands for food by the ever-increasing global population. Two strategies are involved in phytopathogen control: application of chemical pesticides and biological control mechanisms that employ microorganisms against plant pathogens.

12.1.1 Application of Chemical Pesticides

Chemical pesticides (bactericides, fungicides, and insecticides) have been used for plant pathogen control since the early twentieth century. Such practices have contributed significantly to enhancing crop productivity and have played a fundamental role in suppressing pests. However, the negative impact of pesticide use on the environment became a major concern in the mid-1970s. Pesticide use is now considered hazardous because, in addition to incurring high costs, its highly toxic chemicals persist in topsoil, leach into groundwater, contribute to the development of resistance in all classes of pests, produce toxicity in humans and animals, and inhibit root and stem growth (Boesten, 2000; Saxena and Pandey, 2001; Nyporko et al., 2002; Mazid, 2011). In addition, pesticides disrupt soil equilibrium among chemical, physical, and biotic components, which damages soil productivity (Abdel-Raouf et al., 2012). Ibraheem (2007) reported that in Egyptian soil, pesticides have toxic effects on nitrogen (N_2)-fixing cyanobacteria *Anabaena subtropica* and *Anabaena variabilis* by inhibiting photosynthetic reactions. These are the organisms that contribute significantly to the soil fertility and steady input of fixed N_2 (Roger et al., 1986). Negative environmental implications of chemical pesticides and concerns about human health, motivate researchers to assist in the development of alternative crop-protection methods. One such alternative measure involves biological control (Natarajan et al., 2013).

12.1.2 Biological Control of Phytopathogens

Biological control is defined as the reduction of the amount of inoculum or the disease-producing activity of a pathogen by methods that involve one or more non-human organisms (Cook and Baker, 1983; Campbell, 1989). Biological control encompasses several approaches such as mass introduction of antagonists, plant breeding, and specific cultural practices aimed at modifying the microbial balance (Alabouvette et al., 2006). Although this strategy was first proposed in the mid-twentieth century, it is only since 2005 that researchers have begun to focus on the replacement of chemical pesticides with biological control agents. Fungi and bacteria are the chief biological agents that have been investigated for the control of plant pathogens, particularly soil-borne fungi. Viruses, amoeba, nematodes, and arthropods have also been considered as possible biocontrol agents (Whipps and McQuilken, 1993; Kulik, 1995).

Algae are a morphologically and geographically diverse group of organisms. They are the primary producers of the marine environment, which covers 71% of the Earth's surface. They also occur in freshwater lakes and ponds as well as in soil, rocks, ice, snow, plants, and animals (Andersen, 1992). They exist in single-cell, colonial, or organisms grouped together with many cells (multicellular) form. In spite of being ubiquitous, and their biotechnological potential, algae have failed to receive as much attention as fungi and bacteria for the potential biocontrol of phytopathogens.

Algae are a heterogenous group of organisms that consists of two major types: 1. Microalgae, which are found in almost all habitats in which conducive sunlight and moisture are available. They also exist as phytoplankton throughout the oceans. Phytoplanktons include diatoms, dinoflagellates, yellow-brown flagellates, and blue-green algae (cyanobacteria). 2. Macroalgae (seaweeds), which occupy the littoral zone as green algae, red algae, and brown algae (El Gamal, 2010).

In this chapter, we highlight the biocontrol potential of algae (particularly cyanobacteria) as well as seaweeds. We also make recommendations to plant pathologists for the utilization of these microorganisms and their products for the efficient control of plant pathogens.

12.1.3 Cyanobacteria

Cyanobacteria are Gram-negative, oxygenic photosynthetic prokaryotes. They convert light into chemical energy using water as an electron donor, and release oxygen (O_2) as a byproduct during photosynthesis. Early unicellular and filamentous cyanobacteria formed 3.5 billion years ago; the endolithic forms appeared about 1.5 billion years ago (Whilmotte, 1994).

It is widely accepted that cyanobacteria are responsible for the formation of atmospheric O_2 and that they are the antecedents of the present-day chloroplasts, algae and green plants (Miyagishima, 2005; Mulkidjanian et al., 2006). They are found in almost all kinds of environments: aquatic, including freshwater and marine; terrestrial, including deserts, rocks, and mountain soils; and extreme conditions, including hot springs and polar regions; as well as in alkaline and acidic conditions, and symbiotic environments (Thajuddin and Subramanian, 2005; Singh et al., 2008; Abed et al., 2009). They are a morphologically diverse group of organisms that exist in unicellular (e.g., *Synechocystis*), colonial (e.g., *Gloeothece*), trichomatous (e.g., *Oscillatoria* and *Spirulina*), filamentous (e.g., *Lyngbya* and *Phormidium*), heterocystous (*Anabaena*, *Nostoc*, and *Calothrix*), false filamentous (*Scytonema*), and heterotrichous (e.g., *Stigonema*) (Figure 12.1) and have been grouped into three orders (Chroococcales, Chaemosiphonales, and Hormogonales) by Geitler (1932); five orders (Chroococcales, Chaemosiphonales, Pleurocapsales, Nostocales, and Stigonemetales) by Desikachary (1959); and four orders (Chroococcales, Oscillatoriales, Nostocales, and Stigonematales) by Anagnostidis and Komárek (1990). Several cyanobacteria have the ability to fix atmospheric N_2, which leads to the formation of ammonia and a byproduct, hydrogen (H_2) (Burris, 1991), using specialized cells called heterocysts (Wolk, 1996; Thiel and Pratte, 2001). Cyanobacteria possess chlorophyll *a*, phycocyanins, and phycoerythrin as photosynthetic pigments. These pigments give the characteristic blue-green color to cyanobacteria, hence known as blue-green algae.

An immense body of knowledge on the diversity and physiology of cyanobacteria serves as an appropriate base for exploring their biotechnological applications (Thajuddin and Subramanian, 2005; Abed et al., 2009; Thajuddin, 2010). Since 2000, cyanobacteria have been considered to be a promising and rich source of bioactive compounds (Bhadury and Wright, 2004) with antibacterial (Jaki et al., 2000; Ali et al., 2008), antifungal (Kajiyama et al., 1998), antiviral (Patterson et al., 1994), anticancer (Gerwick et al., 1994), and immunosuppressive properties (Koehn et al., 1992). However, certain toxic cyanobacteria or harmful algal blooms excrete bioactive compounds, called cyanotoxins, into surrounding waters. These compounds showed inhibitory effects on the growth, photosynthesis, respiration of other algae (Flores and Wolk, 1986; Gross et al., 1991; Shunmugam et al., 2014) and higher plants (MacKintosh et al., 1990; Smith and Doan, 1999; Lehtimaki et al., 2011; Shunmugam et al., 2013). Cyanotoxins have been classified according to the symptoms they cause in humans and vertebrates: hepatoxin (such as microcystin, nodularin, and cylindrospermopsin), neurotoxins (e.g., anatoxin-a, anatoxin-a[s], and saxitoxin) and irritant-dermal toxins (Sivonen and Jones, 1999; Carmichael, 2001). In addition to bioactive compounds, cyanobacterial extracts are known to contain plant growth regulators such as giberellin, indole 3-acetic acid, and indole 3-butyric acid, which promote seed germination and somatic embryogenesis in plant species such as *Daphne* spp. (Wiszniewska et al., 2013), *Daucua carota* L. (Hellebust, 1974), *Gossypium hirsutum* L. (Gurusaravanan et al., 2013), *Arachis hypogaea* L., and *Moringa oleifera* Lam. (Gayathri et al., 2015).

The inherent properties of cyanobacteria that make them attractive candidates for use in biotechnological applications include photosynthetic efficiency, ease of genetic modification, and inexpensive growth requirements (i.e., sunlight, CO_2, and water along with a few essential mineral nutrients). More than 35 genomes of cyanobacteria have been sequenced and their transcriptomes/

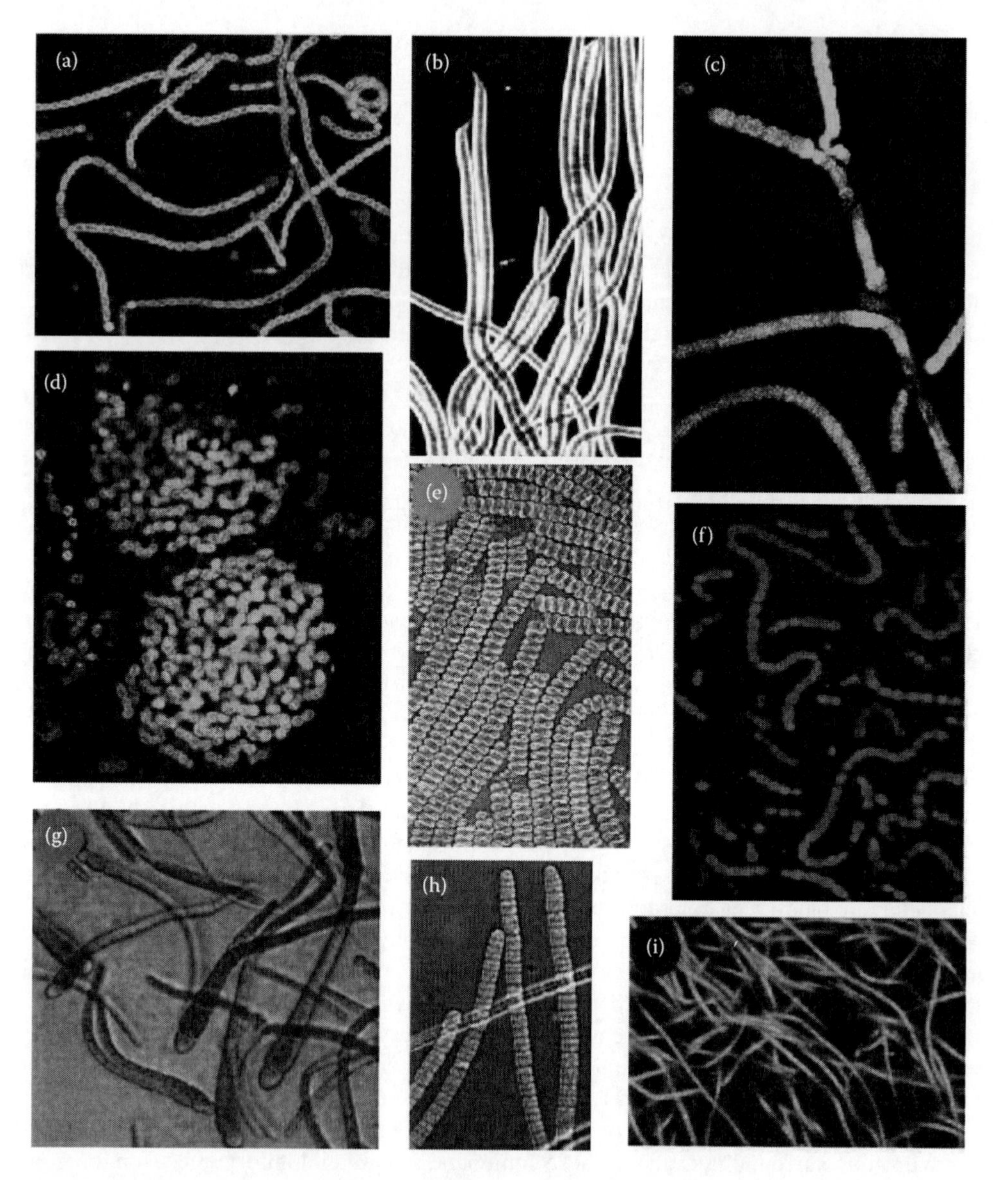

FIGURE 12.1 (a) *Anabaena variabilis*, (b) *Oscillatoria salina*, (c) *Scytonema* sp., (d) *Nostoc punctiforme*, (e) *Spiruling subsalsa*, (f) *Anabaena* sp., (g) *Calothrix* sp., (h) *Oscillatoria tenuis*, and (i) *Phormidium* sp. Figures a, c, d, f, and i-confocal microscopic view. Figure b-dark field view; Figures d, g, and h-bright field view.

metabolomes of them have been characterized, all of which makes these organisms recommended for highly innovative biotechnological approaches (Lu, 2010). Cyanobacteria such as *Spirulina platensis* and *Arthrospira* sp. have been found to contain medically important gamma-linolenic acid, which is converted inside the human body into arachidonic acid and further into prostaglandin E2. This (gamma-linolenic acid) compound lowers the blood pressure and plays an important role in lipid metabolism (Abed et al., 2009). Cyanobacteria also produce commercially important UV-absorbing compounds such as mycosporine-like amino acids and scytonemin (Fleming and Castenholz, 2007). In addition, cyanobacterial pigments such as carotenoids and phycobiliproteins have been extensively utilized in nutraceutical and pharmaceutical industries (Nayak et al., 2007). Phycobilin extract from *Spirulina* sp., which is widely used in the dairy and cosmetic industries, has been shown to contain anti-inflammatory properties (Reddy et al., 2000). Some cyanobacteria have been shown to be potential sources for the large-scale production of vitamins such as B and E. Cyanobacteria are also producers of biotechnologically important thermostable enzymes such as endonucleases (Piechula et al., 2001), prenyl transferase (Ohto et al., 1999), and polyphosphate kinase (Sato et al., 2007). *Spirulina platensis* produce and accumulate biodegradable plastics such as polyhydroxy alkanoates (PHA) under phototrophic and/or mixotrophic growth conditions with acetate (Vincenzini et al., 1990).

Cyanobacteria, along with aerobic organotrophs, constitute ideal consortia for the bioremediation of petroleum products and other organic compounds such as surfactants and herbicides (Radwan and Al-Hasan, 2000; Abed and Köster, 2005). It was reported that hypersaline *Phormidium* strains were able to effectively degrade polycyclic aromatic hydrocarbons such as naphthalene and anthracene (Kumar et al., 2009). The immobilized and free filaments of genus *Phormidium* (*Phormidium* sp. NTMS02) and *Oscillatoria* sp. NTMS01 reported to remove metals such as lead (Pd) (Kumar et al., 2011) and chromium (Cr) (Rajeshwari et al., 2012).

Cyanobacteria are promising organisms for biofuel production when compared with higher plants and eukaryotic algae (Anahas and Muralitharan, 2015). The dominant forms of biofuels currently in use are biohydrogen, bioethanol, and biodiesel. *Synechocystis* PCC 6803 has been employed as model organisms in studies that involve phototrophic biosynthesis of free fatty acids and liquid fuels such as alkyl esters, alkanes, and isoprenoids. Because H_2 has the highest mass-energy density when utilized, it does not generate CO_2 and other pollutants; however, its ability to produce microorganisms make it a suitable candidate for the production of renewable and eco-friendly fossil-fuel alternatives (Edwards et al., 2007).

12.1.4 Macroalgae

Macroalgae (eukaryotic algae) evolved about 1.2 billion years ago when a unicellular eukaryote successfully invaded a photosynthetic cyanobacterium in a process called endosymbiosis. This resulted in a photosynthetic plastid surrounded by two membranes with a greatly reduced cyanobacterial genome (Reyes-Prieto and Bhattacharya, 2007). After the initial occurrence of primary endosymbiosis with cyanobacteria, two evolutionary lineages arose: the green lineage (green algae), characterized by the pigments chlorophyll *a* and *b*; and the red lineage (red algae), characterized by chlorophyll *b* and phycoerythrin (Pulz and Gross, 2004). During the secondary endosymbiosis event, red or green algae were engulfed as plastids into a eukaryotic host; this combination gave rise to heterokant algae, dinoflagellates, cryptophytes, and euglenida (Pulz and Gross, 2004). Algae are mostly photosynthetic, except for a few species that are facultative heterotrophic and some non-photosynthetic species that are obligative heterotrophic (Parker et al., 1961). Algae are found mainly in freshwater or marine environments, but species belonging to about 147 genera were reported to occur in terrestrial habitats (Metting, 1981). To date, approximately 40,000 algal species have been described and around 25,000 species remain unidentified (Guiry, 2012).

These organisms are preferred for the production of biotechnologically important products based on genetic and metabolic engineering approaches because of their combined features of plant systems and bacterial systems, and the availability of a vast variety of unicellular algae for genetic manipulation (Wijffels et al., 2013). Eukaryotic macroalgae has the ability to store secondary metabolites in cellular compartments (e.g., vacuoles) and to efficiently produce and secrete active proteins (in their chloroplasts). These properties make these organisms ideal hosts for biotechnological applications. Moreover, a successful transformation system has been developed that includes techniques to introduce DNA into cells (Kindle et al., 1990), suitable promoters, new selectable marker genes, and expression vectors (Apt and Behrens, 1999) in diatom *Phaeodactylum* and green alga *Chlamydomonas*. Diatoms have the potential to accumulate high amounts of lipids, and improving this quality is a primary aim of biofuel production. Algal biomass in the form of sun-dried or spray-dried powder and tablets are sold mostly for human health food, aquaculture, and animal husbandry. The most commonly used algal biomasses for human consumption are *Chlorella* and *Duneliella*, in addition to cyanobacteria *Spirulina*. Macroalgae have been traditionally used as soil fertilizer, a property that is still being exploited today. They are available as liquid fertilizers, used as surface solidification process against erosion in more arid regions, and have been found to promote germination, plant growth, and flowering. Macroalgae are also popular for the production of economically important products such as polysaccharides (e.g., agar, alginates, and carrageenans) and pigments such as β-carotene, astaxanthin, lutein, zeaxanthin, and canthaxanthin (Pulz and Gross, 2004). Recently, eukaryotic algae were shown to contain antibacterial and anticandidal activity, especially in *Scenedesmus bijugatus* var. *bicellularis* (Ali et al., 2012). In addition, eukaryotic algae are reported to play a pivotal role in the bioremediation of a synthetic dye, Rhodamine B (Baldev et al., 2013).

12.2 ANTIBACTERIAL ACTIVITY OF CYANOBACTERIA

Although the industrial potential of cyanobacterial biocontrol agents is well exploited in Western countries, research in this area in India began only in the last decade. Because cyanobacteria do not produce a particular class of chemicals, their secondary metabolite spectra include 40.2% lipopeptides, 5.65 amino acids, 4.2% fatty acids, 4.2% macrolides, 9.4% amides, and others. The wide spectrum of cyanobacterial secondary metabolites is attributed to their ubiquitous occurrence and long evolutionary history (Singh et al., 2011). To develop new antibiotics, scientists are screening cyanobacterial extracts for antibacterial activity (Skulberg, 2000) and have found them to be potentially active against various bacteria. Unfortunately, very few antibacterial compounds from cyanobacteria have been structurally characterized to date (Table 12.1). Noscomin from *Nostoc commune* showed antibacterial activity against both Gram-positive and Gram-negative bacteria compared with the standard drugs tested (Mundt et al., 2003). Similar antibacterial activity was reported for *Anabaena* extracts by Bhateja et al. (2006). Among the nine ambiguines isolated from *Fisherella* sp., ambiguine-I isonitrile showed more potent antibacterial activity against Gram-positive bacteria than the commercial antibiotics tested (Raveh and Carmeli, 2007). Other chemical compounds belonging to the norbietane family, isolated from the cyanobacterium *Micrococcus lacustris*, showed broad-spectrum resistance to both Gram-positive and Gram-negative bacterial isolates (Gutierrez et al., 2008).

In addition to laboratory research that uses crude extracts of cyanobacterial strains, much attention is being paid to several commercially available cyanobacterial products for activity against

TABLE 12.1
Details of *In Vitro* Antibacterial Activity of Cyanobacteria

Cyanobacteria	Compound	Biological Activity against	References
Nostoc commune	Noscomin	Gram-positive bacteria Gram-negative bacteria	Jaki et al. (1999)
Nostoc sp. CAVN10	Carbamido-cyclophanes	Gram-positive bacteria	Bui et al. (2007)
Fischerella sp.	Ambiguine I isonitrile	Gram-positive and Gram negative bacteria	Raveh and Carmeli (2007)
Micrococcus lacustris	Norbietane diterpenoid	Gram-positive bacteria	Gutierrez et al. (2008)
N. muscorum	Phenolic compound	Gram-positive and Gram-negative bacteria	El-Sheekh et al. (2006)
Fischerella sp.	Hapalindole T	Gram-positive and Gram-negative bacteria	Asthana et al. (2006)
Lyngbya sp.	Pahayokolide A	Gram-positive bacteria	Berry et al. (2004)
Oscillatoria redeki HUB051	Fatty acids	Gram-positive bacteria	Mundt et al. (2003)

phytopathogens. El-Mougy and Abdel-Kader (2013) evaluated the efficacy of two commercial cyanobacteria compounds, Weed-Max and Oligo-Mix, against some soil-borne pathogens. These algal compounds, when supplemented in the growth medium, inhibited the growth of root rot pathogens *Alternaria solani, Fusarium solani, Fusarium oxysporum, Rhizoctonia solani, Sclerotium rolfsii, Sclerotinia sclerotiorum*. These compounds help to reduce root rot disease and improve crop yields when combined with bioagents *Trichoderma harzianum* or *Bacillus subtilis* as integrated soil treatments of vegetable plants such as cucumber, cantaloupe, tomato, and pepper.

12.3 ANTIBACTERIAL ACTIVITY OF MACROALGAE

Marine macroalgae, often known as seaweeds, are a group of algae that are thought to contain biologically active compounds. They have long been used in agriculture, in horticulture to feed livestock, and recently in the form of extracts to promote plant growth (Kulik, 1995). Research on seaweed as a biocontrol agent is quite recent; previous reports on algal biologically active compounds dealt mainly with human pathogenic bacteria, fungi, and food-spoilage microorganisms (Kulik, 1995). Very few reports showed that application of seaweed extracts resulted in direct or indirect protection against pathogens (Subramanian et al., 2011). Details about macroalgal strains, their biological activity, and their target organisms are summarized in Table 12.2.

Lettuce plants (*Lactuca sativa* L.) sprayed with an extract of alga *Ascophyllum nodosum* (L.) during the growing stage reduced plant sensitivity to unspecified disease from 18% (in unsprayed plants) to 12% (Abetz and Young, 1983). Soil application of liquid seaweed extracts to cabbage (*Brassica oleracea* var. *capitata*) reduced the incidence of damping-off disease in seedlings caused by *P. ultimum* (Dixon and Walsh, 2002). Raghavendra et al. (2007) evaluated the effect of a commercial product of seaweed, *Sargassum wightii* (Dravya), on bacterial blight caused by phytopathogen *Xanthamonas campestris* in cotton. The seeds soaked with Dravya (1:500) followed by three foliar sprayings at 10-day intervals (10, 20, and 30 days after soaking) showed not only a reduction in blight appearance but also increased plant height, total number of bolls formed, boll weight, stem growth, chlorophyll content, total phenols, and peroxidase activity.

Kumar et al. (2008) screened 12 different seaweed extracts (*Chaetomorpha antennina, Dictyota dichotoma, Enteromorpha flexuosa, Laurencia obtusa, Gracilaria corticata, G. verrucosa, Grateloupia lithophila, Padina boergesenii, S. wightii, Turbinaria conoides, Halimeda tuna*, and *Ulva lactuca in vitro* against the phytopathogenic bacterium *Pseudomonas syringae*, which causes leaf spot disease in the medicinal plant *Gymnema sylvestre*. The maximum activities were recorded in methanolic extract, followed by ethyl acetate extracts of the brown seaweed *S. wightii*,

TABLE 12.2
Details of *In Vitro* Antibiotic Activity of Macroalgae toward Different Target Organisms

Macroalgae	Biological Activity	Target Organisms	References
S. wightii	Antibacterial	*Xanthomonas oryzae* pv. *oryzae*	Mariadoss (1998)
Enteromorpha flexuosa			
Delisia pulchra	Antibacterial	*Erwinia carotovora*	Manefield et al. (2001)
Gracilaria corticata	Antibacterial	*Pseudomonas aeruginosa*	Choudhury et al. (2005)
S. wightii Greville	Antibacterial	*X. campestris* pv. *Malvacearum*	Raghavendra et al. (2007)
S. wightii	Antibacterial	*P. syringae*	Kumar et al. (2008)
Turbinaria conoides			
Ascophyllum nodosum	Antibacterial	*P. syringae*	Subramanian et al. (2011)
Lessonia trabeculata	Antibacterial	*Botrytis cineria*	Jiménez et al. (2011)
Gracillaria chilensis			

and by ethyl acetate of the brown algae *T. conoides*. Of the 12 seaweeds tested, 70% activity was recorded in brown seaweed, 30% in red seaweed, and 15% in green seaweed. *Sargassum wightii* had already been reported to be an excellent source of antibacterial agents in the control of the bacterial blight of rice caused by *X. oryzae* (Arunkumar and Rengasamy, 2000a,b; Arunkumar et al., 2005). Mariadoss (1998) reported the activity of a methanolic extract of *S. wightii* and chloroform and a methanol extract of *E. flexuosa* against a Gram-negative bacterium, *X. oryzae* pv. *oryzae*, the causal organism of leaf blight of rice.

Hexadecatrienoic acid isolated from *P. tricornutum* displays activity against Gram-positive microbes. High levels of palmitoleic acid and other bioactive fatty acids were also found in the fusiform morphotype of *P. tricornutum*, rather than in the oval morphs of this microalga; that fatty acid is active against various Gram-positive pathogens at micromolar concentrations and its lethal effects start immediately upon exposure (Smith et al., 2010). The antimicrobial activity against a Gram-negative bacterium shown by pressurized (liquid) ethanol extracts from *Haematococcus pluvialis* in its red stage was associated with the presence of short-chain fatty acids, namely butanoic and methyl lactic acids (Santoyo et al., 2009). The exact mechanism of action of fatty acids remains unknown, however. They may act upon multiple cellular targets, but cell membranes are the most probable targets because membrane damage will likely lead to cell leakage and reduction of nutrient uptake, along with inhibiting cellular respiration.

It has been proposed that the induction of systemic resistance (SR) is an effective strategy to protect plants from pathogen attacks. The plants are protected from invading organisms through the activation of genes and biochemical pathways. Such processes are induced by chemicals that are commonly known as elicitors. Marine algae are rich in unique polysaccharides; sometimes, these can be potent elicitors of plant defense responses (Mercier et al., 2001; Cluzet et al., 2004; Sangha et al., 2010). Sangha et al. (2010) reported that a polysaccharide, λ-carrageenan, induced a jasmonic acid (JA)-dependent response by *Arabidopsis* to *S. sclerotiorum*. Similarly, carrageenans of certain red algae elicited a salicylic acid (SA)-dependent response in tobacco against *P. parasitica* var. *nicotianae* (Mercier et al., 2001). Subramanian et al. (2011) studied the mechanism of *A. nodosum*, a brown macroalgae, which induced resistance in *A. thaliana* against *P. syringae* pv. *tomato* DC3000. This study revealed that when *A. nodosum* extracts were applied to *A.* roots, the extracts induced JA-dependent systemic resistance in leaves against *P. syringae* pv. *tomato* DC3000. The brown alga *A. nodosum* is the most widely used seaweed in agriculture- and horticulture-crop protection (Rayorath et al., 2008). *Ascophyllum nodosum* contains laminaran (β-D-(1 → 3) glucan) that elicits plant growth and defense responses by the induction of antimicrobial phytoalexins (Patier et al., 1993).

12.4 CONCLUSION

Although the plant-growth-promoting activities of algae in the form of biofertilizers have been known for centuries, only recently have studies focused on the role of their metabolites in phytopathogen control. In this chapter, the different beneficial roles of algae and cyanobacteria in biological control of phytopathogens were considered. As important components of agriculture and plant ecosystems, algae and cyanobacteria have the ability to improve soil and plant properties. Several of the biological compounds reported herein have shown microbicidal/static activity toward the phytopathogenic microbes tested. Despite these promising results, however, much work remains to be done not only on the exact chemical nature of the bioactive principle but also on the mechanisms of microbicidal/static activity toward targeted phytopathogenic organisms.

ACKNOWLEDGMENTS

The authors are grateful to the Department of Biotechnology (DBT, Govt. of India), for financial support (Project Ref. No.: BT/PR6619/PBD/26/310/14.02.2013) and to the Department of Science and Technology (DST, Government of India) for allowing our use of the laser-scanning confocal microscope purchased through the PURSE Program. The first author, Sumathy Shunmugam, acknowledges Dr. D. S. Kothari Post-Doctoral Scheme of the University Grants Commission (UGC), New Delhi, India for fellowship.

REFERENCES

Abdel-Raouf, N., Al-Homaidan, A.A., Ibraheem, I.B.M. 2012. Agricultural importance of algae. *African Journal of Biotechnology* 11: 11648–11658.

Abed, R.M.M., Dobretsov, S., Sudesh, K. 2009. Applications of cyanobacteria in biotechnology. *Journal of Applied Microbiology* 106: 1–12.

Abed, R.M.M., Köster, J. 2005. The direct role of aerobic heterotrophic bacteria associated with cyanobacteria in the degradation of oil compounds. *International Biodeterioration and Biodegradation* 55: 29–37.

Abetz, P., Young, C.L. 1983. The effect of seaweed extract sprays derived from *Ascophyllum nodosum* on lettuce and cauliflower crops. *Botanica Marina* 26: 487–492.

Alabouvette, C., Olivain, C., Steinberg, C. 2006. Biological control of plant diseases: The European situation. *European Journal of Plant Pathology* 114: 329–341.

Ali, D.M., Kumar, R.P., Shenbagavalli, T., Nivetha, T.M., Ahamed, A.P., Al-Dhabi, N.A., Thajuddin, N. 2012. New reports on anti-bacterial and anti-candidal activities of fatty acid methyl esters (FAME) obtained from *Scenedesmus bijugatus* var. *bicellulari* biomass. *RSC Advances* 2: 11552–11556.

Ali, D.M., Kumar, T.V., Thajuddin, N. 2008. Screening of some selected hypersaline cyanobacterial isolates for biochemical and antibacterial activity. *Indian Hydrobiology* 11: 241–246.

Anagnostidis, K., Komárek, J.A. 1990. Modern approach to the classification systems of cyanophytes. 5-stigonematales. *Algological Studies* 59: 1–73.

Anahas, A.M.P., Muralitharan, G. 2015. Isolation and screening of heterocystous cyanobacterial strains for biodiesel production by evaluating the fuel properties from fatty acid methyl ester (FAME) profiles. *Bioresource Technology* 184: 9–17.

Andersen, R.A. 1992. Diversity of eukaryotic algae. *Biodiversity Conservation* 1: 267–292.

Apt, K.A., Behrens, P.W. 1999. Commercial developments in microalgal biotechnology. *Journal of Phycology* 35: 215–226.

Arunkumar, K., Rengasamy, R. 2000a. Antibacterial activities of seaweed extracts/fractions obtained through a TLC profile against phytopathogenic bacterium *Xanthomonas oryzae* sp *oryzae*. *Botanica Marina* 43: 417–421.

Arunkumar, K., Rengasamy, R. 2000b. Evaluation of antibacterial potential of seaweeds occurring along the coast of Tamilnadu, India against the plant pathogenic bacterium *Xanthomonas oryzae* Pv. *oryzae* (Ishiyama) dye. *Botanica Marina* 43: 409–415.

Arunkumar, K., Selvapalam, N., Rengasamy, R. 2005. The antibacterial compound sulphoglycerolipid 1-0 palmitoyl-3-0 (6′-sulpho-α-quinovopyranosyl)-glycerol from *Sargassum wightii*. *Botanica Marina* 48: 441–445.

Asthana, R.K., Srivastava, A., Singh, A.P., Deepali, Singh, S.P., Nath, G., Srivastava, R., Srivastava, B.S. 2006. Identification of an antimicrobial entity from *Fischerella* sp. colonizing neem tree bark. *Journal of Applied Phycology* 18: 33–39.

Baldev, E., Ali, D.M., Ilavarasi, A., Pandiaraj, D., Sheik Syed Ishack, K.A., Thajuddin, N. 2013. Degradation of synthetic dye, Rhodamine B to environmentally non-toxic products using microalgae. *Colloid Surface B: Biointerfaces* 105: 207–214.

Berry, J., Gantar, M., Gawley, R.E., Wang, M., Rein, K.S. 2004. Pharmacology and toxicology of phayokolide A, a bioactive metabolite from a fresh water species of *Lyngbya* isolated from the Florida everglades. *Comparative Biochemistry and Physiology C* 139: 231–238.

Bhadury, P., Wright, P.C. 2004. Exploitation of marine algae: Biogenic compounds for potential antifouling applications. *Planta* 219: 561–578.

Bhateja, P., Mathur, T., Pandya, M., Fatma, T., Rattan, A. 2006. Activity of blue green microalgae extracts against in vitro generated *Staphylococcus aureus* with reduced susceptibility to vancomycin. *Fitoterpia* 77: 233–235.

Boesten, J.J.T.I. 2000. From laboratory to field: Uses and limitations of pesticide behaviour models for the soil/plant system. *Weed Research* 40: 23–138.

Bui, T.N., Jansen, R., Pham, T.L., Mundt, S. 2007. Carbamidocyclophanes A–E, chlorinated paracyclophanes with cytotoxic and antibiotic activity from the Vietnamese cyanobacterium *Nostoc* sp. *Journal of Natural Products* 70: 499–503.

Burris, R.H. 1991. Nitrogenases. *Journal of Biological Chemistry* 226: 9339–9342.

Campbell, R. 1989. *Biological Control of Microbial Plant Pathogens*. Cambridge, UK: Cambridge Univeristy Press, 218 p.

Carmichael, W.W. 2001. Health effects of toxin producing cyanobacteria: The CyanoHABs. *Human and Ecological Risk Assessment* 7: 1393–1407.

Cluzet, S., Torregrosa, A., Jacquet, C., Lafitte, C., Fournier, L., Mercier, L., Salamagne, S., Briand, X., Esquerré-Tugayé, M.T., Dumas, B. 2004. Gene expression profiling and protection of *Medicago truncatula* against a fungal infection in response to an elicitor from green algae *Ulva* spp. *Plant, Cell and Environment* 27: 917–928.

Cook, R., Baker, K.F. 1983. *The Nature and Practice of Biological Control of Plant Pathogens*. St. Paul, MN: American Phytopathological Society, 539 p.

Desikachary, T.V. 1959. *Cyanophyta*. New Delhi: Indian Council of Agricultural Research, 686 p.

Dixon, G.R., Walsh, U.F. 2002. Suppressing *Pythium ultimum* induced damping-off in cabbage seedlings by biostimulation with proprietary liquid seaweed extracts managing soil-borne pathogens: A sound rhizosphere to improve productivity in intensive horticultural systems. Proceedings of XXVI International Horticultural Congress, Toronto, Canada, pp. 11–17.

Edwards, P.P., Kuznetsov, V.L., David, W.I.F. 2007. Hydrogen energy. *Philosophical Transctions of the Royal Society of London* 365: 1043–1056.

El Gamal, A.A. 2010. Biological importance of marine algae. *Saudi Pharmaceutical Journal* 18: 1–25.

El-Mougy, N.S., Abdel-Kader, M.M. 2013. Effect of commercial cyanobacteria products on the growth and antagonistic ability of some bioagents under laboratory conditions. *Journal of Pathogens* 2013: 11 p., doi:10.1155/2013/838329.

El-Sheekh, M.M., Osman, M.E.H., Dyab, M.A., Amer, M.S. 2006. Production and characterization of antimicrobial active substance from cyanobacterium *Nostoc muscorum*. *Environmental Toxicology* 21: 42–50.

Fleming, E.D., Castenholz, R.W. 2007. Effects of periodic desiccation on the synthesis of the UV-screening compound, scytonemin, in cyanobacteria. *Environmental Microbiology* 9: 1448–1455.

Flores, E., Wolk, C.P. 1986. Production, by filamentous, nitrogen-fixing cyanobacteria, of a bacteriocin and of other antibiotics that kill related strains. *Archives of Microbiology* 145: 215–219.

Gayathri, M., Kumar, P.S., Prabha, A.M.L., Muralitharan, G. 2015. *In vitro* regeneration of *Arachis hypogaea* L. and *Moringa oleifera* Lam. using extracellular phytohormones from *Aphanothece* sp. MBDU 515. *Algae Research* 7: 100–105.

Geitler, L. 1932. *Cyanophyceae. Rabenhorst's Kryptogamen-Flora von Deutschland, Österreich und der Schweiz*. Leipzig, Germany: Akademische Verlagsgesellschaft.

Gerwick, W.H., Roberts, M.A., Proteau, P.J., Chen, J.L. 1994. Screening cultured marine microalgae for anticancer-type activity. *Journal of Applied Phycology* 6: 143–149.

Gross, E.M., Wolk, C.P., Jüttner, F. 1991. Fischerellin, a new allelochemical from the freshwater cyanobacterium *Fischerella muscicola*. *Journal of Phycology* 27: 686–692.

Guiry, M.D. 2012. How many species of algae are there? *Journal of Phycology* 48: 1057–1063.

Gurusaravanan, P., Vinoth, S., Kumar, M.S., Thajuddin, N., Jayabalan, N. 2013. Effect of cyanobacteria extracellular products on high-frequency *in vitro* induction and elongation of *Gossypium hirsutum* L. organs through shoot apex explants. *Journal of Genetic Engineering and Biotechnology* 11: 9–16.

Gutierrez, R.M.P., Flores, A.M., Solis, R.V., Jimenez, J.C. 2008. Two new antibacterial norbietane diterpenoids from cyanobacterium *Micrococcus lacustris*. *Journal of Natural Medicines* 62: 328–331.

Hellebust, J.A. 1974. Extracellular products. In *Algal Physiology and Biochemistry*. Stewart, W.D.P. (ed.). Oxford: Blackwell Scientific Publications, pp. 838–863.

Ibraheem, I.B.M. 2007. Cyanobacteria as alternative biological conditioners for bioremediation of barren soil. *Egyptian Journal of Phycology* 8: 99–116.

Jaki, B., Heilmann, J., Sticher, O. 2000. New antibacterial metabolites from the cyanobacterium *Nostoc commune* (EAWAG 122b). *Journal of Natural Products* 63: 1283–1285.

Jaki, B., Orjala, J., Sticher, O. 1999. A novel extracellular diterpenoid with antibacterial activity from the cyanobacterium. *Nostoc commune*. *Journal of Natural Products* 62: 502–503.

Jiménez, E., Dorta, F., Medina, C., Ramírez, A., Ramírez, I., Peña-Cortés, H. 2011. Anti-phytopathogenic activities of macro-algae extracts. *Mar Drugs* 9: 739–756.

Kajiyama, S., Kanazaki, H., Kawazu, K., Kobayashi, A. 1998. Nostifungicidine, an antifungal lipopeptide from the field-grown terrestrial blue-green algae *Nostoc commune*. *Tetrahedron Lett* 39: 3737–3740.

Kindle, K.L., Richards, K.L., Stern, D.B. 1990. Engineering the chloroplast genome: Techniques and capabilities for chloroplast transformation in *Clamydomonas reinhardtii*. *Proc Natl Acad Sci USA* 88: 1721–1725.

Koehn, F.E., Lomgley, R.E., Reed, J.K. 1992. Microcolins A and B, new immunosuppressive peptide from the blue-green algae *Lyngbya majuscule*. *Journal of Natural Products* 55: 613–619.

Kulik, M.M. 1995. The potential for using cyanobacteria (blue-green algae) and algae in the biological control of plant pathogenic bacteria and fungi. *European Journal of Plant Pathology* 101: 5855–99.

Kumar, M.S., Muralitharan, G., Thajuddin, N. 2009. Screening of a hypersaline cyanobacterium *Phormidium tenue* for the degradation of aromatic hydorcarbons: Naphthalene and anthracene. *Biotechnology Letters* 31: 1863–1866.

Kumar, M.S., Rajeshwari, K., Johnson, S., Thajuddin, N., Gunasekaran, M. 2011. Removal of Pb (II) by immobilized and free filaments of marine *Oscillatoria* sp. NTMS01 and *Phormidium* sp. NTMS02. *Bulletin of Environmental Contamination and Toxicology* 87: 254–259.

Kumar, C.S., Sarada, D.V.L., Rengasamy, R. 2008. Seaweed extracts control the leaf spot disease of the medicinal plant *Gymnema sylvestre*. *Indian Journal of Science and Technology* 1: 1–5.

Lehtimaki, N., Shunmugam, S., Jokela, J., Wahlsten, M., Carmel, D., Keränen, M., Sivonen, K. et al. 2011. Nodularin uptake and induction of oxidative stress in spinach (*Spinachia oleracea*). *Journal of Plant Physiology* 168: 594–600.

Lu, X. 2010. A perspective: Photosynthetic production of fatty acid-based biofuels in genetically engineered cyanobacteria. *Biotechnology Advances* 28: 742–746.

MacKintosh, C., Beattie, K.A., Klumpp, S., Cohen, P., Codd, G.A. 1990. Cyanobacterial microcystin-LR is a potent and specific inhibitor of protein phosphatases 1 and 2A from both mammals and higher plants. *FEBS Letters* 264: 187–192.

Manefield, M., Welch, M., Givskov, M., Salmond, G.P.C., Kjelleberg, S. 2001. Halogenated furanones from the red alga, *Delisea pulchra*, inhibit carbapenem antibiotic synthesis and exoenzyme virulence factor production in the phytopathogen *Erwinia carotovora*. *FEMS Microbiology Letters* 205: 131–138.

Mariadoss, K.L. 1998. Studies on extraction, isolation and characterization of bioactive compounds from seaweeds and their effect on bacterial blight of rice caused by *Xanthomonas oryzae* pv. *oryzae* (Ishiyama) dye. PhD Thesis. University of Madras, Chennai, India.

Mazid, S. 2011. A review on the use of biopesticides in insect pest management. *International Journal of Advanced Science Technology* 1: 169–178.

Mercier, L., Lafitte, C., Borderies, G., Briand, X., Esquerré-Tugayé, M.-T., Fournier, J. 2001. The algal polysaccharide carrageenans can act as an elicitor of plant defence. *New Phytologist* 149: 43–51.

Metting, B. 1981. The systematics and ecology of soil algae. *Botanical Review* 47: 195–311.

Miyagishima, S. 2005. Origin and evolution of the chloroplast division machinery. *Journal of Plant Research* 118: 295–306.

Mulkidjanian, A.Y., Koonin, E.V., Makarova, K.S., Mekhedov, S.L., Sorokin, A., Wolf, Y.I., Dufresne, A. et al. 2006. The cyanobacterial genome core and the origin of photosynthesis. *Proceedings of National Academy of Science USA* 103: 13126–13131.

Mundt, S., Kreitlow, S., Jansen, R. 2003. Fatty acids with antibacterial activity from the cyanobacterium *Oscillatoria redekei* HUB051. *Journal of Applied Phycology* 15: 263–267.

Natarajan, C., Gupta, V., Kumar, K., Prasanna, R. 2013. Molecular characterization of a fungicidal endoglucanase from the cyanobacterium *Calothrix elenkinii*. *Biochem Genet* 51: 766–779.

Nayak, S., Prasanna, R., Prasanna, B.M., Sahoo, D.B. 2007. Analysing diversity among Indian isolates of *Anabaena* (Nostocales, Cyanophyta) using morphological, physiological and biochemical characters. *World Journal of Microbiology and Biotechnology* 23: 1575–1584.

Nyporko, A.Y., Yemets, A.I., Klimkina, L.A., Blume, Y.B. 2002. Sensitivity of *Eleusine indica* callus to trifluralin and amiprophosmethyl in correlation to the binding of these compounds to tubulin. *Russian Journal of Plant Physiology* 49:413–418.

Ohto, P., Ishida, C., Nakane, H., Muramatsu, M., Nishino, T., Obata, S. 1999. A thermophilic cyanobacterium *Synechococcus elongatus* has three different Class I prenyltransferase genes. *Plant Molecular Biology* 40: 1573–1628.

Parker, B.C., Bold, H.C., Deason, T.R. 1961. Facultative heterotrophy in some Chlorococcacean algae. *Science* 133: 761–763.

Patier, P., Yvin, J.C., Kloareg, B., Lienart, Y., Rochas, C. 1993. Seaweed liquid fertilizer from *Ascophyllum nodosum* contains elicitors of plant D-glycanases. *Journal of Applied Phycology* 5: 343–349.

Patterson, G.M.L, Larsen, L.K., Moore, R.E. 1994. Bioactive natural products from blue-green algae. *Journal of Applied Phycology* 6: 151–157.

Piechula, S., Waleron, K., Swiatek, W., Biedrzycka, I., Podhajska, A.J. 2001. Mesophilic cyanobacteria producing thermophilic restriction endonucleases. *FEMS Microbiology Letters* 198: 135–140.

Pulz, O., Gross, W. 2004. Valuable products from biotechnology of microalgae. *Applied Microbiology and Biotechnology* 65: 635–648.

Radwan, S.S., Al-Hasan, R.H. 2000. Oil pollution and cyanobacteria. In *The Ecology of Cyanobacteria*. Whitton, B.A., Potts, M. (eds.). Dordrecht, The Netherlands: Kluwer Academic Publishers, pp. 307–319.

Raghavendra, V.B., Lokesh, S., Prakash, H.S. 2007. Dravya, a product of seaweed extract (*Sargassum wightii*), induces resistance in cotton against *Xanthomonas campestris* pv. *malvacearum*. *Phytoparasitica* 35: 442–449.

Rajeshwari, K., Kumar, M.S., Thajuddin, N. 2012. Adsorption isotherms for Cr (VI) by two immobilized marine cyanobacteria. *Annals of Microbiology* 62: 241–246.

Raveh, A., Carmeli, S. 2007. Antimicrobial ambiguines from the cyanobacterium *Fisherella* sp. collected in Israel. *Journal of Natural Products* 70: 196–201.

Rayorath, P., Khan, W., Palanisamy, R., MacKinnon, S.L., Stefanova, R., Hankins, S.D., Critchley, A.T., Prithiviraj. B. 2008. A brown seaweed, *Ascophyllum nodosum*, induces gibberellic acid (GA3) independent amylase activity in barley. *Journal of Plant Growth Regulation* 27: 370–379.

Reddy, C.M., Bhat, V.B., Kiranmai, G., Reddy, M.N., Reddanna, P., Madyastha, K.M. 2000. Selective inhibition of cyclooxygenase-2 by C-phycocyanin, a biliprotein from *Spirulina platensis*. *Biochemical and Biophysical Research Communications* 3: 599–603.

Reyes-Prieto, A., Bhattacharya, D. 2007. Phylogeny of nuclear-encoded plastid-targeted proteins supports an early divergence of glaucophytes within plantae. *Molecular Biology and Evolution* 24: 2358–61.

Roger, P.A., Ardales, T.S., Watanabe, I. 1986. Chemical composition of cultures and natural samples of N_2-fixing blue-green algae from rice fields. *Biology and Fertility of Soils* 2: 131–146.

Sangha, J.S., Ravichandran, S., Prithiviraj, K., Critchley, A.T., Prithiviraj, B. 2010. Sulfated macroalgal polysaccharides λ-carrageenan and ι-carrageenan differentially alter *Arabidopsis thaliana* resistance to *Sclerotinia sclerotiorum*. *Physiological and Molecular Plant Pathology* 75: 38–45.

Santoyo, S., Rodríguez-Meizoso, I., Cifuentes, A., Jaime, L., García-Blairsy Reina, G., Señorans, F.J., Ibáñez, E. 2009. Green processes based on the extraction with pressurized fluids to obtain potent antimicrobials from *Haematococcus pluvialis* microalgae. *LWT—Food Science and Technology* 42: 1213–1218.

Sato, M., Masuda, Y., Kirimura, K., Kino, K. 2007. Thermostable ATP regeneration system using polyphosphate kinase from *Thermosynechococcus elongatus* BP-1 for D-amino acid dipeptide synthesis. *Journal of Bioscience and Bioengineering* 103: 179–184.

Saxena, S., Pandey, A.K. 2001. Microbial metabolites as eco-friendly agrochemicals for the next millennium. *Applied Microbiology and Biotechnology* 55: 395–403.

Shunmugam, S., Hinttala, R., Lehtimäki, N., Mittinen, M., Uusimaa, J., Majamma, K., Sivonen, K. et al. 2013. *Nodularia spumigena* extract induces upregulation of mitochondrial respiratory chain complexes in spinach (*Spinacia oleracea* L.). *Acta Physiologiae Plantarum* 35: 969–974.

Shunmugam, S., Jokela, J., Wahlsten, M., Battchikova, N., Rehman, A.U., Vass, I., Karonen, M. et al. 2014. Secondary metabolite from *Nostoc* XPORK14A inhibits photosynthesis and growth of *Synechocystis* PCC 6803. *Plant, Cell and Environment* 37: 1371–1381.

Singh, S.M., Singh, P., Thajuddin N. 2008. Biodiversity and distribution of Cyanobacteria at Dronning Maud Land, East Antarctica. *Acta Botanica Malacitana* 33:17–28.

Singh, R.K., Tiwari, S.P., Rai, A.K., Mohapatra, T.M. 2011. Cyanobacteria: An emerging source for drug discovery. *The Journal of Antibiotics* 64: 401–412.

Sivonen, K., Jones, G. 1999. Cyanobacterial toxins. In *Toxic cyanobacteria inwater. A Guide to Their Public Health Consequences, Monitoring and Management, Published on Behalf of the World Health Organization*. Chorus, I., Bartram, J. (eds.). London: E & FN Spon, pp. 41–112.

Skulberg, O.M. 2000. Microalgae as a source of bioactive molecules; experience from cyanophyte research. *Journal of Applied Microbiology* 12: 341–348.

Smith, G.D., Doan, N.T. 1999. Cyanobacterial metabolites with bioactivity against photosynthesis in cyanobacteria, algae and higher plants. *Journal of Applied Phycology* 11: 337–344.

Smith, V.J., Desbois, A.P., Dyrynda, E.A. 2010. Conventional and unconventional antimicrobials from fish, marine invertebrates and micro-algae. *Marine Drugs* 8: 1213–1262.

Subramanian, S., Sangha, J.S., Gray, B.A., Singh, R.P., Hiltz, D., Critchley, A.T., Prithiviraj, B. 2011. Extracts of the marine brown macroalga, *Ascophyllum nodosum*, induce jasmonic acid dependent systemic resistance in *Arabidopsis thaliana* against *Pseudomonas syringae* pv. tomato DC3000 and *Sclerotinia sclerotiorum*. *European Journal of Plant Pathology* 131: 237–248.

Thajuddin, N. 2010. Biotechnological potentials of Cyanobacteria and their industrial applications. In *Industrial Exploitation of Microorganisms*. Maheshwari, D.K., Dubey, R.C., Saravanamuthu, R. (eds.). New Delhi: IK International Publishing House, pp. 70–85.

Thajuddin, N., Subramanian, G. 2005. Cyanobacterial biodiversity and potential applications in biotechnology. *Current Science* 89: 47–57.

Thiel, T., Pratte, B. 2001. Effect on heterocyst differentiation of nitrogen fixation in vegetative cells of the cyanobacterium *Anabaena variabilis* ATCC 29413. *Journal of Bacteriology* 183: 280–286.

Vincenzini, M., Sili, C., Philippis, R., Ena, A., Materassi, R. 1990. Occurrence of poly-b-hydroxybutyrate in *Spirulina* species. *Journal of Bacteriology* 172: 2791–2792.

Whilmotte, A. 1994. Molecular evolution and taxonomy of the cyanobacteria. In *The Molecular Biology of Cyanobacteria*. Bryant, D.A. (ed.). Dordrecht, The Netherlands: Kluwer Academic Publishers, pp. 1–25.

Whipps, J.M., McQuilken, M.P. 1993. Aspects of biocontrol of fungal plant pathogens. In *Exploitation of Microorganisms*. Jones, G. (ed.). . London, UK: Chapman and Hall, pp. 45–79.

Wijffels, R.H., Kruse, O., Hellingwerf, K.J. 2013. Potential of industrial biotechnology with cyanobacteria and eukaryotic microalgae. *Current Opinion in Biotechnology* 24: 405–413.

Wiszniewska, A., Hanus-Fajerska, E., Grabski, K., Tukaj, Z. 2013. Promoting effects of organic medium supplements on the micropropagation of promising ornamental *Daphne* species (Thymelaeaceae). *In Vitro Cellular and Developmental Biology-Plant* 49: 51–59.

Wolk, C.P. 1996. Heterocyst formation. *Annual Review of Genetics* 30: 59–78.

13 Arbuscular Mycorrhizal Fungi-Mediated Control of Phytopathogenic Bacteria

Karunakaran Rojamala, Perumalsamy Priyadharsini, and Thangavelu Muthukumar

CONTENTS

ABSTRACT Bacteria are one of the major causal agents of the plant diseases that create great economic losses in agriculture. An effective biocontrol of bacterial pathogens is essential to reduce the environmental risks posed by chemical control agents. A biocontrol agent for the control of plant diseases should colonize the rhizosphere and reduce the population or activity of the pathogens, but should not leave behind any toxic residues. The natural activities of microorganisms in soil or rhizospheres may contribute to the biocontrol of pathogens and improve the supply of nutrients, thereby enabling plants to maintain their health and fitness and produce higher yields. The various processes of arbuscular mycorrhizal fungal (AMF) symbiosis, which can also alleviate the negative effects of plant pathogens, are the preferred choice for the control of plant pathogens. Nevertheless, the effect of AMF on pathogenic bacteria varies with the host plant. AMF uses several mechanisms to control bacterial plant pathogens; in addition, various factors determine the efficiency of AMF as a disease-control agent. AMF can protect plants from pathogens in both their vegetative and reproductive phases. The ability of AMF to control plant diseases improves with the application of organic amendments. AMF also stimulates the activity of beneficial microorganisms in the rhizosphere that are antagonistic to bacterial plant pathogens. Although studies are limited on the influence of AMF on plant pathogenic bacteria, available evidence suggests that AMF can be used as a successful biocontrol agent.

KEYWORDS: plant pathogenic bacteria, arbuscular mycorrhizal fungi, biocontrol, phytohormones, root exudates

13.1 INTRODUCTION

Plants are highly susceptible to root and soil-borne diseases that cause great losses in yield and quality (Sharma et al., 2004). Control of plant diseases is highly essential for maintainance of the quality and abundance of food, feed, and fiber produced by growers worldwide (Junaid et al., 2013). Losses in yield and reductions in the quality of products caused by the plant pathogens have prompted intensive research on the biology and control of soil-borne pathogens in crop production. Some of the main causative organisms of plant diseases include fungi, nematodes, bacteria, and viruses; the latter two include bacterial pathogens, which cause a number of plant diseases that must be controlled.

Conventional practices to control bacterial plant pathogens include crop rotation, seed certification, development of resistant cultivars, and the uses of antibiotics and fumigation. However, controlling bacterial pathogens is often problematic because these organisms produce long-term persistent survival structures in soil (Grosch et al., 2005). Growers used to depend totally on chemicals or agronomic and horticultural practices to control these pathogens. Although these remedies have resulted in improved productivity and crop quality since the early 1900s, the environmental pollution caused by the excessive use and misuse of agrochemicals has led to strict regulations on their use (Pal and Gardener, 2006). Therefore, many researchers are being prompted to develop alternative strategies that involve the introduction or manipulation of microorganisms to enhance the protection of plants against pathogens (Grosch et al., 2005). Biological control, one of the alternative strategies developed to control plant pathogenesis, refers to the purposeful utilization of an introduced or resident living organism to suppress the activities and populations of one or more pathogens (Pal and Gardener, 2006). This may involve the use of microbial inoculants to suppress a single type or class of plant diseases. Because microbial diversity is a significant natural resource (Kennedy and Smith, 1995), the process of biological control preserves environmental quality by reducing chemical input (Altier, 1994; Barea and Jeffries, 1995).

In biological control, seeds as well as soil can be treated with the same control agent. Among the rhizosphere microorganisms, fungi and bacteria play very important roles in improving plant health and fitness because they are able to improve plant tolerance to biotic stress and also stimulate plant growth. Of the different types of plant–microbe interactions, the most common and widespread association is mycorrhizal symbiosis (Smith and Read, 2008). Of the many types of mycorrhizas, arbuscular mycorrhizal (AM) symbiosis is the most widespread type; it occurs in the plant taxa of about 80% of terrestrial plant families (Smith and Read, 2008). AM fungi (AMF), which compete with plant pathogens for nutrients and space, have been successfully used for biological control (Berg et al., 2006). The negative antagonistic interaction of the AMF with various soil-borne plant pathogens is one of the main reasons that it has potential to be used as a biocontrol agent (Tahat et al., 2010).

13.2 PLANT PATHOGENIC BACTERIA

Plant pathogenic bacteria can be defined as any prokaryote that causes damage to a host plant, regardless of the severity of the interaction (Bradbury, 1986). Bacteria are the major causal agents of diseases in terms of the species involved (Agrios, 2005). Of the 7100 or more bacterial species that have been identified to date, only 150 are phytopathogenic (Buonaurio, 2008). These phytopathogenic bacteria cause diseases in their host plants by using certain specific mechanisms to obtain nutrients for their growth from them (Buonaurio, 2008). Almost all bacterial parasites reside either within a plant or on its surface, or in plant debris or soil (as saprophytes). The characterization of plant pathogenic bacteria is of prime importance, especially in the development of methods to prevent or cure plant diseases that are based on the characteristics of the microbial pathogen. Most plant diseases are caused by Gram-negative bacteria belonging to the phylum Proteobacteria, including families such as Xanthomonadaceae, Pseudomonadaceae, and Enterobacteriaceae. Pathogenic bacteria belong to genera such as *Agrobacterium*, *Acidovorax*, *Burkholderia*, *Clavibacter*, *Erwinia*, *Pantoea*, *Pectobacterium*, *Phytoplasma*, *Pseudomonas*, *Ralstonia*, *Spiroplasma*, *Streptomyces*, *Xanthomonas*, and *Xylella* (Ade and Innes, 2007). Bacterial diseases are most frequent and severe in tropical and subtropical countries, where warmth and humidity provide ideal conditions for bacterial growth.

Damage caused to the host plant by plant pathogenic bacteria includes different kinds of symptoms, irrespective of the severity of the interaction. Most plant pathogenic bacteria cause diseases that are of little or no economic importance (Bradbury, 1986); two diseases caused by bacteria to economically important plants are bacterial wilt and crown gall. When plant pathogenic bacteria cause internal interference in the host plant, various symptoms result (e.g., galls, spots, cankers, and chlorosis). Such symptoms result in reduced plant growth and production. For example, wilt-causing bacteria physically block vascular tissues, which leads to the inhibition of the circulation of water and essential nutrients within the plant. A plant system defends itself against a pathogen both physically and chemically.

13.3 ARBUSCULAR MYCORRHIZAL FUNGI AND PLANT PATHOGENIC BACTERIA

Some microorganisms inhibit mycorrhizal growth in either the pre- or post-colonization state, during which competition for nutrients is not as strong (Miransari, 2009). However, microbes may release toxic substances and modify rhizospheric soil properties in ways that directly affect both the AMF and the plants. Disease-causing plant pathogens can be controlled by manipulating indigenous microbes or by introducing antagonists (Linderman, 1992). Mycorrhizae could affect the numbers as well as the composition of bacterial populations in rhizospheric and hyphospheric regions (Andrade et al., 1997; Linderman, 1988). Association with AMF induces resistivity in host plant against various pathogens (Ismail and Hijri, 2010; Kempel et al., 2010). For example, tomato plants (*Lycopersicon esculentum*) infected with stolbur phytoplasma showed minimum pathogenic effect in mycorrhizal conditions, compared with non-mycorrhizal conditions (Lingua et al., 2002).

Most plant pathogenic bacteria alter the host plants' physiology as well as its biochemical activities, which may be lethal effects. AMF, however, have the ability to reduce the defects in host plants caused by pathogenic bacteria (Halos and Zorilla, 1979). In an experiment under greenhouse conditions on tomato that was inoculated with three different AMF (*Funneliformis mosseae* [=*Glomus mosseae*], *Scutellospora* sp., *Gigaspora margarita*) and the bacterial wilt pathogen *Ralstonia solanacearum*, results suggested that *F. mosseae* induced a higher bioprotective effect against the pathogenic bacteria than the other AMF tested (Tahat, 2009).

Brown et al. (1996) studied the efficacy of AMF in control of bacterial wilt disease in three varieties of tomato and two varieties of eggplant inoculated with an AMF, *F. mosseae*, and a bacterial pathogen, *Pseudomonas solanacearum*. Results suggested that the AMF increased the growth and yield of both crops and also suppressed bacterial wilt disease. Similar results have been reported for *Pseudomonas syringae* pv. *mori* on mulberry (Sharma et al., 1995), *Pseudomonas* on grapevine (Waschkies et al., 1994), and actinomycetes on apple (Otto and Winkler, 1995).

13.4 STEPS INVOLVED IN BIOPROTECTION OF AMF AGAINST PLANT PATHOGENS

AMF-induced host protection against microbial pathogens often exhibits varied responses (Appoloni et al., 2008; Wehner et al., 2010; Sikes, 2010). Association with AMF causes physiological and developmental alterations in the host plant, including improvement in host nutrition, competition for photosynthates and colonization sites, changes in root biomass and architecture, and alterations in other rhizospheric microorganisms. Such changes were reported as AM-induced bioprotection by compensating damage or by stimulating antagonistic activity of rhizosphere microbiota against certain root pathogens (Azcón-Aguilar and Barea, 1997; Barea et al., 2005). However, results from several studies have either failed to explore the exact mechanism of nutrient involvement (Shaul et al., 1999; Fritz et al., 2006) or the regulation of the host defense mechanism by AMF in mycorrhizal-induced resistance (Cordier et al., 1998; Pozo et al., 2002; Pozo and Azcón-Aguilar, 2007). Although AMF bioprotection against plant pathogens has been demonstrated, the mechanisms that underlie it remain unclear (Whipps, 2004; Pozo and Azcón-Aguilar, 2007). This lack of clarity may be due to the involvement of complex parameters, such as a combination of interactions between the partners (plant, AMF, and pathogen); growth conditions, which should be favorable for each partner; and the individual identity that defines the specificity of the system (Pozo and Azcón-Aguilar, 2007). In other words, understanding of each interaction is based on the prerequisite of the mechanism and on the bioprotective effects of the mycorrhizal symbiosis.

Research on the ability of AMF association to reduce damage from plant pathogens (Harrier and Watson, 2004) has mainly focused on root diseases (Linderman, 1992), but it has been suggested that AMF could act as an alternative to inorganic fertilizers and pesticides in cases of chemical resistance and thus contribute substantially to sustainable agriculture. The role of AMF in the biocontrol of plant diseases has been extensively reviewed (Schonbeck, 1979; Dehne, 1982; Bagyaraj, 1984; Smith, 1988; Caron, 1989; Paulitz and Linderman, 1991; Linderman, 1992, 1994; Siddiqui and Mahmood, 1995; Azcón-Aguilar and Barea, 1996; Mukerji, 1999; Siddiqui et al., 1999; Barea et al., 2005; Smith and Read, 2008). The varied AMF and plant pathogen interactions found by these studies include reduction in the damage caused by plant pathogens and selective enhanced resistance or tolerance in hosts. Such protection is not uniformly effective across all pathogens and soils, however, because environmental conditions play important roles in disease incidence and development. As previously stated, AMF–plant pathogen interactions vary not only with the host but also with environmental conditions.

13.5 MECHANISMS INVOLVED IN THE AMF-MEDIATED BIOCONTROL OF PLANT PATHOGENIC BACTERIA

Research on AMF in controlling plant pathogens has been conducted since the 1980s. Suppression of root pathogens and stimulation of saprotrophs and plant-growth-promoting microorganisms by AMFs are significant (Kapoor and Mukerji, 1998). Changes in the population composition of soil microorganisms that are induced by the formation of arbuscular mycorrhiza may favor certain components of the microbiota; in turn, the microbiota may be antagonistic toward root pathogens (Azcón-Aguilar and Barea, 1996). Mechanisms that could account for the protective activity ascribed to AMF include improvement of plant nutrition, compensation for root damage, competition for photosynthates or colonization/infection sites, changes in root-system anatomy or morphology, changes in mycorrhizosphere microbial populations, and activation of plant defense mechanisms (Azcón-Aguilar and Barea, 1996).

13.5.1 Improvement of Plant Nutrition

Root colonization by AMF is often followed by improvement in a plant's mineral nutrient status. Some studies have suggested that AMF may increase the tolerance of a host plant to pathogens by enhancing the host plant's uptake of essential nutrients, especially P, even during plant–pathogen competition for space (Gosling et al., 2006). In addition to P, AMF can enhance the uptake of N, K, Ca, Cu, Mn, S, and Zn (Pacovsky et al., 1986; Smith and Gianinazzi-Pearson, 1988). For example, Tahat (2009) reported that the concentrations of N (41%), P (133%), K (49%), Fe (44%), and Zn (33%) in tomato shoots increased in response to root colonization by *F. mosseae*. AMF increases the nutrient content in shoots due to an extra radical hyphal network in the soil that increases the roots' absorbing area. Host susceptibility to pathogens and tolerance to disease can also be influenced by the nutritional status of the host and the fertility status of the soil (Cook and Baker, 1982). However, enhanced mineral nutrition of mycorrhizal plants failed to affect pathogens in some reports (Graham and Egel, 1988).

13.5.1.1 Tolerance to Pathogen

AMF are known for their ability to enhance plant tolerance to pathogens without excessive yield losses and, in some cases, to enhance the inoculum density of the pathogens. This compensation is apparently related to enhanced photosynthetic capacity (Karajeh and Al-Raddad, 1999; Abdalla and Abdel-Fattah, 2000; Heike et al., 2001) and a delay in senescence, caused by a pathogen, that cancels the positive relationship between disease severity and yield loss (Heike et al., 2001). The incidence of infection by the pathogen remains unaffected by AMF colonization; however, mycorrhizal plants are able to tolerate infection of pathogens better than non-mycorrhizal plants. The efficiency and efficacy of AMF in promoting the growth of plants enables mycorrhizal plants to tolerate pathogens (Hwang, 1988).

The tolerance of a plant to a pathogen varies with the AMF species, although some ineffective species reduce pathogen entry by triggering a defense reaction in the plant (Davis and Menge, 1981). Therefore, the selection of AMF for biological control of plant diseases must proceed with caution. Li et al. (1997) reported that *Glomus macrocarpum* decreased the occurrence of bacterial angular leaf spot of cucumber, although no positive effect on growth or yield was noted. This conclusion indicates that the plant's own tolerance to the pathogen could be the mode of action. In another study, Sampo et al. (2012) showed that the presence and viability of phytoplasmas were restricted to the youngest leaves of mycorrhizal *Chrysanthemum* plants, which suggested the likely involvement of *F. mosseae*-mediated host tolerance to the infection.

13.5.2 Damage Compensation

AMF increase the tolerance of the host to pathogen attack by compensating for the losses in root biomass or function caused by pathogens (Linderman, 1994). This effect illustrates an indirect contribution to biological control through two types of conservation of root-system function: one by AMF hyphae increasing the root-absorbing surface as they grow into the soil, and the other by the maintenance of root-cell activity through arbuscular formation (Gianinazzi-Pearson et al., 1995). The extensive colonization of roots by AMF increased the concentration of chlorophyll *a* and chlorophyll *b* in tomato (Tahat, 2009), which subsequently contributed to increased photosynthetic efficiency in the leaves and enhanced the growth of plants infected by *R. solanacearum*.

13.5.3 Competition for Host Photosynthates

AMF and root pathogens compete for the carbon compounds reaching the roots of the host plant; thus, their growth depends upon the photosynthates of the host plant (Smith, 1987; Linderman, 1994; Smith and Read, 2008). Competition between AMF and pathogen includes the physical exclusion of the pathogen (Davis and Menge, 1981; Hussey and Roncadori, 1982; Smith, 1988). In their study on the effect of *Rhizophagus fasciculatus* (= *Glomus fasiculatum*) on cucumbers, Christensen and Jakobsen (1993) found that AMF probably decreased the amount of plant-root-derived organic matter available for bacterial growth and also increased the spatial variability by competition. The AM plants seemed to be better adapted to compete with the saprophytic soil microflora for common nutrients (e.g., N and P) compared with the non-mycorrhizal plants. Generally, AMF have superior (primary) access to the available photosynthates compared with the pathogens. This higher carbon utilization by AMF may inhibit the growth of the pathogen (Linderman, 1994). Because AMF are solely dependent on the host plant for carbon, 4%–20% of the net photosynthates that are allocated to the host roots may be transferred to the fungus. Limited data support this hypothesis, however (Smith and Read, 2008).

The primary utilization of carbon compounds by AMF within roots seems to correspond to the inhibition of pathogen growth (Azcón-Aguilar and Barea, 1996). Ample evidence has confirmed that the carbon demand of AMF increases as its symbiosis with host plant matures (Lerat et al., 2003). Although the extra-radical hyphae of AMF in soil do not have direct access to photosynthates, the intra-radical hyphae take up glucose and fructose. This intake, however, is mainly confined to the arbuscular and intercellular hyphal interfaces (Douds et al., 2000). Arbuscules that contain cells drain higher amounts of sucrose, as evidenced by the absence of sucrose in these cells; this effect is related to an increased transcription of sucrose catabolism genes, which also regulate the transcription of defense-gene expression (Vierheilig et al., 1995, 2001; Ernst and Siri-Prieto, 2009). Hence, defense-gene expression is regulated by the photosynthates level and its changes in the host plant (Blee and Anderson, 2000).

13.5.4 Competition for Infection/Colonization Sites

The underlying mechanism of the interaction between AMF and soil microorganisms is the physical competition between AMF and rhizosphere microorganisms for space in the root niche (Bansal and Mukerji, 1996). Research by Dehne (1982), who documented AMF and root pathogens colonization of the same host tissues as well as their development in different root cortical cells, suggested rivalry for space between the AMF and pathogens colonizing the roots. Tahat et al. (2012) showed that *F. mosseae*, when inoculated along with *R. solanacearum*, was able to colonize the cortical cells of tomato roots better than the bacterial pathogen. As a result of this colonization, *F. mosseae* was able to protect the host cell from the *R. solanacearum* invasion so totally that no structures belonging to the pathogen were seen. Scanning electron microscope and transmission electron microscope observations of the root cells confirmed these observations (Tahat et al., 2012).

13.5.5 Anatomical and Morphological Changes in the Root System

Colonization by AMF induces remarkable changes in root morphology, specifically the meristematic and nuclear activities of root cells (Atkinson et al., 1994). These changes might affect the development of pathogen infection in the host. The most frequent consequence of AMF colonization is increased branching, which results in a relatively large proportion of higher-order roots that can provide more potential infection sites (Hooker et al., 1994). Berta et al. (1993) showed that the mycorrhizal adventitious roots are generally of larger diameter than non-mycorrhizal roots. Garcia-Garrido and Ocampo (1989) found that the root-length colonization of tomato by AMF was similar in all treatments tested regardless of the presence or absence of the bacterial pathogen *Ps. syringae*. However, there was a depression in the growth of shoots and roots following the inoculation of non-mycorrhizal plants with *Ps. syringae*. It has been reported that the inoculation of *Glomus versiforme* sharply decreased the population of pathogenic bacteria in the xylem of tomato by 81.7%, and in the root surface by 79.3% (Zhu and Yao, 2004). Similarly, in a split-root experiment, *Gl. versiforme* inhibited populations of *R. solanacearum* even though these had been inoculated into different parts of the root system. When half of the root system was inoculated with *R. solanacearum* and the other half was inoculated with or without *Gl. versiforme*, the populations of *R. solanacearum* on the root surface and in the rhizosphere respectively decreased by 13.7% and 43.4%. These results further confirm that inoculation with *Gl. versiforme* could inhibit the *R. solanacearum* population in the root rhizosphere, on the rhizosplane, and even in the host xylem tissue (Genin and Boucher, 2002).

In Tahat et al. (2012)'s examination of the effects of three species of AMF (*F. mosseae*, *Gi. margarita*, and *Scutellaspora* sp.) on tomato growth infected with *R. solanacearum*, inoculation of *F. mosseae* contributed to a significant increase in the number of root tips as well as root surface area, root volume, and root length when compared with the inoculation with *Gi. margarita*. Lingua et al. (2002) showed that AMF colonization reduced the severity of the symptoms induced by stolbur phytoplasma in tomato plants. The morphological parameters (i.e., the fresh weights of root and shoot, plant height, length of internode, number of leaves, and diameter of the adventitious roots) in the infected plants were almost similar to those of the healthy plants in the presence of AMF; in addition, reduced nuclear senescence was observed. The percentages of nuclear populations with different ploidy levels were found to be intermediate between AMF-colonized and phytoplasma-infected plants (Lingua et al., 2002). Increased lignification may protect roots from pathogen invasion along with contributing to the process of elevating phenol metabolism within a host plant (Morandi, 1996).

13.5.6 Microbial Changes in the Mycorrhizosphere

AMF-induced changes in host plant physiology can play a decisive role in root exudation patterns, and consequently cause quantitative and qualitative alterations in microbial populations in the rhizosphere (Azcón-Aguilar and Bago, 1994; Smith et al., 1994). This effect could influence plant growth and health. Changes in soil microorganism population induced by AM formation may lead to stimulation of certain components of the microbiota, which in turn may be antagonistic to root pathogens. Secilia and Bagyaraj (1987) isolated more pathogen–antagonistic actinomycetes from the rhizosphere of AM plants than from non-mycorrhizal plants. The prophylactic ability of some AMF could be exploited in association with coexisting rhizosphere microorganisms used as biological control agents (Linderman, 1994; Barea et al., 1996). AMF, along with plant growth-promoting rhizobacteria, play an important role in the biological control of root pathogens and improved nodulation in legumes (Barea et al., 1996). For example, inoculation of tomato plants with *Gl. versiforme* was found to reduce the population of *R. solanacearum* in the rhizosphere by 26.7% (Zhu and Yao, 2004). Similarly, *F. mosseae* prevented the infection of soybean plants by *Ps. syringae* by suppressing the population density of the pathogen in the rhizosphere (Shalaby and Hanna, 1998).

13.5.7 Activation of Plant Defense Mechanisms

When specific plant defense mechanisms are activated in response to AM colonization, it forms an obvious basis for the protective capacity of AMF. The elicitation of a specific plant defense reaction in response to AM symbiosis could predispose the plant to an early response to attack by a root pathogen (Gianinazzi-Pearson et al., 1994). During their life cycles, plants evolve a number of defense responses that are elicited by various signals, including responses that have been associated with pathogen attack (Huynh et al., 1992). Many of the compounds that are involved in plant defense (Bowles, 1990) have been studied in relationship to AM formation, including phytoalexins, enzymes of the phenol propanoid pathway, chitinases, β-1,3-glucanases, peroxidases, pathogenic-related (PR) proteins, callose, hydroxy-proline-rich glycoproteins, and phenolics (Gianinazzi-Pearson et al., 1994). The role of phenols in suppressing pathogens has been extensively discussed by Sedlarova and Lebeda (2001). The increased phenol concentration in plant tissues after pathogen attack is an important mechanism by which pathogen activity may be limited or its population may be decreased. In tomato, inoculation of *R. solanacearum* and *Gl. versiforme*, either individually or in combination, increased soluble phenol contents by 13.5%–83% (Zhu and Yao, 2004).

13.5.8 Changes in Chemical Constituents

After AMF colonization, the P levels of the host roots are typically enhanced, which modifies the phospholipid composition and therefore the membrane permeability as well; these changes result in reduced leakages of net amounts of sugars, carboxylic acids, and amino acids into the rhizosphere (Ratnayake et al., 1978; Schwab et al., 1983). These alterations also arrest the chemotactic effect of the pathogen upon the plant roots and discourage pathogen entry. Altered levels of IAA, cytokinin (Torelli et al., 2001), and abscisic acid (ABA) (Esch et al., 1994) have been reported in AMF-colonized plants. The level of endogenous IAA was reduced in phytoplasma-infected *Chrysanthemum* plants, but could be partly relieved by *F. mosseae* inoculation (Sampo et al., 2012). IAA as well as cytokininsis seem to be involved in the recovery of phytoplasma-infected periwinkle plants (Ćurković Perica, 2008). Hoshi et al. (2009) reported that the expression of a Candidatus *Phytoplasma asteris* virulence factor in transgenic tobacco plants downregulated auxin signaling and biosynthesis pathway genes, which resulted in altered plant growth. Therefore, it can be speculated that the reduced levels of endogenous IAA in phytoplasma-infected *Chrysanthemum* plants could be partly relieved by *F. mosseae*.

13.5.8.1 Systemic and Induced Resistance

Systemic and induced resistance (SIR) typically refers to the sustained induction of resistance or tolerance to disease in plants by pre-inoculation with a pathogen (Kuc, 1995; Handelsman and Stabb, 1996). The SIR phenomenon in mycorrhizal plants has been demonstrated as localized and systemic resistance to the pathogen (Cordier et al., 1998). An increase in the lignin deposition in plant cell walls after AMF colonization can restrict pathogen spread (Dehne and Schonbeck, 1979). SIR appears to be a promising field that can be used for the effective control of plant diseases. In their studies on the effect of *Gl. versiforme* on *R. solanacearum*, Zhu and Yao (2004) reported that, after inoculation with AMF, SIR to pathogens is weaker than locally induced resistance. After inoculation with *F. mosseae*, phloem cells in *Chrysanthemum carinatum* infected with chrysanthemum yellows showed distorted cell walls, deposition of callose near the sieve pores, and filaments of PR proteins (Sampo et al., 2012).

13.5.8.2 Phytoalexins and Phytoanticipins

Phytoalexins are produced in response to microbial infection (Paxton, 1981), whereas phytoanticipins are stored in plant cells in anticipation of or prior to pathogen attack (Van Etten et al., 1995). Phytoalexins levels elicited by pathogens have been shown to be much higher than those elicited by symbiotic organisms (Wyss et al., 1991). In general, in the presence or absence of pathogens in plant roots, phytoalexins are induced in mycorrhizal plants that neutralize the negative effects of pathogens. Morandi (1996) showed that phytoalexin toxic components were not detected during the first stages of AM formation but could be detected during the later stages of symbiosis.

13.5.8.3 Role of AMF in Phytohormones Production and Induced Resistance in Plants

Phytohormones, the chemical substances secreted by plants, play vital roles in plant growth and development processes, responses to biotic stresses (Bari and Jones, 2009). Most plant pathogen responses are regulated by the coordinated activity of signal transduction pathways (Grant and Lamb, 2006; Adie et al., 2007). Key signaling molecules within plant hormones include salicylic acid (SA), jasmonic acid (JA), and ethylene (ET); during plant–pathogen interaction, these were modified in level and also in their gene expression (Robert-Seilaniantz et al., 2007). SA promotes resistance against pathogens in a biotrophic state, whereas JA and ET act as positive signals in the activation of defenses against necrotrophic pathogens (Thomma et al., 2001; Rojo et al., 2003;

Glazebrook, 2005). Recently, auxins and ABA were also reported to be involved in plant–pathogen interactions (Mauch-Mani and Mauch, 2005; Fan et al., 2009; Kazan and Manners, 2009). For instance, compared with ABA-rich tomato plants, ABA-deficient plants have been reported as more susceptible to pathogens such as *Ps. syringae* (Thaler and Bostock, 2004) and *Erwinia chrysanthemi* (Asselbergh et al., 2008). Phytohormones, which act as signaling molecules, are altered during mycorrhizal formation and growth (Ludwig-Müller et al., 2002). Nevertheless, the changes in phytohormone levels in response to mycorrhization and their interactive role have yet to be clearly understood (Ludwig-Müller, 2000; Hause et al., 2007; Lopez-Raez et al., 2010). Several studies have suggested that AMF association tremendously increases phytohormone levels and numbers of lateral roots in host plants (Barea and Azcón-Aguilar, 1982; Barker and Tagu, 2000; Meixner et al., 2005).

13.5.8.4 Root Exudates

Plant roots are subjected to biotic and abiotic stresses in soil; they respond by secreting chemical substances (exudates) to protect themselves against the negative effects of stress and to substantially influence positive interactions. Root exudates, which are released by the root system into the soil, alter rhizospheric conditions by modifying pH and mineral availability through desorption (McNear, 2013). Root exudates can also influence the growth and interactions of microorganisms in the soil (Rovira, 1969), and have been shown to have active involvement in the regulation of symbiotic and protective interactions between plants and microbes (Jones et al., 2003). Physiological disturbances in root morphology (Groleau-Renaud et al., 1998), soil compaction, and drought conditions (Brimecombe et al., 2000) have all been shown to result in increased secretion of root exudates. Plant–plant interactions are also often regulated by root exudates (Bais et al., 2006; Parepa et al., 2012).

In general, the release of organic compounds mineralizes the acquired nutrients and mediates plant–microbe interactions (Pierret et al., 2007). Root exudates that contain carbon-based compounds mediate between a host and neighboring plants and microbes (Bais et al., 2006; Broeckling et al., 2008) in ways that are determined by various factors of the host plant. Release of certain exudates by roots increases the number of signaling pathways, which positively influences AMF hyphal growth and colonization of roots by AMF (Vierheilig, 2004; Akiyama et al., 2005; Greipsson and Ditommaso, 2006). However, this process may also induce defense genes.

In addition to root exudates, AMF also produce hyphal exudates that can act either as substrates for bacterial growth or as antagonists to bacterial populations. For example, hyphal exudates of *Glomus irregulare* (=*Glomus intraradices*) demonstrated both antagonistic and stimulatory effects on fungi and bacteria, depending on the organism (Filion et al., 1999; Toljander et al., 2007). The effect of AM–bacteria interactions (Bharadwaj et al., 2008a) on the production of exudates and their interactive effects on pathogens is not clearly understood. However, AMF has been shown to enhance the production of secondary metabolites in plants, as biocontrol substances that may not favor the survival of pathogenic organisms. Sedlarova and Lebeda (2001) and Gershenzon (2002) suggested that an increase in the phenol concentration in plant tissue in response to AM symbiosis limits pathogen activity.

13.6 EFFECT OF AMF IN STIMULATING BENEFICIAL MICROORGANISMS THAT ARE ANTAGONISTIC TO BACTERIAL PLANT PATHOGENS

The capacity of AMF to control disease symptoms as well as the intra-radical and rhizosphere proliferation of soil-borne pathogens is quite significant. These are complex processes that are influenced by many mechanisms that act synergistically. Of these, the capacity of the AMF extra-radical network to stimulate beneficial microorganisms seems to be strongly implicated. Various bacteria with high antagonistic capacities against several soil-borne pathogens have been identified within the extra-radical structures or the mycorrhizosphere of several AMF species (Lioussanne, 2010).

Bacteria isolated from surface-decontaminated spores of *Gl. irregulare* and *F. mosseae* extracted from field rhizospheres of *Festuca ovina* and *Leucanthemum vulgare*, when classified within two phylogenetic clusters A, corresponded to *Proteobacteria* and B, *Actinobacteria*, and *Firmicutes* (Bharadwaj et al., 2008a). Further studies showed that the selected bacteria, *Stenotrophomonas maltophilia* (two isolates), *Pseudomonas* spp. (three isolates), *Bacterium subtilis* (one isolate), and *Arthrobacter ilicis* (one isolate) were all antagonistic against *Erwinia carotovora* var. *carotovora* (Ecc), produced siderophores and proteases, and decreased the weight of rotten potato tissues caused by Ecc (Bharadwaj et al., 2008a). The ability of AMF to specifically harbor and then stimulate the rhizobacteria with biocontrol properties suggests that these bacteria would directly reduce pathogen development within the mycorrhizosphere and, consequently, would strongly contribute to the biocontrol mediated by AMF in soil-borne diseases (Lioussanne, 2010).

The AMF *Rhizophagus irregularis* and the rhizobacterium (RB) *Pseudomonas fluorescens* significantly inhibited the development of tomato leaf specks when seedlings were planted and inoculated with AMF and RB in their two-leaf stage (Salem, 2003). The preventive was barely evident in the second leaves, but with the surface increases provided by the third, fourth, and fifth leaves, more specks successively developed in the uninoculated controls. In other words, inoculation with AMF as a single treatment significantly increased the resistance of tomato plants to leaf speck. Moreover, appropriate treatment with RB in the rhizosphere resulted in greater resistance in every successive generation of leaves and combined inoculation with AMF and RB caused very strong resistance in all parts of each experimental plant. In addition, resistance was maintained even in the further generation of leaves. It was concluded that highly effective biological control of the soil-borne pathogen *Ps. syringae* in tomato could be realized only with the application of AMF and RB inoculants (Salem, 2003).

Bacteria can sometimes be found in the cytoplasm of AMF spores (Cruz, 2004; Lumini et al., 2007). The beneficial effects of mycorrhizae in the rhizosphere result from synergistic interactions among all of the rhizosphere microorganisms that are crucial to plant growth (Linderman, 1992). Thus, the relationship between AMF and their associated bacteria may be of great importance to sustainable agriculture (Cruz and Ishii, 2011). Bacteria may also form aggregates (biofilms) on the surfaces of AMF hyphae and spores, which in turn suppress pathogens and affect nutrient biodynamics (Cruz and Ishii, 2011).

13.7 *IN VITRO* STUDIES ON AMF AND AMF-ASSOCIATED BACTERIA

Arbuscular mycorrhizal fungi that interact with bacteria (AMF-associated bacteria or AMB) in the mycorrhizosphere, such as mycelial exudates of *R. irregularis*, have been demonstrated to affect bacterial and fungal pathogenesis (Filion et al., 1999; Ravnskov et al., 1999). AMB isolates *Pseudomonas* FWC30 and FWC70, *Stenotrophomonas* LWC2, and *Bacillus* FWC 42 were selected based on their strong *in vitro* antagonistic abilities against four potato pathogens (Bharadwaj et al., 2008b). The combined effects of *Glomus irregular* and AMB were found to inhibit the growth of *Pectobacterium carotovorum* ssp. *carotovorum*. It is thought that siderophore-mediated competition for nutrients and production of *Pe. carotovorum* ssp. *carotovorum* growth inhibitors could provide the mechanism for the process of inhibition. The role of *R. irregulare* exudates in controlling the pathogen could involve the stimulation of AMB growth as well as production of unidentified metabolites with anti-bacterial effects. Most of the AMB isolates belong to the phylum gamma *Proteobacteria* (Bharadwaj et al., 2008a). The AMB strain FWC70 (*Pseudomonas putida*) isolated from *R. irregulare* stimulated primary and lateral root development, root and shoot length, and the number of leaves per plant in potato, along with an antagonism to the potato pathogen *Pe. carotovorum* ssp. *carotovorum* (Bharadwaj et al., 2008b).

Transformed carrot-root cultures in a two-compartment plate system were used for *in vitro* studies on interactions among *R. irregularis*, 10 isolates of AMB, and one plant pathogen (Bharadwaj et al., 2012). The exudates of *R. irregularis* were found to stimulate the growth of all 10 AMB isolates. This result that *R. irregularis* exudates contain carbohydrates, amino acids, and unidentified compounds that could serve as a substrate to stimulate AMB growth. In terms of AMB effects on bacterial plant pathogens, Bharadwaj et al. (2012) also found that the growth of *Pe. carotovorum* ssp. *carotovorum* was inhibited in the presence of the AMB isolates. These *in vitro* studies suggest mutual growth stimulation between *R. irregularis* and AMB and that together, they seem to provide cumulative resistance to the growth of bacterial pathogens. The cultivated variety of pear, *Pyrus communis* OHF-333 rootstock, is susceptible to pear decline (PD) phytoplasmas. When interactions between the AMF *R. irregularis* obtained from *in vitro* culture and PD phytoplasma ware studied in this root stock, the PD-infected trees were found to be negatively affected by the disease during the first year. In addition, inoculation with *R. irregularis* increased shoot length in the non-PD as well as the PD-infected trees. In the second year, however, results confirmed the beneficial long-term effects of mycorrhizae. Although PD infection was detrimental to the shoot growth, it did not affect the stem girth (Garcia-Chapa et al., 2004). A differential growth of *Clavibacter michigensis* and *Pseudomonas chlororaphic* was attributed to the substances released from *R. irregularis* under *in vitro* culture conditions (Filion et al., 1999).

13.8 LIMITATIONS

One of the most important factors that may affect the functioning of naturally occurring AMF in the context of biological control of plant pathogenic bacteria is the diversity of mycorrhizal fungi at a given site. The benefits of AMF are generally influenced by host type and crop rotation (Johnson et al., 1992; Bever et al., 1996). Additional factors that can influence the biocontrol activities of AMF, especially in agro systems, are mass production for field applications, inoculation techniques, and the competitiveness of the elite mycorrhizal fungi. The main hurdle to exploiting the beneficial effects of AMF in the control of bacterial diseases of plants is the obligate nature of the symbiont. Moreover, the cost of inoculum production can be higher than for other organisms. Indeed, the lack of consistency in efficacious products, along with poor market demands, have contributed significantly to the insignificant and slow progress in this field.

13.9 CONCLUSION AND FUTURE CONSIDERATIONS

Mycorrhizal fungi provide a very effective alternative method of disease control, especially for pathogens that affect below-ground plant parts. AMF have enormous potential to control the plant pathogenic bacteria that cause soil-borne diseases, because root diseases are the most difficult to manage and lead to losses in disturbing proportions. Moreover, mycorrhizal symbiosis substantially influences plant growth under a variety of stressful conditions; their role in biological control of soil/root-borne pathogens is therefore of immense importance in both agricultural systems and forestry (Linderman, 1994). Although it is difficult to reach practical conclusions because of the complexity of microbe–soil–plant systems and the decisive influence of prevailing environmental conditions, it may be possible to find the right combination of factors to exploit the prophylactic ability of AMF. Examples of successful practical applications remain scarce, however (Hooker et al., 1994; Linderman, 1994).

Further research is necessary in this field. Current knowledge suggests that management recommendations for the biological control of target diseases in sustainable agrosystems, particularly with nursery and biological crops, could be made in the future. Appropriate AMF must be used, preferably in association with other pathogen-antagonistic soil microbiota. Possible roles of AMF in the control of plant pathogenic bacteria can also be considered and exploited in plant-breeding programs that aim to select pathogen-resistant cultivars. Applications of microbial inocula might help to improve nutrient supplies to plant and control plant pathogens (Bashan, 1998). The effectiveness of mycorrhizal inoculum, which should be relatively universal for various plants and soils, should also be relatively easy to evaluate on a standard scale (Vessey, 2003). However, investigations on the effects of AMF on plant pathogenic bacteria are rendered complex by varying parameters. A model system should combine interactions among three different partners (the plant, the AMF, and the pathogen) but growth conditions should be the same for each partner; in addition, it must be taken into account that the identities of the partners define the specific nature of the system.

Independent understanding of each interaction (plant/pathogen, root/AM) is an important prerequisite for improving knowledge in this area of research and for identifying the processes that underlie the bioprotective effects of mycorrhizal symbiosis. It has been suggested that the selection of AMF species for a desired activity must be based on their ability to survive, aggressively colonize host roots, and operate efficiently (Bagyaraj, 1994). The number of investigations on the control of plant pathogenic bacteria by AMF is limited to those involving fungi and nematodes. Therefore, understanding the mechanisms that are involved in the AMF-mediated control of bacterial plant diseases should be emphasized.

REFERENCES

Abdalla, M.E., Abdel-Fattah, G.M. 2000. Influence of the endomycorrhizal fungus *Glomus mosseae* on the development of peanut pod rot disease in Egypt. *Mycorrhiza* 10: 29–35.

Ade, J., Innes, R.W. 2007. Resistance to bacterial pathogens in plants. *Encyclopedia of Life Sciences* 1–6.

Adie, B., Chico, J.M., Rubio-Somoza, I., Solano, R. 2007. Modulation of plant defenses by ethylene. *Journal of Plant Growth Regulation* 26: 160–177.

Agrios, G.N. 2005. *Plant Pathology*, 5th edn., San Diego, CA: Elsevier-Academic Press, p. 922.

Akiyama, K., Matsuzaki, K., Hayashi, H. 2005. Plant sesquiterpenes induce hyphal branching in arbuscular mycorrhizal fungi. *Nature* 435: 824–827.

Altier, M.A. 1994. Sustainable agriculture. *Encyclopedia Agricultural Science* 4: 239–247.

Andrade, G., Mlhara, K.L., Linderman, R., Bethlenfalway, G.B. 1997. Bacteria from rhizosphere and hyphosphere soils of different arbuscular mycorhhizal fungi. *Plant and Soil* 192: 71–79.

Appoloni, S., Lekberg, Y., Tercek, M.K., Zabinski, C.A., Redecker, D. 2008. Molecular community analysis of arbuscular mycorrhizal fungi in roots of geothermal soils in Yellowstone National Park (USA). *Microbial Ecology* 56: 649–659.

Asselbergh, B., Achuo, A.E., Höfte, M., van Gijsegem, A.F. 2008. Abscisic acid deficiency leads to rapid activation of tomato defense responses upon infection with *Erwinia chrysanthemi*. *Molecular Plant Pathology* 9: 11–24.

Atkinson, D., Berta, G., Hooker, J.E. 1994. Impact of mycorrhizal colonization on root architecture, root longevity and the formation of growth regulators. In *Impact of Arbuscular Mycorrhizas on Sustainable Agriculture and Natural Ecosystems*. Gianinazzi, S., Schuepp, H. (eds.). Berlin: Birkhauser Basal, pp. 89–99.

Azcón-Aguilar, C., Bago, B. 1994. Physiological characteristics of the host plant promoting and undisturbed functioning of the mycorrhizal symbiosis. In *Impact of Arbuscular Mycorrhizas on Sustainable Agriculture and Natural Ecosystems*. Gianinazzi, S., Schuepp, H. (eds.). Berlin: Birkhauser Basal, pp. 47–60.

Azcón-Aguilar, C., Barea, J.M. 1996. Arbuscular mycorrhizas and biological control of soil-borne plant pathogens—An overview of the mechanisms involved. *Mycorrhiza* 6: 457–464.

Azcón-Aguilar, C., Barea, J.M. 1997. Applying mycorrhiza biotechnology to horticulture significance and potentials. *Scientia Horticulturae* 68: 1–24.

Bagyaraj, D.J. 1984. Biological interactions with VA mycorrhizal fungi. In *VA Mycorrhiza*. Powell, C.L., Bagyaraj, D.J. (eds.). Boca Raton, FL: CRC Press, pp. 131–153.

Bagyaraj, D.J. 1994. Vesicular-arbuscular mycorrhiza: Application in agriculture. In *Techniques for Mycorrhizal Research, Methods in Microbiology*. Noris, J.R., Read, D.J., Varma, A.K. (eds.). London: Academic Press, pp. 819–833.

Bais, H.P., Weir, T.L., Perry, L.G., Gilory, S., Vivanco, J.M. 2006. The role of root exudates in rhizosphere interactions with plants and other organisms. *Annual Review of Plant Biology* 57: 233–266.

Bansal, M., Mukerji, K.G. 1996. Root exudaties in rhizophere biology. In *Concepts in Applied Microbiology and Biotechnology*. Mukerji, K.G., Singh, V.P. (eds.). New Delhi: Aditya Books, pp. 97–100.

Barea, J.M., Azcón-Aguilar C. 1982. Production of plant growth regulating substances by the VA mycorrhizal fungus *Glomus mosseae*. *Applied Environmental Microbiology* 43: 810–813.

Barea, J.M., Azcón-Aguilar, C., Azcón, R. 1996. Interactions between Mycorrhizal fungi and rhizosphere microorganisms within the context of sustainable soil–plant systems. In *Multitrophic Interactions in Terrestrial Systems*. Gange, A.C., Brown V.K. (eds.). Oxford: Blackwell, pp. 65–77.

Barea, J.M., Jeffries, P. 1995. Arbuscular mycorrhizas in sustainable soil plant systems. In *Mycorrhiza Structure, Function, Molecular Biology and Biotechnology*. Hock, B., Varma, A. (eds.). Heidelberg: Springer, pp. 521–559.

Barea, J.M., Pozo, M.J., Azcón, R., Azcón-Aguilar, C. 2005. Microbial co-operation in the rhizosphere. *Journal of Experimantal Botany* 56: 1761–1778.

Bari, R., Jones, J.D. 2009. Role of plant hormones in plant defense responses. *Plant Molecular Biology* 69: 473–488.

Barker, S.J., Tagu, D. 2000. The role of auxins and cytokinins in mycorrhizal symbioses. *Journal of Plant Growth Regulation* 19: 144–154.

Bashan, Y. 1998. Inoculants for plant growth promoting bacteria in agriculture. *Biotechnology Advances* 16: 729–770.

Berg, G., Opelt, K., Zachow, C., Lottmann, J., Gotz, M., Costa, R., Smalla, K. 2006. Therhizhosphere effect on bacteria antagonistic towards the pathogenic fungus *Verticillium* differs depending on plant species and sight. *FEMS Microbiology Ecology* 56: 250–261.

Berta, G., Fusconi, A., Trotta, A. 1993. VA mycorrhizal infection and the morphology and function of root systems. *Environmental and Experimental Botany* 33: 159–173.

Bever, J.D., Morton, J.B., Antonovics, J., Schultz, P.A. 1996. Host-dependent sporulation and species diversity of arbuscular mycorrhizal fungi in mown grassland. *Journal of Ecology* 84: 71–82.

Bharadwaj, D.P., Alstrom, S., Landquist, P.O. 2012. Interactions among *Glomus irregulare*, arbuscular mycorrhizal spore-associated bacteria and plant pathogens under *in vitro* conditions. *Mycorrhiza* 22: 437–447.

Bharadwaj, D.P., Lundquist, P.O., Alstrom, S. 2008a. Arbuscular mycorrhizal fungal spore-associated bacteria affect mycorrhizal colonization, plant growth and potato pathogens. *Soil Biology and Biochemistry* 40: 2494–2501.

Bharadwaj, D.P., Lundquist, P.O., Persson, P., Alstrom, S. 2008b. Evidence for specificity of cultivable bacteria associated with arbuscular mycorrhizal fungal spores (multitrophic interactions in the rhizosphere). *FEMS Microbiology Ecology* 65: 310–322.

Blee, K.A., Anderson, A.J. 2000. Defense responses in plants to arbuscular mycorrhizal fungi. In *Current Advances in Mycorrhizae Research*. Podila, G.K., Douds, D.D. (eds.). St. Paul, MN: The American Phytopathological Society, pp. 27–44.

Bowles, D.J. 1990. Defense related proteins in higher plants. *Annual Review of Biochemistry* 59: 873–907.

Bradbury, J.F. 1986. *Guide to Plant Pathogenic Bacteria*. Kew, UK: CAB International.

Brimecombe, M.J., De Leij, F.A., Lynch, J.M. 2000. The effect of root exudates on rhizosphere microbial populations. In *The Rhizosphere: Biochemistry and Organic Substances at the Soil–Plant Interface*. Pinton, R., Varanini, Z., Nannipieri, P. (eds.). New York: Marcel Dekker, pp. 95–140.

Broeckling, C.D., Broz, A.K., Bergelson, J., Manter, D.K., Vivanco, J.M. 2008. Root exudates regulates soil fungal community composition and diversity. *Applied and Environmental Microbiology*74: 738–744.

Brown, M.B., Luis, E.M., de Castro, A.M. 1996. VAM fungus as a biological control agent of bacterial wilt caused by *Pseudomonas solanacearum*. National Institute of Molecular Biology and Biotechnology, University of the Philippines, Los Banos (UPLB) College, Laguna.

Buonaurio, R. 2008. Infection and plant defense responses during plant bacterial interactions. *Plant–Microbe Interactions* 8: 169–187.

Caron, M. 1989. Potential use of mycorrhizae in control of soil-borne diseases. *Canadian Journal of Plant Pathology* 11: 177–179.

Christensen, H., Jakobsen, I. 1993. Reduction of bacterial growth by a VAM fungus in the rhizosphere of cucumber (*Cucumis sativus* L.). *Biology and Fertility of Soils* 15: 253–258.

Cook, R.J., Baker, K.F. 1982. *The Nature and Practice of Biological Control of Plant Pathogens*. St. Paul, MN: The American Phytopathological Society, p. 539.

Cordier, C., Pozo, M.J., Barea, J.M., Gianinazzi, S., Gianinazzi-Pearson, V. 1998. Cell defense responses associated with localized and systemic resistance to *Phytophthora* induced in tomato by an arbuscular mycorrhizal fungus. *Molecular Plant-Microbe Interacions* 11: 1017–1028.

Cruz, A.F. 2004. Element storage in spores of *Gigaspora margarita* Becker & Hall measured by electron energy loss spectroscopy (EELS). *Acta Botanica Brasilica* 18: 473–480.

Cruz, A.F., Ishii, T. 2011. Arbuscular mycorrhizal fungal spores host bacteria that affect nutrient biodynamics and biocontrol of soil borne plant pathogens. *Biology Open* 1: 52–57.

Ćurković Perica, M. 2008. Auxin-treatment induces recovery of phytoplasma-infected periwinkle. *Journal of Applied Microbiology* 105: 1826–1834.

Davis, R.M., Menge, J.A. 1981. *Phytophthora parasitica* inoculation and intensity of vesicular-arbuscular mycorrhizae in citrus. *New Phytologist* 87: 705–715.

Dehne, H.W. 1982. Interaction between vesicular-arbuscular mycorrhizal fungi and plant pathogens. *Phytopathology* 72: 1115–1119.

Dehne, H.W., Schonbeck, F. 1979. The influence of endotrophic mycorrhiza on plant diseases. II. Phenol metabolism and lignifications *Fusarium oxysporum*. Untersuchungenzum Einfluss der endotrophen Mycorrhiza auf Pflanzenkrankheiten. II. Phenols toffwechsel und Lignifizierung. *Phytopathologische Zeitschrift* 95: 210–216.

Douds, Jr., D.D., Pfeffer, P.E., Shachar-Hill, Y. 2000. Carbon partitioning, cause and metabolism of arbuscular mycorrhizas. In *Arbuscular Mycorrhizas Physiology and Function*. Kapulnik, Y., Douds, D.D. Jr. (eds.). Dordrecht, The Netherlands: Kluwer, pp. 107–129.

Ernst, O., Siri-Prieto, G. 2009. Impact of perennial pasture and tillage systems on carbon input and soil quality indicators. *Soil and Tillage Research* 105: 260–268.

Esch, B., Hundeshagen, H., Schneider-Poetsch, J., Bothe, H. 1994. Demonstration of abscisic acid in spores and hyphae of the arbuscular-mycorrhizal fungus *Glomus* and in the N_2-fixing cyanobacterium *Anabaena variabilis*. *Plant Science* 99: 9–16.

Fan, J., Hill, L., Crooks, C., Doerner, P., Lamb, C. 2009. Abscisic acid has a key role in modulating diverse plant pathogen interactions. *Plant Physiology*150: 1750–1761.

Filion, M., St-Arnaud, M., Fortin, J.A. 1999. Direct interaction between the arbuscular mycorrhizal fungus *Glomus intraradices* and different rhizosphere microorganisms. *New Phytologist* 141: 525–533.

Fritz, M., Jakobsen, I., Lyngkjaer, M.F., Thordil-Christensen, H., Pons-Kuhnemann, J. 2006. Arbuscular mycorrhiza reduces susceptibility of tomato to *Alternaria solani*. *Mycorrhiza* 16: 413–419.

Garcia-Chapa, M., Batlle, A., Lavina, A., Camprubi, A., Estaun, V., Calvet, C. 2004. Tolerance increased to pear decline *Phytoplasma* in mycorrhizal OF-333 pear stock. *Acta Horticulturae* 657: 437–441.

Garcia-Garrido, J.M., Ocampo, J.A. 1989. Effect of VA-mycorrhizal infection of tomato on damage caused by *Pseudomonas syringae*. *Soil Biology and Biochemistry* 21: 165–167.

Genin, S., Boucher, C. 2002. *Ralstonia solanacearum*: Secrets of a major pathogen unveiled by analysis of its genome. *Molecular Plant Pathology* 3: 111–118.

Gershenzon, J. 2002. Secondary metabolites and plant defence. In *Plant Physiology*. Taiz, L., Zeiger, E. (eds.). Sunderland, MA: Sinauer Associates Incorporation Publications, pp. 283–311.

Gianinazzi-Pearson, V., Gollotte, A., Dumas-Gaudat, E., Franken, T., Gianinazzi, S. 1994. Gene expression and molecular modifications associated with plant responses to infection by arbuscular mycorrhizal fungi. In *Advances in Molecular Genetics of Plant-Microbe Interactions*. Daniels, M., Downic, J.A., Osbourn, A.E. (eds.). Dordrecht, The Netherlands: Kluwer, pp. 179–186.

Gianinazzi-Pearson, V., Gollotte, A., Lherminier, J., Tisserant, B., Franken, P., Dumas-Gaudot, E., Lemoine, M.C. et al. 1995. Cellular and molecular approaches in the characterization of symbiotic events in functional arbuscular mycorrhizal associations. *Canadian Journal of Botany* 73: S526–S532.

Glazebrook, J. 2005. Contrasting mechanisms of defense against biotrophic and necrotrophic pathogens. *Annual Review of Phytopathology* 43: 205–227.

Gosling, P., Hodge, A., Goodlass, G., Bending, G.D. 2006. Arbuscular mycorrhizal fungi and organic forming. *Agriculture, Ecosystems and Environment* 113: 17–35.

Graham, J.H., Egel, D.S. 1988. *Phytophthora* root development on mycorrhizal and phosphorus fertilized non-mycorrhizal sweet orange seedlings. *Plant Disease* 72: 611–614.

Grant, M., Lamb, C. 2006. Systemic immunity. *Current Opinion in Plant Biology* 9: 414–420.

Greipsson, S., Ditommaso, A. 2006. Invasive non-native plants alter the occurrence of arbuscular mycorrhizal fungi and benefit from this association. *Ecological Restoration* 24: 236–241.

Groleau-Renaud, V., Plantureux, S., Guckert, A. 1998. Influence of plant morphology on root exudations of maize subjected to mechanical impedence in hydroponic conditions. *Plant and Soil* 201: 231–239.

Grosch, R., Lottmann, J., Faltin, F., Berg, G. 2005. Use of bacterial antagonists to control diseases caused by *Rhizoctonia solani*. *GensundePflanzen* 57: 199–205.

Halos, P.M., Zorilla, R.A. 1979. VAM increase growth and yield of tomatoes and reduce infection by *Pseudomonas solanacearum*. *Philippines Agriculture* 62: 309–315.

Handelsman, J., Stabb, E.V. 1996. Biocontrol of soil borne plant pathogens. *Plant Cell* 8: 1855–1869.

Harrier, L.A., Watson, C.A. 2004. The potential role of arbuscular mycorrhizal (AM) fungi in the bioprotection of plants against soil-borne pathogens in organic and/or other sustainable farming systems. *Pest Management Science* 60: 149–157.

Hause, B., Mrosk, C., Isayenkov, S., Strack, D. 2007. Jasmonates in arbuscular mycorrhizal interactions. *Phytochemistry* 68: 101–110.

Heike, G., von Alten, H., Poehling, H.M. 2001. Arbuscular mycorrhiza increased the activity of a biotrophic leaf pathogen: Is a compensation possible? *Mycorrhiza* 11: 237–243.

Hooker, J.E., Jaizme-Vega, M., Atkinson, D. 1994. Biocontrol of plant pathogens using arbuscular mycorrhizal fungi. In *Impact of Arbuscular Mycorrhizas on Sustainable Agriculture and Natural Ecosystems*. Gianinazzi, S., Schuepp, H. (eds.). Basel, Switzerland: Birkhauser, pp. 191–200.

Hoshi, A., Oshima, K., Kakizawa, S., Ishii, Y., Ozeki, J., Hashimoto, M., Komatsu, K. et al. 2009. A unique virulence factor for proliferation and dwarfism in plants identified from a phytopathogenic bacterium. *Proceedings of the National Academy of Sciences of the United States of America* 106: 6416–6421.

Hussey, R.S., Roncadori, R.W. 1982. Vesicular-arbuscular mycorrhizae may limit nematode activity and improve plant growth. *Plant Disease* 66: 9–14.

Huynh, Q.K., Hironaka, C.M., Levine, E.B., Smith, C.E., Borgmeyer, J.R., Sha, D.M. 1992. Antifungal proteins from plants. Purification, molecular cloning and antifungal properties of chitinases in maize seeds. *Journal of Biological Chemistry* 26: 6635–6640.

Hwang, S.F. 1988. Effect of VA mycorrhizae and metalaxyl on growth of alfalfa seedlings in soils from fields with "alfalfa sickness" in Alberta. *Plant Disease* 72: 448–452.

Ismail, Y., Hijri, M. 2010. Induced resistance in plants and the role of arbuscular mycorrhizal fungi. In *Mycorrhizal Biotechnology*. Thangadurai, D., Busso, C.A., Hijri, M. (eds.). Enfield, NH: Enfield Science, pp. 77–99.

Johnson, N.C., Copeland, P.J., Croolstol, R.K., Pfleger, F.L. 1992. Mycorrhizae: Possible explanation for yield decline with continuous corn and soybean. *Agronomy Journal* 84: 387–390.

Jones, D.L., Dennis, P.G., Owen, A.G., van Hees, P.A.W. 2003. Organic acid behavior in soils—misconceptions and knowledge gaps. *Plant and Soil* 248: 31–41.

Junaid, J.M., Bhat, N.A.D.T.A., Bhat, A.H., Bhat, M.A. 2013. Commercial biocontrol agents and their mechanism of action in the management of plant pathogens. *International Journal of Modern Plant and Animal Sciences* 1: 39–57.

Kapoor, R., Mukerji, K.G. 1998. Microbial interactions in mycorrhizosphere of *Anethum graveolens* L. *Phytomorphology* 48: 383–389.

Karajeh, M., Al-Raddad, A. 1999. Effect of VA mycorrhizal fungus (*Glomus mosseae* Gerd. & Trappe) on *Verticillium dahlia* Kleb. of olive. *Dirasat Agricultural Science* 26: 338–341.

Kazan, K., Manners, J.M. 2009. Linking development to defense: Auxin in plant–pathogen interactions. *Trends in Plant Science* 14: 373–382.

Kempel, A., Schmidt, A.K., Brandl, R., Schädler, M. (2010). Support from the underground: Induced plant resistance depends on arbuscular mycorrhizal fungi. *Functional Ecology* 24: 293–300.

Kennedy, A.C., Smith, K.L. 1995. Soil microbial diversity and the sustainability of agricultural soils. *Plant and Soil* 170: 75–86.

Kuc, J. 1995. Systemic induced resistance. *Aspects of Applied Biology* 42: 235–242.

Lerat, S., Lapointe, L., Gutjahr, S., Piche, Y., Vierheilig, H. 2003. Carbon partitioning in a split-root system of arbuscular mycorrhizal plants is fungal and plant species dependent. *New Phytologist* 157: 589–595.

Li, S.L., Zhao, S.J., Zhao, L.Z., Li, S.L., Zhao, S.J., Zhao, L.Z. 1997. Effects of VA mycorrhizae on the growth of eggplant and cucumber and control of diseases. *Acta Phytophylactica Sinica* 24: 117–120.

Linderman, R.G. 1988. Mycorrhizal interactions with the rhizosphere microflora: The mycorrhizosphere effect. *Phytopathology* 78: 366–371.

Linderman, R.G. 1992. Vesicular-arbuscular mycorrhizae and soil microbial interactions. In *Mycorrhizae in Sustainable Agriculture*. Bethlenfalway, G.J., Linderman, R.G. (eds.). Madison, WI: American Society of Agronomy, pp. 1–26.

Linderman, R.G. 1994. Role of VAM fungi in biocontrol. In *Mycorrhizae and Plant Health*. Pfleger, F.L., Linderman, R.G. (eds.). St. Paul, MN: The American Phytopathological Society, pp. 1–27.

Lingua, G., D'Agostino, G., Massa, N., Antosiano, M., Berta, G. 2002. Mycorrhiza-induced differential response to a yellows disease in tomato. *Mycorrhiza* 12: 191–198.

Lioussanne, L. 2010. The role of the arbuscular mycorrhiza-associated rhizobacteria in the biocontrol of soil borne phytopathogens. *Spanish Journal of Agricultural Research* 8: S51–S61.

Lopez-Raez, J.A., Verhage, A., Fernandez, I., Garcia, J.M., Azcón-Aguilar, C., Flors, V., Pozo, M.J. 2010. Hormonal and transcriptional profiles highlight common and differential host responses to arbuscular mycorrhizal fungi and the regulation of the oxylipin pathway. *Journal of Experimental Botany* 61: 2589–2601.

Ludwig-Müller, J. 2000. Hormonal balance in plants during colonization by mycorrhizal fungi. In *Arbuscular Mycorrhizas: Physiology and Function*. Kapulnik, Y., Douds, D.D. (eds.). Dordrecht, The Netherlands: Kluwer, pp. 263–285.

Ludwig-Müller, J., Bennett, R.N., Garcia-Garrido, J.M., Piche, Y., Vierheilig, H. 2002. Reduced acid application cannot be attributed to increased glucosinolate levels. *Journal of Plant Physiology* 159: 517–523.

Lumini, E., Bianciotto, V., Jargeat, P., Novero, M., Salvioli, A., Faccio, A., Becard, G., Bonfante, P. 2007. Presymbiotic growth and sporal morphology are affected in the arbuscular mycorrhizal fungus *Gigaspora margarita* cured of its endobacteria. *Cellular Microbiology* 9: 1716–1729.

Mauch-Mani, B., Mauch, F. 2005. The role of abscisic acid in plant–pathogen interactions. *Current Opinion in Plant Biology* 8: 409–414.

McNear Jr., D.H. 2013. The rhizosphere-roots, soil and everything in between. *Nature Education Knowledge* 4: 1.

Meixner, C., Ludwig-Muller, J., Miersch, O., Gresshoff, P., Staehelin, C., Vierheilig, H. 2005. Lack of mycorrhizal autoregulation and phytohormonal changes in the super nodulating soybean mutants 1007. *Planta* 222: 709–715.

Miransari, M. 2009. Contribution of arbuscular mycorrhizal symbiosis to plant growth under different types of soil stress. *Plant Biology* 12: 563–569.

Morandi, D. 1996. Occurrence of phytoalexins and phenolic compounds in endomycorrhizal interactions, and their potential role in biological control. *Plant and Soil* 185: 241–251.

Mukerji, K.G. 1999. Mycorrhiza in control of plant pathogens: Molecular approaches. In *Bio-Technological Approaches in Biocontrol of Plant Pathogens*. Mukerji, K.G., Chamola, B.P., Upadhyay, R.K. (eds.). New York: Kluwer, pp. 135–155.

Otto, G., Winkler, H. 1995. Colonization of rootlets of some species of Rosaceae by actinomycetes, endotrophic mycorrhiza, and endophytic nematodes in a soil conducive to specific cherry replant disease. *Zeitschrift Fur Pflanzenkrankheiten Und Pflanzenschutz* 102: 63–68.

Pacovsky, R.S., Bethlenfalway, G.J., Paul, E.A. 1986. Comparisons between P-fertilized and mycorrhizal plants. *Crop Science* 16: 151–156.

Pal, K.K., Gardener, B.M. 2006. Biological control of plant pathogens. *The Plant Health Instructor*, pp. 1–25. DOI: 10.1094/PHI-A-2006-1117-02.

Parepa, M., Schaffner, U., Bossdorf, O. 2012. Sources and mode of actions of invasive knotweed allelopathy: The effects of leaf litter and trained soil on the germination and growth of native plants. *NeoBiota* 13: 15–30.

Paulitz, T.C., Linderman, R.G. 1991. Mycorrhizal interactions with soil organisms. In *Handbook of Applied Mycology. Soil and Plants*. Vol. 1. Arora, D.K., Rai, B., Mukerji, K.G., Knudson, G.R. (eds.). New York: Marcel Dekker, pp. 77–129.

Paxton, J.D. 1981. Phytoalexins—A working redefinition. *Phytopathologische Zeitschrift* 101: 106–109.

Pierret, A., Doussan, C., Capowiez, Y., Bastardie, F., Pages, L. 2007. Root functional architecture: A framework for modeling the interplay between roots and soil. *Vadose Zone Journal* 6: 269–281.

Pozo, M.J., Azcón-Aguilar, C. 2007. Unravelling mycorrhiza induced resistance. *Current Opinion in Plant Biology* 10: 393–398.

Pozo, M.J., Cordier, C., Dumas-Gaudat, E., Gianinazzi, S., Barea, J.M., Azcón-Aguilar, C. 2002. Localized vs systemic effect of arbuscular mycorrhizal fungi on defense responses to *Phytophthora* infection in tomato plants. *Journal of Experimental Botany* 53: 525–534.

Ratnayake, M., Leonard, R.T., Menge, J.A. 1978. Root exudation in relation to supply of phosphorus and its possible relevance to mycorrhiza formation. *New Phytologist* 81: 543–552.

Ravnskov, S., Nybroe, O., Jakobsen, I. 1999. Influence of an arbuscular mycorrhizal fungus on *Pseudomonas fluorescens* DF57 in rhizosphere and hyphosphere soil. *New Phytologist* 142: 113–122.

Robert-Seilaniantz, A., Navarro, L., Bari, R., Jones, J.D. 2007. Pathological hormone imbalances. *Current Opinion in Plant Biology* 10: 372–379.

Rojo, E., Solano, R., Sanchez-Serrano, J.J. 2003. Interactions between signaling compounds involved in plant defense. *Journal of Plant Growth Regulation* 22: 82–98.

Rovira, A.D. 1969. Plant root exudates. *Botanical Review* 35: 35–57.

Salem, S.F. 2003. Impact of mycorrhizal fungi and other symbiotic microbes as biocontrol agents on soil borne pathogens and some ecophysiological changes in tomato roots. Dissertation submitted to Szentistván University, Gödöllő, Hungary.

Sampo, S., Massa, N., Cantamessa, S., D'Agostino, U., Bosco, D., Marzachi, C., Berta, G. 2012. Effects of two AM fungi on phytoplasma infection in the model plant *Chrysanthemum carinatum*. *Agriculture and Food Science* 21: 39–51.

Schonbeck, F. 1979. Endomycorrhiza in relation to plant disease. In *Soil Borne Plant Pathogens*. Schipper, B., Gams, W. (eds.). New York: Academic Press, pp. 271–280.

Schwab, S.M., Leonard, R.T., Menge, J.A. 1983. Quantitative and qualitative comparison of root exudates of mycorrhizal and non-mycorrhizal plant species. *Canadian Journal of Botany* 62: 1227–1231.

Secilia, J., Bagyaraj, D.J. 1987. Bacteria and actinomycetes associated with pot cultures of vesicular-arbuscular mycorrhizas. *Canadian Journal of Botany* 33: 1069–1073.

Sedlarova, M., Lebeda, A. 2001. Histochemical detection and role of phenolic compounds in the defense response of *Lactuca* spp. to lettuce downy mildew (*Bremialactacae*). *Journal of Phytopathology* 149: 693–697.

Shalaby, A.M., Hanna, M.M. 1998. Preliminary studies on interactions between VA mycorrhizal fungus *Glomus mosseae*, *Bradyrhizobium japonicum* and *Pseudomonas syringae* in soybean plants. *Acta Microbiologica Polonica* 47: 385–391.

Sharma, M.P., Gaur, A., Tanu, U., Sharma, O.P. 2004. Prospects of arbuscular mycorrhiza in sustainable agriculture of root and soil-borne diseases of vegetable crops. In *Disease Management of Fruits and Vegetables*. Mukerji K.G. (ed.). Dordrecht, The Netherlands: Kluwer, pp. 501–539.

Sharma, D.D., Govindaiah, R.S. Katiyar, P.K. Das, L. Janardhan, A.K. Bajpai, P.C. Choudhry, Janardhan, L. 1995. Effect of VA-mycorrhizal fungi on the incidence of major mulberry diseases. *Indian Journal of Sericulture* 34: 34–37.

Shaul, O., Galili, S., Volpin, H., Ginzberg, I., Elad, Y., Chet, I., Kapulnik, Y. 1999. Mycorrhiza-induced changes in disease severity and PR protein expression in tobacco leaves. *Molecular Plant-Microbe Interactions* 12: 1000–1007.

Siddiqui, Z.A., Mahmood, I. 1995. Some observations on the management of the wilt disease complex of pigeonpea by treatment with a vesicular arbuscular fungus and biocontrol agents for nematodes. *Bioresource Technology* 54: 227–230.

Siddiqui, Z.A., Mir, R.A., Mahmood, I. 1999. Effects of *Meloidogyne incognita*, *Fusarium oxysporum* sp. *pisi*, *Rhizobium* sp., and different soil types on the growth chlorophyll and carotenoid pigments of pea. *Israel Journal of Plant Sciences* 47: 251–256.

Sikes, B.A. 2010. When do arbuscular mycorrhizal fungi protect plant roots from pathogens? *Plant Signalling and Behavior* 5: 763–765.

Smith, G.S. 1987. Interactions of nematodes with mycorrhizal fungi. In *Vistas on Nematology*. Veech, J.A., Dickon, D.W. (eds.). Hyattsville, MD: Society of Nematology, pp. 292–300.

Smith, G.S. 1988. The role of phosphorus nutrition in interactions of vesicular-arbuscular mycorrhizal fungi with soil borne nematodes and fungi. *Phytopathology* 78: 371–374.

Smith, S., Gianinazzi-Pearson, V. 1988. Physiological interaction between symbionts in vesicular-arbuscular mycorrhizal plants. *Annual Review of Plant Physiology and Plant Molecular Biology* 39: 221–244.

Smith, S.E., Gianinazzi-Pearson, V., Koide, R., Cainey, J.W.G. 1994. Nutrient transports in mycorrhizas: Structure, physiology and consequences for efficiency of the symbiosis. In *Management of Mycorrhiza in Agriculture, Horticulture and Forestry*. Robson, A.D., Abbott, L.K., Malajczuk, N. (eds.). Dordrecht, The Netherlands: Kluwer, pp. 103–113.

Smith, S.E., Read, D.J. 2008. *Mycorrhizal Symbiosis*, 3rd edn, London: Academic Press.

Tahat, M.M. 2009. Mechanisms involved in the biological control of tomato bacterial wilt caused by *Ralstonia solanacearum* using arbuscular mycorrhizal fungi. Dissertation submitted to Universiti Putra Malaysia, Malaysia.

Tahat, M.M., Sijam, K., Othman, R. 2010. Mycorrhizal fungi as a biocontrol agent. *Plant Physiology Journal* 9: 198–207.

Tahat, M.M., Sijam, K., Othman, R. 2012. Ultrastructural changes of tomatoes (*Lycopersicon esculentum*) root colonized by *Glomus mosseae* and *Ralstonia solancearum*. *African Journal of Biotechnology* 11: 6681–6686.

Thaler, J.S., Bostock, R.M. 2004. Interactions between abscisic acid-mediated responses and plant resistance to pathogens and insects. *Ecology* 85: 48–58.

Thomma, B., Tierens, K.F.M., Penninckx, I., Mauch-Mani, B., Broekaert, W.F., Cammue, B.P.A. 2001. Different micro-organisms differentially induce *Arabidopsis* disease response pathways. *Plant Physiology and Biochemistry* 39: 673–680.

Toljander, J.F., Lindahl, B.D., Paul, L.R., Elfstrand, M., Finlay, R.D. 2007. Influence of arbuscular mycorrhizal mycelial exudates on soil bacterial growth and community structure. *FEMS Microbiology Ecology* 61: 295–304.

Torelli, A., Trotta, A., Acerbi, L., Arcidiacono, G., Berta, G., Branca, C. 2001. IAA and ZR content in leek (*Allium porrum* L.) as influenced by P nutrition and arbuscular mycorrhizae, in relation to plant development. *Plant and Soil* 226: 29–35.

Van Etten, H.D., Mansfield, J.W., Bailey, J.A., Farmer, E.E. 1995. Two classes of plant antibiotics: Phytoalexins versus "phytoanticipins." *Plant Cell* 6: 1191–1192.

Vessey, J.K. 2003. Plant growth promoting rhizobacteria as biofertilizers. *Plant and Soil* 255: 571–586.

Vierheilig, H. 2004. Regulatory mechanisms during the plant-arbuscular mycorrhizal fungus interaction. *Canadian Journal of Botany* 82: 1166–1176.

Vierheilig, H., Alt, M., Lange, J., Gut-Rella, M., Wiemkin, A., Botler, T. 1995. Colonization of transgenic tobacco constitutively expressing pathogenesis related proteins by the vesicular arbuscular mycorrhizal fungi *Glomus mosseae*. *Applied Environmental Microbiology* 61: 3031–3034.

Vierheilig, H., Knoblauch, M., Juergensen, K., van Bel, A.J.E., Grundler, F.M.W., Piché, Y. 2001. Imaging arbuscular mycorrhizal structures in living of *Nicotiana tabacum* by light, epifluorescence and confocal laser scanning microscopy. *Canadian Journal of Botany* 79: 231–237.

Waschkies, C., Schropp, A., Marschner, H. 1994. Relations between grape vine replant disease and root colonization of grape vine (*Vitis* sp.) by fluorescent pseudomonads and endomycorrhizal fungi. *Plant and Soil* 162: 219–227.

Wehner, J., Antunes, P.M., Powell, J.R., Mazukatow, J., Rillig, M.C. 2010. Plant pathogen protection by arbuscular mycorrhizas: A role for fungal diversity? *Pedobiologia* 53: 197–201.

Whipps, J.M. 2004. Prospects and limitations for mycorrhizas in biocontrol of root pathogens. *Canadian Journal of Botany* 82: 1198–1227.

Wyss, P., Boller, T., Wiemken, A. 1991. Phytoalexin response is elicited by a pathogen (*Rhizoctonia solani*) but not by a mycorrhizal fungus (*Glomus mosseae*) in soybean roots. *Experientia* 47: 395–399.

Zhu, H.H., Yao, Q. 2004. Localized and systemic increase of phenols in tomato roots induced by *Glomus versiforme* inhibits *Ralstonia solanacearum*. *Journal of Phytopathology* 152: 537–542.

14 Plant Growth-Promoting Rhizobacteria as a Tool to Combat Plant Pathogenic Bacteria

Periyasamy Panneerselvam, Govindan Selvakumar, Boya Saritha, and Arakalagud Nanjundaiah Ganeshamurthy

CONTENTS

ABSTRACT Plant growth-promoting rhizobacteria (PGPR) are rhizosphere-colonizing bacteria that improve plant growth and development through various mechanisms such as improvement of plant nutrition, protection of plants from various pathogens, induction of plant host-defense mechanisms, and others. In agricultural production systems, plant pathogens—particularly fungi, bacteria, viruses, and mycoplasmas—cause severe economic losses. Traditionally, plant diseases are managed by synthetic chemicals that escalate the cost of production and contaminate the environment with their toxic residues. Therefore, alternative methods for the control of plant pathogens have become vital. Among the many available disease-control strategies, the use of microbes is considered to be green and eco-friendly. Although the bio-control of plant pathogens was initiated in the 1920s, the first commercial bacterial biocontrol agent (*Agrobacterium radiabacter* 84) was not introduced until the early

1970s when it was applied for crown gall disease management. Since then many *Bacillus*-, *Pseudomonas*-, and occasionally *Streptomyces* sp.-based commercial products, have been developed and used for plant protection. However, doubts persist about the use of bio-control agents due to their inconsistent results in field conditions. In this chapter, plant growth-promoting mechanisms, salient traits for selection of a bio-control agent, and integrated approaches for plant bacterial disease management are discussed in detail.

KEYWORDS: plant growth-promoting rhizobacteria, bacterial plant pathogens, biological control

14.1 INTRODUCTION

The rhizosphere comprises the soil surrounding plant roots, where microbial and chemical activities are highly influenced by root exudates. In this region, the microbes become attracted by nutrient exudates from plant roots. The rhizoplane (i.e., the rhizosphere and the surfaces of the roots within it) is intensively colonized by microorganisms compared with non-rhizospheric soils. This phenomenon, known as the rhizosphere effect, was first revealed by Hiltner (1904). The rhizobacterial association with plants has three main effects on plant growth: deleterious, neutral, and growth-promoting or stimulatory. Deleterious effects are seen in reductions of plant growth, biomass, and yield. Bacteria that neither promote nor reduce plant growth are called neutral, and rhizobacteria that improve plant growth are called plant growth-promoting rhizobacteria (PGPR) (Kloepper and Schroth, 1978). Common PGPR include members of the genera *Azotobacter*, *Azospirillum*, *Bacillus*, *Burkholderia*, *Gluconoacetobacter*, *Pseudomonas*, and *Paenibacillus*, as well as some of the members of Enterobacteriaceae, yeast, actinobacteria, and others.

Plant growth-promoting rhizobacteria can affect plant growth by direct or indirect mechanisms (Glick, 1995; Gupta et al., 2000), some of which are probably active at different stages of plant growth. Effects include mineral nutrient solubilization/mobilization, repression of soil-borne pathogens; improvement of plant tolerance to drought, salinity, and metal toxicity; and increased production of phytohormones (Gupta et al., 2000). PGPR also cause cell-wall structural modifications and biochemical/physiological changes that lead to the syntheses of proteins and chemicals that are involved in plant defense mechanisms.

In the agricultural ecosystem, most crop plants are subject to heavy productivity losses due to disease incidence. Although such plant diseases can be managed by applications of chemicals and modifications of cultural practices, complete control is very difficult due to the resurgence of pathogens. At the same time, problems resulting from chemical use and heavy-metal residues encourage eco-friendly alternatives. Alternatives or supplements to inorganic fertilizers and plant protection chemicals are essential for future agricultural production—not only because of the likelihood of price increases, but also because of the need to maintain long-term soil productivity and ecological sustainability. The indiscriminate use of chemicals leads to accumulations of residues and heavy metals that deteriorate soil health, which in turn affects the final quality of produce. Above all, as consumers have become more health-conscious, the quality of food has become a contentious issue. Therefore, there is a need to ensure contamination-free food for the humans and live stock. Although a number of alternatives are already available, PGPR-mediated agriculture is gaining vital significance worldwide and is being accepted in a number of crops as a safe mode of disease control.

The array of efficient microbes present in soil can be utilized for sustainable as well as safe agricultural production that do not damage the environment. The use of microbial inoculants for plant disease management plays a dynamic role in reducing chemical usage besides improving crop yields and maintaining the soil health. In general, the application of PGPR has been hampered by inconsistent performances in field tests, which are usually attributed to inefficiency, poor adaptability, rhizosphere competence (Weller, 1988), and others. Better crop-, location-, and soil-specific microbial inoculants must be developed. Biological control of phytopathogens by PGPR has been studied since the mid-twentieth century but is still not considered to be commercially feasible due to various lacunae. In the case of bacterial plant pathogens, for example, very few commercially successful products are available. The inconsistent performances of PGPR in bacterial disease management are mainly due to improper selection of strains, poor rhizosphere competence, use of single inoculants, routine attempts using only limited bacteria, and other factors such as plant nutrient status, micronutrient deficiencies, and so forth. This chapter contains an overview of PGPR and their roles in bacterial plant disease management.

14.2 PLANT PATHOGENIC BACTERIA

Many bacteria can cause diseases in agriculture, horticulture, and forestry. Based on scientific and economic importance, Mansfield et al. (2012) made a survey and compiled a list of the top ten bacterial plant pathogens (Table 14.1): *Pseudomonas syringae* pathovars, *Ralstonia solanacearum*, *Agrobacterium tumefaciens*, *Xanthomonas oryzae* pv. *oryzae*, *Xanthomonas campestris* pathovars, *Xanthomonas axonopodis* pathovars, *Erwinia amylovora*, *Xylella fastidiosa*, *Dickeya* (*dadantii* and *solani*), and *Pectobacterium carotovorum* (and *Pectobacterium atrosepticum*).

14.3 PGPR AND THEIR MECHANISMS OF BIOCONTROL

Plant growth-promoting rhizobacteria exhibit several mechanisms of biocontrol, most of which involve competition and production of secondary metabolites that directly affect a pathogen. Examples of such metabolites include antibiotics, cell-wall-degrading enzymes, siderophores, and hydrogen cyanide (HCN) (Weller, 1988; Kloepper, 1993; Enebak et al., 1998). PGPR suppress plant pathogenic bacteria by different modes of action (Table 14.2); of these, the most common mechanisms include the secretion of antibiotics, bacteriocins, and siderophores, and the induction of systemic resistance. Combining all of these PGPR-mediated mechanisms may result in strong plant protection against various diseases. For example, bacterial wilt caused by *R. solanacearum* was controlled by rhizobacteria to an extent of 16.66%–83.33% in tomato (Jagadeesh, 2000), whereas inoculation of three strains of fluorescent *Pseudomonas* (RJA112, RBG114, and *Arthrobacter* RBE201) resulted in an 83.33% suppression. These results may be due to the combined or synergistic effects of different bacteria. Different modes of bacterial mechanisms for the suppression of *R. solanacearum* are given in Table 14.3. These results also indicate that antibiotics, siderophores, bacteriocin, organic acid production, and induced systemic resistance (ISR) mediated by different antagonists play important roles in the suppression of *R. solanacearum*.

14.3.1 Antagonistic Activity

Plant growth-promoting rhizobacteria antagonize plant pathogenic bacteria by the secretion of metabolites and antibiotics. Huang et al. (2013) reported that PGPRs isolated from a pathogen-prevalent environment possessed better antagonistic activity. The different types of antimicrobial compounds produced by bacteria include volatiles (HCN, aldehydes, alcohols, ketones, and sulfides), nonvolatile polyketides (diacetyl phloroglucinol [DAPG] and mupirocin), heterocyclic nitrogenous compounds (phenazine derivatives: pyocyanin, phenazine-1-carboxylic acid; PCA, PCN, and hydroxy phenazines) (de Souza et al., 2003), phenylpyrrole antibiotic (pyrrolnitrin) (Ahmad et al., 2008), and lipopeptide antibiotics (iturins, bacillomycin, surfactin, and Zwittermicin A) (Bouizgarne, 2013). Fluorescent pseudomonads and *Bacillus* species are also active in the suppression of plant pathogenic microorganisms. These bacterial antagonists enforce suppression of plant pathogens by the secretion of the abovementioned extracellular inhibitory metabolites at low concentrations. For example, black rot caused by *X. campestris* pv. *campestris* (Xcc) causes severe economic losses in all developmental stages of crucifers, but the lipopeptide-producing *Bacillus* strains actively suppress Xcc during the late growth phase (Mariano et al., 2001).

TABLE 14.1
Common Diseases Caused by Plant Pathogenic Bacteria

Name of the Plant	Causal Agent	Bacteria/Actinobacteria	Name of the Disease	References
Apple	*Erwinia amylovora*	Bacteria	Fire blight disease	Baker (1971)
Bamboo	*Xanthomonas campestris* pv. *vasculorum*	Bacteria	Gumming disease	Bradbury (1973)
Blueberry	*Nocardia vaccinia*	Actinobacteria	Galles disease	Demaree and Smith (1952)
Castor Bean	*Xanthomonas campestris* pv. *Ricini*	Bacteria	Leaf spot disease (Black rot)	Yoshi and Takimoto (1928)
	Pseudomonas solanacearum	Bacteria	Bacterial wilt disease	Smith and Godfrey (1921)
Citrus	*Xanthomonas axonopodis*	Bacteria	Bacterial Spot Disease	Gottwald et al. (2002)
Cotton	*Xanthomonas campestris* pv. *malvacearum*	Bacteria	Bacterial blight disease	Innes and Last (1961)
	Erwinia aroidea	Bacteria	Boll rot disease	Ashworth et al. (1970)
Eucalyptus	*Xanthomonas campestris* pv. *eucalypti*	Bacteria	Dieback disease	Truman (1974)
Grapes	*Agrobacterium tumefaciens*	Bacteria	Crown gall disease	Burr and Otten (1999)
	Xylella fastidiosa	Bacteria	Pierce's disease	Berisha et al. (1998), Nunney et al. (2013)
Guayule	*Erwinia carotovora* f.sp. *Parthenii*	Bacteria	Bacterial rot disease	Starr (1947)
Hemp	*Pseudomonas cannabina*	Bacteria	Bacterial blight disease	Sutic and Dowson (1959)
Jute	*Xanthomonas campestris* pv. *nakataecorchori*	Bacteria	Leaf spot disease	Sabet (1957)
Kenef	*Pseudomonas solanacearum*	Bacteria	Bacterial wilt disease	Joachems and Mass (1922)
Maize	*Corynebacterium nebraskense*	Bacteria	Bacterial wilt disease	Wysong et al. (1973)
	Erwinia carotovora	Bacteria	Bacterial stalk rot disease	Hingorani et al. (1959)
	Pseudomonas alboprecipitans	Bacteria	Bacterial leaf blight disease	Gitaitis et al. (1976)
	Pseudomonas andropogonis	Bacteria	Bacterial stripe disease	Elliot and Smith (1929)
	Pseudomonas coronafacient	Bacteria	Chocolate spot	Ribeiro et al. (1977)
	Pseudomonas syringae	Bacteria	Bacterial leaf spot disease	Kendrick (1926)
Peanut	*Pseudomonas solanacearum*	Bacteria	Bacterial wilt disease	Mcclean (1930)
Pears	*Erwinia amylovora*	Bacteria	Fire blight disease	Baker (1971)

(*Continued*)

TABLE 14.1 (*Continued*)
Common Diseases Caused by Plant Pathogenic Bacteria

Name of the Plant	Causal Agent	Bacteria/Actinobacteria	Name of the Disease	References
Potato	*Dickeya dadantii*	Bacteria	Soft rot disease	Slawiak et al. (2009)
	Pectobacterium caratovorum	Bacteria	Bacterial soft rot	Gardan et al. (2003)
	Actinomyces globisporus	Actinobacteria	Potato scab disease	Krasilnikov (1949)
	Bacillus mesentericus	Actinobacteria	Potato rot disease	Brierley (1928)
Rice	*Xanthomonas campestris* pv. *oryzae*	Bacteria	Bacterial blight disease	Tagami et al. (1963, 1964)
	Xanthomonas campestris pv. *oryzicola*	Bacteria	Bacterial leaf streak disease	Bradbury (1971)
	Xanthomonas oryzae pv. *oryzicola*	Bacteria	Bacterial leaf streak disease	Zou et al. (2012)
	Pseudomonas panici	Bacteria	Bacterial stripe disease	Goto and Ohata (1961)
	Pseudomonas oryzocola	Bacteria	Bacterial sheath rot disease	Tanii et al. (1976)
	Erwinia chrysanthemi	Bacteria	Bacterial foot rot disease	Goto (1979)
	Xanthomonas itoaena	Bacteria	Black rot disease	Tochinai (1932)
Safflower	*Pseudomonas syringae*	Bacteria	Bacterial blight disease	Erwin et al. (1964)
Sesame	*Pseudomonas sesame*	Bacteria	Leaf spot disease	Thomas (1965)
	Xanthomonas sesame	Bacteria	Leaf spot disease	Habish and Hammad (1969)
Soybean	*Pseudomonas glycinae*	Bacteria	Bacterial blight disease	Kennedy and Mew (1971)
	Pseudomonas syringae	Bacteria	Crinkle leaf disease	Tachibana and Shih (1965)
	Xanthomonas phaseoli var. *sojens*	Bacteria	Bacterial pustule disease	Allington and Feaster (1946)
	Pseudomonas tabaci	Bacteria	Wild fire disease	Chamberlain (1956)
	Corynebacterium flaccumfacens	Bacteria	Bacterial wilt disease	Dunleavy (1963)
Sugarcane	*Xanthomonas albilineans*	Bacteria	Leaf scald disease	Persley (1973)
	Pseudomonas	Bacteria	Bacterial sun spot disease	Zummo and Freeman (1975)
	Actinomyces like organism	Actinobacteria	Ratoon stunt disease	Damann and Derrick (1976)

TABLE 14.2
PGPR Strains and Their Role in Suppression of Plant Pathogenic Bacteria

PGPR Bacteria	Pathogenic Bacteria	Mode of Action	References
Erwinia herbicola	*Erwinia amylovora*	Suppressed the pathogen by bacteriocin production	Beer and Vidaver (1978)
Fluorescent pseudomonads	*Erwinia caratovora*	Seed piece inoculation with PGPR reduced pathogen population by 95%–100% controlling soft rot disease	Kloepper (1983)
Pseudomonas fluorescens	*Xanthomonas campestris* pv. *citri*	Prevents citrus canker by siderophore production	Unnamalai and Gnanamanickam (1984)
Agrobacterium radiobacter	*Agrobacterium tumefaciens*	Agrocin-like substance inhibited *Agrobacterium tumefaciens*	Chen and Xiang (1986)
Pseudomonas putidaW4P63	*Erwinia caratovora*	Suppressed soft rot potential of tubers by production of siderophore	Xu and Gross (1986)
Agrobacterium radiobacter K1026	*Agrobacterium* biovar 2	Suppressed crown gall on almond seedlings by Agrocin 84 production	Jones and Kerr (1989)
Agrobacterium radiobacter K84	*Agrobacterium tumefaciens*	Suppressed crown gall of fruit trees by production of Agrocin 84	Jones et al. (1991)
Pseudomonas fluorescens A506	*Erwinia amylovora*	Reduction in population of pathogen in pear flowers due to competition	Wilson and Lindow (1993)
Pseudomonas putida, *Serratia marcescens*, *Flavomonas orizihabitans*, *Bacillus pumilus*	*Pseudomonas syringae* pv. *lachrymans*	Mixed PGPR strains protected cucumber against the pathogen by induced systemic resistance	Wei et al. (1996)
Pseudomonas fluorescens A506	*Erwinia amylovora*	Strain A506 controlled frost and fire blight disease in the presence of antibiotics Streptomycin and Oxytetracyclin	Lindow et al.(1996)
Bacillus pumilis, *Bacillus subtilis*, *Curtobacterium flaccumfaciens*	*Pseudomonas syringae* pv. *Lachrymans*, *Erwinia Tracheiphila*	Seed treatment with a mixed inoculum caused reduction in angular spot and wilt of cucumber	Raupach and Kloepper (1998)
Fluorescent pseudomonads	*Ralstonia solanacearum*	Suppressed wilt of tomato by siderophore production	Jagadeesh et al. (2001)
Pseudomonas fluorescens	*Xanthomonas oryzae* pv. *oryzae*	Resists the rice bacterial blight pathogen by induced systemic resistance	Vidhyasekharan et al. (2001)
Azospirillum braziliense	*Pseudomonas syringae* pv. *tomato*	Prevented bacterial speck disease in tomato by competition	Bashan and de-Bashan (2002)
Serratia 12, *Pseudomonas*, *Bacillus* BB11	*Ralstonia solanacearum*	Mixed inoculum of all three strains suppressed tomato wilt	Guo et al. (2004)
Bacillus cereus, *Bacillus lentimorbus*, *Bacillus pumilus*	*Xanthomonas campestris* pv. *campestris*	Reduced severity of black rot of cabbage	Massomo et al. (2004)
Pseudomonas fluorescens SS101 (Pf.SS101)	*Pseudomonas syringae* pv. *tomato*	Protected *Arabidopsis* through induction of systemic resistance against a virulent bacterial leaf pathogen	Judith et al. (2012)

TABLE 14.3
Antagonistic Bacteria and Their Mode of Action against the Plant Pathogen *Ralstonia solanacearum*

Antagonist	Mechanism(s)	References
Pseudomonas solanacearum B1	Bacteriocin production	Cuppels et al. (1978)
Pseudomonas fluorescens	Induced systemic resistance	Kempe and Sequeira (1983)
Bacillus spp.	Antibiotics production	Anuratha and Gnanamanickam (1990)
Erwinia sp.	Anibiotics production	Fucikovsky et al. (1990)
Pseudomonas cepacia B5	2-Keto gluconic acid production	Aoki et al. (1991)
Pseudomonas glumae	Induction of systemic resistance	Furuya et al. (1991)
Pseudomona fluorescens PF59	Siderophore production	Hartmann et al. (1992)
Pseudomonas solanacearum	Bacteriocin production	Arwiyanto et al. (1994)
Pseudomona fluorescens P.f. G32	Antibiotics, siderophore production	Mulya et al. (1996)
Streptomyces corchorusii	Antibiotics production	El-Abyad et al. (1996)
Bacillus spp.	Antibiotics production	Perez et al. (1997)
Pseudomonas brassicacearum J12	2,4-Diacetylphloroglucinol (2,4-DAPG)	Zhou et al. (2012)

It has been demonstrated that lipopeptide can stimulate ISR in plants, probably by interacting with plant cell membranes and inducing temporary alterations in the plasma membrane (Ongena et al., 2009). Wright et al. (2001) reported that *Pantoea agglomerans* produced the antibiotic pantocin AB, which suppressed the disease caused by *Erwinia herbicola* in apple. Niu et al. (2013) found that the antibiotic polymixin P was suppressive against *E. amylovora* Ea273 and *Erwinia caratovora*, which respectively cause fire blight and soft rot diseases.

Sometimes, bacterially produced antimicrobial compounds may not be efficient for all groups of pathogenic bacteria. Similarly, some studies showed that *in vitro* activities could not be related to *in situ* activity. Numerous investigations have demonstrated that actinobacteria, particularly *Streptomyces*, produce numerous secondary metabolites with antibiotic properties. Currently, 42% of the 23,000 known microbial secondary metabolites are produced by actinobacteria (Bouizgarne, 2013). Actinobacteria, particularly the genus *Streptomyces*, produce 70%–80% of known bioactive natural products (Berdy, 2005). In agricultural production systems, bacterial plant diseases are being extensively managed by antibiotic sprays. Because this method is often not advisable, the possibility of using actinobacteria as bio-inoculants for the management of bacterial diseases should be explored.

14.3.2 Siderophore Production

Siderophores, which are low-molecular-weight, extracellular compounds with a high affinity for ferric iron, are secreted by microorganisms to take up iron from the environment (Hofte, 1993). Their mode of action in the suppression of disease was thought to be solely based on competition with the pathogen for iron (Bakker et al., 1993; Duijff et al., 1997). Under iron-limiting conditions, PGPR produce low-molecular-weight compounds called siderophores to competitively acquire ferric ion (Whipps, 2001). Several fluorescent pseudomonads are known to be highly competitive for iron. Because many pathogens need iron, which must be taken up from the environment, a lack of iron may reduce their pathogenicity. Competition for iron can be enhanced by adding selected bacteria to soil, root systems, tubers, or seeds (Schippers et al., 1987). It has been reported that certain endophytic bacteria isolated from field-grown potato plants can reduce the *in vitro* growth of *Streptomyces scabies* and *X. campestris* through the production of siderophore and antibiotic compounds (Sessitsch et al., 2004). Environmental factors such as pH, iron levels, the presence of other trace elements, and the supplies of carbon, nitrogen, and phosphorus, can also modulate siderophore synthesis (Duffy and Défago, 1999).

14.3.3 Hydrogen Cyanide Production

Hydrogen cyanide, which is produced by many rhizobacteria, is postulated to play a major role in the biological control of pathogens (Défago et al., 1990). Production of HCN by certain strains of fluorescent pseudomonads has been observed in the suppression of soil-borne pathogens (Voisard et al., 1989). The cyanide ion is exhaled as HCN and metabolized to a lesser degree into other compounds. HCN production first inhibits electron transport; next, the energy supply to the cell is disrupted, which leads to the death of the organisms. HCN also inhibits the proper functioning of enzymes (Corbett, 1974) and is known to inhibit the action of cytochrome oxidase (Gehring et al., 1993).

14.3.4 Detoxification of Pathogen Virulence Factors

Detoxification of pathogen virulence factors is another kind of mechanism of biological control. For example, certain biocontrol agents are able to detoxify albicidin toxin produced by *Xanthomonas albilineans* (Basnayake and Birch, 1995; Zhang and Birch, 1997).

14.3.5 Induced Systemic Resistance

Yet another possible mechanism for biological control of plant pathogens is the use of bacterial metabolites that increase a plant's resistance to pathogens by induced systemic resistance (ISR). Resistance that is elicited in plants by the application of chemicals or necrosis-producing pathogens is called systemic acquired resistance (SAR). Pieterse et al. (1998) proposed that ISR and SAR can be differentiated not only by their elicitor molecules but also by the signal transduction pathways that they elicit within the plant. Accordingly, ISR is elicited by rhizobacteria or other nonpathogenic microorganisms, and SAR is elicited by pathogens or chemical compounds.

Several PGPR strains can act as inducers of ISR (Kloepper et al., 1992); in fact, PGPR-mediated ISR may be an alternative to chemical inducers. Plants treated with PGPR provided systemic resistance against a broad spectrum of plant pathogens to reduce the incidence of bacterial disease (Ramamoorthy et al., 2001; Ping and Boland, 2004; Ryu et al., 2004a,b). The expression of ISR is dependent upon the combination of host plant and bacterial strain (Van Loon et al., 1998; Kilic-Ekici and Yuen, 2004). Most reports of PGPR-mediated ISR involve free-living rhizobacterial strains, but ISR activity has also been observed in endophytic bacteria. Volatile organic compounds may be key in this process (Ping and Boland, 2004; Ryu et al., 2004a,b). For example, volatiles released by *Bacillus subtilis* GBO3 and *Bacillus amyloliquefaciens* IN937a were able to activate an ISR pathway in *Arabidopsis* seedlings inoculated with the soft rot pathogen of *Erwinia carotovora* ssp. *carotovora* (Ryan et al., 2001). Jones et al. (2005) reported that the combined effect of PGPRs and SAR compounds was effective in controlling bacterial leaf spot disease in tomato caused by *X. campestris* pv. *vesicatoria*.

A large number of defense enzymes have been reported to be associated with ISR. These include ascorbate peroxidase (APX), β1,3-glucanase, catalase (CAT), chitinase, lipoxygenase (LOX), peroxidase (PO), phenylalanine ammonia lyase (PAL), polyphenol oxidase (PPO), proteinase inhibitors, and superoxide dismutase (SOD) (Koch et al., 1992; Schneider and Ullrich, 1994; Van Loon, 1997). These enzymes also bring about liberation of the molecules that elicit the initial steps in the induction of resistance (Keen and Yoshikawa, 1983; Van Loon et al., 1998).

ISR by PGPR has been noticed in a number of crops, including *Arabidopsis* (Pieterse et al., 1996), brinjal, chilli (Ramamoorthy and Samiyappan, 2001; Bharathi et al., 2004), carnation (Van Peer et al., 1991), cucumber (Wei et al., 1996), mango (Vivekananthana et al., 2004), potato (Doke et al., 1987), radish (Leeman et al., 1996), rice (Vidhyasekaran et al., 1997; Nandakumar et al., 2001), sugarcane (Viswanathan and Samiyappan, 1999), and tomato (Duijff et al., 1997) against a broad spectrum of pathogens, including fungi and bacteria.

14.3.5.1 Accumulation of Cell Wall Components

Inoculation of plants with PGPR, which leads to the strengthening of cell walls, also alters host physiology and metabolic responses, which lead to an enhanced synthesis of plant defense chemicals upon challenge by pathogens (Ramamoorthy et al., 2001; Nowak and Shulaev, 2003). Application of endophytic *Pseudomonas fluorescens* WCS417r on tomato increased the thickening of the outer peripheral and outermost part of the radial side of the first layer of cortical cell walls (Duijff et al., 1997). Grapevine treated with *Burkholderia phytofirmans* PsJN enhanced the phenolic compound accumulation;strengthening of cell walls in the exodermis and several cortical cell layers was also observed during endophytic colonization of the bacterium (Compant et al., 2005). When challenged by a pathogen, PGPR-inoculated plants formed structural barriers (e.g., thickened cell wall papillae) due to the deposition of callose and the accumulation of phenolic compounds at the site of attack (Benhamou et al., 1996). This result indicates that structural changes in plant anatomy due to PGPR inoculation are protective against various phytopathogens.

14.3.5.2 Accumulation of PR Proteins and Defense Related Enzyme

The inoculation of PGPR causes a number of biochemical and physiological changes in plants (Ramamoorthy et al., 2001), including induced accumulations of pathogenesis-related proteins (PR proteins), chitinases, and some peroxidases (Viswanathan and Samiyappan, 1999; Park and Kloepper, 2000; Jeun et al., 2004). However, certain PGPR do not induce PR proteins (Hoffland et al., 1995; Pieterse et al., 1996; van Wees et al., 1999), but instead increase accumulation of other defense enzymes (Chen et al., 2000; Ongena et al., 2000). Seed treatment with PGPR strains (Hynes and Lazarovits, 1989) could increase the level of PR proteins in beans. Similarly, PR proteins were induced in intercellular fluid in the leaves of tobacco plants grown in the presence of *P. fluorescens* CHA0 (Maurhofer et al., 1994). Rice leaves pretreated with *P. fluorescens* and challenge-inoculated with *X. oryzae* pv. *oryzae* showed increases in lignin content, peroxidase activity, and 4-coumarate CoA ligase activity (Vidhyasekaran et al., 2001). Amrita et al. (2013) reported that the isolate *Bacillus pumilus* (BRHS/C1) was most effective in improving the growth of *Listea monopetala* (Roxb.) Pers. by the production of defense-related enzymes such as peroxidase, phenylalanine ammonia lyase, chitinase, and β-1,3-glucanases.

14.3.5.3 Production of Signaling Compounds

Plant–PGPR interaction regulates the salicylic acid (SA), jasmonic acid (JA), and ethylene (ET) pathways in plant systems, which initiate basal resistance against pathogens. These three signals are important in ISR as well as SAR. Both types of induced resistance are effective against a broad spectrum of pathogens. *Serratia marcescens* 90–166 mediates ISR to bacterial pathogens by producing SA, using the salicylate responsive reporter plasmid pUTK21. The strain 90-166 induced disease resistance to *P. syringae* pv. *tabaci* in wild-type *Xanthi*-nc and transgenic NahG-10 tobacco that expressed salicylate hydroxylase (Press et al., 1997). Several genera of bacteria, including pseudomonads, are known to synthesize SA. Root colonization of *A. Arabidopsis thaliana* by the nonpathogenic, rhizosphere-colonizing bacterium *P. fluorescens* WCS417r was found to elicit ISR against *P. syringae* pv. *tomato* (Pst) (Knoester et al., 1999).

14.4 OTHER APPROACHES TO BACTERIAL DISEASE MANAGEMENT IN PLANTS

14.4.1 Plant Nutrient Management and Its Role in Suppression of Phytopathogens

Plant nutrient management is central to plant disease management. Some nutrients (e.g., N, P, K, Mn, Zn, B, Cl, and Si) are very important for imparting resistance/tolerance to plants against many diseases. Excess nitrogen application causes plants to become vulnerable to various diseases, whereas sufficient potassium content decreases the susceptibility of plants against many diseases (Dordas, 2008). Balanced potassium content in plants reduces the incidence of diseases such as bacterial leaf blight, sheath blight, stem rot, black rust in wheat, sugary disease in sorghum, and bacterial leaf blight in cotton (Chase, 1989). Phosphorus, the second-most-important nutrient in most crops, is part of many organic molecules of the cell and is involved in various metabolic processes. Research findings indicate that application of P can reduce bacterial leaf blight in rice and fungal diseases in tobacco, soybean, barley, and sugarcane (Huber and Graham, 1999; Kirkegaard et al., 1999; Reuveni et al., 2000).

Balanced micronutrient contents in plants can reduce the severity of diseases and improve plant resistance (Huber and Watson, 1974; Graham, 1983; Engelhard, 1989; Graham and Webb, 1991; Datnoff et al., 2006) to various diseases. The essential micronutrients are involved in many of the metabolic processes that affect the responses of plants to pathogens (Marschner, 1995). In plant systems, the increase of disease resistance has been shown to be more pronounced at zinc-sufficient conditions. Application of micronutrients (i.e., borax and zinc) significantly reduced banana wilt (Sanjeev and Eswaran, 2008). These findings indicate that balanced integrated nutrient management to stimulate resistant mechanisms in plants is a critical intervention against many diseases. Soils are inhabited by a remarkable diversity of beneficial microorganisms that function as nitrogen fixers, phosphate solubilizers and mobilizers, and micronutrient solubilizers. The potential of these microbes can be explored to counteract both major and micronutrient deficiencies in plants as well as to augment tolerance to various diseases. These goals can possibly be achieved by better understanding of native microbes and preferential rhizosphere engineering.

14.4.2 Mycorrhiza-Associated Bacteria and Their Antagonistic Potential against Phytopathogens

During arbuscular mycorrhizal (AM) fungal colonization of plants, the fungal spores provide shelter for some beneficial bacteria (Panneerselvam et al., 2012). Therefore, the beneficial effects of AM fungi in the rhizosphere are the result of synergistic interactions among the rhizosphere microbes. The relationship between AM fungi and their associated bacteria may therefore be of great importance for sustainable agriculture (Barea et al., 1997). The interaction of AM fungi with bacterial association may give either positive or negative effects (Sylvia et al., 1998), depending on the associated bacteria. This finding clearly indicates that not all mycorrhiza-associated bacteria are always useful for the establishment of AM fungal colonization in crop plants. These associated bacteria have very important roles in promoting mycorrhizal colonization in host plants by the secretion of metabolites that are helpful in the easy proliferation of fungal hyphae and the colonization of host-plant roots (Schrey et al., 2005). Mycorrhiza-associated actinobacteria are also important in plant growth promotion as well as bacterial disease suppression (Poovarasan et al., 2013).

The antibacterial activity of actinobacteria has been reported on numerous pathogenic bacteria such as *E. amylovora*, *Pseudomonas viridiflova*, *Staphylococcus aureus*, *Klebsiella pneumonia*, *Agrobacterium tumefaciens*, and *Clavibacter michiganensis* ssp. *michiganensis* (Miyadoh, 1993; Rizk et al., 2007; Oskay, 2009). Foliar spray of mycorrhiza-associated *Streptomyces fradiae*, *Streptomyces avermitilis*, and *Streptomyces canus* significantly reduced the load of *X. axonopodis* pv. *punicae* on pomegranate leaves. Under greenhouse conditions in whole-plant bioassays, *S. fradiae* and *Streptomyces cinnamonensis* caused 86.6% growth inhibition of the pathogen (Poovarasan et al., 2013). This finding clearly indicates that mycorrhiza-associated actinobacteria can be used as potential biocontrol agents for managing bacterial diseases in crop plants.

14.4.3 Utility of Plant Endophytes in Disease Suppression

A large number of plant endophytic bacteria live within plants and establish close relationships with their hosts as a result of the coevolutionary processes. Endophytes provide a wide range of benefits to plants (Sturz et al., 2000) such as promoting growth (Barka et al., 2002; Kang et al., 2007), suppressing disease severity (Coombs et al., 2004; Kloepper et al., 2004; Senthilkumar et al.,

TABLE 14.4
Endophytic Bacteria That Elicit Induced Systemic Resistance in Plants against Bacterial Plant Diseases

Endophytic Bacteria	Observed Response	References
Serratia marcescens 90–166	Decreased severity of bacterial angular leaf spot of cucumber caused by *Pseudomonas syringae* pv. *lachrymans*	Liu et al. (1995)
	Decreased severity of tobacco wildfire, caused by *Pseudomonas syringae* pv. *tabaci*	Press et al. (1997)
	Reduced symptoms of *Pseudomonas syringae* pv. *tomato*	Ryu et al. (2003)
Pseudomonas fluorescens 89B-61	Decreased severity of angular leaf spot caused by *Pseudomonas syringae* pv. *lachrymans* on cucumber	Wei et al. (1996)
	Decreased severity of tobacco wildfire caused by *Pseudomonas syringae* pv. *tabaci*	Park and Kloepper (2000)
	Reduced symptoms of *Pseudomonas syringae* pv. *tomato* on *Arabidopsis*	Ryu et al. (2003)
Bacillus amyloliquefaciens IN937a	Reduced incidence or severity of vegetable diseases caused by *Ralstonia solanacearum*	Jetiyanon et al. (2003), Jetiyanon and Kloepper (2002)
Bacillus pumilus SE34	Reduced incidence or severity of vegetable diseases by *Ralstonia solanacearum*	Jetiyanon et al. (2003), Jetiyanon and Kloepper (2002)
	Reduced symptoms of *Pseudomonas syringae* pv. *tomato*	Ryu et al. (2003)
Azospirillum sp. B510	Exhibited resistance against diseases caused by the virulent bacterial pathogen *Xanthomonas oryzae in* an SA-independent manner	Yasuda et al. (2009)
Streptomyces lydicus WYEC108 + *Pseudomonas fluorescens* A506	Mixed inoculum reduced bacterial spot disease caused by *Xathomonas* sp. in tomato	Cuppels et al. (2013)

2007), inducing plant defense mechanisms (Bargabus et al., 2002; Mishra et al., 2006; Bakker et al., 2007), and enhancing plant mineral uptake (Malinowski et al., 2000). The roles of endophytes in the control of plant diseases in annual, biennial, and perennial crops have been reported by many researchers (Lodewyckx et al., 2002; Li et al., 2003; Bargabus et al., 2004; Kloepper et al., 2004, Hu et al., 2006; Zhao et al., 2006; Nguyen and Ranamukhaarachchi, 2010; Almoneafy et al., 2012).

Applications in the field have been limited, however, due to inconsistent disease control effects; in turn, these may be due to the improper selection of efficient strains. Feng et al. (2013) reported that the population of endophytes in resistant tomato cultivar at different growth stages was significantly higher than in susceptible cultivars. This finding indicates a positive relationship between populations of endophytic bacteria and plant resistance to bacterial wilt. Similar results were observed in other crops, such as cotton (Wang et al., 1997), tobacco (Bottomly et al., 2004), and pea (Ma et al., 2003). The selection of endophytes from disease-resistant plants may provide efficient bio-control agents for bacterial pathogens. Many research findings have shown that some endophytic strains elicit systemic protection to plants against bacterial diseases; this area of agricultural production requires more exploration (Table 14.4).

14.5 RHIZOSPHERE COMPETENCE AND BIOCONTROL EFFICIENCY

In many instances, the application of PGPR has been hampered by inconsistent performances under field conditions (Thomashow, 1996), usually due to poor rhizosphere competence (Schroth and Hancock, 1981; Weller, 1988). Rhizosphere competence of biocontrol agents comprises effective rhizosphere colonization, with the ability to survive in plant roots over a substantial time period, in the presence of the native microorganisms (Weller, 1988; Parke, 1991; Whipps, 1997). Rhizosphere competence is therefore one of the prerequisites for the selection of efficient biological control agents for controlling soil-borne pathogens. The plant-root surface and surrounding rhizosphere provide rich carbon sinks (Rovira, 1965). The root surface and rhizosphere nutrient-rich environments attract a great diversity of microorganisms, including phytopathogens. Competition for these nutrients is a fundamental mechanism by which PGPR protects plants from pathogens (Duffy, 2001).

In general, PGPR contact root surfaces by the actions of flagella and chemotactic responses (De Weger et al., 1987; De Weert et al., 2002), which may vary from plant to plant. The rhizosphere competence of rhizobacteria is strongly correlated with their ability to use organic acids as carbon sources; the composition and quantity of root exudates also influence the nature of bacterial activity (Loper and Schroth, 1986; Goddard et al., 2001). Root colonization is an important factor because a poor colonizer could cause decreased biocontrol activity (Bouizgarne, 2013). Some studies have indicated that the introduced bacterial population size over the crop-development period decides the efficiency of biocontrol activity against plant pathogens (Bull et al., 1991). In general, motile rhizobacteria may colonize the rhizosphere more profusely than the nonmotile organisms do, which would result in better rhizosphere activity and nutrient transformation (Raj et al., 2005).

The failure of PGPR in managing some important plant diseases is mainly due to poor survival and colonization in the rhizospheres of crop plants, as well as poor ability to tolerate abiotic stresses such as drought, salinity, high temperature, and so forth. For example, the perennial tree rhizosphere is dominated by specific groups of native microorganisms that cannot be neutralized by an alien microbial inoculum. Thus, the isolation of competitive and effective strains suited for different rhizospheres must also be ensured.

14.6 TRAITS OF EFFICIENT BIOCONTROL PGPR

The following points should be considered in the selection of PGPR isolates for the management of plant pathogens:

- Screening of candidate bacteria from pathogen-suppressive soil (Cook and Baker, 1983; Weller, 1988) will help in the selection of efficient strains.
- Organisms should have multiple antagonistic mechanisms (e.g., antibiotics, siderophores, and bacteriocin production) and the ability to mediate ISR in plants.
- The method of *in vitro* screening assay for selection of antagonists should be modified in such a way that it simulates natural conditions. Examples include tuber-slice assay (Rhodes et al., 1987) and seedling bioassay (Randhawa and Schaad, 1985).
- Selected organisms should have wider environmental adaptability (e.g., pH, temperature, and moisture).
- Organisms must have the ability to colonize plant roots (Juhnke et al., 1987) with sufficient population levels.
- Organisms must have good ecological competence and the ability to compete and survive in nature.
- Because the variability in root colonization may bring inconsistent results (Weller et al., 1988), organisms should have good rhizosphere competence. If required, crop-specific inoculants should be developed.
- Motile bacteria are hypothesized to colonize the roots efficiently than nonmotile bacteria; however, in-depth research is needed.
- Better rhizosphere colonization at lower temperatures and pH extremes may help bacteria escape competition from native microbes.
- Root colonization by the applied strain may be affected by the host genotype; similarly, the use of crop-specific microbial strains may enhance root colonization.

- PGPR traits in relation to rhizosphere competence have not been clearly described. We do know, however, that selection of bacteria based on their exo-polysaccharide production, fimbriae, flagella, chemotaxis, osmotolerance, and utilization of the carbohydrates in the root exudates are essential for better colonization.
- The exo-polysaccharide production of bacteria plays an important role in bacterial–plant association (Douglas et al., 1982).
- Fimbriae-producing bacteria aid the adhesion of bacterial cells to plant roots (Haahtela and Korhonen, 1985).
- The ability to tolerate dry soil and low osmotic potential will improve rhizosphere colonization under adverse conditions. Reports have indicated that proline-overproducing bacteria acquire more osmotolerance (Csonka, 1981).
- PGPR strains should be selected based on the rhizosphere matric potential of the crop plants. Howie et al. (1987) reported that a rhizosphere matric potential between −0.3 and −0.7 bars is ideal for bacterial cell growth.
- Microbes that possess cellulolytic activity will be more competent than nonproducers in terms of carbohydrate utilization (Ahmad and Baker, 1987) in the rhizosphere.
- Commercial formulations must preserve microbial activity for long periods of time without deterioration of their shelf life.
- Repeated *in vitro* culturing of PGPR strains may lead to the loss of field efficacy (Schroth et al., 1984).
- The PGPR must include a broad spectrum of activity. If it suppresses only one pathogen, other predominant pathogens will render the treatment ineffective (Weller et al., 1988).
- When used along with other compatible bio-inoculants and organic waste recyclers, a bio-control agent may show better disease control than in singular use.

14.7 CONCLUSION AND FUTURE STRATEGIES

The broad spectrum of PGPR action (antibiotic, siderophore, and bacteriocin production, and the induction of ISR) is reported to contain effective, economical, and practical ways to protect plants from bacterial diseases. Many experimental results have shown that, due to their synergistic behaviors, mixtures of PGPR isolates can express strong plant protection against many diseases (Stockwell et al., 2011). Developing a PGPR consortium would therefore be effective for the management of bacterial diseases in plants. Because most biological control work and commercial formulations to date pertain mainly to the genera *Pseudomonas* and *Bacillus*, researchers should shift their attention to other bacterial groups. When new PGPR formulations are introduced in agriculture, it is very essential to adopt the highest standards of environmental protection, as well as the protection of the health of all higher life forms, based on risk perception. When PGPR strains are selected, special attention must be paid to their rhizosphere competence and colonization, apart from their antagonistic traits. The success of any bio-control agent in field conditions mainly depends on how it reaches sufficient levels at the target sites and whether its formulation can support microbial activity for sufficient time periods.

The development of proper delivery mechanisms based on the cultivation practices for different crops is essential. In general, the free-living rhizobacteria must tolerate more environmental stresses than endophytes must in the effective colonization of plant roots; therefore, the use of plant growth-promoting endophytic bacteria should be investigated. In plant disease management, the unidirectional approach of using only biocontrol organisms must become a multidirectional or integrated approach (i.e., application of bio-control agents along with biofertilizers and other plant growth promoters). For example, because plant nutrient deficiencies reduce plants' innate immunities to various pathogens, the combined use of nutrient solubilizers or mobilizers along with bio-control agents would give better performance owing to the synergistic effects of the applied microbes. The failures of most bio-inoculants are mainly due to poor survival, lack of rhizosphere competence, and inability to tolerate abiotic stresses such as drought, salinity, acidity, and high temperature. Thus, suitable strains should be indentified to address these issues and, eventually, develop effective biological control mechanisms.

REFERENCES

Ahmad, F., Ahmad, I., Khan, M.S. 2008. Screening of free-living rhizospheric bacteria for their multiple plant growth promoting activities. *Microbiological Research* 163: 173–181.

Ahmad, J.S., Baker, R. 1987. Rhizosphere competence of *Trichoderma harzianum*. *Phytopathology* 77: 182–189.

Allington, W.B., Feaster, C.V. 1946. The relation of stomatal behaviour at the time of inoculation to the severity of infection of soybeans by *Xanthomonas phaseoli var. sojense* (Hedges) (starr) Burk. *Phytopathology* 36: 385–386.

Almoneafy, A.A., Xie, G.L., Tian, W.X., Xu, L.H., Zhang, G.Q., Ibrahim, M. 2012. Characterization and evaluation of *Bacillus* isolates for their potential plant growth and biocontrol activities against tomato bacterial wilt. *African Journal of Biotechnology* 11: 7193–7201.

Amrita, A., Usha, C., Bishwanath, C. 2013. Improvement of health status of *Listea monopetala* using plant growth promoting rhizobacteria. *International Journal of Bio-Resource and Stress Management* 4(2): 187–191.

Anuratha, C.S., Gnanamanickam, S.S. 1990. Biological control of bacterial wilt caused by *Pseudomonas solanacearum* in India with antagonistic bacteria. *Plant and Soil* 124: 109–116.

Aoki, M., Uehara, K., Koseki, K., Tsuji, K., Iijima, M., Ono, K., Samejima, T. 1991. An antimicrobial substance produced by *Pseudomonas cepacia* B5 against the bacterial wilt disease pathogen *Pseudomonas solanacearum*. *Agricultural and Biological Chemistry* 55: 715–722.

Arwiyanto, T., Goto, M., Tsuyumu, S. 1994. Biological control of bacterial wilt of tomato by an avirulent strain of *Pseudomonas solanacearum*. *Annals of the Phytopathologicial Society of Japan* 60: 421–430.

Ashworth, L.J., Hildebrand, D.C., Schroth, M.N. 1970. Erwinia-induced internal necrosis of immature cotton bolls. *Phytopathology* 60: 602–607.

Baker, K.F. 1971. Fire Blight of pome fruits: The genesis of the concept that bacteria can be pathogenic to plants. *Hilgardia* 40: 603–633.

Bakker, P.A.H.M., Pierterse, C.M.J., Van Loon, L.C. 2007. Induced systemic resistance by fluorescent *Pseudomonas* spp. *Phytopathology* 97: 239–243.

Bakker, P.A.H.M., Raaijmakers, J.M., Schippers, B. 1993. Role of iron in the suppression of bacterial plant pathogens by fluorescent pseudomonads. In *Iron Chelation in Plants and Soil Microorganisms*. Barton, L.L. (ed.). San Diego: B.C. Hemming in Academic Press, pp. 269–278.

Barea, J.M., Azcon-Aguilar, C., Azcon, R. 1997. Interactions between mycorrhizal fungi and rhizosphere microorganisms within the context of sustainable soil–plant systems. In *Multitrophic Interactions in Terrestrial Systems*. Gange, A.C., Brown, V.K. (eds.). Oxford, UK: Blackwell Science, pp. 65–77.

Bargabus, R.L., Zidack, N.K., Sherwood, J.E., Jacobsen, B.J. 2002. Characterization of systemic resistance in sugar beet elicited by a non pathogenic, phyllosphere colonizing *Bacillus mycoides*, biological control agent. *Physiological and Molecular Plant Pathology* 61: 289–298.

Bargabus, R.L., Zidack, N.K., Sherwood, J.E., Jacobsen, B.J. 2004. Screening for the identification of potential biological control agents that induce systemic acquired resistance in sugar beet. *Biological Control* 30: 342–350.

Barka, E.A., Gognies, S., Nowak, J., Audran, J.C., Belarbi, A. 2002. Inhibitory effect of endophytic bacteria on *Botrytis cinerea* and its influence to promote the grapevine growth. *Biological Control* 24: 135–142.

Bashan, Y., de-Bashan, L. 2002. Protection of tomato seedlings against infection by *Pseudomonas syringae* pv. *tomato* by using the plant growth-promoting bacterium *Azospirillum brasilense*. *Applied and Environmental Microbiology* 68: 2635–2643.

Basnayake, W.V.S., Birch, R. G. 1995. A gene from *Alcaligenes denitrificans* that confers albicidin resistance by reversible antibiotic binding. *Microbiology* 141: 551–560.

Beer, S.V., Vidaver, A.K. 1978. Bacteriocins produced by *Erwinia herbicola* inhibits *Erwinia amylovora*. In *Proceedings of the Third International Congress of Plant Pathology,* Laux, W. (ed.), München, Germany, August 16–22, 1978.

Benhamou, N., Kloepper, J.W., Quadt-Hallmann, A., Tuzun, S. 1996. Induction of defense-related ultrastructural modifications in pea root tissues inoculated with endophytic bacteria. *Plant Physiology* 112: 919–929.

Berdy, J. 2005. Bioactive microbial metabolites. *Journal of Antibiotics* 58: 1–26.

Berisha, B., Chen, Y.D., Zhang, G.I., Xu, B.Y., Chen, T.A. 1998. Isolation of Pierce's disease bacteria from grapevines in Europe. *European Journal of Plant Pathology* 104: 427–433.

Bharathi, R., Vivekananthan, R., Harish, S., Ramanathan, A., Samiyappan, R. 2004. Rhizobacteria-based bio-formulations for the management of fruit rot infection in chillies. *Crop Protection* 2: 835–843.

Bottomly, T.S., Fortnum, B.A., Kurtz, H., Kluepfel, D. 2004. Diversity of endophytic bacteria in healthy and *Ralstonia solanacearum* infected tobacco seedlings. *Agronomy-Phytopathology*, p. 8 (abstract).

Bouizgarne, B. 2013. Bacteria for plant growth promotion and disease management. In *Bacteria in Agrobiology: Disease Management*. Maheshwari, D.K. (ed.). Berlin, Heidelberg: Springer-Verlag, doi: 10.1007/978-3-642-33639-3_2.

Bradbury, J.F. 1971. Nomenclature of the bacterial leaf streak pathogen of rice. *International Journal of Systematic Bacteriology* 21: 72.

Bradbury, J.F. 1973. *Xanthomonas vasculorum*. Descriptions of Fungi and Bacteria. *IMI Descriptions of Fungi and Bacteria* 38: 380.

Brierley, P. 1928. Pathogenicity of *Bacillus mesentericus*, *B. aroideae*, *B. caratovorus* and *B. phytophthorus* to potato tubers. *Phytopathology* 18: 819–838.

Bull, C.T., Weller, D.M., Thomashow, L.S. 1991. Relationship between root colonization and suppression of *Gaeumannomyces graminis* var. *triciti* by *Pseudomonas fluorescens* strain 2–79. *Phytopathology* 81: 954–959.

Burr, T.J., Otten, L. 1999. Crown gall of grape: Biology and disease management. *Annual Review of Phytopathology* 37: 53–80.

Chamberlain, D.W. 1956. Pathogenicity of *Pseudomonas tabaci* on soybean and tobacco. *Phytopathology* 46: 51–52.

Chase, A.R. 1989. Effect of nitrogen and potassium fertilizer rates on severity of *Xanthomonas* blight of *Syngonium podophyllum*. *Plant Disease* 73: 972–975.

Chen, C., Belanger, R.R., Benhamou, N., Paulitz, T.C. 2000. Defense enzymes induced in cucumber roots by treatment with plant growth-promoting rhizobacteria (PGPR) and *Pythium aphanidermatum*. *Physiological and Molecular Plant Pathology* 56: 13–23.

Chen, X.Y., Xiang, W.N. 1986. A strain of *Agrobacterium radiobacter* inhibits growth and gall formation biotype III strains of *A. tumefaciens* from grapevine. *Acta Microbiologica Sinica* 26: 193–199.

Compant, S., Reiter, B., Sessitsch, A., Nowak, J., Clement, C., Ait Barka, E. 2005. Endophytic colonization of *Vitis vinifera* L. by a plant growth promoting bacterium, *Burkholderia* sp. strain Ps JN. *Applied and Environmental Microbiology* 71: 1685–1693.

Cook, R.J., Baker, K.F. 1983. *The Nature and Practice of Biological Control of Plant Pathogens*. St. Paul: American Psychopathological Society, 539 p.

Coombs, J.T., Michelsen, P.P., Franco, C.M.M. 2004. Evaluation of endophytic actinobacteria as antagonists of *Gaeumannomyces graminis* var. *tritici* in wheat. *Biological Control* 29: 359–366.

Corbett, J.R. 1974. Pesticide design. In *The Biochemical Mode of Action of Pesticides*. London: Academic Press, pp. 44–86.

Csonka, L.N. 1981. Proline overproduction results in enhanced osmotolerance in *Salmonella typhimurium*. *Molecular Genetics and Genomics* 182: 82–86.

Cuppels, D.A., Hanson, R.S., Kelman, A. 1978. Isolation and characterization of a bacteriocin produced by *Pseudomonas solanacearum*. *Journal of General Microbiology* 109: 295–303.

Cuppels, D.A., Higham, J., Traquair, J.A. 2013. Efficacy of selected streptomycetes and a streptomycete + pseudomonad combination in the management of selected bacterial and fungal diseases of field tomatoes. *Biological Control* 67(3): 361–372.

Damann, K.E., Derrick, K.S. 1976. Bacterium associated with ratoon stunting disease in Louisiana. *Sugarcane Pathologists' Newsletter* 15–16: 20–22.

Datnoff, L., Elmer, W., Huber, D. 2006. *Mineral Nutrition and Plant Diseases*. St. Paul, MN: APS Press.

De Souza, J.T., de Boer, M., de Waard, P., van Beek, T. A., Raaijmakers, J.M. 2003. Biochemical, genetic, and zoosporicidal properties of cyclic lipopeptide surfactants produced by *Pseudomonas fluorescens*. *Applied and Environmental Microbiology* 69: 7161–7172.

De Weert, S., Vermeiren, H., Mulders, I.H.M., Kuiper, I., Hendrickx, N., Bloemberg, G.V., Vanderleyden, J., de Mot, R., Lugtenberg, B.J.J. 2002. Flagella-driven chemotaxis toward exudate components is an important trait for tomato root colonization by *Pseudomonas fluorescens*. *Molecular Plant-Microbe Interactions* 15: 1173–1180.

De Weger, L.A., van der Vlugt, C.I.M., Wijfjes, A.H.M., Bakker, P.A.H.M., Schippers, B., Lugtenberg, B. 1987. Flagella of a plant-growth-stimulating *Pseudomonas fluorescens* strain are required for colonization of potato roots. *Journal of Bacteriology* 169: 2769–2773.

Défago, G., Berling, C.H., Borger, U., Keel, C., Voisard, C. 1990. Suppression of black rot of tobacco by a *Pseudomonas* strain: Potential applications and mechanisms. In *Biological Control of Soil Borne Plant Pathogen*. Hornby, D., Cook, R.J., Henis, Y. (eds.). Wallingford: CAB International, pp. 93–108.

Demaree, J.B., Smith, N.R. 1952. *Nocardia vaccinii* n. sp. causing galls on blueberry plants. *Phytopathology* 42: 249–252.

Doke, N., Ramirez, A.V., Tomiyama, K. 1987. Systemic induction of resistance in potato plants against *Phytophthora infestans* by local treatment with hyphal wall components of the fungus. *Journal of Phytopathology* 119: 232–239.

Dordas, C. 2008. Role of nutrients in controlling plant diseases in sustainable agriculture: A review. *Agronomy for Sustainable Development* 28: 33–46.

Douglas, C.J., Halperin, W., Nester, E.W. 1982. *Agrobacterium tumefaciens* mutants affected in attachment to plant cells. *Journal of Bacteriology* 152: 1265–1275.

Duffy, B.K. 2001. Competition. In *Encyclopedia of Plant Pathology*. Maloy, O.C., Murray, T.D. (eds.). New York: John Wiley & Sons, pp. 243–244.

Duffy, B.K., Défago, G. 1999. Environmental factors modulating antibiotic and siderophore biosynthesis by *Pseudomonas fluorescens* biocontrol strains. *Applied and Environmental Microbiology* 65: 2429–2438.

Duijff, B.J., Gianinazzi-Pearson, V., Lemanceau, P. 1997. Involvement of the outer membrane lipopolysaccharides in the endophytic colonization of tomato roots by biocontrol *Pseudomonas fluorescens* strain WCS417r. *New Phytologist* 135: 325–334.

Dunleavy, J.M. 1963. A vascular disease of soybeans caused by *Corynebacterium* sp. *Plant Disease Report* 47: 612–613.

El-Abyad, M.S., El-Sayed, M.A., El-Shanshoury, A.R., El-Batanouny, N.H. 1996. Effect of culture conditions on the antimicrobial activities of UV-mutants of *Streptomyces corcharusi* against bean and banana wilt pathogens. *Microbiology Research* 151: 201–211.

Elliot, C., Smith, E.F. 1929. A bacterial stripe disease of sorghum. *Journal of Agricultural Research* 38: 1–22.

Enebak, S.A., Wei, G., Kloepper, J.W. 1998. Effects of plant growth-promoting rhizobacteria on loblolly and slash pine seedlings. *Forest Science* 44: 139–144.

Engelhard, A.W. 1989. *Soil Borne Plant Pathogens Management of Disease with Macro and Microelements*. St. Paul, MN: APS.

Erwin, D.C., Starr, M.P., Desjardins, P.R. 1964. Bacterial leaf spot and stem blight of safflower caused by *Pseudomonas syringe*. *Phytopathology* 54: 1247–1250.

Feng, H., Li, Y., Liu, Q. 2013. Endophytic bacterial communities in tomato plants with differential resistance to *Ralstonia solanacearum*. *African Journal of Microbiology Research* 7(15): 1311–1318.

Fucikovsky, L., Luna, I., Lopez, C. 1990. Bacterial antagonists to *Pseudomonas solanacearum* in potatoes and some other plant pathogens. In *Proceedings of the International Conference on Plant Pathogenic Bacteria*. Akademie Kiado, Budapest, Hungary, pp. 201–206.

Furuya, N., Kushima, Y., Tsuchiya, K. 1991. Protection of tomoto by pretreatment with *Pseudomonas glumae* from infection with *Pseudomonas solanacearum* and its mechanisms. *Annals of the Phytopathological Society of Japan* 57: 363–370.

Gardan, L., Gouy, C., Christen, R., Samson, R. 2003. Elevation of three subspecies of *Pectobacterium carotovorum* to species level: *Pectobacterium atrosepticum* sp. nov., *Pectobacterium betavasculorum* sp. nov. and *Pectobacterium wasabiae* sp. nov. *International Journal of Systematic and Evolutionary Microbiology* 53: 381–391.

Gehring, P.J., Nolan, R.J., Watanabe, P.G. 1993. Solvents, fumigants and related compounds. In *Handbook of Pesticide Toxicology*, Vol. 2. Hayes, W.J. (ed.). San Diego, CA: Academic Press, pp. 646–649.

Gitaitis, R.D., Stall, R.E., Strandberg, J.O. 1976. Over seasoning of *Pseudomonas alboprecipitans Rosens*, causal agent of bacterial leaf blight of corn. *Proceedings of American Phytopathological Society* 3: 257.

Glick, B.R. 1995. The enhancement of plant-growth by free-living bacteria. *Canadian Journal of Microbiology* 41: 109–117.

Goddard, V.J., Bailey, M.J., Darrah, P., Lilley, A.K., Thompson, I.P. 2001. Monitoring temporal and spatial variation in rhizosphere bacterial population diversity: A community approach for the improved selection of rhizosphere competent bacteria. *Plant and Soil* 232: 181–193.

Goto, M. 1979. Bacterial foot rot disease of rice caused by strain of *Erwinia chrysanthemi*. *Phytopathology* 68: 213–216.

Goto, K., Ohata, K.I. 1961. *Bacterial Stripe of Rice*, Vol. 10. Taiwan: College of Agriculture National Taiwan University (special publication), pp. 49–59.

Gottwald, T.R., Graham, J.H., Schubert, J.S. 2002. Citrus canker: The pathogen and its impact. Online. Plant Health Progress. doi:10.1094/PHP-2002-0812-01-RV

Graham, D.R. 1983. Effects of nutrients stress on susceptibility of plants to disease with particular reference to the trace elements. *Advances in Botanical Research* 10: 221–276.

Graham, D.R., Webb, M.J. 1991. Micronutrients and disease resistance and tolerance in plants. In *Micronutrients in Agriculture*. Mortvedt, J.J., Cox, F.R., Shuman, L.M., Welch, R.M. (eds.). 2nd edn. Madison, WI: Soil Science Society of America, pp. 329–370.

Guo, J.H., Qi, H.Y., Guo, Y.H., Ge, H., Gong, L.Y., Zhang, L.X., Sun, P.H. 2004. Biocontrol of tomato wilt by growth-promoting rhizobacteria. *Biological Control* 29: 66–72.

Gupta, A., Gopal, M., Tilak, K.V. 2000. Mechanism of plant growth promotion by rhizobacteria. *Indian Journal of Experimental Biology* 38: 856–862.

Haahtela, K., Korhonen, T.K. 1985. In vitro adhesion of N_2-fixing enteric bacteria to roots of grasses and cereals. *Applied and Environmental Microbiology* 49: 1186–1190.

Habish, H.A., Hammad, A.H. 1969. Seedling infection of sesame by *Xanthomonas sesame* sabet and Dowson and the assessment of resistance to bacterial leaf spot disease. *Phytopathologische Zeitschrift* 64: 32–38.

Hartman, G.L., Hong, W.F., Hayward, A.C. 1992. Potential of biological and chemical control of bacterial wilt. In *Bacterial Wilt*. Hartman, G.L. (ed.). Canberra: A.C.S Hayward in ACIAR, pp. 322–326.

Hiltner, L. 1904. Über neuere Erfahrungen und Probleme aufdem Gebiete der Bodenbakteriologie unter besonderer Berücksichtigung der Gründüngung und Brache. *Arb DLG* 98: 59–78.

Hingorani, M.K., Grant, U.J., Singh, N.H. 1959. *Erwinia carotovora* f.sp. *zeae*, a destructive pathogen of maize in India. *Indian Phytopathology* 12: 151–157.

Hoffland, E., Pieterse, C.M.J., Bik, L.,Van Pelt, J.A. 1995. Induced systemic resistance in radish is not associated with accumulation of pathogenesis-related proteins. *Physiological and Molecular Plant Pathology* 46: 309–320.

Hofte, M. 1993. Classes of microbial siderophores. In *Iron Chelation in Plants and Soil Microorganisms*. Barton, L.L. (ed.). San Diego: BC Hemming in Academic Press, pp. 3–26.

Howie, W.J., Cook, R.J., Weller, D.M. 1987. Effects of soil matric potential and cell motility on wheat root colonization by fluorescent pseudomonads suppressive to take-all. *Phytopathology* 77: 286–292.

Hu, Q.P., Xu, J.G., Liu, H.L., Chen, W.L., Zhu, Z.Y. 2006. Isolation and identification of endogenetic bacteria in tomato stems and screening of antagonistic bacteria to *Pseudomonas solanacearum*. *Acta Botanica Boreali-Occidentalia Sinica* 26: 2039–2043.

Huang, J., Wei, Z., Tan, S., Mei, X., Yin, S., Shen, Q., Xu, Y. 2013. The rhizosphere soil of diseased tomato plants as a source for novel microorganisms to control bacterial wilt. *Applied Soil Ecology* 72: 79–84.

Huber, D.M., Graham, R.D. 1999. The role of nutrition in crop resistance and tolerance to disease. In *Mineral Nutrition of Crops Fundamental Mechanisms and Implications*. Rengel, Z. (ed.). New York: Food Product Press, pp. 205–226.

Huber, D.M., Watson, R.D. 1974. Nitrogen form and plant disease. *Annual Review of Phytopathology* 12: 139–165.

Hynes, R., Lazarovits, G. 1989. Effect of seed treatment with plant growth promoting rhizobacteria on the protein profiles of intercellular fluids from bean and tomato leaves. *Canadian Journal of Plant Pathology* 11: 191.

Innes, N.L., Last, F.T. 1961. Cotton disease symptoms caused by different concentrations of *Xanthomonas malvacearum*. *Emporium of Cotton Growers Review* 38: 27–29.

Jagadeesh, K.S. 2000. Selection of rhizobacteria antagonistic to *Ralstonia solanacearum* causing bacterial wilt in tomato and their biocontrol mechanisms. PhD Thesis. University of Agricultural Sciences, Dharwad.

Jagadeesh, K.S., Kulkarni, J.H., Krishnaraj, P.U. 2001. Evaluation of the role of fluorescent siderophore in the biological control of bacterial wilt in tomato using Tn5 mutants of fluorescent *Pseudomonas* sp. *Current Science* 81: 882–889.

Jetiyanon, K., Fowler, W.D., Kloepper, J.W. 2003. Broad-spectrum protection against several pathogens by PGPR mixtures under field conditions. *Plant Disease* 87: 1390–1394.

Jetiyanon, K., Kloepper, J.W. 2002. Mixtures of plant growth-promoting rhizobacteria for induction of systemic resistance against multiple plant diseases. *Biological Control* 24: 285–292.

Jeun, Y.C., Park, K.S., Kim, C.H., Fowler, W.D., Kloepper, J.W. 2004. Cytological observations of cucumber plants during induced resistance elicited by rhizobacteria. *Biological Control* 29: 34–42.

Joachems, S.C., Maas, J.G.J.A. 1922. Alijmzickte in de *Hibiscus cannabinus* op Sumatra's ooskust Tesymannia. 33: 542–546.

Jones, D.A., Kerr, A. 1989. *Agrobacterium radiobacter* K1026 a genetically engineered derivative of strain K84 for biological control of crown gall. *Plant Disease* 73: 15–18.

Jones, J.B., Lacy, G.H., Bouzar, H., Minsavage, G.V., Stall, R.E., Schaad, N.W. 2005. Bacterial spot-worldwide distribution, importance and review. *Acta Horticulturae* 695: 27–33.

Jones, D.A., Ryder, M.H., Clare, B.G., Farrand, S.K., Kerr, A. 1991. Biological control of crown gall using *Agrobacterium* strains K84 and K1026. In *The Biological Control of Plant Diseases*. Komada, H., Kiritani, K., Bay-Petersen, J. (eds.). FFTC Book Series no. 42. (Food and Fertilizer Technology Center) for the Asian and Pacific Region, Taipei, Taiwan. pp. 161–170.

Judith, E.M., de Vos, R.C.H., Dekkers, E., Pineda, A., Guillod, L., Bouwmeester, K., van Loon, J.J.A. et al. 2012. Metabolic and transcriptomic changes induced in *Arabidopsis* by the rhizobacterium *Pseudomonas fluorescens* SS101. *Plant Physiology* 160: 2173–2188.

Juhnke, M.E., Mathre, D.E., Sands, D.C. 1987. Identification and characterization of rhizosphere-competent bacteria of wheat. *Applied and Environmental Microbiology* 53: 2793–2799.

Kang, S.H., Cho, H.S., Cheong, H., Ryu, C.M., Kim, J.F., Park, S.H. 2007. Two bacterial endophytes eliciting both plant growth promotion and plant defense on pepper (*Capsicum annuum* L.). *Journal of Microbiology and Biotechnology* 17: 96–103.

Keen, N.T., Yoshikawa, M. 1983. β-1,3-Endoglucanase from soybean releases elicitor-active carbohydrates from fungus cell walls. *Plant Physiology* 71: 460–465.

Kempe, J., Sequeira, L. 1983. Biological control of bacterial wilt of potatoes: Attempts to induce resistance by treating tubers with bacteria. *Plant Disease* 67: 499–503.

Kendrick, J.B. 1926. Holcus bacterial spot of *Zeamays* and *Holcus* species. *IOWA Agricultural Experiment Station Research Bulletin* 100: 303–334.

Kennedy, B.W., Mew, T.W. 1971. Relationship of postinoculation humidity to bacterial leaf blight development on soybean. *Phytopathology* 61: 879–880.

Kilic-Ekici, O., Yuen, G.Y. 2004. Comparison of strains of *Lysobacter enzymogenes* and PGPR for induction of resistance against *Bipolaris sorokiniana* in tall fescue. *Biological Control* 30: 446–455.

Kim, D.H., Misaghi, I.J. 1996. Biocontrol performance of two isolates of *Pseudomonas fluorescens* in modified soil atmospheres. *Phytopathology* 86: 1238–1241.

Kirkegaard, J.A., Munns, R., James, R.A., Neate, S.M. 1999. Does water and phosphorus uptake limit leaf growth of *Rhizoctonia*-infected wheat seedlings? *Plant and Soil* 209: 157–166.

Kloepper, J.W. 1983. Effect of seed piece inoculation with plant growth-promoting rhizobacteria on populations of *Erwinia carotovora* on potato roots and daughter tubers. *Phytopathology* 73: 217–219.

Kloepper, J.W. 1993. Plant growth-promoting rhizobacteria as biological control agents. In *Soil Microbial Technologies*. Metting, B. (ed.). New York: Marcel Dekker, pp. 255–274.

Kloepper, J.W., Ryu, C.M., Zhang, S. 2004. Induced systemic resistance and promotion of plant growth by *Bacillus* spp. *Phytopathology* 94: 1259–1266.

Kloepper, J.W., Schroth, M.N. 1978. Plant growth promoting rhizobacteria on radishes. In *Proceedings of the 4th International Conference on Plant Pathogenic Bacteria,* Angers, France, pp. 879–882.

Kloepper, J.W., Tuzun, S., Kuc, J.A. 1992. Proposed definitions related to induced disease resistance. *Biocontrol Science and Technology* 2: 349–351.

Knoester, M., Pieterse, C.M., Bol, J.F., Van Loon, L.C. 1999. Systemic resistance in Arabidopsis induced by rhizobacteria requires ethylene-dependent signaling at the site of application. *Molecular Plant-Microbe Interactions* 12(8): 720–727.

Koch, E., Meier, B.M., Eiben, H.G., Slusarenko, A. 1992. A lipoxygenase from leaves of tomato (*Lycopersicon esculentum* Mill.) is induced in response to plant pathogenic *Pseudomonas. Plant Physiology* 99: 571–576.

Krasilnikov, N.A. 1949. In: *Determination of Bacteria and Actinomycetes*. Moscow and Leningrad: Institute of Microbiology Academy Nauk SSSR, p. 830.

Leeman, M., DenOuden, E.M., VanPelt, J.A., Dirkx, F.P.M., Steijl, H., Bakker, P.A.H.M., Schippers, B. 1996. Iron availability affects induction of systemic resistance to *Fusarium* wilt of radish by *Pseudomonas fluorescens*. *Phytopathology* 86: 149–155.

Li, Q.Q., Luo, K., Lin, W., Peng, H.W., Luo, X.M. 2003. Isolation of tomato endophytic antagonists against *Ralstonia solanacearum. Acta Phytopathology Sinica* 33: 364–367.

Lindow, S.W., McGourty, G., Elkins, R. 1996. Interactions of antibiotics with *Pseudomonas fluorescens* strain A506 in the control of fire blight and frost injury to pear. *Phytopathology* 86: 841–848.

Liu, L., Kloepper, J.W., Tuzun, S. 1995. Induction of systemic resistance in cucumber against bacterial angular leaf spot by plant growth-promoting rhizobacteria. *Phytopathology* 85: 843–847.

Lodewyckx, C., Vangronsfeld, J., Porteous, R., Moore, E.R.B., Taghavi, S., Mergeay, M., Van der Lelie, D. 2002. Endophytic bacteria and their potential applications. *Critical Review of Plant Science* 21: 583–606.

Loper, J.E., Schroth, M.N. 1986. Importance of siderophores in microbial interactions in the rhizosphere. In *Iron Siderophores and Plant Sisease*. Swinburne, T.R. (ed.). New York: Plenum, pp. 85–98.

Ma, W.B., Guinel, F.C., Glick, B.R. 2003. *Rhizobium leguminosarum* biovar *viciae* 1-aminocyclopropane-1-carboxylate deaminase promotes nodulation of pea plants. *Applied and Environmental Microbiology* 69: 4396–4402.

Malinowski, D.P., Alloush, G.A., Belesky, D.P. 2000. Leaf endophyte *Neotyphodium coenophialum* modifies mineral uptake in tall fescue. *Plant and Soil* 227: 115–126.

Mansfield, J.G., Magori, S., Citovsky, S., Sriariyanum, V., Ronald, M., Dow, P., Verdier, M. et al. 2012. Top 10 plant pathogenic bacteria in molecular plant pathology. *Molecular Plant Pathology* 13: 614–629.

Mariano, R.L.R., Silveira, E.B., Assis, S.M.P., Gomes, A.M.A., Oliveira, I.S., Nascimento, A.R.P. 2001. Diagnose e manejo de fitobacterioses de importância no Nordeste Brasileiro. In *Proteção de Plantas na Agricultura Sustentável*. Michereff, S.J., Barros, R. (eds.). Brasil: UFRPE, Recife, pp. 141–169.

Marschner, H. 1995. *Mineral Nutrition of Higher Plants*, 2nd edn. London: Academic, p. 889.

Massomo, S.M.S., Mabagala, R.B., Swai, I.S., Hockenhull, J., Mortense, C.N. 2004. Evaluation of varietal resistance in cabbage against the black rot pathogen, *Xanthomonas campestris* pv. *campestris* in Tanzania. *Crop Protection* 23: 315–325.

Maurhofer, M., Hase, C., Meuwly, P., Metraux, J.P., Défago, G. 1994. Induction of systemic resistance of tobacco to tobacco necrosis virus by the root-colonizing *Pseudomonas fluorescens* strain CHA0: Influence of the *gacA* gene and of pyoverdine production. *Phytopathology* 84: 139–146.

McClean, A.P.D. 1930. The bacterial wilt disease of peanuts. *South African Department of Agricultural Science Bulletin* 87: 14.

Mishra, R.P., Singh, R.K., Jaiswal, H.K., Kumar, V., Maurya, S. 2006. *Rhizobium* mediated induction of phenolics and plant growth promotion in rice (*Oryza sativa* L.). *Current Microbiology* 52: 383–389.

Miyadoh, S. 1993. Research on antibiotic screening in Japan over the last decade: A producing microorganisms approach. *Actinomycetologica* 9: 100–106.

Mulya, K., Watanabe, M., Goto, M. 1996. Suppression of bacterial wilt disease of tomato by root dipping with *Pseudomonas fluorescens* P.f. G32—The role of antibiotics and siderophores. *Annals of the Phytopathological Society of Japan* 62: 134–140.

Nandakumar, R., Babu, S., Viswanathan, R., Raguchander, T., Samiyappan, R. 2001. Induction of systemic resistance in rice against sheath blight disease by *Pseudomonas fluorescens*. *Soil Biology and Biochemistry* 33: 603–612.

Nguyen, M.T., Ranamukhaarachchi, S.L. 2010. Soil-borne antagonists for biological control of bacterial wilt disease caused by *Ralstonia solanacearum* in tomato and pepper. *Journal of Plant Pathology* 92: 395–406.

Niu, B., Vater, J., Rueckert, C., Blom, J., Lehmann, M., Ru, J., Chen, X., Wang, Q., Borriss, R. 2013. Polymyxin P is the active principle in suppressing phytopathogenic *Erwinia* spp. by the biocontrol rhizobacterium *Paenibacillus polymyxa* M-1. *BMC Microbiology* 13: 137.

Nowak, J., Shulaev, V. 2003. Priming for transplant stress resistance in vitro propagation. *In vitro Cellular and Developmental Biology—Plant* 39: 107–124.

Nunney, L., Vickerman, D.B., Bromley, R.E., Russell, S.A., Hartman, J.R., Morano, L.D., Stouthamer, R. 2013. Recent evolutionary radiation and host plant specialization in the *Xylella fastidiosa* subspecies native to the United States. *Applied and Environmental Microbiology* 79(7): 2189. doi: 10.1128/AEM.03208-12.

Ongena, M., Daayf, F., Jacques, P., Thonart, P., Benhamou, N., Paulitz, T.C., Bélanger, R.R. 2000. Systemic induction of phytoalexins in cucumber in response to treatments with fluorescent pseudomonads. *Plant Pathology* 49: 523–530.

Ongena, M., Henry, G., Adam, A., Jourdan, E., Thonart, P. 2009. Plant defense reactions stimulated following perception of Bacillus lipopeptides. In 8th International PGPR Workshop. USA. 43.

Oskay, M. 2009. Antifungal and antibacterial compounds from *Streptomyces* strains. *African Journal of Biotechnology* 8: 3007–3017.

Panneerselvam, P., Sukhada, M., Saritha, B., Upreti, K.K., Poovarasan, S., Ajay, M., Sullamath, V.V. 2012. *Glomus mosseae* associated bacteria an their influence on stimulation of mycorrhizal colonization, sporulation and growth promotion in guava (*Psidium guajava* L.) seedlings. *Biological Agriculture and Horticulture* 28(4): 267–279.

Park, K.S., Kloepper, J.W. 2000. Activation of PR-1a promoter by rhizobacteria which induce systemic resistance in tobacco against *Pseudomonas syringae* pv. *tabaci*. *Biological Control* 18: 2–9.

Parke, J.L. 1991. Root colonization by indigenous and introduced microorganisms. In *The Rhizosphere and Plant Growth*. Keister, D.L., Gregan, P.B. (eds.). Dordrecht, The Netherlands: Kluwer Academic Publishers, pp. 33–42.

Perez, J.C., Diccion, T.C., Olbinado, G., Bugat, E., Balaoing, J., Gayagay, R., Backian, G.S., Galap, J., Baden, C. 1997. Management of bacterial wilt (*Pseudomonas solanacearum tuberosum* Linn.). *Philippine Journal of Crop Science* 22: 73–79.

Persley, G.J. 1973. Naturally occurring alternative hosts of *Xanthomonas albilineans* in Queensland. *Plant Disease Reporter* 57:1040–1042.

Pieterse, C.M.J., Van Wees, S.C.M., Hoffland, E., Van Pelt, J.A., Van Loon, L.C. 1996. Systemic resistance in *Arabidopsis* induced by biocontrol bacteria is independent of salicylic acid accumulation and pathogenesis related gene expression. *The Plant Cell* 8: 1225–1237.

Pieterse, C.M.J., Van Wees, S.C.M., Van Pelt, J.A., Knoester, M., Loon, R., Gerrits, H., Weisbeek, P.J., Van Loon, L.C. 1998. A novel signaling pathway controlling induced systemic resistance in *Arabidopsis*. *The Plant Cell* 10: 1571–1580.

Ping, L., Boland, W. 2004. Signals from the underground: Bacterial volatiles promote growth in *Arabidopsis*. *Trends in Plant Science* 9: 263–269.

Poovarasan, S., Sukhada, M., Panneerselvam, P., Saritha, B., Ajay, M. 2013. Mycorrhizae colonizing actinomycetes promote plant growth and control bacterial blight disease of pomegranate (*Punica granatum* L. cv *Bhagwa*). *Crop Protection* 53: 1–7.

Press, C.M., Wilson, M., Tuzun, S., Kloepper, J.W. 1997. Salicylic acid produced by *Serratia marcescens* 90-166 is not the primary determinant of induced systemic resistance in cucumber or tobacco. *Molecular Plant-Microbe Interactions* 10: 761–768.

Raj, N.S., Shetty, H.S., Reddy, M. S. 2005. Plant growth-promoting rhizobacteria: Potential green alternative for plant productivity. In *PGPR: Biocontrol and Biofertilization*. Siddiqui, Z.A. (ed.). Dordrecht, The Netherlands: Springer, pp. 197–216.

Ramamoorthy, V., Samiyappan, R. 2001. Induction of defence related genes in *Pseudomonas fluorescens* treated chilli plants in response to infection by *Colletotrichum capsici*. *Journal of Mycology and Plant Pathology* 31: 146–155.

Ramamoorthy, V., Viswanathan, R., Ravichander, T., Prakasam, V., Samiyappan, R. 2001. Induction of systemic resistance by plant growth-promoting rhizobacteria in crop plants against pests and diseases. *Crop Protection* 20: 1–11.

Randhawa, P.S., Schaad, N.W. 1985. A seedling bioassay chamber for determining bacterial colonization and antagonism on plant roots. *Phytopathology* 75: 254–259.

Raupach, G.S., Kloepper, J.W. 1998. Mixtures of plant growth promoting rhizobacteria enhance biological control of multiple cucumber pathogens. *Phytopathology* 88: 1158–1164.

Reuveni R., Dor, G., Raviv, M., Reuveni, M., Tuzun, S. 2000. Systemic resistance against *Sphaerotheca fuliginea* in cucumber plants exposed to phosphate in hydroponics system, and its control by foliar spray of mono-potassium phosphate. *Crop Protection* 19: 355–361.

Rhodes, D., Logan, C., Gross, D. 1987. *Phytopathology*. Selection of Pseudomonads spp. inhibitory to potato seed tuber decay. *Phytopathology* 76: 1078.

Ribeiro, R.L.D., Durbin, R.D., Arny, D.C., Uchytil, T.F. 1977. Characterization of bacterium inciting chocolate spot of corn. *Phytopathology* 67: 1427–1431.

Rizk, M., Rahman, T.A., Metwally, H. 2007. Screening of antagonistic activity in different *Streptomyces* species against some pathogenic microorganisms. *Journal of Biological Sciences* 7: 1418–1423.

Rovira, A.D. 1965. Interactions between plant roots and soil microorganisms. *Annual Review of Microbiology* 19: 241–266.

Ryan, P.R., Delhaize, E., Jones, D.L. 2001. Function and mechanism of organic anion exudation from plant roots. *Annual Review of Plant Physiology and Plant Molecular Biology* 52: 527–560.

Ryu, C.M., Farag, M.A., Hu, C.H., Reddy, M.S., Kloepper, J.W., Pare, P.W. 2004a. Bacterial volatiles induce systemic resistance in *Arabidopsis*. *Plant Physiology* 134: 1017–1026.

Ryu, C.M., Hu, C.H., Reddy, M.S., Kloepper, J.W. 2003. Different signaling pathways of induced resistance by rhizobacteria in *Arabidopsis thaliana* against two pathovars of *Pseudomonas syringae*. *New Phytology* 160: 413–420.

Ryu, C.M., Murphy, J.F., Mysore, K.S., Kloepper, J.W. 2004b. Plant growth-promoting rhizobacterial systemically protect *Arabidopsis thaliana* against cucumber mosaic virus by a salicylic acid and NPR1-independent and jasmonic acid-dependent signaling pathway. *The Plant Journal* 39: 381–392.

Sabet, K.A. 1957. Studies in the bacterial diseases of Sudan crops I. Bacterial leaf spot of jute (*Corchorus olitorius* L.). *Annals of Applied Biology* 45: 516–520.

Sanjeev, K., Eswaran, A. 2008. Efficacy of micro nutrients on banana Fusarium Wilt. (*Fusarium oxysporum* f.sp. *cubense*) and it's synergistic action with *Trichoderma viride*. *Notulae Botanicae Horti Agrobotanici Cluj-Napoca* 36(1): 52–54.

Schippers, B., Bakker, A.W., Bakker, P.A.H.M. 1987. Interactions of deleterious and beneficial rhizosphere microorganisms and the effect of cropping practices. *Annual Review of Phytopathology* 25: 339–358.

Schmidt, E.L. 1979. Initiation of plant root microbe interactions. *Annual Review of Microbiology* 33: 355–376.

Schneider, S., Ullrich, W.R. 1994. Differential induction of resistance and enhanced enzyme activities in cucumber and tobacco caused by treatment with various abiotic and biotic inducers. *Physiological and Molecular Plant Pathology* 45: 291–304.

Schrey, S.D., Schellhammer, M., Ecke, M., Hampp, R., Tarkka, M.T. 2005. Mycorrhiza helper bacterium *Streptomyces* AcH 505 induces differential gene expression in the ectomycorrhizal fungus *Amanita muscaria*. *New Phytology* 168: 205–216.

Schroth, M.N., Hancock, J.G. 1981. Selected topics in biological control. *Annual Review of Microbiology* 35: 453–476.

Schroth, M.N., Loper, J.E., Hildebrand, D.C. 1984. Bacteria as biocontrol agents of plant disease. In *Current Perspectives in Microbial Ecology*, Vol. 52. Klug, M.J., Reddy, C.A. (eds.). Washington, DC: American Society for Microbiology, pp. 362–369.

Senthilkumar, M., Govindasamy, V., Annapurna, K. 2007. Role of antibiosis in suppression of charcoal rot disease by soybean endophyte *Paenibacillus* sp. HKA 15. *Current Microbiology* 55: 25–29.

Sessitsch, A., Reiter, B., Berg, G. 2004. Endophytic bacterial communities of field-grown potato plants and their plant growth-promoting and antagonistic abilities. *Canadian Journal of Microbiology* 50: 239–249.

Slawiak, M., Lojkowska, E., Van der Wolf, J.M. 2009. First report of bacterial soft rot on potato caused by *Dickeya* sp. (syn. *Erwinia chrysanthemi*) in Poland. *Plant Pathology* 58: 794.

Smith, E.F., Godfrey, G.H. 1921. Bacterial wilt of Castor bean (*Ricinus communis* L.). *Journal of Agricultural Research* 21: 255–261.

Starr, M.P. 1947. The causal agent of bacterial root and stem disease of guayule. *Phytopathology* 37: 291–300.

Stockwell, V.O., Johnson, K.B., Sugar, D., Loper, J.E. 2011. Mechanistically compatible mixtures of bacterial antagonists improve biological control of fire blight of pear. *Biological Control* 101(1): 113–123.

Sturz, A.V., Christie, B.R., Nowak, J. 2000. Bacterial endophytes: Potential role in developing sustainable systems of crop production. *Critical Reviews in Plant Sciences* 19: 1–30.

Sutic, D., Dowson, W.J. 1959. An investigation of a serious disease of hemp (*Cannabis sativa* L.) in jugoslavia. *Phytopathologische Zeitschrift* 34: 307–314.

Sylvia, D.M., Fuhrmann, J.J., Hartel, P.G., Zuberer, D. 1998. *Principles and Applications of Soil Microbiology*. Englewood Cliffs, NJ: Prentice Publishers.

Tachibana, H., Shih, M. 1965. A leaf-crinkling bacterium of soybeans. *Plant Disease Report* 49: 396–397.

Tagami, Y., Kuhara, S., Kurita, T. 1963. Epidemiological studies on the bacterial leaf blight of rice, *Xanthomonas oryzae* (Uyeda et Ishiyama) Dowson. 1. The overwintering of the pathogen. *Bulletin of the Kyushu Agricultural Experimental Station* 9: 89–122.

Tagami, Y., Kuhara, S., Kurita, T., Fujii, H., Sekiya, N., Sato, T. 1964. Epidemiological studies on the bacterial leaf blight of rice, *Xanthomonas oryzae* (Uyeda et Ishiyama) Dowson. II. Successive change in the population of the pathogen in paddy fields during rice growing period. *Bulletin of the Kyushu Agricultural Experimental Station* 10: 23–50.

Tanii, A., Miyajima, K., Akita, T. 1976. The sheath brown rot disease of rice plant and its causal bacterium *Pseudomonas fuscovaginae*. *Annals of the Phytopathological Society of Japan* 42: 540–548.

Thomas, C.A. 1965. Effect of photoperiod and nitrogen on reaction of sesame to *pseudomonas sesame* and *Xanthomonas sesami*. *Plant Disease Report* 49: 119–120.

Thomashow, L.S. 1996. Biological control of plant root pathogens. *Current Opinion in Biotechnology* 7: 343–347.

Tochinai, Y. 1932. The black rot of rice grains caused by *Pseudomonas itoana* n.sp. *Annals of the Phytopathological Society of Japan* 2: 453–457.

Truman, R. 1974. Die-back of *Eucalyptus citriodora* caused by *Xanthomonas eucalypti* sp. *Phytopathology* 64: 143–144.

Unnamalai, N., Gnanamanickam, S.S. 1984. *Pseudomonas fluorescens* is an antagonist to *Xanthomonas citri* (Hasse) dye, the incitant of citrus canker. *Current Science* 53: 703–704.

Van Loon, C. 1997. Induced resistance in plants and the role of pathogenesis related proteins. *European Journal of Plant Pathology* 103: 753–765.

Van Loon, L.C., Bakker, P.A.H., Pieterse, C.M.J. 1998. Systemic resistance induced by rhizosphere bacteria. *Annual Review of Phytopathology* 36: 453–483.

Van Peer, R., Niemann, G.J., Schippers, B. 1991. Induced resistance and phytoalexin accumulation in biological control of *Fusarium* wilt of carnation by *Pseudomonas* sp. strain WCS417r. *Phytopathology* 81: 728–734.

van Wees, S.C.M., Luijendijk, M., Smoorenburg, I., van Loon, L.C., Pieterse, C.M.J. 1999. Rhizobacteria-mediated induced systemic resistance (ISR) in *Arabidopsis* is not associated with a direct effect on expression of known defense-related genes but stimulates the expression of the jasmonate-inducible gene *Atvsp* upon challenge. *Plant Molecular Biology* 41: 537–549.

Vidhyasekaran, P., Kamala, N., Ramanathan, A., Rajappan, K., Paranidharan, V., Velazhahan, R. 2001. Induction of systemic resistance by *Pseudomonas fluorescens* Pf1 against *Xanthomonas oryzae* pv. *oryzae* in rice leaves. *Phytoparasitica* 29(2): 155–166.

Vidhyasekaran, P., Sethuraman, K., Rajappan, K., Vasumathi, K. 1997. Powder formulations of *Pseudomonas fluorescens* to control pigeon wilt. *Biological Control* 8: 166–171.

Viswanathan, R., Samiyappan, R. 1999. Induction of systemic resistance by plant growth-promoting rhizobacteria against red rot disease caused by *Colletotrichum falcatum* went in sugarcane. In *Proceedings of the Sugar Technology Association of India*, Vol. 61. New Delhi, India: Sugar Technology Association, pp. 24–39.

Vivekananthana, R., Ravia, M., SaravanaKumara, D., Kumarb, N., Prakasama, V., Samiyappan, R. 2004. Microbially induced defense related proteins against postharvest anthracnose infection in mango. *Crop Protection* 23: 1061–1067.

Voisard, C., Keel, O., Haas, P., Défago, G. 1989. Cyanide production by *Pseudomonas flurescens* helps to suppress black root rot of tobacco under gnotobiotic condition. *European Microbiological Journal* 8: 351–358.

Wang, Q., Lu, S.Y., Mei, R.H. 1997. One of analysis of endophytic bacteria in vascular tissue of cotton—relationship between the dynamics of bacterial in different resistant varieties and soil fertility and the growth period. *China Journal of Microecology* 9: 48–50.

Wei, G., Kloepper, J.W., Tuzun, S. 1996. Induced systemic resistance to cucumber diseases and increased plant growth by plant growth-promoting rhizobacteria under field conditions. *Phytopathology* 86: 221–224.

Weller, D.M. 1988. Biological control of soilborne plant pathogens in the rhizosphere with bacteria. *Annual Review of Phytopathology* 26: 379–407.

Weller, D.M., Howie, W.J., Cook, R.J. 1988. Relationship between in vitro inhibition of *Gaeumannomyces graminis* var. *tritici* and suppression of take all of wheat by fluorescent pseudomonads. *Phytopathology* 78: 1094–1100.

Weller, D.M., Zhang, B.X., Cook, R.J. 1985. Application of a rapid screening test for selection of bacteria suppressive to take-all of wheat. *Plant Disease* 69: 710–713.

Whipps, J.M. 1997. Developments in the biological control of soil-borne plant pathogens. *Advances in Botanical Research* 26: 1–133.

Whipps, J.M. 2001. Microbial interactions and biocontrol in the rhizosphere. *Journal of Experimental Botany* 52: 487–511.

Wilson, M., Lindow, S.E. 1993. Effect of phenotypic plasticity on epiphytic survival and colonization by *Pseudomonas syringae*. *Appied Environmental Microbiology* 59(2): 410–416.

Wright, S.A.I., Zumoff, C.H., Schneider, L., Beer, S.V. 2001. *Pantoea agglomerans* strain EH318 produces two antibiotics that inhibit *Erwinia amylovora in vitro*. *Applied and Environmental Microbiology* 67: 284–292.

Wysong, D.S., Vidaver, A.K., Stevens, H., Stenberg, D. 1973. Occurrence and spread of an undescribed species of *Corynebacterium* pathogenic on corn in the western corn belt. *Plant Disease Report* 57: 291–294.

Xu, G.W., Gross, D.C. 1986. Field evaluations of the interactions among fluorescent pseudomonads, *Erwinia caratovora* and potato yields. *Phytopathology* 76: 423–430.

Yasuda, M., Isawa, T., Shinozaki, S., Minamisawa, K., Nakashita, H. 2009. Effects of colonization of a bacterial endophyte, *Azospirillum* sp. B510, on disease resistance in rice. *Bioscience/Biotechnology and Biochemistry* 73: 2595–2599.

Yoshi, H., Takimoto, S. 1928. Bacterial leaf blight of castor bean and its pathogen. *Journal of Plant Protection* 15: 12–18.

Zhang, L., Birch, R.G. 1996. Biocontrol of sugar cane leaf scald disease by an isolate of *Pantoea dispersa* which detoxifies albicidin phytotoxins. *Letters in Applied Microbiology* 22: 132–136.

Zhang, L., Birch, R.G. 1997. The gene for albicidin detoxification from *Pantoea dispersa* encodes an esterase and attenuates pathogenicity of *Xanthomonas albilineans* to sugarcane. *Proceedings of the National Academy of Sciences USA* 94: 9984–9989.

Zhao, K., Xiao, C.G., Kong, D.Y. 2006. Controlling effect of endophytic bacteria on *Ralstionia solannacearum* and its antifungal spectrum. *Journal of Southwest University (Natural Science Edition)* 28: 314–318.

Zhou, T., Chen, D., Li, C., Sun, Q., Li, L., Liu, F., Shen, Q., Shen, B. 2012. Isolation and characterization of *Pseudomonas brassicacearum* J12 as an antagonist against *Ralstonia solanacearum* and identification of its antimicrobial components. *Microbiological Research* 167(7): 388–394.
Zou, H., Song, X., Zou, L., Yuan, L., Li, Y., Guo, W., Che, Y. et al. 2012. EcpA, an extracellular protease, is a specific virulence factor required by *Xanthomonas oryzae* pv. *oryzicola* but not by *X. oryzae* pv. *oryzae* in rice. *Microbiology* 158: 2372–2383.
Zummo, N., Freeman, K.C. 1975. Bacterial sun spot, a new disease of sugarcane and sweet sorghum. *Sugarcane Pathologists' Newsletter* 13/14: 15–16.

15 Bacteriophages
Emerging Biocontrol Agents for Plant Pathogenic Bacteria

Duraisamy Nivas, Kanthaiah Kannan, Velu Rajesh Kannan, and Kubilay Kurtulus Bastas

CONTENTS

ABSTRACT Bacterial diseases have serious impacts on agricultural food production. Several methods are available to control these phytopathogens, including antibiotics and copper sprays. However, bacterial resistance to antibiotics makes it risky to continue those strategies and highlights the need for effective disease-control agents. An alternative option is the use of lytic phages to control phytopathogens. Phages are *in vivo* anti-phytopathogenic isolates that can be used for *in vitro* biocontrol. In some circumstances, increased phage disease-control efficiency, protective formulations, appropriate application timings, and integrated management strategies can be achieved. The use of phages in plant disease control is a compelling topic, and several successful studies have been done on phytopathogens. In this chapter, we explore the use of phages in plant-disease control and their roles in the development of effective biocontrol agents.

KEYWORDS: phytopathogens, plant diseases, eco-friendly methods, bacteriophages, biocontrol agents

15.1 INTRODUCTION

The growing human population worldwide, declines in land and water availability, and challenging climatic changes are problematic factors in global food production (Ronald, 2011). However, the major limiting factor in agricultural food production is plant diseases (Balogh et al., 2010). Plant diseases caused by bacteria are a major commercial burden to agricultural production. Bacteria are present in plants as epiphytics or endophytics. Plant pathogenic bacteria colonization and suppression of plants result in the formation of plant diseases (Mukhtar et al., 2010). Phyllosphere bacteria in particular can promote plant growth and both suppress and stimulate the colonization and infection of tissues by plant pathogens (Rasche et al., 2006). The impacts of bacterial plant diseases are reduced by a variety of approaches, which can be quantitative as well as qualitative, to improve food production (Frampton et al., 2012). Traditional disease control strategies involve the implementation of operating practices such as removal of infected plant tissue and appropriate disposal to stop the transmission of pathogens from one site to another.

The most effective plant disease management approach requires an integrated strategy that utilizes one or more of the following components: sound cultural practices, plant genetic resistance, induced resistance, and biological and chemical control agents (Balogh et al., 2010). Emergence of interest in the use of bacteriophages to control plant diseases is relatively recent. This chapter considers the use of phages for the control of plant pathogens. First, we address existing plant disease control methods; next, we discuss the use of phages in plant disease control as well as early developments and modern usages. Finally, we present a comprehensive analysis of plant disease control strategies for phages.

15.2 PLANT DISEASE CONTROL STRATEGIES

Plant diseases, the major economic burden in agriculture food production, may be caused by bacteria or fungi. Most plant diseases, which are caused by bacteria, can cause severe economic damage from crop losses due to spots, mosaic patterns or pustules on leaves and fruits, smelly tuber rots, and plant death. Some bacteria cause hormone-based distortion of leaves and shoots (fasciation); others cause crown gall, which is a proliferation of plant cells that produces swelling at the intersection of stem and soil and on roots (Vidaver and Lambrecht, 2004). Disease control in agriculture, particularly of bacterial diseases, is a challenging problem in agriculture. Management of bacterial disease in plants can be quite difficult due to the lack of effective bactericides, high pathogen variability, gene transfers in pathogens that lead to resistance development, and the lack of chemical-based approaches that are effective and also harmless to the larger environment (Obradovic et al., 2004). Most plant disease control methods are achieved by integrated disease management strategies that combine suitable cultural practices, chemicals such as bactericides or plant activators where applicable, plant genetic resistance, induced resistance, biological/chemical control (copper), and the use of antibiotics such as tetracycline and streptomycin (Obradovic et al., 2005).

Most plant bacterial diseases are difficult to control due to the lack of effective bactericide. Streptomycin, an aminoglycoside antibiotic, has been a major antibiotic in agriculture since the 1950s for the control of phytopathogenic bacteria. However, widespread usage of streptomycin has resulted in the development of streptomycin-resistant strains, which have reduced the drug's efficiency (Thayer and Stall, 1961). Resistance to streptomycin by bacterial pathogens is developed by selection procedures and plasmid-encoded resistance genes. The emergence of streptomycin-resistant strains has led to the loss of control of several pathosystems, including bacterial spot of tomato and pepper (Louws et al., 2001). Although this conclusion remains controversial, it seems clear that agricultural use of antibiotics has strongly influenced the selection of resistance genes and that the movement of these genes into pathogens of clinical relevance has occurred (Sean and Britt, 2013). Copper-based bactericides, which were first introduced in the 1880s, are still used to encounter extensive bacterial plant diseases. However, the evolution of copper-resistant bacterial strains has reduced the disease control efficiency of this approach (Balogh et al., 2010). Copper resistance in many bacterial pathogens includes plasmid-borne resistance and chromosomal resistance (Bender and Cooksey, 1987; Basim et al., 1999). Aside from the impact of bacterial resistance on the environment and sustainable agriculture, the problem reduces the desirability of some agricultural food products. However, these problems have also led to the creation of new eco-friendly biocontrol agents such as plant activators and biocontrol agents.

Systemic acquired resistance (SAR) inducers and plant activators are two alternative disease control strategies to overcome resistant bacterial strains. SAR inducers, which are either composed of protein or synthetic chemical products, induce natural plant defense systems and activate them against the broad spectrum of antibacterial resistance. SAR inducers are successful in controlling several diseases, including bacterial spot of tomato (Obradovic et al., 2004) and bell pepper (Romero et al., 2001), leaf blight of onion (Gent and Schwartz, 2005), and fire blight of apple (Maxson et al., 2002). However, even as they reduce these diseases, such inducers may have harmful effects on plants (Gent and Schwartz, 2005). For example, plant inducers used in the control of citrus canker have proven to be ineffective and to reduce yields (Graham and Leite, 2004). Because conventional plant disease control strategies contain so many complications, extensive efforts have been made to control plant diseases with biologically based strategies (i.e., biocontrol agents). Biocontrol agents, which are an eco-friendly approach to the control of plant pathogens, are created by a number of biological approaches that mostly utilize the bacteria themselves. These approaches include antagonistic bacterial saprophytes, plant growth-promoting rhizobacteria (PGPR), site competition, and avirulent strains produced through induced systemic resistance (ISR) and antagonism (Tanaka et al., 1990). Because all of these approaches have shown different levels of effectiveness and reliability, the use of bacteriophages for disease control is now attracting research attention.

15.3 ENTRY OF PHAGES IN DISEASE CONTROL

Therapeutic use of phages was first reported in 1896 by Ernest Hankin, who observed antibacterial activity against *Vibrio cholera* (the causative agent of cholera); at that time, the disease was one of the deadliest threats faced by humans (Abedon et al., 2011). In 1915, Fredric Twort hypothesized that this antibacterial activity could be due to the virus (phage) itself, but did not pursue this line of thought. Felix D'Herelle, who discovered bacteriophages in 1917 (Hermoso et al., 2007), proposed that the phenomenon was caused by viruses capable of parasitizing bacteria and used a combination of the words "phagein" (Greek for "to eat") and "bacteria" to name them. In 1925, D'Herelle's report on the treatment of plague by anti-plague phage drew attention to the use of bacteriophages in disease control (phage therapy) (D'Herelle et al., 1927). Phages were first applied therapeutically in the treatment of infectious staphylococcal skin disease in humans, which was first identified in 1921 by Richard Bruynoghe and Joseph Maisin. Several similarly promising studies, such as those by D'Herelle and others (Sulakvelidze et al., 2001), were encouraged by these early results. The concept of phage therapy in Western countries was abandoned after the emergence of antibiotics in 1940, but the practice continued in the Soviet Union and is still used in Russia today. The phage therapy concept also fell out of favor in Western countries because of the unreliable and inconsistent results of many phage therapy trials (Haq et al., 2012). When complete understanding of phage biology was obtained, the concept was generally accepted; to date, it has been used successfully in humans, animals, and plants.

After D'Herelle's discovery, phages were proposed as an agent for the control of plant diseases and were evaluated for the control of bacterial diseases in plants (Moore, 1926). Phage therapy has been found to be an effective tool for the control of several phytopathogenic bacteria, including *Xanthomonas* spp. (bacterial spot of peach) (Ceverolo, 1973), *Pseudomonas* spp. (bacterial blotch of mushroom) (Kim et al., 2011), *Erwinia* spp. (fire blight of apple and pear) (Nagy et al., 2012), *Pantoea* spp. (Stewart's wilt of corn) (Thomas, 1935), *Ralsotnia* spp. (bacterial wilt of tobacco) (Fujiwara et al., 2011), *Streptomyces* spp. (common scab) (Goyer, 2005), *Dickeya* spp. (blackleg disease of potato) (Adriaenssens et al., 2012), and *Pectobacterium* spp. (blackleg disease of potato) (Lim et al., 2013).

As a result of the widespread acceptance of this lytic principle, scientists believed that phages could also be important for preventive bacterial growth in soil. As previously stated, interest in phage therapy for the control of phytopathogens has increased in recent years due to the nontoxic nature of this approach, but its ability to infect heavy-metal- as well as antibiotic-resistant bacteria is also highly attractive. Phage applications in the field have been so successful that in 2005 a commercial phage product was registered with the US Environmental Protection Agency (EPA; #67986) by OmniLytics Inc., Salt Lake City, Utah (Jackson and Jones, 2004). The development and commercial application of a wide range of phage-based medical products in the former Soviet republics of Georgia and Russia include sprays for the treatment of various plant diseases. Although the host range, lytic spectrum, and cross-resistance properties of phage strains were initially evaluated to avoid problems caused by lysogeny, only lytic phages are used for plant disease therapy. To overcome bacterial resistance, a phage cocktail was developed from strains with different receptor specificities to pathogenic bacteria as well as to different bacterial species. Phage formulations have been developed for efficient disease treatment and prevention as well as sanitation (Kutter, 1997).

15.4 MODERN USE OF PHAGES IN BIOCONTROL

Decades of research on the use of bacteriophages have contributed several important discoveries to the reevaluation of phage therapy. In ancient times, such exploitation was successfully used to control bacterial plant diseases (Thomas, 1935). In modern times, phages have been found to have several potential advantages in disease control. First, as natural components of the biosphere, bacteriophages can be readily isolated from bacteria that occur in a range of locations, including soil (Ashelford et al., 2003), water (Miernik, 2003), plants (Gill and Abedon, 2003), animals (Sulakvelidze and Barrow, 2005), and humans (Letarov and Kulikov, 2008), as well as hydrothermal vents (Liu et al., 2006). Second, phages are self-replicating and self-limiting; they reproduce only as long as the host bacterium is present in the environment and quickly degrade in its absence (Lang et al., 2007). Third, phages target only the bacterial receptors that are essential for pathogenesis; therefore, the resistant mutants of bacterial strains are attenuated in virulence (Kutter, 1997). Fourth, bacteriophages are not only nontoxic to eukaryotic cells (Greer, 2005) but also are specific or highly discriminatory; the latter means that they eliminate only target bacteria without damaging other, possibly beneficial indigenous flora (Summers, 2001). Fifth, phage preparations are fairly easy and inexpensive to produce and can be stored at 4°C in complete darkness for months without a significant reduction in titer (Greer, 2005). Sixth, phage application can be carried out with standard farm equipment; moreover, because phages are not inhibited by the majority of agrochemicals, they can be tank-mixed with them without significant loss in titer (Iriarte et al., 2007). Although copper-containing bactericides have been shown to inactivate phages, this inhibition was eliminated when the phages were applied at least three days later.

As previously stated, bacteriophage use for plant disease control is a rapidly expanding area and control of several plant diseases has been achieved. A considerable amount of work has been published on the use of phages for bacterial plant diseases. Civerolo and Keil (1969) treated peach foliage of *Xanthomonas pruni* by Xp3-A phage and, under greenhouse conditions with a single application of a single-phage suspension, successfully controlled bacterial spot severity on this foliage. Subsequently, Civerolo (1969) observed improved phage activity at 20°C and 27°C but lower Xp3-A phage disease control efficiency at 37°C.

Borah et al. (2000) treated mung bean bacterial leaf spot caused by *Xanthomonas axonopodis* with a co-application of antagonistic phylloplane bacteria and a bacteriophage that was known to be active against the pathogen. No protection was obtained with antagonistic bacteria alone, but application of the phage confined the pathogen population in plants to leaves; disease control efficacy was recorded as 68% in comparison with untreated plants. Flaherty et al. (2001) isolated 16 phages against a pathogen, screened for host range and lytic ability, selected lytic phages with broad host ranges, and developed the H-mutant phage from some of them. Daily application of phage mixtures in foliar spray can reduce disease severity by 75% (relative to control plants) after 6 weeks.

Extensive research has been done on suppressing bacterial blight caused by *Xanthomonas campestris* with phages. However, Neil et al. (2001) revealed that the lytic bacteriophages of *X. campestris* are widespread in New Zealand soils and could be very easily isolated from the top 2.5 cm. However, for the eight soil samples (out of 11) that yielded these phages, control of bacterial blight was not significantly different than on untreated branches of walnut. Flaherty et al. (2001) effectively controlled bacterial blight in greenhouse and field experiments with a mixture of four phages that were known to be active against two predominant strains of *X. campestris*.

Streptomyces scabiei, the common scab pathogen of potato, was controlled by the phages Stscl and Stsc3 by Goyer (2005). The disease control abilities of phages Stscl and Stsc3 were tested in radish seedlings; those that were grown in the presence of *S. scabiei* weighed significantly less (mass decrease 30%) than the negative controls. When the radish seedlings inoculated with Stscl and Stsc3 phages were observed to grow with no disease symptoms, effective disease control of phages in a plant system *in vitro* was confirmed. Lang et al. (2007) reported that *Xanthomonas* leaf blight of onion (*Allium cepa*), caused by *Xanthomonas axonopodis* pv. *alli*, was controlled by mixtures of bacteriophages.

The persistence of phages has been recorded as 72–96 h in field and greenhouse conditions. During biweekly or weekly application of phages, 26%–50% of disease severity was observed; this result was better than application with copper hydroxide plus mancozeb. Integration of bacteriophage mixtures with acibenzolar-*S*-methyl (ASM) appears to be a promising strategy for managing *Xanthomonas* leaf blight of onion and could reduce grower reliance on conventional copper bactericides.

Bacteriophage stability in soil and at different temperatures was studied in *Ralstonia solanacearum* phages φRSA1, φRSB1, and φRSL1 by Fujiwara et al. (2011). The *R. solanacearum* phages were found to be stable in soil, especially at higher temperatures of 37–50°C, and the stable nature of phages in soil was confirmed by recovery of φRSL1 phages from roots of treated plants and from soil 4 months post-infection. Based on these findings, the researchers proposed that these phages are effective for the control of wilting in tomato.

In most greenhouse and field studies, bacteriophages reduced the severity of bacterial spot of tomato to levels equal to or lower than those obtained with copper bactericides. However, the effectiveness of phages for biological control depends not only upon the susceptibility of the target bacterium but also on environmental factors that affect phage survival (Flaherty et al., 2000). In general, the survival of microorganisms in the phyllosphere depends upon the limited resources that are available in this habitat and the ability of organisms to cope with varied environmental stress conditions such as fluctuating water availability, heat, and osmotic pressure, as well as desiccation. However, phage persistence in plant phyllospheres for extended periods is limited by many factors, including sunlight irradiation, temperature, and desiccation, especially in the UV zone and when copper bactericides are used (Iriarte et al., 2007).

Iriarte et al. (2007) developed formulations that mixed phages and skim milk to extend phage persistence in plant phyllospheres. This protective formulation eliminated the reduction caused by both of these factors. Although phage persistence was dramatically affected by UV light, the other factors were less effective. In particular, formulated phages reduced the deleterious effects of the environmental factors. Integrated management strategy, which is central to plant disease control in modern agriculture, includes combinations of chemical control agents and has achieved biocontrol. As a part of an integrated disease management approach, Obradovic et al. (2004) used combinations of phages with other biological control agents and plant inducers. Phage-based integrated management of tomato bacterial spot is now officially recommended to tomato growers in Florida, and bacteriophage mixtures (e.g., Agriphage) that reduce this pathogen are commercially available (Obradovic et al., 2008).

15.5 PHAGE RESISTANCE

The use of antibiotics for the control of phytopathogens was first implemented in 1950. Today, the antibiotics oxytetracycline and streptomycin are the ones most commonly used on plants. The mechanism of resistance to antibiotics by bacteria results from the mutation in the existing DNA of an organism or the acquisition of a new resistance gene. However, the evolution of resistance mechanisms against antibiotics in nature remains unclear (McManus et al., 2002). Although resistance of plant pathogens to oxytetracycline is rare, the emergence of streptomycin-resistant strains of *Erwinia amylovora*, *Pseudomonas* spp. (Scheck et al., 1996), and *X. campestris* has inhibited the control of several important diseases. Fractions of streptomycin resistance genes in plant-associated bacteria show similarities to those found in bacteria isolated from humans, animals, and soil, and are associated with transfer-proficient elements. The most common vehicles of streptomycin resistance genes in human and plant pathogens are genetically distinct, however. The role of antibiotic use on plants during this antibiotic-resistance crisis is the subject of debate.

Resistances of phages, which are also arising in the bacterial system, reduce disease control as well. However, the mechanism of bacterial resistance to phages is well understood and therefore should be given prime consideration in the development of biocontrol agents. Bacteria have developed resistance against most of the phage infection stages. Briefly, mechanisms of bacterial resistance to the phage include interference in phage adsorption, blockage of DNA entry, abortive infection (Abi), CRISPR/Cas systems, and restriction modification (RM) systems.

15.5.1 Interference in Phage Adsorption

In phage adsorption, interference refers to bacterial insensitivity that is due to the absence, alteration, or masking of phage receptors. Cell-surface receptors are necessary phage attachments to bacterial cells (Hill et al., 1996). Identification of these specific receptors by phages is more important in phage adsorption; mutation in the receptors is the major cause of phage resistance (Frampton et al., 2012). However, development of phage resistance can be beneficial. Mutation in the LPS of the phage-resistant strain *Pectobacterium atrosepticum* resulted in reduced virulence in a potato tuber rot assay (Evans et al., 2010). Likewise, flagella mutation in the φAT1 phage-resistant strain *P. atrosepticum* caused attenuations in their motility and virulence (Evans et al., 2010).

15.5.2 Blocking DNA Entry

In host cells, phage DNA entry is blocked by proteins called superinfection exclusion (Sie) systems. In general, most plant pathogens are Gram-negative. Accordingly, Labrie et al. (2010) revealed that two Sie systems in Gram-negative bacteria, *imm* and *Sp*, caused inhibition in phage DNA injection to cells that prevented subsequent phage infection. *Imm* changes the conformation of the injection site and *Sp* inhibits the activity of the T_4 lysozyme. Resistance mechanisms in plant pathogens have yet to be studied in detail.

15.5.3 Abi Systems

Abortive infection leads to the death of infected cells and inhibits phage reproduction. Abi systems have mostly been identified in dairy bacteria but have also been identified in *P. atrosepticum*, whose Abi system is called ToxIN. ToxIn, which inhibits the infection of multiple phages, is effective in different genera but has been observed only in phytopathogens (Fineran et al., 2008). Although studies on Abi systems began in the mid-twentieth century, their modes of action are still not completely understood, in part because they are complex and in part because of knowledge gaps in phage biology (Labrie et al., 2010).

15.5.4 CRISPR/Cas Systems

Since the first description of CRISPR/Cas locci in 1987, comparative analyses have revealed their presence in an increasing number of complete bacterial and archeal genomes (Goode and Bickerton, 2006). In fact, 40% of the sequenced bacteria have been shown to possess adaptable phage resistance systems. The effects on phage multiplication of clustered, regularly interspaced, short palindromic repeats (CRISPRs) and CRISPR-associated (*cas*) genes have been described. These arrays are transcribed and processed into small RNAs that, with the assistance of Cas proteins, target and degrade spacer complementary viral nucleic acids (Frampton et al., 2012). Experimental analyses of CRISPR/Cas systems in plant pathogens have been conducted on *P. atrosepticum*, *E. amylovora*, and *Xanthomonas oryzae*. *P. atrosepticum*, which has three CRISPR/Cas operons, is expressed *in vivo* and indicates phage resistance that could be active in plant infection (Przybilski et al., 2011).

15.5.5 RM System

Many bacterial genera, if not all, possess RM systems. The principal function of an RM system is thought to be cell protection against invading DNA, including viral DNA. RM systems consist of two actors, restriction endonucleases (REase) and methyltransferases (MTase); their functions are generally performed by separate enzymes. These RM systems have been classified into four groups (types I, II, III, and IV). The functions of RM systems are mostly defined as the inhibition of RM enzymes, reduction of recognition sites, occlusion of restriction sites, and base modification (Borra et al., 2012).

In 1989, mixtures of host-range mutant phages (H-mutants) were first used for plant disease control. Production of H-mutants, which was derived from a patented process, was developed to prevent the occurrence of phage-resistant mutants. H-mutant phages have the ability to lyse both wild-type as well as phage-resistant bacterial cells. Additionally, careful choice of the phage cocktail can ensure that resistant bacteria will be attenuated. The concept of the cocktail depends upon the assumption that a phage mixture which uses different receptors is better for biocontrol. To overcome such resistance, the OmniLytics company has developed a management plan that involves monitoring a pathogen and updating its phage mixture if or when bacterial resistance emerges (Frampton et al., 2012).

15.6 DISEASE CONTROL TRAILS BY PHAGES

The initial stage in the development of a phage-based biocontrol agent involves phage isolation. Phages, which can be simply isolated, should be characterized to ensure that they are suitable for disease control. Such testing allows for the preparation of an appropriate phage cocktail and enables the tracking of phages during bioassays and field trails. However, the efficiency of a biocontrol treatment depends on the target agent and its density, as well as contact between host and biocontrol agent. In phage therapy, probability of phage bacterium contact depends on several factors: initial phage concentration, rates of virion decay, phage multiplication ability in the target environment, concentration and location of target bacteria, and presence of adequate water as a medium for phage diffusion (Gill and Abedon, 2003). Successful phage trials have been performed on a range of phytopathogens, including *Xanthomonas* sp., *Pantoea stewartti*, *Agrobacterium tumefaciens*, *S. scabiei*, *Pseudomonas syringae*, and *Erwinia* sp. (Table 15.1). However, a number of factors can contribute to the failure of phage biocontrol trials. Therefore, it is important to check which causative agent produces symptoms of the disease, if it is the targeted bacterial host, and whether phage resistant mutants have the ability to kill the specific host.

Phages are used for plant pathogen control in the rhizosphere as well as the phyllosphere region. In the rhizosphere, several factors influence the success of disease control. In addition, the rate of

TABLE 15.1
Phages in the Control of Plant Diseases

Pathogen	Disease	Host	References
Xanthomonas campestris	Black rot	Crucifers	Tseng et al. (1990)
Pantoea stewartti	Stewart's wilt	Corn	Thomas (1935)
Agrobacterium tumefaciens	Crown gall	Tomato	Stonier et al. (1967)
Xanthomonas phaseoli	Common blight	Beans	Vidaver and Schuster (1969)
Xanthomonas oryzae	Bacterial blight	Rice	Kuo et al. (1971)
Xanthomonas pruni	Bacterial spot	Peach	Ceverolo (1973)
Erwinia carotovora	Soft rot	Potato	Eayre et al. (1995)
Erwinia ananas	Brown spot	Honeydew	Eayre et al. (1995)
Xanthomonas axonopodis	Leaf spot	Mungbean	Borah et al. (2000)
Xanthomonas campestris pv. *pelargonii*	Bacterial blight	Geranium	Flaherty et al. (2001)
Xanthomonas campestris pv. *juglandis*	Blight	Walnut	Neil et al. (2001)
Streptomyces scabiei	Common scab	Potato	Goyer (2005)
Xanthomonas axonopodis pv. *Allii*	Leaf blight	Onion	Lang et al. (2007)
Pseudomonas syringae	Bacterial speck	Tomato	Prior et al. (2007)
Pseudomonas tolaassi	Brown blotch	Mushroom	Kim et al. (2011)
Ralstonia solanacearum	Bacterial wilt	Tobacco	Fujiwara et al. (2011)
Erwinia amylovora	Fire blight	Apple	Nagy et al. (2012)
Dickeya solani	Blackleg	Potato	Adriaenssens et al. (2012)
Pectobacterium carotovorum	Blackleg	Potato	Lim et al. (2013)
Xanthomonas citri	Citrus canker	Citrus	Balogh et al. (2013)

phage diffusion in soil is low, because phage diffusion depends upon the available free water in soil (Gill and Abedon, 2003).

Phages can become trapped in biofilms, reversibly adsorb to soil clay particles, and become inactivated by low soil pH. External and environmental factors can also cause variable results. For example, plants growing in greenhouses are in stable environments, whereas plants in fields are exposed to varied environmental conditions that may vary between geographic locations. Phage survival and persistence are adversely affected by environmental factors such as exposure to high temperatures, high and low pH, desiccation, rain, and UV light (Frampton et al., 2012). The UV-A and UV-B spectra of sunlight have been determined to be the most destructive environmental factors. Methods that have been explored for increasing the efficacy of control in the phyllosphere include enhancing phage longitivity, using protective formulations, propagating phages in the field, and applying treatments in the evening or early morning. Several methods that address these challenges are available and will be discussed in the following section.

15.6.1 Protective Formulations

In order to reduce the possibility of UV-A and UV-B damage, protective formulations have been developed. Some published results have included the characterization of various materials to determine whether they could extend the lives of viruses after exposure to various environmental factors. These materials were used to extend the residual activities of the viruses by providing protection against sunlight, UV light, and rain leaching. Tested protective materials included sucrose, Congo red, lignin, casein, casecrete, skim milk, and pregelatinized corn flour (Balogh, 2002). Several of these materials were evaluated for their ability to extend phage persistence on tomato foliage. Balogh (2002) identified three formulations that had pronounced effects on phage persistence in tomato: 0.5% pregelatinized corn flour and 0.5% sucrose (PCF); 0.5% casecrete NH-400 (a water-soluble casein polymer), 0.25% pregelatinized corn flour, and 0.5% sucrose (casecrete); and 0.75%

skim milk powder and 0.5% sucrose (skim milk). These formulations showed increased phage persistence from several hours to one to days under field conditions, compared with phage preparations applied without the addition of any protective formulations. The use of these formulations resulted in significant control of bacterial spot in greenhouse and field conditions (Balogh, 2003). In trials with phage against *X. axonopodis* pv. *citri* and *citrumelo*, skim milk inhibited the phage action but phage persistence on the plant surfaces was longer (Balogh et al., 2008).

Despite these and other formulation are enhancements, phages persist in plant phyllospheres and thereby increase disease development. It is likely that proteins and sugars in the formulations provided nutritional sources for pathogens and also facilitated the entry of phages into plant leaf surfaces by breaking down surface tension and forming continuous an aqueous layer in the phyllosphere (Jones et al., 2007). To improve phage efficacy, biologically inert formulations should be identified.

15.6.2 Application Timing

Protective formulations are effective against some environmental factors, but additional approaches should be developed to minimize phage exposure to sunlight during phage application. Application timing, which is relevant to the arrival of pathogens, influences the efficacy of disease control in some cases. Balogh et al. (2008), who investigated the effect of the time of day of phage application, demonstrated that evening applications increased phage survival. Ceverolo and Keil (1969) achieved a marked reduction in bacterial spot of peach only if the phage treatment was applied 1 h or 1 d before pathogen inoculation. Disease severity decreased slightly when phage was applied 1 h after inoculation, but no effect was observed when the phage was applied 1 d after inoculation. Balogh et al. (2003) achieved 27% disease reduction in evening applications, which provided better control of tomato bacterial spot than daytime applications did. Schnabel et al. (1999) successfully achieved significant reduction in fire blight of *E. amylovora* on apple blossoms when phage mixture application and pathogen inoculation were done at the same time. However, disease reduction was not significant when the phage was applied 1 d before inoculation. Phages have been found to survive considerably longer when applied close to sunset rather than in the morning or afternoon. Moreover, the timing of field application of phage to times of day when sunlight irradiation is minimal significantly improved disease control efficacy (Iriarte et al., 2007). The frequency of phage application also appears to be specific to the particular phage phytopathogen–plant system; the best results have varied between daily and weekly application (Flaherty et al., 2000).

15.6.3 Phage Propagation in Field

A number of factors, including protective formulations and timing of application, must be considered for maximizing the efficacy of phage therapy and the reliability of disease control. Phage propagation in the target environment is also an important factor in successful phage therapy. To maintain a high phage population in the field, phage-sensitive bacterial host cells were used to increase phage population on plant surfaces. Phage-sensitive bacterial hosts that are successfully maintained in plant surfaces also increase phage populations in the phyllosphere. Nonpathogenic epiphytes could also be involved in this process, because establish easily in the phyllosphere and may be antagonistic against pathogenic bacteria.

Epiphytes that are closely related to pathogenic bacteria are sensitive to the same phages. A pathogenic strain of *Xanthomonas perforans* with less disease severity than the wild-type was tested by Balogh (2006) in a greenhouse study but was able to establish itself among a tomato canopy-supported high population of its phage. In the absence of a host, phage populations disappeared from the tomato foliage six days after application, whereas in the presence of a host phage numbers recorded a week after application were comparable to the original population.

Tanaka et al. (1990) developed a strategy for bacterial wilt disease control by co-application of phage and the phage-sensitive avirulent strain of the pathogen *R. solanacearum*. A nonpathogenic or attenuated strain of the phytopathogenic strains loses its disease-causing ability and will be utilized by a sensitive host for phage propagation in the field as well. Pathogenically attenuated mutants of plant pathogens, which have better colonizing ability in plants, thereby exert more competition and antagonism toward virulent strains while also acting as effective phage propagators. However, they may also cause some disease themselves. In this case, balance between the positive and negative effects of the approach should be carefully evaluated before implementation (Balogh et al., 2010).

15.6.4 Integrated Disease Control Strategy

Researchers have begun to examine the effects of combining phages with existing or new control measures. In the disease triangle model, disease development is affected by three factors: the host plant, the disease-causing ability of the pathogen, and the environment (Balogh et al., 2010). Several methods have been used for the incorporation of phage application, which is a major component of integrated management of plant disease control, and various combinations of biocontrol agents have been tested to develop sustainable strategies for reducing losses caused by plant disease. These combinations have included strains of PGPR, bacterial antagonists, phages, and SAR inducers (Obradovic et al., 2004). Tanaka et al. (1990) successfully reduced the incidence of tobacco bacterial wilt disease caused by *R. solanacearum* through combinations of sensitive *R. solanacearum* and a phage that was active against both virulent and avirulent strains. Tomato has been regarded as a model plant for the development of an integrated management strategy that combines phage and other alternative disease control strategies (Jones et al., 2007), and the severity of some diseases in tomato has been reduced.

One such strategy, for bacterial spot of tomato, involves combining phages with plant activators (Louws et al., 2001), PGPR (Ji et al., 2006), or antagonistic bacteria (Byrne et al., 2005). Combinations of phages with plant activators stimulate plant resistance against pathogens and reduce pathogen populations in plant systems. These combinations both help disease control and create hostile environments for disease development (Obradovic et al., 2005). Flaherty et al. (2000) used SAR inducers for the control of tomato bacterial spot caused by *X. campestris* pv. *vesicatoria*, which had been inconsistent with phage treatments. When phage was applied to plants that had been treated with SAR (ASM inducers), disease control significantly increased. Another combination of harpin protein with phage showed significantly better disease control than the application of harpin alone. Although plants in this experiment developed HR-like necrotic spots at high concentrations of ASM, the combination of ASM with phage resulted in the absence or elimination of HR-like necrotic spots. In another study, control of bacterial spot in tomato could be achieved by integrated management strategy; specifially, the disease severity was greatly reduced by the combination of phage with copper mancozeb. These treatments produce a significant reduction in bacterial spot in comparison to treatment with copper mancozeb alone (Momol). Integrated management strategy is now widely used in Florida greenhouse and production fields as a part of the standard integrated management control of bacterial spot of tomato.

15.7 OTHER PHAGE-BASED BIOCONTROL TECHNOLOGIES

Phage endolysins or lysins, which are highly efficient molecules that can target the integrity of the cell wall, are designed to attack one of the four major bonds in the peptidoglycan. Movement of this enzyme was controlled by a second phage gene product holin from the lytic system (Fischetti, 2010). Today, control of most plant diseases can be achieved by the use of whole phages only, but such use of lysins has not been explored in detail. Phages CMP1 and CN77 have been identified as active against the plant pathogenic bacterium *Clavibacter michiganensis* strain, which is controlled by the endolysins of CMP1 and CN77 due to its resistance to chemical control agents. Both endolysins are specific for *C. michiganensis* (Wittmann et al., 2011). Compared with the application of whole-phage preparations, the development of lysin-based therapies is more technically challenging. For example, generation of lysin was achieved by the proper understanding of phage identification, characterization, cloning, and purification.

15.8 CONCLUSION

Because plant bacterial diseases have serious impact on agricultural food production, approaches such as antibiotic and spray applications have been developed. Bacteriophage use, another tool for the control of phytopathogens, has gained acceptance in recent decades. Since the discovery of phages in the twentieth century, their use has been extensively evaluated against all kinds of bacterial diseases, including plant diseases. Although the concept of phage therapy is novel, it is also ancient. As the most abundant entities on Earth, the advantages of phages for disease control include self-replication, host specificity, nontoxicity to eukaryotes, and nondisruption of normal flora. Compared with other chemical-based disease control agents, phages are recognized as most eco-friendly. Moreover, phages are easy to isolate, very inexpensive to produce, and can be stored at –20°C without titer reduction.

The use of phages in plant disease control is challenging due to bacterial mutation; pathogens will form resistance to bacteriophages. Therefore, using a single phage for disease control is not an ideal approach. However, when phage cocktails are used to increase disease control efficiency, results are considerable and bacterial resistance can be overcome. Limiting environmental factors include sunlight, UV light, temperature, desiccation, and rain leaching. The effect of these environmental factors was reduced by the use of some protective formulations with phages to increase their survival and persistence in plant systems. Application timing, which is also a limiting factor in phagic disease control, can be resolved by applying phages only during daylight. Application of phages to fields during early morning and evening hours shows better results and also minimizes reductions in efficiency that can be caused by application timing. When phages are included in integrated management strategies (i.e., when they are co-applied with plant activators, SAR inducers, and some phage-sensitive bacterial strains), they will increase phage persistence in plant phyllospheres. The success of phage therapy in plant diseases depends not only on the specific pathogens involved, but also upon the complete understanding of phage biology.

REFERENCES

Abedon, S.T., Abedon, C.T., Thomas, A., Mazure, H. 2011. Bacteriophage prehistory. Is or is not Hankin, 1896, a phage reference? *Bacteriophage* 1(3): 174–178.

Adriaenssens, E.M., Vaerenbergh, J.V., Vandenheuvel, D., Dunon, V., Ceyssens, P.J., Proft, M.D., Kropinski, A.M. et al. 2012. T4-related bacteriophage LIMEstone isolates for the control of soft rot on potato caused by "*Dickeya solani*". *PLoS One* 7(3): 1–10.

Ashelford, K.E., Day, M.J., Fry, J.C. 2003. Elevated abundance of bacteriophage infecting bacteria in soil. *Applied and Environmental Microbiology* 69(1): 285–289.

Balogh, B. 2002. Strategies for improving the efficacy of bacteriophages for controlling bacterial spot of tomato. MS thesis, University of Florida, Gainesville, FL.

Balogh, B. 2006. Characterization and use of bacteriophages associated with citrus bacterial pathogens for disease control. PhD thesis. University of Florida, Gainesville, FL.

Balogh, B., Canteros, B.I., Stall, R.E., Jones, J.B. 2008. Control of citrus canker and citrus bacterial spot with bacteriophages. *Plant Disease* 92(7): 1048–1052.

Balogh, B., Dickstein, E.R., Jones, J.B., Canteros, B.I. 2013. Narrow host range phages associated with citrus canker lesions in Florida and Argentina. *European Journal of Plant Pathology* 135: 253–264.

Balogh, B., Jones, J.B., Iriarte, F.B., Momol, M.T. 2010. Phage therapy for plant disease control. *Current Pharmaceutical Biotechnology* 11: 48–57.

Balogh, B., Jones, J.B., Momol, M.T., Olson, S.M., Obradovic, A., Jackson, L.E. 2003. Improved efficacy of newly formulated bacteriophages for management of bacterial spot on tomato. *Plant Disease* 87: 949–954.

Basim H., Stall, R.E., Minsavage, G.V., Jones, J.B. 1999. Chromosomal gene transfer by conjugation in the plant pathogen *Xanthomonas axonopodis* pv. *vesicatoria*. *Phytopathology* 89: 1044–1049.

Bender, C.L., Cooksey, D.A. 1987. Molecular cloning of copper resistance genes from *Pseudomonas syringae* pv. *tomato*. *Journal of Bacteriology* 169: 470–474.

Borah, P.K., Jindal, J.K., Verma, V.P. 2000. Biological management of bacterial leaf spot of mung bean caused by *Xanthomonas axonopodis* pv. *vignaeradiatae*. *Indian Phytopathology* 53(4): 384–394.

Borra, J.M., González, S., Larrea, G.L. 2012. *The Origin of the Bacterial Immune Response*. Landes Bioscience and Springer Science + Business Media, US.

Bruynoghe, R., Maisin, J. 1921. Essais de therapeutique au moyen du bacteriophage. *Compets Rendus SoccieteBiologie*. 85: 1120–1121.

Byrne, J.M., Dianese, A.C., Ji, P., Campbell, H.L., Cuppels, D.A. 2005. Biological control of bacterial spot of tomato under field conditions at several locations in North America. *Biological Control* 32: 408–418.

Ceverolo, E.L. 1973. Characterization of *Xanthomonas pruni* bacteriophages to bacterial spot disease in prunus. *Phytopathology*. 63: 1279–1284.

Ciancio A., Mukerji, K.G. 2008. Integrated management of disease caused by fungi, phytoplasma and bacteria. In Obradovic, A., Jones, J.B., Balogh, B., Momol, M.T. (eds.). Springer, pp. 211–221.

Civerolo, E.L., Keil, H.L. 1969. Inhibition of bacterial spot of peach foliage by *Xanthomonas pruni* bacteriophage. *Phytopathology* 59: 1966–1967.

D'Herelle, F., Malone, R.H., Lahiri, M.N. 1927. Studies on Asiatic cholera. *Indian Medical Research Mem* 14: 1.

Eayre, C.G., Bartz, J.A., Concelmo, D.E. 1995. Bacteriophages of *Erwinia carotovora* and *Erwinia ananas* isolated from freshwater lakes. *Plant Disease* 79(8): 801–804.

Evans, T.J., Ind, A., Komitopoulou, E., Salmond, G.P.C. 2010. Phage-selected lipopolysaccharide mutants of *Pectobacterium atrosepticum* exhibit different impacts on virulence. *Journal of Applied Microbiology* 109: 505–514.

Fineran, P.C., Blower, T.R., Foulds, I.J., Hempherys, D.P., Lilley, K.S., Salmond, G.P.C. 2008. The phage abortive infection system, ToxIN, functions as a protein–RNA toxin–antitoxin pair. *Proceedings of National Academic of Sciences USA* 106: 894–899.

Fischetti, V.A. 2010. Bacteriophage endolysins: A novel anti-infective to control Gram-positive pathogens. *International Journal of Medical Microbiology* 300: 357–362.

Flaherty, J.E., Harbaugh, B.K., Jones, J.B., Somodi, G.C., Jackson, L.E. 2001. H-mutant bacteriophages as a potential biocontrol of bacterial blight of geranium. *HortScience* 36(1): 98–100.

Flaherty, J.E., Jones, J.B., Harbaugh, B.K., Smoodi, G.C., Jackson, L.E. 2000. Control of bacterial spot on tomato in the greenhouse and field with H-mutant bacteriophages. *HortScience* 35(5): 882–884.

Frampton, R.A., Pitman, A.R., Fineran, P.C. 2012. Advances in bacteriophage mediated control of plant pathogens. *International Journal of Microbiology* Vol 2012, Article ID 326452. 1–11.

Fujiwara, A., Fujisawa, M., Hamasaki, R., Kawasaki, T., Fujie, M., Yamada, T. 2011. Biocontrol of *Ralstonia solanacearum* by treatment with lytic bacteriophages. *Applied and Environmental Microbiology* 77(12): 4155–4162.

Gent, D.H., Schwartz, H.F. 2005. Management of *Xanthomonas* leaf blight of onion with a plant activator, biological control agents, and copper bactericides. *Plant Disease* 89: 631–639.

Gill, J.J., Abedon, S.T. 2003. Bacteriophage Ecology and Plants. APSnet Feature, November. Available at: http://www.apsnet.org/online/feature/phages/.

Godde, J.S., Bickerton, A. 2006. The repetitive DNA elements called CRISPRs and their associated genes: Evidence of horizontal transfer among prokaryotes. *Journal of Molecular Evolution* 62: 718–729.

Goyer, C. 2005. Isolation and characterization of phages Stsc1 and Stsc3 infecting *Streptomyces scabiei* and their potential as biocontrol agents. *Canadian Journal of Plant Pathology* 27: 210–216.

Graham, J.H., Leite, J.R.P. 2004. Lack of control of citrus canker by induced systemic resistance compounds. *Plant Disease* 88: 745–750.

Greer, G.G. 2005. Bacteriophage control of foodborne bacteria. *Journal of Food Protection* 68: 1102–1111.

Haq, I.U., Chaudhry, W.N., Akhtar, M.N., Andleeb, S., Qadri, I. 2012. Bacteriophages and their implications on future biotechnology: A review. *Virology Journal* 9 (9): 1–9.

Hankin, M.E. 1896. L'action bactericide des eaux de la Jumna et du Gange sur le vibrion du cholera. *Annales de L'Institut Pasteur (Paris)* 10(5): 11–23.

Hermoso, J.A., Garcia, J.L., Garcia, P. 2007. Taking aim on bacterial pathogens: From phage therapy to enzybiotics. *Current Opinion in Microbiology* 10(5): 461–472.

Hill, C., Garvey, P., Fitzgerald, G.F. 1996. Bacteriophage-host interactions and resistance mechanisms, analysis of the conjugative bacteriophage resistance plasmid pNP40. *Le Lait* 76: 67–79.

Iriarte, F.B., Balogh, B., Momol, M.T., Jones, J.B. 2007. Factors affecting survival of bacteriophage on tomato leaf surfaces. *Applied and Environmental Microbiology* 73(6): 1704–1711.

Jackson, L.E., Jones, J.B. 2004. Bacteriophage: A viable bacteria control solution. Omnilytics White Paper, pp. 1–10.

Ji, P., Campbell, H.L., Klopper, J.W., Jones, J.B., Suslow, T.V., Wilson, M. 2006. Integrated biological control of bacterial speck and spot of tomato under field conditions using foliar biological control agents and plant growth-promoting rhizobacteria. *Biological Control* 36: 358–367.

Jones, J.B., Jackson, L.E., Balogh, B., Obradovic, A., Iriarte, F.B., Momol, M. 2007. Bacteriophages for plant disease control. *Annual Reviews of Phytopathology* 45: 245–262.

Kim, M.H., Park, S.W., Kim, Y.K. 2011. Bacteriophages of *Pseudomonas tolaassi* for the biological control of brown blotch disease. *Journal of the Korean Society for Applied Biological Chemistry* 54(1): 99–104.

Kuo, T., Cheng, L., Yang, C., Yang, S. 1971. Bacterial leaf blight of rice plant IV. Effect of bacteriophage on the infectivity of *Xanthomonas oryzae*. *Botanical Bulletin of Academia Sinica* 12: 1–9.

Kutter, E. 1997. *Phage Therapy: Bacteriophages as Antibiotics Olympia*. Washington: http://www.evergreen.edu/phage/phagetherapy/phagetherapy.htm (accessed April 2014).

Labrie, S.J., Samson, J.E., Moineau, S. 2010. Bacteriophage resistance mechanisms. *Nature Reviews Microbiology* 8: 317–327.

Lang, J.M., Gent, D.H., Schwartz, H.F. 2007. Management of *Xanthomonas* leaf blight of onion with bacteriophages and a plant activator. *Plant Disease* 91(7): 871–878.

Letarov, A., Kulikov, E. 2008. The bacteriophages in human- and animal body-associated microbial communities. *Journal of Applied Microbiology* 107: 1–13.

Lim, J.A., Jee, S., Lee, D.H., Roh, E., Jung, K., Oh, C., Heu, S. 2013. Biocontrol of *Pectobacterium carotovorum* subsp. *carotovorum* using bacteriophage PP1. *Journal of Microbial Biotechnology* 23(8): 1147–1153.

Liu, B., Wu, S., Song, Q., Zhang, X., Xie, L. 2006. Two novel bacteriophages of thermophilic bacteria isolated from deep-sea hydrothermal fields. *Current Microbiology* 53: 163–166.

Louws, F.J., Wilson, M., Campbell, H.L., Cuppels, D.A., Jones, J.B. 2001. Field control of bacterial spot and bacterial speck of tomato using a plant activator. *Plant Disease* 85: 481–488.

Maxson-Stein, K., He, S.-Y., Hammerschmidt, R., Jones, A.L. 2002. Effect of treating apple trees with acibenzolar-*S*-methyl on fire blight and expression of pathogenesis-related protein genes. *Plant Disease* 86: 785–790.

McManus, P.S., Stockwell, V.O., Sundin, G.W., Jones, A.L. 2002. Antibiotic use in plant agriculture. *Annual Review of Phytopathology* 40: 443–465.

Miernik, A. 2003. Occurrence of bacteria and coli bacteriophages as potential indicators of fecal pollution of vistula river and zegrze reservoir. *Polish Journal of Environmental Studies* 13(1): 79–84.

Moore, E.S. 1926. D'Herelle's bacteriophage in relation to plant parasites. *South African Journal of Science* 23: 306.

Mukhtar, I., Mushtaq, S., Ali, A., Khokhar, I. 2010. Epiphytic and endophytic phyllosphere microflora of *Cassytha filiformis* L. and its hosts. *Ecoprint* 17: 1–8.

Nagy, J.K., Kiraly, L., Schwariczinger, I. 2012. Phage therapy for plant disease control with a focus on fire blight. *Central European Journal of Biology* 7(1): 1–12.

Neil, D.L., Romero, S., Kandula, J., Stark, C., Stewart, A., Larsen, S. 2001. Bacteriophages: A potential biocontrol agent against walnut blight (*Xanthomonas campestris* pv. *juglandis*). *New Zealand Plant Protection* 54: 220–224.

Obradovic, A., Jones, J.B., Momel, M.T., Olson, S.M. 2004. Management of tomato bacterial spot in the field by foliar applications of bacteriophages and SAR inducers. *Plant Disease* 88: 736–740.

Obradovic, A., Jones, J.B., Momel, M.T., Olson, S.M., Jackson, L.E., Balogh, B., Guven, K., Iriarte, F.B. 2005. Integration of biological control agents and systemic acquired resistance inducers against bacterial spot on tomato. *Plant Disease* 89(7): 712–716.

Obradovic, A., Jones, J.B., Balogh, B., Momol, M.T. 2008. Integrated management of tomato bacterial spot. In *Integrated Management of Disease Caused by Fungi, Phytoplasma and Bacteria,* Ciancio A., Mukerji, K.G. (eds.). Netherlands: Springer, pp. 211–221.

Prior, S.E., Andrews, A.J., Nordeen, R.O. 2007. Characterization of bacteriophages of *Pseudomonas syringae* pv. *tomato*. *Journal of the Arkansas Academy of Science* 61: 84–89.

Przybilski, R., Richter, C., Gristwood, T., Clulow, J.S., Vercoe, R.B., Fineran, P.C. 2011. Csy4 is responsible for CRISPR RNA processing in *Pectobacterium atrosepticum*. *RNA Biology* 8(3): 517–528.

Rasche, F., Trondl, R., Naglreiter, C., Reichenauer, T.G., Sessitsch, A. 2006. Chilling and cultivar type affect the diversity of bacterial endophytes colonizing sweet pepper (*Capsicum anuum* L.). *Canadian Journal of Microbiology* 52: 1036–1045.

Romero, A.M., Kousik, C.S., Ritchie, D.F. 2001. Resistance to bacterial spot in bell pepper induced by acibenzolar-*S*-methyl. *Plant Disease* 85: 189–194.

Ronald, R. 2011. Plant genetics, sustainable agriculture and global food security. *Genetics* 188(1): 11–20.

Scheck, H.J., Pscheidt, J.W., Moore, L.W. 1996. Copper and *Streptomycin* resistance in strains of *Pseudomonas syringae* from Pacific Northwest nurseries. *Plant Disease* 80(9): 1034–1039.

Schnabel, E.L., Fernando, W.G.D., Meyer, M.P., Jones, A.L., Jackson, L.E. 1999. Bacteriophage of *Erwinia amylovora* and their potential for biocontrol. *Acta Horticulture* 489: 649–654.

Sean, M., Britt, K. 2013. Exploring the risks of phage application in the environment. *Frontiers in Microbiology* 4: 1–8.

Stonier, T., McSharry, J., Speitel, T. 1967. *Agrobacterium tumefaciens* Conn IV. Bacteriophage PB21 and its inhibitory effect on tumor induction. *Journal of Virology* 73: 268–273.

Sulakvelidze, A., Alavidze, Z., Morris, J.R. 2001. Bacteriophage therapy. *Antimicrobial Agents and Chemotherapy* 45(3): 649–659.

Sulakvelidze, A., Barrow, P. 2005. Phage therapy in animals and agribusiness, in bacteriophages: Biology and applications. In *Bacteriophage: Biology and Applications*. Kutter, E., Sulakvelidze, A. (eds.). Boca Raton, FL: CRC Press, pp. 335–380.

Summers, W.C. 2001. Bacteriophage therapy. *Annual Reviews Microbiology* 55: 437–451.

Tanaka, H., Negishi, H., Maeda, H. 1990. Control of tobacco bacterial wilt by an avirulent strain of *Pseudomonas solanacearunm* M4S and its bacteriophage. *Annals of Phytopathological Society of Japan* 56: 243–246.

Thayer, P.L., Stall, R.E. 1961. A survey of *Xanthomonas vesicatoria* resistance to streptomycin. *Proceedings of Florida Horticulture Society* 75: 163–165.

Thomas, R.C. 1935. A bacteriophage in relation to Stewart's disease of corn. *Phytopathology* 25: 371–372.

Tseng, Y.H., Lo, M.C., Lin, K.C., Pan, C.C., Chang, R.Y. 1990. Characterization of filamentous bacteriophage bLf from *Xanthomonas campestris* pv. *campestris*. *Journal of General Virology* 71: 1881–1884.

Twort, F.W. 1915. An investigation on the nature of ultramicroscopic viruses. *Lancet* 2: 1241–1243.

Vidaver, A.K., Lambrecht, P.A. 2004. Bacteria as plant pathogens. *The Plant Health Instructor*. doi: 10.1094/PHI-I-0809-01, http://www.aspnet.org/edcenter/intropp/pathogengroups/pages/bacteria.aspx.

Vidaver, A.K., Schuster, M.L. 1969. Characterization of *Xanthomonas phaseoli* bacteriophages. *Journal of Virology* 4(3): 300–308.

Wittmann, J., Gartemann, K.H., Eichenlaub, R., Dreiseikelmann, B. 2011. Genomic and molecular analysis of phage CMP1 from *Clavibacter michiganensis* subspecies *michiganensis*. *Bacteriophage* 1(1): 6–14.

16 Role of Defense Enzymes in the Control of Plant Pathogenic Bacteria

Muthukrishnan Sathiyabama

CONTENTS

ABSTRACT Plants respond to pathogenic bacteria by activating a variety of active and passive defense mechanisms. These defense responses, which serve to limit the growth and spread of pathogens within the plants, include hypersensitive programmed cell death, cross-linking and reinforcement of cell walls, biosynthesis of phytoalexins, metabolism of phenolic compounds, and transcriptional induction of defense genes, including pathogenesis-related (PR) genes. In addition to the PR proteins, plants produce defense enzymes such as peroxidase (PO) and polyphenol oxidase (PPO), which are catalysts in the formation of lignin. In this chapter, progress in utilizing defense enzymes in the control of plant pathogenic bacteria is summarized and discussed.

KEYWORDS: chitinase, peroxidase, polyphenol oxidase, phytopathogenic bacteria, induced defense response

16.1 INTRODUCTION

Phytopathogenic bacteria are responsible for great losses in economically important crops (Agrios, 1988). They incite diseases in plants by penetrating into host tissues through natural openings such as hydathodes, stomata, lenticels, and stigma, as well as through wounds. They multiply rapidly inside plant tissue under favorable conditions and cause many serious crop diseases. Disease symptoms that result from bacterial pathogens include wilt, galls, specks, spots, cankers, and chlorosis. Most of the bacterial pathogens of plants are Gram-negative bacteria in the genera *Pseudomonas*, *Erwinia*, *Ralstonia*, *Xanthomonas*, and *Agrobacterium*. Certain Gram-positive bacteria (*Streptomyces* spp. and *Clavibacter* spp.) also cause diseases in plants.

Although much research has attempted to resolve and prevent the damage caused by bacteria, the lack of effective control of some severe diseases poses major difficulty (Ferre et al., 2006). Plant protection against such pathogens is mainly based on copper derivatives and antibiotics. However, these compounds are regarded as environmental contaminants, and resistant strains of plant pathogenic bacteria have been reported in several crops (Loper et al., 1991; Sundin and Bender, 1993). The best approach to controlling the growth of phytopathogenic bacteria in plants involves inducing the plants' own innate immunity.

Plants have evolved several different defense mechanisms to prevent bacterial infection. Plant bacterial pathogens (and, in general, other pathogens) reveal themselves to the host immune system through molecules called pathogen-associated molecular patterns (PAMPs) and through microbe-associated molecular patterns such as flagellin or bacterial lipopolysaccharides (LPS). During the initial contact between a plant and a bacterial pathogen, the plant can detect bacterial PAMPs. This perception activates signal-transduction cascades that turn on basal defenses, which include callose deposition to reinforce cell walls, production of reactive oxygen species and ethylene, and transcriptional induction of a large suite of defense genes, including pathogenesis-related (PR) genes. These responses are triggered by plant extracellular receptors, known as pattern recognition receptors, that specialize in the recognition of PAMPs. These basal defenses are generally sufficient to halt the growth of pathogenic microbes and prevent their establishment.

Chitinases, peroxidases (POs), and polyphenol oxidases (PPOs) are defense enzymes that are induced in plants during pathogenesis and participate in phytopathogen (Chen et al., 2000). This chapter contains discussions of the roles of defense enzymes induced by pathogenic/non-pathogenic bacteria, as well as the hypothetical defense mechanisms by which plants directly counteract phytopathogenic bacteria.

16.2 CHITINASES

Chitinases (poly (1,4-*N*-acetyl-β-D-glucosaminide) glycanohydrolase, E.C.3.2.1.14) are abundant proteins found in a wide variety of seed-producing plants. Although the physiological function of chitinases is unknown, correlative evidence shows that they are defense proteins with antimicrobial activity. Some chitinases also display lysozymal activity (Mauch et al., 1988; Majeau et al., 1990) and thus may be involved in conferring resistance to bacterial pathogens. Chitinases fall into four broad classes as proposed by Shinshi et al. (1990) and Collinge et al. (1993). Class I is basic and contains a cystein-rich *N*-terminal domain with putative chitin-binding properties. Class II is acidic, lacks the *N*-terminal cystein-rich domain, and has a high sequence similarity to class I chitinases. Class III members have no serological relationship to class I and class II but have high amino acid sequence homology to the bifunctional chitinase-lysozyme of *Hevea brasiliensis*. Class IV includes several chitinases that have structural similarities but sequence dissimilarities to class I. The majority of class III chitinases are classified as such on the basis of homology to lysozymes with chitinase activity. The first such chitinase described was a basic 30.3 kDa bifunctional lysozyme from *Parthenocissus quinquefolia* (Bernasconi et al., 1987) that showed strong homology to a lysozyme from *H. brasiliensis*. Chitinase hydrolyzes a polymer of *N*-acetyl glucosamine (NAG), whereas lysozyme hydrolyzes a polymer of alternating NAG and *N*-acetylmuramic acid residues; the latter is a major constituent of bacterial cell walls (Graham and Sticklen, 1994).

Class I chitinases have been purified from the leaves, fruits, and stems of various plants. They are usually located in the vacuole (Keefe et al., 1990), although they may be extracellular. Class I chitinases account for the majority of chitinolytic activity in plant material in which both acidic and basic isoforms are found. Class II chitinases, which consist of a monomeric catalytic domain that has strong homology to the catalytic domain of class I chitinases, are generally acidic and extracellular and can be detected in the apoplastic washing fluid.

A 28-kDa acidic cucumber chitinase was found (Metraux et al., 1988, 1989) that showed strong homology to *Pa. quinquefolia* bifunctional enzyme and accumulated in the intercellular leaf fluid in response to infections by *Pseudoperonospora cubensis* and *Pseudoperonospora lachymans* (Metraux et al., 1988). Increased chitinase activity in leaf extracts of French beans infected with avirulent races of *Pseudomonas syringae* pv. *phaseolicola* was reported by Voisey and Slusarenko (1989). Meins and Ahl (1989) reported the localized accumulation of chitinase and chitinase mRNA in tobacco plants infected with *Ps. syringae*. Recent applied research has begun to focus on the production of transgenic crop plants with increased expression of chitinase, in the hope of producing bacterial disease-resistant varieties.

A symbiotic bacterium, *Rhizobium meliloti*, produces nodulation (Nod) factors that induce nodulation in alfalfa. Because these nod factors contain chitin (Lerouge et al., 1990), it is possible that chitinase also plays a role in this bacterium's symbiosis.

16.3 PEROXIDASES

Peroxidases or PO (E.C.1.11.1.17), a group of heme-containing glycosylated proteins, are known to be activated in response to pathogen attacks. Several roles have been attributed to plant PO in host–pathogen interactions (Chittoor et al., 1999; Do et al., 2003; Saikia et al., 2006), where such enzymes have been implicated in the oxidation of phenols (Schmid and Feucht, 1980), lignifications (Saparrat and Guiller, 2005), plant protection (Hammerschmidt et al., 1982), and elongation of plant cells (Goldberg et al., 1986). POs can create a physical barrier to limit pathogen invasion in host tissues by catalyzing the cross-linking of cell wall components in response to pathogen infection. The correlation between increased PO activity during incompatible interactions and reinforcement of cell walls with phenol compounds was also reported (Reimers et al., 1992). In cotton cultivars, the association of PO activity with resistance to bacterial blight was suggested (Martinez et al., 1996). Apoplastic cationic isoforms of PO generated oxygen during the hypersensitive response (HR) *to Xanthomonas campestris* pv. *malvacearu*m suggested PO involvement in the oxidative burst (Bolwell et al., 2002). The oxidative burst in plants infected by incompatible pathogens is a key early event in the expression of resistance (Wojtasjeck, 1997). In rice infected by *Xanthomonas oryzae,* the activity of the PO gene was shown to be correlated with the thickening of vessel secondary cell walls and probably associated with an increase in lignifications (Hilaire et al., 2001). Change in antioxidant enzyme activity by *Xanthomonas* infection was observed in sweet orange and in tomato (Kumar et al., 2011; Chandrashekar and Umesh, 2012). Tobacco plants containing overexpressions of PO genes showed increased levels of resistance to bacterial wild fire caused by *Ps. syringae* pv. *tabaci* (Faize et al., 2012).

16.4 POLYPHENOL OXIDASE

Polyphenol oxidases or PPO (catechol oxidase; E.C.1.10.3.2) are nuclear-encoded enzymes of almost ubiquitous distribution in plants (Mayer and Harel, 1979; Mayer, 1987). This enzyme uses molecular oxygen to oxidize common ortho-diphenolic compounds (e.g., caffeic acid and catechol) to their respective quinones. The quinonoid products of PPO are highly reactive molecules that can covalently modify and cross-link a variety of cellular nucleophiles via. a 1,4 addition mechanism, and the resultant formation of melanin-like black or brown condensation polymers. A role for PPO in plant defenses against pathogens has been postulated from the earliest days of PPO research. This hypothesis has been supported by many correlative studies, for example, Mayer's (2006) finding of the upregulation of PPO in pathogen-challenged plants. Many PPOs have been predicted to contain a proteolytic processing site near the C-terminus of the polypeptide (Marusek et al., 2006). Pathogen-induced PPO activity continues to be reported for a variety of plant taxa, including monocots and dicots (Chen et al., 2000; Deborah et al., 2001). Correlations of high PPO levels in cultivars with a high pathogen resistance continue to provide support for a pathogen defense role of PPO (Raj et al., 2006).

Several research groups have also attempted to correlate the protective effects of rhizosphere bacteria with the induction of defense enzymes including PPO, with mixed success (Chen et al.,

2000; Ramamoorthy et al., 2002). An issue that complicates PPO function has been the separate localization of its phenolic substrates in plant vacuoles; this separation means that the cell would have to be broken in order for PPO to oxidize phenols. Such an effect is most likely to occur after pest or pathogen challenge. Possible roles of PPO in plant defense are commonly discussed, although direct evidence has only recently become available.

Polyphenol oxidases are known to have broad substrate specificities. Therefore, an enzyme from any given source may be capable of oxidizing a variety of simple ortho-diphenolics, such as caffeic acid and its conjugates, catechol derivatives, or dihyroxyphenylalanine. However, enzymes from different plant species exhibit distinct preference profiles. Flavonoids with ortho-dihyroxy phenolic rings have been found to be PPO substrates that support catechin, epicatechin, and myricetin (Guyot et al., 1996; Jiménez and García-Carmona, 1999). Furthermore, monophenolase activity has been described for some PPOs (Wuyts et al., 2006). Such enzymes may hydroxylate a monophenol such as tyrosine, which can undergo further oxidation by PPO activity to quinine.

Although the range of substrates accepted by isolated PPOs can be readily defined *in vitro*, much less is known about the substrates that are utilized in planta or during defense reactions. Caffeic acid esters (e.g., chlorogenic acid) are both excellent PPO substrates and very common plant metabolites. In tomato and coffee, chlorogenic acid has been identified as the most likely *in vivo* PPO substrate (Li and Steffens, 2002; Melo et al., 2006). In *Populus tremuloides*, catechol is postulated to be released by the breakdown of abundant phenolic glycosides; it would therefore be available as a substrate in damaged tissues (Haruta et al., 2001).

Based on browning reactions to reactive PPO-generated quinines, PPO has often been suggested to function as a defense against pest and pathogens.

In *Solanum berthaultii*, high PPO levels have been found in glandular trichomes; in addition, tricome breakage by small-bodied insects such as aphids leads to rapid PPO-mediated oxidation and polymerization of phenolics, which ultimately entrap the invading insects (Kowalski et al., 1992).

Transgenic plants with modified PPO expression possess a unique system by which to assess the involvement of PPO in plant disease resistance. When challenged with the bacterial pathogen *Ps. syringae* pv. tomato, PPO-overexpressing plants showed reduced bacterial growth, whereas PPO antisense-suppressed lines supported greater bacterial numbers (Li and Steffens, 2002; Thipyapong et al., 2004). These studies demonstrate the importance of PPO in pathogen defense. Li and Steffens (2002) suggested several possibilities, including general toxicity of PPO-generated quinines to pathogens and plant cells, which would accelerate cell death; alkylation and reduced bioavailability of cellular proteins to the pathogen; cross-linking of quinines with protein or other phenolics, which would form a physical barrier to pathogens in the cell wall; and quinine redox cycling that leads to H_2O_2 and other reactive oxygen species.

16.5 INDUCED RESISTANCE

Plants are continually exposed to a vast number of potentially phytopathogenic bacteria, against which they try to defend themselves through a multilayered system of passive and active mechanisms. One such mechanism is induced resistance (IR), which can be broadly subdivided into systemic acquired resistance (SAR) and induced systemic resistance (ISR). SAR develops locally or systemically in response to pathogen infection or treatment with certain chemicals (Walters et al., 2005). ISR is persistent and presents complex components in different locations within the plant; these components are responsible for the activities of various compounds.

IR is a state of enhanced defensive capacity that a plant develops when approximately stimulated. At sites not directly exposed to inducers, expressions of IR can be local or systemic. Inducers may be chemical activators, extracts of cells of living organisms, or microorganisms (Romeiro et al., 2010). It is known that susceptible plants have genetic information for efficient mechanisms of resistance to diseases and that these mechanisms can be systemically expressed for long periods of time prior to inoculation with avirulent pathogens, microbial components, or chemical substances

(Kuc, 1995). Structural and biochemical defenses induced in planta by bacterial infection can also contribute to disease resistance. The most-studied induced biochemical defenses are associated with hypersensitive response (HR); these are responsible for restricting bacterial migration and for blocking bacterial multiplication in planta. Induced HR is generated by the plasma membrane NADPH oxidase and by apoplastic POs (Soylu et al., 2005). Early increases in antimicrobial C6-volatiles derived by the lipoxygenase pathway were reported in bean and pepper plants respectively undergoing HR induced by avirulent strains of *Pseudomonas savastanoi* pv. *phaseolicola* and *X. campestris* pv. *vesicatoria* (Croft et al., 1993; Buonaurio and Servili, 1999).

Transcripts of extracellular pepper class II basic chitinase are highly expressed in pepper plants in response to *X. campestris* pv. *vesicatoria* (Hong et al., 2000). It is well known that chitinases have a bifunctional activity of both chitinase and lysozyme, and therefore may hydrolyze bacterial cell walls (Graham and Sticklen, 1994). Transgenic *Arabidopsis* plants overexpressing a chitinase II gene showed protection against *Ps. syringae* pv. *tomato* (Hong and Hwang, 2006). Algam et al. (2010) reported that application of *Paenibacillus polymyxa* increased the activity of chitinase and glucanase in tomato plants, which result in the activation of a more systemic resistance against wilt pathogen *Ralstonia solanacearum*. ISR by inducing plant-growth-promoting rhizobacteria (PGPR) have been demonstrated to enhance plants' own defense capacities by priming for the potential expression of defense genes (Kim et al., 2004; Tjamos et al., 2005).

16.6 ENDOPHYTIC BACTERIA-MEDIATED INDUCTION

Endophytic bacteria are common inhabitants of the internal tissues of various plant species (Strobel et al., 2004). They survive internally without causing harmful effects to the plant, and some can provide beneficial effects that favor phytohormone synthesis, resistance induction, and biological control of pathogens (Ryan et al., 2008) through ISR (Kumari and Srivastava, 1999). They also secrete metabolites/lytic enzymes that can interfere with pathogen growth and/or activities. Expression and secretion of these enzymes by different microbes can sometimes result in the direct suppression of plant pathogen activities.

The first indication that endophytic bacteria could elicit ISR dates to 1991 (Wei et al., 1991). Greenhouse screening of endophytic *Bacillus* spp. that elicited ISR in tomato plants against *R. solanacearum* has been reported (Jetiyanon and Kloepper, 2002).

Modes of action in disease suppression by these bacteria include antibiosis, production of lytic enzymes, and ISR (Bakker et al., 2007). Rajendran et al. (2006) reported that application of endophytic bacteria such as *Bacillus* and *Pseudomonas* to cotton plants induced the expression of PO, PPO, and chitinase and reduced the disease severity of *Xanthomonas axonopodis* pv. *malvacearum*, the causal agent of bacterial blight in cotton. Higher levels of PO have been correlated with enhanced ISR in several plants (Kandan et al., 2002). Tomato plants sprayed with protein fractions of endophytic *Bacillus* isolates promoted the increase of PO and PPO. The increase of these enzymes, which is associated with cell wall reinforcement, and local and systemic resistance (Anterola and Lewis, 2002), resulted in a reduction of bacterial spot caused by *Xanthomonas vesicatoria* on tomato plants (Romeiro et al., 2010; Roberto-Lanna-Filho et al., 2013).

16.7 PGPR-MEDIATED INDUCTION

Plant growth-promoting rhizobacteria are free-living or root-associated bacteria that can increase plant growth and productivity. Metabolic changes involved in the defense mechanisms of plants are correlated with changes in the activity of key enzymes in primary and secondary metabolism. The production of enzymes related to pathogenesis (PR proteins) by strains of rhizobacteria is considered the largest property of the antagonistic strains (Saikia et al., 2004). Chitinases, lipoxygenases, POs, and glucanases are among these enzymes; increases in their activity and accumulation depend mainly on the inducing agent but also on the genotype of the plant, physiological conditions, and the pathogen (Tuzun, 2001). Depending on the pathosystems studied, varieties of substances are produced by rhizobacteria that have been linked to the activation of mechanisms of disease suppression in plants which reduce damage caused by phytopathogens. Specific strains of non-pathogenic PGPR are known to suppress diseases by ISR in plants (Van peer et al., 1991; Wei et al., 1991; Kloepper et al., 1992; van Loon et al., 1998). Alstrom (1991) reported that seed bacterization with rhizosphere *Pseudomonads* induced disease resistance in common bean against the halo blight bacterial pathogen *Ps. syringae* pv. *phaseolicola*.

Induction of systemic resistance in cucumber against bacterial angular leaf spot by *Pseudomonas putida* and *Serrtia marcens* have been reported (Liu et al., 1995). Jetiyanon et al. (1997) reported that increased PO levels as early plant defense response reactions were associated with PGPR-mediated ISR. Efforts have been made to use PGPR such as *Pseudomonas fluorescens* in the management of bacterial leaf blight of rice (Vidyasekaran et al., 2001). Silva et al. (2004) found that the application of rhizobacteria to tomato plants induced the activity of lipooxygenase (LOX), phenylalanine ammonia lyase (PAL), PO and protected the plants from *Ps. syringae* pv. *tomato*, which interprets rhizobacterium-induced ISR within them. Phenylalanine ammonia lyase and POX are important in the biosynthesis of phenolics, phytoalexnins, and lignin, the key factors responsible for disease resistance (Daayf et al., 1997). The increased level of defense-related enzymes during ISR are known to be crucial in host resistance (Chen et al., 2000; Ramamoorthy et al., 2002). Park and Kloepper (2000) reported the activation of PR-1a promoter by rhizobacteria, which initiated ISR in tobacco against *Ps. syringae* pv. *tabaci*. Application of *Ps. fluorescens* to tomato plants resulted in increased production of PO and PPO, and reduced bacterial wilt disease (Seleim et al., 2014).

Bacillus spp. has been reported to significantly reduce the incidence of disease in a diversity of hosts (Kloepper et al., 2004). Elicitation of ISR against bacterial pathogens by these strains has been demonstrated in greenhouse and field trials on tomato, bell pepper, muskmelon, watermelon, sugar beet, tobacco, and cucumber (Kloepper et al., 2004). ISR PGPRs have also been demonstrated to enhance plant defense capacity by priming for the potential expression of defense genes (Kim et al., 2004; Tjamos et al., 2005). Application of *Bacillus amyloliquefaciens* elicited ISR against *Erwinia carotovora* in *Arabidopsis* (Ryu et al., 2004). Ran et al. (2005) reported that the application of *Pseudomonas* spp. elicited ISR in *Eucalyptus arophylla* against bacterial wilt caused by *R. solanacearum*. Bacterial spot disease caused by *Xanthomonas axonopodis* pv. *vesicatoria* was successfully controlled by *Bacillus* strains isolated from rhizosphere soil of pepper (Mustafa et al., 2008). Romeiro et al. (2010) showed the presence of ISR in tomato with a protein synthesized by the rhizobacterium *Bacillus cereus*.

Rice plants treated with PGPR showed increased synthesis of defense enzymes such as PAL, PO, and PPO, and suppressed the growth of bacterial blight pathogen (*X. oryzae*) in the host (Chithrashree et al., 2011). Kamal et al. (2012) reported that application of *Ps. fluorescens* to tomato plants induced PO and PPO in leaves, and also reduced the disease severity of bacterial wilt caused by *R. solanacearum*. Salicylic-acid-producing rhizobacterial strain *Serratia marcescens* induced resistance to wild fire caused by *Ps. syringae* pv. *tabaci* in tobacco and has been associated with the accumulation of PR genes (Press et al., 1997).

16.8 ROLE OF BACTERIAL DETERMINANTS IN ISR

A molecule termed "bacterial determinant" by van Loon et al. (1998), because it is probably of bacterial origin, gives rise to the signal, which in turn triggers IR. LPS, a component of the cell surfaces of Gram-negative bacteria, act as bacterial determinants and trigger the defense-related ISR signaling pathway (Dow et al., 2000). Several investigators have focused on LPS as ISR elicitors in plants (Romeiro and Kimura, 1997; Dow et al., 2000; Mishina and Zeier, 2007). In tobacco, LPS from *R. solanacearum* induced the synthesis of PR-like polypeptides (Leach et al., 1983). The LPS from *X. campestris* induced transcripts for a defense-related β-1,3 glucanase in turnip (Newman et al., 1995). The increased levels of defense-related enzymes during ISR are known to be crucial to host resistance (Chen et al., 2000; Ramamoorthy et al., 2002).

Bacterial lipopeptides (LPs) such as surfactins, iturins, and fengycins control bacterial plant pathogens (Ongena and Jacques, 2008). The surfactin-type LPs can interact with plant cells as a bacterial determinant for tuning on an immune response through the stimulation of the ISR phenomenon (Ongena and Jacques, 2008). Production of surfactin by *Bacillus* reduces the infection caused by *Ps. syringae* on *Arabidopsis* plants (Bais et al., 2004). In bean and tomato, a role for surfactin in plant defense induction was demonstrated by a similar protective activity directly from the producing strain (Ongena et al., 2007). This macroscopic disease reduction was related to metabolic changes associated with plant defense responses (Ongena et al., 2007). In tomato, LPs that overproduce *Bacillus* isolates stimulate LOX activities. Treatment of *Arabidopsis* plants with flg22, a peptide representing the elicitor active epitope of flagellin, induces the expression of numerous defense-related genes and triggers resistance to pathogenic bacteria (Zipfel et al., 2004).

16.9 SYNTHETIC PEPTIDES AND CHEMICAL-MEDIATED DEFENSE RESPONSE

Endogenous antimicrobial peptides have emerged as good candidates for the protection of crops against pathogenic bacteria (Hancock and Lehrer, 1998; Zasloff, 2002). These peptides have been found in a variety of sources, including mammals, amphibians, insects, and plants, and are known to have an important role in host defense systems and innate immunities (Ganz and Lehrer, 1998; Garcia-Olmedo et al., 1998). Natural antimicrobial peptides exhibit a broad spectrum of activity against bacteria, fungi, and so forth. They are lytic, and have synergistic activity with conventional antibiotics. When Ferre et al. (2006) synthesized short peptides and tested them against *Erwinia amylovora*, *Ps. syringae*, and *X. vesicatoria*, bactericidal activity at micromolar concentrations was observed. These peptides can be considered as potential lead compounds for the development of new antimicrobial agents for use in plant protection, either as components of pesticides or by expression in transgenic plants.

Chemical inducers of disease resistance are essential for the integration of the SAR concept in plant protection strategies. Acibenzolar-*S*-methyl (ASM) is one of the nontoxic synthetic resistance inducers used against plant pathogens. This compound is known to elicit SAR against fungal and bacterial diseases of several plants, including tobacco and cucumber (Lawton et al., 1996) and apple (Brisset et al., 2000; Baysal et al., 2003). It was also effective in controlling bacterial black spot of mango caused by *X. campestris* pv. *mangiferaindicae* (Boshoff et al., 1998). Apple plants treated with ASM showed increased PO levels and IR to fire blight.

Plant activators such as benzothiadiazole or ASM have shown activity against a number of bacterial diseases including bacterial spot (*X. axonopodis* pv. *vesicatoria*) and bacterial speck (*Ps. syringae* pv. *tomato*) of tomato and pepper (Louws et al., 2001) and fire blight of apples *E. amylovora* (Momol et al., 1999). Application of ASM to tomato seedlings induced PO and chitinase, and protected tomato plants from bacterial canker caused by *Clavibacter michiganensis* (Baysal et al., 2003). Babu et al. (2003) reported treatment of rice seedlings with ASM-induced resistance to bacterial leaf blight. Tomato plants sprayed with ASM-induced PO showed resistance to *X. vesicatoria* (Riberiro Junior et al., 2004; Cavalcanti et al., 2006). According to Silva et al. (2007), PO activity involves a variety of processes that are related to plant defense, including hypersensitivity reaction, lignifications, and curing. Reinforcement of cell walls by lignin and phenolic compounds, one of the reactions that is elicited by the defense systems of plants, increases resistance to enzymatic degradation by pathogens and acts as a mechanical barrier to the entry of toxins as well as physical penetration and restriction of tissue colonization by pathogens (Resende et al., 2007). Application of ASM to tomato plants induced changes in the activities of PPO, β-glucosidase, and PO, and protected tomato plants against the wilt pathogen of *R. solanacearum* (Kamal et al., 2012).

16.10 CONCLUSION

The risk of bacterial plant disease epidemics is increasing worldwide. Therefore, development of new control strategies is very much needed. Basic research on plant bacterium interactions, especially at the molecular level, can substantially contribute to attaining this goal. Rapid progress has recently been achieved in the study of plant defense mechanisms against pathogens. However, further efforts are necessary to understand the mechanisms that are responsible for the induction of defense enzymes and for the restriction of bacterial growth during resistance.

REFERENCES

Agrios, G.N. 1998. *Plant Pathology*, 4th edn. San Diego, CA: Academic Press.

Algam, S.A.E., Xie, G., Li, B. 2010. Effects of *Paenibacillus* strains and chitosan on plant growth promotion and control of *Ralstonia* wilt in tomato. *Journal of Plant Pathology* 92: 593–600.

Alstrom, S. 1991. Induction of disease resistance in common bean susceptible to halo blight bacterial pathogen after seed bacterization with rhizosphere pseudomonads. *Journal of General Applied Microbiology* 37: 495–498.

Anterola, A.M., Lewis, N.G. 2002. Trends in lignin modification: A comprehensive analysis of the effects of genetic manipulations/mutations on lignifications and vascular integrity. *Phytochemistry* 61: 221–294.

Babu, R.M., Sajeena, A., Samuneeswari, A.V. 2003. Induction of bacterial blight (*Xanthomonas oryzae* pv. *oryzae*) resistance in rice by treatment with acibenzolar-*S*-methyl. *Annals of Applied Biology* 143: 333–340.

Bais, H.P., Fall., R., Vivanco, J.M. 2004. Biocontrol of *Bacillus subtilis* against infection of *Arabidopsis* roots by *Pseudomonas syringae* is facilitated by biofilm formation and surfactin production. *Plant Physiology* 134: 307–319.

Bakker, P.A.H.M., Pieterse, C.M.J., van Loon, L.C. 2007. Induced systemic resistance by fluorescent *Pseudomonas* spp. *Phytopathology* 97: 239–243.

Baysal, O., Soyulu, E.M., Soyulu, S. 2003. Induction of defense related enzymes and resistance by the plant activator acibenzolar-*S*-methyl in tomato seedlings against bacterial canker caused by *Clavibacter michiganensis* ssp. *michiganenesis*. *Plant Pathology* 52: 747–753.

Bernasconi, P., Locher, R., Pilet, P.E. 1987. Purification and N-terminal amino-acid sequence of a basic lysozyme from *Parthenocissus quinquefolia* cultured in vitro. *Biochimica et Biophysica Acta* 915: 254–260.

Bolwell, G.P., Bindschedler, L.V., Blee, K.A. 2002. The apoplastic oxidative burst in response to biotic stress in plants: A three-component system. *Journal of Experimental Botany* 53: 1367–1376.

Boshoff, M., Kotze, J.M., Korsten, L. 1998. Control of bacterial black spot of mango. *South African Mango Growers Association* 18: 35–38.

Brisset, M.N., Cesbron, S., Thompson, V., Paulin, J.P. 2000. ASM induces the accumulation of defense-related enzymes in apple and protects from fire blight. *European Journal of Plant Pathology* 106: 529–536.

Buonaurio, R., Servili, M. 1999. Involvement of lipoxygenase, lipoxygenase pathway volatiles, and lipid peroxidation during the hypersensitive reaction of pepper leaves to *Xanthomonas campestris* pv. *vesicatoraia*. *Physiological and Molecular Plant Pathology* 54: 155–169.

Cavalcanti, F.R., Resende, M.L.V., Lima, J.P.M.S. 2006. Activities of antioxidant enzymes and photosynthetic responses in tomato pre-treated by plant activators and inoculated by *Xanthomonas vesicatoria*. *Physiological and Molecular Plant Pathology* 68: 198–208.

Chandrashekar, S., Umesh, S. 2012. Induction of antioxidant enzymes associated with bacterial spot pathogenesis in tomato. *International Journal of Agricuture and Veterinary Science* 2: 22–34.

Chen, C., Belanger, R.R., Benhamou, N., Paulitz, T. 2000. Defense enzymes induced in cucumber roots by treatment with plant-growth promoting rhizobacteria (PGPR) and *Pythium aphanidermatum*. *Physiological and Molecular Plant Pathology* 56: 13–23.

Chithrashree, R., Udayasankar, A.C., Nayaka, S.C. 2011. Plant growth promoting rhizobacteria mediate induced systemic resistance in rice against bacterial leaf blight caused by *Xanthomonas oryzae* pv. *oryzae*. *Biological Control* 59: 114–122.

Chittoor, J.M., Leach, J.E., White, E.F. 1999. Differential induction of a peroxidise during defense against pathogens. In *Molecular Biology Intelligence Unit: Pathogeneis-Related Proteins in Plants*, Datta, S.K., Muthukrishnan, S.M. (eds.). New York: CRC Press, pp. 171–193.

Collinge, D.B., Kragh, K.M., Mikkelsen, J.D. 1993. Plant chitinases. *Plant Journal* 3: 31–40.

Croft, K.P.C., Juttner, F., Slusarenko, A.J. 1993. Volatile products of the lipoxygenase pathway evolved from *Phaseolus vulgaris* (L) leaves inoculated with *Pseudomonas syringae* pv. *phaseolicola*. *Plant Physiology* 101: 13–24.

Daayf, F., Schmitt, A., Belanger, R.R. 1997. Evidence of phytoalexins in cucumber leaves infected with powdery mildew following treatment with leaf extracts of *Reynoutria sachalinensis*. *Plant Physiology* 113: 719–727.

Deborah, S.D., Palanisamy, A., Vidhyasekaran, P., Velazhahan, R. 2001. Time-course study of the induction of defense enzymes, phenolics and lignin in rice in response to infection by pathogen and non-pathogen. *Journal of Plant Diseases and Protection* 108: 204–216.

Do, H.M., Hong, J.K., Jung, H.W. 2003. Expression of peroxidise-like genes, H_2O_2 production and peroxidise activity during the hypersensitive responses to *Xanthomonas campestris* pv. *vesicatoria* in *Capsicum annum*. *Molecular Plant Microbe Interaction* 16: 196–205.

Dow, M., Newman, M.A., Roepenack, E.V. 2000. The induction and modulation of plant defence responses by bacterial lipopolysaccharides. *Annual Review of Phytopathology* 38: 241–261.

Faize, M., Burgos, L., Faize, L., Petri, C., Barba-Espin, G., Diaz-Vivancos, P., Clemente-Moreno, M.J., Alburquerque, N., Hernandez, J.A. 2012. Modulation of tobacco bacterial disease resistance using cytosolic ascorbate peroxidase and Cu, Zn-superoxide dismutase. *Plant Pathology* 61: 858–866.

Ferre, R., Badosa, E., Feliu, L. 2006. Inhibition of plant-pathogenic bacteria by short synthetic cecropin A-melittin hybrid peptides. *Applied and Environmental Microbiology* 72: 3302–3308.

Ganz, T., Lehrer, R. I. 1998. Antimicrobial peptides of vertebrates. *Current Opinion in Immunology* 10: 41–44.

Garcia-Olmedo, F., Molina, A., Alamillo, J.M. 1998. Plant defense peptides. *Biopolymers* 47: 479–491.

Goldberg, R., Imberty, A., Liberman, M., Prat, R. 1986. Relationships between peroxidase activities and cell wall plasticity. In: *Molecular and Physiological Aspects of Plant Peroxidases*, Greppin, H., Peneland, C., Gaspar, T (eds.). Geneva: University of Geneva, pp. 208–220.

Graham, L.S., Sticklen, M.B. 1994. Plant chitinase. *Canadian Journal of Botany* 72: 1057–1083.

Guyot, S., Vercauteren, J., Chynier, V. 1996. Structural determination of colourless and yellow dimmers resulting from catechin coupling catalysed by grape polyphenoloxidase. *Phytochemistry* 42: 1279–1288.

Hammerschmidt, R., Nucckles, R., Kuc, J. 1982. Association of enhanced peroxidase activity with induced systemic resistance of cucumber to *Colletotrichum largenarium*. *Physiological Plant Pathology* 20: 73–82.

Hancock, R.E.W., Lehrer, R. 1998. Cationic polypeptides: A new source of antibiotics. *Trends in Biotechnology* 16: 82–88.

Haruta, M., Pedersen, J.A., Constabel, C.P. 2001. Polyphenol oxidase and herbivore defence in trembling aspen (*Populus tremuloides*): cDNA cloning, expression, and potential substrates. *Physiologia Plantarum* 112: 552–558.

Hilaire, E., Young, S.A., Willard, L.H. 2001. Vascular defense responses in rice: Peroxidise accumulation in xylem parenchyma cells and xylem wall thickening. *Molecular Plant-Microbe Interaction* 14: 1411–1419.

Hong, J.K., Hwang, B.K. 2006. Promoter activation of pepper class II basic chitinase gene, CAChi2, and enhanced bacterial disease resistance and osmotic stress tolerance in the CAChi2-overexpressing *Arabidopsis*. *Planta* 223: 433–448.

Hong, J.K., Jung, H.W., Kim, Y.J., Hwang, B.K. 2000. Pepper gene encoding a basic class II chitinase is inducible by pathogen and ethephon. *Plant Science* 159: 39–49.

Jetiyanon, K., Kloepper, J.W. 2002. Mixtures of plant growth promoting rhizobacteria for induction of systemic resistance against multiple plant diseases. *Biological Control* 24: 285–291.

Jetiyanon, K., Tuzun, S., Kloepper, J.W. 1997. Lignification, peroxidise and super oxide dismutases as early plant defence reactions associated with PGPR-mediated induced systemic resistance. In *Plant Growth-Promoting Rhizobacteria—Present Status and Future Prospects*. Ogoshi, A., Kobayashi, K., Homma, Y., Kodama, F., Kondo, N., Akino, S. (eds.). Sapporo: Nakanishi Printing, pp. 265–268.

Jiménez, M., García-Carmona, F. 1999. Myricetin, an antioxidant flavonol, is a substrate of polyphenol oxidase. *Journal of the Science of Food and Agriculture* 79: 1993–2000.

Kamal, A.M.A., Ibrahim, Y.E., Balabel, N.M. 2012. Induction of disease defensive enzymes in response to treatment with acibenzolar-*S*-methyl (ASM) and *Pseudomonas fluorescens* Pf2 and inoculation with *Ralstonia solanacearum* race 3, biovar 2 (phylotype II). *Journal of Phytopathology* 160: 382–389.

Kandan, A.M., Ramiah, R., Radja commare, A. 2002. Induction of phenyl propanoid metabolism by *Pseudomonas fluorescens* against tomato spotted wilt virus in tomato. *Folia Microbiologica* 47: 121–129.

Keefe, D., Hinz, U., Meins, F. 1990. The effect of ethylene on the cell-type specific and intracellular localization of β-1,3 glucanase and chitinase in tobacco leaves. *Planta* 182: 43–51.

Kim, M.S., Kim, Y.C., Cho, B.H. 2004. Gene expression analysis in cucumber leaves primed by root colonization with *Pseudomonas chlororaphis* O6 upon challenge inoculation with *Corynespora cassiicola*. *Plant Biology* 6: 105–108.

Kloepper, J.W., Reddy, M.S., Rodriguez-Kabana, R. 2004. Application of rhizobacteria in transplant production and yield enhancement. *Acta Horticulturae* 631: 219–229.

Kloepper, J.W., Tuzun, S., Kuc, J.A. 1992. Proposed definitions related to induced disease resistance. *Biocontrol Science and Technology* 2: 349–351.

Kowalski, S.P., Eannetta, N.T., Hirzel, A.T., Steffens, J.C. 1992. Purification and characterization of polyphenol oxidase from glandular trichomes of *Solanum berthaultii. Plant Physiology* 100: 677–684.

Kuc, J. 1995. Induced systemic resistance—An overview. In *Induced Resistance to Disease in Plants*. Hammerschmidt, R., Kuc, J. (eds.). Dordecht: Kluwer, pp. 169–175.

Kumar, N., Ebel, R.C., Roberts, P.D. 2011. SOD activity in *Xanthomonas axonopodis* pv. *citri* infected leaves of kumquat. *Journal of Horticultural Science and Biotechnology* 86: 62–68.

Kumari, V., Srivastava, J.S. 1999. Molecular and biochemical aspects of rhizobacterial ecology with special emphasis on biological control. *World Journal of Microbiology and Biotechnology* 15: 535–543.

Lawton, K., Freidrick, M., Hunt, K. 1996. Benzothiadozole induces disease resistance in *Arabidopsis* by activation of systemic acquired resistance signal transduction pathway. *The Plant Journal* 10: 71–82.

Leach, J.E., Sherwood, J., Fulton, R.W., Sequeira, L. 1983. Comparison of soluble proteins associated with disease resistance induced by bacterial lipopolysaccharide and viral necrosis. *Physiological Plant Pathology* 23: 377–385.

Lerouge, P., Roche, P., Faucher, C., Maillet, R., Truchet, G., Prome, J.C., Denarie, J. 1990. Symbiotic host-specificity of *Rhizobium melliloti* is determined by a sulphated and acylated glycosamine oligosaccharide signal. *Nature* 344: 781–784.

Li, L., Steffens, J.C. 2002. Overexpression of polyphenol oxidase in transgenic tomato plants results in enhanced bacterial disease resistance. *Planta* 215: 239–247.

Liu, L., Kloepper, J.W., Tuzun, S. 1995. Induction of systemic resistance in cucumber against bacterial angular leaf spot by plant growth-promoting rhizobacteria. *Phytopathology* 85: 843–847.

Loper, J.E., Henkels, M.D., Roberts, R.G. 1991. Evaluation of Streptomycin, oxytetracycline and copper resistance of *Erwinia amylovora* isolated from pear orchards in Washington state. *Plant Disease* 75: 287–290.

Louws, M., Wilson., H.L., Campbell, D.A. 2001. Field control of bacterial spot and bacterial speck of tomato using a plant activator. *Plant Disease* 85: 481–488.

Majeau, N., Trudel, J., Asselin, A. 1990. Diversity of cucumber chitinase isoforms and characterization of one seed basic chitinase with lysozyme activity. *Plant Science* 68: 9–16.

Martinez, C., Geiger, J.P., Bresson, E. 1996. Isoperoxidase are associated with resistance of cotton to *Xanthomonas campestris* pv. *malvacearum* (race 18). In *Plant Peroxidase: Biochemistry and Physiology* Obinger, O., Burner, U., Eberman, R., Penel, C., Greppin, H. (eds.). Switzerland: University of Geneva, and Austria: University of Vienna pp. 327–332.

Marusek, C.M., Trobaugh, N.M., Flurkey, W.H., Inlow, J.K. 2006. Comparative analysis of polyphenol oxidase from plant and fungal species. *Journal of Inorganic Biochemistry* 100: 108–123.

Mauch, F., Hadwiger, L.A., Boller, T. 1988. Antifungal hydrolases in pea tissue: II. Inhibition of fungal growth by combinations of chitinase and β-1,3 glucanases differentially regulated during development and in response to fungal infection. *Plant Physiology* 87: 325–333.

Mayer, A.M. 1987. Polyphenol oxidases in plants—Recent progress. *Phytochemistry* 26: 11–20.

Mayer, A.M. 2006. Polyphenol oxidases in plants and fungi: Going places? A review. *Phytochemistry* 67: 2318–2331.

Mayer, A.M., Harel, E. 1979. Polyphenol oxidase in plants. *Phytochemistry* 18: 193–215.

Meins, F., Ahl, P. 1989. Induction of chitinase and beta-1,3 glucanase in tobacco plants infected with *Pseudomonas tabaci* and *Phytophthora parasitica* var. *nicotianae. Plant Science* 61: 155–161.

Melo, G.A., Shimizu, M.M., Mazzafera, P. 2006. Polyphenol oxidase activity in coffee leaves and its role in resistance against coffee leaf miner and coffee leaf rust. *Phytochemistry* 67: 277–285.

Metraux, J.P., Burkhart, W., Boyer, M. 1989. Isolation of a complementary DNA encoding a chitinase with structural homology to a bifunctional lysozyme/chitinase. *Proceedings of National Academy of Science USA* 86: 896–900.

Metraux, J.P., Streit, L., Staub, T. 1988. A pathogenesis-related protein in cucumber plants in response to viral, bacterial and fungal infections. *Physiological and Molecular Plant Pathology* 33: 1–9.

Mishina, T.E., Zeier, J. 2007. Pathogen-associated molecular pattern recognition rather than development of tissue necrosis contributes to bacterial induction of systemic acquired resistance in *Arabidopsis. Plant Journal* 50: 500–513.

Momol, M.T., Norelli, J.L., Aldwinckle, H.S. 1999. Evaluation of biological control agents, systemic acquired resistance inducers and bactericides for the control of fire blight on apple blossom. *Acta Horticulturae* 489: 553–557.

Mustafa, M., Yesim, A., Ozden, C. 2008. Biological control of bacterial spot disease of pepper with *Bacillus* strains. *Turkish Journal of Agriculture and Forestry* 32: 381–390.

Newman, M.A., Daniels, M.J., Dow, J.M. 1995. Lipopolysaccharide from *Xanthomonas campestris* induces defense related gene expression in *Brassica campestris*. *Molecular Plant-Microbe Interactions* 8: 778–780.
Ongena, M., Jacques, P. 2008. *Bacillus* lipopeptides: Versatile weapons for plant disease biocontrol. *Trends in Microbiology* 16: 115–125.
Ongena, M., Jourdan, E., Adam, A., Paquot, M. 2007. Surfactin and fengycin lipopeptides of *Bacillus subtilis* as elicitors of induced systemic resistance in plants. *Environmental Microbiology* 9: 1084–1090.
Park, K.S., Kloepper, J.W. 2000. Activation of PR-1a promoter that induce systemic resistance in tobacco against *Pseudomonas syringae* pv. *tabci*. *Biological Control* 18: 2–9.
Press, C.M., Wilson, M., Tuzun, S., Kloepper, J.W. 1997. Salicylic acid produced by *Serratia marcescens* is not the primary determinant of induced systemic resistance in cucumber or tobacco. *Molecular Plant-Microbe Interactions* 10: 761–768.
Raj, S.N., Sarosh, B.R., Shetty, H.S. 2006. Induction and accumulation of polyphenol oxidase activities as implicated in development of resistance against pearl millet downy mildew disease. *Functional Plant Biology* 33: 563–571.
Rajendran, L., Saravanakumar, D., Raguchandar, T., Samiyappan, R. 2006. Endophytic bacterial induction of defense enzymes against bacterial blight of cotton. *Phytopathologia Mediterranea* 45: 203–214.
Ramamoorthy, V., Raghuchander, T., Samiyappan, R. 2002. Induction of defense related proteins in tomato roots treated with *Pseudomonas fluorescens* Pf1 and *Fusarium oxysporum* f. sp. *lycopersici*. *Plant and Soil* 239: 55–68.
Ran, L.X., Li, Z.N., Wu, G.J. 2005. Induction of systemic resistance against bacterial wilt in *Eucalyptus urophylla* by fluorescent *Pseudomonas* spp. *European Journal of Plant Pathology* 113: 59–70.
Reimers, P.J., Guo, A., Leach, J.E. 1992. Increased activity of a cationic peroxidase associated with an incompatible interaction between *Xanthomonas oryzae* pv. *oryzae* and rice (*Oryzae sativa*). *Plant Physiology* 99: 1044–1050.
Resende, M.L.V., Costa, J.C.B., Cavalcanti, F.R. 2007. Selecao de extratos vegetais para inducao de resistancia e ativacao de respostas de defesa em cacaueiro contra a vassuaro-de-bruxa. *Fitopathologia Brasileria* 32: 213–221.
Riberiro Jr., P.M., Perira, R.B., Zacarino, A.B. 2004. Lignificacao induzida por extratos aquosos naturalise producos comercias em tomatorieo infectado por *Xanthomonas campestris* pv. *vesicatoria*. In *Proceeding of the Congresso Brazialiaro de Fitopathologia Gramado*, Brazil, pp. 261–261.
Roberto, L.-F., Souza, R.M., Megalhaes, M.M. 2013. Induced defence responses in tomato against bacterial spot by proteins synthesized by endophytic bacteria. *Tropical Plant Pathology* 38: 295–302.
Romeiro, R.S., Kimura, O. 1997. Induced resistance in pepper leaves infiltrated with purified elicitors from *Xanthomonas campestris* pv. *vesicatoria*. *Journal of Phytopathology* 145: 495–498.
Romeiro, R.S., Lanna-Filho, R., Vieira-Junior, J.R. 2010. Evidence that the biocontrol agent *Bacillus cereus* synthesizes protein that can elicit increased resistance of tomato leaves to *Corynespora cassiicola*. *Tropical Plant Pathology* 35: 11–15.
Ryan, R.P., Germaine, K., Franks, A., Ryan, D.J., Dowling, D.N. 2008. Bacterial endophytes: Recent developments and applications. *FEMS Microbiology Letters* 278: 1–9.
Ryu, C.M., Murphy, J.F., Mysore, K.S., Kloepper, J.W. 2004. Bacterial volatiles induce systemic resistance in *Arabidopsis*. *Plant Physiology* 134: 1017–1026.
Saikia, R., Kumar, R., Arora, D.K. 2006. *Pseudomonas aeruginosa* inducing rice resistance against *Rhizoctonia solani*: Production of salicylic acid and peroxidases. *Folia Micobiologica* 51: 375–380.
Saikia, R., Kumar, R., Singh, T., Srivatsava, A.K. 2004. Induction of defense related enzymes and pathogenesis related proteins in *Pseudomonas fluorescens*-treated chickpea in response to infection by *Fusarium oxysporum* f. sp. *ciceri*. *Mycobiology* 32: 47–52.
Saparrat, M.C.N., Guiller, F. 2005. Lignolitic ability and potential biotechnology applications of the South American fungus *Pleurotus lacioniatocrenatus*. *Folia Microbiologica* 50: 155–160.
Schmid, P.S., Feucht, W.1980. Tissue-specific oxidation browning of polyphenols by peroxidase in cherry shoots. *Gartenbauwisenschaft* 45: 68–73.
Seleim, M., Abo-Elyosi, A., Mohamed, A., Al-Mazoky, A. 2014. Peroxidase and polyphenol oxidase activities as biochemical markers for biocontrol efficacy in the control of tomato bacterial wilt. *Journal of Plant Physiology and Pathology* 2: 1–6.
Shinshi, H., Neuhaus, J.-M., Ryals, J., Meins, F. 1990. Structure of a tobacco endochitinase gene: Evidence that different chitinase genes can arise by transposition of sequences encoding a cysteine rich domain. *Plant Molecular Biology* 14: 357–368.
Silva, H.S., Rumetro, R.S., Cakker, R. 2004. Induction of systemic resistance by *Bacillus cereus* against tomato foliar diseases under field conditions. *Journal of Phytopathology* 152: 371–375.

Silva, V.N., Silva, L.E.S.F., Martinez, C.R. 2007. Estripes de Paenibacillus promovem a nodulacao especifica na simbiose Bradyrhizobium-caupi. *Acta Scientiarum Agronomy* 29: 331–338.

Soylu, S., Brown, I., Mansfield, J.W. 2005. Cellular reactions in *Arabidopsis* following challenge by strains of *Pseudomonas syringae*: From basal resistance to compatibility. *Physiological and Molecular Plant Pathology* 66: 232–243.

Strobel, G., Daisy, B., Castillo, U., Harper, J. 2004. Natural products from endophytic microorganisms. *Journal of Natural Products* 67: 257–268.

Sundin, G.W., Bender, C.L. 1993. Ecological and genetic analysis of copper and streptomycin resistance in *Pseudomonas syringae* pv. *syringae*. *Applied and Environmental Microbiology* 59: 1018–1024.

Thipyapong, P., Hunt, M.D., Steffens, J.C. 2004. Antisense downregulation of polyphenol oxidase results in enhanced disease susceptibility. *Planta* 220: 105–117.

Tjamos, S.E., Flemetakis, E., Papalomatas, E.J., Katinakis, P. 2005. Induction of resistance to *Verticillium dahlia* in *Arabidopsis thaliana* by the biocontrol agent K-165 and pathogenesis-related proteins gene expressions. *Molecular Plant-Microbe Interactions* 18: 555–561.

Tuzun, S. 2001. The relationship between pathogen-induced systemic resistance (ISR) and multigenic (horizontal) resistance in plants. *European Journal Plant Pathology* 107: 85–93.

Van Loon, L.C., Bakker, P.A.H.M., Pieterse, M.J. 1998. Systemic resistance induced by rhizosphere bacteria. *Annual Review of Phytopathology* 81: 453–483.

Van Peer, R., Neiman, G.J., Schippers, B. 1991. Induced resistance and phytoalexin accumulation in biological control of Fusarium wilt of carnation by *Pseudomonas* sp. strain WCS5147. *Phytopathology* 81: 728–734.

Vidyasekaran. P., Kamala, N., Ramanathan, A. 2001. Induction of systemic resistance by *Pseudomonas fluorescens* Pf1 against *Xanthomonas oryzae* pv. *oryzae* in rice leaves. *Phytoparasitica* 29: 155–166.

Voisey, C.R., Slusarenko, A.J. 1989. Chitinase mRNA and enzyme activity in *Phaseolus vulgaris* (L.) increase more rapidly in response to avirulent than to virulent cells of *Pseudomonas syringae* pv. *phasolicola*. *Physiological Molecular Plant Pathology* 35: 403–412.

Walters, D., Walsh, D., Newton, A., Lyon, G. 2005. Induced disease resistance for plant disease control: Maximizing the efficacy of resistance elicitors. *Phytopathology* 95: 1368–1373.

Wei, G., Kloepper, J.W., Tuzun, S. 1991. Induction of systemic resistance of cucumber to *Colletotrichum orbiculare* by select strains of plant growth promoting rhizobacteria. *Phytopathology* 81: 151512.

Wojtasjeck, P. 1997. Oxidative burst, an early plant response to pathogen infection. *Biochemistry Journal* 322: 681–692.

Wuyts, N., De Waele, D., Swennen, R. 2006. Extraction and partial characterization of polyphenol oxidase from banana (*Musa acuminate* Grande naine) roots. *Plant Physiology and Biochemistry* 44: 308–314.

Zasloff, M. 2002. Antimicrobial peptides of multicellular organisms. *Nature* 415: 389–395.

Zipfel, C., Robatzek, S., Navarro, L., Edward, J., Jones, G., Felix, G., Boller, T. 2004. Bacterial disease resistance in *Arabidopsis* through flagellin perception. *Nature* 428: 764–767.

17 Plant Pathogenic Bacteria Control through Seed Application

Yesim Aysan and Sumer Horuz

CONTENTS

ABSTRACT Seeds, the initial source of planting, have had great significance in agriculture for centuries. The value of this role has been recognized not only by producers but also by scientists, who have had high input into the production of healthy plants with high performance and yield. To achieve these goals and also to enhance profitability in agriculture, seeds must be healthy both externally and internally. Since the science of seed pathology has gained more attention, control of seed-borne pathogens and diseases has been attained via integrated disease management strategies. One of these is seed treatments, in which mechanical, biological, physical, or chemical treatments are applied for the avoidance of seed infections. Seed treatments can reduce or eliminate pathogen populations in/on seeds without affecting germination performance. No treatment can completely clean seeds of infections, but treatment can reduce the pathogen inoculum. All measures for seed control through application should be combined in integrated management programs.

KEYWORDS: seed, pathogen, bacteria, treatment, control

17.1 INTRODUCTION

Seed treatments are used on many crop seeds for a variety of purposes. Generally, mechanical, biological, physical, or chemical methods are used to eradicate pathogens not only on/in seeds but also on/in seedlings, bulbs, and other propagation materials, or to reduce inoculum concentrations in order to grow healthy plants. The goal of treatment is to prevent seed infestations on seeds or soil microorganisms. Seed treatments are possibly the cheapest and frequently the safest methods for direct control of plant diseases (Agarwall and Sinclair, 1987; Neergaard, 1988; Maude, 1996).

Seed treatments have been used for centuries and have been widely used on a commercial basis for decades. However, seed treatments that may be profitable for one crop may not be economically reasonable for another. The expense of a seed treatment that may be appropriate for a given crop is related to both the cost of the seed and its intended use. Some seeds (e.g., hybrid flower seeds) are extremely expensive and are produced in small quantities. Such seeds are seldom treated because the risk of damage to them exceeds the expected benefits. Moreover, pesticide registrations are extremely unlikely for such small usage (Taylor and Harman, 1990).

Since 2000 the use of seed treatment has rapidly accelerated and evolved, driven by diverse factors. Increased global trade and possibly climate change have promoted the emergence or re-emergence of diseases and pests in new locales. For a variety of reasons, crop production practices have moved toward tactics that increase the risks for seed decay, seedling disease, and early-season insect attack (Broders et al., 2007a,b; Butzen and Doerge, 2007; Bradley, 2008). These factors and others have resulted in a doubling of the value of the global seed treatment market between 2002 and 2008; in 1995, annual revenue had already exceeded USD $2 billion (McGee, 1995).

The central role of seeds in agriculture has always been recognized, but the importance of this role has been greatly heightened since 1900 and especially since 2000. Several developments have catalyzed the elevation of seeds to the status of most valuable agricultural input (Munkvold, 2009). For decades, in order to grow healthy plants with high yields, producers and especially scientists focused on seeds as the initial source materials in planting. To achieve these goals, seed treatments have been used for decades as seed treatments. Finally, seed treatments in commercial agriculture are performed to enhance profitability. Several disease management practices are currently in use as seed treatments. Their main categories are mechanical, biological, physical, and chemical.

17.2 MECHANICAL TECHNIQUES

Seed lots can contain soil fragments, plant debris, and pathogen particles. Seed color, form, and morphology may alter owing to pathogen infections in seed lots. Seed lots therefore must be cleaned mechanically, using techniques such as storage or sieving in controlled conditions to help reduce pathogen inoculum. Contamination by *Xanthomonas campestris* pv. *malvacearum* on mild, worthless, and pale cotton seeds has been eliminated by floating the seeds (Brinkerhoff and Hunter, 1963). Seed extraction helped to reduce the inoculum incidence of *Clavibacter michiganensis* subsp. *michiganensis* on tomato seeds (Procter and Fry, 1965; Thyr et al., 1973).

17.3 BIOLOGICAL TECHNIQUES

The principle consists of reducing the inoculum of a pathogen or microorganism and/or its ability to cause destructive diseases through one or more biological control microorganisms. Advances in biopesticide agents are attributable, in part, to research on the mechanisms of control, particularly in the soil environment or rhizosphere (Cook, 1993; Harman and Nelson, 1994). Numerous reports have been made of potentially valuable biological control microorganisms, some of which are supplied as seed treatments, but the developmental process to bring these into commercial practice is long and arduous (Cook, 1993). The pursuit of alternatives to chemical pesticides and an increasing interest in the so-called organic (i.e., bio-friendly) production methods since 1990 have stimulated scientific development of biological control agents. Advances have been achieved through a greater understanding of the control mechanisms used by these agents, especially in soil (McGee, 1996).

Cotton was the first large-scale agronomic crop treated with biological control agents for the suppression of seedling diseases that originate in the rhizosphere. Much of the cotton planted in the United States is treated with one or more biological control agents (Brannen and Kennedy, 1996). Generally, it is stated that in control strategies used, biological control is much more variable and less effective than pesticides. The primary reason for this discrepancy is that biological agents used in biological control do not have the ability to survive or compete with other living organisms in the ecological environment (Harman, 1991).

Several biocontrol bacteria and fungi belonging to diverse genera such as *Bacillus*, *Penicillium*, *Enterobacter*, *Pseudomonas*, *Serratia*, *Chaetomium*, and *Trichoderma* are widely used as seed treatments (Agarwall and Sinclair, 1987; Gordon-Lennox et al., 1987; Harman and Nelson, 1994). Of the biological control agents patented by early 1999, 84% were bacteria and 16% were fungi. The bacteria included species of *Streptomyces*, *Pseudomonas*, *Bacillus*, and *Enterobacter*. Species of *Pseudomonas* and *Bacillus* made up the vast majority of these products. Fungi products consisted of various species of *Phomopsis*, *Ectomycorrhizae*, *Trichoderma*, *Cladosporium*, and *Gliocladium* (Duvert, 1999).

Some biological control microorganisms have come into commercial use. Quantum 4000™, which includes the bacterium *Bacillus subtilis*, is sold in the United States (Connick et al., 1990). Mycostop, which contains *Streptomyces griseoviridis*, is a commercial formulation used in Bulgaria, Finland, and Hungary (Mohammadi, 1992). *Pseudomonas cepacia* and *Agrobacterium radiobacter* are known to participate in biological management as seed or plant material treatments (Rhodes and Powell, 1994). System 3, which contains the bacterium *Bacillus subtilis* BG03, and Rhizo-Plus, which contains *Bacillus subtilis* FZB24, are used in biological control as seed treatments. It was stated that the biocontrol agent *Pantoae agglomerans* LRC8311 as a seed treatment inhibited both disease severity and incidence of bacterial wilt of bean caused by *Curtobacterium flaccumfaciens* pv. *flaccumfaciens* (Hsieh et al., 2005a; Huang et al., 2007).

A number of modes of action used by these microorganisms lead to seed and seedling protection. These can be loosely categorized into the areas of antagonism, antibiosis, competition, and mycoparasitism (Mukhopadhyay, 1994). Each of these modes of action has advantages and disadvantages that affect performance. The key disadvantage is that any single mode of action works against a very narrow spectrum of pathogens. In addition, the majority of biological control agent products to date have focused on one disease. Mixtures of organisms with different modes of action or combinations of chemicals and biological control agents might enhance a treatment's spectrum of activity; however, knowledge and understanding of the interactions of such mixtures remain insufficient.

Several types of microbes promote plant growth. The beneficial effects on plant growth by such bacteria, which are generally designated as PGPR (plant growth promoting rhizobacteria), can be direct or indirect. Direct plant growth promotions include biofertilization, stimulation of root growth, rhizoremediation, and plant stress control. Mechanisms of biological control by which rhizobacteria can promote plant growth indirectly (i.e., by reducing the level of disease) include antibiosis, induction of systemic resistance, and competition for nutrients and niches. Most researchers in this area focus on PGPR (Lugtenberg and Kamilova, 2009).

Biocontrol agents differ from the pesticides that are applied to seeds in that they are living organisms with constant reproduction features. In order to increase biological efficiency of these agents, genetically superior strains should be preferred; in addition, efficient seed-application systems must be developed to provide convenient niches and reduce competition with microflora (Harman, 1991; Rhodes and Powell, 1994). A biological agent must decisively compete with a plant pathogen, be affordable and easily applicable for users, and be storable in the long term. Biological agents should stay alive at room temperature for more than a year and effect on wide lands in a wide host plants.

17.4 PHYSICAL TECHNIQUES

The physical method of thermotherapy is based on distinct heat treatments that are meant to reduce pathogen infections in/on seeds or other propagation materials. This commonly used archaic method was quite popular before the advent of chemical treatments (Hermansen and Jorgensen, 1983). Its main principle is the elimination of seed-borne pathogens using various types of heat treatments.

Less-moist seeds are more resistant to such approaches, whereas seeds with high moisture can deceive vigor ability even at low temperatures (Baker, 1962a,b). Because it has been broadly demonstrated that the ability to transport heat by hot water is higher than by saturated water vapor, the duration of both exposure and temperature must be increased when seeds are exposed to steam or dry hot air. Pathogen decrease due to heat treatments has been attributed to protein deterioration, lipid secretion, hormones, stock material consumption, and oxidation (Baker, 1962a,b; Sykes, 1965).

The humidity content, dormancy, age, vigor, and physical properties of seeds, as well as the inoculum type, its concentration, and environmental conditions are the main factors that affect the practicability of physical methods (Baker, 1962a,b). The disadvantages of physical methods are a reduction of seed germination, cracks on seed test, under-performance compared with chemicals, as-yet-unknown mechanisms, failure to eradicate inoculum deep in seeds, and difficulties in carrying out treatment.

17.4.1 Hot-Water Treatments

Hot-water treatments can reduce the incidence of seed-borne pathogens but can cause loss of seed viability as well (Pyndji et al., 1987; Strandberg and White, 1989). The ability of hot water to eliminate low infections of epidemiologically significant bacteria is also limited (Williams, 1980). Appropriate cascade processes in hot-water treatment are as follows: (i) the heat level is increased promptly to suppress the pathogen; (ii) the duration of the treatment eliminates the pathogen but does not affect seed viability; (iii) cessation of treatment is instant and timely; and (iv) the seeds are allowed to dry after treatment. Hot-water treatments are effective to eliminate seed-borne pathogens, but subsequent seed germination may be adversely affected. Challenges in hot-water treatment approaches include setting the heat level such that no harm is caused to the seeds, and deciding the appropriate time duration for treatment. The majority of hot-water treatments involve seed exposure at around 50°C for a few minutes. Fatmi et al. (1991) reported that germination considerably decreased in tomato seeds that had been subjected to 56°C water. Ivey and Miller (2005) demonstrated that hot-water treatment in tomato seeds increasingly reduced a pathogen inoculum and that the quality or quantity of the seedlings from the treated tomato seeds was increased as well. Ozaktan (1991) suggested that hot-water (54–56°C)-treated tomato seeds inhibited *Clavibacter michiganensis* subsp. *michiganensis*. Recommended hot-water treatments for some bacterial pathogens are listed in Table 17.1.

17.4.2 Dry-Heat Treatment

The heat spread that is achieved with dry heat, which involves the formation of a hot-air system, is five times less effective than the heat spread achieved in hot-water treatments (Baker, 1962a,b).

TABLE 17.1
Controlled Seed-Borne Bacterial Pathogens with Hot-Water Treatments and Their Use

Source	Bacterial Pathogen	Recommended Hot-Water Treatment
Pea	*Pseudomonas syringae* pv. *pisi*	55–60°C/15 min
Cowpea	*Xanthomonas campestris* pv. *phaseoli*	52°C/20 min
Wheat	*Xanthomonas campestris* pv. *translucens*	45°C/20 min
Tomato	*Clavibacter michiganensis* subsp. *michiganensis*	53°C/60 min, 50–56°C/25–30 min
Crucifers	*Xanthomonas campestris* pv. *campestris*	50–52°C/30 min
Tobacco	*Erwinia carotovora* pv. *carotovora*	50°C/12 min
Wallflower	*Xanthomonas campestris* pv. *incanae*	54–55°C/10 min

TABLE 17.2
Controlled Seed-Borne Bacterial Pathogens with Dry-Heat Treatments and Their Use

Source	Bacterial Pathogen	Recommended Dry-Heat Treatment
Barley	*Xanthomonas campestris* pv. *translucens*	72°C/4 d
		72°C/7 d
Pea	*Pseudomonas syringae* pv. *pisi*	65°C/1d
Rice	*Pseudomonas fuscovaginae*	65°C/6 d
Rice	*Pseudomonas glumae*	65°C/2 d
Tomato	*Clavibacter michiganensis* subsp. *michiganensis*	80°C/1 h
		76–78°C/2 d
Bean	*Pseudomonas syringae* pv. *phaseolicola*	60°C/1 d
		50°C/3 d
Cucumber	*Pseudomonas syringae* pv. *lachrymans*	70°C/3 d

This technique is utilized to avoid spontaneous or artificial bacterial contamination and to reduce the pathogen inoculum in/on seeds. Dill seeds, for example, are exposed to 65–70°C air for 3 d. However, seed germination can be damaged by this method (Maude, 1996). Hyacinth bulbs gradually exposed to hot air at 26–35°C for 10 d, 20–30°C for another 10 d, and then stored at 80°C, showed reduction in *Pseudomonas gladioli* contaminations. Recommended dry-heat treatments for control of some seed-borne bacterial pathogens are listed in Table 17.2.

17.4.3 Vapor-Heat Treatment

Seed-borne pathogens are eliminated by exposing seeds to vapor heat under a certain pressure. This technique, which is also used for soils and other propagation materials, has similarities to hot-water and dry-heat treatments. Some researchers claim that vapor heat is much safer than hot water although it is less efficient than dry heat. With this method, seeds can rapidly dry; tests are not damaged; and it has been reported seed germination is not heavily affected (Agarwall and Sinclair, 1987; Maude, 1996). However, it has also been reported that vapor-heat treatment is less satisfactory in the control of seed-borne bacterial pathogens and may reduce seed germination (Maude, 1996). Seeds are exposed to vapor at 50–60°C for 20–30 min. In some cases, vapor of seeds must be raised before the treatment for the penetration of the heat into the seeds (Baker, 1962a,b). Despite its abovementioned limitations, vapor heat treatment is routinely used for some bacterial pathogens (Table 17.3).

17.4.4 Radiation Treatment

Radiation treatments may be applied by sound, ultrasound, radio, microwave, infrared, visible light, ultraviolet, X, gamma, and cathode-ray spectra. In tropical or subtropical countries, solar-system LEDs are used to produce radiation for the control of soil-borne pathogens and weeds (Katan, 1981; Katan and De Vay, 1991). Keeping cowpea seeds in water for 4–5 h and spreading them on a floor

TABLE 17.3
Controlled Seed-Borne Bacterial Pathogens with Vapor-Heat Treatments and Their Use

Source	Bacterial Pathogen	Recommended Vapor Heat Treatment
Tomato	*Clavibacter michiganensis* subsp. *michiganensis*	56°C/30 min
Bean	*Pseudomonas syringae* pv. *phaseolicola*	55–60°C/30–60 min
Crucifers	*Xanthomonas campestris* pv. *campestris*	54°C/30 min
Nasturtium	*Rhodococcus fascians*	51°C/30 min

to dry in sunlight reduced contamination of *Xanthomonas campestris* pv. *vignicola* (Jindal et al., 1989). Laser and microwave radiation reduced *Erwinia carotovora* subsp. *carotovora* infection in tobacco seeds and *Pseudomonas syringae* pv. *glycinea* in soybean seeds (Krasnova, 1963; Hankin and Sands, 1977).

17.4.5 Fermentation

This technique is an old one used for tomato seeds, and is still used in non-commercial production to eliminate *C. michiganensis* subsp. *michiganensis* infections. Tomato seeds are kept at 20°C for 4 d, during which a white-type *Bacillus* spp. antagonistic bacteria grows that can kill bacteria on seed coats.

17.5 CHEMICAL TECHNIQUES

Chemical, physical, and biological seed treatments have changed since 1990 as a result of new fungicide chemistry. Economic and environmental considerations now require that chemical products be applied at low and efficient dosages (McGee, 1981). Chemical techniques are widely used to eliminate pathogens and eradicate pathogen inoculum density on/in seeds. Chemical treatments are commonly applied to reduce fungal infections, but have had limited success on bacterial and viral infections (McGee, 1995).

Antibiotics encompass a chemically heterogeneous group of organic, low-molecular-weight compounds, produced by microorganisms, that are deleterious to the growth or metabolic activities of other microorganisms. Antibiotics generally act on several vital processes in microbes, including cell-wall biosynthesis and the syntheses of DNA, RNA, and protein (Raaijmakers and Mazzola, 2012). In the 1950s, plant pathologists quickly recognized the potential of antibiotics for the treatment of plant diseases, especially those caused by bacteria; approximately 40 antibiotics of bacterial or fungal origin were screened for plant disease control in that decade alone (Dekker, 1957; Goodman, 1959; Tomli, 1994). Streptomycin, an aminoglycoside antibiotic, is used to control *Pectobacterium* spp. (formerly *Erwinia* spp.), which causes soft rot diseases in cut flowers and potato-seed pieces; *Pseudomonas cichorii*, which causes bacterial blight of celery; various pathovars of *P. syringae*, which cause fruit-spotting or blossom-blast symptoms on apple, pear, and related landscape trees; *X. campestris* pv. *vesicatoria*, which causes bacterial spot of pepper and tomato; and *Agrobacterium tumefaciens*, which causes crown gall of rose.

The emergence of streptomycin-resistant (SmR) plant pathogens has complicated the control of bacterial diseases in plants. In the United States, streptomycin is permitted on tomato and pepper for control of *X. campestris* pv. *vesicatoria*; however, it is rarely used for this purpose because resistant strains, first discovered in Florida in the early 1960s (Stall and Thayer, 1962), are now widespread (Minsavage et al., 1990). Antibiotic quantities and application periods are crucial (Maude, 1996). Bacterial disease control approaches that include antibiotic treatment are not preferred, owing to antibiotics' inadequate ability to penetrate seeds sufficiently to fully eradicate the inoculum source, and also to their phytotoxic actions and influences in seed germination (Klisiewcz and Pound, 1961; Ralph, 1977; Humaydan et al., 1980; Harman et al., 1987).

Antibiotics applied in polyethylene glycol (PEG) reduced infection by *X. campestris* pv. *phaseoli* in bean seeds, but were phytotoxic (Liang et al., 1992). Other chemical components, including sodium hypochlorite (NaOCl) (Strider, 1979), hydrochloric acid (HCl) (Maude, 1996; Sahin and Miller, 1997; Tireng-Karut and Aysan, 2011), and sodium hydroxide (NaOH) and sodium phosphate (Na_3PO_4) (Yorganci et al., 1993; Karsavuran et al., 2005) may prevent seed contamination by bacterial pathogens without adverse effects on seed viability.

HCl treatment of tomato seeds reduced the incidence of *C. michiganensis* subsp. *michiganensis*, but only in combination with other substances (Thyr et al., 1973; Dhanvantari, 1989; Fatmi et al., 1991; Pradhanang and Collier, 2009). Hopkins et al. (2003) suggested that wet seed treatment involving 30 min exposure to peroxyacetic acid at 1600 μg/mL, followed by seed drying at low humidity in a 40°C drying oven for 48 h, was effective in the elimination of bacterial fruit blotch disease caused by *Acidovorax citrulli*.

Some chemicals, including bronopol, mancozeb, hydroxychinolin sulfate, and thiram, are used to control seed-borne bacterial pathogens (Saygili et al., 2012). Mirik et al. (2005) suggested that bronopol, hydroxychinolin sulfate, and HCl significantly reduced *Xanthomonas axonopodis* pv. *vesicatoria* incidence on pepper seeds. Bronopol and hydroxychinolin sulfate were used to control *X. axonopodis* pv. *malvacearum* on cotton seeds.

Seed-applied crop-protection chemicals are an important and growing facet of crop management, but their full potential is yet to be realized. Because several aphid species and other insects that can be controlled by seed-applied insecticides are vectors of plant pathogens, the use of seed-applied insecticide has contributed to some reductions in disease transmission. In maize, Stewart's wilt caused by the bacterium *Pantoea stewartii* is an important quarantine pathogen that can be seed-transmitted. In addition, seed parent plants that are infected with this bacteria early in their development are more likely to produce infected seeds (Block et al., 1998 and 1999), and sweet corn plants infected early in their development are more likely to die or suffer yield loss through leaf blighting (Suparyono, 1989; Pataky et al., 1995). Therefore, seedling protection against corn flea beetle, the insect vector of *Pantoea stewartii*, is an important management component of both maize seed production and sweet corn production. Seed-applied neonicotinoid insecticides were shown in several studies to effectively prevent feeding by the corn flea beetle and to significantly reduce transmission of *P. stewartii* (Munkvold et al., 1996; Pataky et al., 2000; Kuhar et al., 2002; Andersch and Schwarz, 2003; Pataky et al., 2005). In cantaloupe, seed-applied imidacloprid reduced the severity of bacterial wilt, caused by *Erwinia tracheiphila*, through control of its vector, the striped cucumber beetle (Fleischer et al., 1998).

17.6 SEED TREATMENT METHODS USED IN ORGANIC FARMING

In the 1950s, rapid industrial and population growth worldwide resulted in increased hunger, which led scientists to utilize suitable agricultural lands for the development of chemical-free (i.e., organic) crops of high quality. This utilization was also prompted by widespread and intensive use of pesticides and fertilizers, risks posed by chemical residues, and disruptions of the chemical and physical structures of soils as well as the balances of mineral nutrition in soils (Zengin, 2007). Organic farming is now practiced in many places because it is environmentally healthy, and friendly to humans and other living organisms. Organic farming systems rely on approved practices for the control of plant diseases. Approved practices widely used by organic (and some conventional farmers) include the uses of disease resistant/tolerant cultivars as well as disease-reducing cultural strategies such as crop rotation and sanitation (Raudales and Brian, 2008).

Microbial biopesticides represent an important option for the management of plant diseases. The United States Environmental Protection Agency (EPA) defines biopesticides as certain types of pesticides derived from natural materials such as animals, plants, bacteria, and certain minerals. Based on its active ingredient, a biopesticide are categorized as microbial or biochemical, or as a plant-incorporated protectant (PIP). In general, biopesticides are less toxic, more target-specific, and/or decompose faster after application compared with conventional pesticides. All of these features support the idea that biopesticide application can result in less pollution than conventional chemical pesticides. One disadvantage is that the user of biopesticides must have a thorough understanding of the pest control needs of his or her crop. This is because most biopesticides have more limited target ranges than chemical pesticides.

To assist growers in choosing an appropriate microbial biopesticide the list in Table 17.1 was developed. It includes microbial biopesticides registered with the US EPA and certified for use in organic agriculture by the Organic Materials Review Institute (OMRI). When tomato seeds used in organic farming were subjected to seed treatments to eliminate *Clavibacter michiganensis* subsp.

michiganensis infections, NaOCl, vinegar, hot water, and lactic acid inhibited the pathogen inoculum without adverse effects on seed germination (Tireng-Karut and Aysan, 2011).

17.7 CONCLUSION

The objective of seed pathology is the production and dissemination of high-quality, disease-free seed that maximizes potential crop productivity and value. The use of seed treatments is increasing daily. Aside from the developments discussed above, continued growth in the small but important market in biological seed treatments (especially PGPRs) is expected. Such treatments include microorganisms and non-pesticidal chemicals that enhance plant defense responses against multiple pests/pathogens. Seed treatments with biological control agents provide an alternative to the use of chemical pesticides. Goals for biological seed treatment include control of soil-borne pathogens and their colonization, protection of subterranean plant parts (rhizosphere competence), and increased plant growth. Fungi and bacteria have been shown to accomplish these goals. However, results have been variable and success is not always achieved. In order to have healthy seeds, seed treatments should be used before growing to avoid diseases that can reduce crop quality and productivity.

REFERENCES

Agarwall, V.K., Sinclair, J.B. 1987. *Principles of Seed Pathology*, Vol. 1 (XII, 176 p.) and Vol. 2 (XII, 168 p.) Boca Raton, Florida: CRC Press Inc.

Andersch, W., Schwarz, M. 2003. Clothianidin seed treatment (Poncho®)—The new technology for control of corn rootworms and secondary pests in US-corn production. *Pflanzenschutz Nachr. Bayer* 56: 147–172.

Baker, K.F. 1962a. Principles of heat treatment of soil and planting material. *J. Inst. Agric. Sci* 28: 118–126.

Baker, K.F. 1962b. Thermotherapy of planting material. *Phytopathology*, 52: 1244–1255.

Block, C.C., McGee, D.C., Hill, J.H. 1998. Seed transmission of *Pantoea stewartii* in field and sweet corn. *Plant Dis.*82: 775–780.

Block, C.C., McGee, D.C., Hill, J.H. 1999. Relationship between late season Stewart's bacterial wilt and seed infection in maize. *Plant Dis.* 83: 527–530.

Bradley, C.A. 2008. Effect of fungicide seed treatments on stand establishment, seedling disease, and yield of soybean in North Dakota. *Plant Dis.* 92: 120–125.

Brannen, P.M., Kennedy, D.S. 1996. Biological seed treatments as a component in maintaining seedling health. *Proceedings Beltwide Cotton Conference* TN, USA, January 9–12, 1: 244–247.

Brinkerhoff, L.A., Hunter, R.E. 1963. Internally infected seed as source of inoculum for the primary cycle of bacterial blight of cotton. *Phytopathology* 53: 1397–1401.

Broders, K.D., Lipps, P.E., Paul, P.A., Dorrance, A.E. 2007a. Characterization of *Pythium* spp. associated with corn and soybean seed and seedling disease in Ohio. *Plant Dis.* 91: 727–735.

Broders, K.D., Lipps, P.E., Paul, P.A., Dorrance, A.E. 2007b. Evaluation of *Fusarium graminearum* associated with corn and soybean seed and seedling disease in Ohio. *Plant Dis.* 91: 1155–1160.

Butzen, S, Doerge, T. 2007. Pioneer seed treatment with Dynasty® fungicide. *Crop Insights* 15: 11.

Connick, W.J., Lewis, J.A., Quimby, P.C. 1990. Formulation of biocontrol agents for use in plant pathology. In *New Directions in Biological Control: Alternatives for Suppressing Agricultural Pests and Diseases.* Baker, R., Dunn, P. (ed.). *Proc. UCLA*, Frisco-Colorado, 20–27 Jan. 1989, Alan R. Liss Pubs., pp. 345–372.

Cook, R.J. 1993. Making greater use of introduced microorganisms for biological control of plant pathogens. *Annu. Rev. Phytopathol* 31: 53–80.

Dekker, J. 1957. Internal disinfection of peas infected by means of the antibiotics Rimocidin and Pimaricin and some aspects of the parasitism of the fungus. *Tijdsch. Plantziek* 63: 65–114.

Dhanvantari, B.N. 1989. Effect of seed extraction methods and seed treatments on control of tomato bacterial canker. *Can. J. Plant Pathol.* 11: 400–408.

Duvert, P. 1999. Global Seed Treatment Working Group.

Fatmi, M., Schaad, N.W., Bolkan, H.A. 1991. Seed treatments for eradicating *Clavibacter michiganensis* subsp. *michiganensis* from naturally infected tomato seeds. *Plant Dis.* 75: 383–385.

Fleischer, S.J., Orzolek, M.D., DeMackiewic, D., Otjen, L. 1998. Imidacloprid effects on *Acalymma vittata* (Coleoptera: Chrysomelidae) and bacterial wilt in cantaloupe. *J. Econ. Entomol* 91: 940–944.

Goodman, R.N. 1959. The influence of antibiotics on plants and plant disease control. In *Antibiotics: Their Chemistry and Non-Medical Uses*. Goldberg, H.S. (ed.). Princeton: van Nostrand, pp. 322–448.

Gordon-Lennox, G., Walther, D., Gindrat, D. 1987. Utilisation diantagonistes pour lienrobage des semences: EfficacitE et mode diaction contre les agents de la fonte des semis. *EPPO Bull.* 17: 631–637.

Hankin, I., Sands, D.C. 1977. Microwave treatment of tobacco seed to eliminate bacteria on the seed surface. *Phytopathology* 67: 794–795.

Harman, G.E. 1991. Seed treatments for biological control of plant disease. *Crop Prot.* 10: 166–171.

Harman, G.E., Nelson, E.B. 1994. Mechanism of protection of seed and seedlings by biological seed treatments: Implications for practical disease control. In *Seed Treatment: Progress and Prospects*. Martin, T.J. (ed.). BCPC Monograph No: 57, England: Thornton Heath Inc., pp. 283–292.

Harman, G.E., Norton, J.M., Stasz, T.E. 1987. Nyolate seed treatment of Brassica spp. to eradicate or reduce black rot caused by *Xanthomonas campestris* pv. *campestris*. *Plant Dis.* 71: 27–30.

Hermansen, J.E., Jorgensen, J. 1983. Historical aspects of the control of seedborne cereal disease in Denmark. *Seed Sci. Technol.* 1: 1005.

Hopkins, D.L., Thompson, C.M., Hilgren, J., Lovic, B. 2003. Wet seed treatment with peroxyacetic acid for the control of bacterial fruit blotch and other seedborne diseases of watermelon. *Plant Dis.* 87: 1495–1499.

Hsieh, T.F., Huang, H.C., Erickson, R.S. 2005a. Biological control of bacterial wilt of bean using a bacterial endophyte, *Pantoae agglomerans*. *J. Phytopathol.* 153(10): 608–617.

Huang, H.C., Erickson, R.C., Hsieh, T.F. 2007. Control of bacterial wilt of bean (*Curtobacterium flaccumfaciens* pv. *flaccumfaciens*) by seed treatment with *Rhizobium leguminosarum*. *Crop Prot.* 26(7): 1055–1061.

Humaydan, H.S., Harman, G.E., Nedrow, B.L., Dinitto, L.V. 1980. Eradication of *Xanthomonas campestris*, the causal agent of black rot, from Brassica seeds with antibiotics and sodium hypochlorite. *Phytopathology* 70: 127–131.

Ivey, M.L.L., Miller, S.A. 2005. Evaluation of hot water seed treatment for the control of Bacterial Leaf Spot and Bacterial Canker on fresh market and processing tomatoes. *Acta Hortic.* 695: 197–204.

Jindal, K.K., Thind, B.S., Soni, B.S. 1989. Physical and chemical agents for the control of *Xanthomonas campestris* pv. *vignicola* from cowpea seeds. *Seed Sci. Technol.* 17: 371–382.

Karsavuran, Y., Saygili, H., Tosun, N., Turkusay, H. 2005. Tohumla tasinan hastalik ve zararlilarin mücadelesi. In *Tohum bilimi ve teknolojisi*, Vol. 2. Eser, B., Saygili, H., Gokcol, A., Ilker, E. (eds.). Izmir: Meta Basim, pp. 567–598.

Katan, J. 1981. Solar heating (solarisation) for control of soilborne pests. *Ann. Rev Phytopathol.* 19: 211–236.

Katan, J., De Vay, J.F. 1991. *Soil Solarisation*. Boca Raton, FL: CRC Press, pp. 267.

Klisiewcz, J.M., Pound, G.S. 1961. Studies on control of black rot of Crucifers by seed treating seed with antibiotics. *Phytopathology* 51: 495–500.

Krasnova, M.V. 1963. The effect of some physical factors on the causal agent of bacterioses in soybean seeds. *J. Microbiol. (Kiev)* 25(5): 50–52.

Kuhar, T.P., Stivers-Young, L.J., Hoffman, M.P., Taylor, A.G. 2002. Control of corn flea beetle and Stewart's wilt in sweet corn with imidacloprid and thiamethoxam seed treatments. *Crop Prot.* 21: 25–31.

Liang, L.Z., Halloin, J.M., Saettler, A.W. 1992. Use of polyethylene glycol and glycerol as carriers of antibiotics for reduction of *Xanthomonas campestris* pv. *phaseoli* in navy bean seeds. *Plant Dis* 76: 875–79.

Lugtenberg, B., Kamilova, F. 2009. Plant-growth-promoting Rhizobacteria. *Annu. Rev. Microbiol* 63: 541–556.

Maude, R.B. 1996. *Seed Borne Disease and their Control Principles and Practise*, Vol. XVII. Wallinford, England: CAB International, pp. 280.

McGee, D.C. 1981. Seed pathology: Its place in modern seed production. *Plant Dis.* 65: 638–642.

McGee, D.C. 1995. Epidemiological approach to disease management through seed technology. *Annu. Rev. Phytopathol.* 33: 645–666.

McGee, D.C. 1996. Advances in seed treatment technology. Technical Report No. 11 Asian Pacific Seed Association.

Minsavage, G.V., Canteros, B.I., Stall, R.E. 1990. Plasmid-mediated resistance to streptomycin in *Xanthomonas campestris* pv. *vesicatoria*. *Phytopathology* 80: 719–723.

Mirik, M., Aysan, Y., Karatas, A., Cinar, O. 2005. Fiziksel ve kimyasal uygulama görmüş Biber Tohumlarında *Xanthomonas axanopodis* pv *vesicatoria*'nın aranması. Türkiye 2. *Tohumculuk Kongresi*, 361 (özet), 9–11 Kasım 2005, Adana.

Mohammadi, O. 1992. Mycostop biofungicide-present status. In *Biological Control of Plant Diseases*. Tjamos, E.C., Papavizas, G.C., Cook, R.J. (eds.). New York: Plenum Press, pp. 207–210.

Mukhopadhyay, A.N. 1994. Biocontrol of soil borne fungal pathogens—Current status future prospects and potential limitations. *Indian Phytopathol* 47: 119–126.

Munkvold, G.P. 2009. Seed pathology progress in academia and industry. *Annu. Rev. Phytopathol.* 47: 285–311.
Munkvold, G.P., McGee, D.C., Iles, A. 1996. Effects of imidacloprid seed treatment of corn on foliar feeding and *Erwinia stewartii* transmission by the corn flea beetle. *Plant Dis.*80: 747–749.
Neergaard, P. 1988. *Seed Pathology*, Vols. I and II. Hong Kong: MacMillian Press, XXV. pp.1191.
Ozaktan, H. 1991. Domates bakteriyel solgunluğu (*Clavibacter michiganensis* subsp. *michiganensis*) ile savaşım olanakları üzerine arastirmalar. Ege Universitesi, Fen Bilimleri Enst., Doktora Tezi, Izmir, pp. 98.
Pataky, J.K., Hawk, J.A., Weldekidan, T., Fallah Moghaddam, P. 1995. Incidence and severity of Stewart's bacterial wilt on sequential plantings of resistant and susceptible sweet corn hybrids. *Plant Dis.* 79: 1202–1207.
Pataky, J.K., Michener, P.M., Freeman, N.D., Weinzierl, R.A., Teyker, R.H. 2000. Control of Stewart's wilt in sweet corn with seed treatment insecticides. *Plant Dis.* 84: 1104–1108.
Pataky, J.K., Michener, P.M., Freeman, N.D., Whalen, J.M., Hawk, J.A. 2005. Rates of seed treatment insecticides and control of Stewart's wilt in sweet corn. *Plant Dis.* 89: 262–268.
Pradhanang, P.M., Collier, G. 2009. How effective is hydrochloric acid treatment to control *Clavibacter michiganensis* subsp. *michiganensis* contamination in tomato seed. *Acta Horticulturae* 808: 81–85.
Procter, C.H., Fry, P.R. 1965. Seed transmission of tobacco mosaic virus in tomato. *N.Z.J. Agric.* Press. 8: 367–369.
Pyndji, M.M., Sinclair, I.B., Singh, T. 1987. Soybean seed thermotherapy with heated vegetable oils. *Plant Dis.* 71: 213–216.
Raaijmakers, Jos M., Mazzola, M. 2012. Diversity and natural functions of antibiotics produced by beneficial and plant pathogenic bacteria. *Annu. Rev. Phytopathol.* 50: 403–424.
Ralph, W. 1977. Problem in testing and control of seedborne bacterial pathogens a critical evaluation. *Seed Sci. Technol.* 5: 735–752.
Raudales, R.E. McSpadden Gardener, B.B. 2008. Microbial Biopesticides for the Control of Plant Diseases in Organic Farming. Fact Sheet Agriculture and Natural Resources. HYG-3310-08. Sahin, F., Miller, S.A. 1997. Identification of the bacterial leaf spot pathogen of lettuce, *Xanthomonas campestris* pv. *vitians*, in Ohio, and assessment of cultivar resistance and seed treatment. *Plant Dis* 81: 1443–1446.
Rhodes, D.J., Powell, K.A. 1994. Biological seed treatment—The development process. In *Seed Treatment: Progress and Prospects*. Martin, T.J. (ed.). BCPC Monograph no: 57, England: Thornton Heath Inc., pp. 303–310.
Saygili, H., Sahin, F., Aysan, Y. 2012. Bitki bakteri hastalıkları. In *Bitki Bakteri Hastalıklarıyla Kimyasal Mücadele.* Saygili, H., Sahin, F., Aysan, Y. (eds.). Izmir: Meta Basim, pp. 23.
Stall, R.E., Thayer, P.L. 1962. Streptomycin resistance of the bacterial spot pathogen and control with streptomycin. *Plant Dis. Rep* 46: 389–392.
Strandberg, J.O., White, J.M. 1989. Response of carrot seeds to heat treatments. *J. Am. Soc. Hortic. Sci* 114: 766–769.
Strider, D.L. 1979. Eradication of *Xanthomonas nigromaculans* f.sp. *Zinnia* in Zinnia seed with sodium hypochlorite. *Pl. Dis. Reptr* 63: 869–873.
Suparyono, Pataky, J.K. 1989. Influence of host resistance and growth stage at the time of inoculation on Stewart's wilt and Goss's wilt development and sweet corn hybrid yield. *Plant Dis* 73: 339–345.
Sykes, G. 1965. *Disinfection and Sterilization*, 2nd Edn. London: E. and FN Spon, pp. 212.
Taylor, A.G., Harman, G.E. 1990. Concepts and technologies of selected seed treatments. *Annual Rev. Phytopathol* 28: 321–339.
Thyr, B.D., Webb, R.E., Jaworski, C.A., Ratcliffe, T.J. 1973. Tomato bacterial cancer: Control by seed treatment. *Pl. Dis. Reptr* 57: 974–977.
Tireng Karut, Ş., Aysan, Y. 2011. Organik Tarımda Domates Bakteriyel Kanser ve Solgunluk Hastalığı Etmenine (*Clavibacter michiganensis* subsp. *michiganensis*) Karşı Kullanılabilecek Tohum Uygulamaları. Türkiye IV. Bitki Koruma Kongresi, 28-30 Haziran 2011, Kahramanmaraş. sayfa: 316.
Tomli, C. 1994. *The Pesticide Manual*, 10th Edn. Farnham, England: BCPC Publications.
Williams, P.H. 1980. Black rot: A continuing threat to world crucifers. *Plant Dis.* 64: 736–742.
Yorganci, I., Erkan, S., Ozaktan, H., Eser, B. 1993. Biber, domates, patlıcan tohumlarında viral ve bakteriyel etmenlerin tanılanması ve inaktifleştirilmeleri üzerinde arastirmalar. TUBITAK TOAG-822 nolu proje, VIII 32s.
Zengin. 2007. Organik Tarim. Hasad Yayincilik, 136p.

18 Plant Pathogenic Bacteria Control through Foliar Application

Mustafa Mirik and Cansu Öksel

CONTENTS

ABSTRACT Biological control of pathogens by introduced microorganisms has been studied since the mid-1940s, but has not been considered commercially feasible until recently. The surfaces of plant parts provide a habitat for epiphytic microorganisms, including pathogen and saprophytic organisms, as biological agents. The saprophytic organisms help to reduce the incidence of foliar disease. The activity of both saprophytes and pathogens on leaves is dependent on the microclimatological conditions at the plant surface as well as in the chemical environment. To be successful, the antagonist should be able to multiply on and colonize the plant surface. Biological agents act through various mechanisms, including substrate competition and niche exclusion, siderophores, antibiotics, and induced resistance, according to the relationship between biocontrol agents and pathogens or other microorganisms in the environment. Some studies have shown that biological agents such as *Pseudomonas flourescens*, *Pseudomonas putida*, and *Bacillus subtilis* reduced the severity of foliage bacterial diseases by 25%–92.5%.

KEYWORDS: bacterial agents, foliage bacterial pathogens, bioformulation, *Pseudomonas*, *Bacillus*

18.1 INTRODUCTION

Biological control of plant diseases refers to means of controlling disease or reducing the amount or effect of pathogens that rely upon biological mechanisms, or upon organisms other than man (Campbell, 1989). It also refers to cultural practices (e.g., crop rotation and soil amendments) that affect pathogenic microorganisms. A more narrow definition refers only to the artificial introduction of living microorganisms (biopesticides) into a plant or its environment to control pathogens. Biopesticides are an alternative or supplementary way of reducing the use of chemical pesticides in agriculture (Höfte and Altier, 2010).

Campbell (1989) divided the historical development of biological control into three periods: prehistoric, when traditional methods were predominant; the era of rapid and continuous acquisition of knowledge; and the current era of biological control (Bora and Özaktan, 1998). Although these periods are considered to be distinct, they are linked because knowledge and experience (which we think of as data) are passed down through successive periods. The first and second periods obtained data that are being used today for the preparation of suitable formulations and commercial applications of biological agents. Nearly 30 biological preparations were used commercially in 2000, and this number continues to rise. Biopreparation production is intended to develop the most effective and appropriate formulas to target pathogens *in vitro* and *in vivo*. Shelf-life determination, side effects, and licensure are determined according to the results of current research.

The overuse of chemical pesticides to cure or prevent plant diseases has caused soil pollution and has had harmful effects on human beings. To reduce the use of these chemicals, one possibility is to utilize the activity of microorganisms. It is desirable to replace chemical pesticides with materials that possess the following criteria: high specificity against the target plant pathogens, easy degradability after effective usage, and low mass-production cost. Products that are known as biological control agents (BCA) or biological pesticides fulfill these criteria. Ideally, the microorganisms upon which such products are based are stable, nondominant species that are able to maintain their effectiveness (Shoda, 2000).

The surfaces of aerial plants parts provide habitats for epiphytic microorganisms, many of which are capable of influencing the growth of pathogens. These saprophytic organisms are important in reducing incidences of foliar diseases in field crops. In this chapter, control of pathogens by saprophytic organisms will be emphasized but the use of virulent or avirulent strains of pathogens will not be considered (Blakeman and Fokkema, 1982). To devise ways of enhancing the antagonistic effects of foliar microorganisms against pathogens, it is necessary to understand something of the nature of their specialized microhabitat, usually referred to as the phyllosphere (Last, 1955; Ruinen, 1961) or phylloplane (Preece and Dickinson, 1971; Dickinson and Preece, 1976; Blakeman, 1981).

The activity of both saprophytes and pathogens on leaves is dependent on the microclimatological conditions on the plant surface as well as on the chemical environment (Blakeman, 1972). Various morphological features on the surfaces of leaves can have significant influence. The irregular provision of surface water on leaves results in intermittent growth of microorganisms, particularly bacteria and filamentous fungi, and poses problems for survival during dry periods. Temperatures on leaves can fluctuate widely over a 24 h period; by contrast, indoor environments are much more stable habitats for microbial growth.

Despite the low nutrient content of water films on leaves, yeast and bacteria can form large populations (Blakeman, 1972; Dickinson and Preece, 1976). Nutrients on leaves are primarily derived from leaf tissues as a result of cellular leakage but can be supplemented, usually later in the growing season, by pollen and aphid honeydew deposits (Fokkema, 1981). Leaves of certain plants may secrete phenolic or terpenoid substances onto their own surfaces, which inhibit microbial growth (Blakeman and Atkinson, 1981).

Morphological features on leaf surfaces affect wettability (Matta, 1971). For example, a dense covering of trichomes or crystalline deposits of epicuticular wax increase the surface's water repellency; this in turn restricts microbial growth, and inhibits the spread of pathogen inoculum and the leaching of substances from the leaf. Surfaces of leaves contain numerous microsites that favor both pathogen and saprophyte growth. These sites are frequently located along veins, especially in the depressions between the anticlinal walls of epidermal cells and at the bases of trichomes. At these sites, nutrient leakage is enhanced and microorganisms may be protected from extremes of desiccation by higher relative humidities and longer water-retention periods (Blakeman and Fokkema, 1982).

Bacteria may form resident populations on leaves. Most such bacteria are Gram-negative, often chromogenic, and include the following genera: *Erwinia*, *Pseudomonas*, *Xanthomonas*, and *Flavobacterium*. Gram-positive bacteria such as *Lactobacillus*, *Bacillus*, and *Corynebacterium* are isolated less frequently. In addition to saprophytic bacteria, pathogenic bacteria such as *Pseudomonas syringae* pv. *syringae*, *P. syringae* pv. *morsprunorum*, *P. syringae* pv. *glycinea*, *Erwinia amylovora*, and *Erwinia caratovora* can live in a nonpathogenic epiphytic phase on foliar surfaces (Blakeman and Brodie, 1977). Bacteria often tend to be more abundant on leaves early in the growing season.

The most commonly reported mechanisms of biocontrol by bacterial agents include the production of antibiotics, hydrogen cyanide, lytic exoenzymes (Thomashow and Weller, 1996), and cyclic lipopeptides (Raaijmakers et al., 2006), as well as competition for nutrients and niches (Kamilova et al., 2005), competition for iron mediated by siderophores, competition for carbon (Thomashow and Weller, 1996), and induced systemic resistance or ISR (De Vleesschauwer and Höfte, 2009). The focus of this chapter is on leaf applications to control bacterial diseases, and the effective mechanisms used by bacterial antagonists.

18.2 BACTERIAL BIOLOGICAL AGENTS

Bacteria that are shown or thought to have biocontrol potential are found in many genera, including *Actinoplanes* (Filonow and Lockwood, 1979; Sutherland and Lockwood, 1984), *Agrobacterium*, *Azotobacter*, *Cellulomonas*, *Pasteuria* (Weller, 1988), *Alcaligenes* (Elad and Chet, 1987), *Amorphosporangium* (Filonow and Lockwood, 1979), *Arthrobacter* (Mitchell and Hurwitz, 1965), *Bacillus* (Capper and Campbell, 1986), *Enterobacter* (Kwok et al., 1987), *Erwinia* (Sneh et al., 1984), *Flavobacterium* (Chen et al., 1987), *Hafnia* (Sneh, 1981), *Micromonospora* (Filonow and Lockwood, 1979), *Pseudomonas* (Genesan and Ginanamanickam, 1987), *Rhizobium* and *Bradyrhizobium* (Tu, 1980), *Serratia* (Sneh et al., 1985), *Streptomyces* (Merriman et al., 1974), and *Xanthomonas* (Chen et al., 1987). Obviously, biocontrol agents are not limited to a specific bacterial group. Many of these biocontrol agents are effective under field conditions.

Antagonists should be sought in areas where, despite the presence of a susceptible host, the disease caused by a given pathogen does not occur, has declined, or cannot develop; such areas are preferable to ones where the disease occurs (Baker and Cook, 1974; Blakeman and Fokkema, 1982). Biological control is more successful in the rhizosphere than in the phyllosphere. Indeed, not one of seven commercial products for biocontrol tested by Lewis and Papavizas (1991) was directed against foliar pathogens. Differences in the success rates of biological control methods have been attributed to the environment (Andrews, 1992).

The use of biocontrol on aerial plant surfaces is much less developed than treatments to the soil environment. One isolate of *Bacillus subtilis* in soil provided control of bean rust at a level equivalent to spray application of *B. subtilis* and a fungicide; this trial was carried out on plants in a greenhouse prior to inoculation with spores of the plant pathogen (Baker et al., 1983; Pusey and Wilson, 1984). *B. subtilis* and its related species (i.e., rod-shaped, spore-forming, Gram-positive bacteria) have long been established as industrial bacteria for the production of secretion enzymes such as amylase, protease, pullulanase, chitinase, xylenase, and lipase, among others (Westers et al., 2004). The commercial production of this enzyme represents about 60% of the commercially produced industrial enzymes. *Bacillus* spp. have good potential as biocontrol agents because they have developed strategies for survival in unfavorable environments (Morikawa, 2006).

Commercial strains of *B. subtilis* have been marketed as biocontrol agents for fungal diseases in crops. The commercial biofungicide Serenade, which contains a *B. subtilis* strain, is reported to be effective against a variety of pathogenic bacteria, including *Erwinia*, *Pseudomonas*, and *Xanthomonas* strains. The mechanism of these antibacterial effects is uncertain; however, it is known that *B. subtilis* can produce a variety of antibacterial agents, including a broad spectrum of lipopeptides (e.g., surfactin) that are potent biosurfactants and are important for maintaining the aerial structure of biofilms (Branda et al., 2001). Asako and Shoda (1996) and Morikawa (2006) showed that the half-life of surfactin in soil is longer than that of iturin A. Bais et al. (2004) reported that the biocontrol of *P. syringae* by *B. subtilis* 6051 is related to surfactin formation on the surface of the root (Bais et al., 2004).

Pseudomonas strains, which are some of the most active and dominant bacteria in the rhizosphere, have been intensively investigated as biocontrol agents (Geels and Schippers, 1983; Laville et al., 1992). Most of this research attention has been focused on *Pseudomonas fluorescens*, *Pseudomonas putida*, and *Pseudomonas cepacia* for practical applications. *Pseudomonas* strains produce several kinds of antibiotics that are closely related to the suppression of plant diseases, including pyrrolnitrin, pyoluteorin, and phenezine-1-carboxylate (Weller, 1988; Kraus and Loper, 1995). Siderophores are extracellular, low-molecular-weight compounds that have high affinity for ferric ions and participate in biocontrol (Leong, 1986). Some *Pseudomonas* strains were confirmed to suppress the activities of plant pathogens by the production of siderophore or pseudobactin (Kloepper et al., 1980, 1986). Bacterial siderophores are also known to have functions that help plant growth, such as supplying iron via an iron-chelating function called the plant growth-promoting rhizobacteria (PGPR) effect (Leong, 1986; Loper, 1988; Crowley et al., 1991), which eventually leads to a decrease in disease occurrence. A siderophore-producing *P. fluorescens* induced increased emergence of cotton seedlings in soil infested with *Pythium ultimum*.

18.3 MECHANISM OF BIOCONTROL AGENTS

Biological control has four main mechanisms, which function according to the relationship between biocontrol agents and pathogens or other microorganisms in the environment. These mechanisms are: substrate competition and niche exclusion, siderophores, antibiotics, and induced resistance (Weller, 1988).

18.3.1 Substrate Competition and Niche Exclusion

Competition for nutrients supplied by roots and by seed exudates is probably significant in most interactions between bacteria and pathogens at the root; in addition, such competition is responsible in part for the observed biocontrol by introduced bacteria (Weller, 1988). Nutrient elements such as as carbon and iron are very important in biological control. Large populations of bacteria established on planting material become a partial sink for nutrients in the rhizosphere and phyllosphere, thereby reducing the amount of carbon and nitrogen available to plants. Microorganisms that are already present or added for the purpose of biological control can colonize plant roots in soil; in addition, microorganisms consume root exudates secreted by plants as a carbon source. Root secretions that include complex carbohydrates such as cellulose, hemicellulose, and pectin can only be disintegrated by some specialized microorganisms. Therefore, the biological agents used of carbon must be more competitive than pathogens (Weller, 1988).

Suslow (1982) suggested that niche exclusion is a potentially important mechanism of antagonism of deleterious by PGPR (Suslow and Schroth, 1982). Certain areas on the root appear to be favored for colonization by many bacteria. Inoculating planting material with biological agents presumably prevents or reduces pathogen establishment at these sites. Although *Agrobacterium* 84 suppressed *Agrobacterium tumefaciens*, mainly by the production of Agrocin 84 (Kerr, 1987), physical blockage of infection sites may also contribute to biocontrol (Cooksey and Moore, 1982; Du Plessis et al., 1985).

18.3.2 Siderophores

Siderophores are low-molecular-weight, high-affinity iron (III) chelators that transport iron into bacterial cells (Neilands, 1981; Leong, 1986). When grown under low-iron conditions, fluorescent pseudomonads produce yellow-green fluorescent siderophores (pyoverdine and pseudobactin types) and membrane receptor proteins that specifically recognize and take up the siderophore–iron complex (Weller, 1988). Iron is important for the relationship between pathogens and antagonist microorganisms. Microorganisms benefit from the protein pigments in siderophores for their requirement of iron. Kloepper et al. (1980) were the first to demonstrate the importance of siderophore production as a mechanism of biological control. Siderophores sequester the limited supply of iron (III) in the rhizosphere by biological agents, limit its availability to pathogens, and ultimately suppress pathogen growth (Schroth and Hancock, 1981, 1982).

The availability of iron (III) in soil declines logarithmically as soil pH increases. Therefore, siderophore-mediated suppression should be greater in neutral and alkaline soils than in acid soils (Baker, 1985; Baker et al., 1986; Weller, 1988). Pathogens are thought to be sensitive to suppression by siderophores because they produce no siderophores of their own; they are unable to use siderophores produced by antagonists or other microorganisms in their immediate environment; they produce too few siderophores or the siderophores they produce have lower affinity for iron than those of the antagonists; or they produce one or more siderophores that can be used by the antagonist, but they are unable to use the antagonist's siderophores (Schroth and Hancock, 1981, 1982; Scher and Baker, 1982; Buyer and Leong, 1986; Leong, 1986; Magazin et al., 1986; Weller, 1988). Several reviews have summarized the evidence supporting a role for siderophores in biocontrol (Baker, 1985; Baker et al., 1986; Leong, 1986). The most convincing evidence is the fact that mutants that lack siderophores are less suppressive than their parental strains to pathogens in the rhizosphere (Kloepper et al., 1980; Kloepper and Schroth, 1981; Baker et al., 1986, 1987; Becker and Cook, 1988; Loper, 1988).

18.3.3 Antibiotics

The most important criterion of organism selection for biocontrol agents is the ability to produce antimicrobial metabolites. Antibiotics are heavily involved in disease suppression by some bacteria (Weller, 1988). The first antibiotic defined in biological control, phenazine, is produced by strains of *P. fluorescens* 2-79 and *Pseudomonas aureofaciens* 30-84 that are suppressive to *Puccinia graminis* on wheat (Weller and Cook, 1983). The antibiotic Agrocin 84, which is produced by *Agrobacterium radiobacter* K-84 strains, is used in biological control against grown gall caused by *A. tumefaciens*. The antibiotics 2,4-diacetylphloroglucinol (DAPG), pyoluteorin, pyrrolnitrin, and other phenazine derivates have been described in the biocontrol *Pseudomonas* spp. as the main cause of this antagonistic activity (Thomashow and Weller, 1996; De La Fuente et al., 2004; Weller et al., 2007).

18.3.4 Induced Resistance

All plants possess active defense mechanisms against attack, but these mechanisms can fail when the plant is infected by virulent pathogens that avoid triggering or suppress resistance reactions, or evade the effects of activated defenses. If defense mechanisms are triggered by a stimulus prior to pathogen infection, disease can be reduced. Induced resistance (IR) is a state of enhanced defensive capacity expressed by a plant after appropriate stimulation (Kuc, 1982, 1995; Van Loon et al., 1998). IR occurs naturally as a result of a limited infection by a pathogen, particularly when the plant develops a hypersensitive reaction (HR). Although tissue necrosis contributes to the level of IR attained, the activation of defense mechanisms that limit primary infection appears sufficient to elicit IR (Hammerschmidt and Kuc, 1995). Induced resistance can be triggered by chemicals, nonpathogens, avirulent forms of pathogens, incompatible races of pathogens, or virulent pathogens when infection is stalled due to environmental conditions.

Generally, IR is systemic because the defensive capacity is increased not only in the primary infected plant parts, but also in noninfected, spatially separated tissues. Because of this systemic character, induced resistance is commonly referred to as systemic acquired resistance (SAR) (Ross, 1961a; Ryals et al., 1996; Sticher et al., 1997). However, induced resistance is not always expressed systemically. Localized acquired resistance (LAR) occurs when only tissues that have been exposed to the primary invader become more resistant (Ross, 1961b). sAR and LAR are similar in that they are effective against various types of pathogens. A signal that results in enhanced defensive capacity throughout the plant in SAR appears to be lacking LAR (Van Loon et al., 1998).

SAR is characterized by an accumulation of salicylic acid (SA) and pathogenesis-related proteins (PRPs) (Ward et al. 1991; Uknes et al., 1992; Kessman et al., 1994; Ryals et al., 1996; Sticher et al., 1997). The accumulation of SA occurs locally and also systemically, at lower levels, concomitantly to the development of SAR (Van Loon et al., 1998). Exogenous application of SA also induces SAR in several plant species (Van Loon and Antoniw, 1982; Gaffney et al., 1993; Ryals et al., 1996). Both pathogen- and SA-induced resistance are associated with the induction of several families of PRPs. The induction of PRPs is invariably linked to necrotizing infection, which induces SAR, and has been taken as a marker of the induced state (Ward et al. 1991; Uknes et al., 1992; Kessman et al., 1994). Although SA may be transported from the originally infected leaves, it does not appear to be the primary long-distance signal for systemic induction (Van Loon, 1997).

To date, the nature of this signal has not been established (Vernooij et al., 1994), but it is known that the level of SAR is modulated by ethylene and jasmonic acid or JA (Van Loon and Antoniw, 1982; Xu et al., 1994; Wasternack and Parthier, 1997; Sticher et al., 1997; Knoester, 1998). These results suggest that induction and expression of SAR are regulated through the interplay of several signaling compounds. *Pseudomonas fluorescens* CHAo produces both antibiotics and siderophores (Ahl et al., 1986), but the suppression by this bacterium of black root rot of tobacco (Stutz et al., 1986) appears to be mediated mainly by the production of hydrogen cyanide (Weller, 1988). Kempe and Sequeira (1983) suggested that resistance to a virulent strain of *Ralstonia solanacearum* was induced in potato by the treatment of seed pieces with avirulent or incompatible strains of *Ral. solanacearum* or *P. flourescens* (Weller, 1988). Biological agents may have more than one mechanism to suppress a pathogen, and the relative importance of a particular mechanism may vary with the physical or chemical conditions of the rhizosphere and phyllosphere (Weller et al., 1988).

18.4 EFFECT OF LEAF MORPHOLOGY FOR BIOLOGICAL AGENTS

The phyllosphere microbial community is an open system. Comparisons of the leaf and root habitats show that the leaf is a much more dynamic environment. On a percentage area or unit weight basis, colonization of the root tip and the phylloplane appear similar within about one order of magnitude. The root, unlike the leaf, is surrounded by a medium that can be manipulated for biocontrol (Andrews, 1992).

18.4.1 Surface Features of Leaves and Roots

Leaf topography comprises hills, peaks, valleys, and craters (Cutter, 1976; Juniper and Jeffree, 1983; Allen et al., 1991). The most important of these features are i) veins, which vary in density, pattern, and elevation above the leaf lamina; ii) trichomes and emergences (Juniper and Jeffree, 1983; Wagner, 1991; Andrews, 1992), which (if present) can vary from simple, single-celled, non-glandular hairs to complex, multicellular, glandular structures that may secrete sugar (nectaries), oils, resins, salts, or terpenes, or contain pigment (Juniper and Jeffree, 1983; Andrews, 1992); iii) epidermal cell contours, which may be more or less convex or papillose; and iv) stomata and hydathodes, which may be present or absent. Stomata may be above, below, or at the same plane as the unspecialized epidermal cells, and may have simple or complex antechambers and substomatal cavities (Juniper and Jeffree, 1983; Andrews, 1992). Another important feature, hydathodes, serve to discharge guttation water and dissolved solutes; they vary in complexity and may or may not be associated with the secretory tissue (Esau, 1965). The last of these features, cuticle, is present in most terrestrial plants (Juniper and Jeffree, 1983) but it varies in thickness, morphology, and chemical composition (Esau, 1965; Juniper and Jeffree, 1983; Allen et al., 1991; Andrews, 1992).

18.4.2 Microclimate

All that separates the leaf surface from the elements is a fragile boundary layer of air that is no more than a few millimeters thick (Burrage, 1971, 1976; Lerouge et al., 1990). The phylloplane's environment, which is more variable, includes temperature, relative humidity, dew, rain, wind, and radiation. Moreover, substantial variations in time and space may occur on a scale appropriate to microorganisms. For instance, surface temperatures across the surface of leaf can differ by several degrees (Burrage, 1971; Andrews, 1992), due to differences in diffusional cooling and boundary layer thickness. Accordingly, the water potential of phylloplane microbes will be at some intermediate value between those of the boundary layer and the leaf, and will be continuously variable (Lerouge et al., 1990; Andrews, 1992).

18.4.3 Nutrients

Nutrients on the phylloplane, which originate endogenously or exogenously, include diverse carbohydrates, amino acids, organic acids, sugar alcohols, mineral trace elements, vitamins, and hormones, as well as antimicrobial compounds such as phenols and terpenoids (Tukey, 1970, 1971; Blakeman and Atkinson, 1981; Gaber and Hutchinson, 1988; Fiala et al., 1990; Andrews, 1992). Nutrients are important not only because of their direct role as microbial substrates but also because of their potential indirect effects (James and Gutterson, 1986; Weller and Thomashow, 1990) on the synthesis of antibiotics and siderophores in the phylloplane (Andrews, 1992).

Endogenous nutrients are removed from leaves either by the leaching action of rain, dew, and fog (Tukey, 1970, 1971), or by active exudation via guttation through hydathodes (Esau, 1965; Tukey, 1970, 1971; Frossard, 1981). In general, the loss of nutrients from leaves of terrestrial plants is limited (Fokkema and Schippers, 1986) relative to losses from aquatic plants (Lucas, 1947; Khailov and Burlakova, 1969) and roots (Andrews, 1992). Exudate concentrations vary quantitatively and qualitatively with host, leaf position, leaf surface, age of plant, light, temperature, fertility, pH, leaching medium, and leaf injury (Morris and Rouse, 1985; Derridj et al., 1989; Fiala et al., 1990).

The nutritional environment of the rhizoplane differs from the leaf in several significant respects. Most root exudates originate from young tissue in the root-tip region (Foster et al., 1983; Curl and Truelove, 1986; Lynch, 1990; Keister and Cregan, 1991) and are profuse compared with phylloplane exudates (Tukey, 1970, 1971; Fiala et al., 1990). In general, loss of nutrients from the leaves of terrestrial plants is limited (Andrews, 1992).

18.4.4 Colonization

Potential sources of colonists for phylloplane are soil, seeds, air, buds, and overwintering shoots (Andrews, 1992). Under controlled conditions of high relative humidity, bacteria can reach leaves from seed/soil surfaces (Leben, 1961; Andrews, 1992). Bacteria that originate in soil can survive and grow on leaves, but decline more rapidly under the fluctuating environmental conditions there than phylloplane isolates do (O'Brien and Lindow, 1989). Kempf and Wolf (1989) showed that antagonistic *Erwinia herbicola* from treated seed spread upward and was detectable up to the second 2 cm unit above the ground 10 d after planting, but not at 32 d. Based on the colonization patterns on wheat leaves of different ages in the same field, Fokkema et al. (1979) inferred that no direct transfer of yeasts from older to younger plant parts had occurred, and were unsuccessful in their attempt to manipulate the phylloplane community by seed inoculants.

The general sequence of the colonization of the phylloplane is a function of available inocula, the environment (including nutritional status and micro/macroclimate, as discussed above), and host phenology. There appears to be an early preponderance of bacteria, followed by a sharp increase in yeasts, and eventually a rise in the filamentous fungi (Blakeman, 1985). The colonization pattern on individual leaves is localized and heterogeneous. Preferred colonization sites are along veins and in grooves above the anticlinal walls of epidermal cells (Leben, 1965; Diem, 1974; Andrews, 1992), possibly because of the localized concentrations of nutrients, protection from erosion, trapping and deposition phenomena, and retention of water films.

The colonization process of the rhizoplane differs in many respects from that of the phylloplane. Bacteria are the major root colonists (followed by actinomycetes, filamentous fungi, and yeasts) and tip region is the center of microbial activity (Curl and Truelove, 1986; Fokkema and Schippers, 1986; Bowen, 1991; Hawes and Brigham, 1992). Bacterial populations from the rhizospheres of root systems (as opposed to individual roots) appear to approximate a long-normal distribution (Loper et al., 1984). Colonization can be viewed as a two-stage process in which seed-borne bacteria are initially covered on the root until the carrying capacity of the system is reached (Howie et al., 1987). Water infiltration can actually facilitate this process (Parke, 1991), but its role in fostering colonization of new phylloplane tissues would seem to be negative or at best indirect.

18.5 FORMULATION OF BCAs

Among the strategies tested to enhance the survival of bacteria was the addition of carbon sources to culture medium (i.e., peat) at the time of planting; it was assumed that the carbon sources would be selectively utilized by the biocontrol agent, but not by the pathogen. Even if the populations of bacteria declined below an active threshold, it was assumed that they could recover by selective use of the carbon source. Although it was found that strain 63-28 could utilize mannitol, sorbitol, and trehalose, these carbon sources did not enhance biocontrol in pot experiments. Adjuvants that were also added to the culture media or peat to enhance survival included glycerol, paraffin oil, starch, and proline, but none were successful. When bacteria were adsorbed onto different substrates added to peat, including vermiculite, kaolin, talc, and xanthan gum, strain 63-28 survived best when added to vermiculite that was then air-dried. However, when the bacteria-treated vermiculite was added back to the peat, survival was not enhanced compared with peat alone. Nor did osmotically stressing or conditioning the bacteria during culturing with polyethylene glycol enhance survival. Freeze-drying the bacteria in trehalose or sucrose resulted in the best survival, with high bacterial densities maintained for 6 months to 1 year in dry peat at room temperature. However, when the freeze-dried bacteria were added to peat at 45% moisture, survival declined over a 3-month period. A moisture level of 45% was enough to increase the metabolic activity of the bacteria, which led to the population decline (Paulitz and Belanger, 2001).

Pseudomonas flourescens 63-28 was extensively tested in greenhouse conditions for the control of root diseases (Seresinhe et al., 1997; Hill and Peng, 1999). Because *Pseudomonas* and other Gram-negative bacteria are sensitive to drying, the goal was to develop a formulation that would survive in peat for 6 months to 1 year. Strain 63-28 survived best in peat at 100%–150% moisture (v/v), but poorly at 45% and 25% moisture. Even at the optimum moisture levels, populations declined below the critical threshold of 10^6/g after 1–2 months (Paulitz and Belanger, 2001).

18.6 BIOLOGICAL CONTROL OF FOLIAGE PATHOGENS

Bacillus subtilis QST713, known as Serenade (AgraQuest Inc., Davis, CA, USA) is the latest product based on strain QST713 of *B. subtilis*. It is currently available as a wettable powder; registration is pending for an aqueous suspension formulation (Rhapsody). The product is advertized to have a spectrum of activity that includes more than 40 plant diseases. The bacterium is presumed to work through a number of modes of action, including competition, parasitism, antibiosis, and induction of SAR (Paulitz and Belanger, 2001).

Bora and Ozaktan (2000) tested fluorescent *Pseudomonas* for the biological control of pathogens *Papulaspora byssina*, *Cladobotryum dendroides*, *Pseudomonas tolaasii*, and *Mycogone perniciosa*. They found that *P. flourescens* M15 and *P. putida* 109 controlled brown plaster disease at respective rates of 86.6% and 92.5%. Two applications with *P. flourescens* M4/2i, *P. flourescens* M5/3, and *P. putida* 39.a respectively reduced the severity of the disease caused by *P. tolaasii* by 87.34%–90.5%, 71.3%–71.7%, and 67.5%–72.6%. In the case of *C. dendroides*, fluorescent *Pseudomonas* strains *P. flourescens* M.15 and *P. putida* 39.a respectively provided 292.5 and 240.0 g/tray with higher production than the positive controls.

Lactic acid bacteria (LAB) can be a source of BCA of fire blight disease. Several species of LAB that inhabit plants are currently used as food biopreservatives because of their antagonistic properties against bacteria; in addition, they are considered as generally safe. Candidates for BCA were selected from a large collection of LAB strains obtained from plant environments. Strains were first chosen based on the consistency of their suppressive effects against *E. amylovora* infections in detached plant organs (flowers, fruits, and leaves). *Lactobacillus plantarum* PC40, PM411, TC54, and TC92 were effective against *E. amylovora* in most of the experiments. Moreover, PM411, TC54, and TC92 had strong antagonistic activity against *E. amylovora* as well as other target bacteria, and presented genes involved in plantaricin biosynthesis (*plnJ*, *plnK*, *plnL*, *plnR*, and *plnEF*). These strains efficiently colonized pear and apple flowers; maintained stable populations for at least 1 week under high RH conditions; and survived at low RH conditions. They were effective in preventing fire blight on pear flowers, fruits, and leaves, as well as in the whole plants and in a semi-field blossom assay. This study confirmed the potential of certain strains of *L. plantarum* to be used as an active ingredient of microbial biopesticides for fire blight control that could be eventually extended to other plant bacterial diseases (Rosello et al., 2013).

Gerami et al. (2013) tested epiphytic bacterial isolates against *E. amylovora*, and evaluated them under laboratory and orchard conditions. *Pseudomonas fluorescens* E-10, *P. agglomerans* Abp(2), *P. putida* E-11, and *S. marcescens* Kgh(1) checked antagonistic activity against *E. amylovora* on pear blossom. In all evaluation procedures, biocontrol effects of *P. agglomerans* Abp(2) and *P. fluorescens* E-10 were more promising than those afforded by *P. putida* E-11 and *S. marcescens* Kgh(1). Their selected bacteria were applied twice, at 20% and 80% of blooming. Results showed 46.9% disease occurrence in control plants, while the antagonistic bacteria reduced the disease by 23%–60%. In the orchard trial, the most effective antagonist was *P. agglomerans* Abp(2), and the least effective was *S. marcescens* Kgh(1).

Bacillus thuringiensis suppresses the QS-dependent virulence of the plant pathogen *Erwinia carotovora* through a new form of microbial antagonism and signal interference. *E. carotovora* produces and responds to acyl-homoserine lactone (AHL) QS signals to regulate antibiotic production and the expression of virulence genes, whereas *B. thuringiensis* possesses AHL-lactonase, which is a potent AHL-degrading enzyme (Zhang and Dong, 2004; Dong et al., 2004). *Bacillus thuringiensis* does not seem to interfere with the normal growth of *E. carotovora*; however, it abolishes the accumulation of the AHL signal when the two are cocultured. In plants, *B. thuringiensis* significantly decreases the incidence of *E. carotovora* infection and the symptom development of potato soft rot caused by the pathogen. Biocontrol efficiency should correlate with the ability of bacterial strains to produce AHL-lactose (Morikawa, 2006).

Investigation of the control of *Streptomyces scabies* in potatoes by *Bacillus subtilis* revealed that more antibiotics were produced when the bacteria were grown in a water extract of soybean (Weinhold and Bowman, 1968). Furthermore, when this material was added to soil, buildup of potato scab was prevented, presumably because the soybean substrate supported the antibiotic production in soil by *B. subtilis* and thereby enhanced its ability to control the specific disease. 2,3-Dihydroxybenzoyl-glycine (2,3-DHBG), a siderophore produced by the Gram-positive *B. subtilis* (Ito and Neilands, 1958; Leong, 1986), is also produced by strains of *B. subtilis*, which showed a wide suppressive spectrum on plant pathogens by producing antibiotics. Gram-negative *Escherichia coli* produces a siderophore, entreobactin (Armstrong et al., 1996). *B. subtilis* produced not only antibiotics that suppress plant pathogens, but also siderophores; the regulation of these products by the gene *lpa-14* indicated the possibility of enhanced biocontrol effectiveness by the manipulation of the gene (Shoda, 2000).

This sample also effectively reduced the occurrence of bacterial wilt of tomato caused by *P. solanacearum*. Previously, *B. subtilis* was considered ineffective against *P. solanacearum* (Podile et al., 1985). Integrated biological control of stem canker and black scurf of potato caused by *Rhizoctonia solani* and potato scab caused by *S. scabies* by *B. subtilis* were demonstrated in both the greenhouse and field experiments. In the greenhouse, the occurrence of these diseases was reduced by 60%–70%; a similar effect was observed in the field (Schmiedeknecht et al., 1998). This synergistic phenomenon, which is involved in integrated control that includes the use of fungicides and biocontrol agents, may be more efficient and longer-lasting than control achieved through biocontrol agents or fungicides alone (Shoda, 2000).

When a talc-based formulation of *P. agglomerans* Eh-24 was sprayed on pear trees that were naturally infected with *E. amylovora*, it reduced the percentage of blighted blossom in pear orchards by 63%–76%. Copper oxychloride + maneb mixture was less effective than the bioformulation in reducing the incidence of blossom infection by *E. amylovora* in each pear orchard. In addition, *P. agglomerans* Eh-24 labeled with Str^{R+} was applied at 30% and 100% bloom to monitor its colonization and population dynamics on pear blossoms. In this case, the population size of this strain increased from 2×10^4 to 1.3×10^6 CFU per blossom over 18 d (Ozaktan and Bora, 2006).

Carrer et al. (2008) evaluated the efficiency of actinobacteria *Nocardioides thermolilacinus* SON-17 as a biocontrol agent for tomato diseases after delivery to the phylloplane of tomato plants. In greenhouse assays, when tomato plants were sprayed with a propagule suspension of SON-17 and inoculated with challenging pathogens four d later, the agent was able to reduce disease severity caused by *Alternaria solani*, *S. solani*, *C. cassiicola*, *X. campestris* pv. *vesicatoria*, and *P. syringae* pv. *tomato*. In field trials, weekly spray-applications of actinomycete reduced the severity of the early blight in comparison with the controls. Laux et al. (2003) tested three bacterial antagonists (*P. agglomerans* Pa21889, *B. subtilis* BsBD170, and *Rahnella aquatilis* Ra39) against blossom blight of apple caused by *E. amylovora* under both artificial and natural infection conditions, in comparison with streptomycin (Plantomycin) application. All antagonistic strains gave significant disease reductions (43%–81%).

A total of 167 epiphytic bacterial strains, including *E. herbicola*, fluorescent pseudomonads, and some Gram-positive bacteria, were obtained from the different organs of healthy pear trees in the Aegean region. Eleven bacterial strains that were known to suppress the development of the disease in immature pear fruit assay (IPFA) were used in the pear-blossom assay. *Erwinia herbicola* 78-B/2, 24, and 1B significantly reduced the percentage of the blighted blossoms, by 82.07%–98.03%, compared with the untreated control. Other formulations (e.g., lyophilized, talc-based, and whey-based wettable formulations) of *E. herbicola* Eh-1B and Eh-24, which were the most promising in *in vivo* tests against *E. Amylovora*, were developed as well. The viability of *E. herbicola* cells in lyophilized formulations was successfully protected for at least 180 d at 24°C and 10°C. In the talc-based and whey-based formulations, the bacteria survived up to 180 d in storage at 10°C, although a decline was observed after 120 d. When the bioformulations were stored at 24°C, the shelf lives of *E. herbicola* strains in talc-based and whey-based formulations were estimated to be 60 d.

Bioformulations were also tested for their ability to suppress the development of *E. amylovora* on pear fruits and blossoms. The bioformulations that were tested on pear fruits suppressed the development of fire blight by 71.43%–100%, compared with the application of pathogen alone. In the blossom assays, different bioformulations of *E. herbicola* strains reduced the percentage of the blighted blossoms by 31.47%–84.32%, compared with the application of pathogen alone. Treatments with talc-based formulations of Eh-1B and Eh-24 were more effective in reducing fruit and blossom blight than the lyophilized and whey-based formulations were (Ozaktan et al. 1999).

The integration of foliar bacterial BCA and PGPR was investigated to determine whether biological control of bacterial speck of tomato caused by *P. syringae* pv. *tomato*, and bacterial spot of tomato caused by *X. campestris* pv. *vesicatoria* and *X. vesicatoria*, could be improved. Three foliar BCA and two selected PGPR strains were employed in pairwise combinations. The foliar BCAs had previously demonstrated moderate control of bacterial speck or bacterial spot when applied as foliar sprays. The foliar BCA *P. syringae* Cit7, which was the most effective of the three foliar BCAs, provided significant suppression of bacterial speck in all field trials and bacterial spot in two out of three field trials. Eight of 50 PGPR strains tested significantly suppressed bacterial speck in three experiments. Of the eight strains, *P. fluorescens* 89B-61, *Bacillus pumilus* SE34, and *Bacillus pasteurii* M-38 provided the greatest disease suppression; the mean reduction in lesion numbers exceeded 60%. *Burkholderia gladioli* IN26, which also reduced disease significantly, was the most consistent. Other strains that provided significant disease suppression in all three trials were *Bacillus cereus* 83-6, *Stenotrophomonas maltophilia* IN287, *B. cereus* M-22, and *Bacillus amyloliquefaciens* IN937a.

PGPR SE34 and 89B-61 were selected and used in combination with foliar application of the BCA *P. syringae* Cit7 to evaluate their efficacy in control of bacterial speck on greenhouse tomatoes. In two independent trials, PGPR 89B-61 provided similar levels of disease reduction (76.3% and 52.8%) for the foliar BCA Cit7 (72.0% and 56.0%). In one of the two trials, combined use of 89B-61 and Cit7 provided the largest disease reduction (82.2%), which was significantly higher than that of either Cit7 or 89B-61 applied alone. The disease was significantly reduced by PGPR SE34 in one of the two experiments; however, disease reduction was not significantly greater with the combined use of SE34 and Cit7. *Pseudomonas fluorescens* strain A506 was tested in one experiment, alone or in combination with the two PGPR strains, but it did not provide significant disease reduction or enhance the efficacy of the PGPR strains (Ji et al., 2006).

Rahnella aquatilis strains were inoculated into tomato leaves (foliar application) before challenge by the pathogen, or mixed with the soil (soil application) before sowing. Seeds were also soaked in a bacterial suspension of *Rah. aquatilis* (seed application), or the seedling roots were dipped in *Rah. aquatilis* suspension (root application). In the four experiments, neither *Rah. aquatilis* strains nor seedlings treated with sterile distilled water induced disease symptoms when inoculated into the leaves of 3-week-old tomato seedlings. In contrast to this result, disease symptoms at a rate of 60% were observed after inoculation of *X. campestris* pv. *vesicatoria* into the leaves of control seedlings (which originated from nonbacterized seeds and soil). Seedlings that had undergone root application showed disease symptoms at a rate of 50%. In the foliar application, pre-inoculation treatment with *Rah. aquatilis* 17 reduced infection up to 50%, and pre-inoculation treatment with *Rah. aquatilis* 55 reduced the infection up to 58%. When applied to seeds, both *Rah. aquatilis* strains reduced infection up to 33%.

Mixing bacterial suspensions of *Rah. aquatilis* strains into the soil resulted in less significant reductions. Seedlings treated with *Rah. aquatilis* 17 and then *Xanthomonas* showed 16.6% reduction; seedlings treated with *Rah. aquatilis* 55 and then *Xanthomonas* showed 25% reduction. When seedling roots were dipped in *Rah. aquatilis* suspension, however, infection was reduced up to 40%; moreover, seedlings treated with *Rah. aquatilis* 17 and then with *Xanthomonas* showed 50% reduction. Seedlings treated with *Rah. aquatilis* 55 and then with *Xanthomonas* showed 58% reduction. The application of *Rah. aquatilis* strains onto tomato seedlings at foliage, seed, soil, and root resulted in differences in the fresh and dry weights of the seedlings as well as in the levels of disease reduction.

Foliar application of either *Rah. aquatilis* strain (17 or 55) was most effective in disease reduction (El-Hendawy et al., 2005).

Wilson et al. (2002) noted that even though there is potential for biological control of bacterial speck in tomato (due to the relatively moderate levels of disease control that have been achieved), foliar BCA is likely to be only one component of an integrated disease management program. The most effective BCA, bacterial strain *P. syringae* Cit7, achieved 15%–50% reduction in foliar disease severity with a mean of 28% reduction when used in different tomato-growing regions of North America, under different climatic conditions, on different cultivars, with different horticultural practices (including fungicide regimes) and different users.

18.7 CONCLUSION AND FUTURE PERSPECTIVES

Biological control is the culmination of complex interactions among host, pathogen(s), antagonist, and environment. Identifying important traits allows the efficient selection of new strains. Traits can also be altered to make a strain more effective. Additional research is needed on the physical and chemical factors of foliage that influence both colonization and the expression traits that are important for antagonism in the phyllosphere. By identifying these factors, it may be possible to manipulate them on plants, in the field as well as in greenhouses, to enhance foliage colonization. In order to develop bacterial biocontrol agents for commercial use, the consistency of their performance must be improved. Research on the identification of bacterial traits that function in plant colonization and pathogen antagonism is critically important, and molecular genetics offers the best approach to such studies. Identifying important traits allows more-efficient selection of new strains. Finally, more research is needed on the formulation and delivery of the bacteria. The challenge is to develop inexpensive, easily applied preparations that remain viable under less-than-optimal conditions. It must be kept in mind that growers cannot be expected to buy new equipment or to substantially modify equipment or farming practices to accommodate a biological treatment.

ACKNOWLEDGMENT

The authors thank Cengiz Karagöz, English Instructor at Namik Kemal University Foreign Languages Vocational School, for his editorial advice.

REFERENCES

Ahl, P., Voisard, C., Défag, G. 1986. Iron bound-siderophores, cyanic acid, and antibiotics involved in suppressing of *Thielaviopsis basicola* bay a *Pseudomonas fluorescens* strains. *Journal of Phytopathology* 116: 121–134.

Allen, EA., Hoch, H.C., Steadman, J.R., Stavely, R.J. 1991. Influence of leaf surface features on spore deposition and the epiphytic growth of phytopathogenic fungi. In *Microbial Ecology of Leaves*. Andrews, J.H., Hirano, S.S. (eds.). New York: Springer, pp. 87–110.

Andrews, J.H. 1992. Biological control in the phyllosphere. *Annual Review of Phytopathology* 30: 603–635.

Armstrong, S.K., Pettis, G.S., Forrester, L.J., Mcintosh, M.A. 1996. The *Escherichia coli* enterobactin biosynthesis gene, *entD:* Nucleotide sequence and membrane localization of its protein product. *Molecular Microbiology* 3: 923–936.

Asako, O., Shoda, M. 1996. Biocontrol of *Rhizoctonia solani* damping off of tomato with *Bacillus subtilis* RB14. *Applied and Environmental Microbiology* 62: 4081–4085.

Bais, H.P., Fall, R., Vivanco, J.M. 2004. Biocontrol of *Bacillus subtilis* against infection of *Arabidopsis* roots by *Pseudomonas syringae* is facilitated by biofilm formation and surfactin productin. *Plant Physiology* 134: 307–319.

Baker, C.J., Stavely, J.R., Thomas, C.A., Ssser, M., MacFali, J.S. 1983. Inhibitory effect of *Bacillus subtilis* on *Uromyces phaeoli* and on development of rust pustules on bean leaves. *Phytopathology* 73: 1148–1152.

Baker, K.F., Cook, R.J. 1974. *Biological Control of Plant Pathogens*. St. Paul: American Phytopathological Society, 433 p.
Baker, R. 1985. Biological control of plant pathogens: Definitions. In *Biological Control in Agricultural IPM Systems*. Hoy, M.A., Herzog, D.C. (eds.), Orlando: Academic Press, 606 p.
Baker, R., Elad, Y., Sneh, B. 1986. Physical, biological and host factors in iron competition in soils. In *Iron, Siderophores, and Plant Diseases*. Swinburne, T.R. (ed.). New York: Plenum, pp. 77–84.
Becker, O., Cook, R.J. 1988. Role of siderophores in suppression of *Pythium* species and production of increased growth response of wheat by fluorescent pseudomonads. *Phytopathology* 78: 778–782.
Blakeman, J.P. 1972. Effect of plant age on inhibition of *Botrytis cinerea* spores by bacteria on beetroot leaves. *Physiological Plant Pathology* 2: 143–152.
Blakeman, J.P. 1981. *Microbial Ecology of the Phylloplane*. London, New York: Academic Press. p. 520.
Blakeman, J.P. 1985. Ecological succession of leaf surface microorganisms in relation to biological control. In *Biological Control on the Phylloplane*. Windels, C.E., Lindow, S.E. (eds.). St. Paul, MN: American Phytopathological Society, pp. 6–7, 189 p.
Blakeman, J.P., Atkinson, P. 1981. Antimicrobial substances associated with the aerial surfaces of plants. In *Microbial Ecology of the Phylloplane*. Blakeman, J.P. (ed.). London, New York: Academic Press, pp. 245–264.
Blakeman, J.P., Brodie, I.D.S. 1977. Competition for nutrients between epiphytic microorganisms and germination of spores of plant pathogens on beetroot leaves. *Physiological Plant Pathology* 10: 29–42.
Blakeman, J.P., Fokkema, N.J. 1982. Potential for biological control of plant diseases on the phylloplane. *Annual Review of Phytopathology*. 20: 167–192.
Bora, T., Özaktan, H. 1998. *Bitki hastalıklarıyla biyolojik savaş*. Prizma Matbaası. İzmir, p. 205.
Bora, T., Ozaktan, H. 2000. Biological control of some important mushroom disease in Turkey by fluorescent pseudomonads. *Science and Cultivation of Edible Fungi* 1–2: 689–693.
Bowen, G.D. 1991. Microbial dynamics in the rhizosphere: Possible strategies in managing rhizosphere populations. In *The Rhizosphere and Plant Growth*. Keister, D.L., Cregan, P.B. (eds.). Kluwer Academic Publisher, pp. 25–32.
Branda, S.S., Gonzales-Pastor, J.E., Ben-Yehuda, S., Losick, R., Kolter, R. 2001. Fruiting body formation by *Bacillus subtilis*. *Proceedings of National Academy of Science USA* 98: 11621–11626.
Burrage, S.W. 1971. The micro-climate at the leaf surface. In *Ecology of Leaf Surface Microorganisms*. Preece, T.E., Dickinson, C.H. (eds.), London, UK: Academic Press, pp. 91–101.
Burrage, S.W. 1976. Aerial microclimate around plant surfaces, In: *Microbiology of Aerial Plant Surfaces*. Dickinson, C.H., Preece, T.F. (eds.), London: Academic Press, pp. 173–184.
Buyer, J.S., Leong, J. 1986. Iron transport-mediated antagonism between plant growth-promoting and plant deleterious *Pseudomonas* strains. *Journal of Biological Chemistry* 261: 791–794.
Campbell, R. 1989. *Biological Control of Microbial Plant Pathogens*. Cambridge: University Press.
Capper, A.L., Campbell, R. 1986. The effect of artificially inoculated antagonistic bacteria on the prevalence of take all disease of wheat in field experiments. *Journal of Applied Bacteriology* 60: 155–160.
Carrer, R., Romeiro, R.S., Garcia, F.A.O. 2008. Biocontrol of foliar disease of tomato plants by *Nocardioides thermolilacinus*. *Tropical Plant Pathology* 33(6): 457–460.
Chen, W., Hoitink, H.A.J., Schmitthenner, A.F. 1987. Factors affecting suppression of *Pythium* damping-off in container media amended with compost. *Phytopathology* 77: 755–760.
Cooksey, D.A., Moore, L.W. 1982. Biological control of crown gall with an agrocin mutant of *Agrobacterium radiobacter*. *Phytopathology* 72: 919–921.
Crowley, D.E., Wang, Y.C., Reid, C.P.P., Szaniszlo, P.J. 1991. Mechanisms of iron acquisition from siderophores by microorganisms and plants. *Plant Soil* 130: 179–198.
Curl, E.A., Truelove, B. 1986. *The Rhizosphere*. New York: Springer-Verlag.
Cutter, E.G. 1976. Aspects of the structure and development of the aerial surfaces of higher plants. In: *Microbiology of Aerial Plant Surfaces*. Dickinson, C.H., Preece T.F. (eds.). London, UK: Academic Press, pp. 1–40.
De La Fuente, L., Thomashow, L., Weller, D., Bajsa, N., Quagliotto, L., Chernin, L., Arias, A. 2004. *Pseudomonas fluorescens* UP61 isolated from birds foot trefoil rhizosphere produces multiple antibiotics and exerts a broad spectrum of biocontrol activity. *European Journal of Plant Pathology* 110: 671–681.
De Vleesschauwer, D., Höfte, M. 2009. Rhizobacteria-induced systemic resistance. *Advances in Botanical Research* 51: 223–281.
Delcambe, L., Peypoux, F., Besson, F., Guinand, M., Michel, G. 1977. Structure of iturin and iturin-like substance. *Biochemical Society Transactions* 5: 1122–1124.

Derridj, S., Gregoire, V., Boutin, J.P., Fiala, V. 1989. Plant growth stages in the interspecific oviposition preference of the European corn borer and relations with chemicals present on the leaf surfaces. *Entomologia Experimentalis et Applicata* 53: 267–276.

Dickinson, C.H., Preece, T.F. 1976. *Microbiology of Aerial Plant Surface*. London: Academic Press, p. 669.

Diem, H.G. 1974. Microorganisms of the leaf surface: Estimation of the mycoflora of the barley phyllosphere. *Journal of General Microbiology* 80: 77–83.

Dong, Y.H., Zhang, X.F., Xu, J.L., Zhang, L.H. 2004. Insecticidal by *Bacillus thuringiensis* silences *Erwinia carotovora* virulence by a new form of microbial antagonism, signal interference. *Applied and Environmental Microbiology* 70: 954–960.

Du Plessis, H.J., Hattingh, M.J., Van Vuuren, H.J.J. 1985. Biological control of crown gall in South Africa by *Agrobacterium radiobacter* strain K84. *Plant Disease* 69: 302–305.

Elad, Y., Chet, I. 1987. Possible role of competition for nutrients in biocontrol of *Pythium* damping off by bacteria. *Phytopathology* 77: 190–195.

El-Hendawy, H.H., Osman, M.E., Sorour, N.M. 2005. Biological control of bacterial spot of tomato caused by *Xanthomonas campestris* pv. *vesicatoria* by *Rahnella aquatilis. Microbiological Research* 160: 343–352.

Esau, K. 1965. *Plant Anatomy*. New York: Wiley.

Fiala, V., Glad, C., Martin, M., Jolivet, E., Derridj, S. 1990. Occurrence of soluble carbohydrates on the phylloplane of maize (*Zea mays* L.): Variations on relation to leaf heterogeneity and position on the plant. *New Phytology* 115: 609–615.

Filonow, A.B., Lockwood, J.I. 1979. Evaluation of several actinomycetes and the fungus *Hyphochytrium catenoides* as biocontrol agent for *Phytophthora* root rot of soybean. *Plant Disease* 69: 1033–1036.

Fokkema, N.J. 1981. Fungal leaf saprophytes, beneficial or detrimental? In *Microbial Ecology of the Phylloplane*. Blakeman J.P. (ed.). London: Academic Press, pp. 453–454.

Fokkema, N.J., den Houter, J.G., Kosterman, Y.J.C., Nelis, A.L. 1979. Manipulation of yeasts on field-grown wheat leaves and their antagonistic effect on *Cochliobolus sativus* and *Septoria nodorum. Transactions of the British Mycological Society* 72: 19–29.

Fokkema, N.J., Schippers, B. 1986. Phyllosphere versus rhizosphere as environments for saprophytic colonization. In *Microbiology of the Phyllosphere*. Fokkema, N.J, van den Heuvel, J. (eds.). Cambridge: Cambridge University Press, pp. 137–159.

Foster, R.C., Roviraand, A.D., Cook, T.W. 1983. *Ultrastructure of the Root-Soil İnterface*. St. Paul: American Phytopathology Society, p. 157.

Frossard, R. 1981. Effect of guttation fluids on growth of microorganisms on leaves. In *Microbial Ecology of the Phylloplane*. Belkeman, J.P. (ed.). London: Academic Press, pp. 213–226.

Gaber, B.A., Hutchinson, T.C. 1988. Chemical changes in simulated raindrops following contact with leaves of four boreal forest species. *Canadian Journal of Botany* 66: 2445–2451.

Gaffney, T., Friedrich, L., Vernooij, B., Negmtto, D., Nye, G., Uknes, S., Ward, E., Kessmann, H., Ryals, J. 1993. Requirement of salicylic acid for the induction of systemic acquired resistance. *Science* 261: 754–756.

Geels, F.P., Schippers, B. 1983. Reduction of yield depressions in high frequency potato cropping soil after seed tuber treatments with antagonistic fluorescent *Pseudomonas* spp. *Phytopathology Z* 108: 207–214.

Genesan, P., Ginanamanickam, S.S. 1987. Biological control of *Sclerotium rolfsii* Sacc. in peanut by inoculation with *Pseudomonas flourescens. Soil Biology and Biochemistry* 19: 35–38.

Gerami, E., Hassanzadeh, N,, Abdollahi, H, Ghasemi, A., Heydari, A. 2013. Evaluatıon of some bacterial antagonists for biological control of fire blight disease. *Journal of Plant Pathology* 95(1): 127–134.

Hammerschmidt, R., Kuc, J. 1995. *Induce Resistance to Disease in Plants*. Dordrecht: Kluwer.

Hawes, M.C., Brigham L.A. 1992. Impact of root border cells on microbial populations in the rhizosphere. *Advances in Plant Pathology* 8: 119–148.

Hill, D.J., Peng, G. 1999. Evaluation of AtEze for suppression of fusarium wilt of chrysanthemum. *Canadian Journal of Plant Pathology* 21: 194–195.

Höfte, M., Altier, N. 2010. Fluorescent pseudomonads as biological agents for sustainable agricultural systems. *Research in Microbiology* 161: 464–471.

Howie, W.J., Cook, R.J., Weller, D.M. 1987. Effects of soil matric potential and cell motility on wheat root colonization by fluorescent pseudomonads suppressive to take-all. *Phytopathology* 77: 286–292.

Ito, T., Neilands, J.B. 1958. Products of "low iron fermentation" with *Bacillus subtilis:* Isolation, characterization and synthesis of 2,3-dihydroxybenzoylglycine. *Journal of American Chemical Society* 80: 4645–4647.

James, D.W. Jr., Gutterson, N.I. 1986. Multiple antibiotics produced by *Pseudomonas fluorescens* HV37 and their differential regulation by glucose. *Applied and Environmental Microbiology* 52: 1183–1189.

Ji, P., Campbell, H.L., Kloepper, J.W., Jones, J.B., Suslow, T.V., Wilson, M. 2006. Integrated biological control of bacterial speck and spot of tomato under field conditions using foliar biological control agents and plant growth promoting rhizobacteria. *Biological Control* 36: 358–367.
Juniper, B.E., Jeffree, C.E. 1983. *Plant Surfaces*. London: Edward Arnold, p. 93.
Kamilova, F., Validov, S., Azarova, T., Mulders, I., Lugtenberg, B. 2005. Enrichment for enhanced competitive plant root tip colonizers selects for a new class of biocontrol bacteria. *Environmental Microbiology* 7: 1809–1817.
Keister, D.L., Cregan, P.B. 1991. *The Rhizosphere and Plant Growth*. Dordrecht: Kluwer, 386 p.
Kempe, J., Sequeira, L. 1983. Biological control of bacterial wilt of potatoes: Attempts to induce resistance by treating tubers with bacteria. *Plant Disease* 67: 499–505.
Kempf, H.J., Wolf, G. 1989. *Erwinia herbicola* as a biocontrol agents of *Fusarium culmorum* and *Puccinia recondita* f. sp. *tritici* on wheat. *Phytopathology* 79: 619–627.
Kerr, A. 1987. The impact of molecular genetics on plant pathology. *Annual Review of Phytopathology* 25: 87–110.
Kessman, H., Stamb, T., Hofmann, C., Maetzke, T., Herzog, J. 1994. Induction of systemic acquired disease resistance in plants by chemicals. *Annual Review of Phytopathology* 32: 439–459.
Khailov, K.M., Burlakova, Z.P. 1969. Release of dissolved organic matter by marine seaweeds and distribution of their total organic production to inshore communities. *Limnology and Oceanography* 14: 521–527.
Kloepper, J M., Laverry, R., Lozano, J.C., 1986. Observation on the effect of inoculating cassava plants with *Pseudomonas fluorescens*. *Journal of Phytopathology* 117: 17–25.
Kloepper, J.W., Leong, J., Teintze, M., Scroth, M.N. 1980. Enhanced plant growth by siderophores produced by plant growth promoting rhizobacteria. *Nature* 286: 865–886.
Kloepper, J.W., Schroth, M.N. 1981. Relationship of *in vitro* antibiosis of plant growth-promoting rhizobacteria to plant growth and the displacement of root microflora. *Phytopathology* 71: 1020–1024.
Knoester, M. 1998. The involvement of ethylene in plant disease resistance. PhD thesis, Utrecht University. The Netherlands.
Kraus, J., Loper, J.E. 1995. Characterization of a genomic region required for production of the antibiotics pyoluteorin by the biological control agent *Pseudomonas fluorescens* Pf-5. *Applied and Environmental Microbiology* 61: 849–854.
Kuc, J. 1982. Plant immunization mechanisms and practical implications. In *Active Defense Mechanism in Plants*. Wood, R.K.S. (ed.). New York: Plenum Press, pp. 157–178.
Kuc, J. 1995. Phytoalexins, stress metabolism and disease resistance in plants. *Annual Review of Phytopathology* 33: 157–178.
Kwok, O.C.H., Fahy, P.C., Hoittink, H.A.J., Kuter, G.A. 1987. Interactions between bacteria and *Trichoderma* in suppression of *Rhizoctonia* damping-off in bark compost media. *Phytopathology* 77: 1206–1212.
Last, F.T. 1955. Seasonal incidence of *Sporobolomyces* on cereal leaves. *Transactions of the British Mycological Society* 38: 221–239.
Laux, P., Wesche, J., Zeller, W. 2003. Field experiments on biological control of fire blight by bacterial antagonists. *Z. PflKrankh. PflSchutz* 110(4): 401–407.
Laville, J., Voisard, C., Keel, C., Manrhofer, M., Defago, G., Haas, D. 1992. Global control in *Pseudomonas fluorescens* mediating antibiotic synthesis and suppression of black root rot of tobacco. *Proceedings of National Academy of Science USA* 89: 1562–1566.
Leben, C. 1961. Microorganisms on cucumber seedlings. *Phytopathology* 51: 553–557.
Leben, C. 1965. Influence of humidity on the migration of bacteria on cucumber seedlings. *Canadian Journal of Microbiology* 11: 671–676.
Leong, J. 1986. Siderophores: Their biochemistry and possible role in the biocontrol of plant pathogens. *Annual Review of Phytopathology* 24: 187–209.
Lerouge, P., Roche, P., Faucher, C., Maillet, F., Truchet, G. 1990. Symbiotic host-specificity of *Rhizobium meliloti* is determined by a sulphated and acylated glucosamine oligosaccharide signal. *Nature* 344: 781–784.
Lewis, J.A., Papavizas, G.C. 1991. Biocontrol of plant diseases: The approach for tomorrow. *Crop Protection* 10: 95–105.
Loper, J.E. 1988. Role of fluorescent siderophore production in biological control of *Pythium ultimum* by a *Pseudomonas fluorescens* strain. *Phytopathology* 78: 166–172.
Loper, J.E., Suslow, T.V., Schroth, M.N., 1984. Lognormal distribution of bacterial populations in the rhizosphere. *Phytopathology* 74: 1454–460.
Lucas, L.E. 1947. The ecological effects of external metabolites. *Biological Review of Cambridge Philosophical Society* 22: 270–295.

Lynch, J.M. 1990. *The Rhizosphere. Bacterial Populations from the Rhizosphere of Root Systems (as Opposed to Individual Roots) Appear to Approximate a Long-Normal Distribution*. New York: Wiley.

Magazin, M.D., Moores, J.C., Leong, J. 1986. Cloning of the gene coding for the outer membrane receptor protein for ferric pseudobactin, a siderophore from a plant growth promoting *Pseudomonas* strain. *Journal of Biological Chemistry* 261: 795–799.

Matta, A. 1971. Microbial penetration and immunization of uncongenial host plants. *Annual Review of Phytopathology* 9: 387–410.

Merriman, P.R., Price, R.D., Kollmorgen, J.F., Piggott, T., Ridge, E.H. 1974. Effect of seed inoculation with *Bacillus subtilis* and *Streptomyces grisens* on the growth of cereals and carrots. *Australian Journal of Agricultural Research* 25: 219–226.

Mitchell, R., Hurwitz, E. 1965. Supression of *Pythium debaryanum* by lytic rhizosphere bacteria. *Phytopathology* 55: 156–158.

Morikawa, M. 2006. Beneficial biofilm formation by industrial bacteria *Bacillus subtilis* and related species. *Journal of Bioscience and Bioengineering* 101(1): 1–8.

Morris, C.E., Rouse, D.I. 1985. Role of nutrients in regulating epiphytic bacterial populations. In *Biological Control on the Phylloplane*. Windel, C.E., Lindow, S.E. (eds.), St. Paul, MN: American Phytopathological Society, pp. 63–82.

Neilands, J.B. 1981. Microbial iron compounds. *Annual Review of Biochemistry* 50: 715–731.

O'Brien, R.D., Lindow, S.E. 1989. Effect of plant species and environmental conditions on epiphytic population sizes of *Pseudomonas syringae* and other bacteria. *Phytopathology* 79: 619–627.

Ozaktan, H., Bora, T. 2006. Studies on biological control of fire blight with some antagonistic bacteria. In *Proceedings of the Tenth International Workshop on Fire Blight, Acta Holticulture*, Italy, Vol. 704, pp. 337–340.

Ozaktan, H., Bora, T., Sukan, F.V., Sukan, S. Sargin, S. 1999. Studies on determination of antagonistic potential and biopreparation of some bacteria against the fireblight pathogen. *Proceedıngs of the Eighth International Workshop on Fire Blight, Acta Holticulture*, Turkey, Vol. 489, pp. 663–668.

Parke, J.L. 1991. Root colonization by indigenous and introduced microorganisms. In *The Rhizosphere and Plant Growth*. Keister, D.L., Cregan, P.B. (eds.). Dordrecht: Kluwer, pp. 33–42.

Paulitz, T.C., Belanger, R.R. 2001. Biological control in greenhouse systems. *Annual Review of Phytopathology* 39: 103–133.

Podile, A.R., Prasad, G.S., Dube, H.C. 1985. *Bacillus subtilis* as antagonist to vascular wilt pathogens. *Current Science* 54: 864–865.

Preece, T.F., Dickinson, C.H. 1971. *Ecology of Leaf Surface Micro-Organisms*. London: Academic Press, 640 p.

Pusey, P.L., Wilson, C.L. 1984. Postharvest biological control of stone fruit brown rot by *Bacillus subtilis*. *Plant Disease* 68: 753–756.

Raaijmakers, J.M., de Brujin, I., de Kock, M.J.D. 2006. Cyclic lipopeptide production by plant-associated *Pseudomonas* spp.: Diversity, activity, biosynthesis and regulation. *Molecular Plant-Microbe Interaction* 19: 699–710.

Rosello, G., Bonaterra, A., Frances, J., Montesinos, L., Badosa, E., Montesinos, E. 2013. Biological control of fire blight of apple and pear with antagonistic *Lactobacillus plantarum*. *European Journal of Plant Pathology* 137(3): 621–633.

Ross, A.F. 1961a. Localized acquired resistance to plant virus infection in hypersensitive hosts. *Virology* 14: 329–339.

Ross, A.F. 1961b. Systemic acquired resistance induced by localized virus infection in plants. *Virology* 14: 340–358.

Ruinen, J. 1961. The phyllosphere. I. An ecologically neglected milieu. *Plant and Soil* 15: 81–109.

Ryals, J.A., Neuenschwander, U.H., Willits, M.G., Molina, A., Steiner, H.Y. 1996. Systemic acquired resistance. *Plant Cell* 9: 425–439.

Scher, F.M., Baker, R. 1982. Effect of *Pseudomonas putida* and a synthetic iron chelator on induction of soil suppressiveness to *Fusarium* wilt pathogens. *Phytopathology* 72: 1567–1573.

Schmiedeknecht, G., Bochow, H., Junge, H. 1998. Use of *Bacillus subtilis* as biocontrol agent. II. Biological control of potato diseases. *Z. PflKrankh. PflSchutz*. 105: 376–386.

Schroth, M.N., Hancock, J.G. 1981. Selected topics in biological control. *Annual Review of Phytopathology* 35: 453–476.

Schroth, M.N., Hancock, J.G. 1982. Disease-supressive soil and root-colonizing bacteria. *Science* 216: 1376–1381.

Seresinhe, N., Reyes, A.A., Brown, G.L. 1997. Suppression of *Rhizoctonia* stem rot on poinsettias with *Pseudomonas aureofaciens* strain 63-28. *Canadian Journal of Phytopathology* 19: 116.

Shoda, M. 2000. Bacterial control of plant diseases. *Journal of Bioscience and Bioengineering* 89: 515–521.

Siala, A., Gray, T.R. 1974. Growth of *Bacillus subtilis* and spore germination in soil observed by a fluorescent-antibody technique. *Journal of General Microbiology* 81: 191–198.

Sneh, B. 1981. Use of rhizosphere chitinolytic bacteria for biological control of *Fusarium oxysporum* f. sp. *dianthi* in carnation. *Phytopathology Z* 100: 251–256.

Sneh, B., Dupler, M., Elad, Y., Baker, R. 1984. Clamydospore germination of *Fusarium oxysporum* f. sp. *cucumerinum* as effected by fluorescent and lytic bacteria from *Fusarium* suppressive soil. *Phytopathology* 74: 1115–1124.

Sneh, B., Agami, O., Baker, R. 1985. Biological control of *Fusarium* wilt in carnation with *Serratia liguefaciens* and *Hafnia alvei* isolated from rhizosphere of carnation. *Phytopathology Z* 113: 271–271.

Sticher, L., Mauch-Mani, B., Metraux, J.P. 1997. Systemic acquired resistance. *Annual Review of Phytopathology* 35: 235–270.

Stutz, E.W., Defago, G., Kern, H. 1986. Naturally occurring fluorescent pseudomonads involved in suppression of black root rot of tobacco. *Phytopathology* 76: 181–185.

Suslow, T.V. 1982. Role of root-colonizing bacteria in plant growth. In *Phytopathogenic Prokaryotes*. Mount, M.S., Lacy, G.H. (eds.). London: Academic Press, pp. 187–223.

Suslow, T.V., Schroth, M.N. 1982. Role of deleterious rhizobacteria is minor pathogens in reducing crop growth. *Phytopathology* 72: 111–115.

Sutherland, E.D., Lockwood, J.L. 1984. Hyperparasitism of oospores of some Peronosporales by *Actinoplanes missouriensis* and *Humicola fuscoatra* and other Actinomycetes and fungi. *Canadian Journal of Plant Pathology* 6: 139–145.

Thomashow, L.S., Weller, D.M. 1996. Current concepts in the use of introduced bacteria for biological disease control: Mechanism and antifungal metabolites. In *Plant Microbe Interactions*. Vol. 1, Stacey, G., Keen, N. (eds.). London: Chapman and Hall, pp. 187–236.

Tu, J.C. 1980. Incidence of root rot and over wintering of alfalfa as influenced by rhizobia. *Phytopathology Z* 97: 97–108.

Tukey, H.B. Jr. 1970. The leaching of substances from plants. *Annual Review of Physiology* 21: 305–324.

Tukey, H.B. Jr. 1971. Leaching of substances from plants. In *Ecology of Leaf Surface Micro-Organisms*. Preece, T.F., Dickinson, C.H. (eds.). London: Academic Press, pp. 67–80.

Uknes, S., Mauch-Mani, B., Moyer, M., Potters, S., Williams, S. 1992. Acquired resistance in *Arabidopsis*. *Plant Cell* 4: 645–656.

Van Loon, L.C. 1997. İnduced resistance in plants and the role of pathogenesis-related proteins. *European Journal of Plant Pathology* 103: 753–765.

Van Loon, L.C., Antoniw, J.F. 1982. Comparison of the effects of salicylic acid and ethephon with virus-induced hypersensitivity and acquired resistance in tobacco. *The Netherland Journal of Plant Pathology* 88: 237–256.

Van Loon, L.C., Bakker, P.A.H.M., Pieterse, C.M.J. 1998. Systemic resistance induced by rhizosphere bacteria. *Annual Review of Phytopathology* 36: 453–483.

Vernooij, B., Friedrich, L., Morse, A., Reist, R., Kolditz-Jawhar, R., Ward, E., Uknes, S., Kessmann, H., Ryals, J. 1994. Salicylic acid is not the translocated signal responsible for inducing systemic acquired resistance but is required in signal transduction. *Plant Cell* 6: 959–965.

Wagner, G.J. 1991. Secreting glandular trichomes: More than just hairs. *Plant Physiology* 96: 675–799.

Ward, E.R., Uknes, S.J., Williams, S.C., Dincher, S.S., Wiederhold, D.L., Alexander, D.C., Ahl Goy, P., Métraux, J.P., Ryals, J.A. 1991. Coordinate gene activity in response to agents that induce systemic acquired resistance. *Plant Cell* 3: 1085–1094.

Wasternack, C., Parthier, B. 1997. Jasmonate-signalled plant gene expression. *Trends in Plant Science* 2: 302–307.

Weinhold, A.R., Bowman, T. 1968. Selective inhibition of potato scab pathogen by antagonistic bacteria and substrate influence on antibiotic production. *Plant and Soil* 27: 12–24.

Weller, D.M. 1988. Biological control of soilborne plant pathogens in the rhizosphere with bacteria. *Annual Review of Phytopathology* 26, 379–407.

Weller, D.M., Cook, R.J. 1983. Suppression of take-all of wheat by seed treatments with fluorescent pseudomonads. *Phytopathology* 73: 463–469.

Weller, D.M., Howie, W.J., Cook, R.J. 1988. Relationship between *in vitro* inhibition of *Gaeumannomyces graminis* var. *tritici* and suppression of take-all of wheat by fluorescent pseudomonads. *Phytopathology* 78: 1094–1100.

Weller, D.M., Landa, B.B., Mavrodi, O.V., Schroeder, K.L., De La Fuente, L., Blouin Bankhead, S., Allende Molar, R., Bonsall, R.F. 2007. Role of 2,4-diacetylphloroglucinol-producing fluorescent *Pseudomonas* spp. In the defense of plant roots. *Plant Biology* 9: 4–20.

Weller, D.M., Thomashow, L.S. 1990. Antibiotics: Evidence for their production and sites where they are produced. In *New Direction in Biological Control*. Baker, R., Dunn, P. (eds.). New York: AR Liss. Inc., pp. 703–711.

Westers, L., Westers, H., Quax, W.J. 2004. *Bacillus subtilis* as cell factory for pharmaceutical proteins; A biotechnological approach to optimize the host organism. *Biochimica et Biophysica Acta* 1694: 299–310.

Wilson, M., Campbell, H.L., Ji, P., Jones, J.B., Cuppels, D.A. 2002. Biological control of bacterial speck of tomato under field conditions at several locations in North America. *Phytopathology* 92(12): 1984–1992.

Xu, Y., Chang, P.F.L., Liu, D., Narasimhan, M.L., Raghothama, K.G., Hasegawa, P.M., Bressan, R.A. 1994. Plant defense genes are synergistically induced by ethylene and methyl jasmonate. *Plant Cell* 6: 1077–1085.

Zhang, L.H., Dong, Y.H. 2004. Quorum sensing and signal interference: Diverse implications. *Molecular Microbiology*. 56: 604–614.

19 Modern Trends of Plant Pathogenic Bacteria Control

Kubilay Kurtulus Bastas and Velu Rajesh Kannan

CONTENTS

ABSTRACT Conventional control methods for plant pathogenic bacteria include avoidant, exclusive, eradicative, protective, resistant, and therapeutic applications. Sustainable agriculture has become a norm for modern agriculture. As environmental and ecological issues continue to impact agriculture, all technologies developed for crop production must be economically feasible, ecologically sound, environmentally safe, and socially acceptable. Non-chemical methods for control of crop diseases, such as pathogen-free seeds, disease resistance, crop rotation, plant extracts, organic amendments, and biological control are considered to be less harmful than synthetic chemical pesticides and, therefore, offer great potential for application in conventional agriculture, organic farming, and/or soilless culture. Inoculations with plant-growth-promoting rhizobacteria (PGPR) were effective in controlling bacterial and multiple diseases caused by different pathogens. A number of chemical elicitors of SAR and ISR may be produced upon inoculation by PGPR strains, including salicylic acid, siderophores, lipopolysaccharides, 2,3-butanediol, and other volatile substances. Mycorrhizal fungi can limit subsequent infections by some bacterial pathogens. The use of bacteriophages as an effective phage therapy strategy faces significant challenges for controlling plant diseases in the phyllosphere. A new approach to protect plants against bacterial diseases is based on a type of interference with the communication system known as quorum sensing (QS), which is used by several phytopathogenic bacteria to regulate expression of virulence genes according to population density. Traditional bactericides used too frequently and in excessive amounts on plants to manage these disease cause serious damage to agricultural and natural ecosystems. Therefore, there is a need to curtail pesticide use and reduce the environmental impacts of pesticides. In this connection, recognition of the importance of spices and their derivatives (extracts, essential oils, decoctions, hydrosols) in crop protection is increasing under the concept of integrated pest and disease management. A new generation of chemical compounds is being produced by either chemical synthesis or biotechnological processes. This new strategy includes the use of bioactive products, called plant activators, that induce

SAR in plants to many pathogens, including bacteria. R genes encode putative receptors that respond to the products of Avr genes expressed by the pathogen during infection, which are used in conventional resistance breeding programs. Transgenic plants have increased resistance to many plant pathogenic bacteria. Antimicrobial peptides have received attention as candidates for plant protection products. There is no single method of satisfactory control for plant diseases, especially those caused by bacteria. An integrative approach in accordance with the dynamics of agro-ecosystem management is not only the best strategy; it is the key to success in the control of crop diseases and the achievement of sustainable crop production.

KEYWORDS: plant pathogenic bacteria, PGPR, mycorrhizae, bacteriophage, quorum sensing

19.1 INTRODUCTION

In the twenty-first century, high crop yields are achieved through heavy use of synthetic pesticides and fertilizers, and through new cultivars grown in monoculture. Conventional control methods for plant pathogenic bacteria contain avoidant, exclusive, eradicative, protective, resistant and therapeutic applications. These prevent disease according to the time of year or the location of an area; prevention of the introduction of inoculum; elimination, destruction, or inactivation of the inoculum; utilization of cultivars that are resistant to or tolerant of the bacteria; and the cure of plants that are already infected. The success of plant disease control by synthetic chemicals created a general perception that chemical control could provide a permanent solution to disease problems in modern agriculture (Huang and Wu, 2009).

The term "sustainable agriculture" has become a norm in modern agriculture. As environmental and ecological issues continue to impact agriculture, all technologies developed for crop production must be economically feasible, ecologically sound, environmentally safe, and socially acceptable. Numerous non-chemical methods for the control of crop diseases, such as pathogen-free seeds, disease-resistant varieties, crop rotation, application of plant extracts, organic amendments, and biological control are considered less harmful than synthetic chemical pesticides and, therefore, offer great potential for application in conventional agriculture, organic farming, and/or soilless culture. No single method can provide satisfactory control of crop diseases. Integration of all effective and eco-friendly measures in accordance with the dynamics of agro-ecosystem management would be the best strategy for the efficient control of diseases in crops. In this era of energy conservation and environmental protection, research on energy-saving and environmentally sound methods for the sustainable management of crop diseases is both a priority and a challenge. In the environmental era, all practical solutions to plant disease control must be based on environmental safety, natural resource conservation, and biodiversity maintenance (Huang, 2000).

Numerous disease-control methods are considered more environmentally friendly than chemical control; therefore, they have potential for plant disease management in sustainable agriculture. Use of pathogen-free seeds, bulbs, or seedlings is a prerequisite for the sustainable control of crop diseases because numerous plant pathogens are transmitted to fields or other places via infected or contaminated seeds, bulbs, or seedlings. The development of efficient methods (e.g., molecular techniques or selective media for rapid detection of bacterial pathogens) is of high priority. Such methods are particularly useful in seed certification and plant quarantine. Moreover, because different seed-borne pathogens may occur on the same host crop, developing a multiplex PCR to detect several pathogens would be more efficient and economically feasible than a PCR to detect a single pathogen.

19.2 BIOLOGICAL CONTROL

Plants are surrounded by diverse types of mesofauna and microbial organisms, some of which can contribute to biological control of plant diseases. The microbes that probably contribute most to disease control can be classified as competitive saprophytes, facultative plant symbionts, and facultative hyperparasites. These can generally survive on dead plant material, but they are able to colonize and express biocontrol activities as they grow on plant tissues. Knowledge of the molecular basis of plant–microbe interactions is now being applied to the search for less-aggressive crop-protection methods based on the use of beneficial microorganisms and on stimulation of plant defense responses. Bacteria can establish intimate symbiotic associations with plants or live as epiphytes (they are capable of living [i.e. multiplying] on plant surface without causing disease symptoms) or endophytes (residing within plant cells or tissiues without causing disease symptoms) in more subtle relationships with a plant. The production of pathogens that express different levels of virulence, produce specific antipest molecules, or trigger the activation of a particular mode of action would provide powerful tools to investigate the evolutionary principles and applications of specific biocontrol agents. However, it is imperative that scientists who are developing opportunistic pathogens continue to rigorously weigh the advantages and hazards posed by these new biocontrol agents and provide expertise to regulatory authorities. In addition to microbes, phytopathogen-specific phages have biocontrol potential. Their efficacy in relation to plant disease must be tested before phage preparations are scaled up; this endeavor requires knowledge about phage characteristics and lifestyles (Ackermann et al., 2004).

The terms “biological control” and its abbreviated synonym, “biocontrol,” are used in different fields of biology. The occurrence of pesticide-resistant pathogens and the potential adverse effects of pesticides on the environment require the use of alternative low-risk strategies such as biocontrol agents for disease control. Currently, biocontrol agents are not used extensively in integrated crop disease-management programs. However, the use of beneficial microorganisms is considered to be one of the most promising methods for crop-management practices that are both safer and more rational. Since the mid-1960s, antagonistic bacilli, and fluorescent pseudomonads have been used as biocontrol agents for management of plant diseases. Although fluorescent pseudomonads were identified as biocontrol agents for commercial development in numerous reports, very few pseudomonads-based products showed high value and high commercial returns (Selosse et al., 2004). Systemic acquired resistance (SAR)- or induced systemic resistance (ISR)-stimulator bacterial biocontrol agents have been developed. These include *Bacillus mycoides* (Bargabus et al., 2002), *Bacillus pumilus* (Bargabus et al., 2004), *Bacillus subtilis* (Ryu et al., 2004), *Pseudomonas fluorescens* (Maurhofer et al., 1994; Leeman et al., 1995), *Pseudomonas putida* (Meziane et al., 2005), *Serratia marcescens* (Press et al., 2001), *Actinoplanes* (Sutherland and Lockwood, 1984), *Agrobacterium* and *Azotobacter* (Weller, 1988), *Alcaligenes* (Elad and Chet, 1987), *Amorphosporangium* (Filonow ve Lockwood, 1979), *Enterobacter* (Kwok et al., 1987), *Flavobacterium* (Chen et al., 1987), *Rhizobium ve Bradyrhizobium* (Tu, 1980), *Streptomyces* (Merriman et al., 1974), and *Pantoae agglomerans* (Huang et al., 2007).

Huang et al. (2007) have developed a granulate biofumigant named PBGG, using *Pseudomonas boreopolis*, *Brassica* seed pomace, glycerin, and sodium alginate. Application of 1.0% (w/w) of PBGG to soil infested with *Streptomyces padanus* and *S. xantholiticus* were reported as effective (Chung et al., 2005). Granulated PBGG can be a useful alternative to methyl bromide in the management of soil-borne pathogens (Huang et al., 2007).

19.2.1 Plant Growth-Promoting Rhizobacteria (PGPR)

Rhizobacteria can suppress pathogen growth by competing for nutrients and space, limiting the available Fe^{3+} supply by producing siderophores, producing lytic enzymes, and by antibiosis (Jing et al., 2007). Potential PGPRs are selected for their ability to inhibit the growth of various phytopathogens or miscellaneous rhizospheric bacteria and fungi *in vitro*. Pure cultures of antagonistic rhizobacterial strains are then screened in greenhouse trials by treatment of soil with a bacterial suspension (10^8 CFU mL^{-1}) and measurement of the effect on plants (Baysal et al., 2008). How to elucidate the reasons for one biocontrol agent's superiority over another in controlling plant pathogens can be challenging.

Chernin (2011) showed that certain PGPR are capable of producing volatile compounds that can reduce the signal molecules of pathogens. Specifically, volatile organic compounds produced by strains of *Pseudomonas fluorescens* and *Serratia plymuthica* could disrupt quorum sensing (QS) in a number of plant pathogens, including *Agrobacterium*, *Chromobacterium*, *Pectobacterium*, and *Pseudomonas*. Because PGPRs are used as agricultural inputs in many crops, quorum-quenching (QQ) could serve as a new disease management strategy.

Individual bacteria sense the presence of other members by secreting chemical molecules into the environment. Sufficiently high amounts of these molecules would thus suggest a high population and trigger certain population-dependent activities. In bioluminous bacteria, for example, light production is triggered at high population density. The release of volatile organic compounds (VOCs) is another important mechanism for the elicitation of plant growth by rhizobacteria. Some PGPR strains, including *B. subtilis* GB03, *B. amyloliquefaciens* IN937a, and *E. cloacae* JM22 have been recorded as releasing a blend of volatile components, particularly 2,3-butanediol and acetoin (Ryu et al., 2004). The VOCs produced by rhizobacterial strains can act as signaling molecules to mediate plant–microbe interactions. Volatiles produced by PGPR that colonize roots are generated at sufficient concentrations to trigger plant responses (Ryu et al., 2004). Identified low-molecular-weight plant volatiles such as terpenes, jasmonates, and green leaf components have been shown to be potent signal molecules for living organisms (Farmer, 2001).

Some bacteria confer beneficial traits on the host plant during interaction but are unable to invade tissues and infect them. Isolation of PGPR and biocontrol agents is a time-consuming and ultimately inefficient process that is dependent on efficient screening methods. Very often it requires the analysis of thousands of isolates but yields only a few useful ones because not all isolates of a given species are active. For this reason, a strong research effort is needed in order to find biochemical or molecular markers that are specifically associated with the best performance in beneficial bacteria. PGPR are inhabitants of the rhizosphere, the volume of soil under the immediate influence of a plant's root system. In the rhizosphere, secretion of organic compounds by the plant favors a large, active microbial population. Inoculation of plants with PGPR, mainly of the genera *Pseudomonas*, *Serratia*, *Azospirillum*, and *Bacillus*, enhances root-system and whole-plant growth, and often limits the growth of certain soil-borne plant pathogens (Kloepper, 1991).

Studies have characterized the determinants and pathways of IR stimulated by biocontrol agents and other nonpathogenic microbes. The first of these pathways, systemic acquired resistance (SAR), is mediated by salicylic acid (SA), a compound that is frequently produced following pathogen infection. It also typically leads to the expression of pathogenesis-related (PR) proteins, which include a variety of enzymes that may act directly to lyse invading cells, reinforce cell-wall boundaries to resist infection, or induce localized cell death. A second phenotype, induced systemic resistance (ISR), is mediated by jasmonic acid (JA) and/or ethylene, which are produced after applications of some nonpathogenic rhizobacteria. Interestingly, the SA- and JA-dependent defense pathways can be mutually antagonistic, and some bacterial pathogens take advantage of this conflict to overcome SAR. For example, pathogenic strains of *Pseudomonas syringae* produce coronatine, which is similar to JA, to overcome the SA-mediated pathway (He et al., 2004). Only if induction can be controlled (because it is overwhelmed or synergistically interacts with endogenous signals) will host resistance be increased (Kunst et al., 1997; Pierson and Ishimaru, 2000).

The mechanisms of plant growth stimulation are strongly dependent on the characteristics of both the bacteria and the host plant. In some cases, a relationship has been found among the synthesis of plant-growth regulators such as IAA, siderophores that chelate iron, and the biocontrol of soil-borne plant pathogens or plant-deleterious microorganisms, or even with the induction of systemic defense responses. Biocontrol agents are found either in the aerial plant part or in the root system, which is colonized by an extremely abundant microbiota.

In several instances, inoculations with PGPR were effective in controlling multiple diseases caused by different pathogens, including anthracnose (*Colletotrichum lagenarium*), angular leaf spot (*Pseudomonas syringae* pv. *lachrymans*), and bacterial wilt (*Erwinia tracheiphila*). A number of chemical elicitors of SAR and ISR may be produced by PGPR strains upon inoculation, including salicylic acid (SA), siderophores, lipopolysaccharides, 2,3-butanediol, and other volatile substances (Van Loon et al., 1998; Ongena et al., 2004; Ryu et al., 2004). Again, such molecules may have multiple functions, which would blur the lines between direct and indirect antagonisms. More generally, a substantial number of microbial products have been identified as elicitors of host defenses, which indicates that host defenses are probably continually stimulated over the course of a plant's life cycle (Vallad and Goodman, 2004).

19.2.2 Mycorrhizal Activity

The microbes that are more recalcitrant to *in vitro* culturing have been intensively studied. These include mycorrhizal fungi (e.g., *Pisolithus* and *Glomus* spp.) that can limit subsequent infections of plant pathogens. Because multiple infections can and do take place in field-grown plants, weakly virulent pathogens can contribute to the suppression of more virulent pathogens via the induction of host defenses. Because various epiphytes and endophytes may contribute to biological control, the ubiquity of mycorrhizae deserves special consideration. Mycorrhizae, which are formed as the result of mutualist symbioses between fungi and plants, occur on most plant species. Because they are formed at an early stage of plant development, they represent nearly ubiquitous root colonists that assist plants with the uptake of nutrients (especially phosphorus and micronutrients) (Biermann and Linderman, 1983; Morton and Benny, 1990). During colonization, mycorrhizal fungi can prevent root infections by reducing access sites and stimulating host defenses (Linderman, 1994). Arbuscular mycorrhizal fungi (AMF) protect a host plant against root-infecting pathogenic bacteria. The damage caused by *Pseudomonas syringae* on tomato may be significantly reduced when the plants are well colonized by mycorrhizae (Garcia-Garrido and Ocampo, 1989). The mechanisms involved in these interactions include physical protection, chemical interactions, and indirect effects (Fitter and Garbaye, 1994). The other mechanisms used by AMF to indirectly suppress plant pathogens include enhancing the availability of nutrition to plants, increasing lignification to cause morphological changes in the root, and changing the chemical composition of plant tissues with antifungal chitinase, isoflavonoids, etc. (Morris and Ward 1992). AMF may also alleviate abiotic stress and change the microbial composition of the mycorrhizosphere (Linderman, 1994).

Successful mycorrhizal relationships depend upon AMF species, pathogen species, nutrients, and the synergistic and antagonistic effects of other soil microorganisms to AMFs. Although the specifics of such factors may be unknown, results against fire blight (*Erwinia amylovora*) were more successful with *Glomus intraradices* than with an extensively used copper compound. Along with the economical and environmental advantages, gaining a new approach for fire blight control should encourage more detailed studies on AMFs (Bastas et al., 2006). Today, many mycorrhizae fungi preparations (individual or mix) are available for plant health and growth.

19.2.3 Bacteriophages

The use of bacteriophages as an effective phage therapy strategy faces significant challenges for controlling plant diseases in the phyllosphere. A number of factors must be taken into account when considering phage therapy for bacterial plant pathogens. Given that effective mitigation requires high populations of phage be present in close proximity to the pathogen at critical times in the disease cycle, the single biggest impediment to the efficacy of bacteriophages is their inability to persist on plant surfaces over time due to environmental factors. Successful results were obtained by the use of phage therapy in plant pathosystems such as *Xanthomonas campestris* pv. *campestris* (Bergamin and Kimati, 1981), *Erwinia carotovora* subsp. *carotovora* (Ravensdale et al., 2007), *Xanthomonas citri* subsp. *citri* and *Xanthomonas fuscans* subsp. *citrumelonis* (Balogh et al., 2008), *Xanthomonas axonopodis* pv. *vignaeradiatae* (Borah et al., 2000), *Pseudomonas tolaasii* (Munsch and Olivier, 1995), *Xanthomonas axonopodis* pv. *allii* (Gent and Schwartz, 2005), *Xanthomonas campestris* pv. *vesicatoria* (Bergamin and Kimati, 1981), *Erwinia amylovora* (Schnabel and Jones, 2001; Svircev et al., 2005), *Streptomyces scabies* (McKenna et al., 2001), *Xanthomonas arboricola* pv. *pruni* (Civerolo and Kiel, 1969; Zaccardelli et al., 1992), *Ralstonia solanacearum* (Tanaka et al., 1990), *Xanthomonas campestris* pv. *vesicatoria* (Flaherty et al., 2000; Balogh et al., 2003), *Rhizobium radiobacter* (formerly *Agrobacterium tumefaciens*) (Stanier et al., 1967; Boyd et al., 1971), and *Xanthomonas campestris* pv. *juglandis* (McNeil et al., 2001).

19.2.4 Quorum Sensing (QS)

Quorum sensing is the population-density-dependent regulation of gene expression in bacteria. Bacterial pathogens and symbionts depend substantially on QS to colonize and infect their hosts, because it enables the individual bacterial cells in a local population to coordinate the expression of certain genes, which helps them to act somewhat like a multicellular organism. QS, which is particularly important for the ability of pathogenic bacteria to successfully infect plant hosts, is also important to the regulation of gene expression and behaviors in bacterial symbionts of plants. Genome-level transcriptional studies are needed to gain a better perspective of the full range of functions that are altered in plants that respond to N-acylhomoserine lactones (AHLs). Chemical identification and synthesis are central to analyzing the biochemical mode of action of mimics, determining their effects on relevant bacteria, identifying the enzymatic pathways of mimic synthesis, and exploring the potential for the use of plant QS mimics to improve the health of plants (Bauer and Mathesius, 2004). QS is crucial to aspects of bacterial physiology, including regulation of rhizospheric competence factors (e.g., antibiotic production), horizontal gene transfer, and control of functions that are directly or indirectly related to plant–microbe interactions (Whitehead et al., 2001).

Abolishment of the production of QS signals, a process known as quorum quenching (QQ), results in significantly defective biofilm formation and thereby reduces the ability of a pathogen to colonize a host (i.e., biocontrol). The mechanisms and functions of QQ have been evaluated in order to reveal its possible applications in the control of plant diseases and promotion of plant health. In this new concept, rhizobacterial volatiles are an important alternative to antibiotics in the biocontrol of various plant pathogens because they are capable of inhibiting the QS network mediated by AHL signal molecules. The potential of QQ to develop novel biocontrol strategies for plant pathogens has been recognized (Dong et al., 2007). These studies clearly indicate that QQ can be used for the biocontrol of pathogenic microorganisms by targeting the QS pathway.

As previously mentioned, QS is critical for the infection and invasion processes of many phytopathogens. QS via AHLs is used by many proteobacteria to regulate the expression of virulence-associated factors that orchestrate their temporal and spatial production within a plant. Many phytopathogenic bacteria use AHL-QS in order to spatially and temporally express virulence-associated factors in planta (Cha et al., 1998; Von Bodman et al., 2003; Venturi and Fuqua, 2013). Many AHL-producing bacteria colonize the rhizospheres of plant roots (Elasri et al., 2001). Most of these bacteria (e.g., PGPR) are beneficial to the plant, which means they can promote its growth and/or protect it from microbial pathogens (Lugtenberg and Kamilova, 2009). Mechanisms of disease suppression of microbial pathogens by PGPR strains include niche exclusion by competition for nutrients, production of antimicrobial compounds, and induction of ISR during which the inducing PGPR and pathogen do not undergo any type of direct interaction. ISR differs from pathogen-induced and salicylate-mediated SAR (van Loon et al., 1998; Zamioudis and Pieterse, 2012; Venturi and Fuqua, 2013). ISR that follows colonization of PGPR strains leads to the enhanced expression of various signaling pathways, including jasmonate and ethylene.

The AHLs affect gene expression by altering the levels of many proteins, including those involved in hormonal and defense functions. Many phytopathogenic bacteria use AHL-QS to temporally express virulence-associated factors in the plant after successful colonization. Experiments have been performed with transgenic plants that produce AHLs by expressing a bacterial luxI-type gene by studying plant susceptibility after challenge with a bacterial pathogen. Plant mimics that act as antagonists of AHL-QS might lead to pathogen confusion and decreased pathogenicity. More progress with respect to AHL mimics has been achieved using algae, especially for compounds that inhibit AHL-QS systems (Manefield et al., 2002; Teplitski et al., 2004). A compound that inhibits AHL-QS has also been identified from leguminous plants; it has been identified as L-canavanine (an arginine analog) (Keshavan et al., 2005). All of these inhibiting compounds have been implicated in the enhancement of the proteolytic degradation of QS LuxR receptors, and therefore to affect the QS response (Koch et al., 2005). Research continues to focus on identifying QS inhibitors, because these are believed to have considerable applications in agriculture for controlling bacterial populations.

A new approach to protect plants against bacterial diseases is based on interference with the communication system of QS, used by several phytopathogenic bacteria to regulate expression of virulence genes according to population density (Cui and Harling, 2005). The enzyme, AiiA, isolated from bacterial strain *Bacillus* sp. 240B1, was found to degrade the QS-signaling molecule of the soft rot pathogen *Pectobacterium carotovorum* (formerly *Erwinia carotovora*), and thereby to render the bacteria incapable of infecting the host. Transgenic expression of AiiA in planta was subsequently demonstrated to provide significant enhancement of resistance against soft rot in potato (Dong et al., 2001). The strategy looks technically very promising because the microbial target is likely to be strongly conserved.

19.2.5 Using Essential Oil and Extracts of Medicinal and Aromatic Plants

Rapid and effective control of plant disease and microbial contamination in crop cultivation is generally achieved by the use of synthetic pesticides and antibiotics. However, such control methods of the plant diseases prevent bacterial multiplication but are not always adequate control methods for seed-borne inoculums. These chemicals are also associated with an undesirable effect on the environment due to their slow biodegradation in the environment and some toxic residues in the products for mammalian health. This issue is still in the center of debate although the use of antibiotics is forbidden in many countries of the world. In addition, the risk of developing resistance by microorganisms and the high cost-benefit ratio are other disadvantages of synthetic pesticide uses. Therefore, there has been a growing interest in research concerning the alternative pesticides and antimicrobial active compounds, including the plant extracts and essential oils (Isman, 2000; McManus et al., 2002).

Several medicinal plants and microorganisms are rich sources of bioactive compounds. Although these sources are primarily used for discovering human medicines, they also hold promise for the identication of antimicrobial compounds that target plant pathogens. Many plants and plant products have been reported to possess antimicrobial properties (Grange and Ahmed, 1988). The demand for medicinal plants in pharmaceutical industry has increased considerably. Plant extracts and essential oils have been shown to have antimicrobial effects on bacterial pathogens; moreover, the presence of antibacterial compounds in higher plants has long been recognized as an important factor in disease control. Such compounds have also shown potential in agriculture; however, their plant toxicity is yet to be assessed (Balestra et al., 1998; De Castro, 2001; Varvaro et al., 2001 and 2002; Lo Cantore et al., 2004; Iacobellis et al., 2005).

19.3 CHEMICAL CONTROL OF PLANT PATHOGENIC BACTERIA

Traditional bactericides are bronopol, copperhydroxide, copper sulfate, copper oxychloride, cresol, dichlorophen, dipyrithione, dodicin, fenaminosulf, formaldehyde, 8-hydroxyquinolinesulfate, kasugamycin, nitrapyrin, octhilinone, oxolinicacid, oxytetracycline, probenazole, streptomycin, tecloftalam, and thiomersal. These chemicals, which are excessively used on plants to manage diseases, cause serious damage to agricultural and natural ecosystems. Therefore, there is a need to curtail the use of pesticides and reduce their environmental impacts. In this context, the importance of spices and their derivatives (extracts, essential oils, decoctions, hydrosols) in crop protection is increasingly being recognized under the concept of integrated pest and disease management (IPDM) (Ragsdale, 2000). Under this concept, all possible modes of plant pests and disease control methods are integrated to minimize the excessive use of synthetic pesticides (Beg and Ahmad, 2002). Exploitation of naturally available plant chemicals that retard the reproduction of undesirable microorganisms, which would be more realistic and ecologically sound for plant protection, will be prominent in the future development of commercial pesticides, with emphasis on the management of plant diseases in general but particularly on bacterial diseases (Gottlieb et al., 2002).

It is known that many plant pathogenic bacteria have acquired resistance to synthetic pesticides (White et al., 2002). For instance, pathovars of *Xanthomonas campestris* have developed resistance to antibiotics such as kanamycin, ampicillin, penicillin, and streptomycin (Bender et al., 1990; McManus et al., 2002). The increasing incidence of pesticide resistance is further fueling the need for a new generation of pesticides that are eco-friendly. A green plant, which represents a reservoir of effective novel chemotherapeutants with different modes of action, can act as a valuable source of natural pesticides for resistant pathogens (Newman et al., 2003; Gibbons, 2005). In fact, the popularity of botanical pesticides is increasing to the point that some plant products are in global use as green pesticides. However, although the body of scientific literature that documents the bioactivity of plant derivatives to different pests continues to expand, very few botanicals are currently used in agriculture (Dubey et al., 2008). Many reports have been made on the uses of several plant byproducts on several human pathogenic bacteria and fungi, but far fewer have been made on the management of phytopathogenic bacteria. Plant-based antimicrobials have enormous therapeutic potential because they can accomplish their purpose with fewer of the side effects that are often associated with synthetic antimicrobials.

The new generation of chemical compounds is produced either by chemical synthesis or by biotechnological processes. Strategies may include the use of bioactive products, also called plant activators, that induce systemic acquired resistance (SAR) in plants to many pathogens, including bacteria (Sticher et al., 1997). Plant activators have no direct antimicrobial activity but can elicit plants to initiate preinfection defense reactions such as the accumulation of pathogenesis-related (PR) proteins and ultrastructural changes (Benhamou and Belanger, 1998; Inbar et al., 1998; Louws et al., 2001). Several of these compounds that are currently on the market in some countries include derivatives of acetylsalicylic acid. One such compounde, benzotiadiazole (BTH), was evaluated as a viable alternative to copper-based bactericides for field management of *P. syringae* pv. *tomato* and *Xanthomonas campestris* pv. *vesicatoria*, particularly where copper-resistant populations predominate (Agrawal et al., 1999; Louws et al., 2001).

Harpin protein (Messenger™) isolated from *Erwinia amylovora* was found to be effective in controlling certain bacterial and viral pathogens for which no other alternative control methods are available (United States Environmental Protection Agency, 2000). Messenger™ was found to induce resistance in tomato, orange, grape, and strawberry plants to several fruit rot pathogens (Qiu and Wei, 2000). Messenger™ also had significant effects on many common citrus diseases (Remick and Wei, 2000). ISR 2000, another plant activator, includes 71% *Lactobacillus acidophilus*, 12.7% yeast extract, 10% plant extract, and 0.2% benzoic acid. It can induce SR and enhance crop viability and vigor (Anonymous, 2001). The most intriguing example of an evolving alternative biocontrol agent is the registration and commercialization of the proteins encoded by the *hrp* genes of phytopathogenic bacteria for use in crop protection and as plant enhancers.

An alternative and promising approach to control fire blight involves growth-regulating acylcyclohexanediones (e.g., prohexadione-Ca). The most notable effect of prohexadione-Ca is the control of shoot elongation growth by inhibiting the formation of growth-active gibberellins from inactive precursors (Evans et al., 1999; Rademacher, 2000). Prohexadione-Ca-treated apple, pear, and quince trees were significantly less infected by *E. amylovora* (Fernando and Jones 1999; Momol et al., 1999; Maxson and Jones, 2002; Aldwinckle et al., 2002; Deckers and Schoofs, 2002; Buban et al., 2002; Bazzi et al., 2003; Buban et al., 2003; Norelli et al., 2003; Bastas and Maden, 2004, 2008; Bastas et al., 2010). In addition to its successful use on pome fruits, prohexadione-Ca treatment was also successful in other important crops such as strawberry, grapevine, kiwi fruit, cranberry, rose, and peach. Evidently, prohexadione-Ca-triggered formation of 3-deoxyflavonoids with phytoalexin-like activity can control a wide range of fungal and bacterial diseases in several crop plant species. Initial phytopathological investigations have shown that application of prohexadione-Ca significantly lowers progression of *Ralstonia solanacearum* and reduces leaf lesions of *Pseudomonas syringae* pv. *tomato* and *Xanthomonas campestris* pv. *vesicatoria* on greenhouse tomato plants (Bazzi et al., 2003).

19.4 MANAGEMENT OF PLANT DISEASES USING GENETIC ENGINEERING

Plants use two major immune systems for self-defense from pathogens. In one system, conserved pathogen-associated molecular patterns (PAMP) are recognized by plants through pattern recognition receptors (PRRs) that help activate defense responses. Pathogens inactivate PAMP-triggered immunity by delivering virulent disease-causing effector proteins into cells. Plants counteract these effectors by recognizing them through resistance (R) proteins, a process that results in a more rapid and robust defense system called effector-triggered immunity. Plant defenses include cell-wall reinforcement, hypersensitive response (HR), and expression of many defense genes, such as pathogenesis-related (PR) genes. Introgression of resistant genes from wild relatives by traditional breeding methods is inefficient, however, because of difficulty in sexual hybridization and potential loss of desirable traits. In addition, in most cases the gene-for-gene type R genes are rapidly overcome by new virulent pathogen strains.

Many studies on the basis of the specificity of plant–microbe interactions have employed plant pathogenic bacteria of the genera *Pseudomonas*, *Erwinia*, and *Xanthomonas*. A group of genes that has been implicated consists of HR and pathogenicity (*hrp*) genes, which control the capacity of bacteria to develop HR in non-host plants (Lindgren, 1997). The first confirmation of the role of *hrp* genes was provided by the discovery of harpins, which are proteinaceous elicitors of HR, in *Pseudomonas syringae* and *Erwinia amylovora* (Wei et al., 1992). A second group of genes, the avirulence (*Avr*) genes, code for most of the virulence-associated proteins introduced into the host cell by the type III secretion system, which is controlled by the HRP system, and trigger programmed plant-defense responses such as HR (De Wit, 1997). Studies have also been carried out with *Rhizobium radiobacter*, a member of the family Rhizobiaceae, that causes tumors in several plants. *R. radiobacter* has been taken as a model for inter-kingdom genetic exchange in plants because the ultimate cause of disease is the transfer of a T-DNA region of the Ti plasmid to a plant cell, its integration into chromosomes, and the expression of its encoded plant-regulator genes (Ream, 1989). After the T-DNA is cleaved from the Ti plasmid, the resultant single strands are coated with Vir proteins and secreted to the host-plant cell by means of a type IV secretory pathway. In fact, the plant parasite is a DNA-based genetic element that multiplies synchronously with the plant-cell genome.

Certain signal transduction components are used in *R* gene resistance and in some non-host resistances as well (Parker et al., 1996; Peart et al., 2002). Therefore, it might be possible to identify effective resistance genes against crop pathogens from model species and transfer them into crops. It will be of great interest to determine whether non-host resistance results from the natural pyramiding of *R* genes and/or from the use of *R* genes that recognize virulence factors that are essential for the pathogen to cause disease. Note that non-host resistance might result from several mechanisms (Heath, 2000) and it is possible that genetic dissection of non-host resistance will provide unanticipated tools for engineered resistance. *R* genes have been used in resistance breeding programs for decades, with varying degrees of success. Recent molecular research on R proteins and downstream signal-transduction networks have provided exciting insights that will enhance the use of *R* genes for disease control. Definition of conserved structural motifs in R proteins has facilitated the cloning of useful *R* genes, including several that are functional in multiple crop species and/or provide resistance to relatively wide ranges of pathogens. Numerous signal-transduction components in the defense network have been defined, and several are being exploited as switches by which resistance can be activated against diverse pathogens (Garelik, 2002).

These genes have the advantage of conferring complete resistance against specific pathogen races. Most *R* genes encode the so-called nucleotide binding site, which consists of leucine-rich repeat (NBLRR) proteins that activate downstream defenses to combat disease when the pathogen has a specific avirulence-gene (*Avr*) that corresponds to the specific *R* gene in the plant (McHale et al., 2006). However, resistance obtained by introgression of these types of gene generally has a drawback; namely, that pathogen populations eventually adapt to their presence and overcome them (Hovmoller et al., 1997; McDonald and Linde, 2003). In other words, when the *Avr* gene in the pathogen is inactivated by a mutation, the plant's resistance is no longer functional. Because *Avr* genes often encode effector proteins that have evolved to function in pathogenicity, strong balancing selection persists in natural plant and pathogen populations for polymorphism at the genetic loci in both host and pathogen. This means that many *Avr* alleles are present in natural pathogen populations. However, genotypes that carry a virulent allele of any *Avr* gene locus will eventually migrate to and invade the resistant plant population, which leads to reduced efficiency of the specific resistance gene. These types of resistance genes operate at the recognition stage of an interaction and generally against biotrophic pathogens; in such cases, the expression of resistance is often associated with HR, a form of programmed cell death (PCD).

The *R* genes encode putative receptors that respond to the products of avirulence or *Avr* genes that are expressed by the pathogen during infection. In many cases, when transferred to a previously susceptible plant of the same species, a single *R* gene can provide complete resistance to one or more strains of particular pathogen. For this reason, *R* genes have been used for decades in conventional resistance breeding programs. The strong phenotypes and natural variability at *R* loci have also been exploited by molecular geneticists to clone *R* genes and investigate their molecular modes of action. *R* gene-mediated resistance has several attractive features for disease control. When induced in a timely manner, concerted responses can efficiently halt pathogen growth with minimal collateral damage to the plant. No input is required from the farmer and there are no adverse environmental effects. Unfortunately, however, *R* genes are often quickly defeated by co-evolving pathogens (Pink, 2002). In addition, many *R* genes recognize only a limited number of pathogen strains and therefore do not provide broad-spectrum resistance. But some transgenic uses of single *R* genes have been proven durable. For example, the pepper gene *Bs2* provided longstanding resistance against bacterial spot disease caused by *Xanthomonas campestris*. The *Bs2*, cloned from pepper, was shown to encode a NB-LRR protein (Tai et al., 1999).

In some cases, *R* genes can provide effective protection against pathogens that are transformed into new species and even into new genera; moreover, this protection can be broad spectrum—that is, independent of pathogen race and even species (Rommens et al., 1995; Oldroyd and Staskawicz, 1998; Tai et al., 1999). Rxo1, an *R* gene derived from maize (*Zea mays*), a non-host of the rice bacterial pathogen *Xanthomonas oryzae* pv. *oryzicola*, was successfully transformed into rice (*Oryza sativa*) and shown to confer resistance against *X. oryzae* pv. *oryzicola* (Zhao et al., 2005). Thus, there is potential for the use of *R* genes as transgenes across natural breeding barriers. However, interspecies differences may radically influence *R* gene function (Ayliffe et al., 2004) and therefore the use of *R* genes from closely related species is preferable. The transgenic approach, which circumvents tedious back-crossing, has successfully been accomplished in rice for the *R* gene *Xa21*; in this case it conferred broad but nevertheless race-specific resistance to bacterial leaf blight disease (Wang et al., 2007). *Xa21* was subsequently transformed into a restorer line for hybrid rice and was shown to provide resistance without compromising elite traits (Zhai et al., 2002). Field tests of *Xa21* transgenic rice in the Philippines, China, and India have shown satisfactory results (Datta, 2004). However, deregulation of transgenic *Xa21* rice for large-scale cultivation is pending. It should be noted that conventional breeding assisted by the use of molecular marker techniques has already provided hybrids that contain *Xa21*, pyramided with other resistance genes (Joseph et al., 2004; Zhang et al., 2006) and has thereby created a competitive alternative to the transgenic approach.

Related studies on induced resistance have led to two different strategies for the development of transgenic disease-resistant plants. One of these, a first-generation strategy that is analogous to the strategies used in GM crops to control insect pests, involves the use of single-gene products that have direct inhibitory effects on the pathogen. Second-generation strategies (i.e., more recent studies) are based on an understanding of the mechanisms that regulate disease resistance in plants, such as the *R* genes. Research on plant-defense mechanisms in the 1970s and 1980s rapidly revealed that various defense proteins (i.e., the PR or pathogenesis-related proteins), certain small peptides and a wealth of secondary metabolites possess direct antimicrobial activities (Hammerschmidt, 1999; Broekaert et al., 2000; Field et al., 2006; van Loon et al., 2006; Castro and Fontes, 2007). Topical application of these proteins, which trigger several plant responses to stress, has shown several beneficial effects (Wei et al., 1992). Transgenic plants have also been produced with engineered genes that encode HR elicitors (e.g., harpins) or that overexpress *R* genes or PR proteins. These transgenic plants show increased resistance to many plant pathogenic bacteria and fungi.

Antimicrobial peptides (AMPs) have received attention as candidates for plant-protection products. Bacteriocins, a type of protein and peptide secreted by major groups of bacteria, kill closely related species. Examples of large bacteriocins that inhibit plant pathogenic bacteria have been reported from bacteria associated with plants (Ishimaru et al., 1988; Pham et al., 2004; Parret et al., 2005). Cyclopeptides are secondary metabolites reported from bacteria. They include gramicidins (*Bacillus brevis*), iturins (*Bacillus amyloliquefaciens*), syringomycins (*Pseudomonas syringae* and *Pseudomonas viridiflava*), agrastatins and fengycins (*Bacillus subtilis*), amphisins and viscosins (*Pseudomonas fluorescens*), putisolvins (*Pseudomonas putida*), tolaasins (*Pseudomonas tolaasi*), corpeptins (*Pseudomonas corrugata*), and syringopeptins (*Pseudomonas syringae*). Lipidic cyclopeptides, which are produced by several plant-associated and soil-inhabiting bacteria, have antibacterial, cytotoxic, or surfactant properties (Raaijmakers et al., 2006). Syringomicins and syringopeptins are virulence factors in *Pseudomonas syringae*, but also inhibit Gram-positive bacteria (Grgurina et al., 2005). Pseudopeptides of interest in plant disease control, which are produced by bacteria, have few peptide bonds and complex aminoacid modifications. Pantocines are derivatives from alanine that inhibit transaminase-catalyzed aminoacid biosynthesis in *Erwinia amylovora*, and are produced by strains of *Pantoea agglomerans* (Brady et al., 1999). The analogue of cecropin B, MB39 inhibits *Pectobacterium carotovorum* subsp. *betavasculorum*, *C. michiganensis*, and some pathovars of *Pseudomonas syringae* (Alan and Earle, 2002). Synthetic peptides have been designed by scientific-commercial firms. The strains ESF12, BP76, and BPC194 have antibacterial activity against *E. amylovora* (Powell et al., 1995; Ferre et al., 2006; Monroc et al., 2006a, b).

19.5 INTEGRATION OF CONTROL METHODS

For most crops, there is no single method for the satisfactory control of plant diseases, especially those caused by bacteria. Integrated approaches are the best strategy for effective management of the diseases of greenhouse and field crops. Strategies for the sustainable management of crop diseases include judicious use of naturally toxic substance from plants, soil management techniques such as bio-fumigation and organic soil amendment, crop management techniques that utilize pathogen-free seeds or other planting materials, development of disease-resistant cultivars, application of biological control agents, and cultural practices such as field sanitation and crop rotation. Future success hinges on further integration of these control strategies in crop production systems such as conventional agriculture, organic farming, and soilless cultures. Integrated disease management approaches require a thorough understanding of the ecology of each cropping system, including the crop, the pathogen(s), and the antagonist(s), as well as the surrounding environments. Integration of these effective measures in accordance with the dynamics of agro-ecosystem management is the key to success in the control of crop diseases and the achievement of sustainable crop production.

Because energy conservation and environmental protection are two of the major issues for human beings in the twenty-first century, research on the development of energy-saving and environmentally friendly methods for sustainable disease management in agriculture is both a challenge and a priority.

Given the efforts since the mid-1990s to develop biotechnological approaches for introducing disease resistance into crops, and the lack of concrete results in terms of the numbers of transgenic crops currently in use, one must ask which circumstances would be appropriate for attempts to create disease-resistant plants by genetic engineering. The effectiveness of plant breeding has improved dramatically in recent years through the development of molecular marker technologies, which are particularly beneficial for disease-resistance breeding that minimizes costly (and potentially harmful) phenotypic screening.

The latest technologies that directly incorporate genes into genomes, a process commonly referred to as genetic modification or genetic engineering, are introducing new traits into biocontrol agents and producing biologically based products (e.g., microbial fungicides) that can be used to interfere with pathogen activities. Registered bioagents are generally labeled with short reentry intervals and pre-harvest intervals, which give greater flexibility to growers. When living microorganisms that have prominent effects on target pathogens are introduced, they may also augment natural beneficial populations to further reduce the damage caused by pathogens and to increase plant fitness (Baysal and Tör, 2014). The use of advanced analytical tools and techniques, including transcriptomics, proteomics, and metabolomics, will continue to provide new insights into biologics, their mode of action, and their impact on rhizospheric microbiota. It seems possible that in the near future, inhibitory compounds may be mass-produced by microorganisms with the required properties and used as replacements for the pesticides currently in use.

Developments of compounds suitable for agricultural use as pesticide ingredients have several constraints. These are mainly due to the intrinsic toxicity and low stability of some of the compounds, the necessity of developing suitable formulations, and the need for inexpensive products in plant protection. Therefore, future research should focus upon developing less toxic and more stable compounds, and upon decreasing production costs by improving preparative synthesis and biotechnological procedures that employ microbial systems or transgenic crops as plant factories.

19.6 CONCLUSION

As environmental and ecological issues continue to impact agriculture, technologies developed for crop production must be economically feasible, ecologically sound, environmentally safe, and socially acceptable. Knowledge of the molecular basis of plant–microbe interactions is now being applied to the search for less-aggressive crop protection methods that are based on the utilization of beneficial microorganisms and on stimulation of plant defense responses. The emergence of pesticide-resistant pathogens and the potentially adverse effects of pesticides on the environment require the use of alternative low-risk strategies (e.g., biological control agents) for disease control. Because PGPRs are used as agricultural inputs in many crops, QQ could serve as a new disease-management strategy. Recent studies have characterized the determinants of IR, including SAR and ISR pathways, and has shown that IR can be stimulated by biological control agents and other nonpathogenic microbes. The latter include mycorrhizae, which form as the result of mutualist symbioses between fungi and plants and occur on most plant species. The mechanisms involved in these interactions include physical protection, chemical interactions, and indirect effects. The use of bacteriophages as an effective phage therapy strategy faces significant challenges in the control of plant diseases in the phyllosphere. Quorum sensing is crucial to bacterial physiology functions; these include regulation of rhizospheric competence factors (e.g., antibiotic production), horizontal gene transfer, and control of functions that are directly or indirectly related to plant–microbe interactions.

Plant extracts and essential oils have been shown to have antimicrobial effects on bacterial pathogens. In addition, the presence of antibacterial compounds in higher plants has long been recognized as an important factor in disease control. Strategies for the sustainable management of crop diseases include judicious use of naturally toxic substances from plants, soil management techniques such as biofumigation and organic soil amendment, crop management techniques such as the use of pathogen-free seeds or other planting materials, the use of disease-resistant cultivars, the application of biological control agents, and cultural practices such as field sanitation and crop rotation. The new generation of chemical compound strategies includes the use of bioactive products (plant activators) that induce SAR in plants to many pathogens, including bacteria. Transgenic plants have been produced with engineered genes that encode HR elicitors, such as harpins, or which overexpress *R* genes or PR proteins. The effectiveness of plant breeding has improved dramatically in recent years through the development of molecular marker technologies, which are particularly beneficial for disease resistance breeding because phenotypic screening can be minimized. Future areas of interest include the development of less toxic and more stable compounds as well as the decrease of production costs. The latter can be accomplished by improving preparative synthesis and biotechnological procedures that use microbial systems or transgenic crops as plant factories.

REFERENCES

Ackermann, H.W., Tremblay, D., Moineau, S. 2004. Long-term bacteriophage preservation. *World Fed. Cult. Collec.* News 38: 35–40.

Agrawal, A.A., Tuzun, S., Bent, E. (eds.). 1999. *Inducible Plant Defenses Against Pathogens and Herbivores. Biochemistry, Ecology and Agriculture.* St. Paul, Minnesota: American Phytopathological Society.

Alan, A.R., Earle, E.D. 2002. Sensitivity of bacterial and fungal plant pathogens to the lytic peptides, MSI-99, Magainin II, and Cecropin B. *Mol Plant-Microbe Interact* 15: 701–708.

Aldwinckle, H., Bhaskara, R.M.V., Norelli, J. 2002. Evaluation of control of fire blight infection of apple blossoms and shoots with SAR inducers, biological agents, a growth regulator, copper compounds and other materials. *Acta Hortic* 590: 325–331.

Anonymous. 2001. *Product and Trial Information.* USA: Dossier. Improcrop.

Ayliffe, M.A. et al. 2004. Aberrant mRNA processing of the maize Rp1-D rust resistance gene in wheat and barley. *Mol. Plant-Microbe Interact.* 17: 853–864.

Balestra, G.M., Antonelli, M., Varvaro, L. 1998. Effectiveness of natural products for *in vitro* and *in vivo* control of epiphytic populations of *Pseudomonas syringae* pv. *tomato* on tomato plants. *J. Plant Pathol.* 80: 251.

Balogh, B., Canteros, B.I., Stall, R.E., Jones, J.B. 2008. Control of citrus canker and citrus bacterial spot with bacteriophages. *Plant Dis.* 92: 1048–1052.

Balogh, B., Jones, J.B., Momol, M.T., Olson, S.M., Obradovic, A., Jackson, L.E. 2003. Improved efficacy of newly formulated bacteriophages for management of bacterial spot on tomato. *Plant Dis.* 87: 949–954.

Bargabus, R.L., Zidack, N.K., Sherwood, J.W., Jacobsen, B.J. 2002. Characterization of systemic resistance in sugar beet elicited by a non-pathogenic, phyllosphere colonizing *Bacillus mycoides*, biological control agent. *Physiol. Mol. Plant Pathol.* 61: 289–298.

Bargabus, R.L., Zidack, N.K., Sherwood, J.W., Jacobsen, B.J. 2004. Screening for the identification of potential biological control agents that induce systemic acquired resistance in sugar beet. *Biological Contr* 30: 342–350.

Bastas, K.K., Akay, A., Maden, S. 2006. A new approach to fire blight control: Mycorrhiza. *HortScience* 41(5): 1309–1312.

Bastas, K.K., Maden, S. 2004. Researches on the control of fire blight (*E. amylovora*) with prohexadione-Ca (BAS 125 10 W) and benzothiadiazole + metalaxyl (Bion MX 44 WG). Selcuk Univ. *J. Agric. Fac* 18(33): 49–58.

Bastas, K.K., Maden, S. 2008. Evaluation of host resistance inducers and conventional products for fire blight management in loquat and quince. *Phytoprotection* 88: 93–101.

Bastas, K.K., Maden, S., Katircioglu, Y.Z., Boyraz, N. 2010. Determination of application time for chemical control of fire blight disease in pear varieties. *J. Agri. Sci.* 16(3): 150–161.

Bauer, W.D., Mathesius, U. 2004. Plant responses to bacterial quorum sensing signals. *Curr. Opin.Plant Biol.* 7: 429–433.

Baysal, O., Calıskan, M., Yesilova, O. 2008. An inhibitory effect of a new *Bacillus subtilis* strain (EU07) against *Fusarium oxysporum* f. sp. *radicis lycopersici*. *Physiol Mol Plant Pathol* 73: 25–32.

Baysal, O., Tör, M. 2014. Smart biologics for crop protection in agricultural systems. *Turk. J. Agric. For* 38: 723–731.

Bazzi, C. et al. 2003. Control of pathogen incidence in pome fruits and other horticultural crop plants with prohexadione-Ca. *Eur. J. Hortic. Sci* 68: 108–114.

Beg, A.Z., Ahmad, I. 2002. *In vitro* fungitoxicity of the essential oil of *Syzygium aromaticum*. *World J. Microbiol. Biotechnol.* 18: 313–315.

Bender, C.L., Malvick, D.K., Conway, K.E., George, S., Pratt, P. 1990. Characterization of pXV10A, a copper resistance plasmid in *Xanthomonas campestris* pv. *vesicatoria*. *Appl. Environ. Microbiol.* 56: 170–175.

Benhamou, N., Belanger, R.R. 1998. Benzothiadiasole-mediated induced resistance to *Fusarium oxysporum* f.sp. *radicis-lycopersici* in tomato. *Plant Physiol* 118: 1203–1212.

Bergamin, F.A. Kimati, H. 1981. Studies on a bacteriophage isolated from *Xanthomonas campestris* 1. Isolation and morphology. *Summa Phytopathol* 7(3–4): 35–43.

Biermann, B., Linderman, R.G. 1983. Use of vesicular-arbuscular mycorrhizal roots, intraradical vesicles and extraradical vesicles as inoculum. *New Phytol* 95: 97–105.

Borah, P.K., Jindal, J.K., Verma, J.P. 2000. Integrated management of bacterial leaf spot of ungbean with bacteriophages of Xav and chemicals. *J. Mycol. Plant Pathol* 30(1): 19–21.

Boyd, R.J., Hildebrant, A.C., Allen, O.N. 1971. Retardation of crown gall enlargement after bacteriophage treatment. *Plant Dis. Rep* 55: 145–148.

Brady, S.F., Wright, S.A., Lee, J.C., Sutton, A.E., Zumoff, C.H., Wodzinski, R.S., Beer, S.V., Clardy, J. 1999. Pantocin B, an antibiotic from *Erwinia herbicola* discovered by heterologous expression of cloned genes. *J Am Chem Soc* 121: 11912–11913.

Broekaert, W.F., Terras, F.R.G., Cammue, B.P.A. 2000. Induced and preformed antimicrobial proteins. In Slusarenko, A. J., Fraser, R.S.S., van Loon, L.C. (eds.). *Mechanisms of Resistance to Plant Diseases*. Kluwer: Dordrecht, pp. 371–477.

Buban, T., Sallai, P., Obzsut-Truskovszky, E., Hertelendy, L. 2002. Trials with applying chemical agents other than bactericides to control fire blight in pear orchards. *Acta Hortic* 590: 263–267.

Buban, T., Földes, L., Kormany, A., Hauptmann, S., Stammler, G., Rademacher, W. 2003. Prohexadione-Ca in apple trees: Control of shoot growth and reduction of fire blight incidence in blossoms and shoots. *J. Appl. Bot* 77: 95–102.

Castro, M.S., Fontes, W. 2007. Plant defense and antimicrobial peptides. *Protein Pept. Lett.* 12: 11–16.

Cha, C., Gao, P., Chen, Y.C., Shaw, P.D., Farrand, S.K. 1998. Production of acyl-homoserine lactone quorum sensing signals by gram-negative plant-associated bacteria. *Mol. Plant Microbe Interact* 11: 1119–1129.

Chen, W., Hoitink, H.A.J., Schmitthenner, A.F. 1987. Factors affecting suppression of Pythium damping-off in container media amended with compost. *Phytopathology* 77: 755–760.

Chernin, L. 2011. Quorum-sensing signals as mediators of PGPRs' beneficial traits. In *Bacteria in Agrobiology: Plant Nutrient Management*. Vol. 3. Maheshwari, D.K. (ed.). Berlin, Germany: Springer-Verlag, pp. 209–236.

Chung, W.C., Huang, J.W., Huang, H.C. 2005. Formulation of a soil biofugicide for control of damping-off of Chinese cabbage (*Brassica chinensis*) caused by *Rhizoctonia solani*. *Biol. Contr* 32: 287–294.

Civerolo, E.L., Kiel, H.L. 1969. Inhibition of bacterial spot of peach foliage by *Xanthomonas pruni* bacteriophage. *Phytopathology* 59: 1966–1967.

Cui, X., Harling, R. 2005. N-acyl-homoserine lactone-mediated quorum sensing blockage, a novel strategy for attenuating pathogenicity of Gram-negative bacterial plant pathogens. *Eur. J. Plant Pathol.* 111: 327–339.

Datta, S.K. 2004. Rice biotechnology: A need for developing countries. *AgBioForum* 7: 31–35.

De Castro, S.L. 2001. Propolis: Biological and pharmacological activities. Therapeutic uses of this bee-product. *Annu. Rev. Biol. Sci* 3: 49–83.

De Wit, P.J.G.M. 1997. Pathogen avirulence and plant resistance: A key role for recognition. *Trends Plant Sci* 2: 452–458.

Deckers, T., Schoofs, H. 2002. Host susceptibility as a factor in control strategies of fire blight in European pear growing. *Acta Hortic* 590: 127–138.

Dong, Y.H., Wang, L.H., Zhang, L.H. 2007. Quorum-quenching microbial infections: Mechanisms and implications. *Phil Trans R Soc B* 362: 1201–1211.

Dong, Y.-H., Wang, L., Xu, J.-L., Zhang, H.-B., Zhang, X.F., Zhang, L.H. 2001. Quenching quorum-sensing-dependent bacterial infection by an N-acyl homoserine lactonase. *Nature* 411: 813–817.

Dubey, N.K., Srivastava, B., Kumar, A. 2008. Current status of plant products as botanical pesticides in storage pest management. *J. Biopestic.* 1(2): 182–186.

Elad, Y., Chet, I. 1987. Possible role of competition for nutrients in biocontrol of Pythium damping-off by bacteria. *Phytopathology* 77: 190–195.

Elasri, M. et al. 2001. Acyl-homoserine lactone production is more common among plant-associated *Pseudomonas* spp. than among soilborne *Pseudomonas* spp. *Appl. Environ. Microbiol* 67: 1198–1209.

Evans, J.R., Evans, R.R., Regusci, C.L., Rademacher, W. 1999. Mode of action, metabolism and uptake of BAS 125W, prohexadione-calcium. *HortScience* 34(7): 1200–1201.

Farmer, E.E. 2001. Surface-to-air signals. *Nature* 411: 854–856.

Fernando, W.G.D., Jones, A.L. 1999. Prohexadione-Ca: A tool for reducing secondary fire blight infections. *Acta Hortic* 489: 597–600.

Ferre, R., Badosa, E., Feliu, L., Planas, M., Montesinos, E., Bardaji, E. 2006. Inhibition of plant-pathogenic bacteria by short synthetic cecropin A-melittin hybrid peptides. *Appl. Environ Microbiol* 72: 3302–3308.

Field, B., Jordan, F., Osbourn, A. 2006. First encounters–Deployment of defence-related natural products by plants. *New Phytol.* 172: 193–207.

Filonow, A.B., Lockwood, J.I. 1979. Evaluation of several actinomycetes and the fungus *Hyphochytrium catenoides* as biocontrol agent for Phytophthora root rot of soybean. *Plant. Dis.* 69: 1033–1036.

Fitter, A.H., Garbaye, J. 1994. Interactions between mycorrhizal fungi and other soil microorganisms. *Plant Soil* 159: 123–132.

Flaherty, J.E., Jones, J.B., Harbaugh, B.K., Somodi, G.C., Jackson, L.E. 2000. Control of bacterial spot on tomato in the greenhouse and field with h-mutant bacteriophages. *Hortscience* 35(5): 882–884.

Garcia-Garrido, J.M., Ocampo, J.A. 1989. Effect of VA mycorrhizal infection of tomato on damage caused by *Pseudomonas syringae*. *Soil Biol. Biochem* 21: 165–167.

Garelik, G. 2002. Taking the bite out of potato blight. *Science* 298: 1702–1704.

Gent, D.H., Schwartz, H.F. 2005. Management of xanthomonas leaf blight of onion with a plant activator, biological control agents, and copper bactericides. *Plant Dis.* 89: 631–639.

Gibbons, S. 2005. Plants as a source of bacterial resistance modulators and anti-infective agents. *Phytochemistry Reviews* 4: 63–78.

Gottlieb, O.R., Borin, M.R., Brito, N.R. 2002. Integration of ethnobotany and phytochemistry: Dream or reality? *Phytochemistry* 60: 145–152.

Grange, M., Ahmed, S. 1988. *Handbook of Plants with Pest Control Properties*. New York: John Wiley & Sons.

Grgurina, I., Bensaci, M., Pocsfalvi, G., Mannina, L., Cruciani, O., Fiore, A., Fogliano, V., Sorensen, K.N., Takemoto, J.Y. 2005. Novel cyclic lipodepsipeptide from *Pseudomonas syringae* pv. *lachrymans* strain 508 and syringopeptin antimicrobial activities. *Ant Agents Chemother* 49: 5037–5045.

Hammerschmidt, R. 1999. Phytoalexins: What have we learned after 60 years. *Annu. Rev. Phytopathol* 37: 285–306.

He, P., Chintamanani, S., Chen, Z., Zhu, L., Kunkel, B.N., Alfano, J.R., Tang, X., Zhou, J.M. 2004. Activation of a COI1-dependent pathway in Arabidopsis by *Pseudomonas syringae* type III effectors and coronatine. *Plant J* 37: 589–602.

Heath, M.C. 2000. Nonhost resistance and nonspecific plant defenses. *Curr. Opin. Plant Biol* 3: 315–319.

Hovmoller, M.S., Ostergard, H., Munk, L. 1997. Modelling virulence dynamics of airborne plant pathogens in relation to selection by host resistance. In *The Gene-for-Gene Relationship in Plant–Parasite Interactions*. Crute, I.R., Holub, E., Burdon, J.J. (eds.). Wallingford, UK: CABI International, pp. 173–190.

Huang, H.C. 2000. Crop protection: Current progress and prospects for the new millennium. *J. Hebei Agric. Sci* 4: 34–48.

Huang, H.C., Wu, M.T. 2009. Plant disease management in the era of energy conservation. *Plant Pathol. Bull* 18: 1–12.

Huang, J.W., Chung, W.C., Huang, H.C., Shiau, J.H. 2007. A granulate biofumigant for control of soilborne diseases in crops. (Republic of China Patent Certificate No. I 276402, March 21, 2007; Expiration date: March 9, 2024.

Iacobellis, N.S., Lo Cantore, P., Capasso, F., Senatore, F. 2005. Antibacterial activity of *Cuminum cyminum* L. and *Carum carvi* L. essential oils. *J. Agric. Food Chem* 53: 57–61.

Inbar, M., Doostdar, H., Sonoda, R.M., Leibee, G.L., Mayer, R.T. 1998. Elisitors of plant defensive systems reduce insect densities and disease incidence. *J. Chem. Ecol* 24: 135–149.

Ishimaru, C., Klos, E.J., Brubaker, R.R. 1988. Multiple antibiotic production by *Erwinia herbicola*. *Phytopathology* 78: 746–750.

Isman, M.B. 2000. Plant essential oils for pest and disease management. *Crop Protect* 19: 603–608.

Jing, Y.D., He, Z.L., Yang, X.E. 2007. Role of soil rhizobacteria in phytoremediation of heavy metal contaminated soils. *J. Zhejiang Univ. Sci* 8: 192–207.

Joseph, M. et al. 2004. Combining bacterial blight resistance and Basmati quality characteristics by phenotypic and molecular marker-assisted selection in rice. *Molecular Breeding* 13: 377–387.

Keshavan, N.D., Chowdhary, P.K., Haines, D.C., Gonzalez, J.E. 2005. L-Canavanine made by *Medicago sativa* interferes with quorum sensing in *Sinorhizobium meliloti*. *J. Bacteriol* 187: 8427–8436.

Kloepper, J.W. 1991. Plant growth-promoting rhizobacteria as biological control agents of soilborne diseases. In *The Biological Control of Plant Diseases*. Bay-Petersen, J. (ed.). FFTC book series no. 42. Taipei, Taiwan: Food and Fertilizer Technology Center.

Koch, B., Liljefors, T., Persson, T., Nielsen, J., Kjelleberg, S., Givskov, M. 2005. The LuxR receptor: The sites of interaction with quorum-sensing signals and inhibitors. *Microbiology* 151: 3589–3602.

Kunst, F. et al. 1997. The complete genome of the gram-positive bacterium *Bacillus subtilis*. *Nature* 390: 249–256.

Kwok, O.C., Fahy, H.P.C., Hoittink, H.A.J., Kuter, G.A. 1987. Interactions between bacteria and *Trichoderma* in suppression of Rhizoctonia damping-off in bark compost media. *Phytopathology* 77: 1206–1212.

Leeman, M., Van Pelt, J.A., Den Ouden, F.M., Heinbroek, M., Bakker, P.A.H.M. 1995. Induction of systemic resistance by *Pseudomonas fluorescens* in radish cultivars differing in susceptibility to Fusarium wilt, using novel bioassay. *Eur. J. Plant Pathol* 101: 655–664.

Linderman, R.G. 1994. Role of AM fungi in biocontrol. Pages 1–25 In *Mycorrhizae and Plant Health*. Pfleger, F.L. and Linderman, R.G. (eds.). St. Paul, MN: APS Press.

Lindgren, P.B. 1997. The role of hrp genes during plant-bacterial interactions. *Annu. Rev. Phytopathol* 35: 129–152.

Lo Cantore, P., Iacobellis, N.S., De Marco, A., Capasso, F., Senatore, F. 2004. Antibacterial activity of Coriandrum sativum L. and *Foeniculum vulgare* Miller var. *vulgare* (Miller) essential oils. *J. Agric. Food Chem* 52: 7862–7866.

Louws, F.J., Wilson, M., Campell, H.L., Cuppels, D.A., Jones, J.B., Shoemaker, P.B., Sahin, F., Miller, S. 2001. Field control of bacterial spot and bacterial speck of tomato using plant activator. *Plant Dis.* 85: 481–488.

Lugtenberg, B., Kamilova, F. 2009. Plant-growth-promoting rhizobacteria. *Annu Rev Microbiol* 63: 541–555.

Manefield, M. et al. 2002. Halogenated furanones inhibit quorum sensing through accelerated LuxR turnover. *Microbiology* 148: 1119–1127.

Maurhofer, M., Hase, C., Meuwly, P., Metraux, J.P., Defago, G. 1994. Induction of systemic resistance to tobacco necrosis virus by the root-colonizing *Pseudomonas fluorescens* strain CHA0: Influence of the gacA gene and of pyoverdine production. *Phytopathology* 84: 139–146.

Maxson, K.L., Jones, A.L. 2002. Management of fire blight with gibberellin inhibitors and SAR inducers. *Acta Hortic* 590: 217–223.

McDonald, B.A., Linde, C. 2003. The population genetics of plant pathogens and breeding strategies for durable resistance. *Euphytica* 124: 163–180.

McHale, L., Tan, X.P., Koehl, P., Michelmore, R.W. 2006. Plant NBS-LRR proteins: Adaptable guards. Genome Biology 7: http://genomebiology.com/2006–7/4/212/abstract

McKenna, F., El-Tarabily, K.A., Hardy, G.E.S.T., Dell, B. 2001. Novel *in vivo* use of a polyvalent *Streptomyces* phage to disinfest *Streptomyces scabies*-infected seed potatoes. *Plant Pathol* 50(6): 666–675.

McManus, P.S., Stockwell, V.O., Sundin, G.W., Jones, A.L. 2002. Antibiotic use in plant agriculture. *Annu Rev Phytopathol* 40: 443–465.

McNeil, D.L., Romero, S., Kandula, J., Stark, C., Stewart, A., Larsen, S. 2001. Bacteriphages: A potential biocontrol agent against walnut blight (*Xanthomonas campestris* pv. *juglandis*). *N.Z. Plant Prot* 54: 220–224.

Merriman, P.R., Price, R.D., Kollmorgen, J.F., Piggott, T., Ridge E.H. 1974. Effect of seed inoculation with *Bacillus subtilis* and *Streptomyces grisens* on the growth of cereals and carrots. *Aust. J. Agric. Res* 25: 219–226.

Meziane, H., Van der Sluis, I., Van Loon, L.C., Hofte, M., Bakker, P.A.H.M. 2005. Determinants of *Pseudomonas putida* WCS358 involved in inducing systemic resistance in plants. *Mol. Plant Pathol* 6: 177–185.

Momol, M.T., Ugine, J.D., Norelli, J.L., Aldwinckle, H.S. 1999. The effect of prohexadione-Ca SAR inducers and calcium on the control of shoot blight caused by *E. amylovora* on apple. *Acta Hortic* 489: 601–605.

Monroc, S., Badosa, E., Besalu, E., Planas, M., Bardaji, E., Montesinos, E., Feliu, L. 2006a. Improvement of cyclic decapeptides against plant pathogenic bacteria using a combinatorial chemistry approach. *Peptides* 27: 2575–2584.

Monroc, S., Badosa, E., Feliu, L., Planas, M., Montesinos, E., Bardaji, E. 2006b. *De novo* designed cyclic cationic peptides as inhibitors of plant pathogenic bacteria. *Peptides* 27: 2567–2574.

Morris, P.F., Ward, E.W.R. 1992. Chemoattraction of zoospores of the plant soybean pathogen, *Phytophthora sojae*, by isoflavones. *Physiol. Mol. Plant Pathol* 40: 17–22.

Morton, J.B., Benny, G.L. 1990. Revised classification of arbuscular mycorrhizal fungi (zygomycetes): A new order glomales, two new suborders, glomineae and gigasporineae and gigasporaceae, with an amendation of glomaceae. *Mycotaxon* 37: 471–491.

Munsch, P., Olivier, J.M. 1995. Biocontrol of bacterial blotch of the cultivated mushroom with lytic phages: Some practical considerations. In *Science and Cultivation of Edible Fungi, Proceedings of the 14th International Congress*. Elliott, T.J. (ed.). The Netherlands: Balkema, AA., Rotterdam, Vol. II, pp. 595–602.

Newman, D.J., Cragg, G.M., Snader, K.M. 2003. Natural products as sources of new drugs over the period. 1981–2002. *J. Nat. Prod.* 66: 1022–1037.

Norelli, J.L., Jones, A.L., Aldwinckle, H.S. 2003. Fire blight management in the twenty-first century. *Plant Dis.* 87: 756–765.

Oldroyd, G.E.D., Staskawicz, B.J. 1998. Genetically engineered broad spectrum disease resistance in tomato. *Proceedings of the National Academy of Sciences of the United States of America* 95: 10300–10305.

Ongena, M., Duby, F., Rossignol, F., Fouconnier, M.L., Dommes, J., Thonart, P. 2004. Stimulation of the lipoxygenase pathway is associated with systemic resistance induced in bean by a nonpathogenic Pseudomonas strain. *Mol. Plant-Microbe Interact* 17: 1009–1018.

Parker, J.E. et al. 1996. Characterization of eds1, a mutation in Arabidopsis suppressing resistance to Peronospora parasitica specified by several different RPP genes. *Plant Cell* 8: 2033–2046.

Parret, A.H.A., Temmerman, K., De Mot, R. 2005. Novel lectin-like bacteriocins of biocontrol strain Pseudomonas fluorescens Pf-5. *Appl Environ Microbiol* 71: 5197–5207.

Peart, J.R. et al. 2002. Ubiquitin ligase-associated protein SGT1 is required for host and nonhost disease resistance in plants. *Proc. Natl. Acad. Sci. U. S. A* 99: 10865–10869.

Pham, H.T., Riu, K.Z., Jang, K.M., Cho, S.K., Cho, M. 2004. Bactericidal activity of glycinecin A, a bacteriocin derived from *Xanthomonas campestris* pv. *glycines*, on phytopathogenic *Xanthomonas campestris* pv. *vesicatoria* cells. *Appl. Environ. Microbiol* 70: 4486–4490.

Pierson, L.S., Ishimaru, C.A. 2000. Genomics of plant-associated bacteria: A glimpse of the future that has become reality. APSnet August Feature Story. www.apsnet.org/online/feature/Genomics/Top.html

Pink, D.A.C. 2002. Strategies using genes for non-durable resistance. *Euphytica* 1: 227–236. Tai, T.H. et al. 1999. Expression of the Bs2 pepper gene confers resistance to bacterial spot disease in tomato. *Proc. Natl. Acad. Sci. U. S. A* 96: 14153–14158.

Powell, W.A., Catranis, C.M., Maynard, C.A. 1995. Synthetic antimicrobial peptide design. *Mol Plant-Microbe Interact* 8: 792–794.

Press, C.M., Loper, J.E., Kloepper, J.W. 2001. Role of iron in rhizobacteria mediated induced systemic resistance of cucumber. *Phytopathology* 91: 593–598.

Qiu, D., Wei, Z.M. 2000. Effects of Messenger on gray mold and other fruit rot diseases. *Phytopathology* 90: 62.

Raaijmakers, J.M., de Bruijn, I., de Kock, M.J.D. 2006. Cyclic lipopeptide production by plant-associated Pseudomonas ssp: Diversity, activity, biosynthesis, and regulation. *Mol Plant-Microbe Interact* 19: 699–710.

Rademacher, W. 2000. Growth retardants: Effects on gibberellin biosynthesis and other metabolic pathways. *Annu. Rev. Plant Physiol. Plant Mol. Biol* 51: 501–531.

Ragsdale, N.N. 2000. The impact of the food quality protection act on the future of plant disease management. *Annu. Rev. phytopathol.* 38: 577–596.

Ravensdale, M., Blom, T.J., Gracia-Garza, J.A., Svircev, A.M., Smith, R.J. 2007. Bacteriophages and the control of *Erwinia carotovora* subsp *carotovora*. *Can. J. Plant Pathol* 29: 121–130.

Ream, W. 1989. *Agrobacterium tumefaciens* and interkingdom genetic exchange. *Annu Rev Phytopathol* 27: 583–618.

Rommens, C.M.T., Salmeron, J.M., Oldroyd, G.E.D., Staskawicz, B.J. 1995. Interspecific transfer and functional expression of the tomato disease resistance gene Pto. *Plant Cell* 7: 1537–1544.

Ryu, C.M., Farag, M.A., Hu, C.H., Reddy, M.S., Kloepper, J.W., Pare, P.W. 2004. Bacterial volatiles induce systemic resistance in Arabidopsis. *Plant Physiol* 134: 1017–1026.

Schnabel, E.L., Jones, A.L. 2001. Isolation and characterization of five *Erwinia amylovora* bacteriophages and assessment of phage resistance in strains of *Erwinia amylovora*. *Appl. Environ. Microbiol* 67(1): 59–64.

Selosse, M.A., Baudoin, E., Vandenkoornhuyse, P. 2004. Symbiotic microorganisms, a key for ecological success and protection of plants. *CR. Biol* 327: 639–648.

Stanier, T., McSharry, J., Speitel, T. 1967. *Agrobacterium tumefaciens* conn IV. Bacteriophage PB21 and its inhibitory effect on tumor induction. *J. Virol* 1(2): 268–273.

Sticher, L., Mauchmani, B., Mauchmani, B., Metraux, J.P. 1997. Systemic acquired resistance. *Annu. Rev. Phytopath* 35: 235–270.

Sutherland, E.D., Lockwood, J.L. 1984. Hyperparasitism of oospores of some Peronosporales by *Actinoplanes missouriensis* and *Humicola fuscoatra* and other Actinomycetes and fungi. *Can. J. Plant. Pathol* 6: 139–145.

Svircev, A.M., Lehman, S.M., Kim, W.S., Barszcz, E., Schneider, K.E., Castle, A.J. 2005. Control of the fire blight pathogen with bacteriophages. In *Proceedings of the 1st International Symposium on Biological Control of Bacterial Plant Diseases*. Zeller, W., Ullrich, C., (eds.). Berlin, Germany: Die Deutsche Bibliothek - CIPEinheitsaufnahme, pp. 259–261.

Tai, T.H. et al. 1999. Expression of the Bs2 pepper gene confers resistance to bacterial spot disease in tomato. *Proceedings of the National Academy of Sciences* 96: 14153–14158.

Tanaka, H., Negishi, H., Maeda, H. 1990. Control of tobacco bacterial wilt by an avirulent strain of *Pseudomonas solanacearunm* M4S and its bacteriophage. *Ann. Phytopathol. Soc. Jpn* 56: 243–246.

Teplitski, M. et al. 2004. *Chlamydomonas reinhardtii* secretes compounds that mimic bacterial signals and interfere with quorum sensing regulation in bacteria. *Plant Physiol* 134: 137–146.

Tu, J.C. 1980. Incidence of root rot and over wintering of alfalfa as influenced by rhizobia. *Phytopathol. Z* 97: 97–108.

Vallad, G.E., Goodman, R.M. 2004. Systemic acquired resistance and induced systemic resistance in conventional agriculture: Review and interpretation. *Crop Sci* 44: 1920–1934.

Van Loon, L.C., Bakker, P.A.H.M., Pieterse, C.M.J. 1998. Systemic resistance induced by rhizosphere bacteria. *Annu. Rev. Phytopathol* 36: 453–483.

Van Loon, L.C., Rep, M., Pieterse, C.M.J. 2006. Significance of inducible defense-related proteins in infected plants. *Annual Review of Phytopathology* 44: 135–162.

Varvaro, L., Antonelli, M., Balestra, G.M., Fabi, A., Scermino, D. 2001. Control of phytopathogenic bacteria in organic agriculture: Cases of study. *J. Plant Pathol* 83: 244.

Varvaro, L., Antonelli, M., Balestra, G.M., Fabi, A., Scermino, D., Vuono, G. 2002. Investigations on the bactericidal activity of some natural products. In *Proc. Int. Cong. Biol. Products: Which Guarantees for the Consumers?* Milan, Italy.

Venturi, V., Fuqua, C. 2013. Chemical signaling between plants and plant-pathogenic bacteria. *Annu. Rev. Phytopathol* 2013. 51: 17–37.

Von Bodman, S.B., Bauer, W.D., Coplin, D.L. 2003. Quorum sensing in plant-pathogenic bacteria. *Annu. Rev. Phytopathol* 41: 455–482.

Wang, G-L., Song, W.Y., Ruan, D.L., Sideris, S., Ronald, P.C. 2007. The cloned gene, Xa21, confers resistance to multiple *Xanthomonas oryzae* pv. *oryzae* isolates in transgenic plants. *Molecular Plant-Microbe Interactions* 9: 855.

Wei, Z.M., Wei, Z., Laby, R.J., Zumoff, C.H., Bauer, D.W., He, S.Y., Collmer, A., Beer, S.V. 1992. Harpin, elicitor of the hypersensitive response produced by the plant pathogen *Erwinia amylovora*. *Science* 257: 85–88.

Weller, D.M. 1988. Biological control of soilborne plant pathogens in the rhizosphere with bacteria. *Annu. Rev. Phytopathol* 26: 379–407.

White, D.G., Zhao, S., Simjee, S., Wagner, D.D., McDermott, P.F. 2002. Antimicrobial resistance of food-borne pathogens. *Microbes and Infections* 4: 405–412.

Whitehead, N.A., Barnard, A., Slater, M.L., Simpson, H.N.J.L., Salmond, G.P.C. 2001. Quorum-sensing in Gram-negative bacteria. *FEMS Microbiol Rev* 25: 365–404.

Zaccardelli, M., Saccardi, A., Gambin, E., Mazzuchi, U. 1992. *Xanthomonas campestris* pv. *pruni* bacteriophages on peach trees and their potential use for biological control. *Phytopathol. Mediterr* 31: 133–140.

Zamioudis, C., Pieterse, C.M. 2012. Modulation of host immunity by beneficial microbes. *Mol. Plant-Microbe Interact* 25: 139–150.

Zhai, W.X. et al. 2002. Breeding bacterial blight-resistant hybrid rice with the cloned bacterial blight resistance gene Xa21. *Mol. Breed.* 8: 285–293.

Zhang, J., Li, X., Jiang, G., Xu, Y., He, Y. 2006. Pyramiding of Xa7 and Xa21 for the improvement of disease resistance to bacterial blight in hybrid rice. *Plant Breed* 125: 600–605.

Zhao, B.Y., Lin, X.H., Poland, J., Trick, H., Leach, J., Hulbert, S. 2005. From the cover: A maize resistance gene functions against bacterial streak disease in rice. *Proceedings of the National Academy of Sciences* 102: 15383–15388.

20 Scientific and Economic Impact of Plant Pathogenic Bacteria

Velu Rajesh Kannan, Kubilay Kurtulus Bastas, and Rajendran Sangeetha Devi

CONTENTS

ABSTRACT Plant pathogenic bacteria occupy a wide host range in almost all the regions of Earth and in almost all types of plant systems. This group of pathogens, which mainly belong to the Xanthomonadaceae, Pseudomonadaceae, and Enterobacteriaceae families, target all types of plants that can supply appropriate food and shelter on their surfaces as well as in their tissue regions. The genera most often represented are *Erwinia*, *Pectobacterium*, *Pantoea*, *Agrobacterium*, *Pseudomonas*, *Ralstonia*, *Burkholderia*, *Acidovorax*, *Xanthomonas*, *Clavibacter*, *Streptomyces*, *Xylella*, *Spiroplasma*, and *Phytoplasma*. Phytopathogenic bacteria often cause hormonal imbalances in infected plants that result in stunting, overgrowth, galls, root branches, defoliation, resetting, leaf epinasty, and others. These problems alter the nutrition levels of affected plants on qualitative and quantitative levels; in addition, they negatively impact global food supplies and economics. Worst of all, phytopathogen-infested plants spread toxic molecules to herbivores and carnivores through the natural food chain, which can collapse entire ecosystems.

KEYWORDS: plant pathogenic bacteria, disease impacts, opportunistic rhizosphere pathogens, synthesizing toxic molecules, protection strategies

20.1 INTRODUCTION

The diversity of plant pathogens, which are found all over the globe, includes bacteria, fungi, oomycetes, and nematodes. Of the 7100 classified bacterial species, roughly 150 cause diseases to plants by obtaining nutrients from them for their own growth; in addition, they use specific mechanisms to secrete proteins and other molecules to cellular compartments of their hosts to modulate plant defense circuitry and enable parasitic colonization (Abramovitch et al., 2006; Birch et al., 2006; Chisholm et al., 2006; Kamoun, 2006, 2007; Block et al., 2008; Davis et al., 2008; Misas-Villamil and Van der Hoorn, 2008). Bacterial diseases are most frequent and severe in tropical and subtropical places, where warm and humid conditions are ideal for bacterial growth. Indeed, consistent crop losses are annually recorded in such countries.

The problem of plant diseases, particularly in developing and Third World countries, is exacerbated by the paucity of resources devoted to their study. In part, this lack may be a result of governmental error but it also arises from the difficulty of quantifying plant diseases and relating these quantifications to the failure of crops to reach manageable yields. The inability to supply such hard data to authorities indicates that plant pathology continues to be grossly underfunded in comparison to its impact. As a result, more than 800 million people may not have adequate food on any given day and at least 10% of global food production is lost to plant disease (James, 1998; FAO, 2000; Christou and Twyman, 2004).

The vital mission of the discipline of plant pathology, which comprises microbiology, plant science, and agronomy, is to control plant diseases. Plant pathologists are highly aware that their data is directly related to the causes of food shortages and to impairments to food production caused by plant pathogens. They must also keep in mind that contamination by microbial agents within food supplies may vary, and that such contamination may influence the normal properties of food and food-related raw materials according to numerous factors; including the ability to cultivate crops in certain areas or regions, issues of distribution, consumer awareness, and others. This chapter contains a detailed consideration of the scientific and economic impact of plant pathogenic bacterial disease.

20.2 PLANT–PATHOGENIC BACTERIA INTERACTIONS

The ecological coordination of structures within interactions among organisms has immense influence up to kingdom levels. Obviously, this means that plant–microbe interactions are both ubiquitous and essential, whether these associations produce positive and beneficial results or negative ones. The plant–pathogen interaction process on molecular levels involves secreted molecules, known as effectors; the understanding of these is critical for a complete mechanistic picture of the processes that underlie host colonization and pathogenicity levels. The study of all of these subjects has significantly increased our knowledge of how plant diseases are caused by a diversity of plant bacterial pathogens, how they select their host targets, and how and where molecules produced both by bacteria and plants interact and affect the outcomes of their interactions.

The effectors of bacterial pathogens are related to proteins and other small molecules that alter host-cell structure, metabolic function, and behavior. These alterations facilitate infection (virulence factors and toxins), trigger defense responses (avirulence factors and elicitors), or both (Huitema et al., 2004; Kamoun, 2006, 2007). The concept of extended phenotype (i.e., genes whose effects reach beyond the cells in which they reside), which was introduced by Richard Dawkins (1999) in his seminal work, views effectors as parasite genes that have phenotypic expression in host bodies and their behaviors. Indeed, effectors are the products of genes that reside in pathogen genomes, but their functions occur at the interface with host plants or even inside plant cells (Kamoun, 2006, 2007).

Effectors may occur in the molecular forms of toxins, degradative enzymes, and others. Although many effectors interfere the host's inherent immunity, some may alter host plant behavior and morphology. The best example of the latter, coronatine, was shown by Melotto et al. (2006) to trigger stomatal reopening in *Arabidopsis* and thereby to facilitate bacterial entry into the plant apoplast. Interactions between pathogenic bacteria and plants can be either incompatible or compatible. Incompatible interactions, which occur when the bacterium encounters a non-host plant (non-host resistance) or a resistant host plant (cultivar-specific resistance), often evoke hypersensitive response (HR) in the plant. During HR, rapid, programmed death of plant cells occurs at the initial site of infection. During compatible interactions, bacteria infect susceptible host plants and cause disease symptoms that may be evident throughout the plants. Other notable effects of HR are the pathogenicity (*hrp*) gene cluster that encodes the type III secretion system, effector trafficking, and host targets for defense suppression (Huynh et al., 1989; Li et al., 2002; Kvitko et al., 2009; Zhang et al., 2010; Jovanovic et al., 2011).

20.3 TYPES OF PLANT PATHOGENIC BACTERIA

Both wild and cultivated plant species are subject to diseases. About 80,000 plant diseases have been recorded throughout the world; the earliest known records of their occurrence were found in fossils about 25,000 years old (Chu et al., 1989). Plant diseases have affected even developed nations of which the best example is the potato late blight in Ireland (1845–1860) (Cox and Large, 1960), a million people died of hunger and more than a million struggled to evacuate. The species of bacteria that are most pathogenic to plants belong to numerous genera (Leyns et al., 1984; Strange, 2003), including *Erwinia*, *Pectobacterium*, *Pantoea*, *Agrobacterium*, *Pseudomonas*, *Ralstonia*, *Burkholderia*, *Acidovorax*, *Xanthomonas*, *Clavibacter*, *Streptomyces*, *Xylella*, *Spiroplasma*, and *Phytoplasma*.

20.3.1 Prominent Plant Pathogenic Bacteria

Plant pathologists have listed 10 most virulent bacterial plant pathogens based on their pathogenesis, economic impact, and molecular aspects (Table 20.1). The survey was compiled from 458 votes by members of the international plant pathology community.

Because each group is able to cause multiple diseases that are damaging and/or fatal to plants, these pathovars continue to have a huge impact on the scientific understanding of microbial pathogenicity. For example, studies on the molecular biology of the virulence of and the plant defenses against *P. syringae* have contributed new insights to microbial pathogenicity, including the fact that the virus can also affect human beings (Mansfield et al., 2012). Other notable examples of the

TABLE 20.1
Top 10 Plant Pathogenic Bacteria

Rank	Bacterial Pathogens
1	*Pseudomonas syringae* pathovars
2	*Ralstonia solanacearum*
3	*Agrobacteruim tumefaciens*
4	*Xanthomonas oryzae* pv. Oryzae
5	*Xanthomonas campestris* pathovars
6	*Xanthomonas axonopodis* pv manihotis
7	*Erwinia amylovora*
8	*Xylella fastidiosa*
9	*Dickeya* (dadantii and solani)
10	*Pectobacterium carotovorum* (and *P. atrosepticum*)

Source: Adapted from Mansfield, J. et al. 2012. *Molecular Plant Pathology* 13: 614–629.

hypersensitive response are the pathogenicity (hrp) gene cluster encoding the type III secretion system, effector trafficking, and host targets for defense suppression (Huynh et al., 1989; Li et al., 2002; Kvitko et al., 2009; Zhang et al., 2010; Zhang et al., 2010; Jovanovic et al., 2011). *Ralstonia solanacearum*, which has a very broad host range and can infect 200 plant species in more than 50 families, is the causal agent of potato brown rot; bacterial wilts of tobacco, eggplant, and some ornamentals; and Moko disease of banana. Although the direct economic impact of *R. solanacearum* is difficult to quantify, this pathogen is extremely dangerous because of its wide geographical distribution and host range (Mansfield et al., 2012). In potato alone, this pathogen is responsible worldwide for annual losses that are estimated at US$ 1 billion (Elphinstone, 2005).

Agrobacterium is the first organism found to be capable of inter-kingdom gene transfer. Since this discovery, a great deal has been learned about the molecular mechanisms that underlie *A. tumefaciens*-mediated genetic transformation, a highly complex process that is regulated by numerous bacterial and host factors (Gelvin, 2010; Pitzschke and Hirt, 2010; Mansfield et al., 2012). Novel mechanisms within the interactions of *A. tumefaciens* with its hosts are still being discovered (Mansfield et al., 2012). Scientific milestones have also been reached through the study of species of *Xanthomonas*, which cause at least 350 different plant diseases (Leyns et al., 1984) that lead to soft rots and tremendous associated economic losses. In addition, it is inevitable for some of the longstanding translational breakthrough practices such as treatment of leukemias (Mansfield et al., 2012). *Xanthomonas* spp. has many pathovars that cause economically important diseases on various economically significant host plants (Rademaker et al., 2005; Young et al., 2008).

Erwinia amylovora is of great historical importance to phytobacteriologists. Shortly after the pioneering work of Pasteur and Koch in the nineteenth century on bacterial pathogens of humans and animals, it was the first pathogenic bacterium confirmed to cause disease in plants (Griffith et al., 2003). *E. amylovora* is justifiably referred to as the "premier phytopathogenic bacterium" (Mansfield et al., 2012). Remarkably, its 3.90 Mb genome is the smallest of the plant pathogenic bacteria that have been sequenced (Sebaihia et al., 2010). This small size is consistent with its lack of the plant-cell-degrading tools that are common to most other phytopathogenic bacteria, such as cell-wall-degrading enzymes and low-molecular toxins (Mansfield et al., 2012).

Xylella spp., a nutritional pathogenic bacterium associated with several important plant diseases, causes Pierce disease of grapevine, citrus variegated chlorosis, and almond leaf scorch disease. Other host species for this bacterium are elm, oak, oleander, maple, sycamore, coffee, peach, mulberry, plum, periwinkle, pear, and pecan. This single species includes numerous strains that have been well characterized as pathotypes; cross-infection among different host and strains has also been observed, but without the development of disease symptoms (Mansfield et al., 2012).

In 1995, *Erwinia chrysanthemi* was transferred to a new genus, *Dickey* spp., whose members all cause economically important diseases on numerous plant hosts worldwide, including 10 monocot and 16 dicot families (Samson et al., 2005; Ma et al., 2007). *Dickeya dadantii* and *Dickeya solani* exist mainly in tropical and subtropical environments, however, where they are able to colonize a wide host range that includes *Saintpaulia* and potato. The area of study includes plant defenses and pathogen responses to defense (Li et al., 2009; Segond et al., 2009; Yang et al., 2010), pathogenesis in the pea aphid, (Costechareyre et al., 2010), and interactions between phytopathogens and human pathogens (Yamazaki et al., 2011; Mansfield et al., 2012).

The erwinias are taxonomically closely related to *Pectobacterium* spp. *Pectobacterium carotovorum* is geographically widely distributed, whereas *Pca* (Phenazine-1-carboxylic acid) is largely confined to cooler climates (Pérombelon, 2002; Toth et al., 2003). This soft rot pectobacteria was an important model pathogen in the early days of the genetic analysis of phytopathogenesis. Moreover, analysis of the roles of plant cell-wall-degrading enzymes in virulence led to the discovery of enzyme-secretion systems and the realization that protein secretion systems are operated by common mechanisms in molecular pathogenesis across plant and animal pathogens (Wharam et al., 1995; Evans et al., 2009). In addition, *Pectobacterium atrosepticum* was the first enterobacterial phytopathogen to be genomically sequenced, a process that uncovered traits such as its type IV and type VI secretion systems, its production of new secondary metabolite toxins, and its nitrogen fixation capability (Bell et al., 2004; Liu et al., 2008; Mattinen et al., 2008). It has also been determined that this enterobacterial plant pathogen is taxonomically related to animal pathogens (Mansfield et al., 2012). The agricultural impacts of *Pectobacterium* spp. have enormous translational significance for biological control. For example, periplasmic L-asparaginase from *P. carotovorum*, which causes soft rot, is used clinically in the treatment of acute lymphocytic leukemias; in addition, some related recombinant *Erwinia* spp. have been considered as possible tools for the biotechnological manufacture of vitamin C (Robert-Baudouy, 1991; Mansfield et al., 2012).

Plant pathogenic bacteria cause diseases and symptoms such as galls and overgrowths, wilts, leaf spots, specks, blights, soft rots, scabs, and cankers. Some produce toxins or inject special types of proteins that lead to host-cell death, or they produce enzymes that break down key structural components of plant cells and their walls. One example is the production of enzymes by soft rot bacteria that degrade the pectin layer that holds plant cells together. Still others colonize water-conducting xylem vessels, which causes plants to wilt and die (Ellis et al., 2008). In due course, plant pathogenic bacteria that produce toxins and/or other chemicals induce metabolic abnormalities in affected plants that may be spread to humans and animals when they consume such plants or their products (Ellis et al., 2008).

20.4 PATHOGENIC BACTERIAL IMPACT IN PLANTS

20.4.1 Growth and Development

Pathogenesis is the step-by-step development of a disease. It may include prolonged events that lead to the production of a disease, such as serial changes in the structures and functions of cells in tissues and organs. These changes may be caused by microbial, bacterial, or chemical means, or by physical agents. The disease mechanism(s) by which an etiological factor causes disease can also be used to describe the trajectory of the disease (i.e., acute, chronic, or recurrent). Biomolecular weapons have been developed; these include enzymes, toxins, growth regulators, and polysaccharides that are identical to the ones secreted by pathogens as they carry out their activities.

Plant pathogenic bacteria are known to secrete either single or multiple enzymes, such as cutinases, cellulases, pectinases, and lignases, and to do so in specific pathogen–host combinations. Pathogenic bacteria may secrete enzymes continually or upon contact with the host plant surface; in either case, the enzymes penetrate into living parenchyma tissues and cause degradation of middle lamella. Bacteria-producing cutinase may be linked with pathogenic virulence. Pectin-degrading enzymes are involved in a wide range of plant diseases, particularly the soft rot diseases caused by *Erwinia carotovora*. Cellulolytic enzymes are vitally involved in softening and disintegrating cell walls, spreading pathogens, and the collapse of cells and tissues. They also participate indirectly in disease development by releasing soluble sugars that may be used as nutrients by pathogens, and possibly by releasing material into a plant's vascular system that interferes with internal water transport.

Plant growth regulators play instrumental roles in plant pathogenesis by controlling plant growth through the hormonal activities of naturally synthesized compounds such as auxins, gibberellins, cytokinins, and ethylene. Plant pathogenic bacteria often cause an imbalance in a plant's hormonal system by disturbing its hormone levels. Common symptoms of infected plants are related to growth regulations; these include stunting, overgrowing, galling, root branching, defoliation, resetting, leaf epinasty, and others. Levels of bacterial infection in plants are directly related to auxin levels; examples include southern bacterial wilt caused by *Rolstonia solanacearum* and crown gall caused by *Agrobacterium tumefaciens*, in which the causal bacteria alter and increase the auxin levels within the plant to promote disease.

The programmed death of plant cells is another important event that shows the impact of bacteria in infected plants; often the interaction between a plant and a bacterial pathogen often results in plant cell death. This event can occur when a pathogen unsuccessfully parasitizes the host as well as when the pathogen successfully causes disease (Maher et al., 1994). For example, some pathogens oblige the development of parasites that depend on living cells, but other pathogens may benefit from the release of nutrients from dead cells. The foliar bacterial pathogens *Xanthomonads* and *Pseudomonads*, which multiply between plant cells in the apoplasm, cause the death of the host-leaf cells and are assumed to gain additional access to plant tissue through wound sites (Tsuji et al., 1991). However, foliar bacteria pathogens may persist longer in tolerant tissue. If this is so, when infected leaves senesce and fall into soil they may provide a larger reservoir for bacteria. The growth and development of plant pathogenic bacteria are more than demonstrations of integrated biotic and abiotic signals from the environment, understanding them is essential for the study of the population biology of pathogen effectors and their co-evolving host genes (Jones and Dangl, 2006).

20.4.2 Economical Impact

The impact of plant bacterial diseases is generally more significant on global levels, but such impact varies according to the establishment and patterns of the epidemics, which in turn are strongly influenced by prevailing specific area climatic conditions and the species and/or cultivars of specific host plants (Stefani, 2010). For example, Dunegan observed in 1932 that 25%–75% of peach fruits in neglected peach orchards could show lesions. Since that time, bacterial spot has been detected on different stone fruit species in different areas of the world. Although *Xanthomonas arboricola* pv. *pruni* (Xap) can affect all cultivated *Prunus* species and their hybrids, the most severe epidemics have been reported on Japanese plum (*Prunus salicina*), *P. japonica* and its hybrids, and *P. persica* and its hybrids (Ritchie, 1995; Stefani, 2010). These pathogens are now widespread in China, South Africa, and Uruguay; Argentina, Australia, Brazil, Bulgaria, Canada, India, Japan, Korea, Mexico, Moldova, Pakistan, Romania, Russia, Ukraine, and the United States have reported local outbreaks (EPPO/OEPP, 2006; Stefani, 2010).

From the mid-1970s through the 1980s, bacterial canker of stone fruits was a severe and recurrent disease in three Italian regions (Romagna, Friuli, and Veneto), where it affected mostly plum and peach, the most important fruit crops in Italy (Bazzi and Mazzucchi, 1980; Stefani et al., 1989). Previously, in 1968, the U.S. Agricultural Research Service and the U.S. Department of Agriculture had jointly decided to introduce the Calita variety into Italy because of its outstanding fruit quality, regular production, and firmness under Italian conditions (Nicotra, 1978), despite a forecast of possible stone-fruit economic losses of 3.1 million Australian dollars in years favorable to the disease. The decision was aided by inconsistent epidemic patterns that could have been attributed to differential pathogenicity features of bacterial strains, differences in infection rates across stone fruit species and cultivars, and differences in cropping conditions such as irrigation, fertilization, pruning time, and frequency. The economic impact of this particular disease largely manifests in three major variables: reduced quality and marketability of fruits, reduced orchard productivity, and increased costs in nursery production (Stefani, 2010).

The phytopathogenic *Agrobacterium* spp., which has worldwide impact, is mainly found in nurseries of fruit trees, nuts, grapevines, vegetables, and ornamentals. It also causes major economic losses in vineyards and fruit orchards (Pulawska, 2010). In nurseries, regular losses of up to 30% occur because the infected plants are unfit for the market and must be destroyed. The main reason for the losses in orchards is that the tumors girdle the modified roots or affect plant morphology such that transport of water and nutrients is impaired; in turn, these effects lead to reduced growth, lack of vigor, and poor productivity or even premature plant death. Important crops that are affected by *Agrobacterium* include stone fruits such as cherry, peach, apricot, and plum (Ali et al., 2010), as well as apple, pear, and grape (Ali et al., 2010; Rouhrazi and Rahimian, 2012); nuts such as almond, pecan, and walnut (Pulawska, 2010); vegetables such as tomato and sweet pepper (Cubero et al., 2006); and ornamentals such as rose, chrysanthemum, and aster (Aysan and Sahin, 2003; Cubero et al., 2006).

Rhodococcus fascians infection has caused high economic damage in ornamental plants by rendering their flowers commercially unfit (Putnam and Miller, 2007; Depuydt et al., 2008). Historically, incidences of the disease have been reported at nurseries of tropaeolum, petunia, dahlia, chrysanthemum, pelargonium, lily, fresia, brassica, impatiens, hebe, and carnation (Vantomme et al., 1982; Ulrychová and Petrů, 1983; Cooksey and Keim, 1983; Zutra et al., 1994). Other horticultural and greenhouse crops such as strawberry, lettuce, melon, and spinach are vulnerable as well (Faivre-Amiot, 1967). Little information on current worldwide losses is available, however, because nurseries are reluctant to provide such data (Tarkowski and Vereecke, 2014).

The impact of the important plant pathogenic bacterial agent *P. syringae* is steadily increasing due to the reemergence of old diseases such as bacterial speck of tomato and bacterial cancer on kiwi, and the emergence of new diseases such as bleeding cancer of horse chestnut (Mansfield et al., 2012). The worldwide spread of these diseases is most likely caused by the sowing of contaminated seeds for annual crops and by the use of contaminated vegetative materials to propagate woody plants. In fruit trees, systemic *P. syringae* infection and eventual death of young plants is a perennial problem in nurseries; in addition, rapid declines in the productivity of older orchards results from canker development, girdling, and death of scaffold branches (Kennelly et al., 2007; Marcelletti et al., 2011). It is evident that *P. syringae* pv. *actinidiae* is responsible for sudden, severe, and repeated occurrences of bacterial canker in kiwi production facilities worldwide, in kiwigreen (*Actinida deliciosa*) as well as kiwigold (*Actinida chinensis*), and that these occurrences cause serious economic losses (Tarkowski and Vereecke, 2014).

20.4.3 Nutritional Impact

The wide-ranging Xanthomonadaceae, Pseudomonadaceae, and Enterobacteriaceae families are targeted to all types of plants in favorable conditions, from which they obtain food and shelter on surfaces as well as within tissue regions. Except for a few rare cases, phytopathogenic bacteria provoke diseases in plants by penetrating host tissues through natural openings such as stomata, hydathodes, lenticels, nectarthodes, and stigma, or through wounds. Bacteria colonize the apoplast (i.e., the intercellular spaces or xylem vessels) where they respectively cause parenchymatous and vascular or parenchymatous vascular diseases. Some bacterial species also have the capacity to survive as epiphytes on plant surfaces (e.g., phylloplane, rhizoplane, carpoplane). From an epidemiological point of view, this poses ability a particular danger because these bacteria are quick to infect plants in favorable conditions. Once they are inside plant tissues, bacteria may implement two main attack strategies to exploit the host plant's nutrients: biotrophy, in which the plant cells are kept alive as long as possible and bacteria extract nutrients from live cells, and necrotrophy, in which bacteria kill plant cells and extract nutrients from dead cells.

Biotic stresses, including biotrophic and necrotrophic bacteria, are major constraints in production. Bacterial canker is the usual name for diseases caused by the abovementioned pathogens because of homologies in symptoms; however, although the disease caused by the persicae is known as bacterial dieback. Generally speaking, all pathogens may attack all aerial organs of a host. The most characteristic symptoms are cankers and necroses that develop on trunks and branches, where they are often located around spurs, wounds, and branch junctions. In early infections on branches, the tissue is sunken, water-soaked, and slightly brown-colored. It gradually darkens and finally becomes reddish-brown or black. Cankers and necroses can be associated with orange-brown gummosis, in which small round lesions of various sizes appear on leaves; these are initially light brown initially but change to dark brown, and can be surrounded by a yellowish halo. The necrotized tissue often drops out of the leaves (the so-called shot-hole symptoms). These pathogens can also attack blossoms, which become brown, shriveled, and often drop before fully opening. Sunken, brown-black, irregular, or regular necroses on immature fruits of susceptible cultivars of sweet and sour cherry are well known.

20.5 OPPORTUNISTIC CONTAMINATION OF PATHOGENIC BACTERIA

Globally, the Food and Agriculture Organization (FAO) plays a key role in addressing the challenges of food security, including the deprivations caused by plant diseases (FAO, 2000). Throughout the advanced world, national government agencies have provided assistance to agricultural development in developing countries, sometimes with specific attention to plant disease issues. At a voluntary level, the International Society for Plant Pathology (ISPP) is one of the many scientific organizations that are deeply concerned about global food security and the contributions that science can make to enhance it. Through ISPP, the society addresses these issues through a task force on global food security (http://www.isppweb.org/food security background.asp).

20.5.1 Cross-Kingdom Plant Pathogenic Bacteria

The ways that phytopathogens utilize specific elements to help breach reinforced cell walls and manipulate plant physiology to facilitate the disease process are similar to the ways that human pathogens use determinants to exploit mammalian physiology and to overcome highly developed adaptive immune responses. Studies have highlighted the ability of seemingly dedicated human pathogens to cause plant diseases as well as the ability of specified plant pathogens to produce human diseases. Such microbes supply systems that are of great interest in the study of the evolution of cross-kingdom pathogenicity, as well as the benefits and compromises of exploiting multiple hosts with significantly different morphologies and physiologies (Kirzinger et al., 2011).

Plant pathogens have evolved a repertoire of pathogenicity factors that invade plant host cells, facilitate disease development, and eventually promote pathogen dispersal (Panstruga and Dodds, 2009). Interestingly, some plant pathogens cause disease in humans although they are frequently isolated from human infections in the nosocomial environment. The bacterial genus *Pantoea* comprises seven species that are known to cause plant disease. These include *Pantoea agglomerans*, *Pantoea ananatis*, *Pantoea citrea*, *Pantoea dispersa*, *Pantoea punctata*, *Pantoea stewartii*, and *Pantoea terra* (Coutinho et al., 2002; Koutsoudis et al., 2006; Marín-Cevada et al., 2006; Cruz et al., 2007; Masyahit et al., 2009).

Pantoea agglomerans causes crown and root gall disease of gypsophila and beet (Coutinho et al., 2002; Koutsoudis et al., 2006; Cruz et al., 2007; Masyahit et al., 2009), *P. ananatis* causes bacterial blight and dieback of *Eucalyptus* (Coutinho et al., 2002), brown stalk rot of maize (Goszczynska et al., 2007), and stem necrosis of rice (Cother et al., 2004). *P. citrea* is the causal agent of pink disease in pineapple (Marín-Cevada et al., 2006). Species of *Pantoea*, however, have also been revealed as pathogenic to humans. Now classified as an opportunistic human pathogen, *P. agglomerans* is implicated in United States and Canadian outbreaks of septicemia caused by contaminated closures on infusion fluid bottles (Maki et al., 1976). *Pantoea agglomerans* has been associated with contaminations of intravenous fluids, parenteral nutrition, blood products, propofol, and transference tubes that have caused illness and even death to humans (Matsaniotis et al., 1984; Alvarez et al., 1995; Bennett et al., 1995; Habsah et al., 2005; Bicudo et al., 2007); *P. agglomerans* has also been obtained from joint fluids of patients with synovitis, osteomyelitis, and arthritis (Cruz et al., 2007), where infection often occurs following injuries with wood slivers, plant thorns, or wooden splinters (Flatauer and Khan, 1978; de Champs et al., 2000; Kratz et al., 2003; Cruz et al., 2007). Several studies have conclusively shown that these human pathogenic bacterial species are also capable of colonizing and causing diseases in a wide variety of plant hosts (Table 20.2).

20.5.2 Opportunistic Pathogens in the Rhizosphere

The rhizosphere provides essential nutrients for the existence of all types of microorganisms by producing biochemicals that, with the support of host plants, continuously multiply. This region is also an upright opportunist to mature plants as well to animal pathogens. For *Salmonella enterica*

TABLE 20.2
Cross-Kingdom Plant Pathogenic Bacteria

Pathogen	Plant Host/Niche	Human Disease/Condition
Enterobacter cloacae	Macadamia, dragon fruit, orchids, papaya	Respiratory/skin/urinary infection, septicemia
Enterococcus faecalis	*Arabidopsis thaliana*	Urinary/abdominal/cutaneous infections, septicemia
Burkholderia ambifaria	Soil, maize roots	Unknown
Burkholderia cenocepacia	Soil, maize roots, onion	Septicemia
Burkholderia cepacia	Soil, rice, maize, wheat, onion	Septicemia
Burkholderia gladioli	Onion, gladiolus, iris, rice	Septicemia
Burkholderia glathei	Soil	Unknown
Burkholderia glumae	Rice	Chronic granulomatous disease
Burkholderia mallei	Soil	Melioidosis/Glanders
Burkholderia plantarii	Rice, gladiolus, iris	Melioidosis
Burkholderia pseudomallei	Tomato	Melioidosis/Glanders
Burkholderia pyrrocinia	Soil	Melioidosis/Glanders
Pantoea agglomerans	Crown/root gall	Arthritis/septicemia
Pantoea ananatis	Eucalyptus, maize, rice	Septicemia
Pantoea citrea	Pineapple	Septicemia
Pantoea dispersa	Seeds	Septicaemia
Pantoea punctata	Japanese mandarin oranges	Unknown
Pantoea septica	Unknown	Septicemia
Pantoea stewartii	Maize	Unknown
Pantoea terrea	Japanese mandarin oranges	Unknown
Salmonella enterica	Tomato, *Arabidopsis thaliana*	Gastroenteritis/typhoid fever
Serratia marcescens	Squash, pumpkin	Septicemia, urinary tract infection

Source: Adapted from Kirzinger, M.W.B., Nadarasah, G., Stavrinides, J. 2011. *Genes* 2: 980–997.

serovar *typhimurium* and *E. coli* O157:H7, among others, the plant environment is a niche for pathogens that, via plant vectors, cause diseases only in debilitated or immune-compromised humans. These so-called opportunistic or facultative human pathogens have been associated with significant case fatality ratios in Europe and Northern America, and their impact on human health has increased substantially since 1980 (Berg et al., 2005; Teplitski et al., 2011). Various wild and cultivated plant species, particularly *Bacillus cepacia*, *Pseudomonas aeruginosa*, and *Salmonella maltophilia*, have been reported to host opportunistic human pathogens in their rhizospheres (Berg et al., 2005). Other bacterial species that cause skin, wound, and urinary tract infections (e.g., *B. cereus*, *Proteus vulgaris*) can also be found in rhizosphere environments (Berg et al., 2005).

Although many studies have highlighted the presence of opportunistic human pathogens in the rhizosphere (Table 20.3), relatively little is known about their virulence to their clinical counterparts. In a recent study on *P. aeruginosa* PaBP35, a strain was isolated from the aerial shoots of black pepper plants grown in a remote rainforest in southern India (Kumar et al., 2013). These polyphasic approaches show that PaBP35 is a singleton, among a large collection of *P. aeruginosa* strains, that clusters distantly from typical clinical isolates; however, subsequent analyses revealed that PaBP35 was resistant to multiple antibiotics; grew at temperatures of up to 41°C; produced rhamnolipids, hydrogen cyanide, and phenazine antibiotics; displayed cytotoxicity on mammalian cells; and caused infection in an acute murine airway infection model (Kumar et al., 2013).

TABLE 20.3
Occurrence of Potentially Human Pathogenic Species in the Rhizosphere of Diverse Plants

Origin of Rhizosphere	Pathogenic Bacteria	References
Oilseeds rape (*Brassica napus*)	*Aeromonas salmonicida, Bacillus cereus, Burkholderia cepacia, Chromobacterium violaceum, Chryseomonas luteola, Chryseobacterium indologenes, Cytophaga johnsonnae, Enterobacter intermedius, Pantoea agglomerans, Proteus vulgaris, Pseudomonas aeruginosa, Salmonella typhimurium, Serratia grimesii, Serratia liquefaciens, Serratia proteamaculans, Serratia rubidaea, Sphingomonas paucimobilis, Stenotrophomonas maltophilia*	Berg et al. (1996) Berg et al. (2002) Graner et al. (2003)
Potato (*Solanum tuberosum*)	*Achromobacter xylosoxidans, Alcaligenes faecalis, Bacillus cereus, Burkholderia cepacia, Chromobacterium violaceum, Cytophaga johnsonae, Enterobacter amnigenus, Enterobacter cloacae, Enterobacter intermedius, Flavimonas oryzihabitans, Francisella philomiragia, Janthinobacterium lividum, Kluyvera cryorescens, Ochrobactrum anthropic, Pantoea agglomerans, Proteus vulgaris, Pseudomonas aeruginosa, Serrtia grimesii, Sphingobactrium spiritivorum, Sphingomonas paucimobilis, Staphylococcus epidermis, Staphylococcus pasteuri, Staphylococcus xylosus, Stenotrophomonas maltophilia*	Lottmann et al. (1999) Gupta et al. (2001) Lottmann and Berg (2001) Berg et al. (2002) Reiter et al. (2002) Krechel et al. (2002) Sessitsch et al. (2004) Berg et al. (2005)
Strawberry (*Fragaria xananassa*)	*Acinetobacter baumannii, Acinetobacter calcoaceticus, Burkholderia cepacia, Pantoea agglomerans, Proteus vulgaris, Salmonella typhimurium, Serratia grimesii, Serratia proteamaculans, Staphylococcus epidermis*	Berg et al. (2002) Berg et al. (2005)
Alfalfa (*Medicago sativa*)	*Stenotrophomonas maltophilia, Flavobacterium johnsonae*	Schwieger and Tebbe (2000)
Goosefoots (*Chenopodium album*)	*Stenotrophomonas maltophilia, Flavobacterium johnsonae*	Schwieger and Tebbe (2000)
Sunflower (*Helianthus annuus*)	*Burkholderia cepacia, Stenotrophomonas maltophilia, Flavobacterium odoratum*	Hebbar et al. (1991)
Maize (*Zea mays*)	*Burkholderia cepacia, Klebsiella pneumoniae, Serratia liquefaciens, Sphingomonas paucimobilis, Stenotrophomonas maltophilia*	Lambert et al. (1987) Dalmastri et al. (1999) Chelius and Triplett (2000)
Rice (*Oryza sativa*)	*Aeromonas veronii, Alcaligenes xylosoxidans, Enterobacter cloacae, Ochrobactrum anthropic, Pseudomonas aeruginosa, Sarratia marcescens*	Mehnaz et al. (2001) Gyaneshwar et al. (2001) Tripathi et al. (2002)
Wheat (*Triticum sativum*)	*Burkholderia cepacia, Enterobacter agglomerans, Ochrobactrum anthropi, Ochrobactrum tritici, Pseudomonas aeruginosa, Salmonella typhimurium, Staphylococcus aureus, Stenotrophomonas maltophilia, Streptococcus pyogenes*	Morales et al. (1996) Germida and Siciliano (2001)

Source: Adapted from Berg, G., Eberl, L., Hartmann, A. 2005. *Environmental Microbiology* 7: 1673–1685.

Interestingly, the mechanisms involved in rhizosphere colonization and the antimicrobial activity of human pathogenic bacteria appear to be similar to the mechanisms involved in the virulence within and colonization of human tissues (Berg et al., 2005; van Baarlen et al., 2007; Holden et al., 2009). For example, *B. cepacia* strains caused disease in a lung-infection model and was also virulent in alfalfa (Bernier et al., 2003), and *Salmonella enterica* pathogenicity genes as well as genes related to carbon utilization were differentially regulated in the presence of lettuce root exudates (Klerks et al., 2007). The combined results of chemotaxis assays postulate that root exudates trigger chemotaxis in *S. enterica* and switch on genes that play a role in adherence (Klerks et al., 2007). Several other genes and traits, including fimbriae, adhesins, and capsule production, have been identified as participating in the attachment of human pathogens to plant surfaces. Further discussion on this topic can be found in several comprehensive studies (van Baarlen et al., 2007; Tyler and Triplett, 2008; Holden et al., 2009; Teplitski et al., 2011).

Plant colonization by human pathogens is made possible when, following attachment, human pathogenic bacteria (in particular, Enterobacteriaceae) invade root tissue (Warriner and Namvar, 2010). In contrast to their invasion of animal hosts, enteric bacteria appear to reside mostly in the apoplastic spaces of plant hosts (Holden et al., 2009). Several studies have indicated that human pathogenic bacteria enter the root tissue at sites of lateral root emergence. This entry has been observed for *Salmonella* and *E. coli* O157: H7 in roots of *Arabidopsis* and lettuce, and for *Klebsiella pneumoniae* in multiple plant species (Tyler and Triplett, 2008). Both endophytic and systemic colonization of barley by the food-borne pathogen *S. enterica* serovar *typhimurium* were demonstrated by Kutter et al. (2006). Damaged roots provide easy access points for opportunistic human pathogens, which results in invasion and endophytic colonization.

P. aeruginosa PaBP35 is able to colonize the shoots of black pepper stem cuttings at relatively high densities (>105 CFU per gram of shoot segment) after 8 min of root treatment. But the invasive and endophytic behavior of strain PaBP35 is not limited to black pepper; it has also been demonstrated in tomato seedlings. Interestingly, PaBP35 established significantly higher population densities in the roots and shoots of tomato seedlings than *P. aeruginosa* PA01 did; this result suggests some level of specificity in endophytic colonization (Kumar et al., 2013).

Many studies have focused on pathogen infections within edible plants which can be traced to bacteria that are pathogenic to humans. For example, fruits and vegetables infected with soft rot pathogens showed significant increases in *Salmonella* and *E. coli* O157:H7, which can be fatal to humans (Teplitski et al., 2011). Subsequent loss of cell-tissue integrity, concomitant release of nutrients from infected plant tissue, and degradation of plant tissue by macerating plant pathogens may also lead to an increase in pH that could be beneficial to enteric pathogens (Holden et al., 2009; Teplitski et al., 2011). Plant pathogens may also suppress plant defenses that otherwise would have limited the invasion and endophytic colonization of bacteria that cause disease in humans (Iniguez et al., 2005). Teplitski et al. (2011) postulated that controlling plant pathogens would reduce the predisposition of plant produce to colonization by human-associated pathogens.

Many plant-associated organisms are known to alter the morphology of their host plant, resulting in malformations that either create a protective niche or enhance dispersal. Classic examples include: rhizobial nodules (Oldroyd and Downie, 2008), galls induced by *Agrobacterium* spp. and other bacteria (Chapulowicz et al., 2006), and witches' broom and other developmental alterations caused by pathogens such as phytoplasmas (Hogenhout et al., 2008).

20.6 PLANT PATHOGENIC BACTERIAL INFECTION ON NON-HOST ORGANISMS

A diversity of plant pathogenic bacteria secrete proteins and other molecules to different cellular compartments of their hosts and sometimes to non-target hosts; these molecules modulate host plant defense circuitry and functions. All of these pathogens secrete proteins and small molecules that alter host-cell structures and functions. These alterations either facilitate infection (which is described in terms of virulence factors and toxins) or trigger defense responses (which are described in terms of avirulence factors and elicitors), or both (Huitema et al., 2004; Kamoun 2006, 2007). The interactions of bacteria with plants can be either incompatible or compatible. Incompatible interactions occur when the bacterium encounters a non-host plant (a non-target host) or a resistant host plant (a cultivar-specific host) and such interactions are frequently associated with HR (i.e., rapid, programmed plant-cell death that occurs at the site of infection). Compatible interactions occur when the bacterium infects one or more susceptible host plants and causes disease symptoms.

The severity of disease symptoms caused by bacteria in plants is determined by the production of a number of virulence factors, including phytotoxins, plant cell-wall-degrading enzymes, extracellular polysaccharides, and phytohormones. In general, plant pathogenic bacteria produce a wide spectrum of non-host specific phytotoxins. These toxic compounds cause symptoms in many plants whether or not the plants can be infected by the toxin-producing bacterium (Table 20.4). On the basis of the symptoms they induce in plants, phytotoxin pathogens have been grouped as necrosis-inducing and chlorosis-inducing (Buonaurio, 2008). During their lifetimes, plants are continually exposed to a vast number of potential phytopathogenic bacteria, against which they try to defend themselves through a multilayered system of passive and active defense mechanisms. When defense responses are effective, plants are considered to be resistant. Most plant species are resistant to most species of potential bacterial invaders, a phenomenon termed non-host or species resistance. It is likely that during evolution, non-host resistance has been overcome by individual phytopathogenic strains of a given bacterial species through the acquisition of virulence factors, which enabled them to either evade or suppress plant defense mechanisms (Nürnberger et al., 2004; Buonaurio, 2008).

20.6.1 Synthesis of Toxic Molecules by Plant Pathogenic Bacteria and Their Impacts

The major outcome interactions between a host plant and a pathogenic bacterium is the development of learned and specific virulence factors; these give selective advantages to the organism that produces the toxic molecules (Alouf and Freer, 1999). Our understanding of the mechanisms of bacterial toxic molecular or phytotoxic activity has increased enormously (Cossart et al., 1996; Schiavo and van der Goot, 2001). Phytotoxins are produced by plant pathogens or in host–pathogens interactions that directly injure plant cells and influence the course of disease development or symptoms. Bacterial pathogens also produce a number of secondary metabolites that are toxic to plant cells as well as herbivorous and carnivorous consumers of infected plants.

Studies of phytotoxins are carried out more often when there is visible evidence of such activity. But some phytotoxins may instead act by changing metabolic processes in the host in such a way that the deleterious activity is manifested only at biochemical levels (Bender et al., 1999). Phytopathogenic prokaryotes affect the ability of a bacterium to induce HR in non-host plants and pathogenicity in host plants, and to grow within or on the surface of plants (Gopalan et al., 1996). They may also be involved in biosynthesis of a type III secretion pathway that is similar in both plant and animal pathogens and is used to secrete virulence proteins (Salmond, 1994). Evidence is mounting that mechanisms which function in clinical pathogens of animals, such as the type III secretion systems in *Salmonella*, *Shigella*, and *Yersinia*, are similar to those in phytopathogenic species (Roine et al., 1997).

Phytopathogenic bacteria of the *Pseudomonas* genus, especially *Pseudomonas syringae*, produce a wide spectrum (Table 20.5) of non-host specific phytotoxins (i.e., toxic compounds that cause symptoms in many plants whether or not the plants can actually be infected). Phytotoxins of *Pseudomonas* spp. have been grouped as necrosis-inducing and chlorosis-inducing. *Pseudomonas syringae* pv. *syringae*, which causes many diseases and types of symptoms in herbaceous and woody plants, produces necrosis-inducing phytotoxins such as lipodepsipeptides. Based on their amino acid chain length, these are usually divided in two groups: mycins (e.g., syringomycins) and peptins (e.g., syringopeptins) (Gross et al., 1998). Because all analyzed strains of *P. syringae* pv. *syringae* produce both syringomycins and syringopeptins, interrelated roles have been suggested for them in plant–pathogen interactions (Bender et al., 1999).

P. syringae phytotoxins can contribute to systemic movement of bacteria in planta (Patil et al., 1974), lesion size (Bender et al., 1987; Xu and Gross, 1988), and pathogen multiplication in its host (Feys et al., 1994; Mittal and Davis, 1995). The primary symptom elicited by coronatine is a diffuse chlorosis that can be induced in a wide variety of plant species (Gnanamanickam et al., 1982). Interestingly, the reaction of *Arabidopsis thaliana* to exogenously applied coronatine is atypical;

TABLE 20.4
Non-Host Specific Plant Pathogenic Bacteria Producing Phytotoxins

Host/Diseases	Bacteria Pathogens	Phytotoxins	References
Infect ryegrass, soybean, crucifers, *Prunus* spp. tomato	*Pseudomonas syringae* pv. *atropurpurea, glycinea, maculicola, morsprunorum, tomato*	Coronatine	Mitchell (1982), Wiebe and Campbell (1993)
Chlorosis in bean	*Pseudomonas syringae* pv. *tabaci, coronafaciens, garcae*	Tabtoxin	Uchytil and Durbin (1980), Gasson (1980)
Halo blight of bean and canker of kiwifruit	*Pseudomonas syringae* pv. *phaseolicola, actinidae*	Phaseolotoxin	Mitchell (1976), Bender et al. (1999)
Apical necrosis in mango, blister bark in apple, blossom blast in pear, brown spot in bean plant leaf	*Pseudomonas syringae* pv. *syringae*	Mangotoxin, syringomycins, syringopeptins	Latorre and Jones (1979), Mansvelt and Hattingh (1986), Ballio et al. (1991), Arrebola et al. (2003)
Bacterial canker in peach	*Pseudomonas syringae* pv. *persicae*	Persicomycins	Barzic and Guittet (1996)
Bacterial canker in stone fruit	*Pseudomonas syringae* pv. *syringae, aptata, atrofaciens*	Syringomycins	Ballio et al. (1994a)
Chlorosis of developing shoot tissues in sunflower	*Pseudomonas syringae* pv. *tageti*	Tagetitoxin	Rhodehamel and Durbin (1989)
Bacterial canker in stone fruit	*Pseudomonas syringae* pv. *morsprunorum*	Coronatine	Latorre et al. (1979)
Bacterial leaf spot on cruciferous plants	*Pseudomonas syringae* pv. *maculicola*	Coronatine	Cuppels and Ainsworth (1995)
Soft rot disease in diverse crops (cabbage, potato, onion, radish, etc.)	*Pectobacterium carotovorum* subsp. *Carotovorum*	Carotovoricin, Carocin	Roh et al. (2010)
Soft rot disease in potato	*Pectobacterium atrosepticum*	Coronafacic acid	Bell et al. (2004)
Wilt disease in tomato, pepper, eggplant, Irish potato	*Ralstonia solanacearum*	Infection promoting effector proteins (T3Es)	Vasse et al. (2000)
Bacterial blight in rice	*Xanthomonas oryzae* pv. *oryzae*	Phenylacetic acid, *trans*-3-methyl-thio-acrylic acid, 3-methyl-thio-propionic acid	Lee et al. (2005)
Blight disease	*Xanthomonas campestris* pv. manihotis	3-Methyl-thio-propionic acid	Perreaux et al. (1986)
Cassava bacterial blight	*Xanthomonas axonopodis* pv. *manihotis*	3-Methyl-thio-propionic acid, carboxylic acid	Perreaux et al. (1986) Ewbank (1992)
Fire blight of apple and pear	*Erwinia amylovora*	Amylovorin or fire blight toxin	Goodman et al. (1974)
Soft rot disease	*Dickeya dadantii*	Cyt toxin	Costechareyre et al. (2012)
Tuber soft rots in potato	*Dickeya solani*	Oocydin, Zeamine	Garlant et al. (2013)
Redroot pigweed	*Xanthomonas campestris* pv. *retroflexus*	Organic acids, Cyclo (proline-phenylalanine), Polysaccharide	Mingzhi et al. (2007)

TABLE 20.5
Phytotoxins Produced by *Pseudomonas* spp

Producing Organism	Toxin	References
Pseudomonas syringae pv. *atropurpurea, glycinea, maculicola, morsprunorum, tomato*	Coronatine	Ichihara et al. (1977)
Pseudomonas corrugata	Corpeptin	Emanuele et al. (1998)
Pseudomonas fuscovaginae	Fuscopeptin	Ballio et al. (1996)
Pseudomonas syringae pv. *persicae*	Persicomycins	Barzic and Guittet (1996)
Pseudomonas syringae pv. *actinidiae, phaseolicola*	Phaseolotoxin	Mitchell (1976)
Pseudomonas andropogonis	Rhizobitoxine	Mitchell et al. (1986)
Pseudomonas syringae pv. *syringae, aptata, atrofaciens*	Syringomycins	Ballio et al. (1994a,b), Fukuchi et al. (1992), Vassilev et al. (1996)
Pseudomonas fuscovaginae	Syringomycins	Batoko et al. (1998)
Pseudomonas syringae pv. *syringae*	Syringopeptins	Ballio et al. (1991)
Pseudomonas syringae pv. *tabaci, coronafaciens, garcae*	Tabtoxin	Stewart (1971)
Pseudomonas syringae pv. *tagetis*	Tagetitoxin	Rhodehamel and Durbin (1989)
Pseudomonas tolaasii	Tolaasin	Rainey et al. (1991)
Pseudomonas marginalis (*P. fluorescens*)	Viscosin	Laycock et al. (1991)

Source: Adapted from Bender, C., Alarcón-Chaidez, F., Gross, D.C. 1999. *Microbiology and Molecular Biology Reviews* 63: 266–292.

instead of chlorosis, anthocyanins accumulate at the site of inoculation and the tissue develops a strong purple hue (Bent et al., 1992). Coronatine is also known to induce hypertrophy, inhibit root elongation, and stimulate ethylene production (Volksch et al., 1989; Kenyon and Turner, 1992). Visual assessment of phytotoxin production in planta can be subjective, however, because plants exhibit a limited number of visible reactions to phytotoxins. Some phytotoxins may act by changing metabolic processes in the host such that deleterious activity might be manifested only at the biochemical level (Bender et al., 1999).

Phaseolotoxin, a non-host-specific phytotoxin produced by *P. syringae* pv. *Phaseolicola*, has an inhibitory effect on the proliferation of leukemia and on pancreatic cell lines (Bachmann et al., 2004); in fact, interference by ornithine decarboxylase has long been considered a promising therapeutic approach against proliferative diseases, including various malignancies (Auvinen et al., 1992; Davidson et al., 1999). *P. syringae* pv. *syringae* produces toxic cyclic lipodepsipeptide action that has been observed in bilayer lipid membranes and human red blood cells (Feigin et al., 1997; Blasko et al., 1998); it also induces lysis of human red blood cells and tobacco protoplasts, presumably due to the colloid osmotic shock triggered by the ion flux through pores (Menestrina et al., 1994).

Dalla Serra et al. (1999) compared the hemolytic effect of cyclic lipodepsipeptides on human and rabbit red blood cells and found that the most effective is syringomycin E, followed by syringotoxin; the weakest are syringopeptins. Hutchison and Gross (1997) observed that syringopeptins are 85% as effective as syringomycin E on horse erythrocytes. Contradictory results were reported by Lavermicocca et al. (1997), whose findings in sheep red blood cells indicated that syringopeptins have the highest lytic activities; syringomycin E is less effective; and the least active is syringotoxin. These findings were in agreement with Sorensen et al. (1996). Although the structural modification of syringomycin E may improve the therapeutic index of this potent peptide, it should be noted that an HeLa cytotoxicity assay cannot predict its toxicity in humans and animals. However, it has been found that the pore inactivation of syringotoxin in bilayer lipid membrane is due to a decrease in the number of pores and not to changes in their conductance schedule (Szabò et al., 2002).

20.7 PROTECTION STRATEGIES OF PLANT PATHOGENIC BACTERIA

The risk of occurrence of bacterial plant disease epidemics is rising with the worldwide exchanges of plant materials also increase; moreover, the current strategies for control are often unsatisfactory. Development of new control strategies is of the utmost urgency; basic research on plant pathogenic bacterium interactions, especially at a molecular level, can substantially contribute to the attainment of this goal (Buonaurio, 2008). Bacterial spot of tomato (*Lycopersicon esculentum* Mill.) caused by *Xanthomonas campestris* pv. *vesicatoria* (Doidge), Dye (*Xanthomonas axonopodis* pv. *vesicatoria* (Vauterin et al., 1995) are most serious diseases in many areas (Ward and O'Garro, 1992; Uys et al., 1996). The disease affects stems, leaves, and fruits (Sherf and MacNab, 1986; Agrios, 1997) and causes significant losses when environmental conditions are suitable for the bacterial plant pathogen (Pohronezny and Volin, 1983).

Numerous strategies have been used for controlling this disease, including sanitation, the usage of pathogen-free seed, other cultural practices (Sherf and MacNab, 1986; Jones et al., 1991), as well as the use of tomato cultivars that are resistant to *X. campestris* pv. *vesicatoria* (Mew and Natural, 1993; Bouzar et al., 1999). Chemical control with copper and streptomycin sprays has also been attempted (Thayer and Stall, 1961; Conover and Gerhold, 1981; Jones and Jones, 1985). Disadvantages of chemical applications, including chemical residues on fruit, high costs, and emergence of resistant bacterial strains have been reported (Jones and Jones, 1985; Stall et al., 1986; Ritchie and Dittapongitch, 1991).

Gilbert et al. (2010) reported that the lack of chemicals for restraining bacterial diseases in orchards creates a need for alternative methods such as approaches based on prevention, biological control, and plant resistance. These approaches require accurate knowledge of the identity, ecology, and pathogenicity of the *P. syringae* strains encountered in the field. More information is needed to improve our knowledge of the ecology and epidemiology of plant pathogens; to identify the most damaging organisms; to better differentiate, classify, and detect pathogens; and to allow for selection of correct reference strains when breeding stone fruit cultivars for resistance.

20.8 CONCLUSION

Plant pathogenic bacterial impact on global agriculture is increasing. The human population, which has more than doubled since year 1960s, is projected to exceed nine billion by 2050. The issues of global food security are among the most persistent in international politics. The disproportionate global distribution of food is influenced by social and economic factors, expanding urbanization, bioenergy production, and the marginalization of land fit for agriculture. It is clear that food security will require an increase in sustainable agricultural production, particularly the production of cereals such as wheat, rice, and maize (http://faostat.fao.org).

Possible solutions include the exclusion, elimination, or reduction of pathogen inocula and development through good cultural practices, intercropping and rotation, judicious use of pesticides, breeding programs that exploit the gene pools of related species of plants, understanding and combating virulence mechanisms of pathogens, biological control, post-harvest protection, and improvement of plant performance through biotechnology. At the biological level, there is clearly much to be done to expand our knowledge of plant pathogens and our ability to deal with them. At the political level, plant pathologists should persistently draw these opportunities to the attention of politicians and public and private investors, because such people are able to affect the challenging demands of food security and plant nutrient balance without contaminated molecules (including biologically synthesized toxic residuals) which are vital to the support of human life, especially in developing countries. Appropriate funding from public and private sectors should be allocated in order to efficiently reduce or eradicate the plant bacterial diseases that destroy the most important crops.

REFERENCES

Abramovitch, R.B., Anderson, J.C., Martin, G.B. 2006. Bacterial elicitation and evasion of plant innate immunity. *Nature Reviews Molecular Cell Biology* 7: 601–611.

Agrios, G.N. 1997. *Plant Pathology*, 4th Edn. New York: Academic Press.

Ali, H., Ahmed, K., A. Hussain, Imran. 2010. Incidence and severity of crown gall disease of cherry, apple and apricot plants caused by *Agrobacterium tumefaciens* in Nagar Valley of Gilgit-Baltistan, Pakistan. *Pakistan Journal of Nutrition* 9: 577–581.

Alouf, J.E., Freer, J.H. 1999. *The Comprehensive Sourcebook of Bacterial Protein Toxins*. London, UK: Academic Press.

Alvarez, F.E., Rogge, K.J., Tarrand, J., Lichtiger, B. 1995. Bacterial contamination of cellular blood components. A retrospective review at a large cancer center. *Annals of Clinical and Laboratory Science* 25: 283–290.

Arrebola, E., Cazorla, F.M., Duran, V.E., Rivera, E., Olea, F., Codina, J.C., PerezGarcia, A., de Vicente, A. 2003. Mangotoxin: A novel antimetabolite toxin produced by *Pseudomonas syringae* inhibiting ornithine/arginine biosynthesis. *Physiological and Molecular Plant Pathology* 63: 117–127.

Auvinen, M., Paasinen, A., Andersson, L.C., Holtta, E. 1992. Ornithine decarboxylase is critical for cell transformation. *Nature* 360: 355–358.

Aysan, Y., Sahin, F. 2003. An outbreak of crown gall disease on rose caused by *Agrobacterium tumefaciens* in Turkey. *Plant Pathology* 52: 780.

Bachmann, A.S., Xu, R., Ratnapala, L., Patil, S.S. 2004. Inhibitory effects of phaseolotoxin on proliferation of leukemia cells HL-60, K-562 and L1210 and pancreatic cells RIN-m5F. *Leukemia Research* 28(3): 301–306.

Ballio, A., Barra, D., Bossa, F., Collina, A., Grgurina, I., Marino, G., Moneti, G. et al. 1991. Syringopeptins new phytotoxic lipodepsipeptides from *Pseudomonas syringae* pv. *syringae*. *FEBS Letters* 291: 109–112.

Ballio, A., Collina, A., Di Nola, A., Manetti, C., Paci, M., Segre, A. 1994a. Determination of structure and conformation in solution of syringotoxin, a lipodepsipeptide from *Pseudomonas syringae* pv. *syringae* by 2D NMR and molecular dynamics. *Structural Chemistry* 5: 43–50.

Ballio, A., Bossa, F., Di Giorgio, D., Ferranti, P., Paci, M., Pucci, P., Scaloni, A., Segre, A., Strobel, G.A. 1994b. Novel bioactive lipodepsipeptides from *Pseudomonas syringae*: The pseudomycins. *FEBS Letters* 355: 96–100.

Ballio, A., Bossa, F., Camoni, L., Di Giorgio, D., Flamand, M.-C., Maraite, H., Nitti, G., Pucci, P., Scaloni, A. 1996. Structure of fuscopeptins, phytotoxic metabolites of *Pseudomonas fuscovaginae*. *FEBS Letters* 381: 213–216.

Barzic, M.R., Guittet, E. 1996. Structure and activity of persicomycins, toxins produced by *Pseudomonas syringae* pv. *persicae*, *Prunus persica* isolate. *European Journal of Biochemistry* 239: 702–709.

Batoko, H., de Kerchove d'Exaerde, A., Kinet, J.M., Bouharmont, J., Gage, R.A., Maraite, H., Boutry, M. 1998. Modulation of plant plasma membrane H1-ATPase by phytotoxic lipodepsipeptides produced by the plant pathogen *Pseudomonas fuscovaginae*. *Biochimica et Biophysica Acta* 17: 216–226.

Bazzi, C., Mazzucchi, U. 1980. Epidemie di *Xanthomonas pruni* su susino *Xanthomonas pruni*. *Informatore Fitopatologico* 30(5): 11–17.

Bell, K., Sebaihia, M., Pritchard, L., Holden, M., Holeva, M.C., Thomson, N.R., Bentley, S.D. et al. 2004. Genome sequence of the enterobacterial phytopathogen, *Erwinia carotovora* subsp. *atroseptica* and characterisation of novel virulence factors. *Proceedings of the National Academy of Sciences*. USA 101: 11105–11110.

Bender, C.L., Stone, H.E., Sims, J.J., Cooksey, D.A. 1987. Reduced pathogen fitness of *Pseudomonas syringae* pv. tomato Tn5 mutants defective in coronatine production. *Physiological and Molecular Plant Pathology* 30: 272–283.

Bender, C., Alarcón-Chaidez, F., Gross, D.C. 1999. *Pseudomonas syringae* phytotoxins: Mode of action, regulation, and biosynthesis by peptide and polyketide synthetases. *Microbiology and Molecular Biology Reviews* 63: 266–292.

Bennett, S.N., McNeil, M.M., Bland, L.A., Arduino, M.J., Villarino, M.E., Perrotta, D.M., Burwen, D.R., Welbel, S.F., Pegues, D.A., Stroud, L. 1995. Postoperative infections traced to contamination of an intravenous anesthetic, propofol. *The New England Journal of Medicine* 333: 147–154.

Bent, A.F., Innes, R.W., Ecker, J.R., Staskawicz, B.J. 1992. Disease development in ethylene-insensitive *Arabidopsis thaliana* infected with virulent and avirulent *Pseudomonas* and *Xanthomonas* pathogens. *Molecular Plant-Microbe Interactions* 5: 372–378.

Berg, G., Eberl, L., Hartmann, A. 2005. The rhizosphere as a reservoir for opportunistic human pathogenic bacteria. *Environmental Microbiology* 7: 1673–1685.

Berg, G., Marten, P., Ballin, G. 1996. *Stenotrophomonas maltophilia* in the rhizosphere of oilseed rape—Occurrence, characterization and interaction with phytopathogenic fungi. *Microbiological Research* 151: 19–27.

Berg, G., Roskot, N., Steidle, A., Eberl, L., Zock, A., Smalla, K. 2002. Plant-dependent genotypic and phenotypic diversity of antagonistic rhizobacteria isolated from different *Verticillium* host plants. *Applied and Environmental Microbiology* 68: 3328–3338.

Bernier, S.P., Silo-Suh, L., Woods, D.E., Ohman, D.E., Sokol, P.A. 2003. Comparative analysis of plant and animal models for characterization of *Burkholderia cepacia* virulence. *Infection and Immunity* 71: 5306–5313.

Bicudo, E.L., Macedo, V.O., Carrara, M.A., Castro, F.F., Rage, R.I. 2007. Nosocomial outbreak of *Pantoea agglomerans* in a pediatric urgent care center. *Brazilian Journal of Infectious Diseases* 11: 281–284.

Birch, P.R., Rehmany, A.P., Pritchard, L., Kamoun, S., Beynon, J.L. 2006. Trafficking arms: Oomycete effectors enter host plant cells. *Trends in Microbiology* 14: 8–11.

Blasko, K., Schagina, L.V., Agner, G., Kaulin, Y.A., Takemoto, J.Y. 1998. Membrane sterol composition modulates the pore forming activity of syringomycin E in human red blood cells. *Biochimica et Biophysica Acta* 1373: 163–169.

Block, A., Li, G., Fu, Z.Q., Alfano, J.R. 2008. Phytopathogen type III effector weaponry and their plant targets. *Current Opinion in Plant Biology* 11: 396–403.

Bouzar, H., Jones, J.B., Stall, R.E., Louws, F.J., Schneider, M., Rademaker, J.L.W., de Bruijn, F.J., Jackson, L.E. 1999. Multiphasic analysis of *Xanthomonads* causing bacterial spot disease on tomato and pepper in the Caribbean and Central America: Evidence for common lineages within and between countries. *Phytopathology* 89: 328–335.

Buonaurio, R. 2008. Infection and plant defense responses during plant-bacterial interaction, In: *Plant-Microbe Interactions*. Ait Barka, E., Clement, C. (eds.). India: Research Signpost, pp. 169–197.

Chapulowicz, L., Barash, I., Schwartz, M., Aloni, R., Manulis, S. 2006. Comparative anatomy of gall development on *Gypsophila paniculata* induced by bacteria with different mechanisms of pathogenicity. *Planta* 224: 429–437.

Chelius, M.K., Triplett, E.W. 2000. Immunolocalization of dinitrogenase reductase produced by *Klebsiella pneumonia* in association with Zea mays L. *Applied and Environmental Microbiology* 66: 783–787.

Chisholm, S.T., Coaker, G., Day, B., Staskawicz, B.J. 2006. Host-microbe interactions: Shaping the evolution of the plant immune response. *Cell* 124: 803–814.

Christou, P., Twyman, R.M. 2004. The potential of genetically enhanced plants to address food insecurity. *Nutrition Research Reviews* 17: 23–42.

Chu, P.W.G., Helms, K., Martin, R.R. 1989. Subterranean clover distortion disease: A new virus-like disease of subterranean clover and lucerne with unusual features. In *Proceedings of the 7th Australasian Plant Pathology Society Conference*, Brisbane, 3–7 July, p. 96.

Conover, R.A., Gerhold, N.R. 1981. Mixture of copper and maneb or mancozeb for control of bacterial spot of tomato and their compatibility for control of fungus diseases. *Proceedings of the Florida State Horticultural Society* 94: 154–156.

Cooksey, D.A., Keim, R. 1983. Association of *Corynebacterium fascians* with fasciation disease of Impatiens and Hebe in California. *Plant Disease* 67: 1389–1389.

Cossart, P., Boquet, P., Normark, S., Rappuoli, R. 1996. Cellular microbiology emerging. *Science* 271: 315–316.

Costechareyre, D., Dridi, B., Rahbe, Y., Condemine, G. 2010. Cyt toxin expression reveals an inverse regulation of insect and plant virulence factors of *Dickeya dadantii. Environmental Microbiology* 12: 3290–3301.

Costechareyre, D., Balmand, S., Condemine, G., Rahbe, Y. 2012. *Dickeya dadantii*, a plant pathogenic bacterium producing Cyt-Like entomotoxins causes septicemia in the pea aphid *Acyrthosiphon pisum PLoS One* 7(1): e30702.

Cother, E.J., Reinke, R., McKenzie, C., Lanoiselet, V.M., Noble, D.H. 2004. An unusual stem necrosis of rice caused by Pantoea ananas and the first record of this pathogen on rice in *Australasian. Plant Pathology* 33: 495–503.

Coutinho, T.A., Preisig, O., Mergaert, J., Cnockaert, M.C., Riedel, K.H., Swings, J., Wingfield, M.J. 2002. Bacterial blight and dieback of *Eucalyptus* species, hybrids, and clones in South Africa. *Plant Disease* 86: 20–25.

Cox, A.E., Large, E.C. 1960. Potato blight epidemics throughout the world. United States Department of Agriculture. *Agricultural Handbook* 174: 1–230.

Cruz, A.T., Cazacu, A.C., Allen, C.H. 2007. *Pantoea agglomerans*, a plant pathogen causing human disease. *Journal of Clinical Microbiology* 45: 1989–1992.

Cubero J, Lastra, B., Salcedo, C.I., Piquer, J., López, M.M. 2006. Systemic movement of *Agrobacterium tumefaciens* in several plant species. *Journal of Applied Microbiology* 101: 412–421.

Cuppels, D.C., Ainsworth, T. 1995. Molecular and physiological characterization of *Pseudomonas syringae* pv. tomato and *Pseudomonas syringae* pv. *maculicola* strains that produce the phytotoxin coronatine. *Applied and Environmental Microbiology* 61: 3530–3536.

Dalla Serra, M., Bernhart, I., Nordera, P., Di Giorgio, D., Ballio, A., Menestrina, G. 1999. Conductive properties and gating of channels formed by syringopeptin 25A, a bioactive lipodepsipeptide from *Pseudomonas syringae* pv. *syringae*, in planar lipid membranes. *Molecular Plant-Microbe Interactions* 12: 401–409.

Dalmastri, C., Chiarini, L., Cantale, C., Bevivino, A., Tabacchioni, S. 1999. Soil type and maize cultivar affect the genetic diversity of maize root-associated *Burkholderia cepacia* populations. *Microbial Ecology* 38: 273–284.

Davidson, N.E., Hahm, H.A., McCloskey, D.E., Woster, P.M., Casero, R.A. 1999. Clinical aspects of cell death in breast cancer: The polyamine pathway as a new target for treatment. *Endocrine-Related Cancer* 6: 69–73.

Davis, E.L., Hussey, R.S., Mitchum, M.G., Baum, T.J. 2008. Parasitism proteins in nematode-plant interactions. *Current Opinion in Plant Biology* 11: 360–366.

Dawkins, R. 1999. *The Extended Phenotype: The Long Reach of the Gene*. Revised edition. New York: Oxford University Press.

de Champs, C., Le Seaux, S., Dubost, J.J., Boisgard, S., Sauvezie, B., Sirot, J. 2000. Isolation of *Pantoea agglomerans* in two cases of septic mono-arthritis after plant thorn and wood sliver injuries. *Journal of Clinical Microbiology* 38: 460–461.

Depuydt, S., Putnam, M., Holsters, M., Vereecke, D. 2008. *Rhodococcus fascians*, an emerging threat for ornamental crops. In: *Floriculture, Ornamental and Plant Biotechnology: Advances and Topical Issues*. da Silva, J.A. Teixeira (ed.). volume V, 1st Edn. London: Global Science Books. p. 480–489.

Dunegan, J.C. 1932. The bacterial spot disease of the peach and other stone fruits. Technical Bulletin No. 273. Washington, DC, USA: US Department of Agriculture. pp. 53.

Ellis, S.D., Boehm, M.J., Coplin, D. 2008. OSU Extension Factsheet—Introduction to Plant Disease Series PP401.06 entitled "Bacterial Diseases of Plants". (Available online at: http://ohioline.osu.edu/hyg-fact/3000/index.html).

Elphinstone, J.G. 2005. The current bacterial wilt situation: A global overview. In: *Bacterial Wilt Disease and the Ralstonia solanacearum Species Complex*. Allen, C., Prior, P., Hayward, A.C. (eds.). St Paul, MN: APS Press. pp. 9–28.

Emanuele, M.C., Scaloni, A., Lavermicocca, P., Iacobellis, N. S., Camoni, L., Di Giorgio, D., Pucci, P., Paci, M., Segre, A., Ballio, A. 1998. Corpeptins, new bioactive lipodepsipeptides from cultures of *Pseudomonas corrugata*. *FEBS Letters* 433: 317–320.

EPPO/OEPP. 2006. Data sheets on quarantine pests. *Xanthomonas arboricola* pv. *pruni*. http://www.eppo.org/QUARANTINE/listA2.htm.

Evans, T.J., Perez-Mendoza, D., Monson, R., Stickland, H. G., Salmond, G.P.C. 2009. Secretion systems of the enterobacterial phytopathogen, *Erwinia*. In: *Bacterial Secreted Proteins*. Wooldridge, K. (ed.). Norfolk: Caister Academic Press, pp. 479–503.

Ewbank, E. 1992. Etude du catabolisme d'acides aminés conduisant à la formation de phytotoxines par *Xanthomonas campestris* pv. *manihotis* (agent pathogFne de la bactériose du manioc). PhD thesis, Université Catholique de Louvain, Louvain-la-Neuve, Belgium.

FAO. 2000. *The State of Food Insecurity in the World (SOFI)*. Rome, Italy: FAO, UN. www.fao.org/FOCUS/E/SOFI00/sofi001-e.htm.

Faivre-Amiot, A. 1967. Quelques observations sur la présence de *Corynebacterium fascians* (Tilford) Dowson dans les cultures maraichères et florales en France. *Phytiatrie Phytopharmacie* 16: 165–176.

Feigin, A.M., Schagina, L.V., Takemoto, J.Y., Teeter, J.H., Brand, J.G. 1997. The effect of sterol on the sensitivity of membranes to the channel-forming antifungal antibiotic, syringomycin E. *Biochimica et Biophysica Acta* 1324: 102–110.

Feys, B.J.F., Benedetti, C.E., Penfold, C.N., Turner, J.G. 1994. *Arabidopsis* mutants selected for resistance to the phytotoxin coronatine are male sterile, insensitive to methyl jasmonate, and resistant to a bacterial pathogen. *Plant Cell* 6: 751–759.

Flatauer, F.E., Khan, M.A. 1978. Septic arthritis caused by *Enterobacter agglomerans*. *Archives of Internal Medicine* 138: 788.

Fukuchi, N., Isogai, A., Nakayama, J., Takayama, S., Yamashita, S., Suyama, K., Takemoto, J. Y., Suzuki, A. 1992. Structure and stereochemistry of three phytotoxins, syringomycin, syringotoxin and syringostatin, produced by *Pseudomonas syringae* pv. *syringae*. *Journal of the Chemical Society. Perkin Transactions* 1 pp. 1149–1157.

Garlant L, Koshinen, P., Rouhiainene, L., Laine, P., Paulin, L., Auvinene, P., Holm, L., Pirhonen, M. 2013. Genome sequence of *Dickeya solani*, a new soft rot pathogen of potato, suggests its emergence may be related to a novel combination of non-ribosomal peptide/polyketide synthase clusters. *Diversity* 5: 824–842.

Gasson, M.J. 1980. Indicator technique for antimetabolic toxin production by phytopathogenic species of *Pseudomonas*. *Applied and Environmental Microbiology* 39: 25–29.

Gelvin, S.B. 2010. Plant proteins involved in *Agrobacterium*-mediated genetic transformation. *Annual Review of Phytopathology* 48: 45–68.

Germida, J.J., Siciliano, S.D. 2001. Taxonomic diversity of bacteria associated with the roots of modern, recent and ancient wheat cultivars. *Biology and Fertility of Soils* 33: 410–415.

Gilbert, V., Planchon, V., Legros, F., Maraite, H., Bultreys, A. 2010. Pathogenicity and aggressiveness in populations of *Pseudomonas syringae* from Belgian fruit orchards. *European Journal of Plant Pathology* 126: 263–277.

Gnanamanickam, S.S., Starratt, A. N., Ward, E.W.B. 1982. Coronatine production *in vitro* and *in vivo* and its relation to symptom development in bacterial blight of soybean. *Canadian Journal of Botany* 60: 645–650.

Goodman, R.N., Huang, J.-S., Huang, P.-Y. 1974. Host-specific phytotoxic polysaccharide from apple tissue infected by *Erwinia amyluvora*. *Science* 183: 1081–1082.

Gopalan, S., Bauer, D.W., Alfano, J.R., Loniello, A.O., He, S.Y., Collmer, A. 1996. Expression of the *Pseudomonas syringae* avirulence protein AvrB in plant cells alleviates its dependence on the hypersensitive response and pathogenicity (Hrp) secretion system in eliciting genotype-specific hypersensitive cell death. *Plant Cell* 8: 1095–1105.

Goszczynska, T., Botha, W.J., Venter, S.N., Coutinho, T.A. 2007. Isolation and identification of the causal agent of brown stalk rot, a new disease of maize in South Africa. *Plant Disease* 91: 711–718.

Graner, G., Persson, P., Meijer, J., Alstrom, S. 2003. A study on microbial diversity in different cultivars of *Brassica napus* relation to its wilt pathogen, *Verticillium longisporum*. *FEMS Microbiology Letters* 29: 269–276.

Griffith, C.S., Sutton, T.B., Peterson, P.D. 2003. *Fire Blight: The Foundation of Phytobacteriology*. St. Paul, MN: APS Press.

Gross, D.C., Scholz-Schroeder, B.K., Zhang, J.-H., Grgurina, I., Mariotti, F., Della Torre, G., Guenzi, E., Grandi, G. 1998. Characterization of the thiotemplate mechanisms of syringomycin and syringopeptin synthesis by *Pseudomonas syringae* pv. *syringae*, In *Molecular Genetics of Host-Specific Toxins in Plant Disease*. Kohmoto and O. C. Yoder (ed.). Dordrecht, The Netherlands: Kluwer Academic Publishers, pp. 91–98.

Gupta, C.P., Sharma, A., Dubey, R.C., Maheshwari, D.K. 2001. Effect of metal ions on growth of *Pseudomonas aeruginosa* and siderophore and protein production. *Indian Journal of Experimental Biology* 39: 1318–1321.

Gyaneshwar, P., James, E.K., Mathan, N., Reddy, P.M., Reinhold-Hurek, B., Ladha, J.K. 2001. Endophytic colonization of rice by a diazotrophic strain of *Serratia marcescens*. *J Bacteriol* 183: 2634–2645.

Habsah, H., Zeehaida, M., van Rostenberghe, H., Noraida, R., Wan Pauzi, W.I., Fatimah, I., Rosliza, A. R., Nik Sharimah, N.Y., Maimunah, H. 2005. An outbreak of *Pantoea* spp. in a neonatal intensive care unit secondary to contaminated parenteral nutrition. *Journal of Hospital Infection* 61: 213–218.

Hebbar, O., Berge, T., Henlin, S., Singh, S.P. 1991. Bacterial antagonists of sunflower (*Helianthus annuus* L.) fungal pathogens. *Plant Soil* 133: 131–140.

Hogenhout, S.A., Oshima, K., Ammar, D., Kakizawa, S., Kingdom, H. N., Namba, S. 2008. Phytoplasmas: Bacteria that manipulate plants and insects. *Molecular Plant Pathology* 9: 403–423.

Holden, N., Pritchard, L., Toth, I. 2009. Colonization out with the colon: Plants as an alternative environmental reservoir for human pathogenic enterobacteria. *FEMS Microbiology Reviews* 33: 689–703.

Huitema, E., Bos, J.I.B., Tian, M., Win, J., Waugh, M. E., Kamoun, S. 2004. Linking sequence to phenotype in *Phytophthora*-plant interactions. *Trends in Microbiology* 12: 193–200.

Hutchison, M.L., Gross, D.C. 1997. Lipopeptide phytotoxins produced by *Pseudomonas syringae* pv. *syringae*: Comparison of the biosurfactant and ion channel-forming activities of syringopeptin and syringomycin. *Molecular Plant-Microbe Interactions* 10: 347–354.

Huynh, T.V., Dahlbeck, D., Staskawicz, B.J. 1989. Bacterial blight of soybean: Regulation of a pathogen gene determining host cultivar specificity. *Science* 245: 1374–1377.

Ichihara, A., Shiraishi, K., Sato, H., Sakamura, K., Nishiyama, K., Sakai, R., Furusaki, A., Matsumoto, T. 1977. The structure of coronatine. *Journal of the American Chemical Society* 99: 636–637.

Iniguez, A.L., Dong, Y., Carter, H.D., Ahmer, B.M.M., Stone, J.M., Triplett, E.W. 2005. Regulation of enteric endophytic bacterial colonization by plant defenses. *Molecular Plant-Microbe Interactions* 18: 169–178.

James, C. 1998. Global food security. Abstract. *International Congress of Plant Pathology*. 7th Edinburgh, UK, Aug. No. 4.1 GF. http://www. bspp.org.uk/icpp98/4/1GF.html.

Jones, J.B., Jones, J.P. 1985. The effect of bactericides, tank mixing time and spray schedule on bacterial leaf spot of tomato. *Proceedings of the Florida State Horticultural Society* 98: 244–247.

Jones, J.B., Jones, J.P., Stall, R. E., Zitter, T.A. (eds.) 1991. Compendium of tomato diseases. *American Phytopathological Society* St. Paul, MN: APS Press.

Jones, J.D., Dangl, J.L. 2006. The plant immune system. *Nature* 444: 323–329.

Jovanovic, M., James, E.H., Burrows, P. C., Rego, F. G., Buck, M., Schumacher, J. 2011. Regulation of the co-evolved HrpR and HrpS AAA+ proteins required for *Pseudomonas syringae* pathogenicity. *Nature Communications* 2: 177.

Kamoun, S. 2006. A catalogue of the effector secretome of plant pathogenic oomycetes. *Annual Review of Phytopathology* 44: 41–60.

Kamoun, S. 2007. Groovy times: Filamentous pathogen effectors revealed. *Current Opinion in Plant Biology* 10: 358–365.

Kennelly, M.M., Cazorla, F.M., de Vicente, A., Ramos, C. 2007. *Pseudomonas syringae* diseases of fruit trees: Progress toward understanding and control. *Plant Disease* 91: 4–17.

Kenyon, J.S., Turner, J.G. 1992. The stimulation of ethylene synthesis in *Nicotiana tabacum* leaves by the phytotoxin coronatine. *Plant Physiology* 100: 219–224.

Kirzinger, M.W.B., Nadarasah, G., Stavrinides, J. 2011. Insights into cross-kingdom plant pathogenic bacteria. *Genes* 2: 980–997.

Klerks, M.M., Franz, E., Van Gent-Pelzer, M., Zijlstra, C., Van Bruggen, A.H. 2007. Differential interaction of *Salmonella enterica* serovars with lettuce cultivars and plant-microbe factors influencing the colonization efficiency. *International Society for Microbial Ecology* 1: 620–631.

Koutsoudis, M.D., Tsaltas, D., Minogue, T. D., von Bodman, S.B. 2006. Quorum-sensing regulation governs bacterial adhesion, biofilm development, and host colonization in *Pantoea stewartii* subspecies *stewartii*. *Proceedings of the National Academy of Sciences. USA* 103: 5983–5988.

Kratz, A., Greenberg, D., Barki, Y., Cohen, E., Lifshitz, M. 2003. *Pantoea agglomerans* as a cause of septic arthritis after palm tree thorn injury; case report and literature review. *Archives of Disease in Childhood* 88: 542–544.

Krechel, A., Faupel, A., Hallmann, J., Ulrich, A., Berg, G. 2002. Potato-associated bacteria and their antagonistic potential towards plant pathogenic fungi and the plant parasitic nematode *Meloidogyne incognita* (Kofoid & White) Chitwood. *Canadian Journal of Microbiology* 48: 772–786.

Kumar, A., Munder, A., Aravind, R., Eapen, S.J., Tummler, B., Raaijmakers, J.M. 2013. Friend or foe: Genetic and functional characterization of plant endophytic *Pseudomonas aeruginosa*. *Environmental Microbiology* 15: 764–779.

Kutter, S., Hartmann, A., Schmid, M. 2006. Colonization of barley (*Hordeum vulgare*) with *Salmonella enterica* and *Listeria* spp. *FEMS Microbiology Ecology* 56: 262–271.

Kvitko, B.H., Park, D.H., Velásquez, A.C., Wei, C.F., Russell, A.B., Martin, G.B., Schneider, D.J., Collmer, A. 2009. Deletions in the repertoire of *Pseudomonas syringae* pv. tomato DC3000 type III secretion effector genes reveal functional overlap among effectors. *PLoS Pathogens* 5(4): e1000388.

Lambert, B., Frederik, L., Van Rooyen, L., Gossele, F., Papon, Y., Swings, J. 1987. Rhizobacteria of maize and their antifungal activities. *Applied and Environmental Microbiology* 53: 1866–1871.

Latorre, B.A., Jones, A.L. 1979. *Pseudomonas morsprunorum*, the cause of bacterial canker of sour cherry in Michigan, and its epiphytic association with *Pseudomonas syringae*. *Phytopathology* 69: 335–339.

Lavermicocca, P., Iacobellis, N., Simmaco, M., Graniti, A. 1997. Biological properties and spectrum of activity of *Pseudomonas syringae* pv. *Syringae* toxins. *Physiological and Molecular Plant Pathology* 50: 129–140.

Laycock, M.V., Hildebrand, P.D., Thibault, P., Walter, J.A., Wright, J.L.C. 1991. Viscosin, a potent peptidolipid biosurfactant and phytopathogenic mediator produced by a pectolytic strain of *Pseudomonas fluorescens*. *Journal of Agricultural and Food Chemistry* 39: 483–489.

Lee, B.M., Park, Y.J., Park, D.S., Kang, H.W., Kim, J.G., Song, E.S., Park, I.C. et al. 2005. The genome sequence of *Xanthomonas oryzae* pathovar oryzae KACC10331, the bacterial blight pathogen of rice. *Nucleic Acids Research* 33: 577–586.

Leyns, F., De Cleene, M., Swings, J. G., De Ley, J. 1984. The host range of the genus *Xanthomonas. The Botanical Review* 50: 308–356.

Li, C.M., Brown, I., Mansfield, J., Stevens, C., Boureau, T., Romantschuk, M., Taira, S. 2002. The Hrp pilus of *Pseudomonas syringae* elongates from its tip and acts as a conduit for translocation of the effector protein HrpZ. *EMBO Journal* 21: 1909–1915.

Li, Y., Peng, Q., Selimi, D., Wang, Q., Charkowski, A.O., Chen, X., Yang, C.H. 2009. The plant phenolic compound p-coumaric acid represses gene expression in the *Dickeya dadantii* Type III secretion system. *Applied and Environmental Microbiology* 75: 1223–1228.

Liu, H., Coulthurst, S.J., Pritchard, L., Hedley, P.E., Ravensdale, M., Humphris, S., Burr, T. et al. 2008. Quorum sensing coordinates brute force and stealth modes of infection in the plant pathogen *Pectobacterium atrosepticum. PLoS Pathogens* 4: e1000093.

Lottmann, J., Heuer, H., Smalla, K., Berg, G. 1999. Influence of transgenic T4-lysozyme-producing plants on beneficial plant associated bacteria. *FEMS Microbiology Ecology* 29: 365–377.

Lottmann, J., Berg, G. 2001. Phenotypic and genotypic characterization of antagonistic bacteria associated with roots of transgenic and non-transgenic potato plants. *Microbiological Research* 156: 75–82.

Ma, B., Hibbing, M.E., Kim, H.S., Reedy, R.M., Yedidia, I., Breuer, J., Breuer, J. et al. 2007. Host range and molecular phylogenies of the soft rot enterobacterial genera *Pectobacterium* and *Dickeya. Phytopathology* 97: 1150–1163.

Maher, E.A., Bate, N.J., Ni, W., Elkind, Y., Dixon, R.A., Lamb, C.J. 1994. Increased disease susceptibility of transgenic tobacco plants with suppressed levels of preformed phenyl-propanoid products. *Proceedings of the National Academy of Sciences, USA*91: 7803–7806.

Maki, D.G., Rhame, F.S., Mackel, D.C., Bennett, J.V. 1976. Nationwide epidemic of septicemia caused by contaminated intravenous products. I. Epidemiologic and clinical features. *American Journal of Medicine* 60: 471–485.

Mansfield, J., Genin, S., Magori, S., Citovsky, V., Sriariyanum, M., Ronald, P., Dow, M. et al. 2012. Top 10 plant pathogenic bacteria in molecular plant pathology. *Molecular Plant Pathology* 13: 614–629.

Mansvelt, E.L., Hattingh, M.J. 1986. Bacterial blister bark and blight of fruit spurs of apple in south Africa caused by *Pseudomonas syringae* pv. *syringae. Plant Disease* 70: 403–405.

Marcelletti, S., Ferrante, P., Petriccione, M., Firrao, G., Scortichini, M. 2011. *Pseudomonas syringae* pv. *actinidiae* draft genomes comparison reveal strain-specific features involved in adaptation and virulence to *Actinidia* species. *PLoS ONE* 6: e27297.

Marín-Cevada, V., Vargas, V.H., Juárez, M., López, V.G., Zagada, G., Hernández, S., Cruz, A. et al. 2006. Presence of *Pantoea citrea*, causal agent of pink disease, in pineapple fields in Mexico. *Plant Pathology* 55: 294.

Masyahit, M., Sijam, K., Awang, Y., Ghazali, M. 2009. First report on bacterial soft rot disease on dragon fruit (*Hylocereus* spp.) caused by *Enterobacter cloacae* in peninsular Malaysia. *International Journal of Agricultural and Biological Engineering* 11: 659–666.

Matsaniotis, N.S., Syriopoulou, V.P., Theodoridou, M.C., Tzanetou, K.G., Mostrou, G.I. 1984. Enterobacter sepsis in infants and children due to contaminated intravenous fluids. *Infection Control* 5: 471–477.

Mattinen, L., Somervuo, P., Nykyri, J., Nissinen, R., Kouvonen, P., Corthals, G., Auvinen, P., Aittamaa, M., Valkonen, J. P., Pirhonen, M. 2008. Microarray profiling of host-extract induced genes and characterization of the type VI secretion cluster in the potato pathogen *Pectobacterium atrosepticum. Microbiology* 154: 2387–2396.

Mehnaz, S., Mirza, M.S., Haurat, J., Bally, R., Normand, P., Bano, A., Malik, K.A. 2001. Isolation and 16S rRNA sequence analysis of the beneficial bacteria from the rhizosphere of rice. *Canadian Journal of Microbiology* 47: 110–117.

Melotto, H.C., Underwood, W., Koczan, J., Nomura, K., He, S.Y. 2006. Plant stomata function in innate immunity against bacterial invasion. *Cell* 126: 969–980.

Menestrina, G., Schiavo, G., Montecucco, C. 1994. Molecular mechanisms of action of bacterial protein toxins. *Molecular Aspects of Medicine* 15(2): 79–193.

Mew, T.W., Natural, M.P. 1993. Management of *Xanthomonas* diseases. In: *Xanthomonas*. Swings, J.G., Civerolo, E.L. (eds.). London: Chapman & Hall, pp. 341–362.

Mingzhi, L.I., Ling, X.U., Ziling, S.U.N., Yongquan, L.I. 2007. Isolation and characterization of a phytotoxin from *Xanthomonas campestris* pv. *retroflexus. Chinese Journal of Chemical Engineering* 15(5): 639–642.

Misas-Villamil, J.C., van der Hoorn, R.A. 2008. Enzyme-inhibitor interactions at the plant-pathogen interface. *Current Opinion in Plant Biology* 11: 380–388.
Mitchell, R.E. 1976. Isolation and structure of a chlorosis-inducing toxin of *Pseudomonas phaseolicol*a. *Phytochemistry* 15: 1941–1947.
Mitchell, R.E. 1982. Coronatine production by some phytopathogenic Pseudomonads. *Physiological Plant Pathology* 220: 83–89.
Mitchell, R.E., Frey, E.J., Benn, M.H. 1986. Rhizobitoxine and L-threo-hydroxythreonine production by the plant pathogen *Pseudomonas andropogonis*. *Phytochemistry* 25: 2711–2715.
Mittal, S.M., Davis, K.R. 1995. Role of the phytotoxin coronatine in the infection of *Arabidopsis thaliana* by *Pseudomonas syringae* pv. *tomato*. *Molecular Plant-Microbe Interactions* 8: 165–171.
Morales, A., Garland, J.L., Lim, D.V. 1996. Survival of potentially pathogenic human-associated bacteria in the rhizosphere of hydroponically grown wheat. *FEMS Microbiology Ecology* 20: 155–162.
Nicotra, A. 1978. Calita, new plum variety. *Acta Horticulturae* 74: 49–50.
Nürnberger, T., Brunner, F., Kemmerling, B., Piater, L. 2004. Innate immunity in plants and animals: Striking similarities and obvious differences. *Immunological Reviews* 198: 249–266.
Oldroyd, G.E.D., Downie, J.M. 2008. Coordinating nodule morphogenesis with rhizobial infection in legumes. *Annual Review of Plant Biology* 59: 519–546.
Panstruga, R., Dodds, P.N. 2009. Terrific protein traffic: The mystery of effector protein delivery byfilamentous plant pathogens. *Science* 324: 748–750.
Patil, S.S., Hayward, A.C., Emmons, R. 1974. An ultraviolet-induced non toxigenic mutant of *Pseudomonas phaseolicola* of altered pathogenicity. *Phytopathology* 64: 590–595.
Pérombelon, M.C.M. 2002. Potato diseases caused by soft rot erwinias: An overview of pathogenesis. *Plant Pathology* 51: 1–12.
Perreaux, D., Maraite, H., Meyer, J.A. 1986. Detection of 3-methylthiopropionic acid in cassava leaves infected by *Xanthomonas campestris* pv. *manihotis*. *Physiological and Molecular Plant Pathology* 28: 323–328.
Pitzschke, A., Hirt, H. 2010. New insights into an old story: *Agrobacterium* induced tumor formation in plants by plant transformation. *EMBO Journal* 29: 1021–1032.
Pohronezny, K., Volin, R.B. 1983. The effect of bacterial spot on yield and quality of fresh market tomatoes. *HortScience* 18: 69–70.
Pulawska, J. 2010. Crown gall of stone fruits and nuts—Economic significance and diversity of its causal agent tumorigenic *Agrobacterium* spp. *Journal of Plant Pathology* 92: 87–98.
Putnam, M.L., Miller, M.L. 2007. *Rhodococcus fascians* in herbaceous perennials. *Plant Disease* 91: 1064–1076.
Rademaker, J.L., Louws, F.J., Schultz, M.H., Rossbach, U., Vauterin, L., Swings, J., de Bruijn, F.J. 2005. A comprehensive species to strain taxonomic framework for *Xanthomonas*. *Phytopathology* 9: 1098–1111.
Rainey, P.B., Brodey, C.L., Johnstone, K. 1991. Biological properties and spectrum of activity of tolaasin, a lipodepsipeptide toxin produced by the mushroom pathogen *Pseudomonas tolaasii*. *Physiological and Molecular Plant Pathology* 39: 57–70.
Reiter, B., Pfeifer, U., Schwab, H., Sessitsch, A. 2002. Response of endophytic bacterial communities in potato plants to infection with *Erwinia carotovora* subsp. *atroseptica*. *Applied and Environmental Microbiology* 68: 2261–2268.
Rhodehamel, N.H., Durbin, R.D. 1989. Toxin production by strains of *Pseudomonas syringae* pv. *tagetis*. *Physiological and Molecular Plant Pathology* 35: 301–311.
Ritchie, D.F. 1995. Bacterial spot. In: *Compendium of Stone Fruit Diseases*. Ogawa, J.M., Zehr, E.I., Bird, G.W. (eds.). St. Paul, MN, USA: APS Press, pp. 50–52.
Ritchie, D.F., Dittapongitch, V. 1991. Copper- and streptomycin-resistant strains and host differentiated races of *Xanthomonas campestris* pv. *vesicatoria* in North Carolina. *Plant Disease* 75: 733–736.
Robert-Baudouy, J. 1991. Molecular biology of *Erwinia*: From soft-rot to antileukaemics. *Trends in Biotechnology* 9: 325–329.
Roh, E., Park, T.H., Kim, M.I., Lee, S., Ryu, S., Oh, C.S., Rhee, S., Kim, D.H., Park, B.S., Heu, S. 2010. Characterization of a new bacteriocin, Carocin D, from *Pectobacterium carotovorum* subsp. *carotovorum* Pcc21. *Applied and Environmental Microbiology* 76: 7541–7549.
Roine, E., Wei, W., Yuan, J., Nurmiaho-Lassila, E.-L., Kalkkinen, N., Romantschuk, M., He, S.Y. 1997. Hrp pilus: An hrp-dependent bacterial surface appendage produced by *Pseudomonas syringae* pv. tomato DC3000. *Proceedings of the National Academy of Sciences, USA* 94: 3459–3464.
Rouhrazi, K., Rahimian, H. 2012. Genetic diversity of Iranian *Agrobacterium* strains from grapevine. *Annals of Microbiology* 62: 1661–1667.

Salmond, G.P.C. 1994. Secretion of extracellular virulence factors of plant pathogenic bacteria. *Annual Review of Phytopathology* 32: 181–200.

Samson, R., Legendre, J.B., Christen, R., Fischer-Le Saux, M., Achouak, W., Gardan, L. 2005. Transfer of *Pectobacterium chrysanthemi* and *Brenneria paradisiaca* to the genus *Dickeya* gen. nov. as *Dickeya chrysanthemi* comb. nov. and *Dickeya paradisiaca* comb. nov. and delineation of four novel species, *Dickeya dadantii* sp. nov., *Dickeya dianthicola* sp. nov., *Dickeya dieffenbachiae* sp. nov. and *Dickeya zeae* sp. nov. *International Journal of Systematic and Evolutionary Microbiology* 55: 1415–1427.

Schiavo, G., van der Goot, E.G. 2001. The bacterial toxin toolkit. *Nature Reviews Molecular Cell Biology* 2: 530–537.

Schwieger, F., Tebbe, C.C. 2000. Effect of field inoculation with *Sinorhizobium meliloti* L33 on the composition of bacterial communities in rhizospheres of a target plant (*Medicago sativa*) and a non-target plant (*Chenopodium album*)–linking of 16S rRNA gene-based single-strand conformation polymorphism community profiles to the diversity of cultivated bacteria. *Applied and Environmental Microbiology* 66: 3556–3565.

Sebaihia, M., Bocsanczy, A.M., Biehl, B.S., Quail, M.A., Perna, N.T., Glasner, J.D., DeClerck, G.A. et al. 2010. Complete genome sequence of the plant pathogen *Erwinia amylovora* strain ATCC 49946. *Journal of Bacteriology* 192: 2020–2021.

Segond, D., Dellagi, A., Lanquar, V., Rigault, M., Patrit, O., Thomine, S., Expert, D. 2009. NRAMP genes function in *Arabidopsis thaliana* resistance to *Erwinia chrysanthemi* infection. *Plant Journal* 58: 195–207.

Sessitsch, A., Reiter, B., Berg, G. 2004. Endophytic bacterial communities of field-grown potato plants and their plant growth-promoting abilities. *Canadian Journal of Microbiology* 50: 239–249.

Sherf, A.F., MacNab, A. A. 1986. *Vegetable Diseases and Their Control*. New York: Wiley.

Sorensen, K.N., Kim, K.-H., Takemoto, J.Y. 1996. *in vitro* antifungal and fungicidal activities and erythrocyte toxicities of cyclic lipodepsipeptides produced by *Pseudomonas syringae* pv. *syringae*. *Antimicrobial Agents and Chemotherapy* 40: 2710–2713.

Stall, R.E., Loschke, D.C., Jones, J.B. 1986. Linkage of copper resistance and a virulence loci on a selftransmissible plasmid in *Xanthomonas campestris* pv. *vesicatoria. Phytopathology* 76: 240–243.

Stefani E., Bazzi, C., Mazzucchi, U., Colussi, A. 1989. *Xanthomonas campestris* pv. *pruni* in pescheti del Friuli. *Informatore Fitopatologico* 39(7–8): 60–63.

Stefani, E. 2010. Economic significance and control of bacterial spot/canker of stone fruits caused by *Xanthomonas arboricola* pv. *pruni. Journal of Plant Pathology* 92: 99–103.

Stewart, W.W. 1971. Isolation and proof of structure of wildfire toxin. *Nature* 229: 174–178.

Strange, R.N. 2003. *Introduction to Plant Pathology.* New York: John Wiley & Sons.

Szabò Z., Gróf, P., Schagina, L.V., Gurnev, P.A., Takemoto, J.Y., Mátyus, E., Blaskò, K. 2002. Syringotoxin pore formation and inactivation in human red blood cell and model bilayer lipid membranes. *Biochimica et Biophysica Acta* 1567: 143–149.

Tarkowski, P., Vereecke, D. 2014. Threats and opportunities of plant pathogenic bacteria. *Biotechnology Advances* 32: 215–229.

Teplitski, M., Warriner, K., Bartz, J., Schneider, K.R. 2011. Untangling metabolic and communication networks: Interactions of enterics with phytobacteria and their implications in produce safety. *Trends in Microbiology* 19: 121–127.

Thayer, P.L., Stall, R.E. 1961. A survey of *Xanthomonas vesicatoria* resistance to streptomycin. *Proceedings of the Florida State Horticultural Society* 75: 163–165.

Toth, I.K., Bell, K., Holeva, M.C., Birch, P.R.J. 2003. Soft rot erwiniae: From genes to genomes. *Molecular Plant Pathology* 4: 17–30.

Tripathi, A.K., Verma, S.C., Ron, E.Z. 2002. Molecular characterization of a salt-tolerant bacterial community in the rice rhizosphere. *Research in Microbiology* 153: 579–584.

Tsuji, J., Somerville, S., Hammerschmidt, R. 1991. Identification of a gene in *Arabidopsis thaliana* that controls resistance to *Xanthomonas campestris* pv. *campestris*. *Physiological and Molecular Plant Pathology* 38: 57–65.

Tyler, H.L., Triplett, E.W. 2008. Plants as a habitat for beneficial and/or human pathogenic bacteria. *Annual Review of Phytopathology* 46: 53–73.

Uchytil, T.F., Durbin, R.D. 1980. Hydrolysis of tabtoxin by plant and bacterial enzymes. *Experientia* 36: 301–302.

Ulrychová, M., Petrů, E. 1983. Isolation of some strains of *Corynebacterium fascians* (TILFORD) DOWSON in Czechoslovakia. *Journal of Plant Biology* 25: 63–67.

Uys, M.D.R., Thompson, A.H., Holz, G. 1996. Diseases associated with tomato in the main tomato growing of South Africa. *Southern African Society for Horticultural Sciences* 6: 78–81.

van Baarlen, P., van Belkum, A., Thomma, B.P.H.J. 2007. Disease induction by human microbial pathogens in plant model systems: Potential, problems and prospects. *Drug Discovery Today* 12: 167–173.

Vantomme, R., Elia, S., Swings, J., De Ley, J. 1982. *Corynebacterium fascians* (Tilford 1936) Dowson 1942, the causal agent of leafy gall on lily crops in Belgium. *Parasitica* 38: 183–192.

Vasse, J., Genin, S., Frey, P., Boucher, C., Brito, B. 2000. The hrpB and hrpG regulatory genes of *Ralstonia solanacearum* are required for different stages of the tomato root infection process. *Molecular Plant-Microbe Interactions* 13: 259–267.

Vassilev, V., Lavermicocca, P., Di Giorgio, D., Iacobellis, N.S. 1996. Production of syringomycins and syringopeptins by *Pseudomonas syringae* pv. *atrofaciens*. *Plant Pathology* 45: 316–322.

Vauterin, L., Hoste, B., Dersters, K., Swings, J. 1995. Reclassification of *Xanthomonas*. *International Journal of Systematic Bacteriology* 45: 472–489.

Volksch, B., Bublitz, F., Fritsche, W. 1989. Coronatine production by *Pseudomonas syringae* pathovars: Screening method and capacity of product formation. *Journal of Basic Microbiology* 29: 463–468.

Ward, H.P., O'Garro, L.W. 1992. Bacterial spot of pepper and tomato in Barbados. *Plant Disease* 76: 1046–1048.

Warriner, K., Namvar, A. 2010. The tricks learnt by human enteric pathogens from phytopathogens to persist within the plant environment. *Current Opinion in Biotechnology* 21: 131–136.

Wharam, S., Mulholland, V., Salmond, G.P.C. 1995. Conserved virulence factor regulation and secretion systems in bacterial pathogens of plants and animals. *European Journal of Plant Pathology* 101: 1–13.

Wiebe, W.L., Campbell, R.N. 1993. Characterization of *Pseudomonas syringae* pv. *maculicola* and comparison with *Pseudomonas syringae* pv. tomato. *Plant Disease* 77: 414–419.

Xu, G.-W., Gross, D.C. 1988. Evaluation of the role of syringomycin in plant pathogenesis by using Tn5 mutants of *Pseudomonas syringae* pv. *syringae* defective in syringomycin production. *Applied and Environmental Microbiology* 54: 1345–1353.

Yamazaki, A., Li, J., Hutchins, W.C., Wang, L., Ma, J., Ibekwe, A.M., Yang, C.-H. 2011. Commensal effect of pectate lyases secreted from *Dickeya dadantii* on proliferation of *Escherichia coli* O157:H7 EDL933 on lettuce leaves. *Applied and Environmental Microbiology* 77: 156–162.

Yang, S., Peng, Q., Zhang, Q., Zou, L., Li, Y., Robert, C., Pritchard, L. et al. 2010. Genome-wide identification of HrpL-regulated genes in the necrotrophic phytopathogen *Dickeya dadantii* 3937. *PLoS One* 5: e13472.

Young, J.M., Park, D.C., Shearman, H.M., Fargier, E. 2008. A multilocus sequence analysis of the genus *Xanthomonas*. *Systematic and Applied Microbiology* 5: 366–377.

Zhang, J., Li, W., Xiang, T., Liu, Z., Laluk, K., Ding, X., Zou, Y. et al. 2010. Receptor-like cytoplasmic kinases integrate signaling from multiple plant immune receptors and are targeted by a *Pseudomonas syringae* effector. *Cell Host and Microbe* 7: 290–301.

Zutra, D., Cohen, J., Gera, A., Loebenstein, G. 1994. Association of *Rhodococcus (Corynebacterium) fascians* with the stunting-fasciation syndrome of carnation in Israel. *Acta Horticulturae* 377: 319–323.

Index

E

F

J

K

L

M

N

O

P

Q

R

S

T

U

V

W

X